Wilderness Management

Stewardship and Protection of Resources and Values

Fourth Edition

Chad P. Dawson
Professor of Recreation Resources Management,
Former Chair, Department of Forest and Natural Resources Management,
College of Environmental Science and Forestry,
State University of New York,
Syracuse, New York
cpdawson@esf.edu

John C. Hendee
Professor Emeritus and Former Dean, College of Natural Resources,
University of Idaho, Moscow, Idaho,
Vice President, Science and Education, The WILD Foundation
hendeejo@uidaho.edu

International Wilderness Leadership (WILD) Foundation
Boulder, Colorado

Fulcrum Publishing
Golden, Colorado

Copyright © 2009 International Wilderness Leadership (WILD) Foundation

All rights reserved. No part of this book may be reproduced, stored in a retrieval system, or transmitted in any form or by any means, electronic, mechanical, photocopying, recording, or otherwise, without written permission from the publisher.

Library of Congress Cataloging-in-Publication Data
  Dawson, Chad P. (Chad Patrick), 1948-
   Wilderness management : stewardship and protection of resources and values / Chad P. Dawson, John C. Hendee. -- 4th ed.
    p. cm.
   Rev. ed. of: Wilderness management / John C. Hendee, Chad P. Dawson, 2002
   Includes bibliographical references and index.
   ISBN 978-1-55591-682-4 (pbk.)
   1. Wilderness areas--United States--Management. 2. Wilderness areas--Management. 3. Nature conservation--United States. 4. Nature conservation. I. Hendee, John C. II. Hendee, John C. Wilderness management. III. Title.
  QH76.H46 2008
  333.78'2--dc22
              2008041003

Cover photos: "Many ecosystem benefits flow from wilderness" © 2009 by Chad P. Dawson; inset photo of three students © 2008 courtesy of WILDLINK
Design by Patty Maher

Printed in the United States of America by Color House Graphics, Inc.
0 9 8 7 6 5 4 3 2 1

Fulcrum Publishing
4690 Table Mountain Drive, Suite 100
Golden, Colorado 80403
(800) 992-2908 • (303) 277-1623
www.fulcrumbooks.com

Contents

Publisher's Preface by Vance G. Martin,
  President, International Wilderness Leadership (WILD) Foundation ix
Foreword by National Wilderness Coordinators, U.S. Federal Agencies,
  and Other Wilderness Leaders xi
History, Authorship, and Sponsorship of the Book
  by Chad P. Dawson and John C. Hendee xiii
Authors' Preface xv

I. The Setting

Chapter 1. Wilderness Management: Philosophical Direction 1
  Introduction 2
  Wilderness and Other Land Uses 3
  What Is Wilderness? 4
  Historical Origins of the Wilderness Concept 5
  Wilderness and the Early American Scene 5
  Historical Wilderness Themes and Values 6
  Wilderness Management and Wilderness Designation 12
  The Effects of Wilderness Designation Decisions on Management 13
  The Need for Wilderness Management 16
  Alternatives to Management? 18
  Wilderness Philosophy and Wilderness Management 19
  Anthropocentric and Biocentric Perspectives 19
  In Support of Biocentric Management 22
  Summary 25
  Study Questions 26
  Reader Exercise: How Do Your Personal Wilderness Values Rate? 26
  Case Discussion Questions 28
  References 29

Chapter 2. Historical Roots of Wilderness Management
  by Roderick Nash and John C. Hendee 31
  Introduction 31
  The Intellectual Dilemma 31
  Destruction by Popularity—A Threat to Wilderness 32
  Roots of Wilderness Appreciation 36
  Evolution of Wilderness in the National Parks 36
  Evolution of Wilderness in the National Forests 39
  The Wilderness Management Idea Begins 43
  Wilderness Management Evolves 45
  Study Questions 47
  Acknowledgments 48

Case Discussion: Modern-Day American Views on Wilderness 48
Case Discussion Questions 48
References 49

Chapter 3. International Wilderness
by Vance G. Martin and Alan Watson 50
Introduction 51
Forces of Change and Wilderness Protection in the Twenty-first Century 51
Three Classes of Wilderness Protection: A Global View 58
The World Wilderness Congress 83
International Journal of Wilderness 83
Summary 83
Study Questions 84
Acknowledgments 84
Case Discussion: Canadian Wilderness—The Quetico Provincial Park 85
Case Discussion Questions 86
References 86

II. U.S. Legal Authority and Process for Wilderness

Chapter 4. The Wilderness Act: Legal Basis for Wilderness
Protection and Management 89
Introduction 90
The L-20 Regulation 90
The U Regulations 91
Statutory Protection for Wilderness 92
A Brief Legislative History of the Wilderness Act 94
The Wilderness Act of 1964 96
Some Exceptions and Ambiguities in Wilderness Legislation 109
The Evolution of Wilderness Protection 109
Study Questions 110
Acknowledgments 110
Case Discussion: Compare the 1956 Bill to the 1964 Wilderness Act 111
Case Discussion Questions 111
References 111

Chapter 5. Management Implications in the Wilderness Designation
Process, Wilderness Designation Acts, and Wilderness Litigation 113
Introduction 114
Part I: Evolution of the Designation Process 115
Part II: Management Implications in Wilderness Designation Acts 143
Part III: Management Implications from Litigation Involving Wilderness 155
Study Questions 159
Acknowledgments 160
Case Discussion Questions: Olympic National Park Wilderness 160
References 160

Chapter 6. The National Wilderness Preservation System and Complementary
Conservation Areas 163
Introduction 163
The National Wilderness Preservation System 164
Complementary Conservation Areas 168
Summary 176

Study Questions 176
Acknowledgments 177
Case Discussion: Recreational Use on the Colorado River in
    Grand Canyon National Park 177
Case Discussion Questions 178
References 178

III. Wilderness Management and Planning Concepts

Chapter 7. Principles of Wilderness Management 179
Introduction 180
Principle 1: Manage Wilderness as the Most Pristine Extreme on the
    Environmental Modification Spectrum 180
Principle 2: Manage Wilderness Comprehensively, Not as Separate Parts 181
Principle 3: Manage Wilderness, and Sites Within, Following a
    Concept of Nondegradation 182
Principle 4: Manage Human Influences, a Key to Wilderness Protection 183
Principle 5: Manage Wilderness Biocentrically to Produce Human Values
    and Benefits 184
Principle 6: Favor Wilderness-Dependent Activities 185
Principle 7: Guide Wilderness Management Using Written Plans with
    Specific Area Objectives 186
Principle 8: Set Carrying Capacities as Necessary to Prevent Unnatural Change 187
Principle 9: Focus Management on Threatened Sites and Damaging Activities 188
Principle 10: Apply Only the Minimum Tools, Regulations, or Force to Achieve
    Wilderness-Area Objectives 189
Principle 11: Involve the Public as a Key to the Success of
    Wilderness Management 189
Principle 12: Monitor Wilderness Conditions and Experience Opportunities
    to Guide Long-Term Wilderness Stewardship 190
Principle 13: Manage Wilderness in Relation to Its
    Adjacent Lands 191
Summary 193
Study Questions 193
Case Discussion: Assisted Trout Migration into Wilderness? 193
Case Discussion Questions 194
References 194

Chapter 8. Wilderness Management Planning 195
Introduction 196
The Need for Planning 196
Trends and Evolution of Wilderness Planning Direction 197
Wilderness Management Planning in U.S. Federal Agencies 199
Planning Tools and Techniques 203
Summary 214
Study Questions 214
Acknowledgments 214
Case Discussion: Comprehensive Conservation Planning for the
    Kenai National Wildlife Refuge, Alaska 215
Case Discussion Questions 215
References 216

Chapter 9. Managing for Appropriate Wilderness Conditions: The Limits of
Acceptable Change Process 217
  Introduction 218
  Considerations in Managing Visitor Use and Impact 219
  The Carrying-Capacity Concept 219
  The Limits of Acceptable Change (LAC) Concept 222
  Implementing the LAC Process 234
  Direct Visitor Management May Be Necessary When Standards Are Exceeded 246
  Summary 247
  Study Questions 247
  Acknowledgments 247
  Case Discussion: Campsite Indicators on the White Mountain National Forest 248
  Case Discussion Questions 249
  References 249

IV. Wilderness Resources, Values, and Threats to Them

Chapter 10. Wilderness Ecosystems by Jerry F. Franklin and Gregory H. Aplet 251
  Introduction 252
  An Ecosystems Primer 252
  The Nature of Wilderness Ecosystems 257
  Changes in Wilderness Ecosystems 259
  Managing Wilderness Ecosystems 264
  Boundary Effects 264
  Summary 270
  Study Questions 271
  Acknowledgments 271
  Case Discussion: Protecting Hemlocks in the Southeastern U.S. 271
  Case Discussion Questions 272
  References 273

Chapter 11. Fire in Wilderness Ecosystems 275
  Introduction 276
  Fire Occurrence and Behavior 277
  The Natural (Historical) Role of Fire in Wilderness Ecosystems 280
  How Has the Historical Role of Fire Been Modified by Fire Suppression? ... 283
  Objectives of Wilderness Fire Management 285
  Wilderness Fire Policy Alternatives and Their Consequences 289
  Wilderness Fire-Management Planning: Considerations and Constraints ... 291
  History and Evolution of Wilderness Fire-Management Programs 296
  Fire-Dependent Ecosystems in American Wilderness Regions 298
  Trends and Future Needs 302
  Wilderness Fire Programs and Ecosystem Management 303
  Study Questions 303
  Acknowledgments 303
  Case Discussion: Ignition Is One of the Challenges to Managing Fires
    in Wilderness 304
  Case Discussion Questions 304
  References 305

Chapter 12. Wildlife in Wilderness: A North American and International
  Perspective by John C. Hendee and David J. Mattson 308
  Introduction 309

The Wilderness Wildlife Resource 309
Wildlife-Related Problems in Wilderness Stewardship 318
Wilderness Management Objectives and Guidelines for Wildlife 327
Summary 330
Study Questions 331
Acknowledgments 331
Case Discussion: Wolf Reintroduction Controversy 331
Case Discussion Questions 332
References 332

Chapter 13. Potential Threats to Wilderness Resources and Values 335
    Introduction 336
    Wilderness Conditions and Values Affected by Threats 337
    Potential Threats to Wilderness 339
    Monitoring Threats to Wilderness 352
    Summary 352
    Study Questions 353
    Acknowledgments 353
    Case Discussion: Interactions Between Potential Threats to Wilderness 354
    Case Discussion Questions 354
    References 355

V. Wilderness Uses and Their Management

Chapter 14. Wilderness Use and User Trends 357
    Introduction 358
    The Importance of Understanding Wilderness Use and Users 358
    Wilderness-Dependent Uses 358
    Wilderness Use: An Overview 359
    Wilderness Recreational Use 369
    User Characteristics 382
    Factors Affecting Wilderness Use Trends and Management 384
    Wilderness Recreation Use Projections 387
    Summary 389
    Study Questions 390
    Acknowledgments 390
    Case Discussion: Constraints to Wilderness Participation 391
    Case Discussion Questions 391
    References 391

Chapter 15. Ecological Impacts of Wilderness Recreation and Their Management
    by David N. Cole 395
    Introduction 396
    Importance of Wilderness Recreation Impacts 396
    Recreational Activities and Associated Impacts 398
    Managing Campsite Impacts 407
    Managing Trail Impacts 423
    Managing Pack and Saddle Stock 429
    Summary 434
    Study Questions 435
    Acknowledgments 435
    Case Discussion: Baseline Wilderness Data 435
    Case Discussion Questions 436
    References 436

Chapter 16. Wilderness Visitor Management: Stewardship for Quality Experiences
　　by Chad P. Dawson, John C. Hendee, and Rudy M. Schuster 439
　　Introduction 440
　　Visitor Management Considerations 441
　　Elements of Visitor Use Subject to Management 448
　　Managing for Wilderness Experiences 451
　　Wilderness Visitor Management Approaches 454
　　Indirect and Direct Management Techniques 460
　　Other Management Tools and Techniques 475
　　Use Simulation Models 478
　　Summary 480
　　Study Questions 482
　　Acknowledgments 482
　　Case Discussion: Relationships with Wilderness 483
　　Case Discussion Questions 483
　　References 484

VI. The Future
　　Chapter 17. Future Issues and Challenges in Wilderness Stewardship 487
　　　　Wilderness—Past, Present, and Future 487
　　　　A Growing Importance of Wilderness 488
　　　　Trends in Wilderness Use 490
　　　　Management Funding and Resources 491
　　　　An Expanding—but Ultimately Limited—Wilderness System 492
　　　　Conclusion 493
　　　　References 494

Appendix A. The Wilderness Act *495*
Appendix B. Chronological Listing of Wilderness Designation and
　　Wilderness-Related Laws *501*
Glossary *507*
Abbreviations and Acronyms *513*
Index *515*
About the Authors *526*

Publisher's Preface

by Vance G. Martin, President,
International Wilderness Leadership (WILD) Foundation

Wilderness is essential to a well-balanced natural-resource program. We need wild places with naturalness and solitude and where biological evolution can proceed naturally to produce the genetic diversity on which all life and societies are built. These remote and relatively unspoiled areas provide important research opportunities and benchmarks for monitoring changes in the Earth's climate and natural systems; they provide clean water and important habitat for wildlife. Increasingly, wilderness areas are popular for recreation and personal growth and healing that generate economic benefits in addition to social and spiritual values. Wilderness areas are valuable; however, if we do not manage them wisely, their value will decline as surely as if they were never designated.

This book presents state-of-the-art information about wilderness management by leading authorities in the field, building on three previous editions, in 1978, 1990, and 2002 respectively, and the contributions and reviews over the years by dozens of collaborators. The book traces the history, philosophy, legal basis, and processes for designating wilderness and accepted practices for managing the National Wilderness Preservation System. It looks beyond U.S. boundaries at the growing international recognition and application of the wilderness concept, and the need for stewardship and protection of its many values wherever wilderness exists.

The U.S. Forest Service, through the Government Printing Office, published the first edition of *Wilderness Management* in 1978; a revised, second edition was published by North American Press of Fulcrum Publishing in 1990; and a revised, third edition was published by Fulcrum Publishing in 2002.

Wilderness management is a growing focus of four federal agencies: the U.S. Forest Service, National Park Service, Bureau of Land Management, and U.S. Fish and Wildlife Service. They are joined by many state agencies that also manage wilderness or similar land classifications in the United States and, increasingly, by natural-resource agencies in other countries.

This book is meant for everyone who is concerned about wilderness—land managers, scientists, wilderness users, teachers, students, citizens, environmentalists, natural-resource developers, outfitters and guides, consultants, planners, and policy makers in the United States and other countries. We all have a stake in the future of wilderness.

The WILD Foundation

In recognition of the biological, economic, cultural, and spiritual values of wilderness, The International Wilderness Leadership (WILD) Foundation was established in 1974. Its worldwide goals are to increase the understanding, protection, and sustainability of wilderness and wildlands, and to inspire an ecological conscience among current and future leaders. For more information on WILD, contact WILD president Vance G. Martin at vance@wild.org or visit the website at www.wild.org.

Foreword

by National Wilderness Coordinators, U.S. Federal Agencies, and Other Wilderness Leaders

The 1964 U.S. Wilderness Act stands the test of time, virtually unamended in forty-four years. Today, there are more than 700 wilderness units totaling more than 107 million acres in the United States National Wilderness Preservation System (NWPS) under stewardship of the Bureau of Land Management, U.S. Fish and Wildlife Service, U.S. Forest Service, and National Park Service. Twelve states within the United States also manage some land for wilderness purposes. More than a dozen other countries also have wilderness systems designated by law or administrative policy. We are challenged to manage these wild and natural places as the Wilderness Act directs: "to keep them unimpaired for future use and enjoyment as wilderness." Even though political decisions remain about designating additional land as wilderness, responsibilities for stewardship of existing wilderness continue to grow.

The toughest challenge is to keep wilderness wild, and (as stated in the Wilderness Act of 1964) "affected primarily by the forces of nature, with the imprint of man's work substantially unnoticeable." Wilderness is at risk from external threats, such as global warming and invasion by exotic species, and internally, due to recreational and other wilderness uses.

Wilderness stewardship seeks to maintain the wildness of wilderness while at the same time making it accessible for appropriate human use and enjoyment. Management and use of range, minerals, wildlife and fish, water, and other resources in wilderness must be compatible with the wilderness concept. Fire, insects, disease, and other natural forces must be allowed to play their ecological roles within wilderness as long as they do not threaten human life, resources, and properties outside the wilderness boundary. Increasingly, the need to allow and even encourage natural fire regimes is a challenge. As wilderness and land development and global environmental changes continue, wilderness can provide reservoirs of gene pools and natural ecosystems, and be a benchmark against which to compare humankind's impacts on land and ecosystems.

Wilderness is a "special place," somewhere to experience solitude, unconfined recreation, and wild natural surroundings. Each visitor has a special responsibility to not degrade the wilderness experience of other visitors or wilderness resources. Wilderness users need to understand and abide by a Leave No Trace ethic, where each is a "visitor who does not remain." Wilderness management must be "light on the land," striving to use minimum tools, and doing only necessary wilderness tasks.

Thirty years of wilderness progress and action have passed since the first edition of this textbook was published in 1978, eighteen years since the second edition in 1990, and seven years since the third edition in 2002. U.S. wilderness acreage has increased dramatically since the Wilderness Act was passed, and it has become increasingly clear that protecting wilderness requires more than just setting land aside and leaving it alone. Wilderness management, the stewardship and protection of resources and values, is essential. It is the breadth of these wilderness stewardship challenges and opportunities that this book addresses.

We are pleased to jointly introduce this fourth edition of *Wilderness Management: Stewardship and Protection of Resources and Values*. We expect that it will help wilderness stewards and users, resource policy makers, and concerned citizens better care for enduring wilderness resources in the United States and worldwide.

Karen Taylor Goodrich,
Chair, Inter-Agency Wilderness Policy Council,
Associate Director, Visitor and Resource Protections,
National Park Service, U.S. Department of the Interior

Bill Meadows,
President, The Wilderness Society
Washington, D.C. USA

Carl Rountree, Director,
National Landscape Conservation System and
Community Partnerships,
Bureau of Land Management,
U.S. Department of the Interior

Andrew Muir,
Chair, The Wilderness Foundation, South Africa

Greg Siekaniec,
Assistant Director,
National Wildlife Refuge System,
Fish and Wildlife Service,
U.S. Department of the Interior

Vance G. Martin,
President, The WILD Foundation
www.wild.org

Chris Brown,
Director for Wilderness and Wild and Scenic Rivers,
U.S. Forest Service,
U.S. Department of Agriculture

History, Authorship, and Sponsorship of the Book

by Chad P. Dawson and John C. Hendee

This book was first conceived in 1971 at a U.S. Forest Service–sponsored symposium, "Management Implications of Wilderness Research," where managers, researchers, and environmentalists discussed mutual concerns for wilderness management and the need for new and relevant information. The first edition of *Wilderness Management*, coauthored by John C. Hendee, George H. Stankey, and Robert C. Lucas, with many collaborating chapter coauthors and reviewers, was published in 1978; a revised second edition by the same coauthors was published in 1990. A third edition, by John C. Hendee and Chad P. Dawson in 2002, retained relevant material from the first and second editions but was rewritten to incorporate new changes and materials appropriate under the now broader title *Wilderness Management: Stewardship and Protection of Resources and Values*. This fourth edition, with Dawson assuming senior authorship, continues the evolution of content in response to expanded wilderness knowledge, challenges, and its application as a land-use designation in the United States and abroad.

We thank and recognize our colleagues George Stankey and Bob Lucas for their major coauthorship and contributions to the first two editions. Their contributions to wilderness are many and deep and still influence this book. Likewise, we thank the dozens of agency managers, scientists, university and environmental colleagues, and contributing chapter authors, coauthors, and reviewers for their contributions to this and earlier editions. The book, the entire field, and all of us in wilderness work have changed, broadened, and evolved through the collective efforts of many.

The first edition of this book was published by the U.S. Department of Agriculture, U.S. Forest Service, through the Government Printing Office. The International Wilderness Leadership (WILD) Foundation sponsored and financed the second and third editions; the second edition was published by the North American Press of Fulcrum Publishing and the third edition was published by Fulcrum Publishing. These arrangements continued for this fourth edition. WILD retains all net proceeds from book sales for its wilderness leadership programs.

Like the first three editions, this fourth edition is directed toward wilderness stewards everywhere, especially in U.S. federal land management agencies—the U.S. Forest Service, National Park Service, U.S. Fish and Wildlife Service, and Bureau of Land Management—and toward teachers, students, resource managers, citizens, conservationists, scientists, and wilderness users. The book's focus is updated by embracing current literature, experience, research, and emerging events, including major additions of wilderness areas and acreage, substantial legislation, eight World Wilderness Congresses, several national wilderness conferences, fifteen years of the *International Journal of Wilderness*, www.wilderness.net, and other expanding Internet sources for wilderness information. Because wilderness literature has expanded severalfold since the first and second editions, we've tried to be less encyclopedic, citing the most important new studies and classic ones, but presenting now accepted practices and concepts without extensive citations. Each chapter has been revised, reviewed, and edited, new material added, and other material deleted. Some appendices on wilderness-related laws and wilderness-area acreages were removed from this fourth edition to reduce the size of the book and now reside on the WILD Foundation website (www.wild.org), or they are available online at www.wilderness.net.

Our case examples and background material represent the expertise of all the U.S. agencies and, although we recognize specific policy differences among the agencies, we have sought a level of presentation that transcends agency differences. This is not a book constrained by any agency policy—our goal is to inspire and guide creative stewardship of the National Wilderness Preservation System independent of differences in current organizational perspectives, policies, or practices. Although each agency is directed by internal policies and federal guidelines, all operate under the Wilderness Act and all have high aspirations for preserving designated wilderness under their management. The expanded and updated coverage of international wilderness seeks to inspire protective designation, management, and stewardship of wilderness values wherever possible around the globe.

We have sought the most up-to-date coverage of all topics, supplementing our expertise with collaborating authorship and coauthorship of chapters or sections of chapters by colleagues who are experts in these areas. Chapter coauthors include Roderick Nash on "Historical Roots of Wilderness Management," Chapter 2; Vance G. Martin and Alan Watson on "International Wilderness," Chapter 3; Chapter 10 by Gregory Aplet, and Jerry Franklin on "Wilderness Ecosystems"; David Mattson with John C. Hendee on "Wildlife in Wilderness: A North American and International Perspective," Chapter 12; David Cole, Chapter 15, on "Ecological Impacts of Wilderness Recreation and Their Management"; and Rudy M. Schuster with Chad P. Dawson and John C. Hendee on "Wilderness Visitor Management: Stewardship for Quality Experiences" for Chapter 16. Contributions of specific sections of chapters, or noteworthy involvement of colleagues in their preparation, are identified as appropriate.

All parts of the book bear the imprint of the two book coauthors, either writing new material or editing contributed material as necessary for continuity with the rest of the book. Wherever the word *we* appears (outside of direct quotations), it refers to a consensus opinion held by us.

We are grateful to our many colleagues in universities, the U.S. wilderness management agencies, international colleagues, and the conservation community who provided encouragement and constructive review for the first three editions of this book. We have not included a detailed list of reviewers and contributors for previous editions, but our appreciation continues for their participation in the evolution of this book over a span of thirty years. At the end of each chapter, we do acknowledge colleagues who contributed information or provided review for that chapter in this fourth edition.

To our reviewers, we offer sincere gratitude—their review and input were invaluable, but they share no responsibility for our interpretations or statements. We thank the staff at Fulcrum Publishing for their work to help us make this book as good as we can, especially Patty Maher. We give special thanks to Vance G. Martin and The WILD Foundation for three times sponsoring the book's production, and to Bob Baron, president of Fulcrum Publishing, for three times accepting the challenge of publishing this textbook.

Authors' Preface

Wilderness is the topic of a substantial and growing literature. Hundreds of academic and commercial books, book chapters, journal and magazine articles, reports and brochures, and numerous Internet sources describe *wilderness values* (the rationale for preservation, public appeals, political movements), *wilderness places* (histories, geographies, noteworthy areas, biographies of people intimately linked with specific areas), and *wilderness activities* (how, where, and what-to-do and not-to-do guides). Faced with such a bewildering selection of wilderness material, a reader could easily conclude that everything worth knowing about wilderness has already been written. But on the topic of wilderness management—stewardship and protection of resources and values—readings are in shorter supply, but growing. Coverage of wilderness management is widely scattered among short articles, research reports, graduate theses, Internet sources, and government publications and manuals. The goal of this textbook is to complement this extensive literature with a comprehensive and synthesized overview identifying principles and concepts.

Today, wilderness management is more important and complex now that the National Wilderness Preservation System has grown to more than 107 million acres. Although a majority of the wilderness designation decisions in the United States have been made, the wilderness system will still grow, perhaps by another 10 to 20 million acres. But once decisions have been made to designate an area as wilderness, the long-term preservation of those values that originally led to its wilderness designation will depend on management—the stewardship and protection of its resources and values.

The book's seventeen chapters bring together important information, established concepts, and examples pertaining to wilderness designation and stewardship. The information includes management philosophy, principles and concepts, research findings, and management experience from all four of the federal agencies managing wilderness, from state agencies, and from several other countries.

The book's specific objectives are:

1. To describe the evolution of the U.S. National Wilderness Preservation System: its philosophical, historical, and legal origins; its current size, number of areas, and distribution; and its probable future. We also describe state wilderness in the United States and some wilderness systems in other countries.
2. To provide a common reference for managers, students, scientists, educators, and citizens, who must work together to steward the U.S. National Wilderness Preservation System, state wilderness systems, and wilderness in other countries to protect their resources and values.
3. To propose principles and concepts from which management policy and actions to preserve wilderness might be derived and to describe current wilderness management approaches, policies, procedures, and techniques.
4. To introduce readers to pressing wilderness management issues, impacts, the implications of alternative methods of dealing with them, pertinent literature, current problems, solutions, and research.
5. To describe differences and interrelationships between wilderness designation and wilderness management and between management of wilderness and management of contiguous non-wilderness lands.

We recognize that readers harbor diverse views about wilderness designation and management, and we do not expect universal agreement with our treatment of this value-laden topic. We have tried to avoid polarity of opinion about wilderness and have attempted to maintain a broad, conceptual perspective. *Rather than offer stock solutions, we have tried to identify alternatives and their implications, and to cite examples where possible. Where we do advocate specific action, we have tried to clearly state our position and our reasons.* Although public and private agencies and many individuals continue to hold various notions about wilderness and its management, we are gratified by an increasing convergence of views on wilderness stewardship while recognizing that seemingly incongruous political pressures are ever present. We hope this book will stimulate even more discussion and encourage consensus to meet the challenges of wilderness designation, stewardship, and management that face governmental agencies, managers, and citizens in the decades ahead.

Organization of the Fourth Edition

This fourth edition follows the same mega-organization as earlier editions. This framework is designed for a logical flow through the subject matter, but each section and chapter stands alone enough to invite reading and study independent of the rest of the book. *Wilderness Management: Stewardship and Protection of Resources and Values* contains seventeen chapters divided into six sections described as follows.

I. The Setting

Chapter 1, "Wilderness Management: Philosophical Direction," describes the differences and relationships between wilderness designation and management, the need for wilderness, the philosophical and pragmatic bases for wilderness management, and the need for commonsense policies to meet today's needs. The second chapter, with Roderick Nash, professor emeritus of history and environmental studies, University of California, Santa Barbara, and new material by the coauthors, explores the history of the wilderness management idea and the beginning of its acceptance in the United States. Chapter 3 by Vance G. Martin, president of The WILD Foundation, and Alan Watson, of the Aldo Leopold Wilderness Research Institute, reviews international concepts and systems of wilderness preservation, and describes the growing international interest in wilderness and the nations beyond the United States that have established systems or policies for protecting wilderness or comparable areas and values.

II. U.S. Legal Authority and Process for Wilderness

Three chapters explain the enabling legislation and subsequent status of the nation's wilderness system. Chapter 4, "The Wilderness Act: Legal Basis for Wilderness Management," explains the background and direction of the Wilderness Act of 1964. Chapter 5, "Management Implications In the Wilderness Designation Process, Wilderness Designation Acts, and Wilderness Litigation," explains the evolution of wilderness designation process in the federal agencies and reviews important management direction contained in the wilderness designation laws to date and the implications of litigation related to those wilderness laws. Chapter 6 describes the current "National Wilderness Preservation System and Complementary Conservation Areas," such as wild and scenic rivers and state wildernesses, and their interrelationships and potential future.

III. Wilderness Management and Planning Concepts

Three chapters identify broad directions and concepts for managing wilderness and protecting its associated values. Chapter 7 describes general "Principles of Wilderness Management"; Chapter 8 discusses the importance of "Wilderness Management Planning," reviews current planning policies and processes of the various agencies, and offers suggestions for completing plans and making them work; and Chapter 9 reviews the idea of sustainable carrying capacity by "Managing for Appropriate Wilderness Conditions" through a "Limits of Acceptable Change Process."

IV. Wilderness Resources, Values, and Threats to Them

Four chapters explore aspects of wilderness resources that must be managed: ecosystems, fire, wildlife, and current threats to wilderness. Chapter 10, "Wilderness Ecosystems," by Greg Aplet, ecologist for The Wilderness Society, and Jerry Franklin, professor of Forest Resources at the University of Washington, Seattle, describes basic concepts that control wilderness ecosystems (and all ecosystems) with notes on how this relates to today's ecosystem management emphasis for all federal lands. Chapter 11, "Fire in Wilderness Ecosystems," discusses the natural role of fire in wilderness, the need for management to maintain or restore this role, the methods used, and progress toward this goal. Chapter 12, "Wildlife in Wilderness: A North American and International Perspective," with David Mattson of the U.S. Geological Survey, reviews wildlife values, relationships, and problems in North America, and suggests some management objectives and guidelines for wildlife in wilderness. Chapter 13, "Potential Threats to Wilderness Resources and Values," summarizes the nature and degree of nineteen threats to wilderness.

V. Wilderness Uses and Their Management

Three chapters address visitor use and how to manage it. Chapter 14, "Wilderness Use and User Trends," discusses the various uses of wilderness and their current and projected levels of use. Chapter 15, "Ecological Impacts of Recreation and Their Management," by David N. Cole, research biologist at the Aldo Leopold Wilderness Research Institute, Missoula, Montana, discusses the ecological impacts of recreational use and how to manage them, based on his extensive research on the subject and related new evidence. Chapter 16, "Wilderness Visitor Management: Stewardship for Quality Experiences," with Rudy M. Schuster, explains direct and indirect approaches to managing wilderness visitors to produce and protect wilderness experiences.

VI. The Future

A short, concluding Chapter 17 identifies some current issues and potential future problems and opportunities for wilderness. Foremost is the need for continued integration of thinking and actions by wilderness managers, educators, scientists, and citizens on how to address the challenges of protecting wilderness resources and values for the benefit of future generations.

Italics for Emphasis

We identify with italics those key passages that synthesize and/or generalize key concepts and/or relationships. Our goal with this emphasis is to identify key material for browsing readers and for those reviewing material.

Terminology

Throughout the book, unfamiliar terms, or words used in a special way, are explained where they first occur in the text, and subsequent uses are without additional explanation. Readers may refer to the Glossary and Abbreviations and Acronyms for definition of unfamiliar terms and abbreviations.

Before 1964, the term *wilderness* was used to describe humanity's changing perception of unknown areas or lands modified primarily by natural forces. But after 1964, although retaining this historic and familiar meaning, the term has acquired a new, more precise legal definition. In the United States, the word *wilderness* now refers to areas designated by Congress as units of the National Wilderness Preservation System. This legal designation sometimes prompts reference to "Capital 'W' Wilderness" for legally designated wilderness and "little 'w' wilderness" for roadless land that is not yet, but could someday be, designated (i.e., "de jure" versus "de facto" wilderness). For readability and clarity, we have elected to capitalize the word *Wilderness* only when it is used as a proper noun, for example, the Bob Marshall Wilderness, the National Wilderness Preservation System, or the Wilderness Act. All other references to wilderness as a general term or in the legal sense use a lowercase *w*, and this generally applies to discussions of international wilderness.

The term *act* used alone might occasionally raise questions. As it is used in this book, if the context is to a period before January 1975, the term refers only to the Wilderness Act of 1964. After January 1975, the term refers to the general intent or provisions of the Wilderness Act of 1964 *and* the so-called Eastern Wilderness Act of 1975, which extended application of the Wilderness Act to areas in the eastern United States. The term *wilderness designation acts* refers to legislation designating specific areas as wilderness or as wilderness study areas to be evaluated for their suitability to become units of the National Wilderness Preservation System. The term *wilderness acts* refers collectively to all the just mentioned terms. This is usually in some generic reference to congressional direction for the National Wilderness Preservation System that is embodied in the Wilderness Act of 1964, its extension to eastern areas through the so-called Eastern Wilderness Act of 1975, and collectively in all the wilderness laws, some of which often include management direction for specific areas. Appendix B lists all 171 wilderness acts from 1964 through 2007.

Statistical sources for information concerning the National Wilderness Preservation System used in this book include information published by the Aldo

Leopold Wilderness Research Institute and at www.wilderness.net, a website supported by the Aldo Leopold Wilderness Center, the Arthur Carhart National Wilderness Training Center, and the University of Montana's College of Forestry and Conservation's Wilderness Institute. Authoritative sources differ slightly on several statistical characteristics of the National Wilderness Preservation System, including the number of units, their acreage, and the agencies that manage them. Several factors affect these figures and account for the discrepancies: units can be in more than one state; land acquisitions might have changed the acreage; boundaries might have been adjusted; measurements might have been refined; and official mapping might have been completed after publication of the statistics. But these discrepancies, especially concerning unit acreage, are insignificant in assessing total area of the National Wilderness Preservation System, which includes more than 700 units and 107 million acres.

—Chad P. Dawson and John C. Hendee,
December 2008

I. The Setting

Chapter 1
Wilderness Management: Philosophical Direction

Introduction	2
Wilderness and Other Land Uses	3
What Is Wilderness?	4
Historical Origins of the Wilderness Concept	5
Wilderness and the Early American Scene	5
Historical Wilderness Themes and Values	6
Experiential	7
Scientific	8
Symbolic and Spiritual	10
Economics	10
Relating Wilderness Themes and Values to Management	11
Recent Research in Values	12
Wilderness Management and Wilderness Designation	12
The Effects of Wilderness Designation Decisions on Management	13
Interim Management	13
Precedents, Purity, and Nondegradation	14
Designation of Areas with Unusual Management Problems	15
Designation Limiting Uses and Management	16
The Need for Wilderness Management	16
The Management Paradox	16
More Internal and External Threats to Wilderness	17
Past Disruption of Nature	18
Giving Meaning to Wilderness Designation	18
Alternatives to Management?	18
Wilderness Philosophy and Wilderness Management	19
Anthropocentric and Biocentric Perspectives	19
The Anthropocentric Philosophy	20
The Biocentric Philosophy	21
In Support of Biocentric Management	22
Wilderness Naturalness and Solitude	22
Two Qualifications to Biocentrism	24
A Commonsense Policy Is Needed	25
Summary	25
Study Questions	26
Reader Exercise: How Do Your Personal Wilderness Values Rate?	26
Case Discussion Questions: Anthropocentric and Biocentric Values	28
References	29

Introduction

Since 1964, important progress in wilderness preservation has been made in the United States as a result of the combined efforts of conservation organizations, concerned citizens, federal agencies, wilderness users, and the U.S. Congress. *The Wilderness Act of 1964 (U.S. Public Law 88-577) and 171 additional wilderness designation laws (see Appendix B) that have classified more land as wilderness seek to maintain a portion of the nation in a natural and undeveloped condition for future generations.* The U.S. National Wilderness Preservation System (NWPS) includes separate wilderness areas in the national forests, national parks, fish and wildlife refuges, and public lands administered by the Bureau of Land Management (BLM). By the end of 2007, more than 107 million acres in more than 700 areas were protected in the NWPS. States protect another 3 million acres of lands as wilderness under state laws, and several nations besides the United States also protect wilderness—all inspired to some degree by the U.S. Wilderness Act of 1964.

Public interest in wilderness has been strong and consistent over many decades now (Cordell et al. 2003; Scott 2004; Schuster et al. 2005). Wilderness is often treated as an icon of nature and naturalness and portrayed in the visual arts as a primitive and natural landscape with striking physical features like glaciers, mountains, deserts, shorelines, rivers and waterfalls, wildlife, and primeval forests and grasslands. Our interest in nature has been discussed in popular books and reports in recent years. Bird (2007) investigated, through an extensive literature review, the positive linkages between the natural environment, biodiversity, and mental health. The Biophilia Hypothesis of E. O. Wilson has caught the imagination of many with his idea that human development through the ages has resulted in a "human dependence on nature that extends far beyond the simple issues of material and physical sustenance to encompass as well the human craving for aesthetic, intellectual, cognitive, and even spiritual meaning and satisfaction" (Kellert and Wilson 1993, 20). Scientists and educators have expressed concerns that future generations are growing up without that strong connection with nature (Louv 2006).

Our final achievements in wilderness stewardship and preservation are as yet unknown because designating areas as part of the NWPS is only part of the task. As George Marshall, former president of the Sierra Club and brother of wilderness advocate Robert Marshall, noted, "At the same time that wilderness boundaries are being established and protected by Acts of Congress, attention must be given to the quality of wilderness within these boundaries, or we may be preserving empty shells" (Marshall 1969, 14). This is the challenge of wilderness management to formulate and implement programs of stewardship and protection that will achieve the objectives of the NWPS.

Wilderness management is relatively new, having evolved since the early 1960s and passage of the Wilderness Act. Wilderness management is seen as increasingly important because of growing direct pressures from all kinds of wilderness users and indirect human impacts on all lands. In brief, *the goals of wilderness management are to protect a designated wilderness area's naturalness and solitude, that is, its wilderness character.* This is not an easy task while also providing for its use and enjoyment as wilderness by increasing numbers of people engaging in primitive forms of recreation, hunting and fishing in some areas, and by traditional uses such as grazing, mining, and air access in place before wilderness designation. Further, management must accommodate special provisions and exceptions to the Wilderness Act that may be in the wilderness designation legislation for some areas. For example, there are special provisions for wilderness in Alaska, such as snowmobile access and subsistence hunting,

Fig. 1.1. The view southeast from Silver Pass in the John Muir Wilderness, on the Sierra National Forest, California. Once an area is legally designated as wilderness, it joins the National Wilderness Preservation System. Preservation of its wilderness qualities then depends on its management. Photo courtesy U.S. Forest Service.

fishing, and gathering under the Alaska National Interest Lands Conservation Act, or ANILCA (U.S. Public Law 96-487). Increasing wilderness use compromises solitude, and naturalness may be compromised by fire protection, fish stocking, grazing, heavy visitor use, and climate change. *Simply designating an area as wilderness does not ensure its preservation. Enlightened wilderness management and stewardship are needed.*

Probably most of the areas and acreage to be set aside as wilderness in the United States have already been designated, although another 10 to 20 million acres are possible, maybe even likely. Therefore, protecting wilderness values in the United States rests primarily on the careful stewardship of areas already in the NWPS and to a lesser extent on designating more areas. This is a sobering responsibility for natural-resource professionals and wilderness advocates, and future generations will be our critics. Resource professionals are challenged to develop and apply effective and proactive approaches to wilderness management now, rather than when problems are out of control and our efforts can, at best, be reactions to pressing needs. Gaining the support and involvement of wilderness advocates is an important part of the challenge.

Wilderness and Other Land Uses

This book focuses primarily on the management of *legally designated wilderness* areas in the United States on federal lands formally protected by the 1964 Wilderness Act and its extension to lands in the eastern United States by the so-called Eastern Wilderness Act of 1975 (U.S. Public Law 93-622); to public lands administered by the Bureau of Land Management by the Federal Land Policy and Management Act of 1976, or FLPMA (U.S. Public Law 94-579); and, in 1980, to Alaska by the ANILCA (U.S. Public Law 96-487). The NWPS consists of more than 107 million acres of wilderness administered by four federal agencies: one under the Department of Agriculture, the U.S. Forest Service (USFS), and three under the Department of the Interior, the National Park Service (NPS), the U.S. Fish and Wildlife Service (FWS) and the Bureau of Land Management (BLM).

Our goal is to further the protection and appropriate use of legally designated wilderness areas through their proper stewardship and management. We have not addressed the issues of how much wilderness should be set aside, where it should be located, how large an area should be, and so forth. We believe that such issues are primarily political and will be decided largely through the political process. Resource managers can best facilitate that process by helping to define and assess potential land-use alternatives for areas proposed as wilderness. Managers' recommendations are useful when they include their expert knowledge, data, and analyses. Providing factual analysis of the consequences and outcomes of proposed alternatives is a key responsibility for natural-resource managers.

We are fortunate in the United States to have a variety of environments, ranging from paved and densely populated urban areas to uninhabited primeval forests and deserts. That range can be described as being along an environmental modification spectrum (EMS). Providing a full spectrum of natural environments is a desirable response to the broad range of tastes and interests in any society. At one end of the spectrum are cities, the most modified of environments; at the opposite end are the least modified environments and most primeval, wildernesses. Considering landscapes for recreational purposes leads to the parallel concept of a recreation opportunity spectrum (ROS) that includes urban and suburban recreation areas, rural countryside, highly developed campgrounds, intensively managed private lands, publicly owned multiple-use forests, national parks, recreation and scenic areas, roadless wildlands, and wilderness. Through the Wilderness Act, U.S. society has chosen to preserve, unimpaired for future generations, as wilderness, a selection of our least modified environments and landscapes at the primeval end of the spectrum.

Wilderness, as the least modified extreme on the spectrum of land uses, is sensitive to a variety of uses, including those on adjacent public and private lands. Thus, wilderness managers must consider the compatibility of uses within a given wilderness and also respond to influences and interrelationships among uses and management on adjacent lands.

Moreover, because U.S. federal agencies that manage wilderness—the NPS, FWS, BLM, and USFS—also manage lands devoted to many other purposes and activities, wilderness management must be carefully coordinated with other agency and interagency land-management plans and the laws that direct them.

While most of this book focuses on wilderness managed by federal agencies in the United States, the concepts are relevant to wildernesses designated by states (see Chapter 6), and by law, regulation, or policy in other countries (see Chapter 3). Thus, by outlining wilderness stewardship and management philosophies; approaches, ideas, and concepts; and some policies, procedures, actions, and likely consequences based on experience and research, *we anticipate that the book will be useful not only to federal and state wilderness managers in the United States, but also in other countries where wilderness is protected or contemplated.*

What Is Wilderness?

What is wilderness? This crucial question affects all wilderness designation and management decisions. *Legally*, wilderness can be defined narrowly as *an area possessing qualities outlined in section 2(c) of the Wilderness Act of 1964. Socially*, it is whatever people think it is, potentially the entire universe, the terra incognita of people's minds. We can call these two definitions *legal wilderness* and *sociological wilderness*. Deriving a universally accepted definition of sociological wilderness seems unlikely because perceptions of wilderness vary so widely. For example, for some urbanites with scant outdoor experience, wilderness might be perceived as any relatively undeveloped wildland, uncut forest, or woodlot. In addition, as illustrated in Chapter 3, people from different cultures and countries may have widely different perceptions of what qualifies as wilderness.

On the other hand, legal wilderness as defined by the Wilderness Act, section 2(c) is much more precise: "*A wilderness, in contrast with those areas where man and his own works dominate the landscape, is hereby recognized as an area where the earth and its community of life are untrammeled by man, where man himself is a visitor who does not remain.*" This legal definition places wilderness on the "untrammeled" or "primeval" portion of the EMS. Furthermore, it is sanctioned by the traditions of land use in America and rests on ideas espoused decades ago. For example, Aldo Leopold (1921, 719) envisioned wilderness as "a continuous stretch of country preserved in its natural state, open to lawful hunting and fishing, devoid of roads, artificial trails, cottages, or other works of man." Robert Marshall (1930, 141) offered a similar definition:

> I shall use the word wilderness to denote a region which contains no permanent inhabitants, possesses no possibility of conveyance by any mechanical means and is sufficiently spacious that a person in crossing it must have the experience of sleeping out. The dominant attributes of such an area are: First, that it requires any one who exists in it to depend exclusively on his own effort for survival; and second, that it preserves as nearly as possible the primitive environment. This means that all roads, power transportation and settlements are barred. But trails and temporary shelters, which were common long before the advent of the white race, are entirely permissible.

Because this book is primarily about managing legal wilderness on federal lands in the United States, our definition of wilderness mirrors that outlined in the Wilderness Act. This definition prescribes conditions for areas included in the NWPS and indicates the purposes that management programs for these areas are designed to achieve. The Wilderness Act lends both quantitative and qualitative substance to the traditionally elusive question, What is wilderness?

Although recognizing that wilderness has taken on added precision in its legal definition, we should not forget the evolution of the concept. *Wilderness is still largely a social and political phenomenon of twentieth-century North America. Recently, however, its recognition has spread internationally.* Chapter 3 describes in greater detail the status of such lands elsewhere in the world and the cultural uniqueness of wilderness in any country. Even in the United States, the perception of wilderness has evolved from that of a forbidding landscape to that of a valued cultural, economic, social, and environmental resource (Cordell et al. 2003; Cordell, Bergstrom et al. 2005). For example, substantial economic and social values are often tied to the protection of wilderness in particular vicinities that make nearby locations attractive places to live or to visit as a recreationist or tourist (Power 1996a, 1996b; Rudzitis 1996; Holmes and Hecox 2004; Green et al. 2005), to conserve biodiversity (McNeely 2007), and as sources of ecological services such as clean air and water, climate modification, and more (Patterson 2007).

Our intent is not to trace in detail the origins of the word *wilderness* and its cultural evolution, or to annotate the extensive literature about the many values and philosophies of wilderness. Many other books and articles do this; even a complete listing is beyond the scope of this book. However, *readers should understand the origin of the term wilderness and the diversity of human values associated with it. This is necessary to understand the Wilderness Act and to be able to relate it to ways in which these values can be depreciated or enhanced through wilderness management or stewardship.* Thus, the remainder of this chapter briefly reviews the origins of the wilderness concept and basic values that underlie its legal definition, and management or stewardship, because we use these terms interchangeably.

Historical Origins of the Wilderness Concept

A strong religious flavor influenced the early origins of the word *wilderness*. The word appeared in the fourteenth-century English translation of the Bible from Latin and was used as a synonym for uninhabited and arid lands of the Near East (Nash 2001). Lands described as wilderness were virtually uninhabited, desolate and arid, and vast (Outdoor Recreation Resources Review Commission 1962). In such lands, people could not long survive.

The inhospitability of these lands was due to low precipitation, and because climatology was poorly understood, such lands were perceived as evidence of God's displeasure. Wilderness was a cursed land, and when the Lord set out to punish, through act or parable, the wilderness was often the setting—witness the fate of Adam and Eve after being driven from the Garden of Eden. Conversely, the greatest blessing to be bestowed on humanity was to transform the wilderness "to make it blossom like a rose."

The experience of the Israelites reinforced and added another dimension to the Judeo-Christian notion of wilderness. Wilderness was not only the setting for their forty-year-long wanderings, the Lord's punishment for their misdeeds, but a place where they could prove themselves worthy of the Lord and, subsequently, the Promised Land. *The wilderness, thus, was a place where one might purge and cleanse the soul to be fit in the sight of God.* Jesus' forty days in the wilderness, fasting and resisting the temptations of Satan, were preparation for speaking to God (Nash 2001).

Wilderness, then, in early Judeo-Christian thought, was the place of punishment and penitence. However, even if wilderness had been seen as a place to enjoy oneself, early Christians would hardly have allowed themselves the luxury of a pleasure trip into the mountains. The mission of early Christians was to forgo worldly pleasures and seek salvation.

In its origins, Christianity was a highly human-centered religion (White 1967). God created Adam in his own image, and humans stood distinctly apart from nature. However, Christian tenets gradually evolved until it became inappropriate or, at least, unnecessary to insist on the dichotomy of humankind versus nature. The first proponents of this view were regarded as heretics. St. Francis of Assisi, who insisted that animals, too, had souls, was excommunicated. Rather than interpreting natural phenomena such as storms, appearance of islands, and earthquakes as evidence of God's wrath, people came to see nature as a revelation of His handiwork. *Eventually, with the rise of the physical and natural sciences, people began to associate wilderness with inspiration, not terror, and to explain wilderness and other natural phenomena based on science rather than theology.*

This new appreciation of nature was reflected in many ways. For instance, the symmetrical, formal gardens commonly found in the late 1700s, such as the Garden of Versailles, gradually gave way to more informal, pastoral, and natural settings. In art and literature, the wild, turbulent panoramas of the Alps became favorite scenes. Some literary heroes were the persons who knew how to live in harmony with nature (McCloskey 1966). *This gradual evolution of thinking about relationships of humans to their environment represented an important precondition to the recognition of wilderness as a source of human values and to the eventual development of programs for its preservation.*

Wilderness and the Early American Scene

When the first European explorers reached what is now the conterminous forty-eight United States, they found a continent that was inhabited by Native Americans, whose lands, in contrast to the more dense European level of human habitation and cultivation, were considered undeveloped wilderness. In less than 500 years, the 1.9-billion-acre former wild landscape has been reduced by 98 percent. However, concurrent with its diminishing size has been the increase in appreciation of wilderness values. McCloskey (1966) and Roderick Nash (2001) argue that scarcity of wilderness is a necessary precondition for recognizing its value (see Chapter 2), and that what we once were surrounded by, we now dominate. Another hypothesis by Oelschlaeger (1991) is that humans evolved from nature and wildness as a formative source of human existence and that we appreciate and value wilderness as a primordial source of humanity that is expressed in culture and society.

The wilderness was considered a barrier and a threat to sixteenth- and seventeenth-century settlers. It hindered movement and it frequently possessed little that could directly help settlers prosper. However, as settlements by European immigrants rolled across the landscape, a movement was begun to retain some unmodified lands for perpetuity. *Nash (2001) has argued that interest in maintaining these wildlands was motivated in part by the desire to lend a distinctive quality to American culture. Although literary and artistic accomplishments were almost nonexistent in the young nation, it did possess one thing for which there was no European counterpart: wilderness.* Thus, even while strong motivations existed to conquer the wilderness, there were also stirrings of opinions that valued its retention.

One early observer who foresaw the need for long-term protection of the natural environment was George Catlin (1796–1892), a nineteenth-century lawyer, painter, and student of American Indians. Following a series of trips through the northern Great Plains, Catlin concluded that rapid slaughter of the buffalo, the deterioration of Indian cultures as they collided with expanding white culture, and the general disappearance of the primitive landscape were losses American culture could ill afford. Thus, in the early 1800s, he called for establishment of *"a nation's park, containing man and bear, in all the wild and freshness of the nature's beauty!"* (Nash 1982, 101).

Catlin's remarks received little attention, but the seed was planted. In 1858, writing in *Atlantic Monthly*, Henry David Thoreau (1817–1862) asked, "why should not we have our national preserves in which the bear and panther, and some even of the hunter race, may still exist" (as cited in Nash 1982, 102). Thoreau spoke primarily for the view that credited wilderness with the values most important in the molding of humankind, a perspective summarized in the oft-cited statement attributed to him, "in wildness is the preservation of the world." (Nash 1976). He provided a philosophical framework for literary and artistic expressions of feelings about wilderness as a uniquely American asset, the one attribute of the new nation that made it superior to the tired, settled lands of the Old World (Evans 1981). Gradually, this idea became a great force among the leaders of American thought and culture.

More than a century passed between the warnings of Catlin and Thoreau and passage of the Wilderness Act. During that time, numerous advocates argued for the preservation of some of the remaining wilderness landscape, and some important governmental laws indicated increasing recognition of the importance of reserving lands for public purposes. Yellowstone National Park was established in 1872, the first in a long series of important reserves. Earlier, in 1864, the federal government had granted Yosemite valley to the state of California "to hold inalienable for all time" (Ise 1961, n.p.). Although it is almost certain that neither area was intended to remain wilderness (Chapter 2), these actions did set precedents for the federal government to allocate lands for nonconsumptive purposes. Additionally, public policy debates over such issues as roadless areas in national parks transformed into later support for the wilderness movement (Sutter 2002). Moreover, these parks represented official recognition of values and philosophies expressed by people like Thoreau.

Historical Wilderness Themes and Values

The Wilderness Act of 1964 was the product of more than eight years of debate in Congress and multiple legislative proposals (Scott 2004). To fully understand the meaning of that legislation, however, one must reach back into history. The Wilderness Act and the movement leading to it reflect a synthesis of diverse philosophical values, which evolved over many decades (Callicott and Nelson 1998). As McCloskey (1966, 295) notes: "The evolution has blended many political, religious, and cultural meanings into deeply felt personal convictions. Those who administer that law must look to these convictions to understand why the law exists."

The diversity of motives and values among individuals and groups supporting the wilderness movement has been instrumental in its success. We need to understand the appeal of the wilderness idea that inspires such broad endorsement by a majority of Americans and intense commitment by a fervent minority. *The success of wilderness management efforts depends on how clearly we understand these wilderness values and how effectively we protect them.*

The contributions of many wilderness proponents suggest distinct themes around which the wilderness cause has been argued. *Although wilderness may mean something different to each person, four central themes have consistently emerged: experiential, the direct value of the wilderness experience; the value of wilderness as a scientific resource and environmental baseline; the symbolic and spiritual values of wilderness to the nation and the world; and the value of wilderness as a commodity or place that generates direct and indirect economic benefits.* The latter thematic value of wilderness has most recently emerged and is based on the growth of the other values.

Experiential

The wilderness experience is seen as valuable in its own right, and as a factor in forming our national character, part of what makes Americans unique. *Historically, American writers have extolled the closeness to nature, education, freedom, solitude, and simplicity, as well as spiritual, aesthetic, and mystical dimensions of the wilderness experience.*

John Muir, who founded the Sierra Club in 1892, was an articulate and influential early proponent of experiential values of wilderness. Muir, a Scottish immigrant, was raised on a Wisconsin farm. Although a talented inventor, Muir was more intrigued by the ideas he found in science and literature. He was heavily influenced by the writings and philosophies of Thoreau and Ralph Waldo Emerson (Davis 1966–1967). Nevertheless, he found them both wanting in some respects. Muir thought Emerson (1803–1882) had failed to express an appropriate amount of excitement after hiking in the mountains at Yosemite, and he was amused that Thoreau, who repeatedly proclaimed, "In wildness is the preservation of the world," could refer to orchards as forests (as cited in Nash 1982).

To Muir, the essence of wilderness was the freedom, solitude, and beauty of the mountains. These qualities, he felt, could satisfy all human needs. The wilderness experience to Muir was spiritual, the forests were temples, and the trees sang psalms. In the Sierra wilderness, "everything seems equally divine, one smooth, pure wild glow of heaven's love" (Muir 1938, n.p.). The wilderness also offered personal insight; during a raging windstorm in the Sierra, Muir climbed a tree and lashed himself to it to experience nature more closely.

The experiential theme was also reflected in many of the writings of Robert Marshall. Marshall was an extraordinary individual. In a brief but fruitful life (he died at thirty-eight), he accomplished much for wilderness. In 1935, along with Aldo Leopold, he helped found The Wilderness Society, and in 1939 as director of Recreation for the U.S. Forest Service, he formulated the U regulations; both actions strengthened protection of roadless areas before passage of the Wilderness Act in 1964. As an extremely enthusiastic hiker, Marshall routinely hiked thirty-five to forty miles a day, and set a goal of walking thirty miles a day in every state (Edwards 1985). To Marshall, wild scenery was similar to great works of art. In a major paper outlining the future of wilderness, he wrote, "Wilderness furnishes perhaps the best opportunity for pure aesthetic rapture" (Marshall 1930, 145). *Marshall believed the restorative powers of wilderness could help prevent moral deterioration.*

Like Marshall, Aldo Leopold was a USFS employee who helped form that agency's wilderness concept. As a young forester in New Mexico following World War I, he became apprehensive about the expansion of USFS road systems into the backcountry he loved. His prompting led to administrative protection of the Gila River country as a wilderness, the first in the nation. Over the years, his wilderness ideas developed and matured, and his *Sand County Almanac and Sketches Here and There,* published after his death in 1948, is a classic in American conservation literature (Leopold 1949).

In the early 1920s, Leopold argued especially hard for wilderness designation because he wanted to preserve a particular kind of recreational experience, the pack trip. Unless steps were taken to preserve large tracts of land, he stated, the day would come "when a pack train must wind its way up a graveled highway and turn its bell-mare in the pasture of a summer hotel. When that day comes the pack train will be dead, the diamond hitch will be merely rope, and Kit Carson and Jim Bridger will be names in a history lesson" (Leopold 1925, 403).

Leopold pressed for preservation of areas that would provide recreational experiences he thought developed both individual and national character (Leopold 1921). Another young forester of the period, Arthur Carhart, agreed with Leopold's philosophy and pressed for protection of an area surrounding Trappers Lake in Colorado. *To Carhart, recreation was not merely an incidental use of forests; it ranked among the highest of all possible uses because of the moral benefits associated with it*: "Recreation in the open is of the finest grade. The moral benefits are all positive.

The individual with any soul cannot live long in the presence of towering mountains or sweeping plains without getting a little of the high moral standard of Nature infused into his being" (Carhart 1920, 268).

Carhart's immediate superior in Colorado was Carl J. Stahl, a firm supporter of the Trappers Lake proposal. Stahl (1921, 529) also saw forests (and wilderness) as a source for strengthening moral values:

> An appreciation of nature, a stimulation of vigor of the mind and body, and the contentment of soul contributed by association with the forests, go far toward making a useful and contented citizenry. If the American population can be made to feel contented and its effort directed to useful channels, enlistment in the Red organizations of this critical period of unrest can be averted. I can conceive of no more useful purpose the forests can be made to serve.

In the 1920s, contact between humans and nature was seen as a character-building activity, and a counter-force to the perceived evil threat of communism.

Today, the experiential theme of wilderness includes personal growth and healing. Now hundreds of wilderness experience programs use wilderness for environmental education, personal growth, and therapy, such as Outward Bound and the National Outdoor Leadership School (NOLS) (Friese et al. 1998; Russell et al. 2000; see Chapter 14). These programs form an important and growing wilderness industry; they aim at growth in desirable personal qualities (e.g., development of self) such as self-esteem and confidence, independence, outdoor skills, improved group skills (e.g., development of community), and enhanced performance through leadership and team building, spiritual development, and environmental education. Some programs are aimed at disadvantaged groups and individuals and help people in crisis or transition find inspiration and new goals. They provide healing for those dealing with the trauma of domestic instability and chaos or abuse, those adjusting to emotional losses such as death and broken relationships, and those fighting chemical dependency and delinquency.

All of these wilderness experience programs derive from a belief that, in the natural environment (ideally in wilderness), away from social pressures, excessive stimuli, and diversions, we can confront our true and deeper selves, identify our values and priorities, and recover a sense of wholeness. This is the *primal hypothesis,* that the experience of wilderness naturalness and solitude yields important benefits to people, including desirable personal qualities (e.g., development of self), group skills and cohesion (e.g., development of community), and spiritual development (White and Hendee 2000). Such programs reflect one of the central beliefs of the founders of our wilderness system: that the character-building values of wilderness are vital to our society (Scott 2004).

Many recreational experiences take place in designated wilderness, and the kinds of physical challenge, naturalness, and solitude individuals seek are increasingly found only in wilderness. Such programs fulfill Aldo Leopold's dream; he believed that wilderness areas should be places where primitive travel and subsistence skills could be perpetuated (Leopold 1921). Today, such programs challenge wilderness management in its dual role of providing opportunity for experiences aimed at human growth, enrichment, and healing, while also protecting the wilderness resources of naturalness and solitude (Gager et al. 1988).

Scientific

Numerous values arise from the ecological and environmental services provided by wilderness (Cordell, Murphy et al. 2005). For example, due to its generally undisturbed environmental setting, wilderness is an important benchmark source of scientific information about the world around us, how it evolved, how the effects of civilization have altered natural systems, and what the unmodified environment holds for us. *Wilderness is a baseline or control area with which to compare change in other world environments.* Areas where natural processes remain intact are increasingly scarce. Wilderness areas are thus valuable assets: as natural baselines that reveal the extent of impacts elsewhere, as sites where scientists can study natural processes, as gene pools maintaining and reflecting the diversity of nature and providing a gene reservoir we are only now learning to use, and as refuges for certain flora and fauna that cannot survive outside wilderness conditions.

The intricate interrelationships among all organisms have been important to the scientific theme. Muir, for example, saw in wilderness a place where people could feel "part of wild nature, kin to everything" (as cited in Nash 1982, 129). Moreover, it was Muir who expressed the fundamental principle of ecology: whenever you pluck up something, you find everything in the universe attached to it.

Leopold saw scientific value in wilderness as a laboratory for the study of land health. *In 1935, Leopold, Marshall, and others founded The Wilderness Society to promote the protection of lands retaining such naturalness* (Leopold 1941). Although Leopold had long recognized historical and recreational values of wilderness, in 1935 his justifications for such areas predominantly turned to ecological and ethical reasons (Flader 1974).

Because wilderness areas have remained relatively undisturbed over long periods, they are reservoirs of genetic materials that have evolved over eons (Cowan 1968). Such gene pools hold answers to questions yet unasked; once lost, they are impossible to replicate. Similarly, it is important to retain species whose chemical and biological makeup might be useful in the future as the sources of important drugs, for example. Large, undisturbed tracts of wilderness are important sources of genetic diversity and havens of stability in animal and plant populations so essential to retaining genetic pools.

Sometimes the relationship of humans to the world around them can be understood only by analyzing biological systems that have escaped human impact. Wilderness offers an important opportunity to examine ecosystems as they have evolved outside human influence. Understanding such evolution can help prevent errors, at best careless and at worst catastrophic, as we shape and modify the rest of the earth to our purposes. As Leopold once noted, the first principle of intelligent tinkering is to save all the parts, ecologically speaking; a sufficiently large and ecologically representative wilderness system will help that effort.

However, the 1964 Wilderness Act did not directly specify biological diversity or ecological integrity criteria as part of the wilderness designation process; thus, some important physical-biological systems are underrepresented or not included in the current NWPS. The USFS's second Roadless Area Review and Evaluation (RARE II) from 1977 to 1979 specified that the nation's ecosystems, landforms, and wilderness wildlife be important criteria for identifying candidate areas for wilderness designation. An analysis revealed that of the nation's 261 ecotypes, only 157 were included in the NWPS by the late 1980s (Davis 1989). More ecosystem representation was needed in wilderness, a call echoed by the then-chief of the USFS in outlining a wilderness strategy for his agency in 1999 (Dombeck 1999). By 2005, Cordell, Murphy et al. (2005) reported that only nine of the fifty-two categories in Bailey's Province-level Ecosystem classification were not represented somewhere in the NWPS; however, they observed that some of those categories were only represented in a few NWPS areas or the areas were small in size.

Does it really matter if a natural system disappears? In a plea for an end to wilderness destruction, E. O. Wilson (1984, 15) gave this example:

> Natural products have been called the sleeping giants of the pharmaceutical industry. One in every ten plant species contains compounds with some anticancer activity. Among the leading successes from the screening conducted thus far is the rosy periwinkle, a native of the West Indies. It is…the kind of inconspicuous flowering plant that might otherwise have been unknowingly consigned to extinction by the growth of sugarcane plantations and parking lots. But it also happens to produce two alkaloids, vincristine and vinblastine, that achieve 80 percent remission from Hodgkin's disease, a cancer of the lymphatic system, as well as 99 percent remission from acute lymphocytic leukemia.

This is just one example. Worldwide the pharmaceutical industry generates tens of billions of dollars annually from drugs owing their origins to material from wild areas.

Wilderness also provides large tracts of unmodified habitat that some threatened species, for example, the grizzly bear and the timber wolf, *need to survive. Such wilderness offers the opportunity to study these species to ensure their maintenance.* Wilderness tracts have served as laboratories for greatly increasing our knowledge of the biota around us, and studies of wildlife in wilderness settings have substantially enlarged our understanding of these animals (see Chapter 12). Similarly, the presence of extensive tracts of undeveloped (wilderness) land has made possible important baseline research on vegetative communities, fire history, and studies of other natural biological systems that simply could not have occurred without such tracts (see Chapters 10 and 11).

Wilderness also provides an important laboratory for scientists concerned with human behavior. How individuals relate to one another, how they react to stress and challenge, and how natural environments affect behavior are important topics for wilderness research. One series of studies over a ten-year period concluded that restored functioning was a key benefit of wilderness experience (Kaplan and Kaplan 1989), and numerous studies document the values of wilderness experience for personal growth and healing dysfunctional behavior (Russell et al. 2000). Further, studies of wilderness recreation users can provide important insights on the experiential values espoused by Muir, Marshall, and others. *One of the practical goals of wilderness-user studies is to gain insight into the economic and social benefits of wilderness experience and how maintaining or enhancing wilderness naturalness and solitude through management can provide or increase them* (White and Hendee 2000).

Symbolic and Spiritual

In a world characterized by rapid change and complexity, both exciting and frightening, wilderness symbolizes comforting stability and simplicity. The existence of wilderness reflects self-imposed limits on the technological imperative that we must subdue all the earth just because we can. *Wilderness is a symbol of respect for the naturalness and solitude that have been displaced by civilization on most of the earth.*

In 1962, The Outdoor Recreation Resources Review Commission (ORRRC) called on Wallace Stegner, head of the Creative Writing Center at Stanford University, to comment on the significance of wilderness as "an intangible which has altered the American consciousness" (Outdoor Recreation Resouces Review Commission 1962, 34). Stegner's reply argued forcefully for the maintenance of wilderness for the sake of survival:

> Something will have gone out of us as a people if we ever let the remaining wilderness be destroyed; if we permit the last virgin forests to be turned into comic books and plastic cigarette cases; if we drive the few remaining members of the wild species into zoos or to extinction; if we pollute the last clear air and dirty the last clean streams and push our paved roads through the last of the silence, so that never again will Americans be free in their own country from the noise, the exhausts, the stinks of human automotive waste.

Wilderness is needed, Stegner concluded, because it is "a means of reassuring ourselves of our sanity as creatures, a part of the geography of hope" (ORRRC 1962, 34).

The extensive writings of Sigurd Olson reveal many important human values derived from wilderness. In a series of books published over two decades, Olson describes wilderness as a source of inspiration, insight, and personal peace: "The singing wilderness has to do with the calling of the loons. It is concerned with the simple joys, the timelessness and perspective found in a way of life that is close to the past" (Olson 1957, n.p.). While exploring the Knife River in the Quetico-Superior country, Olson found himself nearly overwhelmed by the environment around him: "I was aware of a fusion with the country, an overwhelming sense of completion in which all my hopes and experiences seemed crystallized into one shining vision" (Olson 1963, n.p.).

Spiritual development from wilderness experience is defined in more recent writings and studies as a deep sense of connection to all things, such as the larger universe, a higher power, nature, a feeling of oneness, what is referred to as "connection to other" as opposed to "connection to self" (McDonald and Schreyer 1991; White and Hendee 2000; McDonald 2003). Spiritual development as a wilderness benefit has received little attention and study, in part because spiritual experiences are intensely personal and often inexpressible, and because of the varied personal meanings of spirituality that make it difficult to define them operationally. However, some studies do report findings on spirituality in wilderness. For example, Riley and Hendee (2001), in a study of participants in a commercial wilderness vision-quest program over a ten-year period, found that to "go on a spiritual journey" was a leading motive for participating, and "spirituality connectedness" and "connection to nature" accounted for 26 percent of the reported benefits. Virtually all these vision questers also agreed that being in wilderness with naturalness and solitude, as opposed to a developed recreation area with roads and campgrounds, was essential to gaining the benefits they reported.

The accounts of many prominent wilderness writers in addition to Muir, Stegner, and Olson extol spiritual benefits from wilderness experience, and spiritual benefits are beginning to be documented in studies of wilderness experience. *Spiritual values* of wilderness will be a challenge for scholars and a desire of users in the future as wilderness stands in greater and greater contrast to societies that are ever more impersonal.

Economics

Wilderness, wildlands, and natural protected areas provide valuable ecological services and direct and indirect income from recreation and nature-based tourism, and thus economic benefits for local, regional, and national economies (Bowker et al. 2005; Rosenberger and English 2005). *The diversity of economic values of wilderness is a growing social and political force for protecting and managing wilderness.* Public opinion survey studies consistently report that the majority of the public supports setting aside wilderness areas for present and future generations (Cordell et al. 2003; Scott 2004; Schuster et al. 2005). Most of these people will not set foot in a wilderness, but they value wilderness from several perspectives: *for bequest value—protection of natural ecosystems and national heritage for future generations; existence value—protection of natural processes and conditions just to know that there are places that are relatively unaffected by humans; and option values—maintaining the option to either protect or develop an area in the future. These are often referred to as the "nonuse" values of wilderness.*

Recreationists traveling to and from wilderness areas use retail lodging and service businesses, and buy food and beverages to support their trips and wilderness experiences. Similarly, proximity to wilderness as a setting for travel and tourism can bring new income from providing goods and services to tourists who enjoy the wilderness from a distance but may not actually hike there. Such revenues may be considered in economic development strategies for communities near wilderness or states with numerous wilderness opportunities (Moisey and Yuan 1992; Yuan and Christensen 1992; Bowker et al. 2005; Rosenberger and English 2005).

Local residents can also benefit from wilderness by enjoying a high-quality local environment with natural amenities. Because of the possibility of a higher quality of life, pristine conditions and natural environments raise property values and attract people and businesses to areas that may have few other economic opportunities (Power 1996a).

The designation of wilderness has been perceived by some opponents as having negative economic consequences based on the misconception that wilderness designation prohibits maintenance or expansion of local, extractive industries. Power (1996b, 9) outlines these concerns, reviews relevant economic information, and concludes:

> Wilderness protection does not impoverish communities by locking up resources. Rather, it protects the economic future of those communities by preserving high quality natural environments that are increasing in demand across the nation. Wilderness protection does not threaten the ongoing development of nonmetropolitan economies in any significant way. Rather, it lays part of the long-run basis for their ongoing development by providing attractive places to live, work, and do business. Because of this, the economic problem posed by protecting landscapes is not how to maintain local economic health, but almost the opposite: How to keep the economic activity attracted to areas adjacent to wilderness from undermining the environmental quality that wilderness protection seeks to insure in the first place.

Economic analysis of wilderness benefits is evolving as new approaches are applied and better information becomes available (Holmes and Hecox 2004; Patterson 2007). For example, Russell et al. (1998) analyzed the benefits of a seven-day wilderness experience program for at-risk youth in the Federal Job Corps. They concluded that social benefits exceeded costs based on the reduced early terminations of participants and their subsequently enhanced employability and reduced social costs from the higher welfare, criminal behavior, drug and alcohol abuse that are typical of early dropouts. Loomis and Richardson (2001) identified eight categories of wilderness benefits (on-site and off-site) and estimated their dollar values where possible—ecological services were the largest—concluding that the estimated benefits of the NWPS amount to $3 to $4 billion annually.

Clearly, wilderness is extremely valuable, generating benefits we are just beginning to better document and understand. Wilderness values will only increase in the future as improved methods of estimating benefits are applied and as the supply of wilderness diminishes and demand for its services increases.

Relating Wilderness Themes and Values to Management

The preceding categorized some of the basic wilderness values identified by early and contemporary authors under four broad themes: experiential, scientific, spiritual, and economic. Most observers have also embraced many values. For instance, McCloskey (1966) identified eleven wilderness values that have emerged since the arrival of European settlers in North America. Muir's strong experiential philosophy was backed by an intense scientific curiosity. Similarly,

Marshall recognized both the spiritual and scientific contributions of wilderness. Leopold's ethical and scientific perspective was complemented by a well-developed appreciation for the recreational values of wilderness. Studies of wilderness visitors show that contemporary wilderness users also identify these and many other values in wilderness.

McCloskey's statement (1966) that *to understand the Wilderness Act, we must understand its historical and philosophical origin*, is the key to this discussion. The philosophies and perspectives discussed earlier increase our understanding of what wilderness is, why we have it, and what values it should provide society. These issues are important in applying the Wilderness Act, both in designating wilderness and in its subsequent management. *Once wilderness designation decisions have been made, the extent to which the values espoused by philosophers like Muir, Marshall, and Leopold are realized will depend on wilderness management. It is essential that managers, educators, and citizens be guided in their efforts by a personal philosophy of wilderness that recognizes the values set forth by both early and contemporary philosophers.*

Recent Research in Values

Bergstrom et al. (2004) have identified a connection between wilderness attributes and human values by using a sequential framework based on four levels: (1) wilderness attributes (e.g., naturalness, wildness, geographic); (2) wilderness functions (e.g., biodiversity preserve; recreation setting); (3) wilderness services (e.g., personal health and growth, animal habitat, scientific discovery); and (4) wilderness values (e.g., social). Their research organizes wilderness values into four categories: social, ecological, economic, and ethical and intrinsic. For example, the social values relate to the noneconomic benefits that accrue to society and individuals from direct use (e.g., recreational activities, experiential use, grazing) and indirect use of wilderness (e.g., existence, bequest to future generations, option for future use, human health and well-being). Social values research includes subjects in the fields of sociology, psychology, anthropology, human health, and others relating to the human experience in wilderness (Schuster et al. 2005; Schuster et al. 2007). All of these four values categories have become the subject of more research as scientists attempt to quantify the value of wilderness and wilderness experiences.

Wilderness Management and Wilderness Designation

There is a fundamental difference between wilderness designation and wilderness management. Wilderness designation *includes all processes and activities of government agencies and interested publics to identify areas for potential protection as wilderness and for their designation by Congress under the Wilderness Act*. Wilderness management *includes government and citizen activity, within the constraints of the Wilderness Act, to identify goals and objectives for designated wilderness areas and the planning, implementation, and administration of policies and management actions to achieve them*. Wilderness management *applies concepts, criteria, guidelines, standards, and procedures derived from the physical, biological, social, and management sciences to preserve naturalness, and outstanding opportunities for solitude or primitive, unconfined recreation, in the stewardship and protection of resources and values in designated wilderness areas.*

In general, the wilderness designation process has typically been that: the managing agency of a national forest, national park, wildlife refuge, or BLM-administered land unit reviews its roadless areas, and with public involvement, identifies lands to be studied more intensively for potential wilderness designation. Often intense public interest in a particular area precipitates the process. Then the agency studies these selected areas, again with public involvement, to determine their suitability, the demand and need for their legal designation as wilderness, and the values and developments such as timber, ski resorts, and mining that would have to be given up. Based on this study, a wilderness proposal is submitted by the agency to Congress for review, evaluation, and possible legislation, to designate the area as wilderness. (See Chapter 5 for more detailed information.)

The congressional wilderness study and review process includes formal public hearings and extensive deliberation in committees as alternative bills are proposed, debated, revised, and perhaps finally enacted to legally designate a specific area as wilderness. During this process, interested groups may submit alternative wilderness proposals. For example, a wilderness study on the Mount Baker–Snoqualmie and Okanogan-Wenatchee National Forests in Washington state led to a USFS wilderness proposal to Congress for a 292,000-acre Alpine Lakes Wilderness, with the addition of 82,000 acres if money were to be allocated to purchase private land. However, the Alpine Lakes Preservation Society proposed a larger, 575,000-acre wilderness surrounded by a 437,000-acre national recreation area. The Alpine Lakes Coalition Society, representing industry and recreation vehicle interests, proposed a 216,000-acre wilderness in two separate parcels. Congress considered all three proposals and, after public hearings and congressional debate, finally passed compromise legislation in 1976 (U.S. Public Law 94-357) establishing a 303,508-acre Alpine Lakes Wilderness. An additional 86,000 acres of intermingled private and public land was defined as "intended wilderness," and Congress authorized funds for acquisition of the 43,543 acres of private land needed to complete the full, intended Alpine Lakes Wilderness. The Alpine Lakes Wilderness was completed in 1980, following years of public involvement between many locals groups and the USFS (Knibb 1982), and included 393,000 acres. A protective management unit of 527,000 acres surrounds the wilderness and is managed by the USFS for multiple-use purposes.

The Alpine Lakes example illustrates the political nature of wilderness designation and the competing demands for use of an area. Congress seeks resource-management proposals and advice from federal agencies during the designation process, but agency views are sometimes given no more consideration than the formal proposals of vested-interest groups. The number of areas and acres proposed for designation by different interest groups might vary substantially from one another. Determining public interest in wilderness designation issues is arrived at through political processes, public debate over alternative proposals, and ultimately decided by Congress based on compromise.

Wilderness management involves applying guidelines and principles to achieve established goals and objectives. The guidelines are first to protect the area as wilderness and implement any special legislative provisions and purposes established for the area in its wilderness designation legislation. Management generally is not as political as wilderness designation, although it has become more so because the process of establishing management objectives, policies, and actions must address the desires of competing users. In some wilderness designation acts, Congress gives special direction for handling a specific management issue, for example, grazing guidelines in the 1980 Colorado Wilderness Act (U.S. Public Law 96-560) and its accompanying committee report. In addition, management challenges facing the responsible agency are a direct function of the attributes of the designated areas. The agency must manage "as wilderness" what it inherits through the designation process. If that process yields heavily impacted areas, with high levels of established or incompatible uses, providing "naturalness and solitude" or other wilderness-dependent values will be more difficult and require more intensive management actions.

The Effects of Wilderness Designation Decisions on Management

Wilderness designation, including its often lengthy process, affects wilderness management in several ways: (1) there must be *interim management* before designation and after designation but before development of a management plan; (2) designation can establish precedents, such as *purity,* and *nondegradation;* (3) designation of areas can include *unusual management problems;* and (4) the designation legislation can prescribe special provisions and *limitations on uses and management.*

Interim Management

Areas under study or consideration for wilderness designation must be managed during the review or study period in a way that does not preclude their being designated by Congress as units of the NWPS. In other words, the decision to *study* an area for possible wilderness

Fig. 1.2. Desert tortoise habitat is on sixty-nine BLM Wilderness Areas designated by the California Desert Protection Act of 1994. The tortoise, however, is at risk on desert roads near wilderness boundaries. Photos courtesy Marilyn and John Hendee.

designation is a decision to temporarily *manage* that area as a wilderness. The management of roadless lands to maintain their wilderness character until Congress decides is based on Wilderness Act directives that only Congress has the right to permit or deny any area's designation as wilderness. If interim management (or lack of management) allows the wilderness character of a proposed area to deteriorate before congressional review, it has effectively reduced the options for wilderness designation by Congress.

While wilderness study and the designation process go on, nonconforming uses or inappropriate or excessive levels of use can become well established and protected by their own advocates. Irreversible damage can result from unmanaged use: concentrations of visitors can compact and erode soils and threatened wildlife dependent on wilderness can be further jeopardized. Correction is always far more difficult, expensive, and controversial than prevention, so *an important interim management task is to prevent damage to areas for which wilderness designation is likely, possible, or pending.*

Thus, roadless areas identified as candidates for possible addition to the NWPS are generally managed and used essentially as wilderness, pending designation decisions. This principle is so well established that a major point of conflict in wilderness designation legislation concerns the *release* of roadless areas from any further consideration for wilderness designation: that is, whether roadless areas, having been considered by Congress but not designated, should be *released* from further consideration, thus allowing them to be used for purposes such as logging or developed recreation that could forever preclude their being designated as wilderness. Following wilderness designation there is a period in which interim management continues before the formulation, acceptance, and implementation of an approved wilderness management plan. Unfortunately, this period can exceed a decade or two as state and federal agencies attempt the management-planning process with limited staff, budgets, other heavy workload demands, numerous other plans to write, diverse user-group demands, and potential litigation from special-interest groups. Two decades ago, Reed et al. (1989) reported that more than 60 percent of federally designated wilderness areas lacked a completed and approved management plan, and management capacities are scarcely better today. The longer a wilderness or study area continues under lax or inappropriate interim management, the more users who experience it might conclude that what they experience is what was intended; hence, the more difficult it will be later to more rigorously protect the resources and provide wilderness experiences (Newman and Dawson 1998).

Precedents, Purity, and Nondegradation

There are concerns that designating wilderness areas with substandard qualities of naturalness and solitude, which thus require special management attention, could establish precedents affecting the designation and management of other areas. This is the so-called purity issue. The fear is that if areas with less pristine qualities are designated as wilderness, the standards used in admitting such areas will set a precedent for designating other substandard wilderness or will be used as a rationale for lowering standards in the more pristine areas already admitted (Allin 1982; Costley 1972). Likewise, if an area designated as wilderness requires special management for some existing feature or established use, purists argue that this could establish a precedent requiring similar management in other wilderness areas. Purity has been a major wilderness designation issue because Congress has increasingly considered and designated wilderness areas with impacts and features that some argued should disqualify them for consideration (such as abandoned roads and railroads, cabins, and homesteads). Now, as new wilderness designation proposals emerge for more impacted areas, the concerns about precedents could focus on past management. If one wilderness contains undesirable conditions, such as intensively developed campsites, or allows aerial fish stocking, relocation of nonnative wildlife, or limited use of motorboats and airplanes, will those lower standards apply to other wilderness areas?

The issue of purity in wilderness designation and management will remain, but *fears of setting precedents have been largely quieted by adopting the principle of nondegradation*—a concept widely used in managing air, water, and noise (Mihaley 1972)—to wilderness (also see Chapter 7).

Under nondegradation, management's obligation is to prevent further environmental degradation of individual areas that meet wilderness standards, while managing to upgrade areas that are below minimum wilderness standards. Thus, no one area sets a standard for another, because each in effect is managed to maintain its own baseline conditions of naturalness, solitude, and other wilderness characteristics, as long as they are above some minimum standard. For example, the presence of high use levels requiring tight management controls in the San Gorgonio Wilderness in California does not justify permitting similar conditions to evolve in the relatively pristine Selway-Bitterroot Wilderness in Idaho and Montana. Likewise, practices essential for managing the typically smaller wilderness areas in the East do not establish standards for western areas that are typically larger. Nor do conditions in heavily used areas near large population centers set precedents for more lightly used, remote areas.

Designation of Areas with Unusual Management Problems

Wilderness designation decisions can affect management if an area is designated as wilderness that presents problems for which technical solutions are limited, unavailable, or infeasible. For example, locating boundaries in vulnerable or unwieldy places (e.g., roadsides and lake shorelines) can make wilderness management difficult, as can the inclusion of private land inholdings or popular locations already affected by high recreational use.

The Wilderness Act explicitly recognized that designation criteria can affect management; in section 2(c) (2), the act prescribes the minimum size of a wilderness area: "at least five thousand acres of land or of sufficient size as to make practicable its preservation and use in an unimpaired condition." Below some minimum size, the act is saying, management to preserve an area's wilderness qualities becomes extremely difficult. Maintaining naturalness and solitude in smaller areas is now a challenge in many wilderness areas in the East where, under the so-called Eastern Wilderness Act (U.S. Public Law 93-622), areas of less than 5,000 acres have been considered "of sufficient size" to be designated as wilderness. In addition, highly irregularly shaped boundaries, boundaries that do not follow easily recognized terrain features, and boundaries that adjoin areas of commodity production or urban development could all accentuate management problems. Agencies have generally attempted to keep wilderness boundaries along ridges or other easily recognized topographical features, arguing that such boundaries make management easier. For example, debate between the USFS and environmentalists over the appropriate location of one section of boundary in the San Rafael Primitive Area tied up the wilderness proposal for more than a year, even though the disputed area involved only 2,000 acres of the nearly 150,000-acre total. Managers favored a ridge-top boundary that facilitated fire protection of the wilderness and reduced its size; environmentalists favored a mid-slope boundary that enlarged the area.

Some areas cannot be managed easily as wilderness for other reasons. Marion Lake in Oregon's Mount Jefferson Wilderness illustrates the difficulties that can occur when an accessible and popular area is included in a wilderness. Marion Lake, an easy one-mile hike from the road, had a long history of heavy recreation use before its inclusion in the wilderness system. Nonetheless, environmentalists strongly supported its inclusion, even though a permanent USFS cabin shared space at the lake with scores of private boats stored along the shoreline. The USFS contended it would be extremely difficult to scale back this established use to a level consistent with that required by the Wilderness Act. Congress disagreed, and the area was included in the Mount Jefferson Wilderness. Despite a substantial reduction in boat storage (accomplished in part by the sale of some abandoned boats at auction and the burning of others), Marion Lake continued to attract heavy use (many recreationists came to fish for trout stocked by the state), heavier use than managers considered appropriate in wilderness. To deal with the undesirable impacts and resulting intensive management required at the lake, Oregon senator Mark Hatfield sponsored legislation to remove it from the Mount Jefferson Wilderness. His proposal, however, failed to attract the support necessary for enactment. Now, more than three decades later, with the NWPS much larger, similar situations continue to arise where easy access to popular locations in proposed wilderness poses an interim and subsequent challenge to wilderness management.

The potential for mining in many USFS-managed and BLM-managed wilderness areas is another complication with which wilderness managers must deal. The Wilderness Act provided for continued prospecting (mineral exploration) until 1983, and operation of valid claims after that date, subject to regulations to protect the wilderness. Some wilderness designation laws withdrew areas from mineral exploration or claim operation. In 1982 and 1983, Congress added legislation that blocked the Secretaries of Agriculture and the Interior from any mineral exploration in wilderness. That legislation remained in effect in later years, serving as imminent protection for wilderness study areas (WSAs) and areas recommended for wilderness should they be designated. Mining claims can be staked on BLM study areas until the areas are designated as wilderness. Valid claims with valuable deposits discovered before the cutoff can be activated in the future. Although BLM and USFS regulations attempt to ensure that mining activities conform to the mandates of the act, balancing the legal rights of miners with the preservation of wilderness values presents managers with yet another complicated challenge.

Designation Limiting Uses and Management

Designating an area as wilderness limits the range of uses and management alternatives and techniques. Areas managed as wilderness will offer only part of the ROS the area might otherwise provide. For example, motorized recreational activities such as trail bikes and snowmobiles generally are not allowed. Certain management activities are likewise generally restricted by wilderness designation, such as the use of motorized equipment and mechanical transport, although some exceptions are provided in certain wilderness designation laws.

The issue of what management actions, facilities, and equipment are allowed in wilderness was an important point of debate in the Endangered American Wilderness Act of 1978 (U.S. Public Law 95-237), which sought to designate more than twenty new areas as wilderness and was a forerunner of the Forest Service's second Roadless Area Review and Evaluation (RARE II). Opponents claimed that wilderness designation could mean wildfires could not be suppressed, trails could be closed, and hunting and fishing could be drastically reduced. To clarify the issue of what is permissible in wilderness, Congressman James Weaver of Oregon, a leading proponent of the bill, engaged the new Assistant Secretary of Agriculture, Rupert Cutler, in a colloquy of twelve questions and answers about USFS wilderness management policy at an Interior Committee hearing. That colloquy (Weaver and Cutler 1977) was an important clarification of wilderness management latitude. It exposed several misconceptions about the extreme purity of USFS wilderness management policy, which was much less restrictive than opponents of wilderness claimed. Subsequently the Endangered American Wilderness Act passed, and the RARE II was initiated (see Chapters 2 and 5 for more information about RARE I and RARE II).

The restrictions on recreational and management activity in wilderness make it all the more important for land managers to provide a spectrum of opportunities and activities in areas that are roadless, but *not* designated wilderness: use of motorized vehicles, trail bikes, concentrated recreation use requiring intense management and development, and so on. A broad range of opportunities in the middle of the EMS, with relatively clear standards for recreational and other uses and management, will help meet the public's desire for diverse kinds of off-road recreation and ease the demands on wilderness to accommodate all kinds of roadless recreation activities. Provision for roadless but nonwilderness land-use categories will provide alternatives for areas that cannot be successfully managed as wilderness.

The Need for Wilderness Management

Managing wilderness is a common practice in the past forty to fifty years. The early leaders of the wilderness movement, John Muir, Aldo Leopold, Arthur Carhart, and Robert Marshall, were primarily concerned with saving wilderness from development. They assumed that designating lands as wilderness and prohibiting road construction, logging, and similar uses would ensure the preservation of wilderness, at least for the time being.

"Draw a line around it and leave it alone" pretty well described the prevailing opinion not so long ago. Today, most people concerned about wilderness have concluded that a "no-management" approach will not work with the many pressures and threats confronting wilderness. Thus, *some of the public concern about wilderness is directed at how wildernesses are to be managed to protect their wilderness characteristics and values that led to their designation.*

The Management Paradox

The term wilderness management *is a paradox (Nash 1982). Wilderness is conceptualized to be an area where the influence of modern people is absent (or at least minimized), but the word* management *suggests humans controlling nature.* In most kinds of intensive resource management, humans do alter and control natural processes. Forest management changes the number, size, distribution, and even species of trees, along with its habitat qualities for different fauna.

Many people react negatively when wilderness management is mentioned, because they envision resource and environmental manipulation. Part of the emerging appeal of the term *wilderness stewardship* in lieu of *management* is to soften the implications of the latter. However, *wilderness management does not necessarily require manipulative activity within the designated area. Wilderness management (stewardship) is essentially the management of human use and internal and external influences to preserve an area's naturalness and solitude. It includes everything done to administer an area, from planning in an office to public education and information programs near the wilderness and to management actions and enforcement activities in the field.* Management includes planning and implementing visitor education and, when necessary, rules, regulations, and visitor management to control overuse; minimal facilities such as trails and signs; decisions about access roads to the wilderness boundary; regulations for recreation stock and livestock grazing; wilderness patrols to monitor conditions and use; and public education, including information in maps, brochures, and guidebooks for visitors by wilderness rangers in the field and in schools and other public settings. We will deal in much greater detail with the components of wilderness management in subsequent chapters.

Wilderness management does not need to be heavy-handed on people or on resources. In recent years, the emphasis on educating and informing wilderness users has emerged and grown. *Wilderness managers are guardians, not gardeners or guards.* Managers should not mold nature to suit people; they should manage human use and influences to not impact natural conditions and processes. *The guiding rule is: wilderness managers should only do what is necessary to meet wilderness objectives using the minimum tools, regulation, or force required to achieve those objectives.*

More Internal and External Threats to Wilderness

Why is management so important now if the founders of the wilderness movement apparently were not that concerned about it? *The main reason for management is the overwhelming increase in the number of threats to wilderness compared to the 1930s, in the time of Marshall and Leopold.* For example, soils and vegetation are impacted from increased recreational use far beyond anything experienced a generation or more ago. Some features of the wilderness experience, the quiet and solitude that were once ensured, are now difficult to find in many places and impossible to experience in others. The founders of the wilderness movement did not ignore management because they provided direction for it in the Wilderness Act. But, it was certainly of lower priority than getting wilderness designated. Today, wilderness management is increasingly a concern of managers and the environmental community alike.

More than nineteen types of internal and external threats to wilderness conditions and values have been identified (see Chapter 13). For example, population growth and residential development near wilderness areas stem from many factors: rising incomes, greater mobility, and growing interest in the environment and outdoor recreation; and increased interest in health and physical activity. Heightened interest and increased use of wilderness are logical outgrowths of the intense controversies and publicity over the designation of particular areas. These battles tend to be lengthy, sustained by dedicated local environmentalists and accompanied by continued publicity (Allin 1982; Evans 1981; Frome 1997; Scott 2004). They have occurred throughout the nation. Public recitation of the virtues of and threats to particular areas contributes to the growth of wilderness visitation and appreciation. Inevitably, this interest has been accompanied by a growing realization of the necessity for wilderness management.

Paralleling increased threats to wilderness has been a decline in acreage of undeveloped land that might qualify as wilderness. The amount of land that has wilderness potential is fixed; more cannot be created within any reasonable time span. Designation as wilderness ensures only that development will not occur; it does not create additional acreage or raise its capacity for use. In other words, the wilderness was there, and it was being used as such before it was designated. So one benefit of designation is better management and protection of the existing supply of wilderness.

Fig. 1.3. U.S. Forest Service wilderness manager Greg Hansen (now retired) is an example of federal agency professionals in the U.S. Forest Service, Bureau of Land Management, National Park Service, and U.S. Fish and Wildlife Service who have made a career in wilderness management. Wilderness management includes planning and implementing policies and plans to achieve objectives of the Wilderness Act. Photo courtesy Marilyn Hendee.

Past Disruption of Nature

Wilderness management is needed to restore the dynamics that have been altered by human interference with natural processes. For example, fire prevention and suppression have limited fire as a natural ecological force in some ecosystems for more than fifty years. This constitutes control of nature just as surely as does a dam across a wild river. Before modern fire prevention and suppression, fires occurred much more frequently and over much larger areas, and thereby prevented, for example, unnatural accumulations of fuel that can produce fires more intense than those that might otherwise occur. Fire management to restore natural conditions is needed in many areas, and wilderness management agencies are implementing fire policies to allow fire a more natural role (see Chapter 11). Insect infestations and disease are allowed to run their course in wilderness where they do not threaten resources outside the protected area, and such infestations provide natural demonstrations of ecological processes that are unlikely to be tolerated elsewhere.

Increasingly, as more comprehensive stewardship strategies are adopted for all public wildlands under the label of ecosystem and landscape management, wilderness provides the natural model against which results can be compared and monitored.

Giving Meaning to Wilderness Designation

Wilderness designation loses much of its meaning if subsequent management policies do not define what it will actually accomplish. *Without management, wilderness designation verges on being an empty symbol, a mere name and a boundary line on a map.*

Some people who recognize the need for management want it to wait until all the wilderness designation decisions have been made: that is, "We'll worry about management when all the wilderness designation is all done." We think this is an indefensible stand. Even though a majority of wilderness designation decisions may have been made, many areas are still under consideration. Threats to wilderness (e.g., global climate change, fragmentation of ecosystems, nonnative plant and animal introductions) have greatly increased in the past fifty years and more are expected. The impacts of such threats will be substantial, and in some places have already affected areas and threaten to change the quality and character of available wilderness experiences.

Alternatives to Management?

Is managing wilderness the only alternative? Two policies could obviate that necessity. First, all use could be prohibited. Some problems, like fire management, could still require action, but essentially no use could perhaps justify no management. However, the Wilderness Act and wilderness philosophy make clear that wilderness is to be used and enjoyed by people. Public use is legally mandated. Second, we could just designate areas and forget about management. Any kind of use not clearly illegal under the Wilderness Act could be allowed in an unlimited amount, and environmental damage or changes in the ecosystem could simply be accepted. Under this option, wilderness could vanish from most places. This result would violate the Wilderness Act, which requires the protection and perpetuation, the preservation, of wilderness.

The middle ground between these two extremes requires management. We see no other course of action. This course was endorsed early in the history of designated wilderness by natural-resource leaders reflecting on the new challenge and who largely concluded that managed wilderness is the only possible kind (Cowan 1968; Spurr 1966; Zivnuska 1973). *The real question is not whether to manage but how to manage wilderness.* That is the topic of this book, and the subject of an expanding compendium of experience and research in agency manuals, scientific and popular literature, conference proceedings, and Internet sources. This expanding knowledge continues to greatly strengthen wilderness stewardship. However, the knowledge is scattered; we have synthesized and summarized it for ease of teaching and application.

Wilderness Philosophy and Wilderness Management

Each federal agency having responsibilities under the Wilderness Act has developed policies and guidelines to apply it to lands under their jurisdiction. Such direction is important so that management does not significantly deviate from area to area. However, policies cannot and should not be so detailed as to cover all contingencies because managers need to retain some flexibility to respond to unique conditions occurring in individual areas. There is always a gap between specific policy direction and unanticipated contingencies, the need for broad guidelines on one hand and flexibility on the other, and this gap must be filled by a manager's philosophical perspective. *A deep philosophical understanding of wilderness, grounded in knowledge of its legislated direction and inherent ecological and social values, is a wilderness manager's most important tool.*

Before considering philosophical perspectives, we call attention to a principle discussed in more detail in Chapter 7: *that wilderness is preserved and managed for the benefits and values it provides people.* Many laws, among them the Endangered Species Act, the Wild and Scenic Rivers Act, and the Wilderness Act, have been enacted to protect and preserve the natural environment for its own sake, but under the assumption that retaining natural features provides important human benefits. Even the statement that a feature is unique and worthy of protection is a human judgment based on the belief that such things are valuable for our pleasure, survival, and well-being.

The idea that wilderness is for use and enjoyment by people has clear statutory support. The Wilderness Act specifically notes that wildernesses will be administered "for the use and enjoyment of the American people." But what kinds of use? The act rules out some uses for instance, motorized equipment, with some minor exceptions included in wilderness designation acts, is not permitted. Still, the act permits considerable diversity in styles of use and in the accompanying developments. Simply arguing that wilderness is for public use only states the obvious. It does little to resolve the issues of what kinds of use, how much, where, and when. Understanding different philosophical perspectives about wilderness and its values is important in addressing these questions regarding the act.

Anthropocentric and Biocentric Perspectives

Consider two alternative philosophical notions about wilderness and the basis for its value: *anthropocentric* and *biocentric* (Hendee and Stankey 1973). *The anthropocentric position takes the "use and enjoyment" phrase of the Wilderness Act quite literally.* Under this philosophy wilderness is viewed primarily from a sociological or human-oriented perspective; *the naturalness of the wilderness is less important than maximizing direct human use.* Programs to alter the physical and biological environment to produce desired settings are encouraged: big trees, open vistas, lots of fish and wildlife. Developed facilities to increase recreational use of wilderness are appropriate. In fact, increasing direct human use could increase some types of human values and benefits under this philosophy. The concept of wilderness carrying capacity does not exist, because people's ever-changing adaptation to their environment (and vice versa) continually changes (lowers) standards of crowding and naturalness. The character of wilderness could change to reflect human desires and contemporary standards of naturalness. Wilderness could still be one extreme on the environmental spectrum, but a shifting extreme, not grounded in any definite standards. "Let's open up our wilderness areas and do away with restrictions" could be a slogan for believers in pure anthropocentrism.

In contrast to the anthropocentric perspective, the *biocentric perspective emphasizes maintaining natural systems, if necessary, at the expense of recreational and other human uses* (Hendee and Stankey 1973). The goal of the *biocentric* philosophy is *to permit natural ecological processes to operate as freely as possible, because wilderness values for society ultimately depend on retaining naturalness.* To the extent that naturalness is lost, the experiential, spiritual, and scientific values of wilderness are lessened.

Labeling these perspectives as *biocentric* versus *anthropocentric* might create a false distinction between "wilderness for people's sake" and "wilderness for wilderness' sake." But, as discussed earlier, wilderness is for people. *The important distinction between these philosophies is the extent to which the human benefits of wilderness are seen as dependent on the natural integrity (naturalness and solitude) of the wilderness setting.*

Their respective proponents have argued these alternative perspectives persuasively. Before turning to a more detailed look at each, two points should be emphasized. First, these philosophies represent extreme, polarized concepts about wilderness management and stewardship, and it is unlikely either could be completely followed. However, they do highlight alternative orientations toward wilderness stewardship. Second, it is difficult, if not impossible, to say that either idea is wrong or right. It seems more important to examine the long-range implications of each and to judge the appropriateness of each in light of society's objectives as reflected in the Wilderness Act. With that in mind, let us look at each perspective more closely. *Readers interested in testing their own anthropocentric versus biocentric instincts can find a diagnostic quiz to take in the "Reader Exercise" at the end of this chapter,* based on Clark and Kozacek (1997).

The Anthropocentric Philosophy

Advocates of the anthropocentric approach would have us facilitate direct human use of wilderness. Wilderness managers could emphasize recreation and comfort. They might develop high-standard trail systems; expand stocking fish to most wilderness lakes; increase and upgrade campsite facilities, shelters, toilets, and similar features; and generally increase recreational carrying capacity, aesthetic features, and user comfort and convenience. The perception of wilderness held by the largest number of users could be the most important guideline for managers.

This orientation emphasizes society's demands on wilderness, not its natural condition, and has important implications for users and the environment. Initially

Fig. 1.4. Visitors aboard a cruise ship observe tidewater glaciers in 2.6 million–acre Glacier Bay Wilderness, in Glacier Bay National Park, Alaska. The need for wilderness management is growing, partly because of steadily increasing use, the impacts of that use, and the special permitted uses such as motorized access in Alaskan wilderness areas. Photo courtesy Carol Dawson.

the emphasis on the aesthetic and recreational qualities of wilderness settings could lead to substantial alteration of the environment, particularly vegetation. Given current knowledge and technology, we could engineer the wilderness scene to produce specific environmental conditions. For example, some have argued that a wilderness should represent an early point in our history, perhaps the land as it was at the time the continent was first settled by Europeans. To create such a setting would require sharp interference with natural processes to steer ecological succession in the desired direction. Fire, chemicals, or machines might be needed. Some desirable results, from an anthropocentric point of view, might be increased scenic views from cleared viewpoints, additional forage for stock from intensively managed (perhaps irrigated and fertilized) range, alpine meadows enlarged and maintained by uprooting invading tree seedlings, and more observations of wildlife stimulated by strategically placed salt licks. Traditional forestry, silviculture, and habitat management might be in order. Wilderness managers might be more like gardeners rather than guardians.

However, the anthropocentric approach would mean losing an essential wilderness quality: naturalness. Furthermore, such an approach (at least after a certain point) would be illegal in designated wilderness. The Wilderness Act says quite clearly that wilderness should be a setting where the forces of nature operate free from human influence. Human influence on ecosystems is already pervasive worldwide, ranging from the introduction of atmospheric pollutants to direct recreational impacts. But minimal influence with natural evolution and processes seems to be the clear intent of the Wilderness Act (sec. 2[a]): "In order to insure that an expanding settlement and growing mechanization does not occupy and modify all areas within the United States, [and] to secure for the American people of present and future generations the benefits of an enduring resource of wilderness." Meeting this goal would not be possible under the anthropocentric approach.

An anthropocentric approach might be particularly detrimental to the scientific values of wilderness. The notion of wilderness as a genetic pool, an environmental baseline, and a refuge for the survival of species especially sensitive to human influence would be lost in a wilderness manipulated and altered to fit changing human tastes. Because the loss of naturalness could in many cases be irreversible, there could be incalculable costs in terms of forgone scientific values and benefits.

Styles of recreation tuned to anthropocentric management philosophy might be convenience oriented. Recreational experiences would be a primary goal and would mean increasing access, reducing difficulty and danger, and facilitating use. Conversely, programs hindering or restricting use might be rejected. For example, if wilderness exists primarily for use and enjoyment, programs limiting use (e.g., wilderness permit systems, rationing, rough trails, minimum party size) are bureaucratic hindrances that would be eliminated. Under an anthropocentric philosophy, then, if problems of environmental impact or excessive congestion did arise, the managerial response might be to harden sites so they could tolerate impacts, revegetating, installing more facilities, upgrading trails, and otherwise intensively managing the wilderness to handle increased levels of use. Opportunities for solitude would decline.

Proponents of the anthropocentric philosophy argue that biocentric wilderness regulations discriminate against people who might want to visit wilderness but cannot because of lack of physical ability and, in effect, close off public lands (wilderness) to a large segment of the public. Such proponents might point to the European Alps, where large numbers of people are accommodated through extensive road systems, cog railways, groomed trails, and mountain huts.

The Biocentric Philosophy

The biocentric perspective emphasizes preserving natural processes and encouraging nearly natural conditions and processes within wilderness ecosystems as they existed in the absence of human influence. This requires controlling human influences in the ecosystems from sources such as excess recreational use and eliminating restrictions on normal processes caused by such policies as fire prevention and suppression. Recreational use in wilderness might be consistent with this perspective only to the point that it does not unduly alter natural conditions and processes. *Like the anthropocentric philosophy, a biocentric approach focuses to a lesser extent on human benefits. The important distinction we make between them is that, biocentrically, these benefits are viewed as being dependent on the naturalness of wilderness ecosystems. Thus, protecting naturalness is essential to providing human benefits. It is the experience of wild (untrammeled) nature, and the ecosystem services that flow from it, on which human benefits depend.*

The biocentric approach to wilderness management has specific implications for the environment and users. Over an extended period of time, we might expect to see environmental conditions evolve that reflect historical patterns of ecological succession. The natural processes such as wind, erosion, insects, disease, and fire that have shaped and altered the landscape might continue to operate much as they always have. One consequence could be a wilderness that is aesthetically unattractive. Insect infestation, erosion, fire, forest disease, and similar processes might be allowed to run their course without human interference; as a result, wilderness landscapes would reflect these natural perturbations. This approach could mean that particularly desirable recreational features, such as high mountain meadows as important stock forage areas or areas of high wildlife visibility, might gradually, or dramatically, shift with advancing ecological succession or following natural disturbances.

These management challenges would present little opportunity to apply intensive natural-resource skills such as silviculture and habitat management. Rather, much more emphasis could be placed on monitoring conditions and controlling visitor behavior to preserve dynamic natural processes. Nature would roll the dice to determine ecological outcomes.

Under a biocentric management philosophy, recreational use of a wilderness might be secondary to maintaining the natural order. Management actions to increase and facilitate use, such as high-standard trails and campsite facilities, might not be appropriate. Where recreational use would cause significant impacts, management could modify visitor behavior and limit or disperse use rather than institute methods to absorb greater impact. Moreover, management programs could promote opportunities emphasizing the primitive environment, challenge, and solitude, activities with fewer environmental impacts than those available on nonwilderness lands or emphasized under an anthropocentric philosophy. *A biocentric philosophy requires that visitors take wilderness on its own terms.*

These two wilderness management philosophies, anthropocentric and biocentric, represent polar viewpoints that are especially useful for the sake of discussion *and planning.* A more moderate view between these two extremes has been called ecocentric (Clark and Kozacek 1997), but we call it midcentric, thinking that ecocentric might imply an extreme beyond biocentric.

In Support of Biocentric Management

Which is the most appropriate management philosophy: biocentric or anthropocentric? As noted earlier, the question of which approach is right and which is wrong has no absolute answer. However, an answer must relate to long-term implications and to the legal mandates set forth in the Wilderness Act.

Facilitating maximum use within wilderness will gradually diminish its naturalness and solitude, alter ecological regimes in subtle or drastic ways, and diminish opportunities for experiences dependent on wild and unaltered settings. The result might be to eliminate one extreme on the environmental modification spectrum (EMS) and recreation opportunity spectrum (ROS) and, thus, lose diversity among wildland settings.

We believe that to achieve the legal goals of the NWPS, management should emphasize the natural integrity of wilderness ecosystems as much as possible, and this reflects a more biocentric management philosophy. This position and its implications are most consistent with the mandate of the Wilderness Act and with the intent of the legislative debate that fashioned the act, and with its historical and philosophical foundations that evolved over the past century. *Consider the following arguments supporting a biocentric perspective.*

First, providing for the diversity of recreationists' tastes could ensure that people who prefer a wild and pristine setting would not be displaced in favor of users whose tastes can be met in nonwilderness settings. Some users depend on a pristine wilderness to satisfy their desires for a wilderness experience. As the supply of pristine settings diminishes, management philosophies should strive to maintain an ROS of diverse opportunities to protect all of them and, from our perspective, especially pristine wilderness.

Second, a human ecology view of outdoor recreation suggests that as people gain outdoor experience through activities such as car camping, they may begin to seek more demanding kinds of experiences. For example, young people may choose to expand the boundaries of childhood car camping into wilderness backpacking experience as young adults. Many campers using developed facilities might in the future opt for wilderness. Increased demands on wilderness are virtually certain. Biocentric management will help maintain opportunities at the primitive end of the EMS and ROS to meet this increasing demand.

Third, management responses to increasing use can lead to unanticipated shifts in the kinds of recreational opportunity an area offers. Developments to protect a site can attract a different clientele. For example, campgrounds developed to protect natural qualities of a site such as tent pads, tables, fireplaces, and toilets may attract a new clientele drawn not by nature but by the comfort and convenience of the facilities and socialization with other users. The biocentric approach is challenged not just by recreational use, but also by wilderness designation proposals enacted to include various exceptions and special provisions for mechanized access and use that negatively impact wilderness. Applying the biocentric criteria of naturalness and solitude might minimize such changes in wilderness and limit the growth of inappropriate kinds of use and impacts.

Fourth, a biocentric approach would preserve the greatest range of future options. Management decisions that facilitate increased use through development are often irreversible, and they narrow the range of environmental opportunities by eliminating unmodified areas that are already in short supply. This leads to loss of diversity and loss of important biological and scientific values.

Wilderness Naturalness and Solitude

The Wilderness Act, in section 2(c), defines wilderness, in part, as a place that "has outstanding opportunities for solitude *or* a primitive and unconfined type of recreation" (italics added).

Until recently, the view of this phrasing was generally understood to mean the "or" was intended to clarify the nature of the experiences wilderness should offer and that the descriptive phrase "solitude or a primitive and unconfined type of recreation" is required in the definition of wilderness (Worf 2001; Worf et al. 1972). Proponents of this interpretation claim that it is supported by the legislative history of the Act and ensuing legislation. For example, the late Pennsylvania congressman John Saylor, a leading supporter of the Wilderness Act, described the wilderness experience as composed of various elements, including solitude (Saylor 1962). Many others voiced similar views during debate on the Wilderness Act. Further, in the so-called Eastern Wilderness Act of 1975 (U.S. Public Law 93-622), the statement of policy notes that wildernesses designated by the Act become a part of the NWPS and that management shall "promote and perpetuate the wilderness character of the land and its specific values of solitude, physical and mental challenge, scientific study, inspiration and primitive recreation." Thus, management to provide these experiences, *including solitude,* would seem required.

But, more recently, a new interpretation of section 2(c) of the Act has emerged: that opportunities for either solitude *or* a primitive kind of recreation are required (that is, either one could qualify an area). This interpretation, if accepted, raises new management possibilities, as the following example illustrates.

At the turn of the millennium, following decades of increased wilderness use now concentrating in high-use locations, the USFS proposed a new wilderness recreation strategy that would accept higher levels of use at popular wilderness locations to keep from diverting excessive use and impacts to more pristine, low-use wilderness (Oye 2001). Thus, under this strategy, the interpretation of section 2(c) of the Wilderness Act would seem to be that solitude *or* (rather than *and*) a primitive and unconfined type of recreation is what is required. This interpretation was very controversial. Opponents of the proposed new wilderness recreation strategy claim that a nondegradation mandate is being violated by accepting lower standards of naturalness and solitude at the high-use sites, thereby creating two classes of wilderness (Worf 2001). However, supporters argue that this is needed; they also point to the dramatically increased use at popular wilderness sites and defeat by environmental groups of wilderness management plans to greatly reduce use. "If we lock people out who will fight to save the wilderness?" asked longtime photographer of the Northwest Ira Spring in his statement supporting the proposed new policy (2001, 17). In many ways, the proposed new policy is a fait accompli because use levels at popular wilderness locations are already beyond levels providing solitude. Rolling back use now might not be politically possible, and use will inevitably divert to more lightly used but pristine sites. A statement of the proposed USFS wilderness recreation policy (Oye 2001), preceded by a statement of the situation requiring action (Cole 2001), followed by arguments against the policy (Worf 2001), and supportive of the policy (spring 2001), appear in the April 2001 issue of the *International Journal of Wilderness.* The policy has not been implemented fully across USFS wilderness areas, but it is cited in some recent USFS plans, such as a management plan for the White Mountain National Forest in New Hampshire and Vermont (U.S. Department of Agriculture, Forest Service 2005).

We believe that section 2(c) of the Wilderness Act does require that both solitude and a primitive and unconfined type of recreation be provided. Some flexibility seems possible by approaching the requirement with an NWPS-wide view, that is, every area cannot provide equal amounts of each requirement and some variation is expected.

Obviously, the levels of solitude found within and between areas will vary, perhaps substantially. The wilderness acts do not define a single standard for solitude that all areas must meet. The existing pattern of trails and campsites in a candidate or designated wilderness precludes such a standard anyway. Visitor use intensities will be higher near trailheads and at popular locations, and it seems neither necessary nor possible to manage for some uniform level of interparty contact. The 1964 act calls for provision of "*outstanding opportunities* for solitude" (italics added), and we interpret that to mean there should be places and times within the NWPS and within individual wildernesses where visitors find little or no contact with others. However, this does not mean on every wilderness acre all the time.

We recognize that the wilderness experience is a product of human perception and cannot be precisely described and packaged. However, managers do need some guidelines. To us, the Wilderness Act, in section 2(c) (1), provides them. In answering the question "What distinguishes wilderness from other settings?" we have sought an answer that would be true to congressional intent and offer specific criteria for managers. We think wilderness is an area (1) featuring substantially natural ecological conditions and (2) offering visitors outstanding opportunities for solitude or a primitive and unconfined type of recreation. There will be naturalness everywhere in the wilderness all the time, but at some places in the wilderness, at some times, there will also be high-use sites where the simultaneous provision of naturalness *and* solitude will not be possible. Further, it may not be possible in such situations to roll back use to a level where naturalness and

Fig. 1.5. On Public Lands Day, September 23, 2000, the BLM Needles, California, field office organized a project in which more than forty volunteers (top) gathered to pick up trash from an old mining and ranching homestead in the Old Woman Mountains Wilderness. Such public involvement in wilderness management provides much-needed help to busy managers. Volunteers (middle) take trash to a staging area (bottom) where a horse-drawn wagon is loaded with old wire, irrigation pipe, fencing, cans, and tin and hauled to the trailhead a mile away. Photos courtesy Marilyn Hendee.

solitude can be provided without diverting excessive use to more sensitive pristine areas or prompting political repercussions that might be destructive to the wilderness—such as removing the wilderness designation from popular sites. So judgment is needed. And ironically, it may be by allowing more use at high-use sites (with appropriate environmental safeguards) that a biocentric approach featuring solitude can be applied to the remaining majority of a wilderness area.

Two Qualifications to Biocentrism

The issue of what philosophy will underlie wilderness management is crucial to the future of the NWPS. The wilderness system we have in the future will be a direct product of the philosophy that guides policy and management decisions today. We judge that a predominantly biocentric philosophy is appropriate, necessary, and defensible. The policies and practices of federal agencies tend to reflect such a perspective over most of their wildernesses (some high-use sites excepted), and we hope this direction continues.

We qualify our support for the biocentric approach in two important respects: First, it must be practical, because the biocentric philosophy, as we describe it, represents an ideal, and its implementation will be inhibited by practical constraints. We noted earlier the proposed Forest Service wilderness recreation strategy to cope with high-use sites and arguments for and against such a strategy. Other examples also apply: ideally, biocentric management would allow wilderness fires to burn. However, this is not practical. In many areas, because of several decades of fire suppression, unnaturally large supplies of highly combustible fuels have accumulated. Fires in these areas might become extraordinarily intense, causing catastrophic damage to wilderness resources and/or to resources outside the wilderness boundary endangering people and property (e.g., the Yellowstone forest fires of 1988).

Because the NWPS includes many relatively small tracts of land scattered throughout areas managed for other purposes, nonwilderness considerations will always influence what happens inside the wilderness boundary and vice versa. *Thus, we endorse the biocentric philosophy, while recognizing that its extreme application might be unrealistic and impractical, and create a backlash against it. A rule of reason must temper its application.* Nevertheless, we think that a management and policy orientation that judiciously strives toward the *intent* of biocentricity is proper and feasible.

Our second qualification is that biocentricity and the entire wilderness preservation movement in a broader sense are viable philosophies only if they are accompanied by (1) equitable provision for alternative outdoor recreation opportunities, and (2) comprehensive efforts to humanize the places where we work and live.

The elitist overtones of biocentricity and wilderness concern us. Are we endorsing a philosophy that offers access only to a privileged few at the expense of the majority? That is only true if alternative outdoor recreation opportunities—a full ROS of them—are not provided. In a democratic society, we see little chance that support for wilderness, much less a biocentric philosophy of wilderness management, can survive unless an equitable range of outdoor recreation opportunities are provided.

Our second qualification has even greater long-term significance. As Sebastian de Grazia noted, "Only if you give the city a pleasant and healthful outdoor environment, can you slacken the…drive for the wilderness. Only the city can save the wilderness" (de Grazia 1970, 96). Obviously, citizens who press for wilderness designation and the agencies that manage wilderness have only a limited capacity to change the poverty that plagues our inner cities, the social inequities that divide our people, the haphazard land-use patterns and transportation systems that blight our landscapes, or the pollution that clouds our land, air, and water. However, even though wilderness is our primary focus, we should not lose sight of these broader issues. This wider perspective links wilderness to issues such as energy use, social equity, land-use planning, and federal programs to provide outdoor opportunities for people of the inner city.

A Commonsense Policy Is Needed

While we lean toward a biocentric wilderness management philosophy, we are not extremists. We call for management with a biocentric emphasis but that it be applied with common sense and sensitivity to local conditions; perhaps it might be more appropriate to call it stewardship. We reiterate the idea that *wilderness management should not mold nature to suit people. Rather, it should manage human use and influences so as not to alter natural conditions and processes. Managers should do only what is necessary to meet wilderness objectives and use only the minimum tools, regulation, and enforcement required to achieve those objectives.* In wilderness, people adapt to nature, to naturalness, and to solitude, and that is the source of human benefits from wilderness experience as well as the ecological and nonuse benefits.

The details of management may vary among wildernesses, but we support our broad notions by citing the late Senator Frank Church, floor manager of the Wilderness Act when it passed the Senate, and former chair of the Interior Subcommittee on Public Lands. In a 1977 Wilderness Resource Distinguished Lecture at the University of Idaho, Senator Church (1977, 13) argued:

> It was *not* the intent of Congress that wilderness be administered in so pure a fashion as to needlessly restrict their customary public use and enjoyment. Quite to the contrary, Congress fully intended that wilderness should be managed to allow its use by a wide spectrum of Americans. There is a need for a rule of reason in interpreting the Act, of course, because wilderness values are to be protected. As I stated in 1972, while chairing an oversight hearing of the Subcommittee on Public Lands:…The Wilderness Act was not deliberately contrived to hamstring reasonable and necessary management activities. We intend to permit the managing agencies…latitude…where the purpose is to protect the visitors within the area…[including, for example] minimum sanitation facilities…fire protection necessities…[and] the development of potable water supplies…The issue is not whether necessary management facilities are prohibited; they are not. The test is whether they are necessary.

Thus, *the wilderness management framework intended by Congress was for the agencies to do only what is necessary.* The facilities just mentioned may be required, such as at high-use sites. But, restrictions on use may sometimes be needed to protect especially fragile locations such as pristine, low-use wilderness zones. However, in adopting regulations, common sense is required.

In summary, if purity is an issue in the management of wilderness, let it focus on preserving the natural integrity of the wilderness environment, and not on needless restriction of facilities *necessary* to protect the area while providing for human use and enjoyment. Our hope is that the philosophy that ultimately prevails will emphasize the natural integrity of wilderness ecosystems, solitude, and primitive styles of recreation, with commonsense applications and adjustments as necessary that respond to the needs of individual areas.

Summary

This chapter's objective has been to set the stage for systematic and progressively more detailed discussions related to management of the NWPS in the United States. So far, we have focused on the broadest direction—wilderness management philosophy.

More specifically, this chapter explored the meanings and definition of wilderness in light of some basic, albeit overlapping, themes and values espoused by historical wilderness spokespeople in the United States. We discussed briefly the difference between wilderness designation and wilderness management. We argued the need for wilderness management from several vantage points. Finally, we described two alternative wilderness management philosophies, anthropocentric and biocentric—acknowledging a middle ground as midcentric—and evaluated their applicability in light of objectives of the Wilderness Act. We concluded that the judicious application of a biocentric philosophy of management (stewardship) could result in the most appropriate protection of the NWPS. Stewardship should strive to maintain the historical natural processes that formed the great American wildernesses experienced by our American forebears.

Study Questions

1. How did early European settlers of America view wilderness? How has that view changed in recent history and why?
2. Name four categories of wilderness values or themes recognized by the wilderness movement in this nation. Briefly explain each.
3. Why do managers need to understand the values people hold for wilderness?
4. How does wilderness designation differ from wilderness management?
5. Who decides which lands shall be designated as wilderness?
6. Why is wilderness management more necessary today than in the past?
7. Contrast the *anthropocentric* and *biocentric* philosophies of wilderness management.
8. Which of the two philosophies do you favor for managing wilderness in the United States? Why?
9. What does de Grazia mean by the statement "Only the city can save the wilderness"? Do you agree or disagree? Why?
10. Do the exercise at the end of the chapter, "How Do Your Wilderness Values Rate?" Where does your wilderness management philosophy fall on the anthropocentric versus biocentric spectrum? How do you think it will change by reading this book?

Reader Exercise:
How Do Your Personal Wilderness Values Rate?

A person's personal orientation and philosophy about wilderness and its management can be an important factor in his or her perception of what are appropriate policies and actions for wilderness stewardship. Just recognizing that wilderness visitors and one's colleagues might have a slightly different philosophy contributes to understanding that wilderness stewardship is not a cut-and-dried business; rather, there is a lot of room for interpretation in making the right decision.

Put Your Values to the Test!

How do your personal wilderness values rate? Take the test, adapted from Kendall Clark and Susan Kozacek (1997) in *International Journal of Wilderness* (3[1] 12–13:) and see. It's easy—and you might be interested in seeing where you fit compared to some others who have taken the test. Remember, answer these questions based on your personal values—not what you think the Wilderness Act or an agency's policy requires.

Stop here and answer the thirty-five questions that follow either "yes" or "no." When you have finished the test, count your number of "yes" answers and then continue reading. There you will find the rest of the story and see how your score might compare. We suggest you take the test before seeing how your score might compare with others.

Wilderness Values Questions

1. Do you feel hunting is an appropriate activity in wilderness?

2. Do you feel it is okay to stock native fish in lakes that historically have not had fish?
3. In an area that has established wildlife watering devices (e.g., guzzlers), do you feel it is appropriate to maintain these and leave them in wilderness?
4. Do you feel it is appropriate to control predators in wilderness that are killing a substantial number of livestock?
5. Are low-level aerial-game surveys in wilderness acceptable to you?
6. Do you feel we should be protecting known threatened and endangered species habitat from Prescribed Natural Fires (PNF)?
7. Is it acceptable to you to have Management Ignited Fires (MIF) in a wilderness area?
8. Do you feel it is appropriate to have technologically advanced data-collecting stations in wilderness to monitor temperature, moisture content, wind, and other factors that might allow better information for PNF and MIF?
9. Do you feel we should be suppressing any fires in wilderness?
10. In your opinion, is it okay to maintain historic cabins in wilderness?
11. Do you feel that there is a point when air quality is more important than allowing extended periods of PNF?
12. Is it okay to interpret in a publicly available book historic structures and cultural resources that are in wilderness?
13. Do you feel that cattle or sheep grazing is an appropriate use for wilderness?
14. Do you feel grazing permittees should be allowed to use motorized equipment for maintaining water developments in wilderness where this has been a historical method of maintenance (for example, using a dozer to clean out a dirt stock tank in wilderness)?
15. Do you feel a hazardous tree along a well-used trail should be cut to protect public safety?
16. Do you feel that cutting logs in trails to facilitate passage by pack strings is appropriate in wilderness?
17. Do you feel we should be placing signs by natural caves in wilderness that pose safety hazards?
18. Do you feel it is appropriate for a visitor center to be giving users more information about hazards in wilderness so we can lessen the potential of search-and-rescue operations?
19. Do you feel that signs should be placed at historic structures to warn people of the potential for Hantavirus?
20. Do you feel we should rescue a person with a broken leg (but not in a life-threatening situation) in wilderness with a helicopter?
21. Do you feel it is okay to use llamas or pack goats in wilderness?
22. Do you feel that it is appropriate to leave some established rock-bolt routes for climbers in wilderness areas?
23. Does the value of having the number of users controlled by a permit system outweigh the value of unregulated use and freedom in wilderness (i.e., do you believe permit systems should be used in wilderness)?
24. Do you feel it is okay to allow people to collect crystals in wilderness?
25. Do you feel it is okay to allow people to collect antlers in wilderness?
26. Do you feel that recreation opportunities are the dominant value of wilderness?
27. Do you feel it is okay to have trail signs in wilderness?
28. Do you feel it is okay to put mileage on signs in the wilderness?
29. If a free one were available to you, would you take a cellular phone into wilderness with the intention that it would only be used to help in an emergency situation?
30. Do you feel okay about burying decomposable garbage in wilderness?
31. If you had a well-behaved dog, would you feel okay about taking it with you to the wilderness?
32. Do you think it is appropriate for outfitters to have business operations dependent on wilderness?
33. Do you feel it is okay to film in wilderness a movie about wilderness values?
34. Do you feel it is appropriate to allow a one- or two-week window for chainsaw use to open trails after an intense blow-down event?
35. Do you feel it is okay to apply a mandatory party size or limited permits to promote solitude in wilderness?

Now count your number of "yes" answers and read the following.

Anthropocentric and Biocentric Philosophies

Two contrasting orientations are often used to characterize philosophies of wilderness stewardship: anthropocentric and biocentric. An *anthropocentric* philosophy sees wilderness primarily from a human-oriented perspective. The naturalness of wilderness is less important than facilitating human use and convenience. Programs that might alter the physical and biological environment to produce desired settings are encouraged. A *biocentric* philosophy emphasizes maintaining natural systems at the expense of recreational and other human uses, if necessary, because wilderness values depend on naturalness and solitude. The goal of this philosophy is to permit natural ecological processes to operate as freely as possible.

The wilderness values test is based on thirty-five questions (Clark and Kozacek 1997) that could be answered "yes" or "no," such as question 9: "Do you feel we should be suppressing any fires in wilderness?" In general, a "yes" answer would place a person on the anthropocentric end of the wilderness values scale, and a "no" answer would reflect a biocentric philosophy. The test is scored by tabulating the number of "yes" answers recorded after all questions have been answered. Of course, the questions present choices that are oversimplified compared to the real world, so you have to respond in a generalized way. In addition, you must keep in mind that it is your "personal" wilderness values that are being measured—not the Wilderness Act or an interpretation of agency policy.

The authors of the test reported that most of the managers they tested respond with between fifteen and twenty-five "yes" answers, and they might characterize them as being in the middle of the anthropocentric–biocentric continuum (we call this midcentric). A few responded with fewer than fifteen "yes" answers, which puts them on the biocentric side of the continuum. Moreover, there are some who have more than twenty-five "yes" answers, reflecting an anthropocentric view.

The test has not been applied to populations of wilderness users, but at the interagency wilderness stewardship training session at Eagle Lake, California, in September 1996, the lowest score by several points was nine "yes" answers by a wilderness vision-quest guide who was at the session to participate on a user panel. It would be interesting and valuable to try the test on wilderness user populations in the future.

When the article describing this test was written, Kendall Clark was district ranger on the Eagle Cap Ranger District of the Wallowa-Whitman National Forest in Enterprise, Oregon, and Susan Kozacek was district ranger on the Wilderness Ranger District of the Gila National Forest in Mimbres, New Mexico.

Case Discussion Questions: Anthropocentric and Biocentric Values

1. How do your personal test results compare to those reported by Clark and Kozacek or to the combined results of others in your discussion group? Are you in the anthropocentric, midcentric, or biocentric portions of the range? Does this accurately reflect what you believe are your values about wilderness?
2. How do your personal wilderness values test results compare to your interpretation of the Wilderness Act? Do you perceive the Wilderness Act to be in the anthropocentric, midcentric, or biocentric portions of the range? Defend with specific supporting references to various sections or sentences of the act.
3. Select some readings of a wilderness visionary (Leopold, Marshall, Olson, or others) and try to determine where he or she might be placed along the anthropocentric-to-biocentric continuum. How do you compare to that visionary? Why?
4. Assume the role of a wilderness visitor or user, such as a big-game hunter, long-distance backpacker, local sheep rancher, floatplane pilot and outfitter, or local resident, day hiking and take the test as you imagine he or she would respond. Compare his or her total test result to what you reported in the three previous questions.
5. How do you expect that your wilderness values and philosophy affect your interpretation of the Wilderness Act compared to others? Do others in your discussion group interpret the Wilderness Act the same or differently from you?

References

Allin, Craig W. 1982. *The Politics of Wilderness Preservation*. Westport, CT: Greenwood Press.

Bergstrom, J. C.; Bowker, J. M.; Cordell, H. K. 2005. An organizing framework for wilderness values. In: Cordell, H. K.; Bergstrom, J. C.; Bowker, J. M. eds. *The Multiple Values of Wilderness*. State College, PA: Venture Publishing, pp. 47–55.

Bird, W. 2007. *Natural Thinking*. London, England: Royal Society for the Protection of Birds. www.rspb.org.uk/policy/health; accessed June 11, 2007.

Bowker, J. M.; Harvard, J. E., III,; Beregstrom, J. C.; Cordell, H. K.; English, D. B. K.; Loomis, J. B. 2005. The net economic value of wilderness. In: Cordell, H. K.; Bergstrom, J. C.; Bowker, J. M., eds. *The Multiple Values of Wilderness*. State College, PA: Venture Publishing, pp. 161–180.

Callicott, J. B.; Nelson, M. P. 1998. *The Great Wilderness Debate*. Athens: University of Georgia Press.

Carhart, Arthur H. 1920. Recreation in the forests. *American Forests*. 26: 268–272.

Church, Frank. 1977. Wilderness in a balanced land use framework. *First Annual Wilderness Resource Distinguished Lecture*. University of Idaho Wilderness Research Center, Moscow, Idaho. March 21. [Reprinted as Whither wilderness, *American Forests*. 83(7): 10–12, 38–41.]

Clark, Kendall; Kozacek, Susan. 1997. How do your wilderness values rate? *International Journal of Wilderness*. 3(1): 12–13.

Cole, David. 2001. Wilderness recreation management strategies: Balancing freedom and protection. *International Journal of Wilderness*. 7(1): 12–13.

Cordell, H. K.; Bergstrom, J. C.; Bowker, J. M. 2005. *The Multiple Values of Wilderness*. State College, PA: Venture Publishing.

Cordell, H. K.; Murphy, D.; Riitters, K.; Harvard J. E., III,. 2005. The natural ecological value of wilderness. In: Cordell, H. K.; Bergstrom, J. C.; Bowker, J. M., eds. *The Multiple Values of Wilderness*. State College, PA: Venture Publishing, pp. 205–249.

Cordell, H. K.; Tarrant, M. A.; Green, G. T. 2003. Is the public viewpoint of wilderness shifting? *International Journal of Wilderness*. 9(2): 27–32.

Costley, Richard J. 1972. An enduring resource. *American Forests*. 78(6): 8–11.

Cowan, Ian McTaggert. 1968. Wilderness, concept, function, and management. *The Horace M. Albright Conservation Lectureship*, Vol. 8. Berkeley: University of California, School of Forestry and Conservation. 36p.

Davis, George D. 1989. Preservation of natural diversity: The role of ecosystem representation within wilderness. In: Freilich, Helen R., comp. *Proceedings of the National Wilderness Colloquium*; January 13–14, 1988; Tampa, FL. General Technical Report SE-51. Athens, GA: U.S. Department of Agriculture, Forest Service, Southeastern Forest Experiment Station, pp. 76–82.

Davis, Millard C. 1966-67. The influence of Emerson, Thoreau, and Whitman on the early American naturalists, John Muir and John Barrows. *Living Wilderness*. 39(95): 19–23.

de Grazia, Sebastian. 1970. Some reflections on the history of outdoor recreation. In: Driver, B. L., ed. *Elements of Outdoor Recreation Planning*. Ann Arbor: University of Michigan Press, pp. 89–97.

Dombeck, Mike. 1999. A wilderness agenda and legacy for the U. S. Forest Service. *International Journal of Wilderness*. 5(3): 4–6.

Edwards, Mike. 1985. A short hike with Bob Marshall. *National Geographic*. 167(5): 664–689.

Evans, Brock. 1981. The wilderness idea as a moving force in American cultural and political history. In: *Congressional Record*; April 27, 1981. Washington, DC: U.S. Senate: S4010–S4014.

Flader, Susan L. 1974. *Thinking Like a Mountain: Aldo Leopold and the Evolution of an Ecological Attitude Toward Deer, Wolves, and Forests*. Columbia: University of Missouri Press.

Friese, Gregory T.; Hendee, John C.; Kinziger, Mike. 1998. The Wilderness Experience Program (WEP) industry in the United States: Characteristics and dynamics. *Journal of Experiential Education*. 2(1): 40–45.

Frome, Michael. 1997. *Battle for the Wilderness*. Salt Lake City: University of Utah Press.

Gager, Dan; Hendee, John C.; Kinziger, Mike; Krumpe, Ed. 1988. What managers are saying—and doing—about wilderness experience programs. *Journal of Forestry*. 96(8): 33–37.

Green, G. T.; Tarrant, M. A.; Raychaudhuri, U.; Zhang, Y. 2005. Wilderness in whose backyard? *International Journal of Wilderness*. 11(3): 31–38.

Gudmundsen, Sandra; Loomis, John B. 2005. Tracking the intrinsic value of wilderness. In: Cordell, H. K.; Bergstrom, J. C.; Bowker, J. M., eds. *The Multiple Values of Wilderness*. State College, PA: Venture Publishing, pp. 251–266.

Hendee, John C.; Stankey, George H. 1973. Biocentricity in wilderness management. *BioScience*. 23(9): 535–538.

Hollenhorst, S. J.; Jones, C. D.; Rowe, D. B.; Mehmood, S.; Zhang, Y.; Williamson, J.; Shafer, E. L.; Choi, B.; Choi, E. 2004. Does wilderness impoverish rural regions? *International Journal of Wilderness*. 10(3): 34–39.

Ise, John. 1961. *Our National Park Policy: A Critical History*. Baltimore: Johns Hopkins University Press.

Kaplan, R.; Kaplan, S. 1989. *The Experience of Nature: A Psychological Perspective*. Cambridge, England: Cambridge University Press.

Keeling, P. M. 2007. Beyond the symbolic value of wildness. *International Journal of Wilderness*. 13(1): 19–23.

Kellert, S. R.; Wilson, E. O., eds. 1993. *The Biophilia Hypothesis*. Washington, DC: Island Press.

Knibb, David. 1982. *Backyard Wilderness: The Alpine Lakes Story*. Seattle, WA: Mountaineers Books.

Leopold, Aldo. 1921. The wilderness and its place in forest recreational policy. *Journal of Forestry*. 19(7): 718–721.

———. 1925. Wildernesses as a form of land use. *Journal of Land and Public Utility Economics*. 1(4): 398–404.

———. 1941. Wilderness as a land laboratory. *Living Wilderness*. 6(6): 3.

———. 1949. *A Sand County Almanac and Sketches Here and There*. New York: Oxford University Press.

Loomis, John B.; Richardson, Robert. 2001. Economic values of the U.S. wilderness system: Research evidence to date and questions for the future. *International Journal of Wilderness*. 7(1): 31–34.

Louv, Richard. 2006. *Last Child in the Woods: Saving Our Children from Nature-Deficit Disorder*. Chapel Hill, NC: Algonquin Books.

Marshall, George. 1969. Introduction. In: McCloskey, Maxine E.; Gilligan, James P., eds. *Wilderness and the Quality of Life*. San Francisco: Sierra Club, pp. 13–15.

Marshall, Robert. 1930. The problem of the wilderness. *Scientific Monthly*. 30: 141–148.

McCloskey, Michael. 1966. The Wilderness Act: Its background and meaning. *Oregon Law Review*. 45(4): 288–321.

McDonald, Barbara. 2003. The soul of environmental activists. *International Journal of Wilderness*. 9(2): 14–17.

McDonald, Barbara; Schreyer, R. 1991. Spiritual benefits of leisure participation and leisure settings. In: Driver, B. et al, eds. *Benefits of Leisure*. State College, PA: Venture Publishing, pp. 179–193.

McNeely, J. A. 2007. Wilderness and the conservation of biological diversity. In: Kormos, Cyril F.; Martin, Vance G., eds. *The WILD Planet Project*. Boulder, CO: The WILD Foundation, pp. 21–23.

Mihaley, Marc B. 1972. The Clean Air Act and the concept of nondegradation: *Sierra Club v. Ruckelhaus*. *Ecology Law Review*. 2(4): 801–836.

Moisey, Neil; Yuan, Michael S. 1992. Economic significance and characteristics of select wild land–attracted visitors to Montana. In: Payne, Claire; Bowker, J. M.; Reed, P. C., eds. *The Economic Value of Wilderness: Proceedings of the National Wilderness Conference*; May 8–11, 1991; Jackson, WY. General Technical Report SE-78. Asheville, NC: U.S. Department of Agriculture, Forest Service, Southeastern Forest Experiment Station, pp. 181–189.

Muir, John. 1938. *John of the Mountains: The Unpublished Journals of John Muir*. Wolfe, Linnie Marsh, ed. Boston: Houghton Mifflin.

Nash, Roderick, ed. 1976. *The American Environment: Readings in the History of Conservation*, 2d ed. Reading, MA: Addison-Wesley.

———. 1982. *Wilderness and the American Mind*, 3rd ed. New Haven, CT: Yale University Press.

———. 2001. *Wilderness and the American Mind*, 4th ed. New Haven, CT: Yale University Press.

Newman, P.; Dawson, Chad P. 1998. The interim management dilemma: The high peaks wilderness planning process from 1972 to 1997. In: Watson, A. E.; Aplet, G. H.; Hendee, J. C., comps. *Personal, Societal, and Ecological Values of Wilderness: 6th World Wilderness Congress Proceedings on Research, Management, and Allocation*, Vol. 1; October 24–29, 1997; Bangalore, India. RMRS-P-4. Fort Collins, CO:

U.S. Department of Agriculture, Forest Service, Rocky Mountain Research Station, pp. 139–143.

Oelschlaeger, M. 1991. *The Idea of Wilderness: From Prehistory to the Age of Ecology*. New Haven: Yale University Press.

Olson, Sigurd F. 1957. *The Singing Wilderness*. New York: Alfred A. Knopf.

———. 1963. *Runes of the North*. New York: Alfred A. Knopf.

Outdoor Recreation Resources Review Commission [ORRRC]. 1962. *Wilderness and Recreation, a Report on Resources, Values, and Problems*. Study Report 3. Washington, DC: U.S. Government Printing Office. 352p.

Oye, Garry. 2001. A new recreation strategy for national forest wilderness. *International Journal of Wilderness*. 7(1): 13–14.

Patterson, Trista. 2007. The economic value of wilderness services from and for wilderness. In: Kormos, Cyril F.; Martin, Vance G., eds. *The WILD Planet Project*. Boulder, CO: The WILD Foundation, pp. 24–27.

Power, Thomas M. 1996a. *Environmental Protection and Local Economic Well-Being: The Economic Pursuit of Quality*. New York: M. E. Sharpe.

———. 1996b. Wilderness economics must look through the windshield, not the rearview mirror. *International Journal of Wilderness*. 2(1): 5–9.

Reed, P.; Haas, G.; Beum, F.; Sherrick, L. 1989. Nonrecreational uses of national wilderness preservation systems: A 1988 telephone survey. In: Freilich, Helen R., comp. *Wilderness benchmark 1988: Proceedings of the National Wilderness Colloquium*; January 13–14, 1988; Tampa, FL. General Technical Report SE-51. Asheville, NC: U.S. Department of Agriculture, Forest Service, pp. 220–228.

Riley, Marilyn; Hendee, John C. 2001. Wilderness vision quest clients, motivations and reported benefits from an urban based program 1988–1997. In: Martin, Vance G.; Sarathy, Partha, eds. *Wilderness and Humanity: The Global Issue: 6th World Wilderness Congress Proceedings*; October 24–29, 1998; Bangalore, India. Golden, CO: Fulcrum, pp. 267–276.

Rosenberger, R. S.; English, D. B. K. 2005. Impacts of wilderness on local economic development. In: Cordell, H. K.; Bergstrom, J. C.; Bowker, J. M., eds. *The Multiple Values of Wilderness*. State College, PA: Venture Publishing, pp. 181–204.

Rudzitis, Gundars. 1996. *Wilderness and the Changing American West*. New York: John Wiley.

Russell, Keith C.; Hendee, John C.; Cooke, Steve. 1998. Social and economic benefits of a U.S. wilderness experience program for youth at risk in the Federal Job Corps. *International Journal of Wilderness*. 4(3): 32–38.

Russell, Keith; Hendee, John C.; Phillips-Miller, Dianne. 2000. How wilderness therapy works: An examination of the wilderness therapy process to treat adolescents with behavioral problems and addictions. In: Cole, D. N.; McCool, S. F.; Borrie, W. T.; O'Loughlin, J., comps. *Proceedings: Wilderness Science in a Time of Change Conference*; May 23–28, 1999; Missoula, MT. RMRS-P-15 Vol. 3. Fort Collins, CO: U.S. Department of Agriculture, Forest Service, Rocky Mountain Research Station, pp. 207–217.

Saylor, John. 1962. A report on wilderness. In: *Congressional Record*; May–June 1962. Washington, DC: House of Representatives.

Schroeder, H. W. 2007. Symbolism, experience, and the value of wilderness. *International Journal of Wilderness*. 13(1): 13–18.

Schuster, R. M.; Cordell, H. K.; Phillips, B. 2005. Understanding the cultural, existence, and bequest values of wilderness. *International Journal of Wilderness*. 11(3): 22–25.

Schuster, R. M.; Tarrant, M.; Watson, A. 2005. The social values of wilderness. In: Cordell, H. K.; Bergstrom, J. C.; Bowker, J. M., eds. *The Multiple Values of Wilderness*. State College, PA: Venture Publishing, pp. 113–142.

Scott, Doug. 1984. Securing the wilderness: The visionary role of Howard Zahniser. *Sierra Club Bulletin*. 69(3): 40–42.

———. 2004. *The Enduring Wilderness*. Golden, CO: Fulcrum Publishing.

Spring, Ira. 2001. If we lock people out, who will fight to save the wilderness? *International Journal of Wilderness*. 7(1): 17–19.

Spurr, Stephen H. 1966. Wilderness management. *The Horace M. Albright Conservation Lectureship*, Vol. 6. Berkeley: University of California, School of Forestry and Conservation.

Stahl, C. J. 1921. Where forestry and recreation meet. *Journal of Forestry*. 19(5): 526–529.

Sutter, Paul S. 2002. *Driven Wild: How the Fight Against Automobiles Launched the Modern Wilderness Movement*. Seattle: University of Washington Press.

U.S. Department of Agriculture, U.S. Forest Service. 2005. White Mountain National Forest Land and Resources Management Plan. Laconia, NH: USFS Eastern Region.

U.S. Public Law 88-577. The Wilderness Act of September 3, 1964. 78 Stat. 890.

U.S. Public Law 93-622. (Eastern Wilderness) Act of January 3, 1975. 88 Stat. 2096.

U.S. Public Law 94-579. Federal Land Policy and Management Act (FLPMA) of 1976. 90 Stat. 2743.

U.S. Public Law 95-237. Endangered American Wilderness Act of February 24, 1978. 92 Stat. 40.

U.S. Public Law 96-487. Alaska National Interest Lands Conservation Act (ANILCA) of December 2, 1980. 94 Stat. 2371.

U.S. Public Law 96-560. (Colorado Wilderness) Act of December 22, 1980. 94 Stat. 3266.

Weaver, James W.; Cutler, Rupert. 1977. Wilderness policy: A colloquy between Congressman Weaver and Assistant Secretary Cutler. *Journal of Forestry*. 75(7): 392–394.

White, David; Hendee, John C. 2000. Primal hypotheses: The relationships between naturalness, solitude and the wilderness experience benefits of development of self (DOS), development of community (DOC) and spiritual development (SD). In: McCool, S. F.; Cole, D. N.; Borrie, W. T.; O'Loughlin, J., comps. *2000 Proceedings: Wilderness Science in a Time of Change Conference*, Vol. 3; May 23–27, 1999; Missoula, MT. RMRS-P-15. Fort Collins, CO: U.S. Department of Agriculture, Forest Service, pp. 223–228.

White, Lynn. 1967. The historical roots of our ecological crisis. *Science*. 155(3767): 1203–1207.

The Wilderness Society. 2000. *The Wilderness Act Handbook*, 4th rev. ed. Washington, DC: The Wilderness Society.

Wilson, Edward O. 1984. Million-year histories: Species diversity as an ethical goal. *Wilderness*. 48(165): 12–17.

Worf, William A. 2001. The Forest Service recreation strategy spells doom to the National Wilderness Preservation System. *International Journal of Wilderness*. 7(1): 15–16.

Worf, William A.; Jorgenson, Glen; Lucas, Robert. 1972. *Wilderness Policy Review*. Washington, DC: U.S. Department of Agriculture, Forest Service.

Yuan, Michael S.; Christensen, N. A. 1992. Wilderness-influenced economic impacts on portal communities: the case of Missoula, Montana. In: Payne, Claire; Bowker, J. M.; Reed, P. C., eds. *The Economic Value of Wilderness: Proceedings of the National Wilderness Conference*; May 8–11, 1991; Jackson, WY. General Technical Report SE-78. Asheville, NC: U.S. Department of Agriculture, Forest Service, Southeastern Forest Experiment Station, pp. 191–199.

Zivnuska, John A. 1973. The managed wilderness. *American Forests*. 79(8): 16–19.

Chapter 2
Historical Roots of Wilderness Management

by Roderick Nash and John C. Hendee

Introduction	31
The Intellectual Dilemma	31
Destruction by Popularity—A Threat to Wilderness	32
Roots of Wilderness Appreciation	36
Evolution of Wilderness in the National Parks	36
Evolution of Wilderness in the National Forests	39
The Wilderness Management Idea Begins	43
Wilderness Management Evolves	45
Study Questions	47
Acknowledgments	48
Case Discussion: Modern-Day American Views on Wilderness	48
Case Discussion Questions	48
References	49

Introduction

Management of a designated wilderness can be thought of as a contradiction in terms. It could even be said that any area that is proclaimed wilderness and managed as such is not wilderness by these very acts. The problem is that *wilderness* traditionally means an environment that humans do *not* influence, a place they do *not* control.

Before herding and agriculture began, there was no distinction between wilderness and civilization. As hunters and gatherers, people did not control their environment; they simply lived in it as part of the ecosystem. However, with the beginning of herding, and later of agriculture, *Homo sapiens* began to experiment with environmental modification by domesticating animals and managing plants, soil, and water. In time, totally humanized environments called towns and cities developed. In the process, people indirectly created wilderness by drawing a physical and, even more importantly, a mental distinction between the places they controlled and the places they did not control (Nash 1975).

Etymologically, the word *wilderness* is derived from the Old English wild-deor-ness, the place of untamed beasts (Nash 1970b). Some scholars believe the origin is the concept of "self-willed" land. *Civilization*, conversely, is an environment under human control. Understandably, since the advent of the civilization that created the word *wilderness*, it has stood for the dark, the chaotic, the unknown and fearful, and the back of beyond. It was defined by the absence of the controlling structures of institutions and technologies. Outlaws and brigands of times past, like today's revolutionary guerrillas and marijuana growers, sought wild country for the same reason it attracts some of today's backpackers who seek to escape civilization's nearly omnipresent cloak of control.

The Intellectual Dilemma

The only wilderness true to the etymological roots of the word is that which humans do not influence in any way whatsoever. The more people learn about wilderness, the more they visit it, map it, manage it, and write about it, the less wild it becomes. From such a perspective, even knowledge about a region disqualifies it as wilderness in the true etymological sense. Management of any kind compromises wildness. Even maps, trails, and signs are civilizing, steps toward ordering the environment in the interest of people, toward lessening the amount of the unknown. The association of rangers, wardens, "adopt a wilderness" groups, "save the whatever" organizations, and search-and-rescue teams for a given area obviously detracts from its wildness.

More subtle, but just as foreign to pure wilderness, are sophisticated management techniques. The notions of carrying capacities, use permits and quotas, regulations on behavior, prescribed fire, and fire control gradually erode the "wild" from "wilderness." For many, the wilderness experience is affected when recreational demand makes it necessary to wait years or trust to a lottery for an entrance permit (e.g., private users seeking trips on the Colorado River in Grand Canyon National Park, Arizona, or wilderness hikers to Mount Whitney on the Inyo National Forest, California) or, at the peril of arrest, or citation and fine, to maintain a rigid backcountry travel itinerary so other parties, following a day behind, have places to camp (for example, in the Boundary Waters Canoe Area Wilderness on the Superior National Forest, Minnesota). For many users, this kind of intensive management transforms wilderness into an open-air motel with advance registration and checkout times. The resulting dissatisfaction underscores the need for less restrictive, *light-handed management techniques* that emphasize visitor education, and voluntary rather than enforced restraint. Indeed, as this book describes, such techniques are evolving from experience and research, along with a management ethic to use only the minimum tools to achieve desired results.

The intellectual dilemma posed by a managed wilderness is compounded by the fact that, *in the last analysis, wilderness is a state of mind. Like beauty, wilderness is defined by human perception.* For some individuals, regulations will not be distracting. However, for others, just the *knowledge* that they visit an area by the grace of, and under conditions established by, a government agency detracts from a wilderness experience. Ironically, the success of management in protecting the wilderness experience may decline in proportion to its effectiveness. *Wilderness management in its first few decades has been largely people management. Now more and more focus is being placed on wilderness ecosystems.* The test may be whether we can resist tinkering with them, or as Glover (2000, 4) asks, "Can we stop trying to control Nature?"

It is hard to deny the principle that management is essential today if wilderness is to have any meaning at all. The pure definition of wilderness (no maps, no knowledge, and a totally blank space on the map) is, at least between the 60th parallels, a thing of the past. No one in the United States can ever again have the experience of a Lewis or a Clark, a Jim Bridger, or a John Wesley Powell. It is even wishful thinking to suppose one might today duplicate David Brower's 1930s experience of making a recorded first ascent almost every time he climbed a Sierra peak. The best that can be hoped for, in the American West for instance, is a chance to be in beautiful and comparatively natural country, away from roads, relatively alone, and dependent, in the short run, on one's own resources for comfort and survival.

Thus, a key factor compelling acceptance of managed wilderness is awareness that *without management, contradictory as it may seem, wilderness would surely be impaired or lost* (Nash 1978, 1982, 2001).

Destruction by Popularity— A Threat to Wilderness

The harbingers of wilderness destruction by popularity are certain spectacular areas to which the recreation-minded public flocks in increasing numbers. Mount

Fig. 2.1. In the twentieth century, increasing numbers of Americans began to consider wilderness a resource; they no longer regarded it as an obstacle to be conquered. Early wilderness visitors, such as this man, did not have lightweight equipment, portable stoves, or dehydrated food. Today, technology includes cell phones, global positioning systems, handheld computers, and other innovations that make wilderness more accessible, but to the minds of some, defeat the spirit of the wilderness experience. Photo courtesy U.S. Forest Service.

Whitney, the highest peak in the United States outside Alaska, is a good example. Mount Whitney is part of the Sierra Nevada and lies in the John Muir Wilderness on the Inyo National Forest and in Sequoia Wilderness in national park wilderness. The peak's first recorded ascent was in 1873; in 1973, approximately 14,000 persons made the climb. A dramatic illustration of the changes popularity has brought to Mount Whitney comes from a man who, on August 4, 1949, climbed the peak with his father. Proudly, they signed the summit register, the sixth and seventh individuals to have done so *that year*. On August 11, 1972, this same man climbed Mount Whitney with his son. Upon signing the register, they noted with some shock that they were the 259th and 260th persons on record *that day*! Presumably, there was less pride, and certainly less wilderness, in the experience.

However, even this example pales in comparison to today's visitation to Mount Whitney. It requires a permit year-round and from May 1 to November 1 there is a limited entry. According to Brian Spitek (2007), wilderness manager on the Mount Whitney Ranger District of the Inyo National Forest, in 2007, permits were issued covering 16,315 visitors for the Whitney Portal Trailhead alone. For North Fork of Lone Pine Creek Portal, the mountaineer's route, permits for an additional 3,000 people were issued. An additional 3,882 legally permitted visitors reached the mountain from other routes, such as the John Muir Trail, or the west side through Sequoia–Kings Canyon National Park. Thus, 23,197 visitors had legal permits to climb Mount Whitney during 2007, plus an unknown number of illegal visitors who dodged the permit system.

Additional evidence that popularity can affect the experience and resource comes from the Grand Canyon, where the more-than-200-mile float trip of the Colorado River is perhaps the most intensively supervised wilderness activity in the United States today. Close control by national park officers is facilitated by severely limited access to the river and the expedition-level difficulty of the trip. As a result, an exceptionally complete set of visitation statistics has been compiled (Table 2.1).

In 2007, a new Colorado River Management Plan (U.S. Department of the Interior, National Park Service 2006) went into effect with a strategy to implement a "launch-based" system that is intended to move toward a more even distribution of use through the year and to achieve more equity between commercial and noncommercial visitor use. The 2006 plan capped Colorado River use at 115,500 commercial user days (the same as the 1989 NPS plan), and 114,006 commercial user days were recorded in 2007, which converts to 18,417 commercial visitors. The number of visitor days for private,

Table 2.1. Number of Visitors Floating the Colorado River Through the Grand Canyon of Arizona

Year	Visitors	Year	Visitors	Year	Visitors	Year	Visitors	Year	Visitors
1867	1[1]	1952	19	1962	372	1973	15,219[3]	1983	15,443
1869–1940	73	1953	31	1963–1964	44[2]	1974	14,253	1984	15,952
1941	4	1954	21	1965	547	1975	14,305	1985	18,113
1942	8	1955	70	1966	1,067	1976	13,912	1986	21,168
1943–1946	0	1956	55	1967	2,099	1977	11,830	1987	18,008
1947	4	1957	135	1968	3,609	1978	14,356	1988	22,088
1948	6	1958	80	1969	6,019	1979	14,678	1989–1998	22,000
1949	12	1959	120	1970	9,935	1980	15,142	1999–2000	23,000[4]
1950	7	1960	205	1971	10,385	1981	17,038	2001–2006	22,200[5]
1951	29	1961	255	1972	16,432	1982	16,949	2007	24,735[6]

1. Some contend that James White, a trapper fleeing Indians, floated the Grand Canyon on a log raft two years before the Powell expedition.
2. Travel curtailed by the completion of the Glen Canyon Dam and the resultant disruption of canyon flow.
3. The downturn in visitation after 1972 resulted from a quota system instituted by Grand Canyon National Park. Numbers applying for noncommercial permits continued to rise sharply.
4. Use was capped by the NPS (Jalbert and Trevino 2000). The number of visitors is estimated from cap in 169,000 user days.
5. Use ranged between 21,613 and 22,400 recreation users and between 165,953 and 177,325 user days in 2001–2006.
6. A new Colorado River Management Plan took effect in 2007. Commercial use remained capped by the NPS at 115,500 and noncommercial user days were allowed to increase. The total number of visitor days was 219,444 in 2007, with 114,006 commercial user days and 105,438 noncommercial user days.

Seventeen-Year Wait for a Grand Canyon Float Trip

Jan Pehrson, a sailing photojournalist, in Sausalito, California waited seventeen years to get a noncommercial permit to float the Grand Canyon. A boating person, but not a river expert when she applied in 1988, she was assured there would be plenty of "river-rat" volunteers with equipment and expertise eager to help staff and equip her trip when she received a permit. Until then she had to file a note of "continuing interest" with the National Park Service. Each year she filed the note and watched her position move up in a queue of thousands of applicants to hundreds, until finally she received her permit for May 2006.

The noncommercial permit allowed her to have no more than sixteen persons on the river at one time and required her to pay an NPS fee of $50 per person. No money was to be paid to anyone other than for the shared expenses of the participants, which amounted to about $800 apiece for the 18-day, 226-mile trip. In order to share the rare opportunity with twenty people rather than just sixteen, four individuals hiked out and four new participants hiked in to join the group at Phantom Ranch on day 8, about 87.8 miles downstream from the Lees Ferry launch point. The trip then continued, arriving at the Diamond Creek take-out at mile point 225.7 on day 18.

Selecting trip participants was very important for such a challenging adventure, and Jan bore the full responsibility for this as the permit holder. Volunteer, experienced river runners with all the essential equipment were located by networking through friends and acquaintances, with the rest of the participants selected to create a compatible, multigenerational group of men and women ranging in age from their early twenties to sixty. Camp chores and other work were shared, but on the river individuals filled assigned roles as "lead oar," rowers, and passengers. The group navigated the river in five rafts carrying food and supplies and from one to four persons each, plus an inflatable kayak. No rafts flipped despite some close calls, and when the inflatable kayak flipped there were no injuries.

As to it being a wilderness experience Jan said: "every day, we saw quite a few commercial groups—maybe ten, heading down river under motor power in two to four rafts with people in lifejackets lined up shoulder to shoulder on the seats. You could hear these trips coming. Almost every party we saw was a commercial trip. One misjudged the current and got skewered on a big rock; we pulled them off. I only remember a few noncommercial trips in the eighteen days we were on the river."

When asked about high points of the trip Jan responded: "It was the culmination of nearly a twenty year wait, so the fact that we had a safe trip and good group experience were the highlights. Contributing to it all was experiencing growth as a small community and as individuals, conquering fear, gaining skills and confidence in navigating rapids, working together and depending on one another. We experienced some minor aches and pains, but no sickness. It was an experience of a lifetime with close friends and people we liked and respected. What could be better than that?"

—Jan Pehrson (© 2006)

noncommercial use was increased in 2007 under the new permit system to 105,438 visitor days, which converts to 6,318 visitors. Starting in 2007, the old waiting list system for permits was replaced by a "weighted" lottery system that is based on length of time the applicant has been applying for a permit. Also, any trip cancellations can be used by a replacement applicant and are not lost from use (Jalbert 2008).

From reviewing these statistics (Table 2.1) and realizing that more use occurs in the summer months, it is clear that the quality of wilderness experienced by the early Grand Canyon river runners has declined precipitously. Some argue that, enjoyable as it is, the locale can no longer be considered wilderness. In disgust, they turn to the few remaining wild rivers, perhaps in Alaska and the Canadian northland. However, many others comply, albeit reluctantly, with the strict management policies currently in effect for the Grand Canyon. The logic that persuades them might be illustrated by comparing access to the Grand Canyon and other popular wildernesses with access to playing time on tennis courts.

Tennis players would obviously prefer to play when they wish, for as long as they wish. However, the popularity of the game does not permit this luxury except on private courts, which can be compared to the game reserves of medieval nobility. On public, tax-supported courts (as in publicly supported wilderness areas) demand frequently exceeds available space. Hence management devices are instituted, such as sign-up sheets, time and frequency limitation rules for waiting players, and encouraging doubles games. Court monitors, like wilderness rangers, enforce the regulations.

An alternative response to the tennis problem would be to have no management. Everyone who wanted to play could squeeze onto a court. "Triples" would be common on the popular courts, and in peak-demand periods, a kind of volleyball-with-rackets, with as many as twenty-five on a side, could be played.

From this perspective, acceptance, indeed preference, for management or self-restraint is understandable. Players recognize that tennis is a game that is played by two or four persons. Out of respect for the integrity of the game, and with their own self-interest in mind, players support management. They sign up, wait their turn, and vacate the court at the appointed hour.

Like tennis, wilderness recreation is also a game that cannot be played at any one time and place by more than a few persons. Moreover, it is a game that depends on the existence of a relatively unmodified natural and physical environment. These realizations prompt many to accept management. The resulting regulations may be distasteful, and are clear violations of the traditional sense of wilderness, but they are the best hope of salvaging an approximation of the wilderness experience from the pressures of popularity.

Of course, another response to the increasing popularity of tennis is to construct more courts. Public agencies are sometimes able to accommodate this demand. However, in the case of wilderness, this is difficult if not impossible. A second Grand Canyon is out of the question, and the total amount of wilderness left on the planet is shrinking rapidly. When Congress designates additional wildernesses, it does not really create more wilderness; it merely recognizes and establishes permanent protection for an already established use of the area (see Chapter 14).

Some tennis players have responded to the frustrations of obtaining court time on public facilities by joining private clubs. In a sense, they have found an economic solution to the problem of scarcity. A few wilderness enthusiasts have followed suit, joining hiking and camping clubs located on private land. Indeed, privatization of outdoor recreational resources is a growing trend, and at least one private wilderness area exists in the United States (Johnson 1996).

The basis for any historical discussion of wilderness management recognizes that management is a necessary newcomer to the wilderness movement. The concept of preservation existed long before there was management except for fire control and associated lookouts, trails, and guard cabins. For decades, few even thought about managing wilderness. Perhaps this tendency sprang from the contradiction between the concepts of wilderness and management. Wilderness was not *supposed* to be managed. It was the region that began where management stopped. However, certainly equal in importance was the fact that for years there were few significant overuse or visitor management problems. As late as 1949, when only a dozen persons a season were climbing Mount Whitney and about the same number were running the Colorado River through the Grand Canyon, their control could hardly be regarded as a pressing issue. For a transitory, enchanted moment in American environmental history, wilderness preservation could exist without wilderness management.

Fig. 2.2. Better access, light and efficient recreational equipment, population growth, and pressures of urban life are a few reasons for increased recreational use of wilderness. Here visitors contemplate Green Lake in North Cascades National Park, Washington, after a tough cross-country hike. Photo courtesy John C. Hendee.

Roots of Wilderness Appreciation

The first accounts of wilderness pleasure trips make us realize just how spectacularly empty the country was at that time. Consider, for example, the 625-mile hike that Joseph N. LeConte and three companions made in the southern section of the Sierra in 1890 and recorded by LeConte in his journal (LeConte 1972). The trip is interesting for its parallels and its contrasts with contemporary patterns of wilderness recreation. The four men, all in their early twenties, were students at the University of California, Berkeley. The trip was their summer vacation; they went into the mountains for fun. One is struck, immediately, by the total lack of regulation. There were no permits in 1890, no regulations, no fish and game laws, not even clear maps. The students simply packed up—using burros rather than backpacks—and headed out. Much of the time they had only a general idea of where they were, and a considerable part of their adventure was knowing no one was poised to bail them out of trouble. In 1890, the wilderness was noteworthy for its emptiness. Except for a few miners on the lower Kings River and one of their geology professors conducting experiments above Yosemite Valley, they saw no people.

The Sierra was even emptier twenty years earlier when John Muir ranged through the mountains with a pocketful of bread and a little tea. Even the Sierra Club's organized outings, begun under Muir's leadership in 1901, hardly compromised the Sierra's isolation. In his day, using pack animals restricted wilderness recreation to easier routes and lower passes.

"Off the beaten track" has been transformed by contemporary backpacking, in part due to the evolution of backpacking equipment. Old-timers are astonished at what the outdoor-equipment industry has wrought. In an earlier era, huge bedrolls, heavy tents, and the weight of canned foods sharply limited the places one could visit and the amount of time a party could spend in wilderness. So did the lack of portable, efficient gear for winter camping and rock climbing. Today's light, streamlined equipment has opened the most remote places at all seasons. Then, too, in earlier years, the population of the United States was much smaller. Moreover, compared to contemporary Americans, most of our grandparents had limited mobility (fewer cars, poor roads), less leisure, and greater dedication to the work ethic.

However, the principal reason wild places were empty was that very few Americans cared to visit them. Even as late as World War II, wilderness appreciation was still in its infancy. The explanation lies in the heavy burden of suspicion and fear that wilderness carried as heritage from a pioneering past. Wilderness was not easy to appreciate, as it was something fought since the dawn of civilization. It was a matter of being too close to wilderness, of having too much of it. For appreciation to flourish, wilderness had to become more scarce, and this, in turn, depended on the rise of an urbanized, industrialized society. The United States was at the brink of the transition from a developing to a developed nation in the late nineteenth century. The frontier ended, according to the U.S. census, in 1890, and then large numbers of Americans began to consider wilderness a resource to be enjoyed rather than an adversary to be conquered.

Evolution of Wilderness in the National Parks

Understandably, the first interest in wilderness for recreation was tentative, and only flirted with the wilderness experience. People wanted wilderness, but not too much. They preferred to be on its edge, to look at it, and to have the security and comforts of civilization. This was the context that generated the first wilderness management decisions. For example, the preservation of the world's first national park, Yellowstone, in 1872, had little to do with providing a true wilderness experience for vacationing Americans. The intent of Congress, as stated in the act creating Yellowstone National Park in 1872, was to create a "public park or pleasuring ground for the benefit and enjoyment of the people." Study of the intent of Yellowstone's proponents indicates that they expected the "enjoyment" to be derived from viewing scenic wonders such as geysers, hot springs, and waterfalls from the civilized vantage point afforded by luxurious lodges. Even Nathaniel P. Langford, a leading explorer and publicizer of the first national park, enthusiastically predicted that it would not be long "before the march of civil improvements will reclaim this delightful solitude, and garnish it with all the attractions of cultivated taste and refinement" (Langford 1972, 97). This was entirely consistent with the established pattern of nature tourism of the nineteenth century, which emphasized the edge of wildness, convenient transportation (usually railroads), and lavish hotel accommodations (Runte 1979).

What did visitors to the early national parks expect? The brochures and literature distributed by park promoters invariably featured the attributes of civilization: comfortable coaches, grand lodges, elegantly dressed tourists. Far from enticing the visitor with visions of wilderness camping, the advertisements tried to convince the tourist that there was no need to rough it. Wildness was to be enjoyed, but at a distance. Too wild a park, it was rightly assumed in the late nineteenth and early twentieth centuries, could be a deterrent to tourism. Interest in wilderness was growing, but it had not yet affected recreational desires enough for wilderness management to exist, even as a concept.

As for the roads and hotels that "opened up" Yellowstone and the other early parks and determined their dominant use, park personnel or Congress *did not decide* to feature wilderness as an attraction. It wasn't even an issue; people took conveniences for granted. The language used in the Yellowstone Act made such an interpretation easy. As long as the timber, mineral deposits, natural curiosities, or wonders were preserved in their natural condition, there was no problem with developing the park for mass tourism. Specifically, Old Faithful and Yellowstone Falls were the objects of concern, not the wild backcountry of the park. As long as these "wonders" were kept in public ownership and free from vandalism, the nineteenth-century purposes of the park were fulfilled. Ironically, today a major push of environmentalists is to rid Yellowstone of an estimated 80,000 snowmobile trips each winter, replacing them with passenger vans to ease the noise, exhaust, and wildlife impacts.

Even the most ardent wilderness preservationists of the time, people like Sierra Club president John Muir, accepted this premise of mass tourism development and its management implications. In 1913, Muir *supported* admitting the first private automobiles into Yosemite Valley (Lillard 1968). His reasoning centered on the need to bring people into the parks to build citizen support for the park idea. Along with cars came civilized lifestyles. Hotels, like the posh Ahwahnee completed in 1926 in the scenic heart of Yosemite Valley, contributed to the people's outdoor pleasure as it was defined in the first decades of the twentieth century. In addition, because the national parks were established by law to be "pleasuring grounds," who could object?

Muir bitterly fought economic development of park wilderness. He opposed grazing, mining, logging, and, unsuccessfully, the 1913 decision in favor of hydropower development that inundated Hetch Hetchy Valley in northern Yosemite National Park. However, Muir did not recognize development for recreation, for the public's pleasure, as a comparable threat to wilderness. He died in 1914, long before wilderness advocates understood that management was an unpleasant necessity.

The passage of the National Park Service Act on August 25, 1916, did nothing to change earlier conceptions of the meaning, purpose, and appropriate uses of national parks. Although the legislation stipulated that anything done in the parks must leave the scenery and wildlife "unimpaired," the whole reason for their existence was indisputably public enjoyment. However, pleasure-seeking people could impair nature. The ambiguity inherent in the National Park Service Act has been the source of extensive commentary and still more extensive agony for subsequent park managers (Sax 1980). However, for Americans in 1916, there was considerably less inconsistency in the act. Because wilderness protection and the provision of a wilderness experience were not recognized goals of park management, few questioned developments such as roads and lodges that eroded wildness. Why should they have questions, after all? Hardly anyone went into the park backcountry at this time. Dramatic assertions (probably true) that more than 90 percent of these early visitors saw only 3 percent of the park were not the result of any conscious management policy. They reflected quite accurately the tastes of recreation-minded Americans in the early twentieth century. Most people did not *want* to experience park wilderness. The practical implications for management of these conceptions of park means and purposes can be found in a letter of management direction, May 13, 1918, from Secretary of the Interior Franklin K. Lane to Stephen T. Mather, the first director of the National Park Service (U.S. Department of the Interior, National Park Service 1970).

The letter opens with the standard insistence that the parks be kept in "absolutely unimpaired form," but quickly makes compromises on behalf of public enjoyment. From the standpoint of wilderness preservation, the most damaging aspect of Lane's letter is the assumption that the public should be encouraged to enjoy the parks "in the manner that best satisfies the individual taste." There is no attempt to define what kind of enjoyment is appropriate in a national park, and no effort to distinguish uses that are consistent with the mandate to leave parkland unimpaired. In effect, Lane is saying that the citizen will bring individual preferences to the parks and the parks will fulfill them. The 1918 letter makes clear that "automobiles and motorcycles will be permitted in all of the national parks; in fact, the parks will be kept accessible by any means practicable." The implications of this statement are extraordinary and clearly work against wilderness. So does the secretary's directive to encourage a full range of accommodations from "luxurious hotels" to "free campsites." Nothing is said about low-density, off-road wilderness uses of the parks. Again, the point is that in 1918, wilderness recreation was not considered part of the statutory purpose of national parks. Americans of this period gave little evidence of being disappointed with such a definition and the resulting management policy.

Stephen T. Mather was the ideal director of the National Park Service (NPS) under the explicit and implicit mandates of the early twentieth century. His talent was public relations, and he recognized that national park survival and growth depended on skillful playing of the "number of visitors game" in the political arena. Immediately after passage of the 1916 legislation, Mather launched a vigorous program to boost visitation to national parks. It included a series of publications, the work of Robert Sterling Yard, and initiating or continuing management policies designed to attract and please visitors. Mather and Yard knew that wilderness might not "sell" to their contemporaries. Instead, they cultivated a resort or circus image of parks. Drive-through sequoia trees, cut initially in the 1880s, continued to be a tourist attraction at Yosemite. At Yellowstone, soap was regularly dumped into the geysers to break their surface tension and cause eruptions at times convenient to tourists. At Old Faithful, the symbol of America's national park in this period, colored spotlights from adjacent hotels illuminated night eruptions. During the hour between eruptions, tourists were entertained by radio music.

Yellowstone's famous roadside bears shared top billing with Old Faithful. By explicit direction of Director Mather and his assistant and subsequent director, Horace M. Albright, the bears were regularly fed hotel garbage before grandstands of camera-wielding tourists. In the 1920s, it must be remembered, bear feedings and caged wildlife around hotels did not violate national park purpose; rather, they expressed it. Public enjoyment could easily be stretched to cover such activities.

At Yosemite National Park in the 1870s, the "firefall" replaced the "chicken fall," in which live chickens were tossed over the cliffs. It continued, under Mather and Albright, to dominate the park experience for most tourists. The firefall involved constructing a huge wood fire on the lip of Glacier Point 3,000 feet above the floor of Yosemite Valley. As dusk fell, the crowds gathered. Music played ("Indian Love Call" was a favorite), and at a voice signal, "Let the fire fall!", the burning logs and embers were pushed over the cliff. The potential for forest fire was fully recognized and carefully avoided, but for decades no one even questioned whether the firefall was an appropriate activity for management to sponsor in a national park. No one asked if this was the *kind* of "enjoyment" parks were created to provide. It was not until the late 1960s that changing interpretations of the meaning and purpose of national parks led to limiting road access and replacing the *resort concept* with more wilderness orientation in national parks (Sutter 2002).

There had, however, been earlier indications of wilderness consciousness in national park circles. In 1929, the phrase "original wilderness character" was used in certain versions of the bill establishing Grand Teton National Park (WY). No hotels or new roads were to be permitted in the park. Although stricken from the final text of the bill (Grand Teton National Park Act of 1929), the omitted phrase clearly indicated a desire to emphasize wilderness in the Grand Teton National Park. The first explicit recognition of wilderness in national park legislation appeared five years later in the act establishing Everglades National Park. Section 4 specified that the Florida wetlands would be "permanently preserved as a wilderness." With an eye toward management, the bill went on to say that "no development of the project or plan for the entertainment of visitors shall be undertaken which will interfere with the preservation of the…essential primitive natural conditions now prevailing in this area" (Everglades National Park Act of 1934). Another milestone in the development of the wilderness idea was the 1940 establishment of Sequoia–Kings Canyon National Park (CA), south of Yosemite, as a roadless and hotel-less park. Providing a wilderness experience was, necessarily, its main objective. Indeed, a preliminary version of the establishing legislation even used the name "Kings Canyon Wilderness National Park." The Sierra Club regarded Kings Canyon as compensation, in a sense, for the heavily developed character of Yosemite. Together the two parks would serve a wide spectrum of recreational interests.

Other evidence shows some early awakening of the need to be concerned about management. In 1928 and 1929, George M. Wright of the NPS saw the need for an organizational unit to monitor impacts on wildlife in national park ecosystems. Wright organized a small group of individuals to begin a nationwide systematic survey of the status of wildlife in the parks, with development of a well-defined wildlife policy as its goal. The work of this group (Wright et al. 1932; Wright and Thompson 1934) provided a historic data baseline concerning wildlife in the parks, including such wilderness-dependent species as the wolf and grizzly bear. Wright's work led to the establishment of the Division of Wildlife Research in the NPS, with Wright as its first head, and ultimately the George Wright Society, which today provides for an annual conference and proceedings of research papers by national park scientists.

The expansion of thinking to include park and wilderness management as well as designation was beginning. One of the first written examples of this transformation as it concerned national parks was a 1936 report of Lowell Sumner, a regional wildlife technician.

In his policy recommendations for Sierra parks, Sumner (1936) wondered "how large a crowd can be turned loose in a wilderness without destroying its essential qualities" (n.p.). He realized that for wilderness to exist in the parks, the areas "cannot hope to accommodate unlimited numbers of people" (n.p.). Construction of tourist facilities would have to be restricted. Finally, Sumner's insights extended to the understanding that wilderness managers could also pose a threat to wilderness values. He urged that only "the very simplest maintenance activity" be undertaken in wilderness (Sumner 1936, n.p.). Heavy management, in other words, could be a liability in dealing with an experience featuring solitude, self-reliance, and freedom from the controls normally present in civilization.

Sumner's thinking on these points matured so that six years later he could discuss the adverse effects of pack-stock grazing, fishing, and sheer numbers of visitors on the biological balances of wilderness areas. Then, in one of the first uses of the term, Sumner (1942, 21) urged that use of wilderness be kept "within the carrying capacity or 'recreational saturation point.'" His 1942 definition described carrying capacity as "the maximum degree of the highest type of recreational use which a wilderness can receive, consistent with its long-term preservation." Wilderness managers should "determine in advance the probable maximum permissible use, short of impairment, of all wilderness areas" (Sumner 1942, 22). Here, in 1942, was the basic logic of modern wilderness recreation management.

Evolution of wilderness in the national parks is not complete. While 84 percent of the NPS is protected, most of it is still waiting to be legislatively designated by Congress (Potts 2007). See Chapter 5 for more information on designation of wilderness in the NPS.

Evolution of Wilderness in the National Forests

Although the national parks of the early twentieth century were playing to crowds who had little interest beyond visiting "pleasuring grounds," the USFS took the first steps toward explicitly identifying wilderness as a specific recreational resource and developing appropriate management techniques. While Gifford Pinchot headed the U.S. Division of Forestry (after 1905, called the U.S. Forest Service), the emphasis was on producing commodities, consistent with the 1897 Organic Act, which mentioned only timber, grazing, and watershed protection as uses of the forest reserves (Roth 1984). The forests were to be used, albeit carefully, as a constant source of valuable products. After Pinchot's departure from office in 1910, the meaning of "forest products" was expanded. Some people began to understand that forests were valuable for more than just commodities. Henry Graves, the new chief of the USFS, began to conceive of the national forests as valuable for recreation. Of course, in these early years, *recreation* meant almost every imaginable outdoor activity, but wilderness had a small and growing significance. For instance, in 1910 Graves asked Treadwell Cleveland, Jr., to write an essay on public-recreation facilities for the American Academy of Political and Social Science. The resulting discussion of the use of logging roads, bridges, and trails by the hunter, angler, and picnicker was unprecedented in the history of American forestry. Moreover, Cleveland (1910, 245) made a significant prediction:

> So great is the value of national forest area for recreation, and so certain is this value to increase with the growth of the country and the shrinkage of the wilderness, that even if the forest resources of food and water were not to be required by the civilization of the future, many of the forests ought certainly to be preserved…for recreation use alone.

However, the USFS, like the NPS, was constrained by the anti-wilderness bias of public opinion in this era. Few people wanted to rough it. Recreational development, therefore, consisted of extending forest roads and leasing sites for summer home and hotel construction. Chief Forester Graves was enthusiastic about progress in these areas in his 1912 report; three years later, he obtained permission from the secretary of agriculture to extend summer home leases to thirty years. The result? More permanent structures were built. Wilderness suffered, but at the time few Americans really cared.

In 1918, landscape architect Frank A. Waugh prepared a report for the USFS titled "Recreation Uses on the National Forests" (Waugh 1918). It marked the emergence of full awareness that recreation was an established rationale for national forests. William B. Greeley, who became chief forester in 1920, and his associate forester L. F. Kneipp gave increasing emphasis to this use and even secured budgetary appropriations for recreation beginning in 1922. Greeley, in particular, valued forest scenery, and on several occasions in the early 1920s vetoed tourist-development plans. The most important decision of the era affecting wilderness concerned the spectacular Trappers Lake in Colorado on national forestland. In 1919, a young USFS landscape architect named Arthur H. Carhart was assigned to survey the area for road access and several hundred vacation homes. The plan was entirely in keeping with USFS definitions of recreation, but Carhart was

Fig. 2.3. Arthur Carhart (left) and Aldo Leopold (above) were instrumental in setting aside 574,000 acres of the Gila National Forest, New Mexico (right), in 1924 as the first designated wilderness in the United States. Photos courtesy U.S. Forest Service.

troubled. The beaver of Trappers Lake had been exploited in the 1850s, but otherwise the lake was untouched and reachable only by a tough five-mile trail. Realizing the rarity of such wilderness in the American West, Carhart had misgivings about developing Trappers Lake even for recreational purposes. He was encouraged in this line of thinking by several wealthy hunters who wanted the country to remain pristine. So, after a summer spent not only surveying, but also developing a conviction that the area should be preserved in its pristine condition, Carhart had the courage to recommend doing nothing at all to Trappers Lake. Probably to his surprise, the Denver District Office of the USFS approved the idea. Trappers Lake was left without roads or summer homes (Allin 1982; Baldwin 1972).

Arthur Carhart followed his pro-wilderness recommendation in Colorado with a similar one for the Superior National Forest in Minnesota. Moreover, late in 1919, he met with the young, nontraditional forester Aldo Leopold. The disappearance of large roadless areas in Arizona and New Mexico was evoking in Leopold misgivings similar to Carhart's; Leopold's efforts to retain large sections of country devoid of human influence included a call for a wilderness of at least 500,000 acres for each of the eleven states west of the Great Plains (Leopold 1921; Roth 1984). In 1924, Leopold had the satisfaction of seeing the USFS designate 574,000 acres of the Gila National Forest, New Mexico, as a reserve for wilderness recreation. The efforts of Carhart and Leopold produced the first designation of public lands specifically for wilderness values in America and, indeed, in the world.

The management consequences of establishing the Gila Wilderness Reserve were minimal. A laissez-faire approach prevailed—prohibit building roads and hotels and then leave it alone. In the 1920s, no attempt was made to manage positively for wilderness values, recreational or otherwise. Wilderness was simply set aside.

Chief Forester William B. Greeley exemplified this philosophy in action in the 1920s. He was enthusiastic about creating wilderness reserves on national forests, largely because he feared that the aggressive leadership of Stephen T. Mather, head of the National Park Service, was threatening his own empire (Gilligan 1953). If the USFS did not move to protect its spectacular scenery and develop its recreational resources, there was a good chance that some of its land would be turned over to the NPS. Such considerations unquestionably supported the intentions of some foresters to preserve wilderness simply because it was a good thing to do.

In 1926, Greeley formulated a policy for wilderness. Commercial uses such as grazing and even logging could continue, but campsites, meadows for pack-stock forage, and special scenic spots, as they were called (Gilligan 1953), would be protected. Greeley also instructed Associate Forester L. F. Kneipp to inventory national forest wilderness, the first roadless area review. The result showed 74 areas, each at least 360 square miles, in the 48 states. The chief forester's ideas of management stopped at this point.

In a 1926 communication to his several districts, Greeley explicitly disavowed any intention to regulate the numbers or the behavior of recreational users of wilderness. "I have no sympathy," he declared, "for the viewpoint that people should be kept out of wilderness areas in any large numbers because the presence of human beings destroys the wilderness aspect…Public use and enjoyment are the only justification for having wilderness reserves at all" (as cited in Gilligan 1953, 104). As for the numbers of visitors, "the only limitation should be the natural one set up by the modes of travel possible" (as cited in Gilligan 1953, 104). Clearly, Greeley did not foresee the time when such limitations would be insufficient to keep wilderness from being degraded, ironically, by those who loved it. In the 1920s and 1930s, there was little reason to worry about overuse because the backcountry was still relatively empty.

At a 1926 session of the National Conference on Outdoor Recreation, Aldo Leopold made a strong plea for more systematic planning to protect wilderness (National Conference on Outdoor Recreation 1926). In Associate Forester Kneipp, Leopold found a supporter close to the center of power in Washington. Three years later, on July 12, 1929, Kneipp wrote Forest Service Regulation L-20, to order and consolidate what had until then been piecemeal preservation. This directive, while not law, was only an expression of agency policy, and standardized the term *primitive area* for a decade. Interestingly, the term *wilderness* was not used in L-20 because Kneipp and his colleagues thought the public might be repelled by its connotations (Pomeroy 1957). Kneipp admitted that the term *wilderness* did not apply to regions that had been, and still were being, commercially used.

L-20, with the amendments and instructions that followed it, required field staff to submit definite management plans for each primitive area. (Further discussion of the L-20 Regulation is found in Chapter 4.) These first, extremely vague management instructions amounted to little more than a list of prohibited and permitted activities. Among those permitted activities were virtually the full range of commercial endeavors customarily pursued in national forests. A notable exception was a section of L-20 that established research reserves (after 1930, called experimental forests). These areas, usually small, embraced natural forest ecosystems of scientific importance. Commercial use of the research reserves was prohibited; even recreational use was discouraged. Here, at least, was implicit recognition that recreation could have an impact on the biological integrity of an area.

L-20 raised important management questions about recreational developments in primitive areas. Some USFS officials, with every good intention, responded to the regulation with an aggressive program of trail and shelter construction to compete with the civilized style of developments common in the national parks and attractive to the majority of vacationers in the 1930s.

Kneipp had a different idea of wilderness management. On May 29, 1930, he wrote with some impatience to the field staff: "There should be no need for developing these areas to take care of the large numbers of people who are not capable of exploring wild country without considerable aid" (as cited in Gilligan 1953, 147). Kneipp went on to direct his forest supervisors to stop plans for trail signs, latrines, corrals, and shelters in the wilderness. He recommended that "primitive simplicity" be used as a criterion for development decisions and concluded: "These primitive areas are for the class who seek almost absolute detachment from the evidences of civilization" (as cited in Gilligan 1953, 147).

Such sentiments must have cheered Robert Marshall. The New York–born son of a well-to-do constitutional lawyer, Marshall devoted his short but intense life to wilderness. Professionally, he trained as a forester and plant pathologist. For recreation, he explored the nation's wildest remaining corners, including the Brooks Range in northern Alaska. A prodigious hiker (he regularly covered thirty-five miles a day and occasionally logged more than seventy), Marshall resented any kind of convenience in wilderness, and his management ideas reflected this viewpoint. From his position after 1933 as director of the Forestry Division of the U.S. Office of Indian Affairs, Marshall crusaded to curtail road building in wild places. So-called fire roads particularly offended him. Easy to build, especially when the Great Depression brought thousands of job-hungry men under federal care in work programs like the Civilian Conservation Corps, dirt roads threatened to divide and conquer the last really large wildernesses in the West. Marshall's greatest achievement, really a memorial because he died two months later at thirty-eight, was promulgating the U regulations by the USFS on September 19, 1939. Superseding the L-20 Regulation with respect to more than 14 million acres of wilderness on the national forests, the U regulations tightened protection. (See Chapter 4 for additional details on the U regulations.) In administratively designated *wilderness* and *wild areas* (the term *primitive* was no longer to be used in classifying areas), there would be no roads or other provision for motorized transportation, no lumbering, and no hotels, lodges, or permanent camps

Fig. 2.4. Many people contributed to the wilderness movement in the United States. Robert Marshall (left), chief of the Division of Recreation and Lands in the USFS, led the establishment of the U regulations in 1939 creating wilderness, wild, and roadless areas, the immediate forerunner of today's NWPS. Photo courtesy The Wilderness Society. Lowell Sumner (right) of the NPS helped inventory wilderness conditions in the Sierra Nevada of California in the 1930s and, at that early date, recognized that those fragile lands had a "saturation point" beyond which recreational use could lead to irreversible damage. Photo courtesy National Park Service.

(Baldwin 1972). Very little was said about management, either in the U regulations or in the subsequent instructions for their implementation. Wilderness preservation was a caretaker function with an emphasis on guarding against outside influences. Wilderness inventory and designation, not management, were the preoccupations of the 1930s.

The USFS, through efforts by Carhart, Leopold, and Marshall, was in the forefront of the wilderness struggles. This was possible in large part because of the discretion accorded the USFS in deciding uses of the national forests. This would change under the 1964 Wilderness Act, which provided only Congress would have the power to designate a wilderness and took land-use discretion concerning wilderness away from the agency. As fights over what qualified as wilderness and the designation of many wilderness areas attest, the USFS, interest groups, and Congress have often disagreed over what areas, and what kinds of areas, should be designated as wilderness.

Still, pride of authorship and leadership stemming from its early efforts to establish wilderness runs deep in the USFS. This legacy has complicated the agency's response to expanded proposals and liberalized definitions of wilderness thrust on it by the environmental movement of recent years.

The Wilderness Management Idea Begins

The same emphasis on "drawing a circle around it and letting it alone" characterized citizen conservation groups in the 1930s. The Wilderness Society had its origins in 1934 and 1935 among a small group of men, Marshall and Leopold included, whose declared objective was "holding wild areas *soundproof* as well as *sightproof* from our increasingly mechanized life" (as quoted in Nash 1982, 206). The whole thrust of their effort was to keep adverse influences *out* of wilderness, and not to try to understand and control what was happening *within* its borders.

The first recognition of a management dimension to wilderness preservation began in the 1930s. Marshall's contribution to the "National Plan for American Forestry" (1933), the so-called Copeland Report, contained sections on the overuse of backcountry campsites and the need to educate recreationists in outdoor etiquette, today called minimum-impact or Leave No Trace (LNT) camping.

Further recognition that wilderness required management came in the summer of 1937 when Marshall, the new chief of the Division of Recreation and Lands in the USFS, toured the Sierra with members of the Sierra Club. On the trip, the party visited high country severely damaged by pack-stock grazing and by campers. Discussions begun on the trip led to Marshall's requesting Professor Joel H. Hildebrand, president of the Sierra Club, to organize a committee to advise the USFS about wilderness management. Marshall provided the committee with key questions that revealed the direction of his thinking about wilderness. One question, for example, concerned the feasibility of distributing use by zoning wilderness, in effect, to achieve certain ends. Specifically, Marshall was anxious that "certain areas may still be preserved in what might be termed a super wilderness condition, or, in other words, kept entirely free even from trails, in order that a traveler can have the feeling of being where no one has been before" (Hildebrand 1938, 90).

The Hildebrand Committee replied with a list of trails currently in the Sierra and a recommendation that construction of new trails be sharply limited and, if necessary, kept at a low (that is, primitive) standard. Responding to other questions from Marshall, the Sierra Club advised restricting trail signs, limiting the use and grazing of pack stock, and prohibiting cutting pine boughs for beds. To manage the wilderness and enforce such regulations, the club suggested appointing high-country rangers or guards. Finally, in a significant forecast, both Marshall and the Sierra Club expressed concern that wilderness be made available to the entire public by encouraging use by younger and poorer people. One idea discussed was making burros and camping equipment available on a rental basis.

The Marshall–Sierra Club interchange in 1937 began a new era in wilderness management. It recognized that recreation was only one value associated with wilderness and that, to maintain wild conditions, recreation should be regulated and restricted. Subsequently, a November 1940 article entitled "Certified Outdoorsmen" by J. V. K. Wagar observed that "nature once certified outdoorsmen." According to him, the weak, foolish, and careless just did not return from the wilderness they entered. "But now," he continued, "there is such ease of transportation and so much improvement in equipment that anyone can become a wilderness traveler" (Wagar 1940, 490).

Wagar's point was that many people in the wilderness did not know how to care for themselves or for the country. His suggested remedy was a program conducted by rangers from the NPS and the USFS to certify outdoorsmen. Those attaining the rank of Expert Outdoorsman were safe to live in the woods. Included in their knowledge was the ability to respect and live gently on the land.

In 1940, Wagar did not go so far as to suggest that *only* certified outdoorsmen be admitted to designated wilderness areas, but the implication was clearly present. More than forty years later Roderick Nash proposed a mandatory "wilderness license" in *Backpacker* magazine (Nash 1981). Critics of the idea point out that such a license might further compromise the freedom of the wilderness. The counterargument is that educated (licensed) visitors could be permitted *more* freedom than visitors unfamiliar with minimum-impact wilderness skills. The same logic prevails in the current effort by the agencies to "educate visitors" (but without licensing) as a major wilderness management tool, including certifying instructors and practitioners in LNT practices, so that they can teach other wilderness managers and users (Swain 1996).

Wilderness enthusiasts have long recognized that too many people, even too many qualified outdoor enthusiasts, can spoil a particular place. As early as August 15, 1926, the *New York Herald Tribune* featured a before-and-after cartoon of a mountain lake. In the first frame, a lone horseman approaches the lake, which is surrounded with pines and full of leaping trout; in the second, a solid rank of fishermen surround the lake, and their camps obliterate the scenery. In this case, the extension of a road to the lake was represented as the cause of the change. However, by the 1930s, some Americans understood that, even without roads, wilderness values could be threatened by overuse. If that solitary horseman were joined by fifty other riders and a hundred backpackers, the problem remained much the same.

As a prime practitioner of wilderness recreation, the Sierra Club continued to take keen interest in developing techniques of wilderness management. In 1947, the *Sierra Club Bulletin* featured another article on recreational impact on wilderness. It was coauthored by Lowell Sumner of the NPS and Richard M. Leonard, the chair of the Sierra Club's Outing Committee, which was by then coordinating a number of large high-country trips annually.

Sumner and Leonard focused particularly on the mountain meadows of the Sierra, and their article included a photographic sequence depicting stages in the transformation of a lush meadow grassland into a dustbowl. The cause was excessive recreational use. Discussing the problem under the heading "Saturation of the Wilderness," the authors declared, "We need more than just a concept…We need a comprehensive technique of use that will prevent oversaturation of wilderness and still enable people, in reasonable numbers, to enjoy wilderness" (Leonard and Sumner 1947, 60).

Among the suggested management tools were rotating camping and grazing sites, limiting the length of a permissible stay by one party in one area, and transporting oats rather than relying on natural grasses for pack stock. According to Sumner and Leonard, twenty-four-hour camping limits existed in 1947 in some meadows—the earliest such rules in wilderness management history.

In 1949, the Sierra Club sponsored a High Sierra Wilderness Conference, which grew into a biennial event that flourished for a quarter of a century. At the initial conference, about a hundred federal and state administrators, outing-club representatives, and professional outfitters and guides met to discuss a common concern: wilderness preservation and management. Attendees conceded that the designation and permanent protection of wilderness from outside influences such as roads and commercial development addressed only part of the problem. The other part was the impact of recreation users on wild country. The attendees had the courage to recognize that they were part of the problem.

By the fourth Wilderness Conference in 1955, a full range of wilderness management concerns was being discussed. So was the idea, still a decade away from fruition, for an NWPS; but most commentators recognized that without proper management the designation of wilderness could be meaningless.

In the 1950s and 1960s, the related concerns of designation and management continued to dominate the American discussion of wilderness. Inventory and designation of wild places progressed, as did the protection of established reserves. Notable here was the Echo Park Dam controversy involving Dinosaur National Monument in Colorado and Utah and, many felt, the integrity of all national parks (Nash 1982, 2001). Part of the price of a 1956 decision not to build a dam in Dinosaur was approval of one in Glen Canyon on the Colorado River.

Completion of Glen Canyon Dam in 1963 intensified efforts of dam builders and wilderness protectors when the Grand Canyon itself became the subject of controversy three years later (Nash 1970a). The success of wilderness advocates in defending the Grand Canyon from dams, coming on top of the passage of the Wilderness Act and its establishment of the NWPS in 1964, constituted dramatic evidence of the new political muscle of the preservation idea. However, these successes were limited to the external dimension of wilderness designation. The internal one—how a designated wilderness would be managed—continued to generate problems. The fact was that the NPS had not substantially departed from the tourism development and management assumptions of the Mather-Albright era.

Park management continued to emphasize visitor numbers, conveniences, and viewing spectacular scenic splendors rather than the wilderness experience. This became clear in 1956, when the NPS launched Mission 66. The program responded to rapidly increasing park visitation, but some feared it was the wrong response because its major thrust was further development. More than a billion dollars were poured into it, mostly for constructing roads, visitor centers, and motel-type accommodations. The entire emphasis of Mission 66 was to improve a park's capability for handling *more* tourists, and little was said about wilderness values and wilderness management. The prevailing management philosophy was more appropriate to an amusement park or resort than to a wilderness. The NPS was not alone in this posture. Operation Outdoors, the USFS counterpart to Mission 66, similarly emphasized tourism facilities and conveniences.

The facility and convenience orientation also appeared in backcountry-management practices of the mid-1960s. Trail standards were improved in the interest of easier access. Picnic tables, bulletin boards, fireplace grates, latrines, and corrals for pack stock were often placed at wilderness campsites. In some California areas, wilderness visitors even found rakes for use in tidying up their campsites (Snyder 1966).

However, the tide was beginning to turn toward visitor education, self-sufficiency, and light-handed styles of management. A USFS employee who returned to the John Muir Wilderness in California in 1973 found stoves, latrines, and fences gone. Instead of concentrating visitors at designated campsites, management had opted to encourage dispersed camping. You could find people, but they were just not as visible due to the lack of centralized, formal camp improvement (Koen 1973).

The year 1973 marked the first use of permits to limit the number of visitors to some of the more heavily used wildernesses, for example the San Gorgonio Wilderness in California managed by the USFS and the Colorado River in Grand Canyon National Park. Granted, wilderness permits had existed in Minnesota's Boundary Waters Canoe Area managed by the USFS since 1966 and were required in many wildernesses for record keeping, but they were freely given to every applicant. New to the early 1970s was the idea of using permits to limit use to a predetermined recreational carrying capacity. Use of other wilderness rules and regulations, such as banning wood fires and a pack-it-out policy for litter, also increased. Grand Canyon National Park rules went so far as to require river runners to pack out solid human waste. At a few heavily used wildernesses and backcountry areas, personnel were assigned to coordinate and disperse visitors.

Wilderness Management Evolves

Wilderness management direction has evolved from the initial heavy-handed visitor-control perspectives of the late 1960s and early 1970s to a light-handed, visitor-education approach that emerged in the 1980s and continues today as the first choice of wilderness managers (more detailed coverage is provided in Chapters 7 and 16). Many trends and events facilitated this evolution.

During the late 1970s and early 1980s, volunteer worker programs expanded considerably in national parks and forests, while at the same time recreation management budgets and personnel ceilings were being reduced. Wilderness management research and field experience were revealing the power and wisdom of education and appeals for cooperation as effective methods of managing visitor behavior (see Chapter 16).

The professionalism of wilderness managers increased. Managers eagerly exchanged views and experiences in regional and national wilderness conferences and workshops sponsored by universities, the USFS, and other agencies that included wilderness

Fig. 2.5. Arizona State University students conduct campsite inventory work under the supervision of their instructors, U.S. Forest Service and Bureau of Land Management personnel. Photos courtesy Greg Hansen and Jim Mahoney.

users, scientists, and university professors. The Arthur Carhart National Wilderness Training Center, Missoula, Montana, was established in the early 1990s to provide specific training for field managers and agency executives concerned with wilderness. The Aldo Leopold Wilderness Research Institute was established in the early 1990s to further research supporting wilderness management.

Periodic World Wilderness Congresses provide a forum in diverse places around the world: South Africa (1977, 2001), Scotland (1983), the U.S. (1987, 2005), Australia (1980), Norway (1993), and India (1998). The *International Journal of Wilderness* was established in 1995 and provides regular issues reporting research findings and management, education, and international perspectives. Further evidence of professionalism is that this textbook is now, after thirty years, in its fourth edition.

Increasing public involvement, a trend spurred by the Wilderness Act, has become a way of life for federal agencies; in addition agencies are bound in partnership with pro-wilderness clients. Most recently, the Internet (for example, www.wilderness.net) has become a major forum for citizen concerns and input to management proposals and environmental impact statements.

With these new initiatives and conditions, light-handed and educational approaches to wilderness management have evolved, with some tentative success. For example, in 1983 the Eagle Cap Wilderness on the Wallowa-Whitman National Forest, Oregon, was the first unit of the NWPS to drop a permit system once instituted (Scholz 1983). This move to a light-handed management strategy rested on the premise that the number of wilderness visitors had reached something of a plateau, or, at least, the near-exponential growth in visitation of the late 1960s and early 1970s had slowed. In addition, wilderness camping ethics and skills were improving, due in no small measure to an educational emphasis featuring public and wilderness-user contact by volunteer rangers. Indeed, a survey of Eagle Cap visitors in 1993, repeating many questions asked in a 1965 survey, documented a positive change in visitors' wilderness-behavior norms (Watson et al. 1996).

Wilderness users are concerned about protecting wilderness resources and values and respecting the interests of other visitors in having a wilderness experience. The hope is that the Eagle Cap Wilderness and other wilderness areas, in response to light-handed management and user education, are healing, and that restrictive permits are no longer needed. However, in 1994, a mandatory-permit system was restored to the Eagle Cap Wilderness, but this time mainly to establish use levels, educate visitors, and see that visitors are apprised of wilderness regulations and LNT principles. After three years with the renewed permit system, use levels were determined but the permit system continued through 2007—without restricting the numbers of visitors—because of its value as a wilderness education technique (Carlson 2000).

Literature that confirms the presence of new wilderness attitudes and influence abounds, as do wilderness education programs by organizations such as Outward Bound, the National Outdoor Leadership School, many environmental organizations, youth groups like the Boy Scouts and 4-H, in addition to college and university courses and conferences and agency workshops. Well-trained wilderness users may need fewer rules and less policing (Nash 1981). The emergence of a new generation of visitors who are sensitive to and respectful of the meaning and value of wilderness and who know how to minimize their impacts on wilderness—to "leave no trace"—increase the possibility that light-handed and educational approaches to wilderness management will be successful (Swain 1996).

Wilderness experience is extremely delicate, and wilderness management must be correspondingly deft. In the new millennium, wilderness managers, working on behalf of the public, must make judgments on several difficult questions. One is access to wilderness by commercially guided groups versus private or self-reliant user groups. Should outfitters and their need to show a profit be favored in allocating wilderness recreational opportunity, or do guides actually diminish the essence of the wilderness experience for visitors? What about the increasingly popular use of wilderness for personal growth and wilderness therapy for troubled adolescents? Should such use of wilderness be favored over other commercial or recreational use? To what extent should the Wilderness Act be bent to accommodate, for example, semipermanent outfitter camps and permanently fixed rock-climbing protection? Are certain kinds of nonmotorized recreation (geocaching, for example) less compatible with wilderness preservation than other activities? To what extent should the absence of solitude at popular wilderness locations be tolerated in order to provide more *visitors* a wilderness experience, or to concentrate rather than disperse visitor impacts? Such issues are at the heart of present and future wilderness management.

Another issue involves communications technology. The Wilderness Act specifically addressed keeping land free of mechanized intrusions. Roads, mines, clearcuts, and cottages were the traditional targets of this policy. However, what if the sights and sounds enter the wilderness in your own pack in the form of cellular telephones, geographic positioning units, or laptop or handheld computers (Van Horn 2007)? The current millennial generation—the

Fig. 2.6. Wilderness management has moved into schools and the community. Rich Hanson, wilderness manager from the BLM's Phoenix Field Office, explains the visitor entry permit system to students in the North Maricopa Mountains Wilderness. Photo courtesy Chad P. Dawson.

so-called digital or net generation—is the largest generation in U.S. history. Many of this generation are heavy users of the Internet, GPS, cell phones, e-mail, instant and text messaging, and computer games, and are often wireless in connection. How and where can these technologies be appropriately used for education of wilderness managers and visitors and for communication between them (Dawson 2007)? *Should cyberspace communication in designated wilderness be off-limits just as it is in certain restaurants, places of entertainment, and cathedrals? The essence of wilderness preservation is restraint, and the essence of wilderness experiences is naturalness and solitude. Defining and implementing policies to address new technological possibilities will be a new frontier for wilderness management* (Nash 1996), now more than ever.

Study Questions

1. What is the "intellectual dilemma" of wilderness management?
2. Compare an expedition through the Grand Canyon today with what one might have experienced a century ago.
3. What accounts for the huge increases in wilderness recreation participation since the early twentieth century?
4. Compare and contrast the purposes for establishment of national parks and national forests. How did those purposes influence early notions of wilderness management in the two agencies?
5. How did Robert Marshall's contributions change ideas about wilderness use and management?
6. Who were Arthur Carhart and Aldo Leopold, and how did they contribute to early recognition of wilderness values by the USFS?
7. The Sierra Club is one of the oldest and most influential environmental groups interested in wilderness. How has the Club's position on wilderness management changed over time? How do you think the Sierra Club and other wilderness user groups will further evolve?
8. How has wilderness management evolved since the 1960s and 1970s, and what challenges does it face now?

Acknowledgments

Roderick Frazier Nash, professor of History and Environmental Studies, University of California, Santa Barbara, originally wrote this chapter. Dr. Nash and John C. Hendee revised it for the second, third, and fourth editions.

For contributing current data and information on Mount Whitney use, we thank Brian Spitek, wilderness manager, Mount Whitney Ranger District, Inyo National Forest. For data on visitors to the Colorado River, we thank Linda Jalbert, wilderness planner and coordinator, Grand Canyon National Park.

Case Discussion:
Modern-Day American Views on Wilderness

While the wilderness values of early wilderness leaders and visionaries have been the subject of much literature on the history of the movement, it is the current generation who must manage and sustain the protection of the National Wilderness Preservation System (NWPS). What Americans see as valuable about protecting wilderness areas is important for legislators and policy makers to understand and to take into account regarding both new wilderness area designations and supporting the existing NWPS.

In 2008, Cordell et al. reported on some of the results of a National Survey on Recreation and the Environment (NSRE) that asked Americans sixteen years of age and older what they valued about wilderness protection. The NSRE during fall 2006 through the fall of 2007 asked about the relative importance of thirteen wilderness values on a five-point scale, from not important to extremely important. The percentages of respondents who reported that a value was very or extremely important are as follows:

Wilderness Value	Percentage
Protecting air quality	93.0
Protecting water quality	90.3
Protecting wildlife habitat	87.7
Knowing future generations will have wilderness areas	85.6
Preserving unique wild plants and animals	83.0
Protecting rare and endangered species	82.7
Providing scenic beauty	76.5
Having an option to visit wilderness areas in the future	76.2
Knowing that wilderness areas exist	75.7
Providing recreation opportunities	71.3
Preserving natural areas for science	64.3
Providing spiritual inspiration	57.8
Providing income for tourist industry	39.6

Clearly, modern-day Americans have strong and positive feelings about twelve of these thirteen wilderness values. The study found no statistically significant differences between the responses of residents of metropolitan and rural areas.

Case Discussion Questions

1. How do these modern-day values compare to the implied and expressed values inherent in the roots of the wilderness movement?
2. Would wilderness visionaries like Arthur Carhart, Aldo Leopold, and Bob Marshall agree with these modern-day wilderness values? Why or why not?
3. Do you expect the percentages to increase or decrease on any of these wilderness values in the coming decades as the NSRE is repeated? What do you expect to change and why?

References

Allin, Craig W. 1982. *The Politics of Wilderness Preservation.* Westport, CT: Greenwood Press.

Baldwin, Donald N. 1972. *The Quiet Revolution: The Grass Roots of Today's Wilderness Preservation Movement.* Boulder, CO: Pruett.

Carlson, Tom. 2000. The Eagle Cap Wilderness Permit System: A visitor education tool. *International Journal of Wilderness.* 6(2): 27–28.

Cleveland, Treadwell, Jr. 1910. National Forests as recreation grounds. *Annals of the American Academy of Political and Social Science.* 35(3): 241–247.

Cordell, H. K.; Betz, C.J.; Stephens, B.; Mou, S.; Green, G. T. 2008. How do Americans view wilderness: Part I. http://warnell.forestry.uga.edu/nrrt/nsre/IrisWild1.html; accessed 2/14/08.

Dawson, C. P. 2007. New opportunities for educating future wilderness and wildland managers in a changing technological world. *International Journal of Wilderness.* 13(3): 36–39.

Gilligan, James P. 1953. *The development of policy and administration of Forest Service primitive and wilderness areas in the Western United States.* Ann Arbor: Dissertation. University of Michigan.

Glover, James M. 2000. Spirit of the wilderness: Can we stop trying to control nature? *International Journal of Wilderness.* 6(1): 4–8.

Hildebrand, Joel H. 1938. Maintenance of recreation values in the High Sierra: A report to the United States Forest Service. *Sierra Club Bulletin.* 23(5): 85–96.

Jalbert, Linda. 2008. Correspondence between Linda Jalbert, wilderness planner and coordinator, Grand Canyon National Park and Chad P. Dawson, January 21, 2008.

Johnson, Randy. 1996. Grandfather Mountain: A private U.S. wilderness experiment. *International Journal of Wilderness.* 2(3): 10–13.

Koen, John. 1973. Personal correspondence to John C. Hendee, George Stankey, and Robert Lucas. On file at U.S. Department of Agriculture, Forest Service, Intermountain Research Station, Forestry Sciences Laboratory, Missoula, MT.

Langford, Nathaniel P. 1972. *Discovery of Yellowstone Park: Journal of the Washburn Expedition to the Yellowstone and Firehole Rivers in the Year 1870.* Haines, Aubrey L., ed. Lincoln: University of Nebraska Press. 125p.

LeConte, Joseph N. 1972. *A Summer of Travel in the High Sierra.* Sargent, Shirley, ed. Ashland, OR: Lewis Osborne. 144p.

Leonard, Richard; Sumner, E. Lowell. 1947. Protecting mountain meadows. *Sierra Club Bulletin.* 32(5): 53–62.

Leopold, Aldo. 1921. Wilderness and its place in forest recreational policy. *Journal of Forestry.* 19(7): 718–721.

Lillard, Richard G. 1968. The siege and conquest of a national park. *American West.* 5(1): 28–31, 67, 69–71.

Nash, Roderick. 1970a. *Grand Canyon of the Living Colorado.* Ballantine, NY: Sierra Club.

———. 1970b. Wild-deor-ness, the place of wild beasts. In: McCloskey, Maxine E., ed. *Wilderness: The Edge of Knowledge.* San Francisco: Sierra Club, pp. 34–37.

———. 1975. The creation of wilderness by herding and agriculture. In: *Program/Journal, 14th Biennial Wilderness Conference;* June 5–8, 1975; New York. Audubon, NY: Sierra Club, pp. 51–55.

———. 1978. Wilderness management: A contradiction in terms? Wilderness Resource Distinguished Lecture. Moscow: University of Idaho Wilderness Research Center. [Also published as Wilderness is all in your mind, *Backpacker.* 1979: (7) (Feb.–Mar.): 39–41, 70, 72–75.]

———. 1981. Worldview: Protecting the wilderness from its friends. *Backpacker.* 9(Feb.–Mar.): 15–16.

———. 1982. *Wilderness and the American Mind,* rev. ed. New Haven, CT: Yale University Press.

———. 1996. A wilderness ethic for the age of cyberspace. *International Journal of Wilderness.* 2(3): 4–5.

———. 2001. *Wilderness and the American Mind,* 4th ed. New Haven, CT: Yale University Press.

National Conference on Outdoor Recreation. 1926. S. Doc. 117, 69th Congress, 1st Session.

Pomeroy, Earl. 1957. *In Search of the Golden West: The Tourist in Western America.* New York: Knopf.

Potts, R. 2007. Changing human relationships with wilderness and wildlands. *International Journal of Wilderness.* 13(3): 4–6, 11.

Roth, Dennis. 1984. The national forests and the campaign for wilderness legislation. *Journal of Forest History.* 28(3): 112–125.

Runte, Alfred. 1979. *National Parks: The American Experience.* Lincoln: University of Nebraska Press.

Sax, Joseph. 1980. *Mountains Without Handrails: Reflections on the National Parks.* Ann Arbor: University of Michigan Press.

Scholz, Sue. 1983. The human approach: Wilderness permits not required. *Wallowa County* [OR] *Chieftain.* July 7: 3 (col. 3).

Snyder, Arnold P. 1966. Wilderness management: A growing challenge. *Journal of Forestry.* 64(7): 441–446.

Spitek, Brian. 2007. Correspondence between Brian Spitek, wilderness manager, Mount Whitney Ranger District, Inyo National Forest, and Chad P. Dawson, December 13, 2007.

Sumner, E. Lowell. 1936. Special report on a wildlife study of the High Sierra in Sequoia and Yosemite National Parks and adjacent territory. In-service report. Washington, DC: National Park Service Archives.

———. 1942. The biology of wilderness protection. *Sierra Club Bulletin.* 27(8): 14–22.

Sutter, Paul S. 2002. *Driven Wild: How the Fight Against Automobiles Launched the Modern Wilderness Movement.* Seattle: University of Washington Press.

Swain, Ralph. 1996. Leave No Trace (LNT)—outdoor skills and ethics program. *International Journal of Wilderness.* 2(3): 24–26.

U.S. Department of the Interior, National Park Service. 1970. Administrative policies for natural areas of the National Park System. Washington, DC: National Park Service.

———. 2006. Colorado River Management Plan.

U.S. Public Law. Yellowstone National Park Establishment Act of March 1, 1872.

U.S. Public Law. National Park Service Act of August 25, 1916. 39 Stat. 535.

U.S. Public Law. Grand Teton National Park Act of February 26, 1929. 70 Stat. 1314.

U.S. Public Law. Everglades National Park Act of May 30, 1934. 48 Stat. B16.

U.S. Public Law. Kings Canyon National Park Act of March 4, 1940.

U.S. Public Law. Grand Teton National Park Act of 1950; Repeal of Grand Teton National Park Act of 1929. 64 Stat. 849.

Van Horn, Joe. 2007. GPS. *International Journal of Wilderness.* 13(3): 7–11.

Wagar, J. V. K. 1940. Certified outdoorsmen. *American Forests.* 46(11): 490–492, 524–525.

Watson, Alan; Hendee, John C.; Zaglauer, Hans P. 1996. Human values and codes of behavior: Changes in Oregon's Eagle Cap Wilderness visitors and their attitudes. *National Areas Journal.* 16(2): 89–93.

Waugh, Frank A. 1918. *Recreation Uses on the National Forests.* Washington, DC: Government Printing Office.

Wright, George M.; Dixon, Joseph S.; Thompson, Ben H. 1932. Fauna of the national parks of the United States. *Fauna Series 1.* Washington, DC: U.S. Department of the Interior, National Park Service.

Wright, George M.; Thompson, Ben H. 1934. Fauna of the national parks of the United States. *Fauna Series 2.* Washington, DC: U.S. Department of the Interior, National Park Service.

Chapter 3
International Wilderness

by Vance G. Martin and Alan Watson

Introduction	51
Forces of Change and Wilderness Protection in the Twenty-first Century	51
How Much Wilderness Is Left in the World?	51
Reclaiming Wilderness in Europe	52
Wilderness Recognition in Developing Countries	53
Wilderness-Related International Protection of Nature	54
The World Conservation Union (IUCN)	54
Biosphere Reserves	56
World Heritage Sites	56
Ramsar Sites	57
Implications of International Recognition for Wilderness	58
Three Classes of Wilderness Protection: A Global Review	58
Class I—Legal Protection of Wilderness	59
Finland	59
Australia	60
Canada	62
South Africa	64
Sri Lanka	68
Flathead Indian Nation	68
Russia	69
Class II—Administrative Zoning of Wilderness	71
New Zealand	71
Zimbabwe	74
Tanzania	74
Italy	74
Japan	75
Class III—Wilderness Protection Through Recognition in Conservation Programs	75
Namibia	75
Philippines	75
Protection of Wilderness Values in Other Regions	76
Latin America	76
Nepal	78
Southeast Asia	78
Antarctica	79
Other Initiatives	80
Protecting Biodiversity	80
Wildland Corridors	81
Transfrontier Conservation	81
Water and Marine Wilderness	81
The World Wilderness Congress	83
International Journal of Wilderness	83
Summary	83
Study Questions	84
Acknowledgments	84
Case Discussion: Canadian Wilderness—The Quetico Provincial Park	85
Case Discussion Questions	86
References	86

Introduction

The concept of wilderness—land and water where natural ecological processes operate as free of human influence as possible and a place to learn and exhibit primitive skills with primitive recreation opportunities and solitude—has spread from its American roots to several other nations. Australia, New Zealand, Canada, Finland, Sri Lanka, Russia, and South Africa have legislatively protected wilderness or comparable, strictly protected reserves. Other countries, such as Italy, Zimbabwe, Namibia, and the Philippines, although not enacting wilderness legislation, have declared wilderness zones in parks, municipal watersheds, game reserves, and forests. This gradual globalization of the wilderness concept is evolving in both developed and developing nations, and the American experience is often an important model for these efforts (Martin and McCloskey 1993). Now wilderness managers from North America are also learning about the meaning of wilderness in other cultures through dialogue and proceedings of the World Wilderness Congresses (WWC), the *International Journal of Wilderness (IJW)*, and a growing network of wilderness training exchanges.

The excitement about wilderness was building rapidly worldwide during the 1990s as the value of this land-use designation and its relevance and potential adaptation to other cultures were increasingly recognized. Now, in the twenty-first century, more partnerships are being formed to support wilderness science, to share information on methods and benefits of wilderness protection, and for wilderness stewardship training.

In this chapter, we describe the values of wilderness in the twenty-first century and its spread and adoption throughout the world. We review international approaches to protecting wilderness landscapes and then describe the wilderness designation and protection efforts in some countries and world regions. Finally, we emphasize progress in the 1990s and look at some of the major problems facing the wilderness idea in other countries, particularly the challenge of meeting basic human needs and winning support of local people while protecting wildlands in developing nations.

The meaning of the word wilderness differs around the world. *Wilderness* is often a generic term outside the environmental community and often simply refers to "nature." This is a frequent meaning of the term *wilderness* in international circles, but there is

Fig. 3.1. Around the globe many nations are following the American idea of maintaining certain lands in a wild state, as wilderness or in some protected area category appropriate to their culture. In Africa, wildlife, such as the elephant, are the international and enduring symbol of wildness and wilderness. These charismatic megafauna are part of the natural system and the history and culture that are Africa. Photo courtesy Ulf Doerner.

increasing acceptance of the term to mean those areas legislated or zoned for protection in their natural condition but often accommodating a wider spectrum of human activity than the U.S. definition might allow.

Wilderness areas have major and increasing importance in the world today. Michael Nelson (1998) cites some thirty different reasons for wilderness conservation, including intrinsic value; for dependent indigenous cultures whose livelihood and cultural identity are inextricably linked to wildlands; for biodiversity protection, national identity, and ecotourism; and many more. We will explore some of these reasons in this chapter.

Forces of Change and Wilderness Protection in the Twenty-first Century

How Much Wilderness Is Left in the World?

Answers to this question differ depending on the wilderness criteria used. The first world wilderness inventory was introduced at the fourth WWC in 1987. The inventory was produced by the Sierra Club, with participation by the United Nations Environment Program, the World Bank, and the World Resources Institute, to provide an initial, reconnaissance-level survey of potential wilderness areas remaining in the world (McCloskey and Spalding 1988). *Although the criteria used were fairly crude—areas of 1 million acres (404,858 ha) or more, essentially roadless, and unaffected by permanent habitation or human structures—the survey revealed that approximately one-third of the earth's land surface still remains in a wilderness state.* More than 40 percent of the total was in the high latitudes, but even in Asia and

Fig. 3.2. Wild weather and sudden flooding on rivers are major obstacles to New Zealand wilderness travel. Wilderness visitors are subject to the changes of nature, and travel schedules can be changed for days without warning. Photo courtesy Les Molloy.

Africa, more than 25 percent of the landmass was still undeveloped, including 36 percent of North America (McCloskey and Spalding 1988). This first inventory of the global extent of remaining wilderness lands provided an indication of the opportunities and challenges for international wilderness conservation and set the stage for more precise inventories that followed.

Hannah et al. (1994) produced a GIS map of human disturbance in natural ecosystems using a habitat index with three categories: undisturbed, partially disturbed, and human-dominated. Undisturbed areas included areas with primary vegetation and population densities of fewer than ten people per square kilometer (0.38 square mile). Their analysis found that approximately 52 percent of the planet's land was undisturbed.

The Wildlife Conservation Society's assessment (Sanderson et al. 2002) used four criteria to assess wilderness: population density, land transformation, human access points, and electrical power infrastructure. These factors were combined to generate a Human Influence Index rating, and the resulting map showed that 17 percent of the planet's land surface remained wild in 2002, though that assessment was conducted without including Antarctica.

In 2003, Conservation International conducted another wilderness assessment, identifying areas over 2.47 million acres (1 million ha) having at least 70 percent intact habitat, as well as intact faunal assemblages of mammals, birds, and large predators, and having a population density of fewer than five people per square kilometer (0.38 sq mi) (Mittermeier et al. 2003).

Under these criteria, 46 percent of the planet was still wilderness. When the population density criteria were reduced to fewer than one person per square kilometer (0.38 sq mi), the remaining wilderness decreased to 38 percent.

Reclaiming Wilderness in Europe

Despite the fact that a large portion of the world remains undeveloped, wild, and remote, in many countries there is no wilderness left to protect. In Switzerland, for example, wilderness has long since vanished, despite its rugged mountain landscapes. From border to border, this small nation has been transformed by human activity into a thoroughly modified landscape.

All over Europe, the wildest places have been intensively used for agriculture and industry for thousands of years. Very little area is strictly protected, even within German national parks. Most German national parks do not meet International Union for the Conservation of Nature and Natural Resources (IUCN) international criteria for even the national park category. The first national park in Germany was established in 1969—the Bavarian Forest National Park, which receives about 1.5 million visitors annually (Brüggemann 1997). Logging and development for tourism access were among the initial plans for the park; however, conservation interests prevailed. Today, the principal objective of park management is noninterference in natural processes—maintenance or recovery of wilderness per se is not an objective. This represents a change from the highly managed landscape and intensive forestry traditionally practiced in Germany and most of Europe. The name of this new management approach is translated as "let nature be."

Great Britain provides additional evidence of why the American lead in wilderness preservation cannot be followed in many places, even if the culture would welcome islands of wildness in its midst. The British landscape has been settled continuously for more than 5,000 years. Everywhere the land has been shaped and altered by this long human occupancy. British national parks serve primarily to preserve a rural lifestyle, complete with traditional agricultural practices. *The British situation further demonstrates that, by the time old and intensely developed nations decide to establish a system of parks and reserves, there may be little alternative but to include substantial amounts of civilization. There is simply no opportunity to preserve an extensive wilderness system like that in North America or Australia.*

Nevertheless, the quest for protection of the wildest areas continues. For example, areas within the Highlands of Scotland that appear inhospitable still generate feelings of wildness. However, many of these seemingly inhospitable areas were once inhabited, but were cleared of local people by the English in the early 1800s to establish sport estates and sheep grazing. These areas are not pristine, but many perceive them as wild. They provide solitude and an opportunity to experience wildness, but are they wilderness? They are not, if the measure is biological intactness. Restoration of the original forest cover would be needed to restore wilderness. Thus, an international example of "rewilding" is occurring in the northwest Highlands of Scotland, where the Trees for Life project started in the mid-1980s to restore the original Caledonian Forest. This initiative, led by Alan Watson Featherstone, provides a model with lessons relevant to other countries where wildlands have disappeared (see www.treesforlife.org.uk).

But, there is still a "wild heart" of Europe. The Carpathian Mountains extend from southern Poland southward almost to the Balkans, including a small area of Ukraine, but are largely contained within Romania. According to some studies, there is still a minimum of 864,838 acres (350,000 ha) of original forest, up to 400 years old, in a landscape mosaic that includes meadows lightly managed as hay fields for hundreds of years by subsistence farmers. Only since the fall of communism have the biological richness and wilderness values of the Carpathians become more widely known. The landscape is rich in biodiversity. Bears, wolves, lynx, and stags abound. Encouraged by the European Union, the Romanian government has started a process of land restitution, giving national parklands back to the families of their original owners. Despite prohibitions on logging, clearcuts are evident on many of the restituted parcels. Nowhere else in Europe is the opportunity and timeliness for wilderness protection as great as in the Carpathian Mountains.

Europe is changing. One initiative by the Protected Area Network (PAN) Parks Foundation (PPF), based in Hungary, is to strengthen the wilderness concept in Europe by the foundation's recognition and designation. Most Europeans do not know that there are still exceptional remnants of wilderness on their continent. In the late 1990s the first steps were taken to join conservation and tourism in the most important wilderness areas of Europe. This initiative, implemented by the PAN Parks, recognizes protected areas that meet the highest standards of management for conservation and human use with the PAN Parks quality seal. Tourism is seen as an opportunity rather than as a threat, to give economic value to wilderness areas and to create support for conservation. Protected areas can join the network of PAN Parks if they meet a quality standard, which is defined through a Principles, Criteria, and Indicators system (www.panparks.org/Introduction/Verification/Principles).

A distinguishing feature of PPF, compared to other initiatives in Europe, is the wilderness concept. These criteria require at least 24,710 acres (10,000 ha) of the territory (called a core zone) to be still in a natural state and exclude extractive human uses. This is a pure wilderness zone and represents the most intact and undisturbed expanse of Europe's remaining natural landscapes (Table 3.1). This wilderness concept is used in the marketing of the areas as tourism destinations and as a tool to create opportunities for experiences linked to them (Bobiec 2002). The knowledge that this area is ruled by nature and not by humans can give special and unique meanings when interpreted by guides.

Wilderness Recognition in Developing Countries

Citizens of developing countries with emerging economies have little understanding of wilderness preservation and are more concerned with meeting basic needs such as food, medical care, and education. This was as true when the United States was scarcely settled as it is in developing nations today. For example, Chief Luther Standing Bear of the Oglala Sioux (as quoted in McLuhan 1971, 45) commented in the nineteenth century on the difference between his culture and that of the European settlers who were replacing it: "We did not think of the great open plains, the beautiful rolling hills, and the winding streams with their tangled growth as 'wild.' Only to the white man was nature a 'wilderness' and only to him was the land 'infested' with 'wild' animals and 'savage' people. To us it was tame."

As Standing Bear implies, only cultures that highly manipulate their environment distinguish between wilderness and civilization. The hunter-gatherer, on the other hand, did not technologically

Table 3.1. Certified Wilderness Is One of the Most Significant Achievements of the PAN Parks Concept and Is Part of PAN Parks' Core Purpose

(www.panparks.org/Network/OurParks, accessed April 2008)

Areas	Total Area		Wilderness		Country
	hectares	acres	hectares	acres	
Bieszczady National Park	28,200	69,654	18,425	45,510	Poland
Fulufjället National Park	30,414	94,883	23,048	56,929	Sweden
Oulanka National Park	27,500	67,925	10,000	24,700	Finland
Central Balkan National Park	71,699	177,097	21,019	51,917	Bulgaria
Retezat National Park	38,138	94,201	14,215	35,111	Romania
Paanajärvi National Park	104,371	257,796	100,000	247,000	Russia
Rila National Park	81,046	200,184	16,222	40,068	Bulgaria
Majella National Park	74,095	183,015	16,200	40,014	Italy
TOTAL	463,463	1,144,755	219,129	541,249	

transform wilderness into civilization, and consequently saw no dichotomy between the two. For Standing Bear, the wildland was simply home. It seems that cultures reach a point where their remaining wilderness is jeopardized before wilderness preservation becomes a social concern. In developing nations, it is necessary to develop concern for preservation before their entire wilderness has disappeared. But, understanding and acceptance of the wilderness concept in developing countries are heavily tied to the relevant culture—which is often aimed at development. However, in some places wilderness protection is being linked to economic planning and rural development as "sustainable development," or meeting the needs of the present without compromising the ability of future generations to meet their own needs (World Commission on Environment and Development 1987).

The twenty-first century is bringing rapid environmental, social, and technological change. Understanding and monitoring the influence these changes have on the relationships people have with wilderness are important. Around the world, we need discussion of what can be protected, why it should be protected, what are the threatening forces, and who will benefit from protection of wilderness character. (For a discussion on Africa, see Shroyer et al. 2003.)

Wilderness-Related International Protection of Nature

Many countries with relatively abundant wildland resources have the lowest standards of living and the highest ambitions for material improvement. In these nations, protection of wilderness resources will only succeed if it is part of an overall strategy for peace and development. It will be necessary to raise living standards, so the countries will be less likely to sacrifice natural resources to pay off foreign debt or to win a war. If an integrated approach to development is not implemented, wildland resources will be under greater pressure from people who need those same resources to survive.

It is not defensible for developed countries to insist that developing countries give priority to the retention of their wildlands, thereby denying the legitimate aspirations of their citizens for an improved level of living. However, if it can be demonstrated that retention of wild nature can be a major source of income, as well as bolster the importance of national heritage and health, then it encourages both preserved wildlands and economic progress. Great effort is being made to implement this strategy, as described below.

The World Conservation Union

The World Conservation Union (International Union for the Conservation of Nature, also known as IUCN) was created in 1948 and is composed of sovereign states, governmental agencies, and nongovernmental organizations (NGOs). *As of 2007, IUCN had 1,056 members, consisting of 749 national NGOs, 108 government agencies, 14 state members, 82 international NGOs, and 33 affiliate members.*

IUCN's purpose is to influence, encourage, and help societies throughout the world to conserve the integrity and diversity of nature and to ensure that any use of natural resources is equitable and ecologically sustainable. IUCN works with agencies in individual nations as well as various international bodies, such as the United Nations Environmental Program (UNEP), Food and Agricultural Organization (FAO), and United Nations Educational, Scientific & Cultural Organization (UNESCO).

The IUCN is administrated through its secretariat in Switzerland, has numerous regional offices, and essentially performs its conservation work through six commissions. The commission with the most relevance to wilderness is the World Commission on Protected Areas (WCPA), which undertakes a variety of responsibilities in aiding nations, particularly developing nations, in planning of protected areas. WCPA provides strategic advice to policy makers, guidance to protected area managers, and advice to the public, corporate donors, and governments on the value of protected areas to increase investment. The WCPA serves the United Nations (UN) by publishing the official UN List of National Parks and Protected Areas, a comprehensive listing of all conservation units managed by every UN member country. The IUCN also cofounded the World Conservation Monitoring Centre (WCMC), located at Cambridge in the United Kingdom, which provides detailed computer-based analyses of the status of protected areas around the world. Finally, the WCPA takes primary responsibility for organizing the World Parks Congress held about every ten years in various locations around the world. The most recent one was held in Durban, South Africa, in 2003.

An overview of international protected areas is found in the *2003 UN List of Protected Areas* (Chape et al. 2003), released at the fifth World Parks Congress in Durban, South Africa. It includes world heritage sites, biosphere reserves, and other sanctuaries ranging from the largest national park, in Greenland, spanning 240 million acres (97 million ha), to privately owned sites covering as few as 16 square miles (10 sq km) set aside for conservation. The report lists 102,102 sites, covering an area of 18.8 million square kilometers, and of that total, 17 million square kilometers (11.5 percent of the earth's land surface) is terrestrial. Protected areas now cover more land than that under permanent, arable crops. Two-thirds of the world's protected areas have been assigned an IUCN management category and cover 81 percent of the total area protected.

- Europe has more than 43,000 protected areas listed, followed by North Eurasia (nearly 18,000), North America (more than 13,000), Australia and New Zealand (almost 9,000), eastern and southern Africa (almost 4,390), and western and central Africa (2,600).
- Central America has the largest size with 9 million square miles (14.5 million sq km) of protected areas covering almost 25 percent of the region; North America has 2.8 million square miles (4.5 million sq km), or just over 18 percent of the region's land surface; and South America has nearly 2.5 million square miles (4 million sq km) of protected areas, close to 42 percent of the region.
- An estimated 4,116 marine areas are protected and cover over 1 million square miles (1.6 million sq km); however, this represents less than 0.5 percent of the seas and oceans. Australia and New Zealand have the largest area of marine protection, covering more than 260,400 square miles (420,000 sq km). Europe has the largest number of Marine Protected Areas, more than 800; however, these are small and many offer only limited levels of protection.
- The coastlines of southern and eastern Africa and of South Asia are some of the least protected, and the Indian Ocean, with its wealth of coral reefs, seagrasses, and mangrove forests, is perhaps the most poorly protected ocean.

Wilderness was not always recognized as a conservation category by the IUCN. Following a resolution of the third World Wilderness Congress in Scotland in 1983, a recommendation was passed at the IUCN General Assembly meeting in Madrid in 1984 asking that a specific category of wilderness be added to the IUCN list. Supporting this request was the belief that formal international recognition of wilderness as a conservation category would lend important credibility to wilderness protection in many countries. *Wilderness was included as a Class I protected area when the IUCN conservation categories were finally approved in 1992.* The categories include strict nature reserves (category 1a) and wilderness areas (category 1b).

In a move to reinforce wilderness in the WCPA's Guidelines for Protected Areas, in 2002 a Wilderness Task Force (WTF) was officially established within the WCPA. This Task Force, led by The WILD Foundation, created a permanent institutional voice for wilderness within the WCPA. Cochaired by Vance Martin (president of WILD) and Khulani Mkhize (chief executive officer of Ezemvelo KZN Wildlife, South Africa), the WTF initially led a series of workshops at the World Parks Congress in Durban, 2003; created a web portal for global wilderness professionals connected to IUCN; successfully maintained the integrity of the wilderness 1b category during the 2007–2008 process of revising the Guidelines for Protected Areas; and rewrote the management criteria for the wilderness category.

Biosphere Reserves

Biosphere reserves are not designated wilderness areas, but they illustrate through international recognition that wild, untrammeled areas have value to society. The biosphere reserves program grew out of recognition of the need for an international network of representative protected areas. This need was identified during a general conference of UNESCO in 1970, when that organization initiated the Man and Biosphere (MAB) Program *to conserve natural areas and their genetic materials. The program's goal was to establish biosphere reserves that might provide, on a representative basis, samples of all major types of ecosystems for on-site conservation (as opposed to in zoos, botanical gardens, or laboratories) and provide sites for baseline research activities important to other MAB projects* (IUCN 1979). At its first meeting in 1971, the MAB Coordinating Council defined a variety of geographical and topical issues for investigation.

Biosphere reserves have two principal distinctive characteristics as a category of conservation land use. First, they are representative of the world's major biomes; the emphasis in their selection is on the extent to which they help complete a portrait of the world's ecosystems, rather than on their outstanding or unique qualities. Second, they often contain areas where natural conditions have been degraded by human activity. *In general, most biosphere reserves are composed of a natural or core zone where minimal human impact is present.* Research and education could be undertaken in this zone, but only if they are not manipulative. *In addition, a biosphere reserve might contain a buffer zone in which educational or research activities involving manipulation might be conducted.* Traditional exploitive activities, including timber harvesting, hunting, and fishing, might occur in a controlled manner. *A reclamation or restoration zone could provide for the study and reclaiming of lands and natural processes disrupted by heavy human- or naturally caused alterations. Finally, a stable cultural zone could be managed to protect and study ongoing cultures and land-use practices conducted in harmony with the environment* (International Union for the Conservation of Nature and Natural Resources 1979).

The first biosphere reserves were designated in 1976. As of 2007 there were 507 sites in 102 countries. Such areas offer an important way to improve the scientific basis of resource management. For example, in the La Michilla Biosphere Reserve in Mexico, research has helped demonstrate how carefully regulated hunting could ensure a long-term source of meat while limiting competition with domestic cattle (zu Hulshoff and Gregg 1985).

The key element of a biosphere reserve is the core area. The core is often a primitive wilderness, strictly protected and maintained free of human disruption to conserve a representative example of a freely operating ecosystem in one of the world's major natural regions. The core contains an area in which natural processes continue undisturbed and in which maximum biological diversity is included (zu Hulshoff and Gregg 1985). In the United States, a number of national parks and wildernesses are biosphere reserves, including Yellowstone National Park and the Three Sisters Wilderness in Oregon.

Biosphere reserves are not legislatively designated by governments. They are the result of recognition that they meet an international protocol or accepted set of standards of UNESCO's Man and Biosphere Program (see www.unesco.org/mab).

World Heritage Sites

World heritage sites differ from biosphere reserves in that they have outstanding or superlative natural, cultural, or historic qualities and are recognized under international protocols of the World Heritage Convention. At the second World National Parks Conference in Yellowstone in 1972, Russell Train, then chair of the President's Council on Environmental Quality, proposed creating a world heritage trust to give international recognition to certain national parks. Generally national parks are based on the idea that certain areas are of such significance and value that they should receive national recognition and protection. Train thought that some areas possessed such universal natural, cultural, or historic value that they belonged to the heritage of the entire world (Train 1974). Being named as a world heritage site would convey that recognition.

In 1975, after the twentieth nation signed the Convention for the Protection of the World Cultural and Natural Heritage (UNESCO), it became a binding international convention for all its signatories. At present, more than 150 nations have adhered to the convention. While the convention includes cultural and natural sites, we focus here on natural sites. A property proposed for the world heritage list must meet at least one of the following criteria for a natural landscape based on the convention concerning the "Protection of World Cultural and Natural Heritage" as adopted by the General Conference of UNESCO on November 16, 1972 (see www.unesco.org):

1. Be an outstanding example representing the major stages of the earth's evolutionary history.
2. Be an outstanding example representing significant ongoing geological processes, biological evolution, and man's interaction with his natural environment.
3. Contain superlative natural phenomena, formations, or features, or areas of exceptional natural beauty.
4. Contain the foremost natural habitats where threatened species of animals or plants of outstanding universal value can survive.

As these criteria suggest, the distinguishing quality of a world heritage site, as opposed to a biosphere reserve, is the presence of outstanding or superlative qualities. IUCN and its WCPA are responsible under the convention for advising UNESCO on natural areas nominated for inclusion on the world heritage list (see www.unesco.org). Prominent areas on the list in the United States include Yosemite and Everglades National Parks. As of 2007, there were 851 sites (660 cultural, 166 natural, 25 mixed) in 141 countries. Italy has the most (41), followed by Spain (40), China (35), Germany (32), France (31) and the UK (27). Nearly 50 percent of all sites are in North America or Europe.

Fig. 3.3. World heritage site status is given to areas possessing outstanding natural qualities, exceptional natural beauty, or threatened species of plants or animals. The Daintree area of northern Queensland, Australia, typifies a rain forest setting. Photo courtesy Ralph Lindsey.

Designation as a world heritage site does not directly impose any new management requirements. However, formal recognition of the international quality of the site implies that the constituency for its appreciation and management surpasses that nation's boundaries. Such recognition carries with it important international prestige, so the sovereign authority would feel an obligation to protect the quality of any site having world heritage status. A site whose condition deteriorates may be placed on a "List of World Heritage Sites in Danger" and is eligible for technical and financial assistance from the World Heritage Fund for protection and restoration (see www.unesco.org).

Ramsar Sites

Ramsar sites (named under the 1971 Ramsar Convention on Wetlands of International Importance) protect wetlands of international importance, especially as waterfowl habitat. This agreement, named for the Iranian town on the Caspian where the convention met, is one of the oldest international conservation treaties. The Ramsar program is administered through a secretariat with offices at IUCN headquarters in Switzerland and at the headquarters of the International Waterfowl and Wetlands Research Bureau in the United Kingdom.

As of 2007, there were 156 contracting parties (countries), with 1,676 wetlands sites totaling 370 million acres (150 million ha). Cuba became a contracting party and the Zapata Swamp was listed as Ciénga de Zapata, in December 2001, with an area totaling

1 million acres (425, 000 ha). The U.K. has the most sites (165), followed by Mexico (67) and Australia (64); the U.S. has 22 sites, including Everglades National Park and Okefenokee National Wildlife Refuge.

Wetlands are defined as "areas of marsh, fen, peat land or water, whether natural or artificial, permanent or temporary, with water that is static or flowing, fresh, brackish or salt, including areas of marine water the depths of which at low tides do not exceed six meters" (Archibald 1988, 54). These areas are critical habitat for all waterfowl and are the areas of highest biological diversity in temperate zones. As with biosphere reserves and world heritage sites, those areas on the Ramsar list have no sovereign rights as a result of their listing. However, such international recognition often serves as an additional layer of protection for sensitive areas, as was the case when wetlands in Great Britain, Italy, Pakistan, and South Africa were saved from development by virtue of their actual or potential Ramsar listing.

Wetlands have important wilderness values but may not rank as high as other recreation areas to wilderness visitors because they are physically difficult to visit. In developed countries, the most frequent users of wetlands are duck hunters and bird-watchers. However, it is important to note that, particularly in tropical areas, wetlands provide income, food, and materials for millions of poor rural people who live nearby. The Okavango Delta in Botswana and the Sundarbans on the India-Bangladesh border (Chandy and Euler 2000) are such examples. For more information, see www.ramsar.org.

Implications of International Recognition for Wilderness

The World Bank conveyed international recognition for wilderness in a 1986 policy statement: certain pristine areas yield more benefits to present and future generations if maintained in their natural state. To protect these long-term values, the World Bank announced that it would usually decline to finance development projects in such areas. If development in wildlands is to occur, careful justification is required and compensation (in the form of financing) to preserve an ecologically similar area or some other mitigating measure must be undertaken (World Bank 1986).

International designation of biosphere reserves, world heritage sites, and Ramsar sites typically is imposed over some other type of protected area designation administered by a sovereign government, such as a national park or nature reserve. For example, although the guidelines for establishing biosphere reserves indicate only that they may coincide with other protected areas, in fact almost 80 percent of them do, and some coincide with several protective designations that include wilderness. All the world heritage sites are areas already protected under some national authority. However, international designation adds the protection of international recognition and prestige, thereby enhancing security for the area.

The Biosphere Reserve Project, the World Heritage Convention, and the Ramsar Convention do not specifically mention wilderness. Nevertheless, wilderness values such as naturalness and low human occupancy and impact are important components of each system, at least implicitly. The connection between biosphere reserve and wilderness designation may be particularly important in developing countries where public understanding and support for wilderness are limited, and biosphere reserves may serve to demonstrate the kinds of benefits to society that can be realized through protecting a nation's natural heritage (Halffter 1985). This, in turn, might help build public and institutional support for preserving large tracts of undeveloped area as wilderness, as part of an overall comprehensive program of nature conservation.

In the United States, where a number of biosphere reserves overlap designated wilderness, such as the Noatak Wilderness in the Noatak National Preserve and Aleutian Islands Wildernesses in Alaska, Everglades National Park Wilderness in Florida, and Three Sisters Wilderness in Oregon, the wilderness cores of these biosphere reserves provide a critical baseline for helping determine the extent of human influence on the surrounding biotic environment and are a source for understanding how natural processes can accommodate human impacts. As public awareness grows about the critical role wilderness plays in improving understanding of our environment, the security of such areas should increase.

Three Classes of Wilderness Protection: A Global Review

Following is a summary of de jure (existing in law) and de facto (existing in fact) protection of wilderness areas and values in countries where such protection exists. A useful way to view this spectrum would be in three classes (Martin 1994):

> **Class I wilderness** includes statutory or legal protection of wilderness, usually with management plans. Eight nations including the United States have legislatively designated wilderness: Australia, Canada, Finland, South Africa, Sri Lanka, the quasi-autonomous Flathead Indian Reservation in the United States, and Russia.
>
> **Class II wilderness** includes recognition and protection of wilderness through administrative zoning as found in several countries, such as New Zealand, Zimbabwe, and Italy.
>
> **Class III wilderness** includes wilderness recognition and protection in resource-management programs in which use of the concept *wilderness* is a key component in the overall conservation effort, such as in Namibia and the Philippines.

Class I—Legal Protection of Wilderness
Finland

The inevitability of commercial forestry in the large continuous areas of wildlands in the northern region of Finland in the 1980s stimulated a wilderness movement that began in the 1950s (Sippola 2000, Kajala 2008). *Finland's Legislative Act on Wilderness Reserves was passed in 1991,* providing statutory protection for about 3.75 million acres (1.5 million ha) in twelve units, about 5 percent of the area of Finland and all north of the Arctic Circle (Kajala 2008).

The Finnish word translated as wilderness *(erämaa)* is defined culturally and historically different, from wilderness in the English language. In Finland, the areas protected as wilderness have always been part of people's everyday life for hunting and fishing, whereas in the Anglo-American culture, wilderness is most typically defined as an area outside of human activities and influence (Saarinen 1998; Sippola 2000). Before legislative protection, most of the wilderness lands were not protected at all, though development was limited.

These areas in Finland were legislatively established to preserve their wilderness character to protect the native Sami culture and the traditional subsistence use of the areas, and to enhance possibilities for multiple use of nature. Therefore, all Finnish wilderness areas belong to IUCN category VI: Managed Resource Protected Area, even though many of Finland's wilderness areas remain in a wild state. To be considered for inclusion in the Finnish wilderness system, areas had to be a minimum of 37,000 acres (15,000 ha), over 6 miles wide, and mostly roadless and uninhabited. The smallest Finnish wilderness area, Tsarmitunturi, is 37,000 acres (15,000 ha), and the largest, Kaldoaivi, is 726,500 acres (294,000 ha) (Kajala 2008).

Additional designation of wilderness lands is not anticipated in Finland, although the Finnish wilderness legislation does state that if the national government acquires private inholdings within a designated wilderness reserve, that land area shall be considered to have become part of the wilderness. In addition to national wilderness areas, Finnish Lapland has about 2.5 million acres (1 million ha) of nature protection areas that are valuable for their wilderness character, where human activities are even more restricted than they are in wilderness, or within national parks, where one common management zone was once called a wilderness zone, but recently this term was changed to remote zone. These are typically low-use recreation zones with little regulation in effect (Kajala 2008).

Protecting traditional relationships with nature and indigenous Sami culture is an important aspect of wilderness (Pietikäinen 1995). Roads, mining, and large-scale tourism are forbidden, but small-scale forestry is allowed, with important restrictions, in some parts of five of the twelve wilderness units. Zoning within the wilderness allows for core sanctuary areas where wilderness conservation and appropriate recreation are the highest priority, and buffer zones for low-intensity recreation services and minor sustainable forestry activities (Tynys 1995).

The Parliament-appointed committee that developed a compromise proposal for wilderness classification in Finland decided to allow forestry in part of the most productive, pine-dominated wilderness forests (Sippola 2000) but on only about 3.5 percent of the overall total wilderness land base. Logging has only occurred in one wilderness area, Hammastunturi, and though debate continues, it seems unlikely that logging will occur in other wilderness areas (Kajala 2008).

The wilderness areas are administered by the Finnish Forest and Park Service (Metsähallitus), which governs most state-owned lands in Finland. During preparation of the Finnish wilderness legislation and subsequent land allocations, the Finnish Forest and Park Service actively engaged the public in the planning process, setting new precedents in public land management in Finland. A 1990 survey by Hallikainen (1995) found that the Finnish people support the wilderness concept with expectations of protecting endangered species, ensuring the presence of current landscapes for future populations, and for recreation use.

Sippola (2000) suggested that reindeer husbandry, which provides an important means of livelihood, adopted management practices that were not sustainable. Forestry is partially incompatible with public expectations of protecting the current landscape and future wilderness recreation values (Hallikainen 1998).

Recreation and tourism associated with the wilderness areas and national parks of northern Finland are becoming more important. Saarinen (1998) found that the most important reasons growing numbers of tourists are visiting some northern Finland sites include opportunities to see natural landscapes and wilderness and enjoy peace and quiet. Some of the most popular wilderness areas provide reserved huts, rental cabins, marked hiking trails, and snowmobile routes. However, to protect their pristine character, most areas do not have such facilities and services.

The Finnish people care about wilderness and wildlands and have protected many areas under a variety of designations, including wilderness (Table 3.2).

Australia

The Australian landscape has been described as old, low, dry, and remote. The continent is one of the oldest landmasses on earth and, as a result, has been substantially eroded. With the exception of the Great Dividing Range along the country's eastern shoreline, much of Australia is flat, with an average elevation of only about 1,000 feet (305 m). The country's highest point, Mount Kosciusko, is only 7,200 feet (2,200 m). Australia is arid with a median annual rainfall over half the continent of less than 12 inches (30 cm), and nearly one-third of the continent reports less than 8 inches (20 cm) of rain per year.

The population density of Australia is the lowest of any settled continent, but in another sense Australia is also the most urbanized; 80 percent of the population resides in a narrow crescent running from Brisbane to Adelaide along the southeastern coast. Although a large share of the country is unoccupied, there are extensive areas that show the influence of agriculture and mining.

Each of the country's six states holds principal authority over the management of its natural resources, including park and conservation management. Thus, the spread of the national parks idea and, later, that of wilderness were driven by regional rather than national issues and interests.

In 1932, the term *primitive area* first appeared in an Australian park proposal (Mosley 1978) and was to be a type of zone established within national parks and a separate reserve. Through the 1930s and 1940s, the conservation community struggled to clarify the concept of primitive area, with controversy over the extent to which it might protect an area's scientific and natural-history values. There was strong support for a system of reserves closed to recreationists but open to naturalists and scientific uses.

In 1944, the 3.2-million-acre (1.3-million-ha) Kosciusko National Park in southern New South Wales was designated with the provision that its governing body set aside no more than 10 percent of the park as a primitive area. However, because grazing would be allowed to continue under primitive area designation, scientists and naturalists opposed it, arguing instead that a smaller reserve, open only to scientific permit holders, be established. The controversy deferred action to protect the area's primitive values until 1962, when primitive-area designation was used to halt the construction of hydroelectric power works in the higher elevations of the Snowy Mountains.

A series of inventories of Australia's wilderness resources was begun in the mid-1970s. Helman et al. (1976) proposed a two-part definition of wilderness: a large area of land perceived to be natural, where genetic diversity and natural cycles remain essentially unaltered and dimensional criteria for mapping areas: (1) a minimum core area of 61,750 acres (25,000 ha), (2) a core area free of major indentations, (3) a core area of at least 6 miles (10 km) in width, and (4) a

Table 3.2. Protected Areas and Wilderness in Finland
(Metsähallitus [Finnish Forest and Park Service], 2005)

Protected Area	Number	Total Area (ha)	Total Area (acres)
National Parks	34	881,800	2,178,046
Strict Nature Reserves	17	58,669	144,912
Other Nature Reserves	348	535,000	1,321,450
Privately Owned Nature Reserves	75	8,500	20,995
Wilderness Areas	12	1,490,300	3,681,041

management (buffer) zone surrounding the core area of about 61,750 acres (25,000 ha) or more.

The area inventoried according to these criteria was limited to eastern New South Wales and southeastern Queensland. Twenty areas, totaling more than 2.5 million acres (1 million ha) in the core area, were identified. Of this total, sixteen were all or partly within an established national park while the remainder were in state forests or other public ownership.

The general criteria used in the Helman inventory were adopted for similar inventories conducted in Victoria and Tasmania. In Victoria, twelve areas covering 1.6 million acres (647,520 ha) were identified. For different landscape types the authors modified the Helman criteria somewhat; for example, in semiarid areas the core area needed a minimum 185,000 acres (75,000 ha) with a minimum width of 12 miles (20 km). In Tasmania, the Helman criteria were used, but some smaller areas were included, with thirteen areas totaling 1.7 million acres (688,000 ha) (Russell et al. 1979).

In South Australia, Lesslie and Taylor (1985) used an alternative inventory procedure to study wilderness. They argued that wilderness is a condition existing along a continuum, characterized by two relative attributes—remoteness and primitiveness—and has no absolute boundaries. Moreover, a single indicator cannot measure these two attributes. Remoteness is a function of how close an area is to settlement as well as to access points or routes. Primitiveness is a perceptual as well as objective phenomenon. Consequently, they assessed the extent of wilderness in South Australia according to ratings on four wilderness quality indicators: (1) remoteness from settlement, (2) remoteness from access, (3) aesthetic primitiveness, and (4) biophysical primitiveness. For example, an area more than 2.25 travel days from a settlement or access point was rated very high on the "remoteness from settlement" and "remoteness from access" indicators. At the other extreme, areas less than 0.25 travel day from settlement or access were excluded from the inventory.

The interest in inventorying the extent of the wilderness resource in Australia reflected the growing interest in its protection and management. In 1983, the Council of Nature Conservation Ministers, a group composed of all commonwealth, state, and territory ministers having responsibility for national parks and wildlife, established a working group to examine wilderness management throughout Australia and

Fig. 3.4. The Valley of the Monoliths, part of the 70,000-acre (28,000-ha) wilderness located within the Budawang Range, in Morton National Park, New South Wales, Australia. Although recreational use is permitted here, overnight camping is prohibited because of the highly erodible terrain. Photo courtesy Rob Jung.

to prepare guidelines for establishing and managing wilderness throughout the country that expressed a consistent approach in wilderness protection.

In Australia today, two of the six states, New South Wales (NSW) and South Australia (SA), have wilderness legislation—the Wilderness Act of 1987 in NSW and the Wilderness Protection Act of 1992 in SA (Prest 2008). Most other states and territories (including Queensland, Victoria, Western Australia, Australian Capital Territory, and Northern Territory) have wilderness provisions within their general nature conservation legislation or within their parks and reserves legislation (Prest 2008).

In NSW, under the Wilderness Act of 1987, whether land is already contained within a national park, or is private, or in other public land jurisdiction, it can be protected as wilderness. To be included, lands must not have been substantially modified by humans or are capable of being restored, are of sufficient size to make protection feasible, and are capable of providing opportunities for solitude and self-reliant recreation.

Prest (2008) suggests that the only criteria for inclusion in the SA wilderness system is that the land and its ecosystems have not been affected to a major extent by modern technology and are not seriously affected by exotic animals, plants, or other organisms.

Australia is a federation of separate states that prize and protect their autonomy. Important areas with outstanding wilderness values still remain, and the protection of wilderness continues to be an issue of major public importance. There has been important progress in protecting wilderness in Australia utilizing a variety of legal and administrative approaches (Table 3.3).

Table 3.3. Summary of Australian Wilderness
(Muir 2007)

State or Territory	Wilderness Units	Total Area (ha)	Total Area (acres)
Australian Capital Territory	1 wilderness zone	28,900	71,383
New South Wales	33 wilderness areas	1,950,000	4,816,500
Northern Territory	0	0	0
Queensland	0	0	0
South Australia	8 wilderness areas	184,419	455,515
Tasmania	0	0	0
Victoria	3 wilderness parks and 19 wilderness zones	842,050	2,079,864
Western Australia	4 wilderness zones	225,500	556,985
Commonwealth of Australia	1 wilderness zone	36,800	90,896

There is a strong belief that protection of wilderness in Australia can make a substantial contribution to biodiversity conservation. According to Prest (2008), challenges remain to obtain necessary resources for management, to move the designation process to a faster pace, and to achieve more engagement by Aborigine communities. Contention remains, however, over Aboriginal hunting, expectations for permanent settlements, and off-road use of vehicles within wilderness areas.

Canada

A comprehensive appraisal of wilderness protection in Canada (Hummel 1989) called for a coordinated effort at provincial, territorial, and federal levels to secure legal protection of at least 12 percent of the nation's landscape by the turn of the century. National and provincial opinion polls reflect support for national commitment to such a goal, challenging the nation's political leaders to respond to the public's vision of a comprehensive program of wilderness protection. Canada is the second largest country in the world and has about 8.4 percent of its area protected in some way from development, most of it in wilderness condition (McNamee 2008). Table 3.4 summarizes the extent of Canadian wilderness protection under national and provincial designations.

Several of the ten Canadian provinces—including Ontario, Alberta, Newfoundland and Labrador, Nova Scotia, and British Columbia—have laws explicitly designed to protect wilderness within those provinces. In addition, the National Parks Act was amended in 1988 to specify that the government could declare any part of a national park to be protected as wilderness, and the government was not to authorize activities that could impair the wilderness character of these areas (McNamee 2008).

In public opinion polls, Canadians have expressed strong support for protection of wilderness lands (Rutledge and Vold 1995). For example, there was almost unanimous support for the government of Canada to take actions to protect wilderness, and 83 percent believed it was important for Canada to be seen as an international leader in protecting wilderness (Parks Canada Agency 2000).

Parks Canada, in addition to being responsible for national historical sites, canals, and marine conservation areas, currently administers forty-two national parks and reserves across the nation, totaling 99,445 square miles (about 276,237 sq km). Commercial resource development is prohibited in the parks. The National Parks Act also requires that all land within the park be vested in the federal government; that is, it precludes any type of provincial or private land-ownership

Fig. 3.5. Much of Canada's far north holds outstanding opportunities for wilderness recreation. Here a kayaker enjoys the solitude and natural landscape of Pelly Bay. Photo courtesy Department of Economic Development and Tourism, Northwest Territories, Canada.

Table 3.4. Wilderness Lands by Province in Canada
(McNamee 2008)

Province or Authority	Wilderness Units	Total Area (ha)	Total Area (acres)
British Columbia	1 wilderness area	19,000	46,930
Alberta	3 wilderness areas	100,800	248,976
Newfoundland/Labrador	2 wilderness areas	396,500	979,355
Ontario	9 wilderness parks	4,823,745	11,914,650
Nova Scotia	31 wilderness areas	285,000	703,950
Parks Canada	2 wilderness areas	85,000	209,950

within a national park boundary. One consequence of this requirement is that, with the exception of national parks in the Northern Territories, more than 90 percent of the national park area existing in 1985 was established before 1930 (Dooling 1985). National park policy strives for a system of parks across the country that provides representation for the country's thirty-nine terrestrial natural regions and twenty-six marine regions; however, the requirement of total federal ownership at a time of concern about provincial jurisdiction, coupled with the high costs of land acquisition, makes it difficult to achieve such representation. In 2007, forty-two national parks and reserves existed in twenty-eight terrestrial natural regions, leaving eleven regions yet to be represented (McNamee 2008).

Although Canada's National Park Act was amended in 1988 to pave the way for the Governor's Council to approve the minister's recommendation for declaring wilderness within national parks and provide a more permanent sense of protection, none were approved by 2000.

In 1997, the Canadian government established an expert panel to provide advice on maintaining the ecological integrity of Canada's national parks. In response to that panel's conclusion that all but one national park faced significant threats to ecological integrity, a new Canada National Parks Act emerged in 2000, authorizing the government to declare any part of a park that existed in a natural state or that was capable of returning to a natural state to be a wilderness area. In October of 2000, the federal government finally designated the first legally protected wilderness areas in the four national parks within the Canadian Rocky Mountains World Heritage Site—Banff, Jasper, Kootenay, and Yoho (McNamee 2008).

At the provincial level, Ontario passed a Wilderness Areas Act in 1959 (McNamee 2008). It was a first step, although it was weak because it did not formally close the land to economic or recreational development. McNamee (2008) describes how many of the initial small areas—limited to less than 640 acres (259 hectares)—included by this act later became much larger provincial parks. In 1978, the Ontario government recognized wilderness parks as one of six official categories of Ontario provincial parks, and in 2006, a new Provincial Parks and Conservation Reserves Act repealed the old Wilderness Areas Act and made wilderness parks an official category in the province's protected areas system. In this latest legislation, there is a strong focus on the recreational and spiritual value of wilderness (McNamee 2008).

Quetico Provincial Wilderness Park is an outstanding example of wilderness preservation in Ontario. The wilderness zone comprises all but 630 acres (250 ha) of the park's 1,175,000-acre (476,000-ha) area. The southern boundary of the Quetico adjoins the Boundary Waters Canoe Area Wilderness (BWCAW) in the United States. The two areas form a nearly 2-million-acre (809,400-ha) tract of wilderness canoeing opportunities astride an international boundary, and many management policies of the two areas are coordinated.

The objective of the visitor management program for Quetico is to minimize, through the control of use levels and user activities, the deterioration of the park's biophysical environment and the quality of the user's wilderness experience. The general approach minimizes restrictions on visitors once they reach the park's interior. Daily entry quotas have been established for each ranger station at the park's boundary to help ensure that interior campsite capacities in selected areas are not exceeded. The visitor management program regulates party size, length of stay at selected high-use areas, packing out refuse, and the use of nonburnable but disposable food and beverage containers.

Five large national parks, including Banff, Jasper, and Wood Buffalo, were designated in the late 1800s and early 1900s in Alberta. Lands with wilderness values were protected within each park, albeit without explicit recognition, and 70 percent of the truly wild country was found in Wood Buffalo National Park in Alberta's far north (Alberta Wilderness Association 1985).

In 1959, the Alberta Provincial Legislature passed the Wilderness Provincial Park Act, which in 1962 was renamed the Willmore Wilderness Park Act. Willmore Wilderness Park, 200 miles (322 km) west of Edmonton in the Canadian Rockies, is an area of outstanding scenery, complex vegetation, and diverse wildlife. Many of the peaks in the area are more than 8,000 feet (2,470 m), and the area receives heavy recreational use. The 1959 enabling legislation permits recreational use, but does not limit coal and mineral exploration and, originally, did not prohibit motorized use, although such use was banned beginning in 1962 (Alberta Wilderness Association 1985). Since the area was established, its original size of more than 2,150 square miles (5,600 sq km) has been reduced twice (1963 and 1965) to its present size of approximately 1,775 square miles (4,600 sq km), due in part to pressures for coal exploration and development (Alberta Wilderness Association 1973).

In 1961, the 484-square-mile (1,259-sq-km) White Goat and the 157-square-mile (412-sq-km) Siffleur Wilderness Areas were established under the Forest Reserves Act and later, in 1965, both areas were reestablished under the Public Lands Act. Then in 1967, a fourth wilderness, the 58-square-mile (152-sq-km) Ghost River Wilderness, was declared under another piece of legislation—the Provincial Parks Act. Alberta now had four wilderness areas established by three separate pieces of legislation. The potential confusion introduced by this maze of legislation, coupled with growing public interest in wilderness protection, particularly along Alberta's eastern slopes, resulted in a new, revised Wilderness Areas Act passed in 1971, but it contained some restrictive language still objectionable to wilderness advocates.

The British Columbia (B.C.) Forest Service can establish wilderness areas with a minimum size of 2,470 acres (1,000 ha). In 1989, reflecting changing public attitudes, a Wilderness Management Policy was issued. It was evident that a significant shift in policy had begun. The 1995 B.C. Park Act legislated a requirement to designate a minimum of 10 million hectares of parkland by January 1, 2000. This goal was surpassed, bringing B.C.'s protected lands to nearly 14 percent of its land base. This increase in designated protected areas was partly fueled by public opinion that placed acquiring large wilderness areas as a top priority for B.C. Parks, ahead of operating existing facilities, acquiring beach areas for recreation, and undertaking conservation projects in the parks, according to the 1998–2001 business plan for B.C. Parks.

The Wilderness and Ecological Reserves Act of Newfoundland and Labrador in 1980 empowered the provincial government to set aside areas that were subject to little or no human activity (McNamee 2008), but with hunting, trapping and fishing, and recreational travel included as likely activities. Environmental protection to ensure traditional relationships with the land was a strong motivation in passing this legislation. Since passing this act, the government of Newfoundland and Labrador has only established two wilderness reserves totaling about 4,000 square kilometers, one of which is the Avalon Wilderness Reserve, which protects caribou habitat (McNamee 2008). There is interest in adding more acreage to this protection.

In Nova Scotia, the Wilderness Areas Protection Act of 1998 provided immediate protection of almost 5 percent of the province's total land base (McNamee 2008). The legislation protects thirty-one areas, and such protection is unique in a province that is 70 percent privately owned. The legislation states that the purpose of wilderness areas is to maintain and restore natural processes and biodiversity, natural landscapes and ecosystems, and unique, rare, and vulnerable natural features; and to provide reference points to understand impacts of human activities on nature and opportunities for scientific research, environmental education, and wilderness recreation.

South Africa

The Republic of South Africa (SA) has had specific measures to protect wilderness for nearly forty years. Today, wilderness on federal (state) land can be designated and protected under the National Environmental Management: Protected Areas Act No. 57 of 2003 (NEM:PAA), and the Forest Act of 1968 (as amended in 1971), which only applied to state forests. The Forest Act of 1968 resulted in eleven areas being protected as wilderness on state forests (Table 3.5). Also, each of SA's nine provinces has Nature Conservation Acts, many of which enable protection of wilderness in provincial protected areas (Glavovic 1985).

The colonial Zululand government proclaimed the first protected area on the African continent in 1895 (Bainbridge and Lax 2008). More than 400 areas have been proclaimed since then in South Africa, covering more than 16 million acres (6.73 million ha), which is nearly 6 percent of the nation's total land surface, and about 5 percent of its marine and coastal environments. Eleven of these areas are protected as wilderness areas and contain just under 741,000 acres (300,000 ha) (Bainbridge and Lax 2008).

NEM:PAA defines wilderness as an area retaining an intrinsically wild appearance and character, or capable of being restored to such, which is undeveloped and roadless and without permanent improvements or human habitation. The purposes of wilderness under this act are to protect ecologically viable areas representative of South Africa's biological diversity; to protect or maintain the natural character of the environment, biodiversity, associated natural and cultural resources, and environmental goods and services; to provide outstanding opportunities for solitude; and to control access, which, if allowed, may be only by nonmechanized means (Bainbridge and Lax 2008). The act provides opportunities for designating wildernesses or wilderness zones in all the protected area categories listed in the act, on all forms of land-tenure in the country, in both terrestrial and marine ecosystems (Bainbridge and Lax 2008). The results of this relatively new legislation have not been realized in SA, but the potential for new wilderness area designation is apparent.

Dedicated professionals in wildlife conservation and forestry brought the wilderness concept to SA from the United States. Important progress in wildlife conservation was made by staff members of the Natal (Provincial) Parks Board, in particular Ian Player, and colleagues such as Jim Feely, who in the early 1950s were working in the well-known Zululand reserves such as Umfolozi Game Reserve. As a result of their work, the Parks Board established the first wilderness zone in the Umfolozi Game Reserve. Player's duties kept him in close contact with the wild country along the Black and White Umfolozi Rivers, as well as with Zulu game guards who collaborated in administering the area and in law enforcement operations. Nightly discussions around campfires with his fellow workers helped shape

Fig. 3.6. Dr. Ian Player, noted South African and international wilderness advocate, and his friend, Magqubu Ntombela, famous game guard and Zulu tribal chief. Player was instrumental in establishing part of the Umfolozi Game Reserve as wilderness in 1958 and in founding the Wilderness Leadership School (WLS), a nonprofit group dedicated to introducing leaders, especially young people, to wilderness. The WLS takes people on walking treks in South Africa's wilderness. Photo courtesy Ian Player.

a deep concern for the country's wild character and appreciation of the Zulu people's keen understanding of their natural environment (Player 1998). Wilderness literature from the United States had a significant impact on these men and provided basic principles underlying the wilderness concept.

Initially there was opposition to the concept of wilderness by members of the Parks Board and some sectors of the public because of concern that designation would deny public access by vehicle. However, Player had maintained contact with North Americans such as Howard Zahniser of The Wilderness Society, who were trying to secure passage of the Wilderness Act in the United States. Zahniser kept Player supplied with current literature about the wilderness campaign in America, every word of which was read and reread by Player. Despite continuing opposition from development interests, the (then) Natal Parks Board decided in the mid-1950s to establish two wilderness zones in Zululand—the Lake St. Lucia wilderness in the area now known as the Greater St. Lucia Wetland Park, and the southern portion of the Umfolozi Game Reserve in 1958. These were the first named and zoned wilderness areas in Africa.

Ian Player's impact on wilderness in South Africa and the world is significant. He developed a deep appreciation of the value of close contact with the natural environment through his long exposure to the African wilds. His contact with the Zulu people had also sharpened his appreciation of their intimate knowledge of nature and a lifestyle that harmonized with their environment. In 1957, he undertook a weeklong trek (or

Table 3.5. Legally Designated National Wilderness Areas in South Africa

Area	Province	Number of Areas	Total Area (ha)	Total Area (acres)
Groendal Wilderness (part of the Baviaanskloof Cluster that includes 4 other candidate areas)[1]	Eastern Cape	1	28,900	71,440
Boosmanbos; Cederberg; Doringrivier; Grootwinterhoek	Western Cape	4	115,274	284,957
Wolkberg	Limpopo Province	1	22,000	54,384
Ntendeka; Mdedelelo, Mkhomazi, Mlambonja, and Mzimkulu[2]	KwaZulu-Natal	5	122,995	304,043
Total		11	289,169	714,826

1. The Baviaanskloof Wilderness is a high-priority candidate area that, when legally designated, will be the largest single wilderness area, of some 177,500 ha (438,780 acres).
2. Much of the Drakensberg Mountains has now been declared part of the Ukhahlamba Drakensberg Park, a world heritage site and, as such, administered via the national Department of Environmental Affairs and Tourism which in turn appoints a provincial management authority.

trail, as wilderness trek experiences are known in SA) along the east coast of the country accompanied by six schoolboys to introduce them to wilderness. This trail experience led Player to form an organization to introduce more people to the wilderness and be guided by an individual who understood some of the secrets of nature and the role such settings could play in gaining self-understanding. In 1963, the Wilderness Leadership School (WLS) was formally registered in SA.

In 1974, Player resigned from civil service to devote his full energies to wilderness conservation and the WLS. He expanded the scope of WLS by establishing the International Wilderness Leadership Foundation in the United States (now known as The WILD Foundation), originally to help finance wilderness trail experiences with the WLS for promising and established leaders. More than 35,000 people, ranging from youths to corporate executives, have since participated in the multiracial trails conducted by the WLS. The experience is intended to create an appreciation of natural processes and the values of wilderness settings for personal growth and building human understanding (Player 1987).

Once established in private life as director of the WLS, Player then conceived and organized the first World Wilderness Congress (WWC), held in Johannesburg in 1977. Seven additional congresses have since been convened in various parts of the globe. Player insists that all interests in wilderness conservation participate, and he considers the fact that the World Wilderness Congresses have continued his greatest achievement by bringing international attention to the wilderness debate in other countries—usually against great odds (Draper 1998). According to Ian McCollum (2007), Player is unsurpassed among many conservation leaders in SA and perhaps the world.

The idea of wilderness conservation also evolved among professional foresters in SA in early 1970 under Danie Ackerman, a Yale forestry graduate, who was chief director of the then Department of Forestry. The Forest Act of 1968 was amended in 1971 to provide legal designation of wildernesses on state forestland. This was the first such provision on the African continent.

At that time the Department of Forestry's primary focus was vegetation management, including controlling invasive fauna and flora, restoring degraded areas by using indigenous pioneer species, and maintaining native fire-dependent species, for which prescribed burning was important. In the Drakensberg Mountains, for instance, this involved constructing firebreaks and doing prescribed burns largely on a biennial rotation and protecting fynbos (shrubland) and evergreen forests from fire (Bainbridge 1987).

Management in the Drakensberg wildernesses includes patrol by armed guards to control illegal trespass, transborder traffic and smuggling in and out of adjacent lands, and protecting rock art sites. In the Zululand wildernesses, particular attention is given to combating illegal hunting of large mammals, including elephant and white and black rhinoceros (Bainbridge 1987).

Recent measures to upgrade protecting and managing SA wildernesses include increased public participation and awareness programs. The leading national wilderness organization is the Wilderness Foundation of South Africa, an NGO also founded by Ian Player and now headed by Andrew Muir. It concentrates on advocacy (public awareness), wilderness experience programs for disadvantaged communities, national policy, provincial wilderness coordination, and establishing wilderness on privately owned land. The Wilderness Action Group (WAG), another NGO, promotes appropriate management and use of the official wilderness systems and designation of all suitable wildlands as wildernesses.

Outsiders may perceive that white conservationists and academics largely dominated the wilderness movement in SA by virtue of the apartheid politics of the time. This was true at government executive levels, but in fact members of other races made important contributions. For example, contributions of numerous field staff who worked tirelessly, selflessly, and loyally with their white counterparts should never be ignored or forgotten. Included are outstanding individuals such as Mangosutho Buthelezi, hereditary prime minister of the Zulu people, who made major contributions to nature conservation and subsequently became the minister of Home Affairs of the democratic government, and Magqubu Ntombela, a remarkable and dedicated game guard who became a mentor to Ian Player and an inspiration to all who knew him (Player 1998).

Significant political changes occurred in SA in 1992—a country with 35 million blacks and only 10 million whites and other races. In the new SA, the mainly black African National Congress controls national politics, and jobs at local, provincial, and national levels are increasingly allocated to people of color. To respond to the end of apartheid three wilderness initiatives in SA are noteworthy. First, to address the critical need for wilderness management training for the many new wilderness field staff, wilderness management training courses were initiated through cooperation with and sponsorship of The WILD Foundation based in the United States and local interests coordinated by WAG. Volunteer contributions have been key to this training. Second, to respond to the new SA, then national director of the Wilderness Leadership School, Andrew Muir, created the Opinion Leaders Programme. Initially funded by a grant from the European Union, this continuing program takes small groups of members of Parliament on five-day wilderness treks (Muir 1999). As a result of this initiative, the chairs of the Environmental Portfolio Committee and key members of Parliament experienced wilderness. Consequently, they were motivated to create a Consultative Forum for the Environment. Third, Imbewu, a creative program to give at-risk youth a wilderness experience and vocational training, was conceived by Muir and implemented in a partnership between the WLS and SA National Parks (Muir 1999). Now operated by the Wilderness Foundation, Imbewu (meaning "seed" in two common languages in the country) aims at changing attitudes on environmental issues and South African cultural history. Working from specially constructed, low-impact residential camps in remote areas of several national parks, Imbewu targets

Fig. 3.7. Open tree savanna on the undulating plains of the Kgalagadi Transfrontier Park in South Africa with camelthorn (*Acacia eriloba*) and sourgrass (*Schmidtia kalihariensis*) the dominant species. Photo courtesy Noel van Rooyen.

Fig. 3.8. Two subadult lions engage in play, an important activity for developing survival skills in adulthood, in the Kgalagadi Transfrontier Park, South Africa. Photo courtesy Piet Heymans.

teenaged blacks living mostly in the townships adjacent to major urban areas in SA. Retired, black game guards who formerly worked in the parks deliver Imbewu programs. These "wise men" (and one "wise woman") have profound cultural and practical information about African wilderness and wildlife and train participants in trades and skills valuable to tourism and environmental protection. Imbewu has been considered so successful, and with such pan-Africa potential, that the U.S.-based African Wildlife Foundation received a grant from the U.S. National Park Service to conduct a parallel program in East Africa (Muir 1999).

Sri Lanka

Sri Lanka is the only South or East Asian country with Class I (legally designated) wilderness. Sri Lanka's 1988 Natural Heritage Wilderness Act provides for the conservation and protection of unique ecosystems, genetic resources, and outstanding natural features by regulating the entry to and residence within these areas and the use of their natural resources by local communities. Natural heritage wilderness is created by order of the minister of Lands and Development, but only after consulting the ministers of the Environment, Wildlife Conservation, Fisheries, Agriculture, Cultural Affairs, and Indigenous Medicine. The recommendation must then be approved by the president and confirmed by Parliament. Management of such areas falls under the conservator of Forests and director of Wildlife Conservation (de Alwis 1998).

Two main rationales form the historical basis for the act. First is the historical need to protect wild landscapes for watershed values; this practice originated with the Kingdom of Ceylon to protect its status as an agricultural power some 800 years ago. Singhalese kings declared specific areas as *thahansi kale,* or forbidden forests. These were typically dense, evergreen tropical forests with rivers and streams. Between 1600 and 1800, colonial rulers destroyed most of these areas for tropical hardwood exports. The few that remain, such as the Sinhgangan (Rain) Forest, Peak Wilderness, and Maha Eliya Montane Forest (now called Horton Plains), are dramatically reduced in size and now protected under the 1998 Natural Heritage Wilderness Act.

The second rationale is cultural and religious. Sri Lanka has been a Buddhist nation for more than 2,300 years, with the important exception of a rebellious enclave of Hindu immigrants in the northeast. Thus, Sri Lanka is steeped in the tenets of compassion, protection, and sanctuary for all life. One ancient teaching explained that humanity and all living beings were equal: "Oh great King, the birds of the air and the animals have an equal right to live in this land as thou, the land belongs to the people and all living beings and thou art only the guardian of it." The concept of sanctuary for animals dates back more than 800 years, but as the Singhalese like to point out, in direct contrast to this were the wildlife sanctuaries established by colonial British in 1909 in which game were protected for sportsmen to kill.

Though few areas are yet actually declared and named as wilderness under the act, Sri Lanka's example shows that culture and religion can be an important force for legal wilderness protection.

Flathead Indian Nation

The Confederated Salish and Kootenai Tribes live on 1.2 million acres (518,000 ha) in northern Montana, United States, and are a quasi-autonomous jurisdiction under tribal governance. *The Mission Mountain Tribal Wilderness Area on the Flathead Indian Reservation in northwest Montana is the only wilderness protected under tribal law found on the more than 300 federally recognized Indian reservations in the United States.* These reservations together comprise some 44 million acres (17.8 million ha) of land, most of which is rural, remote, and with large tracts of wildland, some of which could be considered de facto wilderness. Therefore, this first model of wilderness protection by the Flathead Indian Nation is of great importance.

The Confederated Salish and Kootenai Tribes are the descendants of Salish, Pend d'Oreille, and Kootenai Indians who traditionally occupied a 20-million-acre (8.3-million-ha) area stretching from central Montana to eastern Washington and north into Canada. After the Hellgate Treaty of 1855, most of these ancestral lands were ceded to the U.S. government, with 1.2 million acres (518,000 ha) remaining and now known as the Flathead Indian Reservation (McDonald 1995).

Starting in 1936, there have been several attempts to protect the Mission Range, which is central to the Indians' physical and spiritual geography, but the main effort to protect it came in the mid-1970s in response to a proposal by the Bureau of Indian Affairs to log the mountains. Three greatly respected grandmothers (*yayas*) raised the initial protest and led the way for other community leaders to organize a "Save the Mission Mountains Committee." About this time, Thurman Trosper (a tribal member, retired U.S. Forest Service and National Park Service employee, and past president of The Wilderness Society) returned home to the Flathead Reservation and proposed the idea of a tribal wilderness. Following extended community action, the Tribal Council eventually contracted with the University of Montana's Wilderness Institute to draft a boundary and management proposal for a Mission Mountains Tribal Wilderness Area.

Eventually, in 1982, the Tribal Council approved Tribal Ordinance 79A and the Mission Mountains Tribal Wilderness management plan, designating 92,000 acres (38,333 ha) as wilderness. It was the first time that an Indian tribe had taken action to designate wilderness on its own land and to establish policy direction and provide personnel for its protection and management. Further management plans ensued for protecting grizzly habitat and for fire management, and a Wilderness Buffer Zone of 23,000 acres (9,600 ha) was later established. There are important differences between managing wilderness on the Flathead Nation and managing federal lands under the U.S. Wilderness Act. The needs and values of tribal members take precedence over those of nontribal members resident on the reservation or visiting. Recent research by Watson et al. (2007) demonstrates that this tribal wilderness illustrates the wisdom of tribal people in protecting their relationships with this cultural landscape as well as off-site water, wildlife, aesthetic, and other resource use values.

Russia

Hunting reserves of feudal lords and the oak reserves for Peter the Great's navy in the late seventeenth and early eighteenth centuries marked the beginning of Russia's concern with nature preservation and management. After the 1917 revolution and the establishment of the Union of Soviet Socialist Republics (U.S.S.R.) all land was nationalized.

Although a concept exists in the Russian language that is analogous to wilderness—*dikaya mestnost* (wild place)—a different word has been used for designating "forbidden areas," or *zapovedniks*—literally "forbidden areas." The Soviet Union established areas withdrawn from economic use to preserve representative natural ecosystems and typical and rare landscapes; to protect gene pools for wildlife; and for research and educational purposes (Pryde 1977).

The first preserve, Barguzin Zapovednik, was established as a wildlife preserve to protect a sable population by a regional government in 1916 (Ostergren and Shvarts 2000). In the 1920s, the Ilmen'ski Zapovednik was established to protect its unique geological and mineralogical features, making it the first area in the world protected primarily for scientific purposes (Weiner 1988a). The system expanded until 1951, when 128 areas totaled more than 31 million acres (12 million ha). Then, in 1951, about seven-eighths of the zapovednik system was abruptly eliminated, and it was reduced by one-third in 1961 (Weiner 1988b). Since 1961, the system has been slowly rebuilt and in the 1980s the Soviet Union had doubled the size of those reserves to more than 105 million acres (43 million ha) (Pond 1987).

Despite the privatization of industries and homes following the fall of the Soviet Union in 1991, the vast majority of land has remained under federal control. Expansions continued through the 1990s and the early part of this century. Total public control creates, in theory, a promising political framework for all kinds of conservation, including that of wilderness. Twenty-four preserves have been added in Russia since 1991—a nearly 50 percent increase in land area (Ostergren and Shvarts 2000). It has become an important part of one of the most extensive protected area systems in the world.

In 2007, the Russian Federation network of zapovedniks included 101 areas totaling more than 84 million acres (32.7 million ha) (Shestakov and Kats 2008), the majority of which are off-limits to the general public (Table 3.6). Within the Russian Federation there are also 35 national parks, more than 1,500 *zakazniks* (wildlife refuges with species-specific goals, 69 of which are federal reserves), and several thousand minor national monuments that may protect from one to ten acres (0.4 to 4 ha) (Colwell et al. 1997; Sobelev et al. 1995).

There are important differences between the zapovednik system and other Class I wilderness systems: absence of parliamentary or other public legislative control; lack of consistent, enforced management plans; and some use of motorized vehicles. However, there is enough regard for maintaining wilderness values present

Table 3.6. Wilderness Lands (Zapovedniks) in Russia
(Shestakov and Kats 2008)

Period	Number of Areas	Total for Period (ha)	Total for Period (acres)
Pre-1960	28	3,906,519	9,649,102
1960–1980	16	5,775,600	14,265,732
1981–1990	27	11,366,469	28,075,178
Post-1990	30	11,703,700	28,908,139
TOTAL	101	32,752,288	80,898,151

in the Russian zapovedniks that they are included here as legally protected wilderness.

The most significant new threat is an extreme lack of funds for anything more than cursory field management and protection. Many managers and policy makers were concerned about expanding the zapovednik system while budgets decreased and basic management tools (e.g., helicopter support, gasoline, salaries for employees) disappeared (Ostergren 1998). For example, Taymyr Zapovednik was expanded in July 1993 from 1 million acres to 9.6 million acres (0.4 to 3.8 million ha) on a peninsula on Russia's Arctic coast, creating the Great Arctic Reserve, Russia's largest zapovednik (Surlien 1995).

Despite the dominant role of nature conservation in zapovedniks, their value for recreation also emerged. In the 1960s and 1970s, growing pressures on several nature reserves forced authorities to consider that the areas may have value as settings for recreation. The recreational demands contributed in part to the creation of a Soviet National Park System in the 1970s, which grew to thirty-five parks. Some parklands were zoned to include wilderness conditions (e.g., natural conditions, access by foot or horse only, no permanent structures), but unlike zapovedniks the parks also include economic zones for tourism infrastructure and may have had limited logging.

As a result of recreational demands, the Kavkaz Zapovednik in the Caucasus near the Black Sea was opened to camping, climbing, and hiking. In the Stolby Zapovednik near Krasnoyarsk in central Siberia, 11,000 acres (4,500 ha) of the 118,359-acre (47,900-ha) area was also opened to recreation. In the Katun Zapovednik, established in 1992, mountain climbers on Mount Belukha and rafters seeking class 5 whitewater on the Katun River may pay a nightly fee to stay on the fringes of the protected territory. This is wilderness by any standard, primitive recreational opportunities in virgin forests, with elk, wolves, Arguli sheep, and snow leopards. Visitation is attractive because it may provide desperately needed funds or expanded political clout. Growing recreational demands will require greater effort to manage recreational use to protect natural conditions. Despite different cultures and political systems, the inclusion of recreation on zapovedniks, and the increased importance of U.S. wilderness areas for science, have made the two systems appear more similar over the last few decades (Ostergren and Hollenhorst 1999).

Probably the most significant post-Soviet event for protected areas was passage of an "organic act" for all agencies in 1995 and establishment of the Department of Zapovednik Management within the State Committee of Environmental Protection. Historically, zapovedniks were protected by various administrative authorities, even at the local and provincial levels. This legislation in 1995 (which emphasized biodiversity protection and long-term ecological monitoring) provided a new sense of direction and legal strength to protect these critical areas in Russia. In addition to clarifying responsibilities and strengthening the law enforcement authority of employees, the 1995 law expanded the purpose of zapovedniks to include conservation training for professionals, participation in regional environmental impact assessments, and, very controversially environmental education that may include ecotourism. After eighty years of traditional exclusion, it became legal to bring people into the territory of zapovedniks for purposes other than ecological research (Ostergren 2001). Nonetheless, access to the bulk of zapovednik territory is restricted to ecologists, and the areas will remain inviolate to roads, structures, mechanized travel, and any economic use.

Managing designated areas is a greater concern than establishing new preserves. In May 2000, President Vladimir Putin eliminated the State Committee on Environmental Protection and the Federal Forest Service, the home agencies for zapovedniks and national parks, respectively. The protected areas were placed under the Ministry of Natural Resources. Incredibly, after several political and bureaucratic assaults on the status and purpose of protected areas in the winter of 2000–2001, the director of Zapovednik Management, which now included national parks, emerged stronger and with an increased budget for the entire protected area system. The budget in 2001 was still inadequate but the alternative was reduction or complete eradication.

Through trial and tribulation, dedicated staff remained to conduct research and enforce the law on zapovedniks. Industries polluting water resources were successfully sued for damages, and the 1995 regulations permitting employees to confiscate tools for poaching (e.g., guns, snowmobiles) have contributed to zapovednik fiscal solvency. Limited areas are open for ecotourism, and environmental-education programs are giving young people and community members a renewed appreciation for the value and purpose of zapovedniks. The prospects are hopeful for zapovedniks to continue protecting natural areas and maintaining wilderness conditions well into the twenty-first century.

Because Russia covers eleven time zones and a wide array of ecosystems (e.g., tundra, taiga, bogs, deciduous forest, steppe, mountains, deserts), there is plenty of room for additional protection of wilderness lands. Many of the newer zapovedniks have been established along the southern borders of the former Soviet Union where endangered wildlife species are most frequently located, but adequate funding and staffing, coupled with pressures from competing economic interests (e.g., poaching, food gathering), are continuing problems (Pryde 1987, 1997).

Class II—Administrative Zoning of Wilderness

Nations that protect wilderness by administrative zoning (Class II wilderness protection) include New Zealand, Zimbabwe, Italy, and Japan. This protection is not as secure as Class I because administrative rules and regulations are easier to change than statutory protection, and that was one of the main reasons for the public campaign to secure wilderness legislation in the United States.

New Zealand

New Zealand protects a wilderness system through administrative designation or zoning under provisions provided in its Reserve Act of 1977, the National Parks Act of 1980, and the Conservation Act of 1987 (Cessford and Dingwall 1997; Somers 2008). Physical conditions, cultural values, and vigorous wilderness advocacy combine to make New Zealand's wilderness one with extremely high standards, supported by strict management guidelines.

There are eight designated wilderness areas and two recommended wilderness proposals that total around 1.58 million acres (620,000 ha) in New Zealand (Table 3.7). These designated and proposed wilderness areas total about 3 percent of New Zealand's land area, no doubt that small because the Department of Conservation applies very strict criteria to such designations. These New Zealand "wildernesses" reflect standards of physical and social conditions that are among the most rigorous in the world, and exceed the category I(b) criteria specified by IUCN (1998) for the most protected class of wilderness areas, because IUCN criteria make more allowance for minor human modification and habitation.

Possibly 25 percent of New Zealand's conservation lands outside the designated wilderness areas might qualify as the most remote kind of wilderness under IUCN criteria. But these lands also have a solid legislative basis for conservation management in New Zealand, and potential pressures on them for development and recreation use are not great. Eighty-five percent of New Zealand's 4 million residents are urbanized coastal dwellers, most concentrated in the upper half of the North Island. Very few live in remote areas, and unlike most other countries of the world, no one occupies lands that are managed as protected natural areas. All of this is in spite of widespread ecological change in the country, such as through establishment of large forest plantations. The New Zealand recreation opportunity spectrum (ROS) includes "wilderness," "remote," and "back-country walk-in" opportunity classes that largely match wilderness conditions in the United States (Cessford and Reedy 2001).

New Zealand created one of the world's first national parks in 1894 when the volcanoes and lava fields of the central North Island became Tongariro National Park. These lands had been given to the nation in 1887 by its indigenous Maori guardians, who wanted these sacred mountains enshrined in public protected-area status (Cessford and Reedy 2000).

Much of New Zealand is rugged hills or mountains, including twenty-nine peaks over 9,842 feet (3,000 m) in the Southern Alps of the South Island, and two active volcanoes dominate the extensive volcanic zones of the central North Island. These islands are isolated in the temperate southern Pacific Ocean, and apart from the tiny Polynesian Pacific islands, the closest other landmass is Australia, 1,242 miles (2,000 km)

Fig. 3.9. Most high alpine areas have significant wilderness values in New Zealand and internationally. Some have access by aircraft, the only major intrusion, other than people, to otherwise pristine places. Photo courtesy Gordon Cessford.

away. New Zealand's separation from the ancient southern continent of Gondwana some 80 million years ago resulted in unique plants and wildlife with no mammal or marsupial life, some small bats excepted. The ecological niches such species filled elsewhere were occupied in New Zealand by unique birds, insects, and reptiles. In the absence of any competition or predation by mammals, some bird species became flightless, some insect species grew to large sizes, and plants evolved with minimal pressure from animal browsing.

The arrival of humans introduced new ecological impacts against which the native biota had few natural defenses. In less than one thousand years, it is estimated that humans and their accompanying animals have removed 70 percent of the forest cover and driven 32 percent of the indigenous terrestrial bird species to extinction. Moreover, around a thousand species of indigenous plants and animals are now threatened, including thirty-seven of the fifty remaining native terrestrial bird species. Thus, retaining biological uniqueness by maintaining natural biodiversity is a high priority for the management of New Zealand's conservation lands. These conservation lands (including wilderness) cover almost 30 percent of New Zealand, one of the highest proportions of protected areas of any country in the world (Green and Paine 1997; IUCN 1998). Most of these lands are remote and natural, with high wilderness quality by any standards.

Wilderness has a very specific meaning in New Zealand conservation management, relating most often to specifically designated "Wilderness Areas." Early wilderness advocacy was generated by the Federated Mountain Clubs (FMCs) and their associates. In 1939, Lance McCaskill, a prominent conservationist, visited the United States and met Aldo Leopold and, through his influence, introduced the concept of wilderness into New Zealand National Park circles (Cumberland 1981). In 1949, the American conservationist Olaus Murie visited New Zealand and supported the idea of

Table 3.7. Wilderness Areas in New Zealand in 2007 (Cessford 2008)			
Wilderness Areas	**Area ha**	**(acres)**	**Gazetted (legislated)**
Pembroke Fiordland N.P.	18,000	(44,500)	1974. Forested/alpine mountain lands
Glaisnock Fiordland N.P.	125,000	(309,000)	1974. Forested/alpine mountain lands
Raukumara Island	40,000	(99,000)	1988. Remote forested highlands, North
Tasman N.P.	87,000	(215,000)	1988. Forested highlands, Kahurangi
Hooker-Landsborough Alps	41,000	(101,000)	1990. Mountains and valleys in Southern
Olivine Southern Alps	83,000	(205,000)	1997. Mountains, valleys, and glaciers
Adams Alps	47,000	(116,000)	2002. Mountains and valleys in Southern
Paparoa N.P.	31,000	(76,500)	2002. Forested highlands/mountains, Paparoa
Ruakituri Urewera N.P.	24,000	(59,500)	2006. Remote forested highlands, Te
Gazetted area subtotal	496,000	(1,225,500)	1.85 percent of New Zealand land
Proposed Wilderness (2007)	**Estimated Area ha**	**(acres)**	**Area Proposals**
South West/Cameron	182,000	(450,000)	Lowland forest and lakes, Fiordland N.P.
Pegasus Rakiura N.P.	40,000	(99,000)	Lowland Stewart Island forests/coast
Proposed subtotal	222,000	(549,000)	0.8 percent of New Zealand land area
Wilderness Area total	**718,000**	**(1,775,000)**	**2.7 percent of New Zealand land area**

introducing wilderness protection provisions into the pending 1952 national park legislation. This legislation outlined a procedure whereby distinct areas of wilderness could be established in national park settings, and that such areas should be kept and maintained in a state of nature, without any buildings, roads, or tracks, and only foot access permitted.

Protection of areas with wilderness values is also provided for under the Forest Act of 1949, implemented by the New Zealand Forest Service, and the Reserves Act of 1977, implemented by the Department of Lands and Survey. The Forest Act provided for use of state forestlands for wildlife protection and for recreation, particularly as forest parks, consistent with a wilderness philosophy. Wilderness here was considered to be an area to be kept in a state of nature, where structures and roads were prohibited and where access was by foot only (Lucas 1983). However, little actual wilderness protection was achieved under either law.

Although there was general agreement about what wilderness was, there was no consensus or management philosophy. Because the mechanisms for wilderness protection were spread across three different statutes, three different governmental agencies, and two governmental ministries, it was difficult to promote the assessment and protection of wilderness. In 1976, conservation groups, led by the FMC, began to consider options for wilderness protection (Molloy 1983). Working with the three government management agencies, the FMC helped produce a Joint Wilderness Policy in 1980 that outlined the legislative status of wilderness in New Zealand, discussed social and physical criteria for such areas, proposed common definitions with which each agency would work, and identified general guidelines for establishing and managing wilderness.

In 1981, the FMC sponsored New Zealand's first conference on wilderness and introduced ten proposals for wilderness areas spread throughout the nation, ranging from 75,000 to 233,000 acres (30,000 to 94,000 ha) and covering natural landscapes from coastal forest, steep land forest, and tussock grassland to rugged alpine and glaciated areas. Following the conference in late 1981, the minister of Lands and Forests appointed a Wilderness Advisory Group (WAG) to advise on appropriate policies and action.

The WAG reached a consensus based on the earlier Joint Wilderness Policy and provided a succinct statement of objectives for wilderness protection and criteria for wilderness areas, though allowing for flexibility in designation and management. Wilderness would be large enough to take at least two days' foot travel to traverse, have clearly defined topographic boundaries and be adequately buffered from human influences, and have no developments such as huts, tracks (trails), bridges, signs, or mechanized access. The WAG developed specific recommendations for designating ten areas, totaling more than 1.2 million acres (500,000 ha) (Molloy 1983).

Results have come slowly since the first provision for zoning wilderness in the 1952 National Parks Act, the 1977 Reserves Act, a new National Parks Act in 1980, and a new Conservation Act in 1987. In each of these statutes, the purpose of wilderness areas can be summarized as preservation of a natural state, primarily for wilderness experiences. More specifically the Conservation Act of 1987 lists provisions that must apply to any wilderness area: its indigenous natural resources shall be preserved; no building, machinery, or apparatus shall be erected or maintained in it; no livestock, vehicles, or aircraft shall be allowed to be taken into or used in it; and no roads, tracks, or trails shall be constructed in it (Molloy 1997; Cessford and Reedy 2000).

Today there are few recreation-based threats to the integrity of New Zealand's designated wilderness areas. This is due to their extreme remoteness, the sustained difficulty of access, the low demand pressures, the regulations that are in place, and the focus of management on providing more services for front-country visitors to protected natural areas. Major issues that may affect some wilderness areas in the future relate to general ecological sustainability, the intrusive potential of aircraft overflights, and more recently, renewed customary rights for the indigenous Maori people (Barr

Fig. 3.10. Wilderness users need the highest levels of fitness, outdoor skills, and self-sufficiency to access and explore New Zealand's wilderness. Photo courtesy Gordon Cessford.

Fig. 3.11. A bivouac under a rock overhang provides a welcome alternative to wet, cramped tents in New Zealand wilderness, although in this case, only when the river is not in flood. Photo courtesy Gordon Cessford.

1997, 2001). Continued depletion of indigenous species and habitat by introduced animals and plants remains the greatest overriding threat to conservation values. Traditional Maori values for natural areas and their customary use of resources are now being more formally recognized under negotiated settlements of land grievances arising from the 1840 Treaty of Waitangi between the colonial government and the Maori, but to date this has not had important effects on wilderness.

Zimbabwe

Zimbabwe is the first developing country to have proclaimed a wilderness area (Martin 1990). *The Mavuradonna Wilderness Area was designated in Zimbabwe early in 1989 by tribal authority of the Mzaribani District Council on their own communal lands.* This designation was made in conjunction with national policy direction under Zimbabwe's Communal Areas Management Programme for Indigenous Resources (CAMPFIRE), which develops economic, educational, and cultural linkages between natural-resource management and tribal authorities. The minister of State for Tourism officially opened the area, furthering national recognition of the wilderness concept and this area's designation.

The Mavuradonna Wilderness Area is approximately 192 square miles (500 sq km) in the escarpment of the Zambezi Valley; one road bisects the area. The area will be used for a combination of wilderness "trails" (hiking), sport hunting (primarily bow hunting), and wildlife cropping, with a primary purpose of generating economic benefit for local villagers and farmers. The area will produce some income each year under the proposed uses, derived mostly from wildlife harvesting. As a result of the initial wilderness proclamation, other areas in Zimbabwe were under consideration for wilderness designation by other tribal authorities.

There has been continuing resistance from some sectors against designating wilderness in developing nations. Naysayers argue that the wilderness concept is too elitist, exclusionary, and recreation-oriented to be of value in developing countries where higher-priority issues are about food, education, rising population, and poverty. Therefore, the Mavuradonna Wilderness Area is especially important because it signals a shift in attitude within a developing nation toward accepting the wilderness concept; the wilderness designation is linked to rural development with the management emphasis being on economic sustainability of the local people, and it is an example of a designation that is relative to the culture in which it exists (Martin 1990).

The Mavuradonna Wilderness Area was established with little or no public agitation. White conservationists introduced the concept to the tribal authority, and it met little resistance or controversy. Zimbabwe's adoption of the wilderness concept is fairly recent, but it could mean that the concept is further evolving to include and incorporate the needs of more traditional and diverse people, thereby becoming more relevant to developing nations.

Tanzania

More than one-third of Tanzania is protected as national parks, game reserves, forest reserves, and other types of protected areas. Wilderness zoning in national parks is an important part of a tourism strategy in the country to promote low-impact, sustainable tourism development. Most of Tanzania's national parks have wilderness zones. Tanzania's largest national park, Serengeti National Park, has more than 1.28 million acres (517,000 ha) zoned as wilderness. This park's management plan suggests that promoting tourism for the wilderness zone is crucial to achieving the range of park conservation objectives (Kormos and Martin 2008).

Italy

In Italy, a small but determined effort by the Wilderness Associazione Italiana per la Wilderness (AIW), founded by Franco Zunino, who was inspired by participating in the third World Wilderness Congress (1983), ultimately pioneered a new type of wilderness designation by municipal governments in the 1980s (Zunino 2007). The very first Italian wilderness area, the Fosso del Capanno

Wilderness Area, was established in 1988 through a management agreement with a private foundation and was 283 acres (118 ha) in size. The largest wilderness is the Ausoni Wilderness Area, at 10,338 acres (4,230 ha). Most areas are now protected by municipalities, regional forestry authorities, or private landowners. Only one designation is by a national park authority. By the end of 2006, AIW reported that there were 42 wilderness areas covering 71,600 acres (29,000 ha) in 7 regions of Italy and 15 provinces (Zunino 2007). More wilderness designations are expected. The term *wilderness area* is now in common use in Italy, as much a new linguistic innovation as it is a new land-use and planning category in that country.

Although falling short of the type of protection offered by national legislation, the creative approach in Italy demonstrates the important growth of the wilderness concept internationally, and that it can be a useful and politically viable concept even in highly developed European countries.

Japan

The Nature Conservation Law of 1972 provides the foundation for establishing wilderness areas and nature conservation areas in Japan (Caouette 2008). The Ministry of Environment in consultation with government land-management agencies can declare both types of areas. Currently, there are five designated wilderness areas that protect 2,280 acres (5,631 ha) of national forestland. They are intended to be free from human activities and to protect ecosystems that remain in their original condition. These wilderness areas typically allow only research and science activities having specific permits. While some of these areas are adjacent to national parks, they are not components of larger protected areas or systems, as seen in many countries.

Class III–Wilderness Protection Through Recognition in Conservation Programs

Class III protection for wilderness occurs where wilderness is "recognized" as a distinct component of resource-management programs and a larger conservation effort. Namibia and the Philippines offer such protection for wilderness.

Namibia

Namibia is an arid country on the southwest coast of southern Africa. A high proportion of the country is desert or semidesert, and it includes the great Namib Desert. The country has twelve large protected areas, including the world-famous Etosha Pan and the Skeleton Coast National Parks. *Though Namibia does not have any legally designated wilderness, most of these extensive protected areas contain within their borders some of the best-preserved de facto wilderness zones on the subcontinent* (Cooper 2001).

The Wilderness Action Group of Southern Africa, in collaboration with Namibia's Ministry of Environment and Tourism, the U.S. Forest Service, and local donor organizations and NGOs, including The WILD Foundation, has developed a wilderness management training program similar to that practiced in South Africa. Resource managers from Namibia have participated because of their similar large, wild protected areas. Apart from the program's success in technology transfer of wilderness management principles and practices, it prompted the Namibian Ministry to instruct staff members to nominate candidate wilderness zones within the existing protected areas for formal recognition or designation as wilderness zones.

Namibia has important areas of privately owned land that are either partly or completely managed as natural areas for wildlife production, or as private reserves and game farms. The natural communities on these are largely intact, and a full range of large and small wildlife species are present. Many valuable conservation projects, such as a cheetah conservation program, take place on these privately owned areas and make important contributions to the economy through various forms of tourism, including photography and hunting large game (Barnes 1998). Consideration is being given to retain portions of these private lands as private wilderness areas. Some communally owned tribal lands also have extensive wildlands and wildlife populations, and there is growing interest in the sustainable use of these resources (Cooper 1996).

Philippines

In the Philippines, wilderness is recognized under a national act as a type of land classification, but it does not seem to be singled out for special management. The Philippine National Integrated Protected Areas System Act of 1992 established a national system of 200 potential areas (Pollisco and Meniado 1997). The intent is to maintain essential ecological processes and life-support systems, to protect genetic diversity, to

Fig. 3.12. Wilderness in Africa includes some charismatic megafauna—elephant, lion, and rhinoceros—that are considered icons of wilderness. Photos courtesy Ulf Doerner.

ensure sustainable use of resources, and to maintain natural conditions to the greatest extent possible. The act establishes protection of these places as national parks, game refuges, bird sanctuaries, wilderness areas, watershed areas, mangrove reserves, and virgin forests. Eight proposed areas in six regions have been targeted for immediate inventory and intense public participation in planning, funded by a grant from the European Union (Rambaldi 1997).

In northeastern Luzon, the Palanan Wilderness Area is a relatively new example of Class III wilderness, in which use of the term is descriptive rather than part of a formal land-use management plan. It deserves note for several reasons. It is part of the Northern Sierra Madre Natural Park and is considered one of the most important and largest (494,000 acres or 200,000 ha) tract of tropical rain forest in the Philippines. It is also home to the Agta forest people and has been so for more than 2,500 years. Like many of the protected wilderness areas discussed in this chapter, it is also the result of citizen action by a local NGO, the Northern Sierra Madre Wilderness Foundation. Conservation International, the US-based NGO, is very active working with and supporting the local NGO and other private groups in a community-based natural-resource management plan including traditional people's rights, conservation enterprise, sustainable use, and scientific research.

Protection of Wilderness Values in Other Regions

The nations just discussed, along with the United States, are the only ones in which wilderness protection has achieved some type of formal status. Yet, it would be a mistake to presume that wilderness values are not recognized elsewhere or that they are not given some form of protection. The distinction here is subtle; *formal, explicit wilderness protection is still relatively limited around the world, but if we broaden our perspective to embrace programs designed to protect wilderness values, albeit without the explicit labeling of areas as wilderness, we find additional important activity.* The reasons for not labeling such areas as wilderness vary widely; in some places, wilderness is simply not a part of the language. Elsewhere, there is the belief that other designations, such as national parks or nature reserves, adequately protect wilderness values.

The distinguishing qualities of wilderness, as used in this book, includes areas where natural ecological processes operate generally free of human influence and can provide experiences with opportunities for solitude and primitive recreation. In most of the countries just reviewed, this general concept of wilderness prevails. But programs in many other countries protect natural conditions and processes and also the solitude and primitive recreation opportunities associated with wilderness.

Latin America

The Latin American countries do not yet have protected areas meeting the IUCN protected areas class I definition or the protection criteria of the three classes of wilderness defined in this chapter. But, while formal government protection of wilderness is rare or nonexistent in Latin

America, wilderness values survive indirectly from other conservation strategies and actions.

The first designated wilderness area in Latin America was announced at the eighth WWC in 2005. The El Carmen Wilderness Area is 75,000 privately owned acres (30,500 ha) in the core of a large transition area between Mexico and Texas dubbed the El Carmen–Big Bend Conservation Corridor. This 10-million-acre (4-million-ha) mosaic of protected areas (private and public), ranches, and *ejidos* (communal areas) reflects the vision of the remarkable Mexican conservation photographer Patricio Robles Gil and his organization, Agrupación Sierra Madre. The management, rehabilitation, and wildlife reintroduction work in the core area of the complex, including the El Carmen Wilderness Area, is funded and on land owned by CEMEX, a large international corporation. El Carmen is one of the fabled sky islands of the Chihuahuan and Sonoran Deserts of northern Mexico and is an international biodiversity hotspot.

Throughout Latin America, external pressures on wildlands are the major threat, including hydroelectric dams, transmission corridors, oil pipelines and refineries, and road construction. All of these activities rationalize their worth in terms of large economic returns for the country in which they occur. Commercial enterprises such as cattle ranching, agricultural crops, logging, mining, and forestry plantations also penetrate the ecosystems adjacent to and within the wildland areas. Other pressures on wildlands come from subsistence farmers, refugees from war, and land speculators. Of these pressures, subsistence farming is by far the greatest threat to the integrity of wildland tropical forests, and the forests often accommodate the large number of poor, landless people with whom governments simply cannot cope. These nomadic farmers clear land, grow subsistence food crops for a few years until the soil is depleted, then move and repeat the cycle. Although not condoning these practices, government officials can rarely do much to correct them because, without major economic and social restructuring, the only other place for these people is in urban slums.

The best and most intensive rural development possible must occur adjacent to wildland areas to stop the continual encroachment of subsistence farming. Extensive and shifting practices must be converted to intensive and stabilized practices, combined with other efforts to improve the standard of living of the rural, poor people. This is true in Latin America, but no less so in many parts of Africa and southern and Southeast Asia. *A successful wilderness protection strategy will require integrated, sustainable development involving agriculture, forestry, education, sanitation, health, water quality, power production, and more to meet people's needs.* Winning the support of subsistence people living in or near wildlands may also require some allowance for well-managed use of natural resources next to and even within portions of formally designated parks and reserves. The concept of sustainability is a key to acceptance of and support for protected areas in all developing countries.

Many protected areas serve as municipal watersheds. For example, a major portion of the water used in Caracas, Venezuela, comes from the Guatopo National Park, which was established in 1958 to protect catchment areas that served four dam sites. Up to 20,000 liters of water per second can be supplied through this conservation plan made possible only through protecting the park's 250,000 acres (100,000 ha) of rain forest. Canaima National Park, also in Venezuela, is an example of conservation for the purpose of development. The park's size was increased from 2.5 million to 7.5 million acres (1.0 to 3.0 million ha) in 1975 because it was a source of water for a scheduled hydroelectric plant. This large area protects fragile habitats and ensures production of water quality and quantity and power whose values far outweigh any short-lived agricultural uses that may have been possible (MacKinnon et al. 1986).

Problems and resentment can easily arise if developing countries view preservation of wildlands as being solely for the benefit of developed-world tourists and scientists. But, nature tourism and ecotourism can make important economic contributions to developing countries, as demonstrated in Costa Rica, where citizens have come to recognize its value. Most importantly, wildland tourism needs to directly improve the lives of local peoples so they understand the direct relationship between the well-being of the wildlands and prospects for their improved standard of living.

The debt-for-nature swap is an innovative financial mechanism whereby a third party (usually a U.S. conservation organization) purchases the discounted debt of a developing nation from a major bank (debt that the bank wishes to close for various reasons) and then trades it to the debtor country in return for a plan to protect and manage important ecological areas. The first such swap occurred in July 1987, when Conservation International retired $650,000 of Bolivia's $4.5 billion foreign debt (purchased for $100,000) in return for Bolivia's creating and managing a 2.6-million-acre (100-million-ha) buffer zone around the Benai Biosphere Reserve. This first project proved controversial but useful because it illustrated some of the difficulties of debt-for-nature swaps, such as the potential inflationary effect, the need to work closely with an NGO within the developing country, and the potentially disruptive effects of a change in government. However, despite the difficulties, this strategy to conserve wildlands has been performed successfully by several U.S. conservation groups in countries such as Ecuador, Costa Rica, the Philippines, and the Malagasy Republic, usually at the request of the country's government. Debt-for-nature swaps make up a small percentage of all types of debt conversion transactions, but they are a valuable tool for wildland protection. Communication and cooperation among conservation organizations, indigenous peoples, and involved governments are necessary components of their success.

Nepal

Nepal has no designated wilderness areas, but it has great "wilderness allure." Wedged between India and China, Nepal is a small country that includes one of the most spectacular mountain ranges in the world—the Himalayas. Cultural stresses between natives and visitors are common as trekkers and mountaineers from around the world come here to hike and climb. The region's rugged peaks present some of the most challenging opportunities in the world and are centered on the world's highest peak—Mount Everest, which was designated as Sagarmatha National Park in 1976.

In 1950, Nepal opened its borders to foreign visitors for the first time and, up to that point, no Westerners had traveled to the southern base of Mount Everest. Coupled with the Chinese invasion of Tibet, closing the northerly approaches, Nepal became important to the mountaineering community when it opened its borders. In 1953, New Zealand climber Edmund Hillary and his Sherpa guide, Tenzing Norgay, reached the summit of Mount Everest for the first time and attracted worldwide attention. In the 1960s, a British military attaché at Kathmandu, Colonel James O. M. "Jimmy" Roberts, an enthusiastic mountaineer, left the army to promote trips to the area's great peaks. His company, Mountain Travel, was registered as Nepal's first trekking agency, and the great flow of trekkers to Nepal began in earnest. The burgeoning numbers of trekkers and mountaineers were a mixed blessing. Nepal is a poor country and among the least developed nations in the world. The needs of tourists for food, shelter, firewood, and guide services thus represented a major source of income for area residents. However, a rising standard of living carried with it an increased cost of living for locals. The sale of firewood to trekkers accelerated forest cutting. Meeting the needs of tourists transformed local lifestyles. The cutting was difficult to control, in part, because people needed the increased income to finance the purchase of high-cost food and petroleum fuels that became part of their new lifestyle.

The once relatively natural mountain landscape of Nepal, maintained for generations, has changed rapidly, both culturally and ecologically, as tourism has grown. Although from an American perspective we might not call the area wilderness, it was nonetheless an area where natural ecological processes had operated over a long period in close association with indigenous populations. Now it is hoped that the wild qualities of the region can be managed by designating Conservation Areas, such as at Annapurna, to retain some of the area's unique culture and environment.

Southeast Asia

There are no declared wilderness areas in Southeast Asia. In some countries, extensive logging in wildlands, driven by local and international markets, is displacing indigenous people, destroying natural tropical forest, and leading to damaged watersheds and polluted rivers. For example, in the east Malaysian state of Sarawak on Borneo, the rain forests are being logged to supply local markets and consumers in Europe, North America, and Japan. Extensive logging in Sarawak's wildlands has polluted streams and rivers from heavy erosion and milling wastes. Native people, such as the nomadic Penan, have been displaced, and natural flora and fauna are often threatened. Native people are fighting logging practices, which are not sustainable, but complex issues surrounding deforestation—such as national debt, landownership, corruption, and different concepts of development—make it difficult for natives, local and national governments, and international organizations to reach compromises essential to solutions.

In Indonesia, wildland watersheds are important to complex tropical ecosystems. Many indigenous people, such as the Dayaks, Irianese, and Javanese, depend on intact watersheds to support transportation and irrigation. Increased building of hydropower projects and dams, dumping of industrial wastes, and logging of forests degrade watersheds and lead to intensified flooding and erosion. A combination of local and international business, national debt, and domestic land distribution threaten's the biological diversity of these unique wildlands and the indigenous people's economic and social autonomy.

Antarctica

The earth's largest remaining wilderness, Antarctica is a world of superlatives—the coldest, iciest, windiest, and most remote of the world's continents, girded by the stormiest ocean. Equivalent in size to the combined area of the United States and Mexico, it expands in size each winter when the surrounding seas turn to ice.

Antarctica is vital for understanding our planet. Its ancient rocks, once at the core of the Gondwana supercontinent, hold the key to unlocking the earth's geological history. The vast Antarctic ice sheet, on average 6,560 feet (2,000 m) thick, is a window for observing changes in global climates over tens of thousands of years. The ice is a storehouse for 90 percent of the earth's freshwater resources—enough to raise the global sea level by many meters, if all of it were to melt. In this era of global warming, dramatic changes to the ice cap are already being observed. The continent and surrounding seas are a vital influence on the well-being of human life on the earth through the ways they regulate the global atmosphere and the oceans, thus affecting patterns of global weather and ocean circulation.

Antarctica remains mostly untrammeled wilderness—but it is not pristine. Soon after its discovery human exploitation targeted the marine resources and in just a few decades of the nineteenth century, Antarctica's fur seals were brought to the brink of extinction. Then the whaling industry exhausted the population of great whales, one species after the other. Today, whales are protected by a moratorium on commercial harvesting and a hemisphere-wide sanctuary, but full recovery of whale populations might never occur. In contrast, cessation of the seal harvesting allowed Antarctica's fur seals to recover their pre-exploitation populations. Since the late 1960s, new fisheries commenced in the Southern Ocean for the massive numbers of surface-living krill (*Euphausia superba*), and for rock cod and ice fish, and the fisheries experienced classic boom-and-bust episodes when harvest levels were neither ecologically nor economically sustainable.

On land, steadily expanding national research programs have established more than forty scientific stations, composed of buildings, stores, airstrips, and increased ship and air traffic to sustain 4,000 scientists and support staff in Antarctica each year. A small but growing tourist industry brings more than twice that number of people each year into Antarctica. Although Antarctic scientific and tourist activities generally exhibit a high level of environmental awareness, instances of pollution of land, sea, and air from fires, fuel spills, waste dumping, shipwrecks and aircraft crashes occasionally damage the environment. Pollutants such as DDT, polychlorinated biphenyls (PCBs), and other compounds originating from distant industrialized regions are also found. Of even greater concern is global pollution from chlorofluorocarbons and other chemicals that deplete stratospheric ozone over Antarctica, and the impact of "greenhouse" gases on global warming that increase melting of Antarctic glaciers and ice.

Considered together, these impacts and growing risks of significant environmental damage call for careful stewardship and concerted action to ensure that the wilderness values of the Antarctic region are not lost through uncontrolled human activities.

Activities in Antarctica are administered under the Antarctic Treaty System (ATS), a whole complex of arrangements for regulating relations among nations with interests and claims to the Antarctic. At the heart of the ATS is the 1961 Antarctic Treaty, with more than forty contracting parties, of which twenty-six are consultative parties, conducting demonstrable and substantial research programs in the region. The claims of seven nations to sovereignty over parts of the continent were set aside under the treaty. The treaty also demilitarized the continent, outlawed nuclear explosions and nuclear waste disposal, and promoted international cooperation, particularly in science. Other legal instruments were introduced progressively to the ATS to address the conservation of flora and fauna (1964), seals (1972), and marine living resources (1980).

The 1991 Environmental Protocol to the Antarctic Treaty (Madrid Protocol) lays out a farsighted and comprehensive program for protecting and managing the Antarctic environment. It declares Antarctica a natural reserve devoted to peace and science, bans

Fig 3.13. Trekking in the Annapurna Sanctuary near Pokhara, Nepal, provides the opportunity to see the interaction of mountain communities with wilderness environments. Photo courtesy Chad P. Dawson.

mineral extraction, and presents ambitious mandates requiring all treaty parties and private individuals to minimize their impact on Antarctica. With oversight from a Committee on Environmental Protection, supported by the Scientific Committee for Antarctic Research and Council of Managers of National Antarctic Programs (COMNAP), the protocol establishes rules for environmental protection and management. The COMNAP secretariat is located in Australia.

The Madrid Protocol represents a giant leap forward in protecting the wilderness values of Antarctica. It provides for designating special protection for areas acknowledged to have significant natural, historic, scientific, or landscape values, and areas where multiple uses might cause undesirable environmental impact or disruption among conflicting activities. Management plans are required for most protected areas, of which currently more than fifty are established over natural sites such as breeding colonies of seabirds and seals, and seventy historic sites, including stations from the heroic era of Antarctic exploration (Dingwall 1997).

For more than forty years the Antarctic Treaty has proven effective, practical, and responsive to changing global problems and circumstances. In agreeing to the Environmental Protocol, the treaty governments have signaled a strong commitment to protection and sustainable management of Antarctica's resources. But challenges lie ahead with increasing human presence and growing demands for use of its resources.

Other Initiatives
Protecting Biodiversity

Protecting biodiversity directly supports wildland conservation and wilderness because it is in wild areas that diversity flourishes and species evolve as naturally as possible. Recognizing this, *the Convention on Biological Diversity (CBD) was signed at the Earth Summit in Rio de Janeiro, Brazil, in 1992, and it is the world's premier treaty addressing biodiversity loss, with its three mutually reinforcing objectives: conservation, sustainable use, and equitable sharing of benefits from biodiversity* (Faries and Cervigni 1998). The Global Environmental Facility of the World Bank is the financial mechanism of the Convention on Biological Diversity and provides assistance to developing countries that are parties to the Convention. It is difficult to explain the importance of biodiversity to people with subsistence lifestyles, so it is important for wildland managers and environmental educators to actually demonstrate its value with locally relevant projects. The protection of entire biomes preserves the biodiversity of species needed to maintain the productive capacities of ecosystems. Protected areas provide nesting, calving, spawning sites, and habitat for plant and animal species that local people rely on for food, medicine, or for sale to acquire income.

The plight of indigenous people has long been ignored. But positive examples are beginning to emerge in Central and South America, South and Southeast Asia, Australia, and parts of Africa where indigenous peoples' rights are recognized and their well-being integrated into plans for sustainable wildland management. A particularly hopeful sign is formation of *the Native Lands and Wilderness Council*, which first met in 2005 at the eighth WWC, convened and chaired by natives, and focused specifically on case studies of self-management of indigenous wildlands under tribal direction. Managing wildlands owned or sustained by indigenous people is extremely important because many of these native people are wilderness-dependent and live in large wildland areas. In the short term, at least, protecting their lifestyles protects the wilderness.

Much attention has focused on increased protection of the world's tropical, moist forests, which occur in a belt around the middle of the globe. These lands, occupying only about 12 percent of the earth's total land area, contain an abundant diversity of the world's species—40 to 50 percent of the planet's total number of plant and animal species. Yet, the impact of human occupancy and development on these rich reserves is enormous, with huge losses of tropical forest due to logging. The need to protect tropical wildlands cannot be overemphasized, and some form of wilderness protection could play an important role.

Of many examples, a particularly outstanding one was initiated by Conservation International (CI) in 1998, the Tropical Wilderness Protection Fund (TWPF). Building on a very positive experience in creating the 1.6-million-hectares Central Suriname Nature Reserve, CI obtained a grant of $5 million to stimulate similar park and wildland creation projects in other parts of the world. After two years of successful operation, the TWPF was transformed into the Global Conservation Fund (GCF). A generous five-year, $100 million grant from the U.S.-based Gordon and Betty Moore Foundation launched a campaign to protect 371 million acres (150 million ha) of wilderness and an additional 247,000 to 741,000 acres (100,000 to 300,000 ha) of critical habitat in high-biodiversity hotspots—at a projected cost of $6 billion. As of 2002 the GCF had secured 14.8 million acres (5 million ha) in new protected areas, with another 104 million acres (42 million ha) in process of protection.

Wildland Corridors

Conservation biology is concerned with joining protected wild areas and low-impact areas, such as national forests or private forests and ranchlands, to create wildland habitat corridors to ensure migration of wildlife, enhanced biological intactness, and sustainable ecosystem functioning. One of the earliest and ongoing models, started in 1993, is Yellowstone to Yukon (Y2Y), a joint Canadian-U.S. network (www.y2y.net) of more than 800 organizations, institutions, foundations, and conservation-minded individuals who have recognized the value of working together to restore and maintain an unprotected area of the corridor between Yellowstone (in northwestern Wyoming) and the Yukon (northwestern Canada and northeastern Alaska) region. Y2Y opens new possibilities for ensuring the continued presence of North American wildlife and wildlands. Pioneering conservationist photographer Florian Schulz walked and photographed Y2Y and later edited a book (Schulz 2005) that was titled *Yellowstone to Yukon: Freedom to Roam*. In the wake of Y2Y have come similar initiatives, many of them inspired or managed by The Wildlands Project (www.twp.org), such as Heart of the West, Southern Rockies, and Northern Sierra Madre.

Transfrontier Conservation

Some mega–wildland corridors are *transfrontier,* that is, crossing national political boundaries to embrace entire ecosystems. This is not a new concept, and the first such area was established in 1932 between Glacier National Park (Montana, U.S.) and Waterton National Park (Alberta, Canada). A 2005 survey reported in *Transboundary Conservation* (Mittermeier et al. 2005) indicates that there are 188 Internationally Adjoining Protected Areas (IAPA) or complexes involving 818 protected areas in 112 countries and representing approximately 17 percent of the global extent of high-biodiversity wilderness areas. Nearly half of the total land area of IAPAs is located in North America, followed by Africa. The Russian Federation has the greatest number of IAPAs (21), followed by China (14) and Canada (12). Countries contributing the most area to the overall IAPAs are Greenland, the U.S., and Venezuela.

IAPAs, or transfrontier protected areas, involve important issues in political and land management cooperation, the details of which are not easy even across a friendly border such as that between the U.S. and Canada with the Boundary Waters Canoe Area and Canada's Quetico Provincial Wilderness Park, or between Mexico and the U.S. (Texas) where the El Carmen–Big Bend Complex is being developed.

Elsewhere in the world such tough issues may take years to resolve, if at all. The Peace Parks Foundation (www.peaceparks.org), based in South Africa, has popularized this concept and is working to establish a network of "peace parks" throughout Africa. The Peace Parks Foundation is also consulting on proposed projects in the Middle East and in the Demilitarized Zone between North and South Korea. Peace Parks exemplify what the world could be like with concern for wild nature at the center of human dialogue.

Water and Marine Wilderness

At the sixth WWC in Bangalore, India, in 1998, Michael McCloskey presented the results of his most recent international survey—"Wild Rivers of the World" (McCloskey, Michael 2001). In this second of his pioneering, reconnaissance-level, global surveys, McCloskey studied all rivers at least 50 kilometers long. They earned the "wild" distinction if they were free of dams and their effects and had a pollution-free watershed, with no roads along the banks. Rivers that ran through lightly developed territory (small, scattered settlements), though not moderately or heavily developed, were also included. McCloskey's study identified 6,000 river segments in the world that appeared wild under these criteria. Added together, these segments amount to 484,250 miles (781,051 km), averaging 81 miles (130 km) each segment. This amounts to 19.6 percent of all rivers studied, mostly found in developing nations.

North America had the highest percentage of wild rivers—37.9 percent—most of which were in the sub-Arctic and Arctic north above 53 degrees north latitude. South America still had 26.1 percent of its rivers wild, mostly in the Amazon Basin. Three other regions were below average: Eurasia with 17 percent, mostly in eastern Siberia; Australia and Oceania with 10.6 percent; and, surprisingly, Africa with the lowest percentage (7.6). Nearly 60 percent of wild rivers are in just two countries, Russia and Canada. Another 20 percent are in the United States (mostly in Alaska), Brazil, and China.

There are evolving efforts to apply the wilderness concept to water and marine areas. Initially, this was done by incorporating portions of aquatic areas in wilderness designations with surrounding terrestrial areas, the Boundary Waters Canoe Area Wilderness in the United States being the best example. Internationally, an important example of this is the wilderness zone in Lake St. Lucia located in KwaZulu-Natal in South Africa. A low fence in the lake marks areas for nonmotorized use. There is also a terrestrial wilderness zone on the eastern shore between the lake and the Indian Ocean, which was the focus of one of the most dramatic conservation struggles in Africa during most of the 1990s. Conservation NGOs were pitted against the world's largest mining company, Rio Tinto Zinc, and eventually were successful when President Mandela's government proclaimed the Greater St. Lucia Wetland Park in 1998, which was declared South Africa's first world heritage site in 2000.

At the eighth WWC in Alaska in 2005, CI's Mike Smith presented a new inventory of freshwater wilderness that extended the earlier Michael McCloskey (2001) analysis to include lakes and wetlands as well as streams and to identify large areas of the world where surface waters are most likely to remain in natural condition. This study found important areas of freshwater wilderness only at the northern extremities of the continents and in a few small watersheds of South America. The loss of wild water in much of the tropics is associated with the fact that rivers have often served as avenues for human penetration of tropical forests (Sanderson et al. 2002). Although Amazonia and the Congo Basin are considered terrestrial wildernesses (Mittermeier et al. 2003), they are impacted by river channel fragmentation and human management of flow regimes, which affect large areas within catchments. Southeast Asia, which has some of the world's highest levels of freshwater biodiversity, has long been impacted by agriculture and high human population density. It is now subject to high regulation of water flows at all scales. In the Northern Hemisphere, population density and agriculture eliminate wild waters at midlatitudes. Regulation of water flow extends human impacts to even higher latitudes, leaving only a few northward-flowing watersheds with large-scale wilderness conditions. *These studies show that wild freshwater is the rarest form of wilderness and is the form that is disappearing most rapidly.*

There has also been movement toward oceanic or marine wilderness. The second WWC, in Australia, featured discussion of a proposed wilderness zone in the Great Barrier Reef Marine Park. The Australian Park Authority was planning to designate zones with the Capricornia section of the park for wilderness-type use only. Although small in terms of the overall size of the park, it should be emphasized that, in terrestrial terms, the zones are extensive (Kelleher 1989). The U.S. National Oceanic and Atmospheric Administration announced inquiry into the feasibility of oceanic wilderness (Foster and Lemay 1988) at the fourth WWC, in a special symposium focused on the topic.

At the sixth WWC, Maxine McCloskey added considerably to the concept of marine wilderness with an innovative, global survey of potential wilderness areas on the high seas (McCloskey, Maxine 2001). One of former president Clinton's Executive Orders (May 2000) established a new Marine Protected Area in the northwestern Hawaiian Islands. One of its original intentions was to institute "no take" management, a widely accepted criterion for marine wilderness. In 2006 former president Bush signed a proclamation that created the Northwestern Hawaiian Islands Marine National Monument (renamed the Papahānaumokuākea Marine National Monument in March 2007). This national monument placed nearly 140,000 square miles (362,600 sq km) of the northwestern Hawaiian Islands under the highest form of marine environmental protection. The national monument will preserve access for Native Hawaiian cultural activities, provide for carefully regulated educational and scientific activities, enhance visitation in a special area around Midway Island, prohibit unauthorized access to the monument, phase out commercial fishing over a five-year period, and ban other types of resource extraction and dumping of waste. This marine national monument is the largest single area dedicated to conservation in the history of the United States, and the largest protected marine area in the world. It is more than one hundred times larger than Yosemite National Park.

The World Wilderness Congress

The World Wilderness Congress (WWC), a project of The WILD Foundation (www.wild.org), has met on eight occasions (1977, South Africa; 1980, Australia; 1983, Scotland; 1987, the United States; 1993, Norway; 1998, India; 2001, South Africa; and 2005, Alaska) to explore management, cultural, and scientific elements of wilderness protection (Martin 2001). The WWC has provided a continuing international forum for expanding the wilderness concept, and its plenary and technical symposia proceedings document important international perspectives and information on wilderness values, opportunities, management, science, and cultural issues. The WWC has also been adamant, since its inception, that indigenous and traditional communities must be involved in international wildlands conservation.

The WWC has been a focal point for the worldwide evolution of wilderness definition and legislation, and each of the eight WWCs has worked toward a definition of wilderness acceptable to the world community, with each proceedings recording new perspectives. Ultimately a resolution and proposed definition from the fourth World Wilderness Congress (Martin 1988) was submitted and adopted by the IUCN Commission on National Parks and Protected Areas—now called the World Commission on Protected Areas, discussed earlier. The IUCN protected area classification (category 1b) definition for wilderness was subsequently updated to *"A large area of unmodified or slightly modified land, and/or sea, retaining its natural character and influence, without permanent or significant habitation, which is protected and managed so as to preserve its natural condition"* (Kormos 2008, 22).

There is a distinct difference between this evolving international definition and that contained in the 1964 U.S. Wilderness Act, "where the earth and its community of life are untrammeled by man, where man himself is a visitor who does not remain." In developing countries, many wildland areas are either occupied by indigenous people or are under pressure from subsistence farmers, people who engage in shifting cultivation, grazing, or other extractive and potentially damaging practices.

The WWC is a major *conservation project* of The WILD Foundation, with a three- to five-year cycle, and is very different from most conferences. Each new congress is conceived and implemented in and with a different country. Practical conservation and wilderness objectives are established and a structure and process implemented to achieve them. The WWC has an impressive list of these accomplishments, some of which have been mentioned elsewhere in this chapter, such as helping initiate the Global Environmental Facility, establishing the International League of Conservation Photographers, initiating the Native Lands and Wilderness Council, pioneering the marine wilderness concept, establishing wilderness on private lands, creating the first World Wilderness Inventory, creating the first inventory of freshwater wilderness; initiating the first wilderness law, designating wilderness area in Latin America (Mexico), and others.

International Journal of Wilderness

Another project of The WILD Foundation and stimulated by the WWCs is the *International Journal of Wilderness* (*IJW*). Launched in 1995, the *IJW* is sponsored by fifteen wilderness-related organizations, four of them in countries other than the United States. The *IJW* features articles on wilderness protection, allocation and designation, science, management, and use worldwide. It services a growing professional network of wilderness managers, scientists, and others, and is another reflection of the international growth and acceptance of wilderness. As a forum for scientific and professional articles, opinions, and discussion about wilderness around the world, it is abundantly cited in this text. *IJW* is published in three issues per year (see the website www.ijw.org) and includes peer-reviewed professional and science research articles as well as opinion and feature articles about wilderness designation and management concepts and issues.

Summary

The idea of wilderness began in the United States, but it is gaining increasing recognition and support around the world. The American notions about wilderness are an influence, but not a determinant, on how other nations approach the idea. Wilderness as defined in New Zealand, for example, resembles wilderness under the 1964 U.S. Wilderness Act, but is purer in application. Differences of geography, culture, and economics alter the specific ways in which other countries approach designating and managing wilderness.

This chapter describes countries where wilderness is protected by law (Class I), as an administrative zone in a park or other protected area (Class II), or as part of another conservation program designation (Class III). The concept of wilderness varies around the globe in its approach to designation and protection criteria, ecological protection, and indigenous populations, which are major issues regarding wilderness in Australia, South America, the Philippines, and Finland. These are all issues that individual countries must confront and resolve in light of their own social, economic, and cultural circumstances. In developing countries, the need for resolving wilderness and wildland protection with economic and rural development needs is a primary consideration.

International interest in wilderness is growing. Numerous countries around the world have begun to consider how wilderness might fit into their conservation management systems. Even in some countries where long-term occupancy and development have greatly modified the landscape, the possibility of restoring wilderness and natural conditions is being discussed. The wilderness idea reflects an increasing understanding that wilderness settings are a repository of a country's natural heritage and a symbol of that country's international responsibility to protect the environment.

Study Questions

1. What role did wilderness protection efforts in the United States play internationally in protecting wilderness?
2. What are some differences between wilderness protection in the United States and in other countries?
3. Identify and discuss some of the factors that account for differences in progress in wilderness preservation between the United States and Europe.
4. Identify some of the important relationships between wilderness and international conservation designations such as biosphere reserves, world heritage sites, and Ramsar sites.
5. As population increases throughout the world, do pressures for development in lesser-developed countries lead to undesirable changes in ecological conditions? What strategies and actions will be necessary to improve prospects for wilderness and wildland protection in developing nations?
6. Why was there interest in inventorying the freshwater and marine waters of the world for natural and wild conditions? What is the value of protecting some of these areas?
7. Explain the distinctions between Class I, Class II, and Class III wilderness designation presented in this chapter. What nations have Class I wilderness? How are they different in their mechanisms for protection?
8. How are Russian zapovedniks similar to wilderness in the United States? How are they different?
9. What are transfrontier conservation areas?

Acknowledgments

This chapter was updated for the fourth edition by Vance Martin, president of The WILD Foundation, and Alan Watson, Aldo Leopold Wilderness Research Institute, expanding on their coauthorship of the third edition, building on the second edition by George Stankey, Vance Martin, and Roderick Nash and the first edition by Roderick Nash.

The coauthors wish to thank Cyril Kormos, vice president for policy at The WILD Foundation, for his contributions to the section on international wilderness law and policy. We also thank our colleagues for their reviews and contributions to earlier editions, but the interpretation, conclusions, errors, and omissions are the responsibility of the authors. We thank Bill Bainbridge, consultant (retired), Natal Parks Board, South

Africa; Hugh Barr, past president, Federated Mountain Clubs of New Zealand; Ralf Buckley, Griffith University, Australia; Gordon Cessford, Department of Conservation, New Zealand; Paul Dingwall, Department of Conservation, New Zealand; Brian Dych, B.C. Parks, Victoria, British Columbia, Canada; Tom Elliot, Parks Canada, White Horse, Yukon, Canada; Harold Eidsvik (retired), Parks Canada; Peter Helman, private consultant, Sydney, Australia; Steve Hollenhorst, University of Idaho; Evelyn Hurwich, The Antarctica Society, U.S.A.; Liisa Kajala, Finnish Forest and Park Service & Finnish Forest Research Institute, Enontekio, Finland; Alec Marr and Julie McGiness, The Wilderness Society, Australia; Brad Mills, Land Conservation Council of Victoria, British Columbia, Canada; Pers Nilsens, Parks Canada, Ottawa; Dave Ostergren, Northern Arizona University; George Wallace, International School for Forestry and Natural Resources, Colorado State University, Fort Collins.

Case Discussion:
Canadian Wilderness— The Quetico Provincial Park

The Quetico Provincial Park is a 1.18-million-acre (475,782 ha) wilderness area in northwestern Ontario, Canada. The southern boundary is shared with the Boundary Waters Canoe Area Wilderness in the Superior National Forest in the state of Minnesota (U.S.). The Quetico Provincial Park (QPP) was created in 1913 and classified as a wilderness park in 1977, and it has remained remote with difficult access except for two access roads. Canoeists who enter the area to enjoy the more than six hundred lakes and waterways must access the QPP by obtaining a permit via one of six ranger stations that serve more than twenty access points.

Historically, the area was used by Native peoples and later as a travel route for European explorers, fur trappers, and traders. In 1996, the Canadian federal and provincial governments designated the QPP as part of Boundary Waters—Voyageur Waterway and Canadian Heritage River. This designation was a historic and cultural tribute to the voyageurs who paddled these waters as fur traders.

The QPP is in an ecological transition zone between the plains to the west, the boreal forest to the north, and the mixed hardwood and softwood forests to the south. Fire has been a historic natural influence throughout the area. Logging and some mining activity have impacted the forests but were prohibited by 1971. In 1979, all motorized vehicles including motorboats were banned, with the exception of an agreement with part of the Lac La Croix First Nation, which is allowed to operate motorized guideboats with less than a 10-horsepower motor until the year 2015.

The QPP is representative of the wilderness values most often reported by Canadians in justifying their support of wilderness protection: a bequest to future generations and protection from the advancement of civilization. The QPP is one of the first protected areas in the province of Ontario and the second largest such area. International recognition of the importance of protecting this area was noted by the IUCN, which classified the QPP as a category II protected area—a national park designated for ecosystem protection and recreation use.

Recreational opportunities for remote and primitive canoe camping abound, with more than 2,000 unimproved campsites spread over the lakes of the QPP. There is also one campground with 107 sites that is accessible by vehicle at the Dawson Trailhead. The interior recreational use is restricted to daily quotas and regulated by personnel at the ranger stations through a permit process, with group size limits and the prohibition of motorized and mechanized equipment and of nonburnable, disposable food and beverage containers. The interior has both day use and overnight visitors, with an annual average of 125,000 visitors and 122,000 camper nights.

The initial QPP master plan was approved in 1977 and then revised in 1982, 1988, and 1995. In 2007–2008, the QPP was engaged in a review of its management plan. Revision of that plan is expected to include, but is not limited to, consideration of:

1. Zoning (e.g., creation of nature reserve zones, new access zones).
2. Policies for resource stewardship, operations, and development for:
   - Natural resources, including fisheries management, wildlife management, fire management, and vegetation management.
   - Cultural resources—First Nation values (e.g., spiritual sites, pictographs) and historical sites (e.g., Cabin 16, King's Point).
   - Operations—natural heritage education (e.g., interpretation, information services); research

(e.g., wildlife, natural fire, fisheries); recreation management (e.g., carrying capacity and quotas, new recreation activities, and winter uses); Atikokan-Quetico Tourism recommendations; tourism services (e.g., outfitting); and marketing (e.g., partnerships, relationship with Atikokan, Lac La Croix, Thunder Bay).
- Development—economic development (e.g., Lac La Croix and Atikokan); new access points (e.g., locations, facility requirements); roads (e.g., Batchewaung Lake); and other (e.g., backcountry hiking/skiing trails).
3. Implementation priorities for stewardship, operations, and development policies.
4. Social and economic impact analysis (e.g., job creation, infrastructure).
5. Management planning discussions will recognize existing Aboriginal uses and the aspirations of the Lac La Croix First Nation to achieve economic and employment opportunities associated with the park.

Case Discussion Questions

1. The complexities of the issues to be addressed are a challenge to the planning team. What types of representation should there be on the planning team in view of the various constituents and disciplines affected and the considerations being reviewed? Does that change when you think about the mission of the QPP?
2. When you read over the list of considerations before the planning team, does it seem as if the definition of wilderness should enter into what is considered further and what is dropped? Do you think the current definition of wilderness is based on what is currently allowed or prohibited in the QPP? Would each of the five issue categories above, being considered under the QPP plan revision, negatively impact or positively support that definition of wilderness?
3. Some people argue that Canada has such vast wildlands and waterways in the northern area of the nation that parks in the south like QPP should be declassified as wilderness and further developed. How would you argue for maintaining strong protection of wilderness values in QPP? What could international recognition, such as the IUCN category II, do to support your argument?

References

Alberta Wilderness Association. 1973. *Willmore Wilderness Park.* Calgary, Alberta, Canada: Alberta Wilderness Association.

———. 1985. *Wilderness in Alberta: The Need Is Now.* Calgary, Alberta, Canada: Alberta Wilderness Association.

Archibald, George. 1988. The Ramsar convention and wetland protection. In: Martin, Vance G., ed. *For the Conservation of the Earth: Proceedings of the 4th World Wilderness Congress;* September 11–18, 1987; Denver and Estes Park, CO. Golden, CO: Fulcrum, pp. 52–56.

Bainbridge, W. R. 1987. The use of fire to maintain indigenous vegetation in the wilderness systems in Southern Africa. In Krumpe, E. E.; Weingart, P. D., eds. *Designation and Management of Park and Wilderness Reserves: Proceedings of a Symposium at the 4th World Wilderness Congress;* September 14–18, 1987; Denver and Estes Park, CO. Ojai, CA: The WILD Foundation.

Bainbridge, W. R.; Lax, I. 2008. South Africa. In: Kormos, Cyril, ed. *A Handbook on International Wilderness Law and Policy.* Golden, CO: Fulcrum, pp. 241–267.

Barnes, Jonathan I. 1998. Economic value of wilderness in Namibia. *International Journal of Wilderness.* 4(1): 33–38.

Barr, H. 1997. Establishing a wilderness preservation system in New Zealand: a user's perspective. *International Journal of Wilderness.* 2(2): 7–10.

———. 2001. Establishing a wilderness preservation system in New Zealand: A user's perspective. In: Cessford, G. R., ed. *The State of Wilderness in New Zealand.* Science and Research Unit, Department of Conservation, Wellington, pp. 17–22.

Bobiec, A. 2002. Bialowieza Primeval Forest: The largest area of natural deciduous lowland forest in Europe. *International Journal of Wilderness.* 8(3): 33–37.

Brüggemann, Jens. 1997. National parks and protected areas management in Costa Rica and Germany: A comparative analysis. In: Ghimire, K. B.; Pimbert, M. D., eds. *Social Change and Conservation: Environmental Politics and Impacts of National Parks and Protected Areas.* London: Earthscan, pp. 71–96.

Caouette, Brian. 2008. Japan. In: Kormos, Cyril, ed. *A Handbook on International Wilderness Law and Policy.* Golden, CO: Fulcrum, pp. 171–181.

Cessford, G. R.; Dingwall, P. R. 1997. Wilderness and recreation in New Zealand. *International Journal of Wilderness.* 3(4): 39–43.

———. 2001. Wilderness and recreation in New Zealand. In: Cessford, G. R., ed. *The State of Wilderness in New Zealand.* Science and Research Unit, Department of Conservation, Wellington, pp. 35–42.

Cessford, G. R.; Reedy, M. C. 2000. Wilderness status and associated management issues in New Zealand. In: Watson, A. E.; Aplet, G. H.; Hendee, J. C., comps. *Personal, Societal, and Ecological Values of Wilderness: 6th World Wilderness Congress Proceedings on Research, Management and Allocation,* Vol. II; October 24–29, 1998; Bangalore, India. RMRS-P-14. Ogden, UT: U.S. Department of Agriculture, Forest Service, Rocky Mountain Research Station, pp. 185–192.

———. 2001. Wilderness status and associated management issues in New Zealand. In: Cessford, G. R., ed. *The State of Wilderness in New Zealand.* Science and Research Unit, Department of Conservation, Wellington, pp. 43–56.

———. 2008. Personal communication with Vance Martin.

Chandy, Shibi; Euler, David L. 2000. Can community forestry conserve tigers in India? In: Watson, A. E.; Aplet, G. H.; Hendee, J. C., comps. *Personal, Societal, and Ecological Values of Wilderness: 6th World Wilderness Congress Proceedings on Research, Management and Allocation*, Vol. II; October 24–29, 1998; Bangalore, India. RMRS-P-14. Ogden, UT: U.S. Department of Agriculture, Forest Service, Rocky Mountain Research Station, pp. 155–161.

Chape, Stuart; Blyth, S.; Fish, L.; Fox, P.; Spalding, M., comps. 2003. *2003 United Nations List of Protected Areas*. IUCN. Gland, Switzerland; Cambridge, UK; and UN Environmental Program–WCMC, UK.

Colwell, Mark A.; Koroliuk, Andrei Yu; Dubyrin, Alexander V.; Soboleb, Nikolai A. 1997. Russian nature reserves and conservation of biological diversity. *Natural Areas Journal*. 17(1): 56–68.

Cooper, Trygve, ed. 1996. *Proceedings of the International Wilderness Management Symposium—Waterberg Plateau Park, Namibia*. Ojai, CA: The WILD Foundation.

———. 2001. Wilderness in Namibia. In: Martin, Vance, ed. *Wilderness and Humanity: The Global Issue. Proceedings of the 6th World Wilderness Congress*; October 24–29, 1998; Bangalore, India. Golden, CO: Fulcrum, pp. 128–133.

Cumberland, Kenneth B. 1981. *Landmarks*. Surry Hills, New South Wales, Australia: Reader's Digest Proprietary.

de Alwis, Lyn. 1998. Origins, evolution and present status of the protected areas of Sri Lanka. In: Martin, Vance, ed. *Wilderness and Humanity: The Global Issue. Proceedings of the 6th World Wilderness Congress*; October 24–29, 1998; Bangalore, India. Golden, CO: Fulcrum, pp.115–121.

Dingwall, Paul. 1997. Environmental management for Antarctic wilderness. *International Journal of Wilderness*. 3(3): 22–26.

Dooling, Peter J. 1985. Heritage landscapes: Rethinking the Canadian experience. *The Forestry Chronicle*. 61(4): 319–322.

Draper, M. 1998. Zen and the art of Garden Province maintenance: The soft intimacy of hard men in the wilderness of KwaZulu-Natal, South Africa, 1952–1997. *Journal of Southern African Studies*. 24(4): np.

Faries, William; Cervigni, Raffaello. 1998. Achieving sustainable conservation: The global environment facility. *International Journal of Wilderness*. 4(2): 32–33.

Foster, Nancy; Lemay, Michele H. 1988. Ocean wilderness—Myth, challenge or opportunity? In: Martin, Vance G., ed. *For the Conservation of the Earth. Proceedings of the 4th World Wilderness Congress*; September 11–18, 1987; Denver and Estes Park, CO. Golden, CO: Fulcrum, pp. 171–174.

Glavovic, P. D. 1985. The legal status of wilderness: Aspects of the 1984 Wilderness Act. *The South African Law Journal*. 102: 262–267.

Green, M. J. B.; Paine, J. 1997. State of the world's protected areas at the end of the twentieth century. Paper presented at the IUCN World Commission on Protected Areas Symposium on Protected Areas in the 21st Century: From Islands to Networks; November 24–29, 1997; Albany, Australia. Cambridge, England: IUCN Publications Services Unit. www.iucn.org.

Halffter, Gonzalo. 1985. Biosphere reserves: Conservation of nature for man. *Parks*. 10(3): 15–18.

Hallikainen, Ville. 1995. The social wilderness in the minds and culture of the Finnish people. *International Journal of Wilderness*. 1(1): 35–36.

Hannah, L.; Lohse, D.; Hutchinson, C.; Carr, J. L.; Lankerani, A. 1994. A preliminary inventory of human disturbance of world ecosystems. *Ambio*. 23: 246–250.

Helman, Peter M.; Jones, Alan D.; Pigram, John J.; Smith, Jeremy M. B. 1976. *Wilderness in Australia: Eastern New South Wales and Southeastern Queensland*. Armidale, Australia: University of New England, Department of Geography.

Hummel, Monte. 1989. *Endangered Spaces: The Future for Canada's Wilderness*. Toronto, Ontario, Canada: Key Porter Books.

International Union for the Conservation of Nature and Natural Resources (IUCN). 1979. *The Biosphere Reserve and Its Relationship to Other Protected Areas*. 19p.

———. 1998. *1997 United Nations List of Protected Areas*. Cambridge, England: IUCN Publications Services Unit. www.iucn.org.

Kajala, Liisa. 2008. Finland. In: Kormos, Cyril, ed. *A Handbook on International Wilderness Law and Policy*. Golden, CO: Fulcrum, pp. 143–155.

Kelleher, Graeme. 1982 . Management of the Great Barrier Reef. In: Martin, Vance, ed. *Wilderness*. Moray, Scotland: The Findhorn Press, pp. 136–141.

Kormos, Cyril F., ed. 2008. *A Handbook on International Wilderness Law and Policy*. Golden, CO: Fulcrum.

Kormos, Cyril F.; Martin, Vance G. 2008. Southern and Eastern Africa. In: Kormos, Cyril, ed. *A Handbook on International Wilderness Law and Policy*. Golden, CO: Fulcrum, pp. 313–317.

Kruger, Sonja. 2006. The challenge of wilderness fire stewardship in a time of change. *International Journal of Wilderness*. 12(1): 44–47.

Lesslie, R. G.; Taylor, S. G. 1985. The wilderness continuum concept and its implications for Australian wilderness preservation policy. *Biological Conservation*. 32(4): 309–333.

Lucas, P. H. C. 1983. A New Zealand wilderness preservation system or a joint wilderness policy? In: Molloy, Leslie F., ed. *Wilderness Recreation in New Zealand: Proceedings of the FMC 50th Jubilee Conference on Wilderness*; August 22–24, 1981; Rotoiti Lodge, Nelson Lakes National Park, New Zealand. Wellington: Federated Mountain Clubs of New Zealand, pp. 46–52.

MacKinnon, John; MacKinnon, Kathy; Child, Graham; Thorsell, Jim. 1986. *Managing Protected Areas in the Tropics*. Gland, Switzerland: IUCN.

Martin, Vance G., ed. 1988. Resolution of the Congress. In: *For the Conservation of the Earth: Proceedings of the 4th World Wilderness Congress*; September 11–18, 1987; Denver and Estes Park, CO. Golden, CO: Fulcrum.

———. 1990. International wilderness: Adapting to developing nations. In: Lime, David, ed. *Managing America's Enduring Wilderness Resource*. Conference proceedings; September 11–17, 1989; St. Paul: University of Minnesota Extension Service, pp. 252–266.

———. 1994. Wilderness designation as a global trend. In: Martin, Vance; Tyler, N., ed. *Arctic Wilderness: Proceedings of the 5th World Wilderness Congress*; September 1993. Tromso, Norway. Golden, CO: Fulcrum, pp. 8–19.

———. 2001. The World Wilderness Congress. *International Journal of Wilderness*. 7(1): 4–9.

Martin, Vance; McCloskey, Michael. 1993. International laws governing wilderness. *Journal of Forestry*. 91(2): 35.

McCloskey, J. Michael; Spalding, Heather. 1988. A reconnaissance-level inventory of the wilderness remaining in the world. In: Martin, Vance G., ed. *For the Conservation of the Earth: Proceedings of the 4th World Wilderness Congress*; September 11–18, 1987; Denver and Estes Park, CO. Golden, CO: Fulcrum, pp. 18–41.

McCloskey, Maxine. 2001. The high seas: Is there room for wilderness? In: Martin, Vance; Sarathy, M. A. Partha, eds. *Wilderness and Humanity: The Global Issue. Proceedings of the 6th World Wilderness Congress*; October 24–29, 1998; Bangalore, India. Golden, CO: Fulcrum, pp. 193–197.

McCloskey, Michael. 2001. Wild rivers of the world. In: Martin, Vance, ed. *Wilderness and Humanity: The Global Issue. Proceedings of the 6th World Wilderness Congress*; October 24–29, 1998; Bangalore, India. Golden, CO: Fulcrum, pp. 20–25.

McCollum, Ian. 2007. Wilderness warrior. *Africa Geographic*. June: 65–68.

McDonald, Tom. 1995. Mission Mountains Tribal Wilderness Area of the Flathead Indian Reservation. *International Journal of Wilderness*. 1(1): 20–21.

McLuhan, T. C., ed. 1971. *Touch the Earth: A Self-Portrait of Indian Existence*. New York: Outerbridge and Dienstfrey.

McNamee, Kevin. 2008. Canada. In: Kormos, Cyril, ed. *A Handbook on International Wilderness Law and Policy*. Golden, CO: Fulcrum, pp. 91–117.

Metsähallitus (Finnish Forest and Park Service). 2005. Personal communication with Vance Martin.

Mittermeier, Russell A.; Kormos, Cyril F.; Mittermeier, Cristina Goettsch; Gil, Patricio Robles; Sandwith, Trevor; Besancon, Charles. 2005. *Transboundary Conservation: A New Vision for Protected Areas*. Conservation International.

Mittermeier, Russell A.; Mittermeier, Cristina Goettsch; Gil, Patricio Robles; Pilgrim, John. 2003. *Wilderness: Earth's Last Wild Places*. Conservation International and University of Chicago Press.

Molloy, L. F. 1983. Wilderness recreation—the New Zealand experience. In: Molloy, L. F., ed. *Wilderness Recreation in New Zealand: Proceedings of the FMC 50th Jubilee Conference on Wilderness*; August 22–24, 1981; Lake Rotoiti, Nelson Lakes National Park, New Zealand. Wellington, New Zealand: Federated Mountain Clubs of New Zealand, pp. 4–19.

———. 1997. Wilderness in New Zealand: A policy looking for someone to implement it. *International Journal of Wilderness*. 3(2): 11–45.

Mosley, Geoff. 1978. A history of the wilderness reserve idea in Australia. In: Mosley, Geoff, ed. *Australia's Wilderness: Conservation Progress and Plans: Proceedings of the first National Wilderness Conference*; October 21–23, 1977; Canberra, Australia. Melbourne, Australia: Australian Conservation Foundation, pp. 27–33.

Muir, Andrew. 1999. The wilderness leadership school of South Africa's Imbewu and opinion leader programmes. *International Journal of Wilderness*. 5(2): 41–43.

Muir, Keith. 2007. Action toward wilderness protection in Australia. In: Watson, Alan; Sproull, Janet; Dean, Liese, comps. *Science and Stewardship to Protect and Sustain Wilderness Values: Eighth World Wilderness Congress Symposium September 30–October 6, 2005*; Anchorage, AK. Proceedings RMRS-P-49. Fort Collins, CO: U.S. Department of Agriculture, Forest Service, Rocky Mountain Research Station.

Nelson, Michael P. 1998. An amalgamation of wilderness preservation arguments. In: Callicott, J. B.; Nelson, M. P. *The Great New Wilderness Debate*. Athens: University of Georgia Press, pp. 154–200.

Ostergren, David M. 1998. System in peril: A case study of five central Siberian zapovedniki. *International Journal of Wilderness*. 4(3): 12–17.

———. 2001. An organic act after a century of protection: The context, content and implications of the 1995 Russian Federation law on specially protected natural areas. *Natural Resources Journal*. 41(1): 125–152.

Ostergren, David M.; Hollenhorst, Steven J. 1999. Convergence in protected area policy: A comparison of the Russian zapovednik and American wilderness systems. *Society and Natural Resources*. 12(4): 293–313.

Ostergren, David M.; Shvarts, Evgeny. 2000. Russian zapovedniki in 1998: Recent progress and new challenges for Russia's strict nature preserves. In: Watson, A. E.; Aplet, G. H.; Hendee, J. C., comps. *Personal, Societal, and Ecological Values of Wilderness: 6th World Wilderness Congress proceedings on research, management, and allocation*; Vol. II; October 24–29, 1998; Bangalore, India. RMRS-P-14. Ogden, UT: United States Department of Agriculture, Forest Service, Rocky Mountain Research Station, pp. 209–213.

Parks Canada Agency. 2000. Unimpaired for future generations? Protecting ecological integrity with Canada's National Parks. Vol. I. "A call to action." Report of the Panel on the Ecological Integrity of Canada's National Parks. Ottawa, Ontario: National Parks Canada.

Pietikäinen, S. 1995. Finland's wilderness act—a Scandinavian model. In: Martin, Vance; Tyler, N., eds. *Arctic Wilderness: Proceedings of the 5th World Wilderness Congress*; September 1993; Tromso, Norway. Golden, CO: Fulcrum, pp. 181–186.

Player, Ian. 1987. *South African Passage: Diaries of the Wilderness Leadership School*. Golden, CO: Fulcrum.

———. 1998. *Zulu Wilderness: Shadow and Soul*. Golden, CO: Fulcrum.

Pollisco, Wilfrido S.; Meniado, Angelita P. 1997. Implementation of the National Integrated Protected Areas System Act. *Sylvatrop*. 7(1–2): 2–6.

Pond, Elizabeth. 1987. Soviets send mixed signals on their concern about environment. *The Christian Science Monitor*. April 8: 11 (col. 1).

Prest, James. 2008. Australia. In: Kormos, Cyril, ed. *A Handbook on International Wilderness Law and Policy*. Golden, CO: Fulcrum, pp. 57–89.

Pryde, Philip R. 1987. The distribution of endangered fauna in the USSR. *Biological Conservation*. 42(1): 19–37.

———. 1997. Post-Soviet development and status of Russian nature preserves. *Post-Soviet Geography and Economics*. 38(2): 63–80.

Rambaldi, Giacomo. 1997. RRA as a tool in integrating people's participation in protected areas management in the Philippines. *Sylvatrop*. 7(1–2): 28–39.

Russell, J. A.; Matthews, J. H.; Jones, R. 1979. *Wilderness in Tasmania*. Occasional Papers. 10. Hobart, Australia: University of Tasmania, Centre for Environmental Studies.

Rutledge, Ron; Vold, Terje. 1995. Canada's wilderness. *International Journal of Wilderness*. 1(2): 8–13.

Saarinen, Jarkko. 1998. Wilderness, tourism development, and sustainability: Wilderness attitudes and place ethics. In: Watson, A. E.; Aplet, G. H.; Hendee, J. C., comps. *Personal, Societal, and Ecological Values of Wilderness: Proceedings of the 6th World Wilderness Congress proceedings on research, management, and allocation*, Vol. I; October 24–29, 1998; Bangalore, India. Proc. RMRS-P-4. Ogden, UT: USDA, Forest Service, Rocky Mt. Research Station, pp. 29–36.

Sanderson, E. W.; Jaiteh, M; Levy, M. A.; Redford, K. H.; Wannebo, A. V.; Woolmer, G. 2002. The human footprint and the last of the wild. *Bioscience*. 52 (10): 891–904.

Schulz, Florian. 2005. *Yellowstone to Yukon: Freedom to Roam*. Seattle, WA: Mountaineers Books, p. 196.

Shestakov, Alexander; Kats, Dmitry. 2008. The Russian Federation. In: Kormos, Cyril, ed. *A Handbook on International Wilderness Law and Policy*. Golden, CO: Fulcrum, pp. 215–239.

Shroyer, Maretha; Watson, Alan; Muir, Andrew. 2003. Wilderness research in South Africa: Defining priorities at the intersection of qualities, threats, values and stakeholders. *International Journal of Wilderness*. 9(1): 41–45.

Sippola, Anna-Liisa. 2000. Biodiversity in Finnish wilderness areas: Aspects on preserving species and habitats. In: Watson, A. E.; Aplet, G. H.; Hendee, J. C., comps. *Personal, Societal, and Ecological Values of Wilderness: Proceedings of the 6th World Wilderness Congress proceedings on research, management, and allocation*, Vol. II; October 24–29, 1998; Bangalore, India. RMRS-P-14. Ogden, UT: U.S. Department of Agriculture, Forest Service, Rocky Mountain Research Station, pp. 48–56.

Sobolev, N. A.; Shvarts, E. A.; Kreindlin, M. L.; Mokievsky, V. O.; Zubakin, V. A. 1995. Russia's protected areas: A survey and identification of development problems. *Biodiversity and Conservation*. 4: 964–983.

Somers, Jonty. 2008. New Zealand. In: Kormos, Cyril, ed. *A Handbook on International Wilderness Law and Policy*. Golden, CO: Fulcrum, pp. 193–213.

Surlien, Rakel. 1995. The Great Arctic Reserve—Taymyr Peninsula. In: Martin, Vance; Tyler, Nicholas, eds. *Arctic Wilderness: Proceedings of the 5th World Wilderness Congress*; September 1993; Tromso, Norway. Golden, CO: Fulcrum, pp. 247–252.

Train, Russell E. 1974. An idea whose time has come: The World Heritage Trust, a world need and a world opportunity. In: Elliott, Hugh, ed. *Second World Conference on National Parks: Proceedings of a conference*; September 18–27, 1972; Yellowstone and Grand Teton National Parks, WY. Morges, Switzerland: International Union for the Conservation of Nature and Natural Resources, pp. 377–381.

Tynys, Tapio. 1995. Management and planning for wilderness areas in Finland. *International Journal of Wilderness*. 1(1): 37.

U.S. Public Law 88–577. The Wilderness Act of September 3, 1964. 78 Stat. 890.

Watson, Alan; Borrie, William. 2006. Monitoring the relationship between the public and public lands: Application to wilderness stewardship in the U.S. In: Aguirre-Bravo, Celedonio et al. eds. *Monitoring Science and Technology Symposium: Unifying Knowledge for Sustainability in the Western Hemisphere*. September 20–24, 2004; Denver, CO. RMRS-P-42CD. Fort Collins, CO: U.S. Department of Agriculture, Forest Service, Rocky Mountain Research Station.

Watson, Alan; Matt, Roian; Knotek, Katie; Williams, Dan; Yung, Laurie. 2007. Traditional wisdom—protecting wilderness as a cultural landscape. In: Youngbear-Tibbetts, Holly, ed. *Proceedings: Sharing Indigenous Wisdom International Conference*; June 2007; Green Bay, WI: U.S. Department of Agriculture, Forest Service, National Forest Products Laboratory.

Weiner, Douglas R. 1988a. *Models of Nature: Ecology, Conservation, and Cultural Revolution in Soviet Russia*. Bloomington: Indiana University Press.

———. 1988b. The changing face of Soviet conservation. In: Worster, Donald, ed. *The Ends of the Earth: Perspectives on Modern Environmental History*. Cambridge, England: Cambridge University Press, pp. 252–273.

World Bank. 1986. *The World Bank's Operational Policy on Wild Lands: Their Protection and Management in Economic Development*.

World Commission on Environment and Development (WCED). 1987. *Our Common Future*. New York: Oxford University Press.

zu Hulshoff, Bernd von Droste; Gregg, William P., Jr. 1985. Biosphere reserves: Demonstrating the value of conservation in sustaining society. *Parks*. 10(3): 2–5.

Zunino, Franco. 2007. A perspective on wilderness in Europe. *International Journal of Wilderness*. 13(3): 40–43.

II. U.S. Legal Authority and Process for Wilderness

Chapter 4

The Wilderness Act: Legal Basis for Wilderness Protection and Management

Introduction 90
The L-20 Regulation 90
The U Regulations 91
Statutory Protection for Wilderness 92
A Brief Legislative History of the Wilderness Act 94
    The First Wilderness Bill, Introduced in 1956 94
    Changes and Compromises to Pass a Wilderness Act 96
The Wilderness Act of 1964 96
    Section 1—Title 97
    Section 2—Establishes the Wilderness System 97
    Section 3—National Wilderness Preservation System—
        Extent of System 99
    Section 4—Use of Wilderness Areas 100
        Overflights 102
        Fire, Insects, and Disease 102
        Mining 103
        Water Resource Development 104
        Grazing 104
        Boundary Waters Canoe Area 105
        Outfitting and Guiding, Wildlife, and Other Exceptions 106
    Section 5—State and Private Lands Within Wilderness Areas 108
    Section 6—Gifts, Bequests, and Contributions 108
    Section 7—Annual Reports 108
Some Exceptions and Ambiguities in Wilderness Legislation 109
The Evolution of Wilderness Protection 109
Study Questions 110
Acknowledgments 110
Case Discussion: Compare the 1956 Bill to the 1964 Wilderness Act 111
Case Discussion Questions 111
References 111

Introduction

In the United States, wilderness has evolved from a general, ill-defined, and historical concept to a specific and legal land designation. This evolution reflects the history of citizen concern for wilderness and its protection, and various actions taken by land-management agencies in response to these concerns. To fully understand the basis for wilderness protection in the United States, one must understand the history of the protection measures taken over the past seventy years and the reasons for those actions. We believe the same understanding is necessary to understand wilderness protection in other countries.

This chapter examines the early actions taken by the resource-management agencies to preserve wilderness, the reasons for these actions, and their outcomes. We also examine the growth of concern for statutory protection of wilderness that began following World War II and trace the way in which wilderness became a major item on the nation's political agenda. Finally, we discuss the 1964 Wilderness Act, the basic legislation underlying wilderness protection and management in the United States. All ensuing wilderness designation laws have their roots in the Wilderness Act. Some designation acts have contained special provisions that recognize or accommodate local situations, but such provisions are typically accompanied by specific recognition that they do not constitute precedents for other areas. In other cases, Congress has attempted to clarify the intent underlying certain Wilderness Act clauses, either through legislative language or through House or Senate committee reports. In the final analysis, the Wilderness Act has remained a stable base from which decisions to protect and manage wilderness in the United States can be made. Since enacted in 1964, the Wilderness Act has only been revised with specific changes to the management of the Boundary Waters Canoe Area Wilderness.

The L-20 Regulation

As explained in Chapter 2, *the idea of wilderness protection may be a century old, depending on what is considered*. Although Yellowstone and Yosemite National Parks were not explicitly established to preserve wilderness, they nonetheless were major steps in protecting wilderness values. In the early 1900s, the individual efforts of Arthur Carhart and Aldo Leopold, both working within the United States Forest

Fig. 4.1. President Lyndon Johnson, surrounded by cabinet and congressional officials at the White House, signs the Wilderness Act, September 3, 1964. The Wilderness Act (U.S. Public Law 88-577) provides the basic framework governing the establishment and management of wilderness in the United States. Photo courtesy Lyndon Johnson Memorial Library, Austin, Texas.

Service, brought specific recognition of wilderness as a land use. Yet there was much concern about the extent of the nation's remaining wilderness resource; that is, just how much roadless land remained, how much of it, where, and should it be protected from development?

In 1926, the USFS appraised the extent of wilderness remaining in the national forests. A national inventory of all areas greater than 230,000 acres reported that 74 tracts totaling 55 million acres still remained, with the largest tract about 7 million acres. Three years later, this inventory became a basis for *the first systematic program of wilderness protection; that is, Administrative Regulation L-20 promulgated by the USFS in 1929* (Roth 1984).

From a wilderness viewpoint, the L-20 Regulation was not very protective of the resource. It was primarily a list of permitted and prohibited uses. Timber harvesting was permitted in the so-called primitive areas in the belief that, if properly regulated, it might not be incompatible with the primitive-area designation.

For example, logging occurred in about 80,000 acres of the South Fork Primitive Area, one of the three primitive areas that originally comprised what is now the Bob Marshall Wilderness in Montana.

Many uses generally considered incompatible with wilderness today were allowed. Although there were 72 primitive areas with a gross area of 13,402,121 acres in 10 western states, the management plans for these areas allowed road construction in 15 areas, grazing in 62 areas, and logging in 59 areas. In only 4 primitive areas, totaling 297,221 acres, were logging, grazing, and roads absolutely excluded (Gilligan 1954).

Nor was the L-20 Regulation strictly enforced. The broad latitude in L-20 management stemmed from a belief that primitive-area status did not represent a long-term commitment of resources, but was intended to prevent road building and commercial development in areas of scenic and recreational attraction until such time as detailed management plans might be prepared (Outdoor Recreation Resources Review Commission 1962). Some have also suggested that the L-20 Regulation was a USFS strategy to combat the transfer of national forestlands to the National Park Service (Allin 1982; Roth 1984) by giving the USFS a land-use designation to compete with national parks and prevent transfer of such lands to NPS administration. Former USFS chief Richard McArdle (1952–1962) discounts this argument, pointing out that the creation of national forest wilderness has never prevented transfers of national forestland to the NPS. For example, the 226,000-acre Olympic Primitive Area, established by the USFS in 1938, was subsequently transferred to Olympic National Park in Washington (McArdle 1975).

The NPS was definitely concerned about the increasing awareness of recreation among some USFS officials. National Park Service director Stephen Mather frequently challenged the authority of the USFS to develop recreation programs, arguing that recreation was the sole responsibility of his agency (Baldwin 1972, 62–71).

The interagency rivalry was also reflected in the terminology used to identify wilderness. In the 1920s, the USFS called undeveloped areas "wilderness," while the NPS called them "primitive areas." Following USFS promulgation of the L-20 Regulation establishing primitive areas on the national forests, the NPS switched to calling its undeveloped areas wilderness.

The U Regulations

Dissatisfaction with the looseness of the L-20 Regulation led to its replacement with Regulations U-1, U-2, and U-3(a) in 1939 (Gilligan 1954) *as a means of preserving roadless lands on the national forests.* These new regulations were formulated largely under the influence of Robert Marshall, then chief of the Division of Recreation and Lands in the USFS. Marshall was a dynamic proponent of wilderness, and much of his career centered on efforts to strengthen wilderness preservation programs both inside and outside the federal government. Earlier, as director of the Forestry Division of the U.S. Office of Indian Affairs, Marshall was responsible for the designation of sixteen wilderness areas on Native American reservations (Nash 1982). Along with Aldo Leopold, Marshall was instrumental in establishing The Wilderness Society in 1935. Joining the USFS in 1937, Marshall pushed forcefully for expansion of wilderness reserves on the national forests and the U regulations culminated his efforts. Although Marshall's leadership for wilderness cannot be minimized, the role of his assistant in the Division of Recreation and Lands, John H. Sieker, has not been fully recognized or appreciated. Sieker, a forester educated at Yale and Princeton, enjoyed a close professional relationship with Marshall, endorsing his wilderness ideas, a key factor since he was well respected in the USFS. Marshall, while dynamic and energetic, was viewed by many as eccentric (Mitchell 1985). Politically, he was a liberal, advocating civil rights, civil liberties, and the redistribution of wealth, views that gained the enmity of conservative colleagues (Glover and Glover 1986). Moreover, he was a newcomer, having only recently joined the USFS. As an insider, Sieker became an important force in efforts to implement and gain acceptance for the U regulations (McArdle 1975).

The U regulations broadened the purpose of wilderness beyond what was earlier defined by the L-20 Regulation, and three land-use designations were recognized. Regulation U-1 established wilderness areas—tracts of land of not less than 100,000 acres. Acting on the recommendation of the chief of the USFS, the secretary of agriculture could designate such an area as wilderness, and only the secretary could modify or eliminate a wilderness area.

Regulation U-2 defined wild areas—tracts of land between 5,000 and 100,000 acres that could be established, modified, or eliminated by the chief of the

USFS. Thus, in addition to their size, U-1 and U-2 areas were different in terms of who could establish, modify, or eliminate them, but they were managed identically.

Finally, Regulation U-3(a) established roadless areas to be managed principally for recreational use "substantially in their natural condition." Roadless areas larger than 100,000 acres could be established or modified only by the secretary of agriculture; areas less than 100,000 acres could be established or modified by the chief of the USFS. Timber cutting, roads, and other modifications were permissible if provided for in area-management plans. The only areas ever classified under this regulation were three separate tracts in the Superior National Forest in Minnesota that were consolidated in 1958 to form what is now the Boundary Waters Canoe Area Wilderness.

The U regulations provided much more protection for wilderness than the L-20 Regulation they replaced. The U regulations were intended to be permanent, not just an interim measure to halt haphazard development (Allin 1982). They prohibited timber cutting, road construction, and special-use permits for such things as summer homes and hunting camps. Mechanized access, except where well established or in emergencies, was banned. Grazing and water-resource development was allowed. Mining was also allowed to continue, subject to existing mining and leasing laws. The USFS could insist that all such developments take pains to minimize impact on the wilderness (Roth 1984).

Because many of the seventy-six L-20 primitive areas had been established without adequate surveys, each was to be reviewed and reclassified under the new U regulation guidelines. A ninety-day public review period was called for. While under review, all primitive areas were to be managed according to U regulation requirements.

However, progress in reviewing the primitive areas was very slow; between 1939 and the outbreak of World War II, three areas were reclassified as wilderness, six as wild, and three were consolidated into the Bob Marshall Wilderness in Montana. No reclassifications occurred during the war. After the war, reviews resumed, again slowly. By the late 1940s, only 2 million acres had become wilderness (Roth 1984). The slow progress and the loss of acreage in the reclassification process troubled conservationists. In some places, they argued, timbered lower-elevation areas within the old L-20 primitive-area boundaries were lost when the area was reclassified. Although the USFS argued that total primitive-area acreage had remained stable, critics countered that this was because high-elevation areas, devoid of any timber, had been substituted, thus lowering the quality and variety of the wilderness (Roth 1984). For example, Oregon senator Richard L. Neuberger strongly objected to the removal of more than 53,000 acres of timbered land in Oregon's Three Sisters Primitive Area during its reclassification (Hession 1967). The concerns by Neuberger and others helped lead to pressure for a new and more permanent method of wilderness protection, a bill he was to cosponsor.

Statutory Protection for Wilderness

The L-20 Regulation and the U regulations were administrative designations; they were implemented at the discretion of the secretary of agriculture or the chief of the USFS. Because of the agency discretion under an administrative system of wilderness protection, wilderness proponents were concerned that roadless lands were not adequately protected. The answers seemed to lie in a legal, rather than an administrative, approach to wilderness protection in which the requirements and procedures for safeguarding wilderness values were prescribed by statute rather than left to the discretion of agency administrators.

Legal protection for wilderness was not a new idea. In the 1930s, Marshall supported the idea of establishing wilderness areas by congressional action (Nash 1982), as did H. H. Chapman, professor of forestry at Yale University (Chapman 1938), and Secretary of the Interior Harold L. Ickes (Mackintosh 1985).

Subsequently, the need for a legislatively protected wilderness system received additional support from a report issued by the Legislative Reference Service of the Library of Congress in 1949 (Keyser 1949). The report highlighted the widely disjointed programs of wilderness preservation; included opinions from a survey of numerous federal, state, and nonpublic organizations; and reported substantial concern for the future of wilderness and widespread support for wilderness protection as secure as that of the national parks.

In 1953, a dissertation by James P. Gilligan at the University of Michigan provided an appraisal of the condition and quality of the existing system of wilderness and primitive areas. He noted that although 13 million acres had been administratively set aside as wilderness, wild, or primitive areas, the conditions within them often left much to be desired. For example, he reported the existence of 200 miles of public roads, 145,000 acres of private inholdings, up to 500 mining claims, 60 mines, 24 airstrips, pasturage for 140,000 sheep and 25,000 cattle, and nearly 90 dams. Allin (1982) noted that Gilligan's information confirmed the concerns of many conservationists, that wilderness boundaries frequently had been modified, shifting land suitable to economic development outside the preserved area.

In 1962, the Wild Land Research Center at the University of California, with Gilligan as project director, in its report to the Outdoor Recreation Resources Review Commission (1962), recommended the enactment of legislation creating a wilderness system protected by law. Otherwise, the center predicted that the wilderness resource would be gradually lost, based on the following observations (Outdoor Recreation Resources Review Commission 1962):

1. Land-administering agencies could put wilderness to other uses.
2. Agencies lacked full jurisdiction over some land uses in wilderness, such as mining.
3. There was a lack of coordinated control over wilderness uses.
4. There was a lack of distinctive management policy.

One of the strongest advocates of a national wilderness system was Howard Zahniser, executive director of The Wilderness Society (E. Zahniser 1984; H. Zahniser 1964; Scott 1976). His major theme was the need for a persisting program of wilderness preservation—a cohesive program that could eliminate the need for continual, fragmented holding actions against various threats. As early as 1949, he had outlined a wilderness system similar in structure to that eventually proposed in the first wilderness bill in 1956 (Hession 1967). At the Sierra Club's 1951 Biennial Wilderness Conference, Howard Zahniser (1964) remarked the following:

> Let's try to be done with a wilderness preservation program made up of a sequence of overlapping emergencies, threats, and defense campaigns. Let's make a concerted effort for a positive program that will establish an enduring system of areas where we can be at peace and not forever feel that the wilderness is a battleground.

Zahniser and the others who supported legal protection for wilderness *had three main concerns for the law that was to be written. First, they wanted a clear, unambiguous document.* One of the oft-criticized features of the previous administrative approaches to wilderness protection was that they contained many loopholes through which protection could be contraverted. *Second, they wanted to maintain the political coalition that had formed during the earlier efforts to protect Echo Park in Dinosaur National Monument* in Colorado and Utah from flooding. This coalition of groups had effectively resisted efforts in 1955 by the Bureau of Reclamation and Army Corps of Engineers to construct a dam within the monument as part of an ambitious Upper Colorado River Storage Project. Zahniser recognized the importance of maintaining this tested coalition of allies and was careful to circulate early drafts of the wilderness bill to the various organizations to maintain a uniform front of support. And *third, they wanted to minimize opposition to the legislation.* Zahniser and his colleagues sought to assure federal land-managing agencies that existing jurisdictions would be respected and that the proposed bill would not supersede the purposes for which the land was being administered, with the exception of preserving its wilderness character. To commodity interests, Zahniser indicated that the bill would respect existing uses, and the termination of any uses considered nonconforming (e.g., stock use) would be done in a manner equitable to them (Roth 1984).

Difficulties were encountered immediately as both the USFS and the NPS opposed the legislation. The USFS argued that it was not urgently needed. Moreover, efforts to pass it might jeopardize agency efforts to expand and consolidate its existing program of wilderness protection by generating strong opposition to wilderness from within as well as outside government (McArdle 1975). There was also concern that other special interests such as grazing might lobby to secure similar legislative guarantees for their uses of the national forests. Director Conrad L. Wirth of the NPS also opposed it because it was not necessary, and he claimed that it might endanger national park wilderness areas by lumping them with those of other agencies (Allin 1982; Roth 1984).

Nevertheless, Zahniser persisted in his pursuit for improved wilderness protection. His goal was a congressionally established national wilderness system that

Fig. 4.2. Howard Zahniser, executive director of The Wilderness Society, made monumental contributions to establishing a wilderness system and passing the Wilderness Act. Zahniser died in June 1964, before the act was passed. Photo courtesy The Wilderness Society, Washington, DC.

would encompass areas of adequate size and numbers to meet future needs and to provide legal protection to ensure the perpetuation of their primeval character. To meet this objective he and leaders from the Sierra Club, National Parks Association, National Wildlife Federation, and Wildlife Management Institute prepared a draft bill in 1955 at the urging of Senator Hubert Humphrey (MN). *In 1956, Senator Humphrey and nine other senators introduced the first wilderness bill. Representative John Saylor (PA) introduced similar legislation in the House.* The long congressional struggle for legislative establishment of a national wilderness preservation system was under way.

A Brief Legislative History of the Wilderness Act

It took eight years for the final Wilderness Act to be passed by Congress. During that time, sixty-five different wilderness bills were introduced. Eighteen hearings were held across the nation, many thousands of pages of testimony were printed (McCloskey 1966), *and the final bill was substantially changed from the initial version drafted by Zahniser* to the act signed into law by President Lyndon B. Johnson (U.S. Public Law 88-577) September 3, 1964. We discuss some of the issues that delayed passage as well as the major changes between the first draft and the final act; for a more detailed analysis read Allin 1982 and Scott 2004.

The First Wilderness Bill, Introduced in 1956

The first wilderness bill, introduced in 1956, proposed a wilderness system including lands in the National Forest System, National Park System, National Wildlife Refuges and Game Range System, and lands overseen by the Bureau of Indian Affairs. Altogether, about 65 million acres would have been subject to study; as many as 35 to 45 million acres might actually have been designated wilderness (Hession 1967).

All thirty-seven national forest areas classified as wilderness, wild, or roadless under the U regulations were to be automatically included in the system under the first bill. In addition, the forty-four remaining primitive areas were to be temporarily included within the system, and the secretary of agriculture was given nine years to review the status of each and recommend an appropriate classification. Congress would then decide whether to extend permanent protection to each primitive area or exclude it from the system.

The secretary of the interior was directed to review the National Park System and the National Wildlife Refuges and Game Range System, also within nine years, and to recommend areas that should be designated as wilderness. Unlike the national forestlands, no Department of the Interior holdings were automatically included in the proposed wilderness system. Qualified areas under Bureau of Indian Affairs jurisdiction could be included, but only with the consent of the tribal councils. Moreover, protection would exist only so long as the appropriate Native American representatives concurred (Allin 1982). No time limit was placed on designating Native American lands.

The first proposed wilderness bill would have provided comprehensive protection from development. It would have prohibited logging, prospecting, dams, commercial enterprises, roads, motor vehicles, the landing of aircraft, the extension of motor boating to new areas, new mining, and new grazing. Some nonconforming but existing uses, such as grazing, and motorboat and aircraft use would have been respected (Allin 1982).

The first bill would also have established a National Wilderness Preservation Council composed of the heads of the USFS, NPS, U.S. Fish and Wildlife Service (FWS), Bureau of Indian Affairs, Smithsonian Institution, and six citizen representatives. Its functions would have been to receive and review all wilderness reports and recommendations from the secretaries of agriculture and the interior, to transmit these reports to Congress, and to advise Congress and the president during ensuing deliberations on the agency recommendations. The wilderness council was viewed as one means of checking the broad executive discretion of the administrative agencies. The council, if it thought a secretarial report was unsatisfactory, would have been in a good position to influence congressional response.

The function of Congress in this earliest version of the wilderness bill was as a safeguard against any unwise and arbitrary action on the part of any secretary undertaking a measure that "disregarded conservation." This safeguard took the form of a legislative veto that could be used by either the House or Senate. When any secretarial recommendation came before it, Congress had 120 days within which to register its objection. Otherwise, the secretary's recommendation to support or oppose wilderness classification would become effective pursuant to the law. In other words, statutory protection of wilderness would come about in the absence of any *affirmative action* on the part of Congress.

It is important to note that, in this first bill, congressional authority to formulate and enact legislation was delegated to the secretaries of agriculture and the interior; that is, departmental recommendations became law if not vetoed. The idea of a legislative veto was subsequently dropped from the final version, but it had raised the more basic question: who should take the affirmative action in wilderness designations? In 1983, the U.S. Supreme Court ruled legislative vetoes unconstitutional.

This question—one of the crucial issues during the evolution of the wilderness bill—assumed particular importance during consideration of the bill's provisions for granting *temporary wilderness status* to Interior Department holdings identified as suitable for wilderness designation and USFS primitive areas, pending their review for permanent wilderness designation. Argument centered on two questions: (1) should these lands be included in the wilderness system initially (instead of remaining undesignated during their reviews), and (2) should affirmative congressional action—that is, a bill sponsored and introduced, debated, passed, and signed into law like any other legislation—be required to add any area to the wilderness system?

Representative Wayne Aspinall (CO), chair of the House Committee on Interior and Insular Affairs, argued strongly that each new area should be the subject of a separate congressional evaluation and an individual bill. Aspinall and other conservative legislators argued that this particular use of *affirmative congressional action would help* halt the erosion of congressional authority to the executive branch (Mercure and Ross 1970). Wilderness proponents, on the other hand, were concerned that requiring congressional approval of each individual area could prove to be a cumbersome barrier to rapid and equitable wilderness designation.

To force his position requiring affirmative congressional action on all wilderness proposals, Aspinall refused to allow hearings on the wilderness bill until legislation calling for a general review of all federal land-management policies was agreed on. Aspinall believed that the debate over affirmative action centered on the rules governing the withdrawal of wilderness-quality lands from multiple-use status. With more than 5,000 land-use statutes in place, he argued, an overhaul of the nation's land laws was necessary before making any extensive withdrawals for wilderness purposes (Baker 1985). As chair of the Interior and Insular Affairs Committee, Aspinall was able to make good on this threat. Consequently, wilderness proponents agreed to support his proposed legislation to create a Public Land Law Review Commission on the condition that Aspinall report out a wilderness bill that could be debated and amended on the floor of the House. In addition, Senator Clinton Anderson (NM), a major figure in the drive to pass wilderness legislation, agreed to the affirmative congressional action provision, if Aspinall would release the wilderness bill from committee (Baker 1985).

As the wilderness bill neared the end of its long journey through Congress, final refinements were worked out. The San Gorgonio Wild Area in southern California, originally eliminated from the proposed wilderness system to permit construction of a ski area, was restored to the proposed system. USFS authority to declassify existing primitive areas by administrative action was eliminated, thereby giving the legislative branch full authority to control wilderness declassification. While the Senate acceded to the House provision requiring affirmative congressional action, the House agreed to reduce the time period in which new mineral exploration would be allowed in wilderness from twenty-five years to nineteen years. *The House passed the final version of the wilderness bill by a vote of 373 to one; the Senate, by a margin of 73 to 12.* President Lyndon Johnson, who said passage of the bill was "in the highest tradition of our heritage as conservators as well as users of America's bountiful natural endowments," signed the act on September 3, 1964.

Changes and Compromises to Pass a Wilderness Act

Contrasting the Wilderness Act with the original 1956 bill reveals a number of changes—compromises made to secure congressional support. Major changes included the following:

1. *Bureau of Indian Affairs lands were excluded from the bill.* In 1937, nearly 5 million acres of Native American land had been administratively reserved for wilderness purposes, largely through the efforts of Robert Marshall. However, because the tribal councils, rather than the federal government, held title to these lands, wilderness designation could occur only so long as the appropriate Native American representatives concurred.
2. *The National Wilderness Preservation Council was eliminated.* The USFS opposed the council from the beginning, arguing that it created an unnecessary step in the review process (McArdle 1975). Agency opposition, based partly on the six-to-five layperson–to–agency head representation on the council, remained even after the suggested number of laypersons was reduced to three in 1958. Opposition to the advisory council was also founded on the belief that it might very well end up making most of the final wilderness designation decisions, given its influential advisory role to Congress and the president. Zahniser, who had initially insisted on the council, agreed to its removal and counted on the president to check his cabinet officers' recommendations (Hession 1967).
3. *The USFS primitive areas were not included within the initial wilderness system created in 1964, and their designation would require affirmative action by Congress.* Under the original bill, they would have been temporarily included in the system and a secretarial recommendation regarding their designation would have gone into effect in four months in the absence of congressional action to stop it. This affirmative action provision appears to have been a direct result of discussions between President John F. Kennedy and Congressman Aspinall in 1963, and between New Mexico senator Anderson and Congressman Aspinall in 1964. By agreeing to congressional affirmative action for any wilderness designation, the president hoped to gain support from congressional leaders such as Aspinall to move more of his legislative programs through the Congress before the 1964 election (Hession 1967; Roth 1984). Senator Anderson was reluctant to concede on the issue of affirmative congressional action but had concluded that, without it, there would be no wilderness bill at all (Baker 1985).
4. *Prohibitions on uses of wilderness were less restrictive.* As initially conceived, for instance, all new mining in wilderness would have been prohibited on passage of the bill. Yet it became obvious that such a blanket restriction would mean no wilderness bill would pass (Roth 1984). So, as passed, the Act permitted prospecting to continue until December 31, 1983, and mining on claims established before this date would be allowed to continue indefinitely.

Many supporters of wilderness legislation were disappointed with the discrepancy between the act as passed and the original bill proposed by Zahniser and The Wilderness Society (Nash 1982). *The Wilderness Act clearly contained compromises, yet without them, it is unlikely the bill would have ever passed.* The bill enjoyed the support of several senators and congressional representatives, without whose help it would probably have failed. Senators Clinton Anderson (NM), Frank Church (ID), and Hubert Humphrey (MN) and Representatives John Saylor (PA) and Lee Metcalf (MT) were key supporters (Baker 1985; McArdle 1975). The role of Zahniser as a committed citizen advocate and executive director of The Wilderness Society was also crucial (Scott 2004). Together, their efforts were rewarded with the passage of a piece of legislation unique in American as well as international conservation history—U.S. Public Law 88-577, the Wilderness Act.

The Wilderness Act of 1964

The text of the Wilderness Act of 1964 is Appendix A. This act should be required reading for anyone involved in wilderness designation or management. Many of the arguments and much of the confusion surrounding wilderness stewardship stem from the lack of a careful reading and clear understanding of this important document. As the major piece of legislation guiding both wilderness designation and management, the Wilderness Act is the basic reference document for many questions regarding what can or cannot be done in wilderness.

Nevertheless, parts of the Wilderness Act, as is the case with some other landmark legislation, are subject to differing interpretations, depending on one's particular wilderness philosophy and values (e.g., contrasting perspectives on how pure wilderness should be permitted are found in Foote 1973, Worf 1980, and discussed in Allin 1985).

We review the seven sections of the Wilderness Act to highlight important provisions and note ambiguities and contrasting interpretations. In some cases, subsequent wilderness designation laws, or committee reports on those laws, have clarified or elaborated specific wording of the Wilderness Act, such as livestock grazing and insect and disease control. Although the phrases are mentioned in this chapter, along with some examples of additional direction in subsequent legislation, see Chapter 5 for more complete coverage.

The Wilderness Act provides only broad guidelines and directions; detailed guidelines are contained in departmental regulations for implementing the law that are issued by the secretary of agriculture for wilderness on the national forests and by the secretary of the interior for wilderness administered by the National Park Service, U.S. Fish and Wildlife Service, and Bureau of Land Management.

The remainder of the chapter is devoted to explaining the intent and meaning of the law. Additional explanation of the Wilderness Act is found in an excellent handbook published by The Wilderness Society (2004).

The Wilderness Act is organized into seven sections.

Section 1—Title
The act shall be known as the "Wilderness Act."

Section 2—Establishes the Wilderness System
Section 2 provides a broad statement of policy, defines the term *wilderness*, and sets forth some of the conditions and implications of wilderness designation. Section 2(a) clearly states that the establishment and protection of wilderness is a policy of the U.S. Congress, reflecting a belief that because of population pressures, all areas of the nation will be occupied or modified—except those set aside in their natural condition.

Management is specifically referred to in this section, where the act states:

Wilderness areas shall be administered for the use and enjoyment of the American people in such manner as will leave them unimpaired for future use and enjoyment as wilderness, and so as to provide for the protection of these areas, the preservation of their wilderness character, and for the gathering and dissemination of information regarding their use and enjoyment as wilderness.

Note that the phrase "as wilderness" appears several times. Although it is clear that Congress fully intended wilderness to be for people's use and enjoyment, it is also apparent that such *"use and enjoyment" is to be contingent upon the maintenance of these areas "as wilderness,"* a condition the act later defines.

Section 2(a) also specifies that only federal lands will be included in the National Wilderness Preservation System (NWPS) and that no federal lands except those protected by the act or by a subsequent act shall be designated "wilderness." This provision was included to prevent the executive branch of government from designating wilderness, thus reserving that responsibility for Congress. Section 2(a) merely prohibits *official designation* as wilderness by agencies other than Congress. It does *not* prohibit administrative agencies from *managing* lands for wilderness purposes. The USFS, for example, had such authority based on the Multiple-Use, Sustained-Yield Act of 1960. The Wilderness Act specifically indicates that none of its provisions shall interfere with the purposes of the Multiple-Use, Sustained-Yield Act.

The Multiple-Use, Sustained-Yield Act (U.S. Public Law 86-517) was enacted largely at the urging of the USFS as a reaffirmation of the agency's traditional multiple-use philosophy regarding resource management. The act was intended to counter pressures for dominant-use legislation, but it did provide general statutory sanction of wilderness preservation in the national forests, incorporated in the following statement: "The establishment and maintenance of areas of wilderness are consistent with the purposes and provisions of this Act." Robinson (1975) argues that inclusion of this provision helped lead to the withdrawal of USFS opposition to the proposed wilderness legislation (also see discussion in Allin 1982).

The Wilderness Act's provisions originally affected three federal agencies: the USFS in the Department of Agriculture, and the NPS and FWS, both in the Department of the Interior. Before 1964, the wilderness idea had been formally incorporated into USFS planning through the L-20 Regulation and U regulations. The NPS had used a zoning system to protect wilderness values in undeveloped areas more than one-half mile from roads (Outdoor Recreation Resources Review Commission 1962). Before the Wilderness Act, the FWS had not managed any areas specifically for wilderness purposes because habitat enhancement for wildlife often results in substantial modification of areas, thereby conflicting with wilderness values. Nevertheless, the FWS is charged with wilderness responsibilities under the 1964 Wilderness Act, and the first Department of the Interior area to be admitted to the NWPS was the Great Swamp Wildlife Refuge in New Jersey.

In 1976, Congress passed the Federal Land Policy and Management Act (U.S. Public Law 94-579), giving the Bureau of Land Management (BLM) in the Department of the Interior clear authority and direction for management of the public lands under its jurisdiction. FLPMA, as the act is known, instructs the secretary of the interior to review those roadless lands of 5,000 acres or more, as well as all roadless islands administered by BLM, and to make recommendations regarding the suitability of these areas for wilderness designation. This inventory used the wilderness characteristics specified in the Wilderness Act. The results are described in Chapters 5 and 6.

Before 1976, the BLM managed a system of areas for wilderness preservation purposes referred to as *primitive areas* (not to be confused with national forest areas of the same name designated under the L-20 Regulation) and *natural areas*. However, under the terms of the FLPMA, any area classified by BLM as a *primitive* or *natural* area before November 1, 1975, was to be reviewed by the secretary of the interior for its suitability for wilderness designation, and a recommendation made to the president by July 1, 1980.

In addition to specifying which federal lands shall constitute official wilderness, section 2(a) assigns management responsibilities. It specifies that each federal agency charged with jurisdiction of wilderness will continue to manage those areas originally under its jurisdiction after they have been made part of the NWPS. This clause was included so that no new agency would be created. In addition, budget appropriations to the wilderness system as a separate entity or appropriations based solely on the system's existence are prohibited.

The final subsection of section 2 defines wilderness, and it is this section that has probably led to more confusion and debate than any other. It first defines wilderness in an ideal, almost poetic, sense: "A wilderness, in contrast with those areas where man and his own works dominate the landscape, is hereby recognized as an area where the earth and its community of life are untrammeled by man, where man himself is a visitor who does not remain."

Zahniser specifically chose the word *untrammeled*, even though he was warned that it might confuse the definition. Not to be confused with *untrampled*, *untrammeled* means "not subject to human controls and manipulations that hamper the free play of natural forces."

Section 2(c) then goes on to define wilderness in a legal and workable sense as

> an area of undeveloped Federal land retaining its primeval character and influence, without permanent improvements or human habitation…and which (1) generally appears to have been affected primarily by the forces of nature, with the imprint of man's work substantially unnoticeable; (2) has outstanding opportunities for solitude or a primitive and unconfined type of recreation; (3) has at least five thousand acres of land or is of sufficient size as to make practicable its preservation and use in an unimpaired condition; and (4) may also contain ecological, geological, or other features of scientific, educational, scenic, or historical value.

The definition of wilderness in section 2(c) gives important clues to the congressional view of wilderness. Recognizing that the ideal did not exist, they added a working definition based on reality. Wilderness was clearly intended to be an area where humans' impact was minimal and which was predominantly natural and unmodified. At the same time, the act accommodates reality by stating these areas "*generally* appear" to be "*primarily* affected" by nature with man's imprint "*substantially* unnoticeable" (emphasis added).

In the effort to accommodate reality by expressing the necessary conditions for wilderness in these more general terms, some difficult questions were created. For example, if a large tract of land were being considered for wilderness, with one portion in a pristine condition and another portion modified by previous developmental activity, would the criterion "substantially unnoticeable" be based on the aggregate area or on the modified portion? The answer obviously hinges on whether one looks at the whole area in its entirety, or at its individual parts.

Some have argued for keeping the system pure—that including or adding modified areas, even to large contiguous tracts in pristine condition, would compromise the quality of the entire system (Costley 1972). Yet, others have argued that the entire unit of land must be considered and that the inclusion of small areas of modified land is "substantially unnoticeable" when viewed within the larger context. (The principle of nondegradation, discussed in Chapter 7, is useful in resolving this issue.)

The intent of Congress was to establish a system of areas—an ideal—that embodied values espoused by early wilderness proponents, areas with naturalness and solitude. Yet, unreasonably rigid admission standards clearly were not the intent of Congress, either. The act first describes wilderness as an *ideal concept,* but then goes on to define wilderness *as it is to be considered* for the purposes of the act.

The definition of wilderness in section 2(c) cites the importance of "outstanding opportunities for solitude or a primitive and unconfined type of recreation." In recent years, in response to wilderness crowding, some have proposed that Congress meant that opportunities for either solitude *or* a primitive kind of recreation are required; that is, either one could qualify an area for wilderness designation and, thus, only one such quality need be maintained by wilderness management. As discussed in Chapter 1, it is our belief that both are required, broadly speaking—*naturalness and opportunities for solitude and primitive recreation are the distinguishing qualities of wilderness.*

The 5,000-acre minimum is often cited as absolute, but a careful reading of the act clearly shows that it is a suggested guideline. The intent is that the area designated be large enough to permit preservation objectives. In the conference committee that formulated the final wording of the Wilderness Act, the House recommended 5,000 acres as a minimum limit on the size of individual areas, but conferees decided that a statement of intent that a tract of "sufficient size as to make practicable its preservation and use in an unimpaired condition" would be satisfactory (Baker 1985).

Finally, the definition says wilderness *may* include ecological, geological, scenic, and other features. The important point to note is that these values are neither required for an area to be a wilderness nor by themselves sufficient criteria.

In conclusion, *our interpretation of the act's multifaceted definition of wilderness is that the criteria of naturalness and opportunities for solitude and primitive recreation are the distinguishing qualities of designated wilderness and are the principal criteria to guide wilderness management. Naturalness and solitude are used throughout this book in that context.*

Section 3—National Wilderness Preservation System—Extent of System

Section 3, a five-part section of the Wilderness Act, describes the NWPS and the procedures for admitting areas to the system.

Section 3(a) defines the areas that formed the initial core of the NWPS. These included all USFS areas previously classified administratively as wilderness, wild, or canoe. Fifty-four areas were so designated, covering 9.1 million acres. It also instructed the secretary of agriculture to file accurate boundary descriptions of all these areas and make them available to the public. No Department of the Interior land was included in the initial wilderness system.

Section 3(b) instructs the secretary of agriculture, within ten years of the passage of the act, to review all USFS primitive areas for their suitability as wilderness and to make a report on the findings to the president. In turn, the president is to send Congress recommendations to support, oppose, or modify the secretary's proposal. It also established a timetable for the review of the thirty-four primitive areas (a total of 5.4 million acres at the time the act passed), with one-third to be reviewed within the first three years after the act's passage, two-thirds after seven years, and the remaining areas within ten years. Until Congress would take further action, primitive areas were to be administered as they were at the time of the act's passage.

This subsection also describes the president's latitude for enlarging an existing primitive area. At the time of the recommendation to Congress, the president may make an addition to any existing primitive area of not more than 5,000 acres, as long as no single unit of added land exceeds 1,280 acres. Additions beyond the 5,000-acre limit require congressional approval (McCloskey 1966).

Section 3(c) is similar to 3(b) in that it instructs the secretary of the interior to review all roadless areas

Fig. 4.3. The Wilderness Act allows agencies to use certain facilities and do activities essential to managing a specific wilderness. A remote automated weather station (left) is essential to fire management in the Mount Logan Wilderness, on the Bureau of Land Management's Arizona Strip District. USFS staff measure snow (right) to predict water content in the Selway-Bitterroot Wilderness, Montana. Photo on left courtesy Tom Folk, BLM; photo on right courtesy U.S. Forest Service.

of at least 5,000 acres in the NPS and similar size holdings and *every* roadless island within the National Wildlife Refuges and Game Range System, and to submit a report regarding the suitability of these areas for wilderness designation. The islands were specifically cited because, in spite of the small size of many of them, their isolation makes preservation a practicable alternative. A ten-year review period, along with the timetable described in section 3(b), was established; that is, one-third in three years, two-thirds after seven years, and the remainder at the end of ten years.

Section 3(d) provides guidelines for notifying public and local officials of recommendations that the secretaries of agriculture and the interior intend to submit to the president and provides for public hearings on these recommendations.

Finally, section 3(e) describes the procedures for modifying or adjusting any wilderness boundary. No modification of a wilderness boundary by an agency is permitted, even in cases where nonconforming or illegal uses are found. When exact boundary locations are not clear, the enabling legislation or conference or committee reports are used to establish legislative intent.

Section 4—Use of Wilderness Areas

Section 4 is subtitled "Limitation of Use and Activities." It's first subsection a, indicates that the purposes of the Wilderness Act are "within and supplemental" to the purposes for which national forests and units of the national park and wildlife refuge systems are established. It specifically states that the Wilderness Act in no way interferes with a number of specific acts, such as the Multiple-Use, Sustained-Yield Act and the National Park Organic Act. By providing for wilderness preservation under the multiple-use umbrella, section 4(a) removes a major reason for conflict between these two potentially incompatible laws.

The responsibility of each managing agency to maintain the wilderness character of lands under its jurisdiction is reaffirmed in section 4(b). In addition, it states that, "except as otherwise provided in this Act, wilderness areas shall be devoted to the public purposes of recreational, scenic, scientific, educational, conservation, and historical use."

Section 4(c) describes facilities and activities that are not allowed in wildernesses designated by the act. This section must be carefully read, for it opens with the statement, "*Except as specifically provided for in this Act, and subject to existing private rights*" (emphasis added), before going on to catalog prohibited uses. These exceptions are quite substantial.

Subject to the subsection's opening qualification, commercial enterprises and permanent roads are prohibited. Certain other uses (motorized vehicles and equipment, temporary roads, aircraft landings) are prohibited *"except as necessary to meet minimum requirements for the administration of the area for the purpose of this Act* (including measures required in emergencies involving the health and safety of persons within the area)" (emphasis added).

Subsection 4(c) outlines illegal activities and administrative latitude. The section reads, in part, "*Except as specifically provided for in this Act…and, except as necessary to meet minimum requirements for the administration of the area for the purpose of this Act,*" and is followed by a list of prohibited activities such as temporary roads, motor vehicles. The phrase "minimum requirements" seems to mean "essential or necessary." The phrase "except as necessary to meet minimum requirements for the administration of the area for the purpose of this Act" permits administrators to carry out actions otherwise considered inappropriate in wilderness, *if it becomes apparent these actions are the minimum necessary to manage the area as wilderness.* This wording is the basis for the so-called minimum tool rule for wilderness management widely referenced in this book (see Chapter 7).

The basic interpretation of this section is that *administrators must provide evidence that any of their actions are the minimum necessary for managing the area as wilderness.* Simple convenience or cost advantage is not sufficient reason to justify, for example, the use of mechanized equipment. Worf (1980) contends that nonmechanized wilderness maintenance is often cheaper than mechanized maintenance. If, however, the use of mechanized equipment clearly reduces the amount of impact on the wilderness and its use (e.g., by reducing the level and extent of resource impact, and thus being the minimum tool), then administrators would have the flexibility to use such equipment.

McCloskey (1966) also points out a second ambiguity. Does the phrase "for the purpose of this Act" refer solely to the act's objective in section 2(a)—"to secure for the American people of present and future generations the benefits of an enduring resource of wilderness"—an objective limited to preservation, or does it include a broader purpose—preservation *and* compatible human enjoyment? Differing interpretations of the phrase could lead to different administrative actions. The wording of the act indicates that both use of wilderness and its preservation were intended, but it also suggests in section 4(c)(1) that any construction to facilitate use, such as bridges or shelters, must satisfy the criterion of *necessity.* In general, facilities for the convenience and/or comfort of users do not meet this criterion and have not been provided.

Important exceptions to the prohibited uses are described in section 4(c). These exceptions, which are subject to existing private rights, are outlined in section 4(d)(1) and 4(d)(2) and constitute what we call *allowable, but nonconforming uses;* they are legal, but they are generally considered incompatible with the wilderness protection goals of the act. These nonconforming uses reflect some of the compromises that were necessary to pass a Wilderness Act. Both subsections apply only to those areas of national forestland designated as wilderness by the 1964 act. However, in the interests of consistency, subsequent wilderness designation acts have called for the management of newly

Fig. 4.4. A wilderness outfitter leads pack stock and a party on horseback in the Marble Mountain Wilderness, California. The Wilderness Act generally allows established commercial outfitting activities to continue. Photo courtesy U.S. Forest Service.

designated areas to follow the guidelines contained in the 1964 act, subject to any specific exceptions written into that subsequent legislation.

The following uses (nonconforming, but allowed uses) are expressly permitted in section 4(d):

1. Established uses of aircraft and motorboats.
2. Actions taken to control fire, insects, and disease outbreaks.
3. "Any activity, including prospecting, for the purpose of gathering information about mineral or other resources, if carried on in a manner compatible with preservation of the wilderness environment."
4. Continued application of the U.S. mining and mineral leasing laws to wilderness on national forestlands until December 31, 1983.
5. Water resource development (authorized by the president if determined that such use will better serve the national interest than would its denial).
6. Livestock grazing, where established prior to the act.
7. Management of the Boundary Waters Canoe Area under regulations previously laid down by the secretary of agriculture, which were generally less restrictive than those imposed by the Wilderness Act (resolved by subsequent legislation, as we shall discuss shortly).
8. Commercial enterprises necessary for activities that are appropriate in wilderness (e.g., outfitting and guiding).

In summary, section 4(d) of the act recognized that certain "nonconforming" existing uses—specifically aircraft and motorboats—could be permitted to continue in places where they were well established. However, it also noted that the secretary of agriculture might impose such restrictions as deemed necessary to protect the wilderness resource. So, although these "nonconforming" uses are "allowed where established" by the Wilderness Act, they can nonetheless be controlled by the administering agencies.

Overflights

Administrative action might become necessary to limit aircraft within the backcountry of Grand Canyon National Park in Arizona (the area has been proposed for wilderness designation and is managed as wilderness), where more than 117,000 commercial overflights occurred in 2000. Current regulations permit aircraft to fly as low as 500 feet above the surface, but only in "sparsely populated" areas, yet concentrations of users in many backcountry areas probably require even higher minimum altitudes (Anon. 1985).

Fire, Insects, and Disease

Although the secretary of agriculture can undertake measures to control fire, insects, and disease, this authority is *discretionary;* for example, it is not mandatory that fires be controlled.

The provision of section 4(d)(1) regarding actions to control fire, insects, and disease outbreaks has been elaborated on in subsequent legislative history. House Report 95-540, the committee report filed with the Endangered American Wilderness Act (U.S Public Law 95-232) in 1978, emphasizes that any actions to control fire, insect, or disease outbreaks, such as the use of mechanized equipment, the building of fire roads, or the construction of firebreaks, are permissible if judged necessary for the protection of public health or safety (The Wilderness Society 2004).

The discretionary latitude to control pest threats to wilderness values is exemplified in steps taken by the USFS to control outbreaks of the southern pine beetle in several wildernesses in Texas, Arkansas, Mississippi, and Louisiana. A major outbreak of the southern pine beetle coincided with the designation of several wildernesses in Texas in 1984; by 1985 more than 15,000 separate southern pine beetle spots (ten or more trees) had been located, many within wilderness. A variety of management actions were proposed, including the use of synthetic attractants (pheromones), cutting buffer strips around infected areas, and reducing stand density (Billings 1986). Spread of the southern pine beetle can be rapid; in the Four Notch area in Texas, spread rates of 50 feet per day along a 3.5-mile front were reported (Billings and Varner 1986). The problem was confounded by the presence of several nesting sites of the red-cockaded woodpecker, an endangered species that nests only in mature, live pine trees. Harvesting buffer strips proved successful in halting expansion of the southern pine beetle infestations, but controversy remains as to whether this is the most appropriate technique or the *minimum tool necessary* to control insect outbreaks in wilderness. Ultimately the southern pine beetle epidemic subsided naturally, but not before several treatment strategies were applied, some of them in wilderness.

Mining

The discussion of mining in wilderness occupies a substantial proportion of section 4(d)(3). This part of the Wilderness Act refers to wilderness on the national forestlands because most Department of the Interior lands are withdrawn from mineral entry. Until 1976, however, six units of the NPS were open to mineral entry, including Death Valley National Monument (now Park), California and Nevada; Crater Lake National Park, Oregon; Glacier Bay National Monument (now National Park and Preserve), Alaska; Coronado National Memorial, Arizona; Mount McKinley National Park (now Denali National Park and Preserve), Alaska; and Organ Pipe Cactus National Monument, Arizona. In 1976, Congress passed legislation withdrawing these units from further mineral entry and placed existing claims under stringent regulations issued by the secretary of the interior. (See Chapter 14 for more on mining in wilderness.)

Basically, the Wilderness Act specified that prospecting and mining could continue in wilderness on national forestlands until December 31, 1983. After that date, mining could continue only on valid claims existing before that date. In other words, one could not file a claim for mining after December 31, 1983, but could continue mining a claim that existed before that date. A patent conveying both surface and mineral rights may be taken on a valid claim located prior to the Wilderness Act; for a valid claim located after the date of the Wilderness Act, the patent conveys title to mineral rights only.

A claim permits an individual to occupy a site and use it for the purposes of developing the mineral values it contains; however, uses beyond those necessary for development of the mineral values are not allowed and title to the land remains with the U.S. government. A patented claim conveys full ownership of the site, including surface and subsurface values, to the claimant. Both a claim and a patent are considered property rights.

The secretary of agriculture is instructed to issue *reasonable regulations* governing access to claims and related facilities such as transmission lines, roads, and buildings. Generally, these regulations impose more stringent standards for mineral operations in wilderness than on other national forestlands. Under federal regulations, mining operators in national forests are required to prepare an operating plan that describes who is doing the work, where and when it will be done, the nature of the disturbance the work will create, and measures to be taken to protect other resources. Restoration is also called for when the operation ceases. Surface resources, such as timber, may be used if needed for the mining operation and if founded on sound principles of forest management. After January 1, 1984, or the date of any wilderness designated after this date, but subject to valid existing rights, minerals in wilderness were withdrawn from entry.

Fig. 4.5. Mining is permitted in wilderness, under certain restrictions. Shown is a mining operation on the western edge of the Cabinet Mountains Wilderness in western Montana. Photo courtesy U.S. Forest Service.

This portion of the act also instructs the secretary of the interior to develop, in consultation with the secretary of agriculture, a plan for recurrent surveys of the mineral values in any wilderness and to submit these findings to the public, Congress, and the president. A survey of 74 million acres of wilderness and wilderness study areas in the early 1980s revealed that only 2.7 million acres had a high probability of containing significant oil and gas reserves; half of this total was in western Montana (U.S. Department of the Interior, U.S. Geological Survey 1984).

Mining in wilderness is a paradox, and its presence can make sense only when viewed as a necessary political compromise (Matthews et al. 1985). Nevertheless, many view its presence as an internal contradiction within the Wilderness Act.

Mining in wildernesses located in the eastern United States represents a further complication. The titles to most national forestlands in the East were acquired from private ownership, but mineral rights typically remained in private hands. Thus, the agency cannot prevent private mineral development. The scale of this problem is substantial; in 1984, 103 of the 192 designated and potential wilderness areas in the East contained private mineral rights covering nearly 1 million acres (General Accounting Office 1984). Short of acquiring these mineral rights (estimated in the BWCAW alone as nearly $100 million), the USFS can protect the wilderness qualities of these areas only by the conditions it imposes on the mining operation. At the time the Wilderness Act passed, oil and gas development in wilderness was simply not an issue, but since then, concerns with developing domestic sources of supply and rising prices have changed this situation. A large potential mineral supply lies within designated wildernesses in the Rocky Mountains.

Requests for oil and gas development on national forestlands are approved or rejected by the BLM, acting for the Department of the Interior. It is Department of the Interior policy to request a USFS recommendation on any application, and these recommendations typically have governed the Interior Department's decision (Short 1980). In the event a permit is issued by BLM, the actual site development would be administered by the USFS.

Lands under the administration of the BLM were not addressed in the 1964 Wilderness Act. The FLPMA (U.S. Public Law 94-579) provided authority for designating wilderness on the public lands administered by BLM. The BLM review process is described in more detail in Chapter 5. However, during the review of potentially suitable wilderness lands under BLM jurisdiction, the FLPMA instructs the secretary of the interior to manage such lands to not impair their suitability for preservation as wilderness. This "nonimpairment" standard is subject, however, to the continuation of existing mining and mineral leasing "in the manner and degree in which the same was being conducted" at the time the FLPMA passed. A legal opinion issued in 1978 interprets the nonimpairment language of the FLPMA as being a paramount consideration applying to all areas under BLM administration identified as having wilderness characteristics, exempting from the nonimpairment standard only those uses *actually taking place* at the effective date of the FLPMA (Short 1980).

Water Resource Development

Section 4(d) grants authority to the president to authorize programs for developing water resources within wilderness on the national forestlands, if it is determined that permission for such developments would better serve the interests of the United States than their denial. Actions involved here could include reservoir construction, power projects, and transmission lines. Current controversy surrounds the renovation, repair, and reconstruction of existing reservoirs in wilderness, some of them seventy-five to one hundred years old.

The act does not specifically refer to weather modification, a practice of questionable legality. In 1972, the Bonneville Power Administration (BPA) announced intent to seed clouds in western Montana. This might have affected precipitation levels and patterns in the Bob Marshall Wilderness. The USFS, the State of Montana, and the Montana Wilderness Association opposed the plans. Heavy natural precipitation that winter prompted the BPA to drop its plans, and the case became moot. The Bureau of Reclamation issued a position paper arguing that weather modification would not significantly alter natural processes and therefore should be permitted in wilderness (U.S. Department of the Interior, Bureau of Reclamation, 1974). The issue of weather modification was debated again when considering wilderness in southern California but was not provided for in legislation.

Grazing

Grazing is allowed in wilderness on the national forestlands where it was established before the signing of the Wilderness Act. As with the several other nonconforming but allowed uses, the secretary of agriculture is permitted to impose "such reasonable regulations as are deemed necessary." Although the secretary of agriculture cannot prohibit grazing merely because of wilderness, he can restrict or eliminate such use based on principles of range management (McCloskey 1966).

Grazing in wilderness has led to some confusion. In 1980, Congress expressed the concern that certain national forest administrative policies and regulations were acting to discourage or unduly restrict on-the-ground activities necessary for proper grazing management. There was some pressure to amend the Wilderness Act to clarify the intent of Congress with regard to grazing in wilderness. In a House report on the Colorado Wilderness Act (U.S. Public Law 96-560), the committee observed that attempts to draft specific statutory language covering grazing in the entire NWPS would likely not be successful. Instead, *the committee reaffirmed the existing language of the Wilderness Act and included a series of guidelines and specific statements of legislative policy* (The Wilderness Society 2004, 44–46) that included the following (see also Chapters 5 and 14 for more on grazing):

1. There shall be no curtailments of grazing in wilderness areas simply because an area is, or has been, designated as wilderness, nor should wilderness designations be used as an excuse by administrators to slowly "phase out" grazing.
2. The maintenance of supporting facilities, existing in an area prior to its designation as wilderness (including fences, line cabins, water wells and lines, and stock tanks), is permissible in wilderness.
3. The replacement or reconstruction of deteriorated facilities or improvements should not be required to be accomplished using "natural materials," unless the material and labor costs of using natural materials are such that their use would not impose unreasonable additional costs on grazing permittees.
4. The construction of new improvements or replacement of deteriorated facilities in wilderness is permissible if in accordance with these guidelines and management plans governing the area involved.
5. The use of motorized equipment for emergency purposes, such as rescuing sick animals or the placement of feed in emergency situations, is also permissible.

Subsequent legislation establishing wildernesses in Arizona (U.S. Public Law 98-406), Utah (U.S. Public Law 98-428), and Wyoming (U.S. Public Law 98-550) referred to the congressional guidelines on grazing contained in the Colorado Wilderness Act (U.S. Public Law 96-560), incorporating them as statutory directives for these states as well. It is clear that Congress sees grazing as a continuing, legitimate use in wilderness, and the guidelines in the Colorado Wilderness Act are intended to ensure appropriate treatment of grazing throughout the NWPS. Nevertheless, grazing continues to be a controversial issue in many proposals for new wilderness areas and in management of existing wilderness.

Fig. 4.6. Livestock grazing, when established prior to an area's designation as wilderness, may continue. A drift fence in a desert wash in a BLM-administered wilderness in Arizona is a reminder that even in arid regions there are legal grazing leases where cattle feed on ephemeral vegetation. Photo courtesy Marilyn Hendee.

Boundary Waters Canoe Area

Another controversial portion of section 4(d) involved the management of the Boundary Waters Canoe Area Wilderness (BWCAW) on the Superior National Forest in Minnesota. What is now the BWCAW was originally three roadless areas designated under the U-3(a) Regulation: the Superior, Little Indian Sioux, and Caribou Roadless Areas; they were renamed the BWCAW in 1958. Section 4(d) specified that the BWCAW was to continue to be managed in accordance with regulations established by the secretary of agriculture. In general, the primitive character of the area was to be maintained, but certain other uses, including timber harvesting, vegetation manipulation, and road building, in certain locations were to be permitted, as the management of the area as prescribed under the earlier U-3(a) designation was to continue.

Obviously, the battle for preservation of the BWCAW had not been won. More than 500,000 acres in the area were still virgin forest (Heinselman 1973), the largest contiguous tract of virgin forest left in the eastern United States, and that uniqueness led to great concern about the appropriateness of logging in the BWCAW. The history of resolution to conflicts over the apparent contradictions with wilderness in the act's provisions for the BWCAW is too much to cover here, although more details were provided in the second edition

Fig. 4.7. Under the original terms of the Wilderness Act, many nonconforming uses, such as logging, mining, motorboats, and snowmobiles, were permitted in the Boundary Waters Canoe Area Wilderness. Legislative amendments have curtailed or eliminated most such exceptions in the area, thus enhancing the wilderness environment. Photo courtesy U.S. Forest Service.

of this book, and it is the topic of several books and articles (Haight 1974; Proescholdt 1984, 1995, 2000).

Because management of the BWCAW under terms of the Wilderness Act was to be significantly different from other areas, and there were continuing pressures for mining, logging, and mechanized access, environmentalists pushed for separate legislation for the BWCAW that would put it on a more equal footing with other wildernesses. A citizens' advocacy group, Friends of the Boundary Waters Wilderness, was formed to push for the area's protection. Finally, in 1978, *Congress passed U.S. Public Law 95-495 designating the BWCAW; this law constitutes the only amendment to the Wilderness Act as it repealed paragraph 5 of section 4(d) of the act, that portion dealing with the special exceptions regarding management of the BWCAW.* However, its general effect was to strengthen the wilderness provisions regarding the area and to eliminate some of the complex exceptions that applied to it. The U.S. Court of Appeals rejected a 1981 court challenge of the act's constitutionality.

The new law prohibited motorboats within the area except for certain routes and lakes, specified the size of the motors permitted, prohibited snowmobiles except for a few selected routes, terminated all timber sale contracts existing within the wilderness with compensation to the affected companies, and prohibited all future logging in the virgin forest portions of the area. *The BWCAW is the United States' most visited wilderness, the largest wilderness east of the Rocky Mountains and north of the Everglades, and the largest lake and waterway canoe system in the NWPS.* With the adjoining Voyageurs National Park to the west and Canada's Quetico Provincial Park to the north, and La Verendrye Provincial Park to the east, the *BWCAW is part of a 2.5-million-acre international wilderness complex* (Proescholdt 2000).

The BWCAW was the only area in the NWPS that had an airspace reservation. President Harry Truman signed an executive order in 1949 prohibiting flights below an altitude of 4,000 feet, except for emergencies and for official business of the federal, state, or county governments (Andrews 1953). The 1978 law incorporates the terms of that executive order within it.

U.S. Public Law 95-495 added more than 45,000 acres to the BWCAW, bringing its total size to nearly 1.1 million acres, and established within the wilderness a 222,000-acre Boundary Waters Canoe Area Mining Protection Area. Within this area, exploration for and mining of minerals owned by the United States are prohibited, as is exploration or mining that would in any way affect navigable waters. The act also authorized the secretary of agriculture to acquire any minerals or mineral rights owned by persons other than the federal or state government. If necessary to acquire mineral rights, the secretary is authorized to use the secretary's power of eminent domain; however, congressional appropriation of money to secure these private mineral rights would be necessary (Proescholdt 1984).

Outfitting and Guiding, Wildlife, and Other Exceptions

Recognizing that certain commercial services might be necessary to realize fully the recreational values of wilderness, section 4(d) permits activities such as guiding and outfitting, despite the general prohibition of commercial enterprises cited in section 4(c).

After noting that the act does not touch on the question of federal exemption from state water laws, section 4 concludes with a standard disclaimer that the federal government recognizes state jurisdiction over wildlife and fish on national forests.

Language in the House report on the Endangered American Wilderness Act (U.S. Public Law 95-232) offered further guidelines with regard to fishery-management activities within wilderness. The report notes that fisheries' enhancement activities and facilities are permissible, including fish traps, stream barriers, aerial stocking, and protection and propagation of rare species. These guidelines apply throughout the NWPS (The Wilderness Society 2004).

Congress has attempted, in passing subsequent wilderness legislation, to minimize the number of special exceptions to the Wilderness Act, as the need for consistency in wilderness management was a major rationale for the passage of the act. In some areas, special circumstances have led to exceptions to the general prohibitions in the Wilderness Act. For example, in the Central Idaho Wilderness Act (U.S. Public Law 96-312), establishing what is now the Frank Church–River of No Return Wilderness in 1980, aircraft use of landing strips "in regular use" shall be left open. This denies the USFS the discretion to close airstrips that are contained in the Wilderness Act, which reads "use of aircraft…where these uses have already become established, may be permitted to continue." This exception applies only to the Frank Church–River of No Return Wilderness, where nineteen such landing strips are located.

Major exceptions to the general management provisions of the Wilderness Act are also contained in the Alaska National Interest Lands Conservation Act, known as ANILCA (U.S. Public Law 96-487). Before passage of ANILCA in 1980, no designated wilderness existed in Alaska. At least one reason for this was the feeling that Alaska was different and that wilderness management there would have to be different than in the lower forty-eight states. When the legislation was drafted, Congress included several special provisions for the management of wilderness in Alaska but explicitly stated that such provisions applied only to wilderness in Alaska. These exceptions included the use of snowmobiles, motorboats, and aircraft; temporary fishing and hunting camps; and subsistence uses by natives and nonnatives (The Wilderness Society 2004). The ANILCA even allows log salvage along coastlines within wilderness, but the act also provided authority to the USFS to regulate or limit the above uses to protect an area's wilderness character. See Hummel (2007) for an example of the application of ANILCA to the Stikine-LeConte Wilderness in Alaska.

In summary, then, *the direction contained in section 4 of the Wilderness Act primarily concerns the uses considered appropriate in wilderness, thus providing the guiding criterion for wilderness management.* As the preceding discussion suggests, these guidelines are fairly strict. Controversy continues over the application of these guidelines in the allocation of wilderness, as opposed to its management. This debate—the so-called purity argument—has centered on the extent to which previous human impacts can be accepted in wilderness. The USFS, in particular, took the position that evidence of previous human impact normally precluded an area from consideration as wilderness. This position was defended because to justify the substantial opportunity costs associated with wilderness, it was important that any area selected be of highest quality; that is, the area must exhibit the highest degree of "environmental purity" (Costley 1972; Worf 1980).

Some authors (Behan 1971; Foote 1973) have rebutted the purity doctrine. Former senator Frank Church (ID), floor manager for the wilderness bill at the time of its passage, charged that neither the letter nor the intent of the Wilderness Act called for exclusion of areas once impacted by human use, and that agency claims to the contrary were simply evidence of resistance to congressional will (Church 1977). Allin (1982, 1985) has further argued that USFS reliance on the "purity argument" derives from the organization's concern with preserving a greater degree of administrative control over forestlands. The increasing loss of administrative discretion over the management of the national forests led the agency to consider ways in which it could limit the impact of wilderness designation on commodity production.

The admission standards for designating wilderness are dealt with in section 2(c) of the Wilderness Act, not section 4. Preservationists as well as Congress have viewed reliance on section 4 to define wilderness as inappropriate at best and illegal at worst (Allin 1982). Congressional actions in designating wilderness also indicate that a rigid and stringently pure conception of wilderness is not necessary.

For example, in the Mission Mountains Wilderness in Montana and the Agua Tibia Wilderness in California, Congress overruled impacts judged by the USFS to preclude wilderness designation. In the Mission Mountains, about 2,000 acres of the old primitive area were salvage logged in the 1950s after an insect outbreak. The USFS proposed excluding this area, managing it to restore wilderness values, prior to its inclusion in the wilderness system. Congress, however, included the impacted area. In the Agua Tibia Wilderness proposal, the USFS excluded an area containing a road closed to public use but used for fire protection.

However, Congress approved a bill including the road, ruling that such a development was permissible under the administrative exceptions clause of the Wilderness Act, section 4(c).

Not all impacts are acceptable to Congress, of course. Congress excluded a 6,000-acre area containing a 22-mile-long access road to a mining claim in the Emigrant Basin Primitive Area in California from the wilderness bill it approved for this area. Congress instructed the USFS to reexamine the area at a later date in conjunction with the review of some roadless lands contiguous to the nearby Hoover Wilderness. The challenges and problems of wilderness management are closely tied to decisions regarding such questions as what constitutes a "substantially unnoticeable" impact; that is, the greater the latitude in admission standards, the more urgent will be the need for rigorous management to enhance restoration of the wilderness resource and prevent further damage of the impacted area.

Passage of the so-called Eastern Wilderness Act in 1975 (U.S. Public Law 93-622) provides perhaps the strongest statement of congressional intent with regard to the conditions needed to constitute wilderness. This act (discussed in more detail in Chapter 5) explicitly accepted the notion that areas where previous human activity, such as logging and agriculture, had occurred could still be congressionally designated as wilderness. Nelson (1985) notes that House Report 99-262 on a bill to designate new wilderness in Michigan indicated that within areas proposed, two-thirds of the mineral rights were privately owned, motorboat and off-road vehicle use was present, developed camping facilities were located, and one area contained an experimental forest with structures located in it. Such congressional actions clearly repudiate the purity policy as applied to designation (Allin 1982). However, such steps also reflect an evolving perception on the part of the public as to what constitutes wilderness. The criteria of wilderness naturalness and purity standards over time, and in different parts of the country, appear to have been relaxed (Hendee 1986). During the past two decades (1987–2007) many wilderness designation proposals were put forth with special provisions and exceptions that further expand what are nonconforming but allowable wilderness conditions and activities.

Section 5—State and Private Lands Within Wilderness Areas

Section 5 describes the rules and procedures for access to private or state-owned inholdings within national forest wilderness. It provides "such rights as may be necessary to assure adequate access" to such lands. The right of reasonable access is further endorsed in section 1110(b) of ANILCA: "The State or private owner or occupier shall be given by the Secretary such rights as may be necessary to assure adequate and feasible access." These rights are also taken to apply to wilderness lands outside Alaska (The Wilderness Society 2004). Section 5 also indicates that such lands may be exchanged for federally owned property in the same state and of approximately equal value. But, unless the state or private owner relinquishes the mineral interests in the surrounded land, the U.S. government will not transfer the mineral rights of any exchanged land.

The secretary of agriculture is permitted, subject to the appropriation of funds by Congress, to acquire private inholdings within wilderness if the owner agrees to such acquisition and if the acquisition is approved by Congress.

Condemnation is not permitted by the Wilderness Act, an authority that was important to the development of legislation establishing wilderness in the eastern U.S. (see Chapter 5). However, even in the case of the so-called Eastern Wilderness Act, the authority to condemn private land was limited to the sixteen areas designated by that legislation.

Section 6—Gifts, Bequests, and Contributions

Both the secretary of agriculture and the secretary of the interior are authorized to accept gifts in furtherance of the purposes of the act. Such gifts might be land, money, or both. If the gift involves land adjacent to, rather than within, an existing wilderness, Congress must be notified sixty days before acceptance. On acceptance, the land becomes part of the wilderness. If it is consistent with the purposes of the Wilderness Act, a donor may attach stipulations to the gift.

Section 7—Annual Reports

The last section of the Wilderness Act instructs the two secretaries to report jointly to the president at the opening of each session of Congress on the status of the NWPS. These reports provide information on the number of areas and acres in the NWPS, changes in the NWPS, regulations in effect, and the status of areas under consideration. They are available to the public from the Government Printing Office and from senators and representatives. To date, however, they have fallen short of what was proposed for these reports in the Wilderness Act.

Some Exceptions and Ambiguities in Wilderness Legislation

One additional aspect of the Wilderness Act deserves comment. The act's directives and guidelines for area management pertain only to those areas designated by that legislation. So technically, the Wilderness Act imposes use restrictions only on the fifty-four national forest areas designated as wilderness on its passage. It also provides that, pending designation as wilderness, primitive areas were to be managed under regulations in effect at the time the Wilderness Act was passed. Left in some confusion, then, was the status of the Department of the Interior holdings, the wilderness study areas specified in other wilderness designation acts, and all other lands that might be added in the future to the NWPS.

Why Congress did not clearly specify the management direction for these other areas in the 1964 legislation is not entirely clear. The lack of definitiveness does allow Congress latitude for adding special management provisions to the legislation designating each individual area, perhaps in response to an area's unique features, problems, or uses, and this has been done more frequently during the past two decades.

However, wide use of special management provisions could also undermine one of the major reasons a wilderness bill was initially proposed—*consistency* (Aspinall 1964; H. Zahniser 1964). *Consequently, in all subsequent legislation designating wilderness, Congress has affirmed that these new areas shall be managed in accordance with the 1964 act's provisions.* In other words, although the Wilderness Act did not specifically prescribe the management for areas that would be subsequently designated, Congress has extended the 1964 act's provisions to them. *Where special provisions and exceptions have been included in subsequent legislation, they have been restricted to specific areas and not used as a precedent for changes elsewhere.*

The Wilderness Act has many ambiguities, weaknesses, and omissions—more, or less, depending on one's views (e.g., permitting mining or grazing; failure to describe procedures for the review of national forest roadless lands). The general vagueness of many procedures has placed substantial burden on administrative agencies, citizen advocates, and, especially, the courts. Nevertheless, the act takes away much discretion from the management agencies by using "a statutory structure with detailed requirements—such as no vehicles, no structures, and no commerce—along with a strict, overarching mandate to preserve wilderness character" (Dettmann 2008, 4). And, when an agency fails to heed the act's direction, plaintiffs can appeal to the courts for relief and they have provided, numerous times over recent decades, judicial opinions clarifying the act.

Notwithstanding all the unresolved issues, the Wilderness Act represents the principal statutory foundation for wilderness preservation and management in the United States. It defines the broad goals, objectives, policies, and procedures through which an enduring resource of wilderness is to be designated and managed. With one exception—addressing the management of the BWCAW—Congress has not amended the Wilderness Act since its passage in 1964. Nevertheless, alternative interpretations of the act exist, often with alternative implications for wilderness designation and management.

The Evolution of Wilderness Protection

What did the evolution of wilderness protection culminating in the Wilderness Act accomplish beyond the administrative protection already being provided at the time of its passage?

First, the permanency of wilderness was substantially enhanced. The USFS L-20 Regulation afforded little if any permanency—its primitive-area designation was viewed as only an interim measure to halt haphazard development. Longevity of protection was improved by the U regulations, but administrative discretion to choose the level of protection was a major shortcoming. Similarly, although national park or national wildlife refuge designation ensured protection from many kinds of development, it did not necessarily guarantee permanent protection of wilderness values. The Wilderness Act brought increased assurance that such values would be protected.

Second, permitted uses of wilderness were increasingly restricted. The L-20 Regulation contained little in the way of prohibited uses. Logging and other forms of resource development were permitted. The U regulations were developed primarily to exclude some permitted uses—which they did. However, because the U regulations were instituted at the administrative rather than the legislative level, there remained the possibility that certain uses, inconsistent with wilderness, might be permitted. Only through legislation would certain use prohibitions be secure.

Finally, the evolution from the L-20 Regulation and U regulations to the Wilderness Act reflects a change in purpose. The L-20 Regulation was intended to establish a series of areas for the purposes of public education, inspiration, and recreation (ORRRC 1962). The U regulations emphasized the importance of protecting the natural environment (Marshall 1933) and of retaining the primitive quality of these lands, particularly with regard to the style of travel permitted, permitted uses, and procedures for designating or modifying wilderness areas.

Similarly, the management guidelines for national parks and wildlife refuges before passage of the Wilderness Act did not explicitly define the purposes of wilderness preservation. Although a generally low level of development prevailed in many of these areas, the purposes for these areas lacked clear direction.

In the Wilderness Act, we find a new emphasis on the purpose of wilderness. Although public use and enjoyment are clearly provided for, wilderness, as defined by the act, is a landscape where the earth and its community of life are untrameled by humans. Thus, *the Wilderness Act established a national policy and purpose of maintaining a system of areas where natural processes could operate as freely as possible. Recreational use was an appropriate use of these areas, only so long as it was consistent with this purpose.* The evolution of wilderness protection, from the L-20 Regulation to the U regulations to passage of the 1964 Wilderness Act, was a shift toward a more biocentric concept of wilderness, one focused on protection of the natural processes that created these settings.

This evolution of purpose in wilderness preservation is key, in our opinion. Wilderness preservation has evolved from a holding strategy for minimizing unplanned development until more carefully thought-out plans could be formulated to a carefully defined and legally established national system for protecting the ecological integrity of designated areas.

Our endorsement of a biocentric philosophy of wilderness management rests on the belief that framers of the Wilderness Act clearly intended to create a system of areas where people allowed nature's way, as far as possible, to continue unhampered. We share the lawmakers' recurring insistence that human use of these lands must not interfere with the preservation of the area as wilderness. It is from these assumptions regarding purpose that our interpretation of the Wilderness Act flows and that our management recommendations in this book are founded.

Study Questions

1. Identify and discuss some of the factors leading to the pressure to create a congressionally established and protected wilderness system.
2. Discuss the relationship between the L-20, the U regulations, and the Wilderness Act; how did these earlier administrative regulations affect the Wilderness Act? Consider both positive and negative influences.
3. What were some of the critical differences between the wilderness bill proposed in 1956 and the legislation passed in 1964? What were some of the reasons for these differences?
4. In your judgment, what are some of the major problems with the Wilderness Act in terms of how well it protects wilderness values? How can these problems be resolved?
5. What are some of the major changes that have occurred in applying the Wilderness Act since its passage in 1964? What forces have led to such changes? What future changes would you suspect might occur?
6. Why is the Wilderness Act important to wilderness designation? To wilderness management?

Acknowledgments

George H. Stankey was lead author for this chapter in the first and second editions, and his major contributions remain. John C. Hendee, with help from Jay Watson of The Wilderness Society, provided revisions for the third edition and fourth edition.

Case Discussion: Compare the 1956 Bill to the 1964 Wilderness Act

Four major differences were noted between the original 1956 bill and the Wilderness Act of 1964 in the Chapter 4 section title: "Changes and Compromises to Pass a Wilderness Act." Consider how the NWPS might be different today if the 1956 bill had passed as proposed by comparing the four changes to the 1956 bill and the 1964 Wilderness Act as passed by Congress.

Case Discussion Questions:

1. Bob Marshall had long advocated including wilderness designations on Native American lands. For example, the Mission Mountain Wilderness (managed by the USFS) adjoins the Mission Mountain Tribal Wilderness Area (managed by the Confederated Salish and Kootenai Tribes) on the Flathead Indian Reservation. How might the NWPS be different today if tribal lands had been part of the NWPS?

2. The proposed National Wilderness Preservation Council would have included five agency representatives and six laypersons to review the wilderness designation proposals prior to congressional action and provide an advisory role. The Interagency Wilderness Policy Council today coordinates wilderness policy among the four federal agencies, but does not serve the role envisioned by the National Wilderness Preservation Council. What impacts might a National Wilderness Preservation Council have had on wilderness designations in Congress?

3. All mining would have been prohibited on passage of the 1956 bill; however, the 1964 Wilderness Act allowed prospecting to continue until December 31, 1983, and mining claims before this date would be allowed to continue. See the discussion on mining under the Chapter 4 subsection "Section 4—Use of Wilderness Areas." How might the NWPS be different today if all mining had been prohibited from the NWPS?

References

Allin, Craig W. 1982. *The Politics of Wilderness Preservation*. Westport, CT: Greenwood Press.

———. 1985. Hidden agendas in wilderness management. *Parks and Recreation*. 20(5): 62–65.

———. 1987. Wilderness preservation as a bureaucratic tool. In: Foss, Philip O., ed. *Federal Land Policy*. Westport, CT: Greenwood Press, pp. 127–138.

Andrews, Russell P. 1953. *Wilderness Sanctuary*. New York: Bobbs-Merrill.

Anon. 1985. Aerial tours over protected land: Where and why low-flying aircraft are off limits. *FAA General Aviation News*. 24(4): 3–6.

Aspinall, Wayne N. 1964. Underlying assumptions of wilderness legislation as I see them. *Living Wilderness*. 86: 6–9.

Baker, Richard A. 1985. The conservation congress of Anderson and Aspinall, 1963–64. *Journal of Forest History*. 29(3): 104–119.

Baldwin, Donald L. 1972. *The Quiet Revolution: The Grass Roots of Today's Wilderness Preservation Movement*. Boulder, CO: Pruett Publishing.

Behan, R. W. 1971. Wilderness purism—here we go again. *American Forests*. 78(12): 8–11.

Billings, Ronald F. 1986. Coping with forest insect pests in southern wilderness areas, with emphasis on the southern pine beetle. In: Kulhavy, David L.; Conner, Richard N., eds. *Wilderness and Natural Areas in the Eastern United States: A Management Challenge*; May 13–15, 1985; Nacogdoches, TX. Nacogdoches, TX: Stephen F. Austin University, School of Forestry, pp. 120–125.

Billings, Ronald F.; Varner, Forest E. 1986. Why control southern pine beetle infestations in wilderness areas? The Four Notch and Huntsville State Park experiences. In: Kulhavy, David L.; Conner, Richard N., eds. *Wilderness and Natural Areas in the Eastern United States: A Management Challenge*; May 13–15, 1985; Nacogdoches, TX. Nacogdoches, TX: Stephen F. Austin University, School of Forestry, pp. 130–135.

Chapman, H. H. 1938. National parks, national forests, and wilderness areas. *Journal of Forestry*. 36(5): 469–474.

Church, Frank. 1977. Wilderness in a balanced land use framework. Wilderness Resource Distinguished Lectureship. Moscow: University of Idaho, Wilderness Research Center.

Costley, Richard J. 1972. An enduring resource. *American Forests*. 78(6): 8–11.

Dettmann, J. 2008. The need for wilderness litigation. *International Journal of Wilderness*. 14(1): 4–6.

Foote, Jeffrey P. 1973. Wilderness—a question of purity. *Environmental Law*. 3(4): 255–260.

General Accounting Office. 1984. *Private Mineral Rights Complicate the Management of Eastern Wilderness Areas*. RCED-84-101. Washington, DC: U.S. Government Printing Office.

Gilligan, James P. 1953. The development of policy and administration of Forest Service primitive and wilderness areas in the Western United States. Ann Arbor: University of Michigan. Dissertation.

———. 1954. The contradiction of wilderness preservation in a democracy. In: *Proceedings, Society of American Foresters*; October 24–27, 1954; Milwaukee, WI. New York: Society of American Foresters, pp. 119–122.

Glover, James M.; Glover, Regina B. 1986. Robert Marshall—portrait of a liberal forester. *Journal of Forest History*. 30(3): 112–119.

Haight, Kevin. 1974. The Wilderness Act: Ten years after. *Environmental Affairs*. 3(2): 275–326.

Heinselman, Miron L. 1973. Restoring fire to the canoe country. *Naturalist*. 24(4): 21–31.

Hendee, John C. 1986. Wilderness: Important legal, social, philosophical and management perspectives. In: Kulhavy, David L.; Conner, Richard N., eds. *Wilderness and Natural Areas in the Eastern United States: A Management Challenge*; May 13–15, 1985; Nacogdoches, TX. Nacogdoches, TX: Stephen F. Austin University, School of Forestry, pp. 5–1.

Hession, Jack M. 1967. The legislative history of the Wilderness Act. San Diego, CA: San Diego State College. Thesis.

Hummel, M. 2007. The Stikine-LeConte Wilderness in Alaska: Twenty-six years of management under ANILCA. *International Journal of Wilderness*. 13(2): 8–13.

Keyser, C. Frank. 1949. The preservation of wilderness areas—an analysis of opinion on the problem. Washington, DC: Library of Congress, Legislative Reference Service (renamed Congressional Research Service in 1970).

Mackintosh, Barry. 1985. Harold L. Ickes and the National Park Service. *Journal of Forest History*. 29(2): 78–84.

Marshall, Robert. 1933. *The Forest for Recreation*. Senate Document 12. Washington, DC: U.S. Government Printing Office, pp. 463–487.

Matthews, Olen Paul; Haak, Amy; Toffenetti, Kathryn. 1985. Mining and wilderness: Incompatible uses or justifiable compromise? *Environment*. 27(3): 12–17, 30–36.

McArdle, Richard E. 1975. Wilderness politics: Legislation and Forest Service policy. *Forest History*. 19(4): 166–179.

McCloskey, Michael. 1966. The Wilderness Act of 1964: Its background and meaning. *Oregon Law Review*. 45(4): 288–321.

Mercure, Delbert V., Jr; Ross, William M. 1970. The Wilderness Act: A product of Congressional compromise. In: Cooley, Richard A.; Wandesforde-Smith, Geoffrey, eds. *Congress and the Environment*. Seattle: University of Washington Press, pp. 47–64.

Mitchell, John G. 1985. In wildness was the preservation of a smile: An evocation of Robert Marshall. *Wilderness*. 48(169): 10–21.

Nash, Roderick. 1982. *Wilderness and the American Mind*. 2d ed. New Haven, CT: Yale University Press.

Nelson, Thomas C. 1985. Wilderness re-re-defined. *Journal of Forestry*. 83(12): 717–718.

Outdoor Recreation Resources Review Commission. 1962. *Wilderness and Recreation: A Report on Resources, Values, and Problems*. ORRRC Study Report 3. Washington, DC: U.S. Government Printing Office.

Proescholdt, Kevin. 1984. Boundary Waters: More obstacles for a troubled law. *Sierra*. 69(4): 18–21.

———. 1995. *Troubled Waters: The Fight for the Boundary Waters Canoe Area Wilderness*. St. Cloud, MN: North Star Press.

———. 2000. Boundary Waters Wilderness: An international story. *International Journal of Wilderness*. 6(1): 35–37.

Robinson, Glen O. 1975. *The Forest Service*. Baltimore: The Johns Hopkins University Press.

Roth, Dennis. 1984. The national forests and the campaign for wilderness legislation. *Journal of Forest History*. 28(3): 112–125.

Scott, Douglas. 1976. Howard Zahniser: Architect of wilderness. *Sierra Club Bulletin*. 61(9): 16–17.

———. 2004. *The Enduring Wilderness: Protecting Our Natural Heritage Through the Wilderness Act*. Golden, CO: Fulcrum Publishing.

Short, L. Rex. 1980. Wilderness policies and mineral potential on the public lands. *Proceedings of the Annual Institute of the Rocky Mountain Mineral Law Institute*. 26: 39–67.

U.S. Department of the Interior, Bureau of Reclamation. 1974. Position paper on weather modification over wilderness areas and other conservation areas. Washington, DC: U.S. Department of the Interior, Bureau of Reclamation.

U.S. Department of the Interior, U.S. Geological Survey. 1984. Wilderness mineral potential: Assessment of mineral-resource potential in U.S. Forest Service lands studied, 1964–1984. 2 vol. Geological Survey Professional Paper 1300. Washington, DC: U.S. Department of the Interior, Geological Survey.

U.S. Public Law 86-517. The Multiple-Use Sustained-Yield Act of June 12, 1960. 74 Stat. 215.

U.S. Public Law 88-577. The Wilderness Act of September 3, 1964. 78 Stat. 890.

U.S. Public Law 93-622. (Eastern Wilderness) Act of January 3, 1975. 88 Stat. 2096.

U.S. Public Law 94-579. (Wilderness Study) Federal Land Policy and Management Act (FLPMA) of October 21, 1976. 90 Stat. 2743.

U.S. Public Law 95-232. Endangered American Wilderness Act of February 24, 1978. 92 Stat. 40.

U.S. Public Law 95-495. (Boundary Waters Canoe Area Wilderness) Act of October 21, 1978. 92 Stat. 1649.

U.S. Public Law 96-312. Central Idaho Wilderness Act of July 23, 1980. 94 Stat. 948.

U.S. Public Law 96-487. Alaska National Interest Lands Conservation Act (ANILCA) of December 20, 1980. 94 Stat. 2371.

U.S. Public Law 96-560. (Colorado Wilderness) Act of December 22, 1980. 94 Stat. 3265.

U.S. Public Law 98-406. Arizona Wilderness Act of August 28, 1984. 98 Stat. 1485.

U.S. Public Law 98-428. Utah Wilderness Act of September 18, 1984. 98 Stat. 1657.

U.S. Public Law 98-550. Wyoming Wilderness Act of October 30, 1984. 98 Stat. 2807.

The Wilderness Society. 2004. *The Wilderness Act Handbook: 40th anniversary edition*. Washington, DC: The Wilderness Society.

Worf, William A. 1980. Two faces of wilderness—a time for choice. *Idaho Law Review*. 16: 423–437.

Zahniser, Edward. 1984. Howard Zahniser: Father of the Wilderness Act. *National Parks*. 58(1-2): 12–14.

Zahniser, Howard. 1964. How much wilderness can we afford to lose? In: Brower, David R., ed. *Wildlands in Our Civilization;* March 30–31, 1951; Berkeley, CA. San Francisco: Sierra Club, pp. 46–51.

Chapter 5
Management Implications in the Wilderness Designation Process, Wilderness Designation Acts, and Wilderness Litigation

Introduction 114
Part I: Evolution of the Designation Process 115
   Evaluation of Roadless Lands for Wilderness Suitability 115
   Review of National Forest Roadless Lands 115
      The San Rafael Primitive Area 116
      The Gore Range–Eagles Nest Primitive Area 118
   Designating Nonreserved National Forestlands 119
      U.S. Forest Service Inventory of Potential Wilderness Additions 119
      Roadless Area Review and Evaluation (RARE I) 119
      Wilderness in the East: An Eastern Wilderness Act 121
      The Endangered American Wilderness Act 122
   RARE II—Another Look at Forest Service Roadless Areas 122
      The RARE II Inventory 123
      Evaluation of RARE II Areas 124
      Recommendations from RARE II 125
      The Court Intervenes: *California v. Bergland* 125
   U.S. Forest Service Wilderness Designation in the 1980s 126
      Statewide Wilderness Bills—The Prototype of the 1980s 127
      Sufficiency-Release Language 127
      Soft or Hard Release 128
   Designating Department of the Interior Roadless Lands: NPS, FWS 129
      Roadless-Area Review Criteria: NPS and FWS 130
      Wilderness Study Process: NPS and FWS 131
      Olympic National Park Wilderness Review—A Case Study 132
   The Bureau of Land Management 134
      BLM Roadless Area Review Process 135
      Inventory 135
      Wilderness Study 136
      Reporting Wilderness or Nonwilderness Recommendations 139
   Alaska—A Special Case? 139
      Disposition of the Alaskan Public Domain 139
      The Alaska National Interest Lands Conservation Act (ANILCA) 141
   Part I Summary 142
Part II: Management Implications in Wilderness Designation Acts 143
   Benchmark Wilderness Laws 143
      Wilderness Act of 1964 143
      Eastern Wilderness Act of 1975 144
      Federal Land Policy and Management Act of 1976 (FLPMA) 144
      Endangered American Wilderness Act of 1978 144
      Alaska National Interest Lands Conservation Act of 1980 (ANILCA) 145

> Colorado Wilderness Act of 1980 145
> California Desert Protection Act of 1994 (CDPA) 145
> Management Direction and Special Provisions in Wilderness
>   Legislation 146
> Legislative Management Direction for Specific Issues 146
>   Mining 147
>   Motorized Use 148
>   Grazing 148
>   Buffer Zones 148
>   Fish And Wildlife 149
>   Fire, Insect, and Disease Control 149
>   Facilities and Structures 150
> Special Provisions and Congressional Direction 150
>   Sufficiency-Release Language 150
>   Military Overflights 151
>   Miscellaneous Special Provisions 151
> Wilderness Legislation: A View of the Future 152
>   Alaska 152
>   National Park System Wilderness in the Forty-Eight Contiguous
>   States 152
>   National Forest Roadless Areas and Released Lands 152
>   Bureau of Land Management Wilderness Study Areas 153
>   New Wilderness Knowledge and Legislative Direction 153
> Part II Conclusions and Predictions 154
> Part III: Management Implications from Litigation Involving Wilderness 155
>   Litigation and the Wilderness Legislation 155
>   Boundary Waters Canoe Area 156
>   The Extent of Wilderness Litigation 157
>   Litigation Themes and Management Implications 157
>   Part III Summary 158
> Study Questions 159
> Acknowledgments 160
> Case Discussion Questions: Olympic National Park Wilderness 160
> References 160

Introduction

The wilderness designation process, implementing legislation, and litigating over agency decisions can have important management implications. In Chapter 1, we briefly reviewed the relationship between wilderness designation and management, and in Chapter 4 we reviewed the Wilderness Act and its management direction. *In this chapter, we consider how the wilderness designation process, legislation to designate additional wilderness, and litigation over wilderness decisions also influence management. We have organized this chapter into three parts.*

Part I. We review the evolution of the wilderness designation process, focusing on some key events in the evaluation of roadless lands for their wilderness suitability. For example, how roadless lands were reviewed, evaluated, studied, and designated in the national forests, the national parks and fish and wildlife refuges, and lands administered by the Bureau of Land Management influences management by the kinds of areas they designated and decisions in the process. In dealing with the difficult cases and controversies encountered, a wilderness designation process has evolved that now has more room for citizen involvement, and is increasingly clarified by court appeals when necessary.

Part II. We summarize the management implications of wilderness designation laws that set aside wilderness for protection. We focus on seven benchmark laws providing key policy direction and then summarize management direction and special provisions for wilderness uses and activities in other specific wilderness laws. Appendix B is a chronological listing of the 171 wilderness designation and other related wilderness laws through December 31, 2007, with the managing agency and special provisions indicated. The www.wild.org website provides abstracts of all these 171 wilderness designation and related laws, describing their specific management direction or special provisions. The www.wilderness.net website provides text for many of these legislative acts.

Part III. We review the management implications of litigation related to wilderness laws.

While all this information may seem tediously complex to review, it forms the legal foundation of the NWPS and the designation and management of the areas in the system. We have divided the chapter into three parts to make it more readable and usable, and with separate study questions at the end of the chapter. The references for all three parts are combined at the end of the chapter.

Part I: Evolution of the Designation Process

Evaluation of Roadless Lands for Wilderness Suitability

Section 3 of the 1964 Wilderness Act provided procedures for designating wilderness. First, the act provided for an "instant" wilderness system by proclaiming that all the lands administered by the U.S. Forest Service as wilderness, wild, or canoe areas prior to 1964 would be known henceforth as (legally designated) wilderness areas.

Second, the Wilderness Act instructed the secretaries of agriculture and the interior to review roadless lands within their respective jurisdictions and to recommend to the president the suitability of these lands for designation as wilderness. The USFS, through the secretary of agriculture, was instructed to review those national forestlands previously classified administratively as *primitive areas* (a total of 34 areas and 5.4 million acres) within ten years after passage of the act.

Similar instructions were addressed to the secretary of the interior. Within ten years, all roadless areas in the various units of the National Park System and the national wildlife refuges and game ranges in excess of 5,000 acres, as well as all roadless islands, were to be

Fig. 5.1. A visitor in the Absaroka-Beartooth Wilderness in Montana kneels on headwaters of the Boulder River's East Fork. The Wilderness Act of 1964 instructed the secretary of agriculture to review within ten years all administratively designated primitive areas on the national forests to ascertain suitability of wilderness. Photo courtesy Richard Behan.

reviewed by the respective agencies, and recommended their suitability for wilderness to the president. The president would make his recommendations to Congress. Finally, Congress would propose and vote on legislation to designate (or not) each area as wilderness.

However, three problems immediately surfaced for the agencies: (1) *what criteria to use* when making recommendations about the wilderness character of lands and the need for wilderness; (2) *what lands to consider;* and (3) *who should be the primary force in the review process,* allegedly to complete and round out the NWPS: the agencies or interest groups? These issues have permeated the wilderness review process from 1964 until the present, and no agency experienced this debate more dramatically than the USFS.

Review of National Forest Roadless Lands

The USFS's roadless lands review began in late 1964 with a special task force of experienced wilderness managers whose assignment was to write wilderness policy and guidelines in accordance with the act. The task force believed their work would conclude quickly, but they found that the Wilderness Act set only broad philosophical definitions, that little consensus had developed in their own agency on what wilderness was, and that ultimately the input of citizens and congressional

Fig. 5.2. Some of the finest wilderness solitude is in southwestern deserts, where many areas and millions of acres were designated as wilderness in the 1990s. Wilderness designation under the Wilderness Act is a lengthy process requiring resource inventory, public involvement, formulating alternatives, congressional and public debate, passage through Congress, and finally, the president's signature. Photo courtesy Marilyn Hendee.

action, as required by the act, would strongly influence the nature of the wilderness system (Worf 1980).

The task force took a "pure" or "strict constructionist" approach to wilderness policy, believing Congress had the vast and pristine wildernesses of the West in mind for the NWPS. They knew of the many requests for commodity use of wilderness and decided that strict guidelines were needed to prevent the "cheapening" of the NWPS.

From a more practical standpoint, they believed that wilderness and wilderness management would be expensive, in terms of uses forgone and in terms of management to maintain primitive conditions and experiences. Expense would increase sharply if the agency had to remove evidence of previous human use (Costley 1972). The task force also worried that adding areas that had significant human-caused impacts to the wilderness system would be a dangerous precedent that would permit existing, pristine areas to decline to the lower standard of the newly admitted areas. Finally, the task force believed that consistent, clear, and precise guidelines were needed for a system of comparable, high-quality areas to be designated and managed (Worf 1980).

The task force might have expected the environmental groups that had lobbied for the passage of the act to support its strict constructionist approach toward implementing the act, but the environmentalists did so mainly with respect to wilderness management (Roth 1984), and not with respect to adding new areas to the NWPS. *As the USFS developed its guidelines and conducted its review of the thirty-four primitive areas, environmental groups and the public began to demand a greater role in the wilderness designation process, and their definition of wilderness was not always as pure as that of the USFS.* For each primitive area, the agency strongly advocated its position on whether an area should receive wilderness designation within the existing primitive-area boundaries, have larger or smaller wilderness boundaries, or be allocated to other multiple-use purposes because the area was deemed unsuitable for wilderness.

Environmental groups often disagreed with the USFS recommendations because the Wilderness Act stated that only a bill passed by both bodies of Congress and signed by the president could create a wilderness. This made wilderness designation of areas more difficult, and much more political. For the first time, Congress had given itself the power to determine how a piece of national forestland was to be used. In essence, the USFS became an advisor to Congress and citizen environmental groups about wilderness decisions on lands it managed. Thus, the political process would determine the ultimate size of the NWPS.

The review, debate, and final congressional decisions regarding the San Rafael and the Gore Range–Eagles Nest Primitive Areas in California and Colorado illustrate the increasingly political nature of the wilderness designation process and growing influence of environmental groups and the courts. *Thus, although the wilderness designation process was defined by the Wilderness Act, it was a little different, more complicated, and more political than anyone anticipated.*

The San Rafael Primitive Area: Enter Environmental Groups and Congress

The USFS began review of its primitive areas with the 74,990-acre San Rafael Primitive Area in California—an area with little timber or mineral value and known for its grassy balds, scenic rock outcrops, mountain lions, bears, eagles, condors, and Chumash Indian rock paintings. As directed by the Wilderness Act, the USFS examined contiguous roadless lands and proposed a wilderness area of 110,403 acres. Under the USFS proposal, 36,244 acres were to be added to the primitive area and 831 acres containing a fuel break and a lookout tower were to be withdrawn. Little controversy was expected.

At public meetings, environmentalists urged that five other contiguous areas be included in the wilderness proposal. They also opposed deletion of the 831 acres because this area contained favorite camping spots for backpackers and resting sites for California condors. This testimony advocated increasing the size of the wilderness proposal by almost one-third, to about 158,000 acres.

In response to public testimony and management considerations, the USFS then submitted a revised recommendation to designate 142,722 acres as wilderness, an increase of 32,000 acres over its initial proposal and almost double the size of the original primitive area. President Lyndon Johnson transmitted the recommendation, unmodified, to Congress on February 1, 1967, and strongly urged Congress to give it "early and favorable approval."

The temporary peace and goodwill between the USFS and environmentalists began to unravel later in 1967. Disagreements centered on 4,700 acres—an area that environmentalists maintained should be included in the wilderness proposal because it had wilderness character, was a condor flyway, and contained Chumash Indian pictographs. The USFS reported that 2,500 acres of the area contained an administrative site and road and included areas that had been converted from brush to grass for wildlife habitat, livestock forage, and fire control. The remaining 2,200 acres contained the only suitable site for construction of a larger firebreak. In fact, in June of the preceding year, a fire that ignited from the crash of a small airplane burned more than 90,000 acres, 70,000 of which were in the proposed wilderness. USFS officials believed that even under the most liberal interpretation of the Wilderness Act, the existing fuel break with its bulldozer and plow marks could not be considered wilderness.

The local Santa Barbara Sierra Club had already convinced California senator Thomas Kuchel to introduce a bill that was essentially the environmentalists' proposal. It called for a 158,000-acre wilderness and included the 2,200 acres that the USFS wanted to exclude for potential use as a firebreak. This action was early evidence that in questions of wilderness allocation, grassroots environmental organizations would help shape congressional decisions (Roth 1984). Ultimately, environmentalists indicated a willingness to accept a reduction of some 13,000 acres from their original proposal, but they would not give in on the 2,200-acre addition—the area where the USFS wanted to retain the option of building a firebreak.

The 2,200 acres quickly assumed symbolic political importance far beyond their intrinsic worth as wilderness. Debate was vigorous—for precedent was at stake. Environmentalists feared that if Congress accepted the USFS position, it would ever after simply rubber-stamp USFS wilderness proposals. The USFS believed its professionalism was at stake and was further distressed that its authoritative recommendations about fire management were being questioned. Congressional representatives from the district where the San Rafael Area was located supported the USFS proposal because fire danger was such a serious matter. Ultimately, the USFS position was upheld and the 2,200 acres were excluded when, in a special White House ceremony in March 1968, President Johnson signed *the San Rafael Wilderness bill into law* (U.S. Public Law 90-271). It was the first addition to the NWPS. It had passed through the various steps and procedures prescribed by the Wilderness Act; it received considerable debate in Congress; and *it further defined and demonstrated the wilderness designation process as follows:*

1. A preliminary agency recommendation, resting on substantial review (including field surveys and a mineral assessment), was revised following citizen review and input.
2. The revised recommendation was submitted through the secretary of agriculture to the president.
3. The president submitted the proposal to Congress, recommending passage.
4. Congressional committees considered several draft bills and held public hearings.
5. When the House and Senate approved different versions of the bill, a joint conference committee studied the differences and recommended the Senate version (deleting the 2,200 acres from the proposed wilderness addition).
6. After a debate in the full House of Representatives, the House also accepted the conference committee–recommended bill, and Congress then passed it.
7. The president signed the legislation. More than three years had passed between initial field investigation and final passage.

The San Raphael Primitive Area review provided three important insights into the wilderness classification designation: (1) it began to define more precisely the kinds of lands Congress would accept for the NWPS; (2) it identified key players in the wilderness designation process, with grassroots environmental groups wielding much more influence than anticipated; and (3) it demonstrated an intensity of debate over final provisions of wilderness legislation that had not been anticipated.

The Gore Range–Eagles Nest Primitive Area: Enter the Courts

The study of the wilderness suitability of the Gore Range–Eagles Nest Primitive Area introduced another key player to the wilderness designation process—the courts. As the USFS reviewed the Gore Range–Eagles Nest Primitive Area on the White River National Forest in north-central Colorado, the contiguous undeveloped East Meadow Creek drainage quickly became the focus of attention. The area was rolling, timbered high country with small meadows and parklike stands of old-growth Engelmann spruce, lodgepole pine, and fir. It had high wildlife values, including an elk migration and nursing ground. Moreover, it had become a popular backpacking, horse-packing, and big game hunting area for those who lived in or began their trips from nearby Vail, a wealthy ski resort and town built in 1964.

A primitive truck road was constructed through the area in the 1950s for access to combat a bark beetle infestation. In 1962, the USFS drew up a plan for timber sales in the area and in 1964 built an access road to the edge of the area. USFS officials at the ranger district and forest levels recommended against wilderness designation of East Meadow Creek because of the "bug road," the existence of some private inholdings and unpatented mining claims, its location outside the original primitive area, and because established sawmill operators in the area depended on USFS timber (Roth 1984).

The proposed timber sale provided a dilemma. The primitive-area review process had not yet been completed. If the timber harvesting went forward, it would destroy the wilderness character of lands contiguous to a primitive area, effectively making a decision about wilderness designation—a judgment Congress reserved for itself. On the other hand, a decision against the timber sale would be seen as favoring the expansion of the wilderness system—something only Congress can do. Thus, a middle-ground decision was offered that included timber cuts in only six of the proposed fourteen timber-harvesting blocks, with a large buffer zone between the primitive area and the timber sale.

The compromise did not work. Neither environmental groups nor the forest-products industry liked the decision. Both felt it was a decision that only Congress could make. A third group, the citizens of Vail, was especially upset. They argued that their town depended on recreation dollars, that the timber sale was planned in 1962 before their town even existed, and that timber harvesting in the East Meadow Creek would significantly reduce a valuable, recreation-based economic resource.

The citizens of Vail, the Sierra Club, and several Colorado conservation organizations sought an injunction to stop the timber sale (Roth 1984). Their attorney was aware of the landmark Scenic Hudson decision of 1965 in which the judge gave local citizens' groups standing in court and eventually ruled in favor of their suit to halt the construction of a proposed hydroelectric plant. He also noted the language of the Wilderness Act that said, "Nothing herein contained shall limit the President in…recommending the addition of any contiguous area of national forest lands predominantly of wilderness value."

The district court heard the case, and the judge accepted the plaintiffs' arguments. He granted them the right to sue the government and listened to the testimony of the environmental witnesses who vouched for the wilderness character of the area and the USFS's arguments that the bug road disqualified it. On February 17, 1970, the judge ruled in favor of the plaintiffs and permanently enjoined the timber sale. On October 1, 1971, the Tenth Circuit Court upheld the ruling. The USFS continued to believe that a bad interpretation of the Wilderness Act had been made. It believed that the law did not *require* the review of every single contiguous acre of land with wilderness character by the president and Congress, although the act did not *prohibit* such consideration either. The Department of Justice, on behalf of the USFS, appealed to the Supreme Court, which refused to hear the case (Roth 1984).

Grassroots citizen groups had gained a major victory in the "precedent-setting" Gore Range–Eagles Nest Primitive Area review. The judicial decision in effect protected all undeveloped (roadless) lands contiguous to USFS primitive areas until Congress made the final decision; it demonstrated again that wilderness designation decisions would be shaped by actions beyond the USFS and the president; thus it freed the environmentalists to turn their attention to USFS lands that were not contiguous to wilderness or primitive areas, but which were still roadless.

Designating Nonreserved National Forestlands

One major block of lands potentially available for wilderness designation was not specifically mentioned at all in the Wilderness Act. These were the so-called *de facto wilderness* lands in the National Forest System, areas that are in fact wilderness in the general sense of the term—roadless and undeveloped—but that lack explicit, legal designation and thus protection as wilderness.

Although the act itself did not specifically direct the USFS to review lands other than those designated as primitive areas, the Wilderness Act did not preclude such review. *Thus, the agency decision to review all its roadless lands, in addition to lands specifically noted in the Wilderness Act (primitive areas and their contiguous roadless lands), represented a judgment of agency personnel that the act did not prevent such a review.* The agency also recognized that environmentalists would soon be demanding designation of additional wilderness lands not specified in the act. The agency wanted to seize the initiative—to act more than react and to approach wilderness review and evaluation from a multiple-use planning perspective. They would inventory and evaluate the wilderness potential of all roadless areas on the national forests.

U.S. Forest Service Inventory of Potential Wilderness Additions

Three criteria were formulated to guide USFS managers in identifying potential additions to the wilderness system: (1) *suitability*, (2) *availability*, and (3) *need*. The USFS argued that no formula for reaching these important decisions existed, but that the objective in each case was to determine the predominant public value within the meaning of the Multiple-Use, Sustained-Yield Act of 1960 (U.S. Public Law 86-517): "with consideration being given to the relative values of the various resources, and not necessarily the combination of uses that will give the greatest dollar return or the greatest unit output."

1. *Suitability was defined by minimum conditions set forth in the Wilderness Act and by conditions believed to enhance the opportunity for wilderness-dependent experiences,* that is (a) have an absence of roads or other development; (b) be at least 5,000 acres if not adjoining existing wilderness or primitive areas (smaller, undeveloped contiguous areas were also to be inventoried); (c) be free of present or foreseeable nonconforming uses or activities; (d) possess a wide range of subjective values for helping people discover freedom and spiritual renewal; (e) offer outstanding opportunities for challenge and primitive recreation like camping, ski touring, and hiking; (f) include abundant and varied wildlife; and (g) possess outstanding opportunities for formal and informal education and scientific study.

2. *Availability specified that wilderness designation must represent the highest and best use of the land over time.* The tangible and intangible values of wilderness designation had to offset the potential value of all resources that would be forgone, such as the value of timber, minerals, and nonwilderness recreation, and the costs of managing the area as wilderness.

3. *Need meant that the need for wilderness must be compared to that for other resources.* Need was to be determined by consideration of the location, size, type, and capacity of other wildernesses in the general vicinity, by local and national trends in wilderness use, and by the extent to which nonwilderness lands were available to provide dispersed recreation opportunities.

The initial inventory, in this "Roadless Area Review and Evaluation" (RARE) (subsequently called RARE I because it would later be repeated as RARE II), identified 1,449 roadless areas on the national forests with wilderness potential, containing 56 million acres. The chief of the USFS planned to use this inventory, in combination with other data, to complete a list of "study areas" to receive more formal review and evaluation. During the review process, the USFS held some 300 public meetings and received more than 50,000 written and oral comments—at the time the largest public-involvement effort ever undertaken by the federal government (Hendee 1977). However, of the 1,449 areas, only two were in the East and one was in Puerto Rico. The lack of areas in the East was a disappointment to environmentalists and quickly became a source of controversy.

Roadless Area Review and Evaluation (RARE I)

Just how to go about the review and evaluation of potential wilderness additions prompted concern. The USFS wanted to avoid fights over individual areas, or studying all individual area management options, including wilderness, as each roadless tract was considered. Fresh in mind was controversy in 1972 between the USFS and a citizen group over a roadless area near Lincoln, Montana, which resulted in the designation of the 240,000-acre Scapegoat Wilderness—the first USFS area designated under the 1964 act that had not previously been a primitive area (U.S. Public Law 92-395). The long and costly debate over this area demonstrated the need for a comprehensive review of all roadless lands.

Such a review would serve as a first approximation of which roadless lands deserved careful, detailed study as to their potential for wilderness designation and which lands appeared to be more suited for non-wilderness management. The agency also believed that a comprehensive national review and evaluation of all areas would constitute a national Environmental Impact Statement (EIS) under the recently signed National Environmental Policy Act (NEPA) of 1970 (U.S. Public Law 91-190), and thus avoid the need to write an EIS for each roadless area.

The USFS assembled an interdisciplinary team to complete the national Roadless Area Review and Evaluation of the 1,449 potential wildernesses identified in the inventory. The procedure developed and used by the team, called the Roadless Area Review and Evaluation (RARE), became the principal analytical tool used to recommend which of the inventoried areas should receive further intensive study for possible wilderness designation.

The principal objectives of the RARE selection process were:

1. To obtain as much wilderness value as possible relative to the cost and value of the forgone opportunities to produce other goods and services for society.
2. To disperse the future wilderness system as widely as possible over the United States.
3. To represent as many ecosystems as possible so that the scientific and educational purposes of wilderness preservation are best served.
4. To obtain the most wilderness value with the least relative impact on the nation's wood product output.
5. To locate some new wilderness areas closer to densely populated areas so that more people can directly enjoy their benefits.

To evaluate the undeveloped areas, the RARE process used a number of quantitative and qualitative judgments for each area, including:

1. The total gross acres of roadless area.
2. A quality index (QI). Field personnel rated each area on three factors: (a) scenic quality (S), (b) isolation and likely dispersion of visitors within the area (I), and (c) variety of wilderness experiences and activities available (V). Each of these factors was weighted and used to calculate the quality index by the formula: $QI = 4(S) + 3(I) + 3(V)$.
3. An effectiveness index (EI). To derive this measurement, total gross acres were multiplied by the quality index.
4. Total opportunity costs index. This index was composed of the sum of the following:
    a. Estimated costs for studies, establishment, operation, and maintenance.
    b. Cost, if any, of acquiring private land.
    c. Cost of replacing special-use improvements.
    d. Mineral values.
    e. Potential water development values.
    f. Timber values.

The RARE draft EIS proposed 235 areas covering 11 million acres (about one-fifth of the acreage inventoried) as new wilderness study areas (WSAs) to receive intensive study and early consideration for addition to the NWPS (U.S. Department of Agriculture, Forest Service 1973). Until the studies were completed, no management programs could be undertaken that would alter their undeveloped state. More than 7,000 letters and documents were received in response to the draft EIS (Hendee 1977).

Following this analysis, 61 new wilderness study areas were added and 22 were deleted for a net gain of 39 new areas and a revised total (in the final EIS released in October 1973) of 274 areas and 12.3 million acres, an acreage gain of about 10 percent more than in the draft EIS. Of this total, 46 areas and 4.4 million acres had already been officially committed for wilderness study before the announcement of the chief's list.

The four principal variables—the ones on which marginal cases were decided and which the chief and staff repeatedly used in making final decisions: (1) *public input*, including sentiment of involved citizens and organizations and the views of legislators and government agencies; (2) *potential wilderness quality of the roadless areas* as measured by the quality index; (3) *cost effectiveness,* reflecting the value of other resource uses forgone compared to relative wilderness values; and (4) an *overall judgment factor* resting heavily, but not entirely, on the recommendations of local, regional, and national decision makers (U.S. Department of Agriculture, Forest Service 1973). *Many criticisms were leveled at the roadless area inventory and particularly at the RARE process* (Milton 1975), including timing (only eight months elapsed between the initial inventory of new study areas and the final RARE list) and the RARE methodology—such as the subjective nature of the quality index (QI), the overwhelming influence of size of area on the effectiveness index (EI), and the validity of placing a monetary value in the opportunity cost estimate for preservation values many claimed to be nonquantifiable.

Environmentalists believed worthy roadless areas—most notably in the East—were not included among the 1,449 inventoried areas, and therefore were not adequately considered in the RARE process. In addition, the 274 areas selected for wilderness consideration constituted only 19 percent of the total number of areas inventoried and 23 percent of the acreage. Disappointment in the wilderness recommendations led the Sierra Club and others to file a suit in federal court, which was ultimately dismissed, but without prejudice, meaning it could be heard by the court again.

Both the Sierra Club and the USFS hailed the outcome of the lawsuit as a victory. The judge's decision did not prevent the conversion of roadless areas to multiple-use management; it simply stated that a proper EIS must be filed for *each* area and that in the EIS, wilderness must be considered one of the viable management options (Allin 1982).

But, regardless of who won or lost in this case, *the RARE I process had two important benefits for wilderness. First, the inventory provided a reasonably comprehensive list of remaining roadless lands on the national forests,* and this highlighted the need for management guidelines and programs for these lands to protect their wilderness potential.

Second, the RARE process—for all its flaws—systematically reviewed and weighed the relative values of a variety of uses of roadless areas, taking a regional and national perspective. The study set the stage for later, more complete, and defensible evaluations of roadless lands by the USFS, and by other agencies and environmentalists who were keen spectators of the RARE process.

Wilderness in the East: An Eastern Wilderness Act

Even as the RARE I process was reaching its culmination in the early 1970s, the debate over what to do about national forestlands in the East (generally defined as east of the 100th meridian) was heating up. Environmentalists were disappointed with the amount and kind of land that had been protected in the East under the Wilderness Act. They sought protection for areas that, while not as pristine or as vast as areas in the West, still possessed a primitive character. Most of these acres had previously been logged or even farmed and roaded, but time was rapidly healing these wounds.

The USFS held steadfastly to its purity position that only areas that *retained* primeval conditions qualified for wilderness designation and thus pressed for protection of lands in the East under an alternative wild areas system. Wild areas would be distinct from wilderness areas because they would be primarily for recreation enjoyment. But, at the same time, laws were being passed giving wilderness protection to U.S. Fish and Wildlife Service lands in the East; the National Park Service had recently proposed a 75,000-acre wilderness in the backcountry of Shenandoah National Park in Virginia—all lands that had previously been farmed and inhabited had returned to a wild state. And, in 1970, congressional delegations from West Virginia and Alabama introduced bills to designate lands under the Wilderness Act.

During 1972 and 1973, Congress, the USFS, and environmentalists debated the wild areas–versus–wilderness ideas (see Allin 1982; Hendee et al. 1991; Roth 1984) and their implications for how much land would be set aside for protection, and to what standard it would be protected. By 1973, environmental interests were together in support of adding areas in the East to the NWPS. The USFS dropped the idea of a separate wild-areas system in the East and proposed designating sixteen eastern areas as instant wilderness areas and that another thirty-seven be studied for possible wilderness designation.

Senate bill 3433 was introduced into Congress—a bill that called for nineteen instant wildernesses to be established, and forty study areas to be reviewed by the secretary of agriculture within five years according to the procedures outlined in the Wilderness Act. It contained two particularly important clauses: (1) the secretary of agriculture was given the power of condemnation when private (inholding) landowners failed to use their land in a manner compatible with wilderness and were unwilling to sell voluntarily (this was an important authority absent from the 1964 Wilderness Act, and was included because of many private inholdings in eastern national forests); and (2) all lands designated as wildernesses or as wilderness study areas (WSAs) were withdrawn from mineral entry.

The bill quickly passed the Senate and was sent to the House. House Public Lands Subcommittee chairman John Melcher (MT) required that congressional representatives whose districts included the proposed wilderness areas submit written statements of support. This represented a change from past protocol when only oral support was necessary. To the dismay of the environmental community, this resulted in reducing the number of wildernesses in the bill from 19 to 15, and study areas from 40 to 17, and was signed into law (U.S. Public Law 93-622) by President Gerald Ford January 3, 1975 (Roth 1984).

This law (popularly called the "Eastern Wilderness Act") represented a major challenge for management. It stated the desire of Congress to locate wildernesses nearer population centers, and it included areas that are generally smaller in size and show more past evidence of human use than national forestlands previously placed in the NWPS. Some claimed it was explicit repudiation of the USFS purity principle in wilderness designation and management (Allin 1982).

The Endangered American Wilderness Act

Environmentalists had criticized the RARE I process for failing to select several undeveloped areas located near population centers for wilderness study. Indeed, the USFS had eliminated such areas as Lone Peak near Salt Lake City, the Sandia Mountains near Albuquerque, and Pusch Ridge adjacent to Tucson because they were within "sight and sound" of cities. Environmental groups claimed that Nixon administration messages to Congress in 1972 and 1973, the so-called 1975 Eastern Wilderness Act, and their own constituents supported the notion that accessibility to population centers enhanced rather than detracted from wilderness values.

Sierra Club leaders decided that the time was right for an omnibus wilderness bill that would protect several areas that the RARE process had either excluded or had not recommended for future wilderness study. The bill would thus be symbolic of alleged defects in the USFS's RARE I process, would help build grassroots support for wilderness throughout the country, and would demonstrate this support to every congressional representative and senator. The environmental groups convinced Senator Frank Church (ID) and Representative Morris Udall (AZ) to introduce the Endangered American Wilderness bill in Congress in 1976. At first, congressional mail ran against the bill, but as the Sierra Club's campaign geared up, support began to build. It was fortuitous that both of the bill's sponsors were presidential candidates, and the support of presidential candidate Jimmy Carter and his election greatly increased the likelihood of passage of the bill. Passage was further promised when the new Congress was formed and chair of the House Interior Committee passed to Representative Udall.

Ultimately, under the leadership of the new Carter administration, including Rupert Cutler as the new assistant secretary of agriculture over the Forest Service (he was a former assistant executive director of The Wilderness Society), the USFS supported the bill. President Jimmy Carter signed the Endangered American Wilderness Act February 24, 1978 (U.S. Public Law 95-237), creating 1.3 million acres of wilderness in 17 new wilderness areas or additions to wilderness areas in Arizona, California, Colorado, Idaho, New Mexico, Oregon, Utah, Washington, and Wyoming. It was the largest addition to the NWPS accomplished by a single wilderness designation act to that point. In one sense it repudiated the RARE, because it designated some areas that the RARE process had either ignored in their inventory or had not even recommended for wilderness study.

RARE II—Another Look at Forest Service Roadless Areas

RARE II—the second USFS Roadless Area Review and Evaluation—was a critical influence on the wilderness designation process and, thus, on wilderness management. *By the mid-1970s, there was considerable dissatisfaction with the first RARE process because it favored too much or too little wilderness, depending on the view of the various groups.* When Jimmy Carter was elected president in 1976, controversies over the validity and reliability of the first RARE process had slowed the wilderness designation process to a crawl.

RARE II was primarily the brainchild of Rupert Cutler, then assistant secretary of agriculture overseeing the USFS. He thought RARE I was flawed. He had supported the Endangered American Wilderness Act and thereby had helped "liberalize" the USFS definition of wilderness; he wanted a study to help better define what would meet the new minimum-acceptability standard for wilderness. He was also under pressure from commodity interests and the USFS to resolve future wilderness designations.

The basic objective of RARE II was to accelerate the planning process mandated by the recently passed Forest and Rangeland Renewable Resources Planning Act (RPA) of 1974 (U.S. Public Law 93-378) and the National Forest Management Act (NFMA) of 1976 (U.S. Public Law 94-588). *RARE II called for a comprehensive study to identify roadless and undeveloped lands in the National Forest System and to determine their general uses both for wilderness and for resource management and development. The study would produce recommendations on which roadless lands should be designated wilderness, which lands should be released to uses other than wilderness, and which warranted further study for all uses, including wilderness.*

The study was to be a methodical, rational approach to planning in a national context. The goal to identify roadless areas with important nonwilderness values was equal in importance to identifying areas with wilderness potential. *The desired outcome was a balance of land uses that would best meet the nation's needs within the requirements of NEPA.* Considerable public involvement would shape recommendations, and it was Cutler's dream that the study would serve as one large, programmatic EIS; that is, the RARE II study would suffice as the EIS for all areas recommended for wilderness, nonwilderness, or further wilderness study (Roth 1984). But this, of course, was uncharted territory.

The *USFS began RARE II striving to improve on RARE I in two ways: first, to separate the identification or inventory, and evaluation of each roadless area; and second, to establish greater reliability and validity in the inventory process.* In doing this, the USFS wanted to strengthen public confidence in the RARE II process by separating the assessment from the evaluation components. The assessment component—which included area identification and inventory—had wilderness suitability as its primary criterion. *The assessment question was not whether a roadless area should or should not be wilderness, but rather whether it had attributes to meet minimum criteria for wilderness consideration. Judgments about which areas should be recommended for wilderness were left to the evaluation phase.*

The Rare II Inventory

RARE I had reviewed 1,449 areas containing 56 million acres. The roadless status of most of these areas had remained unchanged, and they became part of the RARE II inventory. The public was asked to comment and suggest additions and deletions for the RARE II study, and more than 50,000 comments were received.

Many more areas in the East were included in RARE II than had been previously studied in RARE I because somewhat less restrictive standards were used to select areas in the inventory there. This responded to the limited wilderness opportunities in the East, the need for wilderness close to people, and the faster regenerative capacity of ecosystems in the East. Areas were selected that did not exceed *more than one* of the following: one-half mile of improved road for each 1,000 acres, if road was under USFS jurisdiction; 15 percent of area in nonnatural planted vegetation; 20 percent of area with timber harvested in the past ten years; and an area could contain a few dwellings if the dwellings and access were obscured by natural features. *The final RARE II inventory list included 2,919 roadless areas and encompassed 62 million acres on national forests and national grasslands in 38 states and Puerto Rico* (U.S. Department of Agriculture, Forest Service 1979).

A variety of data were collected on the RARE II areas. Each area's renewable-resource potential was measured through a system called the Development-Opportunity Rating System (DORS). This system compared the benefits and costs of developing a roadless area for renewable multiple-use outputs, including saw timber, grazing, developed recreation, hunting, and fishing. In addition, site-specific information on potential timber, programmed harvest, and grazing was reviewed, and each area's mineral potential was assessed.

The RARE II process also sought to increase ecological diversity within the NWPS, with the goal of obtaining two or more distinct examples of each of the nation's ecosystems (U.S. Department of Agriculture, Forest Service 1979). Using the Bailey-Küchler mapping system, an approach using Bailey's (1976) ecoregion concept and Küchler's (1966) system of potential natural vegetation, the USFS identified 242 distinct ecosystems in the United States and Puerto Rico (later corrected to 261). The ecosystems were distinct in such physical environment factors as climate and soil and in biological environment factors such as vegetation (Davis 1989). The RARE II process gave preference for wilderness recommendation to those areas that contained an ecosystem not currently represented in the NWPS.

In late 1977, a USFS task force developed the RARE II Wilderness Attribute Rating System (WARS) to inventory the wilderness characteristics of each RARE II area. WARS established a procedure for identifying the attributes of wilderness as defined by the Wilderness Act and for rating their condition. The purpose of WARS was to provide a measure of the area's wilderness quality, just as DORS provided a development-opportunity rating system to measure its potential commodity and developmental values.

Table 5.1 lists the critical and supplemental attributes and their respective components that comprise the WARS. The *critical attributes used in WARS* (those mentioned in the Wilderness Act) were *natural integrity* of the area, *apparent naturalness* (amount and nature of perceptible impacts), *solitude opportunity*, and *primitive-recreation opportunity*. WARS also permitted adjustment in natural integrity and apparent naturalness ratings if area boundaries were adjusted to remove serious intrusions. Each of the four critical attributes was rated on a scale of one to seven, ranging from the outstanding presence of the attribute to its virtual absence. Composite wilderness attribute (WARS) scores were determined by adding the ratings of each of the four critical attributes. Thus, each of the four attributes was weighted equally in the overall score.

There were two supplemental area ratings: a supplementary wilderness-attribute rating and a scenic value rating. These ratings reflect section 2(c)(4) of the Wilderness Act, which indicates that an area "may also contain ecological, geological, or other features of scientific, educational, scenic, or historical value." This suggests that these features are not necessary for wilderness suitability, but that their presence in an extraordinary degree enhances wilderness values. These two variables were not part of an area's overall composite wilderness-attribute score but instead were viewed as supplemental information to help make marginal decisions or to identify areas that might be placed in other types of USFS protected-area management.

Evaluation of Rare II Areas

The inventory process of RARE II resulted in a numerical "wilderness-attribute rating" for each area ranging from four to twenty-eight, which indicated the extent to which the area contained wilderness attributes listed

Table 5.1. Wilderness Attributes (WARS) and Their Components

Wilderness Attributes	Components on Which Ratings Are Based
1. Natural integrity	Fourteen possible physical developments or human-caused impacts (e.g., roads, railroad rights-of-way, reservoirs, grazing, air pollution, etc.), scaled as to their presence, effect on natural integrity, size of area impacted, potential separability from rest of area, duration of impact if uncorrected, feasibility of correcting.
2. Apparent naturalness	Uses the same components as natural integrity, but the ratings differ.
3. Outstanding opportunity for solitude	Size of area, topographic screening, vegetative screening, distance from perimeter to core, human intrusions, scaled as to their degree of impact on opportunity for solitude.
4. Primitive-recreation opportunities	Size of area, topographic screening, vegetative screening, distance from perimeter to core, diversity, challenge, absence of facilities, scaled as to their degree of impact on primitive recreation.
5. Supplementary attributes	
a. Ecological	Presence and abundance of endangered or threatened plants and animals or other special ecological features.
b. Geological	Presence and abundance of special geological features.
c. Scenic	Ratings based on visual management system.
d. Cultural features	Presence of any cultural-historical features.

in the Wilderness Act. The question that remained was which of these areas should be recommended for wilderness designation, released for nonwilderness uses, or held for further study? For example, how far down the wilderness-attribute rating scores should the USFS go before the cutoff point between wilderness and nonwilderness was reached?

In the RARE II draft EIS published on June 15, 1978 (U.S. Department of Agriculture, Forest Service 1978), the public was asked to comment on three things: (1) what individual areas should be allocated to wilderness, nonwilderness, or further planning, and why; (2) what approaches should be used in reaching a decision on allocating the total roadless-area inventory; and (3) what decision criteria should be used in developing a proposed course of action. Response was massive: 264,093 inputs carrying 359,414 signatures, seven times as many public comments as received in RARE I (Hendee 1986; Hendee et al. 1980). Most responses expressed a preference for allocation of an individual area, but some also addressed the issue of alternative allocation strategies and final decision criteria.

Ten general alternative allocation strategies, reflecting a range of mixes of wilderness and nonwilderness options, were offered to the public. The potential physical, biological, social, and economic effects of each alternative were quantified and evaluated by the USFS to the extent feasible. Of primary concern were the benefits realized and benefits forgone (opportunity costs), if more or fewer wilderness areas were recommended. This analysis, along with public input, existing laws and regulations, identified public needs, and USFS professional judgment of decision makers, resulted in the USFS's preferred alternative. In essence, the proposal called for the allocation to nonwilderness of areas with high potential for commodity resource output, and allocation to wilderness of those with high wilderness attributes. This responded to public input that expressed favor for economic values and jobs, timber production, accessibility, high scenery and diversity in the wilderness system, and high-quality additions to the NWPS.

Recommendations from Rare II

The final RARE II EIS was published on January 4, 1979, approximately 18 months after initiation of the study (USDA FS 1979). It called for wilderness designation of 624 areas totaling 15,008,838 acres, allocation to nonwilderness of 1,981 areas totaling 36,151,558 acres, and further planning for 314 areas totaling 10,796,508 acres. The document recommended the proposed wilderness areas to the Ninety-sixth Congress for legislative action and indicated that no activities would be permitted that might alter their wilderness qualities. Areas recommended for nonwilderness would be made available to multiple-use activities on April 15, 1979. Areas listed for further planning would remain undeveloped until their status could be reviewed more thoroughly under routine USFS land- and resource-management planning processes. Until a final recommendation was made on these further planning areas, no commercial timber harvesting would be allowed, and exploration and leasing for oil, gas, and energy minerals would be permitted only under rigid stipulations.

When the RARE II results came forward, environmentalists were disappointed. Only about one-fourth of the inventory of national forest roadless areas were being recommended for wilderness, with only one-sixth of the lands allocated to further planning—meaning wilderness was still considered a possibility. And environmentalists were even more critical about the distribution of proposed wildernesses—of the approximately 15 million acres recommended for wilderness, 5 million were on the Tongass National Forest in Alaska. Furthermore, the permanent loss of roadless areas recommended for release to nonwilderness uses was imminent. The final RARE II EIS called for the release to nonwilderness uses of the roadless areas recommended for nonwilderness during the first cycle of forest plans mandated by the NFMA. These "released" areas became the center of controversy in 1979 and in the 1980s.

The Court Intervenes: California v. Bergland

In California, the general sense was that the wilderness issue should not be settled quickly, and RARE II recommendations for the state reflected this feeling. Of the approximately 6 million acres in California considered by RARE II, about 1 million were recommended for wilderness, 2.4 million for nonwilderness, and 2.6 million were placed in the further-study category. Still, there was controversy. For example, the Trinity County Board of Supervisors created a philosophically mixed committee to study the RARE II recommendations for the Shasta-Trinity National Forest. This committee recommended that 48 percent of the 449,000 roadless acres become wilderness, whereas the initial RARE II recommendations called for only about 4 percent. The

USFS informed the board of supervisors that the RARE II recommendations responded to the larger picture—to a variety of national, regional, and local issues. It could not and would not follow a county recommendation. This position did not sit well, either with the county board or with many influential congressional representatives.

The Shasta-Trinity controversy soon became a test of the entire RARE II process. Against the wishes of both The Wilderness Society and the Sierra Club, Huey Johnson, director of the California Resources Agency, sued the USFS and sought a court injunction against the release of forty-six California roadless areas to nonwilderness uses.

Johnson, in *California v. Bergland*, contended that the programmatic RARE II analysis did not meet the requirements of NEPA, and on January 8, 1980, the district court agreed. *The court enjoined the forty-six areas from development and indicated that before there could be any change in the status of these areas, an EIS for each area would have to be prepared.* In October 1982, the Ninth Circuit Court of Appeals upheld the lower court ruling (Roth 1984).

As a result, the assistant secretary of agriculture for Natural Resources and Environment in the new Reagan administration, John Crowell (former general counsel for Louisiana Pacific Timber Co.), directed the USFS to reevaluate all RARE II recommendations within the ongoing land-use planning process. His inclusion of *all* areas caused environmental groups to fidget, knowing that the further analyses might cause areas recommended *for* wilderness by RARE II to be dropped. Congressional representatives and the USFS squirmed at the thought of another time-consuming and expensive RARE study. In addition, Crowell called for implementing any activities currently planned in areas recommended for nonwilderness uses unless there was a specific court injunction against it. To many environmentalists, this seemed to violate the implications of the California court injunction.

U.S. Forest Service Wilderness Designation in the 1980s

By the early 1980s, RARE II had provided a lot of information, recommendations, and some agreement toward resolving the status of roadless areas on the national forests. A comprehensive and careful inventory of USFS roadless areas had been completed, including maps, lists of wilderness attributes, ecosystem representation, extent and content of public involvement, and descriptions of commodity trade-offs. The USFS, the Carter administration, and the Reagan administration supported wilderness designation for about 15 million acres of national forestlands and further study for 11 million acres. Just how legislation to protect these areas would come forward was unclear, and there was uncertainty over the final resolution of the areas released for non-wilderness uses. Suggestions for a third RARE study had little support. At the same time, the Crowell directive to study these lands under the normal USFS land-management planning process seemed to acknowledge that the RARE II process had failed to achieve a prompt resolution of the wilderness-allocation issue as a coordinated national evaluation of roadless lands.

However, how would the RARE II data reflecting its "objective approach" affect wilderness designation? One analysis of wilderness designation decisions in Arizona, Idaho, and Utah concluded that the RARE II information about the renewable and nonrenewable resource potential of areas was relatively unimportant in explaining the eventual designation of areas to wilderness or allocation to nonwilderness or further planning (Mohai and Verbyla 1987). Although public signatures favoring wilderness and WARS ratings were consistently among the best predictors of wilderness designation, many areas that were allocated to nonwilderness did not have high resource potential, whereas many areas with high resource development opportunity were recommended for wilderness (Mohai and Verbyla 1987). *This analysis confirmed that rational, empirical analyses based on hard data were playing only a limited role in the highly politicized environment. Politics were proving even more important than resource analysis in wilderness decisions.*

Three issues dominated wilderness allocations on the national forests in the 1980s. *First,* how to add areas to the NWPS? Should legislation be drawn up to add single RARE II areas or a few areas at a time to the system? Should one omnibus bill be passed that included all or most areas recommended by RARE II? Or, should each state with RARE II areas have its own wilderness bill? *Second,* how permanent would the "release to other uses" be for areas recommended for nonwilderness by RARE II? The timber industry and the USFS preferred permanent or *long-term release,* or *"hard release"* as it came to be called; this would permit long-term planning for management of commodity or nonwilderness recreational use of these lands. The environmental community favored *short-term release, or soft release—the released areas could continue to be considered for wilderness during Forest Service land-management planning and would thus provide opportunities in the future to gain wilderness protection for some of the released land.* A *third* question concerned the adequacy of the RARE II analysis as a programmatic EIS. Would a separate EIS have to be prepared for each area? It seemed so as the *California v. Bergland* court decision indicated that the final RARE II recommendations did not constitute an acceptable EIS for certain areas in California.

Statewide Wilderness Bills—
The Prototype of the 1980s

As a result of RARE II, politicians were much more aware of the wilderness potential of national forest roadless areas throughout the country. With the data from RARE II, they had a much firmer grasp of where existing and potential areas were, the quality and characteristics of those areas, locations where need was the greatest, and trade-offs that could be made to meet commodity and wilderness needs. Thus, they were better prepared to deflect the emotional appeals of single-interest commodity or environmental groups. They were also better prepared to look at the big picture, how trade-offs in one area might be offset in another. In addition, because there were so many areas to consider, the sheer workload of an area-by-area approach for legislation to designate wilderness was daunting. The idea of omnibus legislation to designate a lot of areas at one time was appealing, but not without problems.

The environmental community opposed the omnibus bill approach from the beginning. They had a fifteen-year history of successful influence on the wilderness designation process with a strategy largely based on grassroots organizations that influenced local forest supervisors and congressional representatives. To move the locus of decision to arbitrators of one comprehensive bill in Washington, DC, might have a devastating effect on their membership, and many favorite areas inevitably would be lost in compromise. Many would not even be considered because they were not on the RARE II list. Finally, most discussion of an omnibus bill included some sort of permanent release to commodity use of areas not recommended by RARE II for wilderness or further study. Environmental groups feared this most of all, for it would forever prevent them from seeking wilderness protection for many areas they felt still deserved wilderness consideration.

Soon, individual state congressional delegations, tired of inactivity on an omnibus RARE II bill, feeling pressure from their own constituencies, and wanting to seize the initiative, began to put forward statewide bills—bills that responded to the RARE II recommendations for their state. This approach was a logical extension of the way Congress had done business in the past, when local congressional representatives typically sponsored bills to designate wilderness areas in their district. With the more comprehensive data developed in RARE II, state delegation members could negotiate among themselves to devise a bill that was acceptable to them. Such *statewide bills rapidly became the norm in the 1980s; virtually all national forest wilderness areas added to the NWPS in the 1980s were included in state bills.*

Sufficiency-Release Language

The *California v. Bergland* decision, that an EIS must be prepared for any area whose status would change with release to nonwilderness uses, was, on the face of it, a clear victory for environmentalists. The court decision raised questions about the sufficiency of the entire RARE II process, and this could have set the stage for another national roadless area review (Allin 1982). Yet, the mood of Congress in the late 1970s and early 1980s was not in favor of further delay, study, or expense for more wilderness evaluation. Environmental groups feared a legislative backlash and perhaps a permanent release of nonwilderness RARE II lands. Representative Tom Foley (WA) had introduced legislation in late 1979 that would have mandated immediate nonwilderness management for the 36 million acres recommended for nonwilderness in RARE II and nonwilderness management by January 1, 1985, for any area in the national forests not given statutory wilderness protection by that time. This would have effectively fixed the final amount of wilderness on the national forests at the level set aside by that date (Allin 1982).

The California wilderness bill provided an opportunity for compromise. Representative Phil Burton (CA), Harold "Bizz" Johnson (CA), and John Seiberling (OH), chair of the House Public Lands Subcommittee, convened a meeting of representatives of industry, environmental groups, and the USFS to settle the issues

Fig. 5.3. Blodgett Canyon in the Bitterroot National Forest was identified during RARE II as having important wilderness values. RARE II was a second effort to break the deadlock on identifying and recommending undeveloped areas in the national forests for wilderness designation. Although the process had many criticisms and concerns, it did identify areas where important wilderness values were found. Photo courtesy Don Dodge.

of "sufficiency" and "release." The eventual agreement was to place within each statewide wilderness bill a *sufficiency clause*—a statement that said that the RARE II study was "sufficient" for NEPA purposes. This was to protect against RARE II lawsuits. Implementing this recommendation regarding release language, however, became controversial.

Soft or Hard Release

The compromise worked out by the Burton-Johnson-Seiberling meetings called for (soft) release of lands not recommended for wilderness or further study for one planning cycle (ten to fifteen years) of the Forest Service planning process required by the NFMA of 1976. When the next generation of plans was prepared, the USFS could then study any of these lands still remaining roadless for their wilderness potential. This type of release (for one planning cycle only) became known as "soft release," or later as "Colorado release" when the terminology was included in the 1980 Colorado wilderness bill. Permanent release to nonwilderness uses became known as "hard release."

The compromise began to unravel with the election of Ronald Reagan in 1980. The leaders of the National Forest Products Association had never been happy with the "soft release" compromise. With the Reagan administration in the White House, they once again lobbied vigorously for permanent release. In early 1981, Senator Samuel Hayakawa (CA), with cosponsorship by Senator James McClure (ID), the new chairman of the Energy Committee, and Senator Jesse Helms (NC), new chairman of the Agriculture Committee, introduced a bill that would not create any wilderness but listed deadlines after which RARE II roadless lands would be forever off-limits to wilderness recommendation. This bill at first appeared likely to pass, for it had the support of a popular administration and powerful members of the Republican-controlled Senate. For the next three years, environmental lobbyists worked hard to stop this and similar bills in committees.

However, as time passed, all sides were becoming impatient for a settlement. Grassroots environmental organizations could not understand why release language should hold up the addition of lands that all parties involved had agreed should be placed in the wilderness system. Many felt that by the end of the first generation of management plans the released lands would have been developed anyway, so the negotiations between hard and soft release proponents were largely moot.

With the approach of December 31, 1983—which would mark the end of new mineral exploration and leasing in wilderness under terms of the Wilderness Act—mining companies mounted a campaign to get the Department of the Interior to lift its administrative moratorium on the processing of claims in existing wilderness. Such mineral exploration and leasing were still legal within wilderness, but because of costs, negative public opinion, and strict USFS regulations, no leases had been issued. Secretary of the Interior James Watt broke from this tradition and considered mineral leasing requests in the Bob Marshall Wilderness of Montana and Washakie Wilderness in Wyoming and actually approved a mineral lease in the Capitan Wilderness of New Mexico. The public outcry to this action was loud and reverberated through Congress. Congressional committees passed resolutions condemning the proposed mining activity and ordering it to be stopped, and bills were introduced in Congress to permanently ban wilderness mineral leasing.

The environmental groups acted quickly to take advantage of the revived interest in wilderness from the mining threat. They pointed out that even the Bob Marshall Wilderness, crown jewel of the wilderness system, was vulnerable to the commodity interest groups. Then they mounted a campaign for passage of a big western state wilderness bill with soft release language—a bill that would act as a precedent for other western states.

The tide for the first time in years was running in the environmentalists' favor. In 1983, the timber industry was in a severe recession, had fewer lobbyists in Washington, and was preoccupied with obtaining relief from money-losing federal timber contracts. A final break occurred when Senator Mark Hatfield (OR) began to move a wilderness bill forward in the summer of 1983. His pace quickened in 1984 when the Oregon Natural Resources Council, a grassroots environmental group, filed a lawsuit to stop development of any RARE II lands in Oregon based on Judge Lawrence Karlton's 1980 *California v. Bergland* decision. Such a lawsuit, if successful, would have a serious negative impact on the troubled timber industry as certainty of future access to released lands would be further in doubt. The Hatfield bill, when introduced in Congress in 1984, contained "Colorado" or "soft" release language.

Meanwhile, the House had passed New Hampshire, Vermont, Wisconsin, and North Carolina wilderness bills that included the soft release language. USFS officials expressed concern that an amendment to a forest management plan might, with soft release terminology, be considered a revised plan. Thus, a second-generation plan might occur much before the anticipated ten to fifteen years.

The final negotiations of the release issue were feverish, with USFS chief R. Max Peterson working closely with Senate and House leaders. *The agreement reached was that future state bills would have no fixed time period for release but would simply refer to the forest plans typically lasting ten to fifteen years. Further, during this period, the USFS was not categorically prohibited from managing released land to preserve its wilderness character; it just was not required to do so.* Commodity interests had gained in the final solution by tightening up the definition of a plan revision; this was to ensure that a plan would really run the expected ten to fifteen years, during which released lands might be considered for commodity use, typically timber harvest. In addition, the agreement stipulated that roadless areas of less than 5,000 acres, and those areas examined in old USFS unit plans before the RARE II study and recommended for nonwilderness use, were released during the first generation of forest plans.

Agreement on the release issue opened the door to rapid wilderness designation. In the summer of 1984, eighteen national forest wilderness bills were passed, adding 6.6 million acres to the NWPS. This was the largest addition to wilderness on the national forests in a single session of Congress since passage of the Wilderness Act. *About 13.6 million acres were released to USFS multiple-use management based on their respective land-management plans under the NFMA.*

Designating Department of the Interior Roadless Lands: NPS, FWS

The Wilderness Act of 1964 clearly described the National Park Service (NPS)–and U.S. Fish and Wildlife Service (FWS)–administered lands subject to wilderness consideration. All roadless lands in excess of 5,000 acres and all roadless islands were to be studied. For those areas considered suitable for wilderness designation following agency study, wilderness proposals were to be developed for submission to Congress through the Department of the Interior (USDI), the Office of Management and Budget, and finally the president.

Operating with these guidelines, the Department of the Interior did not have to decide what lands to consider or how to propose them for wilderness designation. However, it did have to struggle with the problem of establishing wilderness criteria: among the available candidate areas, which ones should be recommended for wilderness designation? On what basis? It was eight years after the act passed before official selection criteria and guidelines were established in the Interior Department. In the meantime, a beginning had to be made.

Soon after passage of the act, the Interior Department estimated that 22.5 million acres of NPS land and 24.1 million acres of FWS holdings were subject to wilderness study—and this area grew as new national parks and refuges were created. Eventually, *during the ten-year period following the passage of the Wilderness Act, the NPS identified for wilderness review 63 areas covering more than 28 million acres, while FWS identified 113 areas totaling 29 million acres* (U.S. Congress 1973).

The review was very slow to reach Congress. Only one FWS area, the Great Swamp National Wildlife Refuge in New Jersey, was submitted and designated as wilderness by 1969. Only two NPS areas, Craters of the Moon National Monument in Idaho, with 43,000 acres, and Petrified Forest National Park in Arizona, with 50,000 acres (U.S. Public Law 91-504), were added to the NWPS by 1970, six years after passage of the Wilderness Act.

Just why the NPS wilderness study process moved so slowly is not completely clear. Allin (1982) contends that the NPS had little enthusiasm for wilderness

Fig. 5.4. The Wilderness Act directed the secretary of the interior to review all roadless areas of 5,000 acres or larger in the National Park System and to recommend which should become wilderness. The Hidden Lake area of Glacier National Park, Montana, was recommended to be designated wilderness, and it has since been under consideration for decades. Photo courtesy Richard Behan.

because such designation would limit its management discretion and prohibit future development for its primary clientele, touring motorists. In addition, early on, the NPS seemed to take a very strict stance on wilderness resource requirements. For example, Director George Hartzog made it clear that national parklands administratively designated within the NPS as Class III, "natural environment areas," would not meet the criteria for wilderness but instead would be viewed as transition or buffer zones between developed and wilderness lands.

Environmentalists pointed out that the sights and sounds of civilization that might be found in these threshold zones were now acceptable in wilderness, and that the NPS really was simply trying to retain the option of development for these zones. Subsequent statements by Director Hartzog confirmed these suspicions (Allin 1982). Moreover, the NPS developed a policy to exclude any private inholdings within a roadless area from a wilderness recommendation unless federal acquisition of the area was ensured. Areas with other nonconforming but legally permitted uses like water-monitoring devices and motorboat patrol zones were also eliminated from wilderness consideration.

The absence of an explicit wilderness designation procedure for USDI lands likely explains some of the delay and difficulty encountered by the NPS and FWS. Traditionally, the NPS had zoned roadless and undeveloped tracts in the individual parks' master plans. For example, much of Yellowstone's 2.2 million acres is zoned "roadless" in the park's master plan. The problem the USDI faced in meeting the obligations of the Wilderness Act was much the same as that faced by the USFS in its roadless area reviews—determining specific boundaries of areas for wilderness study.

Roadless Area Review Criteria: NPS and FWS

In June 1972, more than eight years after passage of the Wilderness Act, assistant secretary of the interior for Fish, Wildlife and Parks Nathaniel Reed issued a memo to the directors of the NPS and the FWS defining *criteria to be followed in determining an area's suitability for wilderness designation.* These criteria suggest a response to public opinion in being less pure in their requirements for wilderness recommendations. In particular, the memo specified that *roadless areas should not be excluded from wilderness consideration by the existence of the following conditions:*

1. Existing or proposed use of tools, equipment, structures, or facilities if these are necessary for the health and safety of wilderness users or the protection of the resource.
2. Prior rights or privileges such as grazing or limited commercial services that are proper for realizing recreational or other wilderness purposes.
3. The existence of unimproved roads, structures, installations, or utility lines that can and should be removed upon designation as wilderness.
4. Use of the area for research unless it requires permanent structures or facilities not needed for management.

In addition, the following nonconforming uses or facilities should not, by themselves, exclude an area from a wilderness recommendation so long as subsequent legislation designating the area specifically allows them: small structures like primitive boat docks or shelters, controlled burning, natural-appearing lakes created by water-development projects, hydrologic devices for monitoring water resources, underground utilities such as gas pipelines and transmission lines, and minimum tools and equipment necessary to maintain the water developments and utilities.

Special provisions should also be included in the wilderness legislative proposals for those areas that are currently surrounded by nonqualifying federal lands, but which in the foreseeable future will qualify, and thus permit the secretary of the interior to designate such lands as wilderness. Finally, no portion of a national park or wildlife refuge should be designated as wilderness unless the wilderness designation is compatible with the purposes for which the park or refuge was established.

Wilderness Study Process: NPS and FWS

The NPS and FWS began their wilderness review with the roadless lands that were under their jurisdiction in 1964. To the extent possible, wilderness study proceeded concurrently with the preparation of park or refuge General Management Plans (GMPs). The process typically had four phases: *preliminary study phase, public meetings, final recommendations, and legislative phase*. The *preliminary phase* included the collection and analysis of basic and field data, preliminary recommendations of wilderness boundaries (if wilderness was to be recommended), preparation of first draft of the preliminary wilderness-suitability study and draft EIS, and briefings on the potential draft to be proposed for higher agency and departmental officials and the appropriate congressional delegation(s).

Changes in the proposal sometimes resulted from agency, departmental, and Council on Environmental Quality (CEQ) review. The *public meetings* provided citizens and organizations an opportunity to comment on the draft proposal. Appropriate notices were listed in advance of public meetings and an informational packet on the issues and concerns regarding the draft wilderness proposal was prepared for distribution. Comments were then recorded and analyzed to aid in subsequent decision making.

The *final recommendations phase* involved preparation of the final wilderness recommendation with the benefit of public input, review, and approval of a final report by the agency at agency regional and Washington offices; submittal of wilderness recommendation and final EIS to the assistant secretary for Fish, Wildlife and Parks for approval; and then to the secretary of the interior. The approved proposal with its final EIS then went to the Office of Management and Budget (OMB) and to the CEQ for review and approval. The approved proposal is submitted by the secretary to the president.

During the final or *legislative phase*, the president submits his recommendation regarding the proposal to Congress; typically, the Washington office of the agency prepares the draft legislation and the transmittal letters from the secretary to the president and from the president to each branch of Congress. A bill or bills including the actual or modified wilderness proposal are then introduced in Congress by a House or Senate member. Then debate, negotiation, compromise, and eventual passage by committees and the full Congress would result in wilderness designation. This resembles closely the process for wilderness bills including USFS lands.

On the face of it, this appears reasonably straightforward, and NPS and FWS wilderness proposals typically have not been as controversial as those on national forests. The prospect of commodity development does not exist on these lands so environmental groups first focused their attention on USFS and BLM lands. Still, there have been controversies regarding Assistant Secretary Reed's guidelines, agency implementation of those guidelines, and frequent reluctance by state congressional delegations to carry a legislative proposal forward. As an example, The Wilderness Society (1974–1975) argued that the Interior Department's wilderness guidelines confuse the stringent management criteria in section 4 of the Wilderness Act with the flexible entry criteria in section 2. As a result, they contended Interior Department officials interpret wilderness designation as a decision to cease virtually all management activity unless specific authorization is given in the wilderness legislation for an area. Based on this interpretation, wilderness designation would be rejected for many areas because it would end management activities needed to accomplish objectives of the legislation originally establishing the park or refuge. The Wilderness Society suggested that a management activity need only meet a minimum necessity test. Administrators need to demonstrate only that a management activity is the minimum necessary for proper administration of the area both for the purposes for which the park or refuge was originally established *and* as wilderness. If the management activity meets this test, it does not constitute a sufficient reason to disqualify an area for designation as wilderness. Of course minimum necessary management would be different than park management as usual.

The relationship between the original legislation establishing a park or refuge and subsequent wilderness designation within some of these areas has also created problems, particularly on national wildlife refuges and game ranges. Many of the proposals submitted by the FWS following review of their roadless areas recommended against wilderness designation. For example, about 29,000 acres of the 40,000-acre Red Rock Lakes National Wildlife Refuge in Montana was recommended for wilderness (U.S. Department of the Interior, U.S. Fish and Wildlife Service n.d.). The remaining 11,000 acres were judged not suitable because of existing and planned developments to manage waterfowl, especially the trumpeter swan. Field studies of the 45,000-acre Laguna Atascosa National Wildlife Refuge in Texas revealed that, although a portion of the area did qualify as wilderness, such designation would conflict with the primary objective of the refuge, which is to provide habitat for waterfowl, so wilderness designation was not recommended (U.S. Department of the Interior, U.S. Fish and Wildlife Service 1970). Congress concurred in substance with both recommendations, designating 32,350 acres of the Red Rocks Refuge as wilderness and concurring with the FWS recommendation against wilderness designation at Laguna Atascosa.

Conflicts between legislative objectives exist despite the declaration in section 4(a) of the Wilderness Act that wilderness designation is "within and supplemental to" the purposes for which national forests, national parks, wildlife refuges, and game ranges were established. Where a legitimate conflict exists between the goals of wilderness and those of the park or refuge enabling legislation, the organic legislation generally takes precedence. For example, where wilderness designation might restrict necessary wildlife management practices on a game range, wilderness designation is limited, rejected, or made with special recognition of the potential intrusion.

The Department of the Interior also recognizes what are called "potential wilderness additions." This was originally conceived as a designation for areas *where clearly nonconforming uses were present (e.g., structures), but which would qualify for wilderness designation once the nonconforming use was removed*. In omnibus legislation passed in late 1976 (U.S. Public Law 94-567), eight national park units, containing 53,506 acres, were identified as potential wilderness additions. Most of these areas were so labeled because of grazing, and it is not altogether clear why they could not have been included in the wilderness because grazing is allowed in wilderness when it is a preexisting right. However, under this designation, the secretary of the interior has authority to establish them as designated wilderness when the nonconforming use ceases. For example, in California, 1,800 acres near Point Reyes were added to the Philip Burton Wilderness after removal of power lines subsequent to the adjoining area's earlier designation (U.S. Public Law 94-544).

By 1974, the end of the ten-year study period, the NPS had completed review and submitted wilderness proposals on fifty-six areas. Yet, controversy, an apparent lack of broad-based public support, lack of a concerted push by environmental groups, or reluctant state congressional delegations has caused millions of acres of NPS-recommended wilderness to languish before Congress. For example, Yellowstone National Park has a 2-million-acre wilderness proposal submitted to Congress in 1972, which is still awaiting congressional action. Although wilderness designation has been implemented in Washington state's Olympic National Park (U.S. Public Law 100-668) and Death Valley National Park in California (U.S. Public Law 103-433), proposals including many millions of acres still await congressional action in more than a dozen parks, including large tracts in Great Smoky Mountains, Grand Canyon, and Yellowstone National Parks. Although some may argue that wilderness protection is ensured under national park status, knowledgeable persons argue that important wilderness values are being lost in these backcountry areas without legal designation as wilderness (Crumbo 1996; Watson 1996).

Olympic National Park Wilderness Review— A Case Study

A summary of the wilderness study in Olympic National Park will help illustrate the review procedure prescribed by the Wilderness Act for roadless areas 5,000 acres or larger in the national parks, wildlife refuges, and game ranges.

It has been the policy of the NPS to prepare a GMP (General Management Plan, formerly called a master plan) for each area of the NPS to provide the framework for its overall management, public use, and physical development. To help determine future use, a land-classification plan based on an area's resources is included in the GMP. In classifying land, the NPS uses six land classes developed by the Outdoor Recreation Resources Review Commission (ORRRC 1962). These six classes, modified for applicability to the NPS, include the following:

Class I—High-density recreation areas
Class II—General recreation areas

Class III—Natural environment areas
Class IV—Outstanding natural areas
Class V—Primitive areas
Class VI—Historic and cultural areas

Roadless areas within a park, typically Class V, but sometimes Class III, IV, and VI lands, have usually been managed and preserved in a roadless, natural condition before review and proposal for designation as wilderness. The review of parklands to develop a recommendation for wilderness designation has usually been carried out in conjunction with a major public review and updating of the park GMP, as was the case in Olympic National Park. Along with the review and updating of the park GMP, roadless portions of the 870,200-acre park were formally reviewed in a wilderness study.

During 1972 and early 1973, with the help of an NPS planning team composed of members from the local park staff, the regional office, and the planning division in the NPS Denver Service Center, a new GMP and a preliminary wilderness proposal were prepared for Olympic National Park. The initial draft called for 93 percent of the park to be designated as wilderness. The park contained four roadless units of 5,000 acres or larger, and most of the acreage in three of them was proposed for wilderness. One unit included the majority of the park (816,650 acres); the other two units were elongated strips of land along the Pacific Ocean comprising 13,160 and 5,080 acres, respectively. A fourth unit, the 26,800-acre Mount Angeles Roadless Area, was not proposed for wilderness in order to retain long-range options for alternate-access development. Also excluded from the preliminary proposal were two 20-acre enclaves in the roadless interior intended for permanent hostels to furnish food and lodging for hikers visiting the park in the future.

In August 1973, this preliminary proposal (along with the new management plan and accompanying draft EIS) was released to other agencies and the public. Five hundred people attended public meetings on the GMP and wilderness proposal held in October and November. Altogether, nearly 6,000 persons, agencies, and organizations responded to the NPS's preliminary wilderness proposal. From November 1973 through early spring 1974, the NPS analyzed and evaluated those responses and prepared a final wilderness recommendation and an altered GMP. The final wilderness recommendation eliminated the 20-acre enclaves intended for hostels and recommended them for wilderness, along with most of the previously excluded Mount Angeles unit. With other minor boundary adjustments, the final wilderness recommendation included 862,139 acres, about 96 percent of the park.

This recommendation was submitted to the Washington office in June 1974 and was transmitted from the NPS to the Department of the Interior and the Office of Management and Budget. A recommendation for wilderness designation of the four units in Olympic National Park, along with fifteen other parks and wildlife refuges, was included in a White House communication to Congress.

Early in 1975, Senate bill 1091 called for designating an Olympic National Park Wilderness identical to that proposed in the NPS recommendation. About the same time, a bill designating a slightly larger wilderness (H.R. 5823) was introduced in the House. However, no congressional action was taken on either version.

By 1987, the wilderness proposal for the Olympic National Park backcountry still had not been acted on by Congress. This decade of inactivity demonstrates a point that we and others (e.g., Mohai and Verbyla 1987) *have made throughout this chapter: the wilderness designation process is very political, and much more than a rational planning process carried out by agency bureaucrats.* In this case, the NPS met its legal requirement for wilderness review and submitted a proposal through the White House to Congress. Why would such a delay then occur? Allin (1982) notes that the NPS has often failed to aggressively seek support for wilderness designation, in part because the agency believed its backcountry areas were already being adequately protected by its management practices and because it did not want its management discretion limited. In addition, the environmental community in the late 1970s and 1980s was focusing its attention and wilderness lobbying efforts at preserving the most threatened areas—the multiple-use lands of the USFS (Allin 1982).

Late in 1987, the political and environmental climate in Washington, DC, had evolved enough to nudge the NPS wilderness proposal forward. In 1984, Congress had passed a Washington State Wilderness Act for national forestlands emanating from the RARE II process (U.S. Public Law 98-339); thus, three years later, the state's environmental lobby was free to turn its attention to the NPS. More important, Senator Dan Evans (WA), former governor and a frequent visitor and longtime supporter of the national park backcountry in the state, decided to retire at the end of the 100th Congress. He wanted to leave a legacy of wilderness protection for the wild country of the national parks he loved. In November 1987, he asked the NPS to help him draft a wilderness proposal for Olympic, Mount Rainier, and North Cascades National Parks.

With this encouragement, the NPS worked quickly. On March 15, 1988, the Washington state delegation introduced the Washington Park Wilderness Act of 1988 (S. 2165), proposing 871,730 acres in Olympic National Park and large acreages in Mount Rainier and the North Cascades National Park complex for wilderness, and a companion House bill (H.R. 4146). The Olympic wilderness proposals were similar to the NPS recommendation of 862,139 acres back in 1974. The increase in size of about 9,000 acres was made possible by park boundary adjustments and land acquisition during the intervening fourteen years, most notably the acquisition of land at Lake Ozette and the extension of park ownership of the Pacific beach to the lowest tide line.

Hearings were held on the Senate and House bills during the summer of 1988, and there was general support for wilderness designation from affected agencies, the city of Seattle, the environmental community, and many public interest groups. The environmental lobby did, however, recommend a slightly larger wilderness proposal by calling for inclusion of the roadless portions of Olympic National Park's strip of land along the Queets River, the surface area of Lake Ozette, the north shore of Lake Crescent, and a small, isolated stretch of wild beach along the Pacific Ocean. Compromises were quickly reached. Congress accepted the Queets River and wild beach proposal but rejected the Lake Crescent addition because of inholdings and the Lake Ozette surface due to motorized access across the lake to private dwellings. On November 16, 1988, during the waning days of the 100th Congress, Congress passed U.S. Public Law 100-668, creating 876,669 acres of Olympic National Park as wilderness. The long process of review, study, debate, and wilderness designation of a national park backcountry was complete.

The Bureau of Land Management

The Wilderness Act of 1964 gave the USFS, the NPS, and the FWS the authority to study, protect, and manage "legal" wilderness on lands under their jurisdiction. The act failed to give comparable authority to the BLM, even though this agency managed 473 million acres of federal land, far more than the other agencies. These were the "forgotten lands," vast acreages of the public domain that were never disposed of through the various federal programs to place land in private ownership or never designated for special purposes like national parks or national forests. About 174 million acres were in the contiguous United States, virtually all in the West. Although grazing, mining, and primitive low-quality roads had impacted many of these lands, their wilderness values seemed immense. Estimates in the early 1970s of BLM lands with wilderness potential in the lower forty-eight states ranged from about 50 million acres to as high as 90 million acres. These lands, along with the BLM lands in Alaska, represented by far the largest block of potential additions to the NWPS.

The Federal Land Policy and Management Act (FLPMA) of 1976 (U.S Public Law 94-579) gave the BLM an organic act, a long-needed legislative statement of direction and purpose for the agency. *Two sections of FLPMA called on the BLM to study, make recommendations to the president, and ultimately manage legally designated wilderness. Section 603 instructed the agency to review its roadless lands* of 5,000 acres or more and its roadless islands and make recommendations regarding their suitability or nonsuitability for wilderness designation within fifteen years after passage of FLPMA. The same was to be done for the BLM administratively classified primitive or natural areas that existed before November 1, 1975, with recommendations regarding these areas to go to the president by June 1, 1980. In ten western states, fifty-three BLM primitive or natural areas existed that, with their contiguous roadless lands, amounted to 1.2 million acres. They had been established by administrative designation, possessed qualities comparable to those of legally defined wilderness, and were being managed for wilderness values.

Section 202 of the FLPMA mandated the development, maintenance, and revision of land-use plans, which among other things *gave priority to the designation and protection of areas of critical environmental concern*. Under this authority, the BLM studied—and continues to study—and develop recommendations for wilderness designation. These lands also include roadless parcels less than 5,000 acres in size (and thus not qualifying for study under FLPMA section 603), but which lie adjacent to existing or proposed wilderness or national park areas.

The final legal directive assigning wilderness study initiatives to the BLM was the Alaska National Interest Lands Conservation Act (ANILCA) of 1980 (U.S. Public Law 96-487). This law withdrew BLM roadless lands in Alaska from wilderness review but stated that the secretary of the interior at personal discretion could periodically study and make wilderness recommendations to Congress.

James Watt, President Ronald Reagan's secretary of the interior, issued a memorandum on March 12, 1981, directing that no further wilderness inventory and review be done in Alaska, and this memorandum remains in effect. The one exception to this is a wilderness review of the Central Arctic Management Area specifically mandated by the ANILCA legislation.

BLM Roadless Area Review Process

The BLM's wilderness review process has three major phases: inventory, study, and reporting.

Inventory

The inventory phase, begun in 1978, sought to identify lands with wilderness attributes—lands that met the minimum standards for wilderness as defined in the Wilderness Act. The BLM attempted to accomplish this task quickly and anticipated spending more time in the study and evaluation of areas that made the "first cut." The task was enormous, and the BLM selected a two-step approach. During the first step, called the *initial inventory* and conducted in 1978 and 1979, areas identified by the BLM staff and the public as *not* having wilderness attributes were eliminated from further review. This was done largely by using resource data and maps available in the BLM district offices, and this process reduced the total acreage under consideration to about 50 million acres.

The remaining lands then became the focus of the *intensive inventory*. BLM resource professionals conducted on-the-ground inspections of each area to assess the presence or absence of wilderness characteristics. The public was invited to participate in the field inspections and to review the agency's assessment procedures and recommendations for each area. Public response was considerable: more than 10,000 comments were received from across the country. The BLM intensive roadless area inventory was essentially completed by the end of 1980, and areas found to possess the basic characteristics of wilderness were classified as wilderness study

Fig. 5.5. A large part of Olympic National Park was designated wilderness in 1988, fourteen years after the initial National Park Service proposal. Photo courtesy National Park Service.

areas (WSAs). A total of 861 areas containing about 25 million acres were classified as WSAs.

At the time the BLM was doing its inventory, the focus of environmental groups was on the USFS's RARE II study. As a result, there were fewer challenges to the BLM process than might otherwise have been the case. Still, the 25 million acres of wilderness study areas were far less than the environmental community expected, and there were accusations that the BLM was applying wilderness-attribute standards too strictly. For example, debate raged over what constituted a road and whether tire tracks across the desert should disqualify an area from wilderness consideration. The displeasure intensified in late 1982 and early 1983 when Secretary of the Interior James Watt removed approximately 85 WSAs, or about 1.5 million acres, from further wilderness study. These WSAs either contained *split-estate lands* (where the federal government owned the surface and someone else owned the subsurface estate), were areas of less than 5,000 acres, or were areas that contained more than 5,000 acres but received a high wilderness rating because of adjacent existing or proposed park or wilderness lands. The Sierra Club, having learned valuable lessons from its legal action against the USFS, sued. The U.S. District Court for the Eastern District of California, in the case of *Sierra Club v. Watt*, ruled in favor of the Sierra Club. The court, among other statements, noted that Watt had failed to use good judgment, and it restored the deleted areas to their wilderness-study-area status. *Once again, the courts supported citizen wilderness views opposing an agency.*

Wilderness Study

In February 1982, the BLM published its "Wilderness Study Policy: Policies, Criterion and Guidelines for Conducting Wilderness Studies on Public Lands" (USDI, Bureau of Land Management 1982). This document specified the BLM's national wilderness program policy, listed and described planning criteria to use in the evaluation of the WSAs, prescribed quality standards for analysis and documentation, and recommended a framework for integrating the wilderness study process into the agency's multiple-resource planning system. Through this process, the BLM hoped to complete a high-quality study of its WSAs, including full public involvement, by the end of 1987. This would allow four years for mineral review of those areas recommended for wilderness and for making reports to the president. However, when additional areas had to be reviewed because of the *Sierra Club v. Watt* decision, this schedule proved to be overly optimistic. In addition, the secretary of the interior, following the lead of the secretary of agriculture and the USFS, decided to submit all recommendations on a *statewide* basis, rather than on an individual-area basis. So a revised schedule called for completion of wilderness studies for submission starting in 1989 and continuing on into 1991, with all studies being submitted before the 1991 due date.

The overall goal of the WSA studies was to determine whether each area, or a portion thereof, was more suitable for wilderness or for other uses. The BLM's wilderness program policy provided the philosophical underpinnings for the review. Basically, this philosophy is that wilderness is one of the several multiple uses of public lands, and therefore the study should be more than just a review of the wilderness potential of the WSAs. The inventory phase had already established that these areas possessed wilderness attributes. The study then should assess the values, resources, and existing and potential uses of the WSAs for wilderness *and* for other commodity or recreational outputs. In addition, any recommendation for a WSA should reflect the mix of resource uses and benefits currently provided in the area. The BLM recognizes that wilderness itself provides multiple uses, offering benefits from primitive recreational use, wildlife habitat protection, watershed protection, and protection of cultural and archeological resources. The agency also views wilderness as a long-term allocation and management commitment. *Thus, the capability of the agency to protect wilderness values in the long run was to be a primary consideration in recommendations for a given WSA.*

From the foregoing BLM philosophy of wilderness, *two criteria* were used to direct and justify all BLM wilderness recommendations—both for those judged suitable and those nonsuitable. *The first criterion was the amount and quality of an area's wilderness values and the second was manageability.*

1. Quality of Wilderness Values

The first criterion acknowledged that although all WSAs had at least minimum wilderness attributes, some were of higher quality than others. To evaluate quality, four components or kinds of wilderness values were measured and recorded: (1) quality of the area's mandatory wilderness characteristics; (2) special features, or quality of the area's optional wilderness characteristics; (3) multiple-resource benefits—the benefits to other multiple resource values and uses that wilderness designation could ensure; and (4) diversity in the NWPS.

The Wilderness Act of 1964 provides a list of mandatory attributes to be evaluated—size, naturalness, and opportunities for solitude or primitive recreation. Objective information was gathered to fully document these and other attributes and criteria for each WSA. This information was summarized in matrix classifications and other descriptive methods, but arbitrary and subjective weighting or ranking was not done. The BLM had watched and learned from the Forest Service problems with weighting and ranking in the RARE I study.

The BLM recognized that a WSA needed only to "generally appear" natural and have human imprints "substantially unnoticeable" for it to receive a wilderness "suitable" recommendation. Yet, areas with fewer such impacts seemed to have higher wilderness potential than those with more. Thus, during the field study, a general description, the location, the size of the area, and the overall influence of human imprints in the WSA were documented. In addition, whether the imprint was the result of activities occurring inside or outside of the area was noted, and the potential for separating the imprinted portions from the rest of the area and recommending the remainder for wilderness designation was recorded. The BLM had also learned that sights and sounds of humans outside a WSA do not necessarily preempt its wilderness designation, thereby acknowledging the will of Congress expressed in House Report 95-546 accompanying the previously discussed Endangered American Wilderness Act of 1978 (U.S. Public Law 95-237).

The BLM used a slightly different interpretation of the Wilderness Acts call for the provision of outstanding opportunities for solitude or primitive and unconfined recreation than did the USFS. It defined these two attributes—solitude and primitive recreation—differently and directed that *if an area had one or the other (solitude or primitive recreation), it qualified for wilderness. If an area had both attributes, it would possess greater wilderness potential. Solitude* was defined as the state of being alone or remote from habitation; isolation; or in a lonely, unfrequented, or secluded place. Environmental features can enhance opportunities for solitude. Size and configuration of area, topographic screening, vegetative screening, presence of outside sights and sounds that enhance or diminish wilderness values, and opportunities for users to find a secluded spot were measured and recorded for each study area. *Primitive* and unconfined type of recreation was defined as activities characteristic of dispersed, undeveloped recreation, which do not require facilities or motorized equipment.

The Wilderness Act indicates that a wilderness area "may also contain ecological, geological, or other features of scientific, educational, scenic or historical value." The BLM recognized that these attributes are supplemental and not mandatory for an area to be recommended for wilderness. Still, the agency recognized such attributes enhance wilderness values and, in marginal areas, may affect the final recommendation. Thus, the study guidelines provide for the recording of the abundance of these attributes and the importance of each to the overall wilderness value of the area.

The BLM responded to previous direction and actions affecting wilderness designation by giving special consideration to multiple-resource values and uses, such as wildlife habitat and archeological sites, that would be threatened if the area were not given the protective status of wilderness. Moreover, those WSAs were given greater wilderness value if, protected in a natural state, they had a high likelihood of fostering the return of animals and fishes that were formerly found in the area, or of improving water quality or visual resources within and beyond the boundaries of the WSA.

The final wilderness value component considered was *diversity*—diversity in ecosystems and landforms represented in the NWPS, balancing the distribution of wilderness areas throughout the nation, and, to the extent possible, giving greater consideration to areas located closer to population centers. These initiatives reflect the previous actions of Congress in passing laws to locate wilderness areas closer to the people (Eastern Wilderness Act; Endangered Wilderness Act) and a long-standing goal of the federal wilderness resource agencies to preserve examples of all of the nation's ecosystems. The BLM used the Bailey-Küchler system to map ecosystems, as the USFS did in its RARE II process, and if an area under study contributed to the diversity of natural systems and features in the NWPS, it received additional wilderness value. On the other hand, if its ecosystem or landform duplicated that of existing or proposed areas, it was given a lower ranking. The same was true for geographical distribution; areas located in regions with fewer wilderness areas received higher consideration. Finally, areas located within a five-hour drive of population centers of 100,000 or more were given special consideration.

2. Manageability

In its evaluation of WSAs, the BLM carefully considered whether it could manage and protect wilderness attributes over the long term. *If current or projected resource conditions or uses would make wilderness protection of an area difficult or impossible, its wilderness potential was believed to be substantially lessened.* Five factors or assumptions were used in making judgments about manageability. *First, management problems* anticipated from nonconforming but allowable uses were considered. Such uses included *mining and grazing.* If it were reasonably certain that the nature or intensity of such uses would substantially destroy wilderness values, the study recommendation would be for nonwilderness allocation. *Second, the status of the land* in the WSA was considered. If someone other than the BLM owned the surface or subsurface rights to the land, the BLM believed that wilderness management would be a problem. Thus, if the WSA contained *private inholdings, state lands,* or valid existing *mineral leases or claims,* the wilderness rating of the area was lowered. The same was true if a private individual or corporation owned *subsurface mineral rights. Third,* the BLM evaluated the impact of the guaranteed *right of access to any private inholding.* Such access generally included the existence or further development of roads through the WSA, over which the BLM had little control. Fourth, the agency adopted a policy of *no buffer zones around a WSA*, a policy that reflected the view of Congress when it passed the Endangered American Wilderness Act. Thus, an area would stand alone, and appropriate use and development of commodity resources could occur up to the boundaries of a wilderness area. If such existing or proposed uses would make the management of the wilderness infeasible, the wilderness potential of the area was lowered. Fifth, the BLM acknowledged that there would be *no change in air quality recommendations* because of a wilderness recommendation. If existing or proposed uses of land adjacent to a WSA caused or might cause air quality of the area to drop below Class II status, the area would less likely be recommended as wilderness. *These BLM manageability judgment criteria indicate that the BLM, like the USFS before it, subscribed to purity in wilderness management and recognized an integral link between wilderness designation decisions and ease and quality of management.* The USFS experience, as we have seen, suggests that the BLM's purity position on wilderness designation might be challenged by public opinion and environmental groups.

3. Quality Standards for Wilderness Study

In addition to the wilderness values and manageability criteria, *the BLM published six standards to be used in analysis and documentation of WSAs.* The purposes of the standards were fourfold: to ensure consistency in analysis and reporting across the WSAs, to ensure consideration of wilderness value within the agency's multiple-use mandate, to place the wilderness review within the agency's land-management planning process, and to ensure that resource analyses and recommendations in study reports meet the requirements of necessary EIS or environmental assessments (EA).

The *first standard* required *careful study of the WSA's energy and mineral values,* with a view of reducing the nation's dependence on foreign resources vital to our economy and security. Thus, the presence or potential presence of vital metals or fuels in a WSA figured prominently in its wilderness recommendation. Even after a study recommendation was made favoring wilderness designation, a formal evaluation of the mineral values contained in the WSA would be made by the U.S. Geological Survey/Bureau of Mines before the study report was forwarded to the president.

Standards two and three called for *consideration and reporting on the range of alternative uses for all or part of the WSA,* followed by a *review of the impact on other resource values if the area was recommended for wilderness, or upon wilderness values if the area was not recommended for wilderness.* In making these analyses, BLM planners reviewed the existence of wilderness and commodity resources within the WSA, and outside the WSA on public or private lands but within the same region. Consideration was given to the impact of land-use recommendations on local and regional economies and whether use and development of nonwilderness resources in the area are compatible with management of the area as wilderness.

Standard four described the *public involvement process* to be used in the analyses, evaluations, and recommendations regarding the WSAs. This process, which includes the general public and state and local governments, seeks to identify issues that the public believes should be considered with respect to a particular WSA, and any values and resources in the WSA that would augment BLM's current information base. It is also intended to tap public opinions on whether an area is suitable for wilderness designation or more suitable for other resource uses. Wilderness recommendations would not be based exclusively on a vote-counting majority-rule system, but BLM district managers and state directors would consider public input along with the WSA's multiple resources and social and economic values and uses.

The fifth standard required that consideration be given to *any adverse or favorable social and economic effects that designation of wilderness areas would have on local areas.* This is a standard planning requirement of all BLM unit analyses. Finally, FLPMA and BLM planning regulations required that wilderness study teams document the extent to which any recommendation is *consistent with fully approved resource-related plans of state and local governments.* If a state or local government informs the BLM that all or part of a WSA report is inconsistent with that government's policies, the BLM will respond to this comment, explaining how the inconsistency was resolved and why. If the current state or local plan is generally consistent with the BLM policies and programs, every effort is to be made to mitigate the impacts of inconsistencies caused by a wilderness study report.

Reporting Wilderness or Nonwilderness Recommendations

The wilderness study process ends with the BLM state director's preliminary wilderness recommendation for WSAs in that director's state. The wilderness reporting process then involves the roles of the BLM national director, the secretary of the interior, and the president in acting on the state director's preliminary recommendation. A somewhat different process is used for those areas found *suitable* by the BLM state director for wilderness versus those the director decides are unsuitable. For those areas with a positive wilderness recommendation, a formal mineral survey is requested by the U.S. Geological Survey/Bureau of Mines. On completion of the survey, the state director then reviews his or her initially positive wilderness recommendation in light of the mineral survey results. If no change is called for, the state director submits the preliminary wilderness recommendation along with the mineral report to the BLM director for final acceptance. With the BLM director's concurrence, the recommendation is then submitted to the assistant secretary of interior for Land and Water Resources, then to the secretary of the interior, and ultimately to the president.

However, if the findings of the mineral survey suggest that the preliminary wilderness recommendation is inappropriate (because it would mean a loss of important mineral potential), then the state director returns the proposal to the appropriate district manager for a revised recommendation. The new recommendation with appropriate supporting information then proceeds up through the agency and departmental decision hierarchy. If the BLM national director, on the preliminary recommendation of the state director, also recommends against wilderness, the recommendation proceeds directly to the assistant secretary. Then with concurrence of the assistant secretary, the "unsuitable" proposal moves on to the secretary of the interior and then to the president.

The BLM had been very active during the 1990s in studying and reporting recommendations on millions of acres of roadless lands under their management. Millions of acres appear to have high wilderness value, but many of these lands also contain valuable nonwilderness resources. As of March 2008, 7.7 million acres of BLM land have been designated as wilderness.

Over the next decade, we anticipate much more attention by environmental groups, commodity groups, and the general public to BLM wilderness recommendations, and even on some areas not recommended by the agency. There will be certain controversy, debate, and compromise. The final outcome, yet to be determined, could add up to 25 million acres of wilderness on BLM lands and strongly influence the character and quality of the NWPS.

Fig. 5.6. The Cottonwood Point Wilderness in Arizona, managed by the Bureau of Land Management, is desert, an ecosystem largely missing in units designated wilderness in the 1964 Wilderness Act. The Bureau of Land Management was not included in the 1964 Wilderness Act, but subsequently the 1976 Federal Land Policy and Management Act (FLPMA) directed the BLM to identify areas for consideration as wilderness. BLM wilderness areas substantially increased representation of desert ecosystems in the 1990s. Photo courtesy Tom Folks.

Alaska—A Special Case?

About 90 percent of Alaska is de facto wilderness and the epitome of the last frontier. This vast area (375 million acres) is worth special attention because it contains such rich natural resources, including wilderness. It is also a study in special problems related to the disposition of public lands to Alaskan natives and state and federal land-management agencies.

Disposition of the Alaskan Public Domain

Since its acquisition from Russia in 1867, most of Alaska has remained in federal ownership. However, with the impending passage of the Alaskan Statehood Act in 1958, pressures for the transfer of much of this land to the state to facilitate economic development began to grow. Congress responded by granting the state a generous package of land rights. The state was given full title to the submerged lands of the continental shelf, estimated at 35 to 45 million acres. It was allowed to select 104 million acres from the federal domain along with full title to all mineral rights on these state-selected lands. Moreover, Alaska was to receive 90 percent of the revenues from mineral leasing on all those lands remaining under federal jurisdiction, twice the percentage other states received (Allin 1982).

A major omission in the disposal of lands to the state was any mention of claims by Alaskan natives. A 1961 Supreme Court decision ruled that native claims against selections made by the state could continue, but it also ruled that because the federal government had clearly given the state the right to select, such selections could continue. The potential legal nightmare such rulings created was headed off by the decision of then Secretary of the Interior Stewart Udall to freeze all further land transactions until such time that Congress could settle the native claims issue. Udall's temporary freeze was replaced by Public Land Order Number 4582 halting the selection process until the end of 1970. Alaska governor Walter J. Hickel, arguing that the state's economic future was stymied because 95 percent of Alaska remained in federal ownership, attempted to have the Land Order set aside, but was unsuccessful (Allin 1982).

The issue was further confounded in 1968 with the discovery of vast oil reserves on Alaska's North Slope. To capitalize on these vast reserves, it became immediately apparent that a pipeline from the oil fields to an ice-free port on Alaska's southern coast would need to be constructed. The pipeline, estimated to be 800 miles long, would cross federal land subject to the Public Land Order. The pipeline also represented an intrusion on wilderness comparable to the construction of the railroads across the western United States a century earlier (Allin 1982). A legal resolution of the native claims was badly needed.

After extensive debate in Congress, the Alaska Native Claims Settlement Act (ANCSA) was passed in 1971 (U.S. Public Law 92-203). ANCSA terminated the land freeze imposed by the Land Order. The secretary of the interior was directed to withdraw up to 80 million acres suitable for addition to the four existing conservation systems; that is, national forests, national parks, fish and wildlife refuges, and wild and scenic rivers (these were the so-called [d][2] lands, referring to the section of ANCSA where the instructions for withdrawal were contained). This was to be completed within nine months and lands withdrawn by the secretary would generally not be available for selection by the state or by the natives. The legislation also contained language that allowed the secretary to withdraw an unspecified amount of land "to insure that the public interest in these lands is properly protected" (Allin 1982, 217) (these were referred to as the [d][1] lands). This led some conservationists to recommend that as much as 50 million acres be set aside in addition to the 80 million acres of (d)(2) lands specifically identified.

ANCSA also authorized the transfer of 40 million acres to the ownership of native villages or newly formed native corporations. Each Alaskan village was permitted to withdraw 23,040 acres (36 square miles). In addition, each of the twelve native corporations was allowed to withdraw an amount of land prescribed by formula in the law.

In late 1972, Interior Secretary Rogers C. B. Morton announced his plans for Alaskan land withdrawal. More than 240 million acres were involved; a little more than 80 million acres were withdrawn for addition to existing conservation systems, including nearly 19 million acres in three new national forests, 32 million acres in new national parks, and 32 million acres in new national wildlife refuges. In addition, twenty new units in the National Wild and Scenic River System were recommended. Moreover, 47 million acres were withdrawn under the (d)(1) authority to protect the public interest in these lands. Also, 4.5 million acres were tied up in pipeline corridors mandated by ANCSA and another 112 million acres were withdrawn as a pool from which the natives could select their 40 million acres.

Secretary Morton's recommendations drew fire from almost every quarter. The state was alarmed by the large withdrawal (47 million acres) under the (d)(1) provisions. Wilderness groups were opposed to the recommendation to establish three new national forests in Alaska; they contended that commodity exploitation under multiple-use management would continue on BLM lands remaining after selections were completed and therefore no new national forestlands were needed. They were also concerned that although the USFS had committed itself to the designation of WSAs on these new forests, there was no guarantee of large-scale wilderness protection for them (Allin 1982).

Therefore, despite ANSCA's legislation regarding disposition of the public domain in Alaska, there still remained sharp disagreements over the future management direction of the land and resources in the state. The continuing dispute triggered yet another effort to arrive at a legislative solution to the dilemma. Beginning in 1973 and extending until 1980, probably the most significant congressional and public debate over U.S. natural-resource allocations raged. On the one side, the State of Alaska and its allies argued for the assertion of state rights and more opportunity for economic development. On the other side, conservation organizations argued that the last great opportunity to preserve a major portion of the nation's wild heritage was at hand and that we could not afford to lose it.

Given the large stakes, it is not surprising that Congress found it difficult to reach a satisfactory resolution over the future of the Alaskan lands. Conservationists were highly organized. The heart of this group was the Alaska Coalition, a diverse collection of state and national organizations ranging from the Sierra Club and The Wilderness Society to the United Auto Workers and National Council of Senior Citizens. The State of Alaska led the opposition, along with local developmental and commodity interests.

The inability of Congress to reach a satisfactory compromise on the issue eventually led to an unusual series of events. In 1978, Secretary of the Interior Cecil Andrus announced that the Carter administration was losing patience over the lack of progress and that it was exploring ways in which it might take action on its own behalf to solve the problem. When Congress failed to respond with any positive proposal, the Carter administration made good on its promise. In late 1978, Secretary Andrus withdrew 110 million acres for three years under the authority of the FLPMA. Two weeks later, President Jimmy Carter invoked the Antiquities Act of 1906 to create seventeen new national monuments covering 56 million acres of the 110 million withdrawn by Andrus. In one stroke, President Carter had more than doubled the size of the NPS. In addition, under presidential direction, Secretary Andrus began the process of creating 40 million acres of new national wildlife refuges in Alaska, and Secretary of Agriculture Bob Bergland withdrew 11 million acres of existing national forests to prevent mineral entry and state selection (Allin 1982). The executive branch had seized the initiative from the legislative branch in a dramatic fashion.

However, a legislative solution to the future of resource management in Alaska was still required. After two more years of congressional debate, it became apparent that some form of compromise had to be reached. The issue came to a head with the national elections in 1980. Not only was the Carter administration voted out of office, but a number of key congressional supporters of conservation legislation in Alaska were also defeated. Consequently, environmental groups agreed to accept legislation that, despite some provisions that they did not favor, was nevertheless acceptable. On December 2, 1980, in the closing days of his administration, President Carter signed the Alaska National Interest Lands Conservation Act into law. The new law had profound implications for the future of wilderness.

The Alaska National Interest Lands Conservation Act (ANILCA)

The ANILCA (U.S. Public Law 96-487) set aside 104.3 million acres of national parks, wildlife refuges, wilderness areas, and other conservation units in Alaska. Of these acres, approximately 56 million were placed in the NWPS. Wilderness acreage by agency was 32.4 million for the NPS, 18.6 million for the FWS, and 5.4 million for the USFS. *ANILCA stated that designated wilderness shall generally be managed according to the provisions of the Wilderness Act of 1964, but with several exceptions to recognize unique conditions in Alaska, minimize impacts on established uses, and permit some economic expansion.* Exceptions such as providing public cabins recognized the unusual severity of Alaska weather and the danger of being attacked by bears; provisions made for *subsistence hunting and gathering* recognized the special relationship between rural Alaska residents and the wild resources on which they depend. *ANILCA did state that its special provisions applied only to Alaska wilderness;* they are not to be viewed as precedents to alter wilderness management and future wilderness legislation in the lower forty-eight states (The Wilderness Society 2000).

Further discussion of the management implications of ANILCA is included in Part II of this chapter. A case example for the Stikine-LeConte Wilderness illustrates application of ANILCA to a USFS-managed wilderness in southeast Alaska (U.S. Department of Agriculture, Forest Service 2005; Hummel 2007).

ANILCA upheld the sufficiency of the RARE II EIS for Alaska. However, it did specifically require an investigation of a 2.1-million-acre tract called the Nellie Juan–College Fiord WSA in the Chugach National Forest. In addition, the Chugach National Forest was to proceed with recommendations regarding the wilderness suitability of fourteen areas identified during RARE II for further study. The final EIS prepared by the forest recommended 1.7 million acres (29 percent

Fig. 5.7. The Alaska National Interest Lands Conservation Act (ANILCA) passed in 1980 and ensured wilderness in Alaska. Certain uses, such as cabins and floatplanes shown here in Admiralty Island National Monument Wilderness, were permitted so visitors could cope with adverse weather and bears. Photo courtesy U.S. Forest Service.

of the total forest area) for wilderness. No further study areas are located on the Tongass National Forest and 30 percent of the forest is presently designated wilderness. ANILCA also specifies that unless Congress expressly authorizes it, the USFS shall not conduct any further statewide roadless review and evaluation, thereby precluding the possibility of a RARE III in Alaska. Yet twenty-eight years after passage of ANILCA the Tongass National Forest Management Plan continues to be appealed by environmental groups for its lack of adequate consideration of wilderness potential.

Part I Summary

The Wilderness Act of 1964 established an instant National Wilderness Preservation System, called for the review of remaining roadless lands, and described a general procedure for adding additional areas to the system but gave only broad guidelines on the kinds of lands most suitable for additional wilderness designation. Part I of this chapter traces the processes of the U.S. Forest Service, National Park Service, U.S. Fish and Wildlife Service, and Bureau of Land Management used to review their roadless lands to identify and evaluate the lands most suitable for wilderness designation. Although tedious, this material provides a basis for understanding how today's situation and directions for wilderness designation and management evolved. There were, and continue to be, stresses and strains as the four federal resource agencies with wilderness jurisdiction have responded to evolving definitions and cultural meanings of wilderness. Congress, the courts, and the public have been involved in the process more than ever anticipated.

Important lessons have been learned. Although the wilderness designation process requires a sound technical basis to identify areas and inventory their wilderness attributes and other values, it is largely a political process. Thus, the American public has been the ultimate judge of the disposition of roadless lands. Environmental groups learned this early on, and they have had a far greater influence on wilderness designation than initially expected.

Agency planners and policy makers have struggled to understand and institutionalize the wilderness ideal, to integrate wilderness within their traditional resource programs, and to retain a position of leadership regarding wilderness. There have been and continue to be occasional problems, controversies, and even impasses. When this occurs, the courts increasingly have been the final arbitrator. Gradually, the agencies have found a more even and defensible course. They have done this by taking a comprehensive planning approach, integrating wilderness into legal planning processes, recognizing that wilderness study involves two very different issues (assessment and evaluation), anchoring assessment in wilderness attributes defined in law, keeping in touch with public values when making evaluative judgments, and working closely with individual members of Congress and congressional committees who make the final designation decisions.

The most difficult task facing the wilderness resource agencies remains keeping in touch with the slowly but inevitably evolving cultural definition of wilderness in the United States and applying it to the management of the remaining roadless areas so these values are protected. Such sensitivity seems necessary when recommending which lands should be wilderness, when integrating wilderness and the other multiple uses of public lands, and ultimately when making individual wilderness management decisions. *We have seen the drift of public opinion away from rigidly applied, pure wilderness standards in wilderness designation—and thus ultimately in wilderness management.* In the final analysis, this evolving process has led to, and is leading to, diversity in the NWPS. This, in turn, can lead to broader political support and a resource capable of providing for a variety of human needs.

Part II: Management Implications in Wilderness Designation Acts

Part II reviews all the wilderness legislation that has been passed to date and focuses on the management implications of several benchmark laws and other wilderness designation laws that contain management direction. Wilderness designation laws are acts of legislation passed by Congress to add areas to the NWPS, and some of them contain management direction.

During the forty-three years since passage of the Wilderness Act, from 1964 through 2007, 170 additional wilderness laws were passed by Congress and approved by the president, primarily to add areas to the NWPS (wilderness designation laws). Through all this wilderness legislation, Congress has designated more than 107.4 million acres of wilderness. This wilderness designation legislation has generally reaffirmed the management direction in the Wilderness Act of 1964, but some laws have provided additional interpretation, clarification, or special direction for wilderness management, as well as providing for wilderness study, adjusting wilderness boundaries, or changing a wilderness name. The same law may also provide for other public lands purposes not directly related to wilderness.

Appendix B lists all 171 wilderness laws including the Wilderness Act, and the www.wild.org website provides abstracts of these laws. Because many of these laws have no name in their official title, and are merely named as an "Act of a particular date," we have added in parentheses a descriptive title of what the act does, for example (Colorado Wilderness) Act of December 22, 1980.

For our purposes, we define a wilderness law as any federal statute that designates new wilderness, adds land to and/or deletes land from existing wilderness (including boundary revisions), or provides for wilderness study. The term omnibus *is specifically used to signify a law designating more than one wilderness.* In several cases, wilderness designation, study, or management provisions are secondary purposes for a law, but those wilderness provisions are what we cover here in Part II.

Because of legislation, *the National Wilderness Preservation System has grown from the fifty-four areas and 9.1 million acres established by the Wilderness Act in 1964 to more than 700 wilderness units totaling 107.4 million acres.* Wilderness has been a topic consistently dealt with by Congress since the first wilderness bill was introduced in 1956.

Benchmark Wilderness Laws

Seven benchmark wilderness laws have set standards for wilderness designation and management, and/or have interpreted or clarified the wilderness concept or provided management direction. These laws are important milestones in wilderness legislation; we call them *benchmark* wilderness laws, beginning with the Wilderness Act (Browning et al. 1988).

Wilderness Act of 1964

This act defined wilderness for the purposes of establishing and managing the National Wilderness Preservation System, designated 9.1 million acres of national forestlands as wilderness, and directed wilderness study of national forest primitive areas and roadless areas within the national park and national wildlife refuge systems. A key element of the act was that the agency's future wilderness study and review include public involvement in the form of public notices and hearings. This was a relatively new requirement at the time and began the process of widespread public involvement in wilderness designation decisions that intensified during the Forest Service's roadless area reviews. Public involvement continues today in the national forest planning process and all natural-resource decisions. In addition, the law specified that only Congress had the authority to designate additional wilderness areas in the NWPS.

The Wilderness Act provided management direction by prohibiting certain uses and allowing others. Prohibited were permanent and temporary roads, most commercial enterprises, motorized equipment and mechanical transport, landing of aircraft (where not an established use prior to designation), and structures and installations—with exceptions for necessary administrative purposes and to protect preexisting private rights.

Special provisions of the Wilderness Act allowed for mining on valid claims and mineral development on leases established before December 31, 1983, with mineral prospecting and surveys permitted to provide information on mineral resources; water development projects with presidential approval; livestock grazing; fire, disease, and insect control; and aircraft landings and motorboat use where already established. Also allowed were certain commercial uses deemed compatible with the wilderness concept, such as outfitting and guiding. See The Wilderness Society handbook (2000) and Chapter 4 for additional explanation of the Wilderness Act.

Eastern Wilderness Act of 1975

U.S. Public Law 93-622, the so-called Eastern Wilderness Act of 1975, added sixteen wildernesses in the East, where only four wildernesses on national forests previously existed. The act also designated seventeen "wilderness study areas" to be managed to preserve their wilderness qualities until their suitability for wilderness designation could be evaluated and until Congress had sufficient time and opportunity to designate the selected areas as wilderness.

While debating proposed wilderness bills, Congress addressed the issue of whether lands in the East should be managed in a "wild-areas system" under different standards from the National Wilderness Preservation System (see Part I of this chapter). The argument for a separate system was based mostly on the premise that roadless areas in the East had been severely modified by previous human use and consequently did not qualify for wilderness designation under criteria of the Wilderness Act. This argument failed, and Congress eventually determined that designated roadless areas in the East should be included and managed as part of the National Wilderness Preservation System (U.S. Senate Report 1975, 93-803). Thus, it is correct to refer to such designated areas as "wilderness in the East," and not as "Eastern wilderness." *The legislation is also referred to as the so-called eastern Wilderness Act because it does not have a formal title and does not establish a separate system of eastern wilderness areas.*

This act also authorized the secretary of agriculture to use condemnation to acquire private lands in the seventeen wildernesses designated by the Eastern Wilderness Act. Condemnation authority had not been provided in the Wilderness Act, but wilderness areas designated in the eastern U.S. contained many private inholdings. Congress believed that condemnation authority would help ensure that private owners used these lands in a manner compatible with the wilderness concept, and would provide a means of acquisition if they did not.

Federal Land Policy and Management Act of 1976 (FLPMA)

The Federal Land Policy and Management Act of 1976, or FLPMA (U.S. Public Law 94-579), provided for wilderness study and designation of lands managed by the Bureau of Land Management, a provision not included in the Wilderness Act. This law addressed many aspects of the management of these public lands and directed that roadless areas having wilderness characteristics be inventoried within fifteen years to determine their suitability for wilderness designation. Suitable areas could be recommended to Congress for wilderness designation under the provisions of the Wilderness Act of 1964. The BLM thus joined the Forest Service, National Park Service, and U.S. Fish and Wildlife Service as a partner in wilderness stewardship.

The BLM's wilderness study areas were to be managed to preserve their wilderness characteristics, but existing uses such as mining, mineral leasing, and grazing were permitted to continue subject to regulations set by the secretary of the interior. Once designated as wilderness, the management provisions of the Wilderness Act pertaining to wildernesses on the national forest would generally apply.

Endangered American Wilderness Act of 1978

U.S. Public Law 95-237 was passed in 1978, partly in response to perceived shortcomings of RARE I, the Forest Service's first "post–Wilderness Act roadless area review and evaluation," of remaining national forest roadless areas to identify those suitable for wilderness study or designation. The Endangered Wilderness Act added sixteen areas to the NWPS, primarily including areas either excluded from the RARE I inventory of roadless lands or not recommended for wilderness study or designation after their review. Thus the name, "endangered wilderness." The Forest Service's "purity" requirements for wilderness designation, including its "sights and sounds doctrine" that wilderness should be out of sight and sound of civilization, came under intense congressional scrutiny during committee hearings. The "sights and sounds doctrine" was deemed contrary to Congress's desire to establish wildernesses near large cities (U.S. House of Representatives 1977). Much of the congressional debate focused on RARE I's criteria for recommending only 274 wilderness study areas totaling about 12 million acres from the 56 million acres inventoried in 1,449 roadless areas. Environmentalists claimed that much more of the national forest roadless land qualified for wilderness designation. By passing the Endangered Wilderness Act, Congress further *established that areas previously influenced by people should not be precluded from consideration for wilderness designation, nor should roadless areas near major cities,* as they could provide much-needed, primitive recreation for the nearby population. In such areas, boundaries were even drawn to provide adequate trailheads and facilities for the large number of wilderness visitors that were anticipated. Congressional committees addressed the interpretation of the Wilderness Act concerning certain uses, activities, and management, and endorsed the Forest Service's plan to conduct RARE II, a second comprehensive Roadless Area Review and Evaluation (Weaver and Cutler 1977). Generally, these Congressional committees supported a less stringent view of designation and management criteria than were being applied by the Forest Service at the time.

Alaska National Interest Lands Conservation Act of 1980 (ANILCA)

The Alaska National Interest Lands Conservation Act, or ANILCA (U.S. Public Law 96-487), was passed in 1980 after many years of debate *and provided management direction for the large tracts of federal land in Alaska.* ANILCA was the first act to pass both bodies of Congress that contained RARE II sufficiency-release language (Gorte 1987). The law declared that Congress had performed its own evaluation of RARE II roadless areas in Alaska and had decided to designate some lands as wilderness, to release some lands to nonwilderness multiple uses, and to hold other areas for wilderness study or further planning. *The law added more than 56 million acres to the NWPS, most of it in units of the National Park System and National Wildlife Refuge System. Millions of additional acres were authorized for further wilderness study* within the national parks and wildlife refuges and the Chugach National Forest, while special guidelines were included for wilderness recommendations of public lands in Alaska managed by the BLM.

Congressional committee hearings closely examined the unique natural environments in Alaska, exemplified by the state's vast size and predominately undeveloped condition (U.S. House of Representatives 1979). Subsequently, *Congress provided special provisions in ANILCA for wildernesses in Alaska* to allow certain motorized use and access, along with maintaining existing wilderness cabins and establishing new cabins where administratively authorized. Provisions for subsistence uses of natural resources applicable to other wildlands in Alaska were also included for wilderness, for establishment and maintenance of structures for aquaculture purposes and temporary construction and use of facilities for hunting and fishing, and for modification of existing timber sales contracts that applied to certain newly designated wildernesses on national forestland. The special provisions for wilderness under ANILCA were an attempt to compromise between the wilderness definition and the nonconforming uses (e.g., motorized access) and cultural heritage of the landscape in Alaska (e.g., subsistence activities); however, the legislation left wilderness managers with a complex stewardship situation to navigate (Tanner 2004; U.S. Department of Agriculture, Forest Service 2005; Hummel 2007).

Colorado Wilderness Act of 1980

U.S. Public Law 96-560 was important because Congress referred to the 1977 House of Representatives Committee Report 96-617 for explicit management direction for livestock grazing. This management direction had far-reaching effects, because the committee report required that livestock grazing in all national forest wildernesses should be managed according to the report's management provisions that were offered as interpretation of the Wilderness Act grazing provisions (Wilkinson and Anderson 1987). *The law also prohibited establishing buffer zones around wildernesses; directed a review of fire, disease, and insect control measures and policies in Colorado's wildernesses; and included RARE II sufficiency-release language for roadless areas in Colorado.* The Colorado Wilderness Act, by including reference to the accompanying committee report on management direction, thus brought a degree of closure to congressional debate over certain wilderness management policy and its application that had begun with consideration of the Endangered Wilderness Act several years earlier.

California Desert Protection Act of 1994 (CDPA)

The California Desert Protection Act (U.S. Public Law 103-433), passed in 1994, designated sixty-nine new BLM wilderness areas encompassing 7.6 million acres. It is a milestone for several reasons, including that it increased the total acreage of the NWPS to more than 100 million acres, far exceeding the early expectations of wilderness proponents, and occurred during the thirtieth-anniversary year of the Wilderness Act. The

Fig. 5.8. The California Desert Protection Act (1994) added more than seventy areas and millions of acres to the National Wilderness Preservation System. The Bureau of Land Management administers Kingston Peak (above) in the Kingston Range. Photo courtesy Jay Swetech.

CDPA more than doubled the number of wilderness areas managed by the BLM and increased BLM's wilderness acreage by 70 percent, thus greatly expanding that agency's wilderness responsibilities.

This act transferred more than 3 million acres of land from BLM jurisdiction to NPS jurisdiction for expansion of Death Valley National Park, Joshua Tree National Park, and creation of the Mojave National Preserve. The three NPS units contain 3.9 million acres of designated wilderness. The enlargement was to give all contiguous federal lands of national park caliber full statutory recognition as national parklands.

The California Desert Protection Act continued a trend of recent, prior wilderness legislation, including the Nevada Wilderness Act (U.S. Public Law 101-195) and the Arizona Desert Wilderness Act (U.S. Public Law 101-628), to not restrict or preclude low-level overflights of military aircraft or designate new, special airspace. The airspace above these areas was seen as critical for training, research, and development of the U.S. armed forces, and there is a lack of alternative sites. The act stipulates military overflights are not incompatible with protection and management of the natural, environmental, and cultural resource values of designated wilderness in the California desert. Another key provision was that the act provided for administrative access by mechanized means to all sixty-nine wilderness areas designated, including vehicle access for law enforcement agencies and for wildlife management and restoration by appropriate state agencies. See Watson and Brink (1996) for additional explanation of the California Desert Protection Act.

Management Direction and Special Provisions in Wilderness Legislation

In the Wilderness Act, Congress provided a legal foundation for wilderness management, recognizing that wilderness requires management to protect its wilderness character and values that are defined in the act. As the NWPS evolved, Congress has refused to amend the Wilderness Act of 1964 to clarify management direction or for any other reason, except for a minor amendment made in the Boundary Waters Canoe Area Wilderness Act of 1978 to provide new management direction in that area. Mostly, Congress has clarified management direction or made special provisions for particular areas through legislation and committee reports. For example, the Colorado Wilderness Act (U.S. Public Law 96-560) further defined grazing management through House committee report (U.S. House of Representatives 1979), but the act and the accompanying committee report expressly stated that the provisions did not amend the Wilderness Act.

Legislative Management Direction for Specific Issues

Management direction as defined here includes those provisions in wilderness laws that affirm, interpret, or modify the management direction of the Wilderness Act, for application to a specific wilderness or group of wildernesses in the case of omnibus laws. Following is a review of management direction included in wilderness legislation subsequent to the Wilderness Act of 1964, updating and expanding work first published by Browning et al. (1988), and included in the second and third editions of this book. The laws described under each activity are usually illustrative, but a complete, chronological list of wilderness laws, identifying those with certain management direction and special provisions, is included in Appendix B. An abstract of each law describing its management direction or special provisions can be found at wilder.ng

Mining

Mining in wilderness is an extremely complex legal issue and is a topic in numerous legal books and articles (Coggins and Wilkinson 1987; Loop 1986; Matthews et al. 1985; Wilkinson and Anderson 1987). Following is a general overview of mining provisions in wilderness designation laws, but readers must be aware that other laws and agency authority and regulations complicate the issue.

Mining and mineral leasing are most often addressed in national forest and BLM wildernesses because national parks and wildlife refuges were normally withdrawn from such activities when established. Mineral exploration on national forestlands is even more complex in the East, where subsurface rights are mostly privately owned and the Forest Service must estimate the impacts of exploration and potential mining in designated wildernesses as individual cases arise (U.S. Senate 1984f, Report 98-614). The U.S. Government Accounting Office (1984) reported that in future wilderness legislation, Congress may need to consider the acquisition of private mineral rights in existing wildernesses in the East or permit mining in these areas, and the Forest Service may need to assess and inform Congress of the potential acquisition costs for private mineral rights of potential wildernesses in the East.

The Wilderness Act permitted mineral prospecting and surveys to continue in national forest wilderness if "compatible" and "consistent" with protection of the wilderness resource, to provide information on the mineral resources in such areas. The law provided for mining and mineral leasing on national forest wilderness until December 31, 1983, but mining on valid claims and mineral development on leases established before that date could also continue or begin at a future date (Gorte 1988). Mining on valid claims, and mineral development on leases existing at the time of wilderness designation, may also occur on those wildernesses established after 1983.

Any mining activities in national forest wilderness today are subject to certain requirements and regulations set by the secretary of agriculture to define a valid claim and to protect wilderness characteristics (Loop 1986; Matthews et al. 1985; The Wilderness Society 2000; Wilkinson and Anderson 1987). FLPMA (U.S. Public Law 94-579) directed that mining be administered in BLM wilderness in generally the same manner as in national forest wilderness. Congress typically excludes from wilderness designation those national forest and BLM areas with high mineral potential, but exceptions have been made in some wilderness laws, as the following examples indicate. These exceptions and special provisions were usually compromises that allowed wilderness designation of an area to proceed despite conflict over its mineral potential or mining activity.

The Boundary Waters Canoe Area Wilderness Act established a BWCA mining protection area that prohibited mining of federally owned minerals in the wilderness and on adjacent nonwilderness lands. This act also set certain restrictions on mining of nonfederally owned minerals within the mining protection area. The Central Idaho Wilderness Act of 1980 (U.S. Public Law 96-312) established a special mining-management zone in the Frank Church–River of No Return Wilderness to allow for cobalt exploration and mining. In establishing that wilderness, Congress elected to try to minimize the adverse environmental effects of mining while also protecting critical bighorn sheep habitat rather than excluding the land in question from the wilderness.

Coal deposits were a major concern in the Cranberry Wilderness of West Virginia, and provisions were made in the West Virginia Wilderness Act of 1983 (U.S. Public Law 97-466) to acquire all nonfederally owned mineral interests and to permit exploration activities and drilling in the wilderness to determine the value of nonfederally owned minerals, subject to guidelines set by the secretary of agriculture. Congress also provided special provisions for phosphate leasing and mining in the Florida Wilderness Act of 1984 (U.S. Public Law 98-430), with procedural steps for presidential and congressional approval of the need to mine phosphate. A Senate committee report for the Florida Wilderness Act expressed hope that future restoration technology would be better able to reduce environmental deterioration caused by phosphate mining (U.S. Senate 1984d, Report 98-580). In the Texas Wilderness Act of 1984 (U.S. Public Law 98-574), Congress designated the Indian Mounds Wilderness, despite active oil and gas drilling, thus leaving the Forest Service to mitigate the damage to wilderness qualities caused by such activities (Evans 1986).

Motorized Use

The Wilderness Act provided for private and state government inholders to maintain their existing access rights and to permit minimal motorized use for necessary administrative purposes. The California Desert Protection Act (U.S. Public Law 103-433) makes clear provision for motorized access for administrative purposes, including law enforcement and wildlife management by appropriate state agencies (Watson and Brink 1996). Wilderness designation laws and accompanying committee reports have also provided specific qualifications and instances where the use of planes and helicopters, motorboats, snowmobiles, and other types of motor vehicles will be allowed. For example, motorboat use is specifically allowed for recreational purposes by the Okefenokee Swamp Act (U.S. Public Law 93-429; U.S. House of Representatives 1974), for private access in one wilderness by the Florida Wilderness Act (U.S. Public Law 98-430), and for both purposes by the BWCA Wilderness Act (U.S. Public Law 95-495); the BWCA Wilderness Act also permits snowmobile use for grooming ski trails near resorts and for access to two remote locations in Canada, and the landing of aircraft where previously established is affirmed by specific wording in the Central Idaho Wilderness Act (U.S. Public Law 96-312; Meyer 1999). Helicopter use is permitted to service vault toilets in certain wildernesses by the Endangered Wilderness Act (U.S. Public Law 95-237) and the Utah Wilderness Act (U.S. Public Law 98-428), and other forms of motorized use are mentioned in the Arizona Wilderness Act (U.S. Public Law 98-406), California Wilderness Act (U.S. Public Law 98-425), Utah Wilderness Act (U.S. Public Law 98-428), and Wyoming Wilderness Act (U.S. Public Law 98-550). ANILCA (U.S. Public Law 96-487) provided for a wide array of motorized use, including snowmobiles, motorboats, and aircraft.

Committee reports have also addressed congressional intent and interpretation of motorized use associated with wilderness recreational access (U.S. House of Representatives 1974), fish and wildlife management (U.S. House of Representatives 1979b; U.S. House of Representatives 1976), watershed protection and flood control projects (U.S. House of Representatives 1984; U.S. Senate 1984e), and grazing activities (U.S. House of Representatives 1979a).

Grazing

Livestock grazing and other wilderness management policies in national forest wildernesses were reviewed by congressional committees in the 95th (1978–1979) and 96th (1980–1981) Congresses. The result was, in 1979, U.S. House Committee Report 96-617, which accompanied the Colorado Wilderness Act (U.S. Public Law 96-560) and provided interpretation and clarification of grazing provisions in the Wilderness Act. The Colorado Wilderness Act directed that grazing management in Colorado wildernesses be guided by House Report 96-617, which stressed its interpretation that the Wilderness Act provided for continuation of existing grazing use; the maintenance and construction of supporting facilities, including fences, line cabins, water wells and lines, and stock tanks; and temporary use of motorized equipment to repair facilities and for emergency purposes. As mentioned earlier, these provisions were mandated by the committee report to apply to grazing activities in all national forest wildernesses.

Several subsequent laws, including the Arizona Wilderness Act, Utah Wilderness Act, Wyoming Wilderness Act, Nebraska Wilderness Act, and El Malpais Wilderness Act in New Mexico, also indirectly refer to the guidelines in House Report 96-617. The El Malpais Wilderness Act grazing provisions are important because they apply to a BLM wilderness, and not a national forest wilderness. In addition, several of these laws provide for administrative review of existing grazing policies to ensure that they are consistent with that report. The New Mexico Wilderness Act addressed grazing directly in authorizing additional fencing as provided in the grazing-allotment management plan for livestock in the Cruces Basin Wilderness.

Buffer Zones

In 1980, a congressional committee examined the issue of buffer zones around national forest wildernesses (U.S. House of Representatives 1980, House Report 96-1126) and pointed out that Congress takes great care in determining and establishing wilderness boundaries with the intent that only lands within the boundaries be managed as wilderness. Similarly, a Senate committee report (U.S. Senate 1984c, Senate Report 98-465) stressed that nonwilderness activities should not be restricted or prevented in areas adjacent to wildernesses simply because such activities can be seen or heard from within the wildernesses. Such restrictions applied by the Forest Service were formerly referred to unofficially as the "sights and sounds doctrine," part of the agency's "pure criteria" for wilderness to ensure high standards for designation and management. The New Mexico Wilderness Act (U.S. Public Law 96-550) in 1980 was the first law to prohibit buffer zones, and subsequent laws contain similar provisions, including the Colorado Wilderness Act, Oregon Wilderness Act, Washington Wilderness Act, Arizona Wilderness Act, Utah Wilderness Act, Arkansas Wilderness Act, Wyoming Wilderness Act, Pennsylvania Wilderness Act, Virginia Wilderness Act, and Michigan Wilderness Act.

Fig. 5.9. The first BLM-designated wilderness to become a unit of the National Wilderness Preservation System in 1983 was Bear Trap Canyon Wilderness, in Montana, and it is a unit of the Lee Metcalf Wilderness. Photo courtesy Bureau of Land Management.

Fish and Wildlife

Fish and wildlife protection has long been associated with wilderness designation, and the presence of fish and wildlife is considered an integral part of wilderness for hunting, fishing, viewing, and other recreational purposes (see Chapter 12). Fish and wildlife professionals often argue against wilderness designation, because it is perceived to preclude the use of many modern wildlife-management techniques, and wildlife-management controversies in wilderness are common (Czech and Krausman 1999). Many committee reports stress the importance of preserving and managing fish and wildlife in individual wildernesses.

One House of Representatives committee report (U.S. House of Representatives 1983) addressed the need to balance management activities with the protection of wilderness characteristics. This report stresses that management agencies have authority to maintain water supply facilities, restore natural vegetation, utilize prescribed burning, enhance and restore fish populations, and use motorized equipment to fulfill fish and wildlife management objectives. Congress reaffirmed such intent in ANILCA (U.S. Public Law 96-487) and provided for fisheries research and enhancement, and motorized use and access for subsistence hunting and fishing purposes. The Central Idaho Wilderness Act (U.S. Public Law 96-312) permitted access to manage bighorn sheep along mining roads authorized in the special mining management zone in the Frank Church–River of No Return Wilderness. The Wyoming Wilderness Act (U.S. Public Law 98-550) permitted occasional motorized use to manage bighorn sheep in the Fitzpatrick Wilderness, and the Endangered Wilderness Act (U.S. Public Law 95-237) authorized a fish and game research program to help protect these resources in the Gospel-Hump Wilderness in Idaho and surrounding nonwilderness lands. The California Desert Protection Act (U.S. Public Law 103-433) specifically provides for mechanized access for wildlife management in the sixty-nine wilderness areas it designated (Watson and Brink 1996).

Fire, Insect, and Disease Control

The Wilderness Act specifically permits fire, insect, and disease control measures, but concern has been expressed about limitations that might be placed on control measures in wilderness. One House committee report (U.S. House of Representatives 1977) stressed that control measures include "the use of mechanized equipment, the building of fire roads, fire towers, fire breaks or other pre-suppression techniques where necessary, and other techniques for fire control." Wilderness legislation also affirms that such protective measures can be applied in wilderness. The Endangered Wilderness Act (U.S. Public Law 95-237) stipulated that fire control measures should be properly utilized to protect watersheds in two California wildernesses. The Colorado Wilderness Act (U.S. Public Law 96-560) directed administrative review of current policies for fire, disease, and insect control in Colorado wildernesses to ensure that these policies were consistent with congressional intent and were adequate for protection of adjacent nonwilderness lands.

Facilities and Structures

In several wilderness designation laws, Congress permitted facilities and structures for different purposes. In 1969, two existing dams were included in the Desolation Wilderness established in California (U.S. Public Law 91-82). A congressional committee previously debated whether the use and management of these dams would severely degrade the wilderness or detract from its surroundings (U.S. Senate 1969). Two reservoirs were also included in the Indian Peaks Wilderness in Colorado in 1978 (U.S. House of Representatives 1978). The use of sanitary facilities, such as vault toilets, was specifically provided for in the Endangered Wilderness Act (U.S. Public Law 95-237) and the Utah Wilderness Act (U.S. Public Law 98-428) to protect watersheds in several wildernesses. The Central Idaho Wilderness Act (U.S. Public Law 96-312) permitted construction of water supply facilities in certain areas. The installation and use of "weather modification special equipment," such as snow gauges, and water quality and quantity measuring instruments were endorsed in a House committee report as beneficial to furthering the "scientific, educational, and conservation purposes" of wilderness and to protecting watersheds and "preserving the wilderness character" in some cases (U.S. House of Representatives 1977). The Vermont Wilderness Act (U.S. Public Law 98-322) permitted the maintenance of trails and shelters along the Appalachian Trail and associated trails. "Hydrologic, meteorological, and telecommunications facilities" needed for flood warning and control purposes were permitted in specified wildernesses in the Arizona Wilderness Act (U.S. Public Law 98-406) and Utah Wilderness Act (U.S. Public Law 98-428).

ANILCA (U.S. Public Law 96-487) permitted the construction and maintenance of new and existing cabins in Alaska wildernesses, temporary construction and use of facilities for hunting and fishing, and also provided for fish hatcheries and weirs to improve fisheries. As mentioned earlier, the New Mexico Wilderness Act (U.S. Public Law 96-550) permitted construction of additional fencing for grazing purposes in the Cruces Basin Wilderness.

Special Provisions and Congressional Direction

Special provisions have been included in wilderness laws and congressional committee reports to provide specific guidelines for designation and management based on unique circumstances of local or regional concern. Understanding the relationship among the designating legislation, the Wilderness Act of 1964, and legislative intent is a complex matter that often requires additional management information about the situations that lead to the need for the special provision (Watson et al. 2004). Following is a discussion of some of these additional special provisions.

Sufficiency-Release Language

Through sufficiency-release language in wilderness designation laws, Congress has assumed a partial role in resolving the question of which remaining national forest roadless areas are eligible for further wilderness review and which lands are released for other uses (see Part I of this chapter). Since its first legislative use in 1980, Congress has consistently applied sufficiency-release language on a state-by-state basis. ANILCA was the first law to contain sufficiency-release language, but as such language evolved it was applied in twenty-seven subsequent laws.

Sufficiency language was constructed by Congress to respond to the *California v. Block* lawsuit, which invalidated the RARE II Final Environmental Impact Statement information for California and prevented development on those roadless lands recommended for nonwilderness uses (Baldwin and Gorte 1984; Gorte 1987). *Sufficiency language states Congress's conclusion that the information in the RARE II Environmental Impact Statement for a particular state or section of a state is adequate for Congress's review, and that no further statewide roadless area reviews will be conducted by the Forest Service in that state, nor will there be judicial review of the decision releasing the national forestlands for nonwilderness uses. In other words, roadless area reviews have been "legally and factually sufficient," and land not designated for wilderness, wilderness study, or further planning is "released" for possible nonwilderness use.*

Release language has generally been debated in two forms: "hard" and "soft." In its strictest form, *hard release language would permanently release roadless lands not designated as wilderness or for wilderness study from further wilderness consideration unless authorized by Congress* (Baldwin and Gorte 1984). Hard release language might also require nonwilderness multiple use of released lands, although such interpretation might be argued. Although hard release language has not yet been included in wilderness legislation, various industry groups have advocated permanent or hard release in testimony on wilderness bills.

Soft release language provides that the wilderness option for RARE II nonwilderness roadless areas will not be considered again by the Forest Service during the development of the initial forest plans in accordance with the National Forest Management Act (U.S. Public Law 94-588), but may be considered during the revision of these plans (Baldwin and Gorte 1984; U.S. Senate 1980). Before the initial plans are revised, the Forest Service is permitted to manage lands not designated as wilderness, wilderness study, or for further planning, for nonwilderness multiple use. Moreover, released lands do not necessarily have to be managed to protect their wilderness characteristics. Under provisions in the National Forest Management Act and Forest Service regulations, roadless areas that remain when the forest plans are revised at the end of each subsequent ten-year planning period must then be reconsidered for wilderness designation in the planning process for the next forest plan (The Wilderness Society 2000; U.S. Senate 1980).

From 1980 through 1983, soft release language of this type appeared in six wilderness laws, including ANILCA, the Alaska National Interest Lands Conservation Act, New Mexico Wilderness Act, Colorado Wilderness Act, Charles C. Deam Wilderness Act in Indiana, Paddy Creek Wilderness Act in Missouri, and West Virginia Wilderness Act.

In 1984, continued pressure from development interests led to some alteration and clarification of soft release language that put more responsibility on the forest planning process to protect or develop a released roadless area (Gorte 1987). *The result was compromise release language stating that released lands not designated as wilderness would be managed for nonwilderness multiple use in accordance with the National Forest Management Act, but protection of a roadless area's wilderness characteristics would be allowed if this decision were made in the planning process* (Gorte 1987; U.S. Senate 1984a) Thus, protection of these released lands for potential wilderness consideration was placed on the planning process. Congressional committees also clarified the circumstances under which forest plans could be revised, which further clarified the timing of wilderness reviews (Gorte 1987; U.S. Senate 1984a; U.S. Senate 1984b). Since 1984, most wilderness laws have contained this *compromise* version of soft release language (Gorte and Baldwin 1987).

Military Overflights

Congress took additional steps in the 1990s to ensure wilderness designation does not preclude low-level military overflights or designation of new special airspace within wilderness, with wording in the Nevada Wilderness Protection Act of 1989, Arizona Desert Wilderness Act of 1990, and California Desert Protection Act. Section 4(d)(1) of the Wilderness Act makes special provisions for the use of aircraft where these uses have become established, and they may be permitted to continue based on the opinion and restrictions of the secretary of agriculture. The original intent of this provision was aimed at wilderness airstrips already established in wilderness. More recently Congress's special provisions for military aircraft to continue making low-level training flights through designated wilderness, and tourist overflights as well, have become controversial based on the probable impacts on nature and disrupting opportunities for visitor solitude. General (nonmilitary) airspace reservation below 5,000 feet was provided for by presidential executive order in the Boundary Waters Canoe Area Wilderness, and the BWCA Wilderness Act directed that the earlier presidential directive control airplane flyovers.

Miscellaneous Special Provisions

These include a variety of provisions not failing under other categories. For example, the Sawtooth Wilderness Act and the Hells Canyon Wilderness Act both designated wilderness within national recreation areas and provided condemnation authority to acquire

Fig. 5.10. Bureau of Land Management signs mark a wilderness boundary and closure of an old road to vehicles. The California Desert Protection Act (1994) designated sixty-nine BLM-managed wilderness areas. Wilderness designation closed many old mining and prospecting access roads. Photo courtesy John C. Hendee.

nonfederal lands, while also withdrawing the areas from further mining, subject to valid existing claims. The Alpine Lakes Area Management Act of 1976 authorized a special study of the Enchantment Area within the wilderness to determine the area's best management. A special study to consider including the Indian Peaks Wilderness in Rocky Mountain National Park was authorized by the Indian Peaks Act. The Virginia Wilderness Act called for a combined state and federal air quality study of designated wilderness study areas. The El Malpais and Cebolla Wilderness Act reserved water rights for the wildernesses designated by the act with certain special provisions.

Wilderness Legislation: A View of the Future

Wilderness has been an ongoing topic of legislation since the Wilderness Act was passed in 1964. Since that time a total of 170 additional laws have been passed, setting aside more than 700 areas, totaling 107.4 million acres in forty-four states. Wilderness will continue to be a focus for legislation as long as additional campaigns to designate more wildernesses are waged and seem to have a chance of succeeding. The Wilderness Society has announced an ultimate goal of 200 million acres of designated wilderness in the United States, and citizen efforts to inventory potential wilderness are active in every region of the country. *We believe that wilderness legislation will be in front of the Congress for a long time—as long as any potential wilderness remains that has not been seriously considered for designation. Following are some likely focal points for wilderness consideration.*

Alaska

No legislative action has taken place following wilderness review of 18 million acres in thirteen Alaskan national parks that were designated for wilderness study by ANILCA (Browning et al. 1988). Draft wilderness proposals were submitted by 1988 for public review and comment before wilderness recommendations were made by the agencies. Action is still pending on FWS recommendations for wilderness designation following review of 55 million acres of the FWS refuge system in Alaska. There are large acreages of potential wilderness in Alaska whose future designation is not clear, but future designations of wilderness in Alaska are probable.

National Park System Wilderness in the Forty-eight Contiguous States

Numerous national parks, including Yellowstone, Olympic, North Cascades, Canyonlands, Big Bend, Glacier, Great Smoky Mountains, and Rocky Mountain, contain large roadless areas that have yet to be designated by Congress as wilderness. Almost 8.8 million acres have been recommended to Congress for wilderness designation in these national parks and thirteen other national park units in the forty-eight contiguous states. About 6 million acres in seventeen other national park units are also being studied for their wilderness potential. Thus, legal designation of up to 15 million acres of national park wilderness in the lower forty-eight states is possible.

Tradition and law protect these national park areas from allocation to commodity uses and development activities such as timber harvesting, mining, and livestock grazing. Thus, environmental interests promoting wilderness designation have focused on national forest and BLM areas. It was anticipated that, as remaining national forest roadless areas were either designated as wilderness or released from further wilderness consideration, and as the BLM wilderness review is completed, there would be a renewed effort to complete wilderness designation in these national parks. Although that may yet happen, many are frustrated that decades later it has not yet occurred, and are concerned that in the meantime wilderness values are being lost (Crumbo 1996; J. Watson 1996).

National Forest Roadless Areas and Released Lands

Conflict over wilderness proposals is intense in Idaho, Montana, and several western states, and has the potential to continue for decades in Idaho where proposed national forest plans leave 9 million acres of roadless lands intact and unresolved as to their future (MacCracken et al. 1993). The continuing concern in Idaho and Montana is the impacts of additional wilderness designation on other land uses, such as timber, mining, and motorized recreation, because these states have extensive public lands and already substantial wilderness designations.

By 1987, Congress had passed twenty-eight wilderness designation laws that included soft release language, with about 21 million acres of national forest roadless areas identified in twenty-three states (excluding Alaska, Kentucky, Nebraska, and Michigan) (Gorte and Baldwin 1987). For more than a decade environmentalists fought to protect released roadless areas as these roadless areas were considered in forest planning processes to keep them from being developed or invaded by roads, timber harvest, or other activity that would preclude their wilderness designation in the future. The reconsideration for wilderness designation of at least some of these lands is now a renewed issue, as release provisions under current national forest plans expire, and remaining roadless lands can be reconsidered for their wilderness potential under new forest plans.

Hope for additional wilderness designation was renewed October 13, 1999, when President Bill Clinton declared a moratorium on new road construction on most national forest roadless lands to further protect roadless areas from development (O'Laughlin and Freemuth 2000). At stake were 60 million acres of inventoried and uninventoried roadless areas. Five states have within their borders 39.8 million acres of the 60-million-acre total: Alaska, 14.8 million acres; Idaho, 9.4; Montana, 6.0; Colorado, 5.5; California, 4.1. Seven other western sates have a total of 17.9 million acres, and 25 other states have a total of 17.9 million acres (O'Laughlin and Freemuth 2000). The George W. Bush administration, which took office in 2001, opposed roadless area protection and tried to overturn the moratorium. Although the final outcome is still unclear, it seems likely that substantial roadless-area protection and future wilderness designations of these lands will occur.

Bureau of Land Management Wilderness Study Areas

The Federal Land Policy Management Act of 1976 (U.S. Public Law 94-579) required that the BLM conduct a wilderness inventory of its roadless areas. The inventory identified 795 wilderness study areas totaling nearly 25 million acres. The inventory and recommendations were to be made to Congress by 1991. Tentatively the BLM has recommended 10 million acres for congressional wilderness designation. The initial results of the BLM recommendations have resulted in the Arizona Desert Wilderness Act of 1990 (U.S. Public Law 101-628), a second Colorado Wilderness Act of 1993 (U.S. Public Law 103-77), and the California Desert Protection Act of 1994 (U.S. Public Law 103-433). However, numerous BLM wilderness study areas are still under consideration

Fig. 5.11. Hoodoos in New Mexico's Bisti/De-Na-Zin Wilderness, managed by the Bureau of Land Management, show the variety of landforms and ecosystems within the National Wilderness Preservation System. Photo courtesy Bureau of Land Management.

by Congress for wilderness designation. The BLM has the smallest current and potential amount of wilderness acreage of the four federal agencies; however, it has a dedicated program of wilderness stewardship and the potential to grow, provided some of the long-standing obstacles to designation can be removed or reduced. Leshy (2005) outlines four issues and conflicts that hinder BLM lands from wilderness designation: potential oil and gas development, off-highway vehicle use for recreation on federal lands, rights-of-way across federal lands, and the number and size of private land inholdings in potential BLM wilderness areas.

New Wilderness Knowledge and Legislative Direction

During the past five decades, a large body of wilderness management information has been developed from experience and research, and this new information has also shaped management direction. Agencies have developed and revised regulations, management policies, and plans, and numerous agency training workshops and conferences have been held. The Arthur Carhart National Wilderness Training Center, Missoula, Montana, provides wilderness management training courses for all four wilderness agencies. Although field experience has been an important source of new wilderness knowledge, so have scholarly inquiries. There have been many scientific studies of resource conditions, wilderness use, and effects of management. The Aldo Leopold Wilderness Research Institute, Missoula, Montana, conducts and supports research nationwide and cooperates in wilderness research at many universities.

In addition, publications such as our text, first published in 1979, have been joined by the *International Journal of Wilderness* (www.ijw.org), launched in 1995, which provides an ongoing forum for wilderness research, education, and management information; national conferences on wilderness research and management have been held and their proceedings are also a rich compilation of wilderness knowledge; a wilderness policy council with representatives from all four federal wilderness agencies meets regularly to coordinate their wilderness policies and actions; eight World Wilderness Congresses have been held in six different countries, with accompanying technical symposia and proceedings. Internet sources, such as www.wilderness.net and Wildnet, provide information and host discussion of wilderness issues. A Wilderness Task Force of the World Conservation Union (IUCN) provides a forum for coordinating wilderness information from around the world, providing a broader context for wilderness management in the United States. Clearly, we know more about wilderness and its management now.

In the U.S. another challenge has been to interpret the new knowledge about wilderness management's applicability within the overriding congressional direction in the Wilderness Act and subsequent wilderness designation laws. Devising and implementing wilderness stewardship principles and policies that respond to evolving knowledge, yet adhere to perceived legislative direction, are sometimes difficult, but always essential. *Just because a technique is available to do the job does not mean it can be used in wilderness*, such as mechanized efficient trail-building machines. *In wilderness management, legislative direction is more important than technical capability.*

Although Congress has been very conservative in clarifying or changing management direction in the Wilderness Act, congressional leaders and congressional reports have more frequently addressed management concerns. For example, in 1977, Senator Frank Church of Idaho stressed that managing agencies use a "rule of reason" and common sense to interpret Congress's intent in the Wilderness Act, and "do only what is necessary" to manage wilderness to protect wilderness characteristics and still provide for "human use and enjoyment" (Church 1977). Two years later, in 1979, a House of Representatives report stated that a strict or "pure" interpretation of the Wilderness Act of 1964 is inappropriate for allocating or managing wildernesses, due to individual differences among wildernesses (House of Representatives 1979a). A few years later, a congressional leader expressed the view that if management conflicts between Congress and the managing agencies become too great in the future, Congress may provide more explicit management guidelines and directives in wilderness legislation (McClure 1985).

Such pronouncements of congressional leaders and congressional committee reports, although lacking the force of law, may influence agency policy. They contribute to caution by the agencies in proposing change in management direction and encourage the enlisting of expert and public views before taking action. At the turn of this century, with many wilderness management issues to address, the USFS called for a panel of experts to review wilderness management direction and issues, and a wilderness summit to hear the views of citizens and interest groups (Brown 2000, 2002; Dombeck 1999). At the same time, the Forest Service proposed a new "Wilderness Recreation Strategy" to address high use and impacts at popular wilderness sites, but great controversy about both its wisdom and consistency with direction in the Wilderness Act limits full implementation (see the April 2001 issue of *International Journal of Wilderness* for a series of articles on the Wilderness Recreation Strategy for the National Forests).

We want to note here that understanding the legislative process is necessary to understand how a challenge to wilderness law might be decided. If a wilderness law were to be challenged, perhaps over implementing a management provision, the courts may review pertinent legislative history, including congressional debate preceding the law's enactment and committee reports accompanying the law, to interpret (determine) Congress's intent for the provision. Altogether, the material to be reviewed might include House and Senate committee reports on proposed bills, floor debate in these bodies, joint House-Senate conference reports, and any committee report referred to in the law. For example, Shannon Meyer addressed the legislative history surrounding wilderness airstrip management (1999). More information on litigation as an influence on wilderness management is in Part III of this chapter.

Part II Conclusions and Predictions

Since 1964, wilderness legislation has been a major topic for congressional action with the passage of the

Wilderness Act. A total of 171 laws have designated more than 700 units of the NWPS, totaling 107.4 million acres, with many laws clarifying or specifying wilderness management direction. Special management directions in wilderness designation legislation have generally affirmed the intent of the Wilderness Act, with congressional reports accompanying wilderness laws most often providing the necessary clarification and interpretation.

Wilderness legislation is likely to remain a major topic for congressional action in the decades to come. Additional designations are likely in NPS and FWS units, national forest roadless lands, and BLM wilderness study areas. As national forest planning cycles are completed, there will be renewed opportunity to consider the wilderness potential of the released national forest roadless areas. *We believe that future wilderness legislation will continue the trend of containing language that affirms and/or clarifies management direction in ways that address local concerns.* There will also be occasional legislation to adjust boundaries, change wilderness names, correct inconsistencies, and otherwise adjust to inevitable change. For example, by 1999 there were eleven name changes of areas in the wilderness system; such adjustments and changes are outlined by Landres and Meyer (2000).

Between 1964 and 2000, only 5 of the 132 laws passed had removed acres from wilderness designation (Landres and Meyer 2000). However, there continue to be proposals to remove wilderness designations or make major special provisions and exceptions for some units, as some recent designations demonstrate (e.g., Steens Mountain Wilderness, in Washington, designated under U.S. Public Law 106-399). We and others are concerned that future wilderness laws may include more such special provisions, nonconforming uses, and exceptions that will dilute the quality of the NWPS (Nickas and Proescholdt 2005).

The 107.4 million acres of wilderness already designated would surprise the original authors of the Wilderness Act, who anticipated an NWPS of only 40 to 50 million acres (Church 1977). Yet, 50 million acres or more of wilderness may likely be added to the NWPS from the roadless areas in Alaska, national parks in the forty-eight contiguous states, remaining national forest roadless areas, and the BLM wilderness study areas. The Wilderness Society and other environmental groups have called for a 200-million-acre wilderness system.

For at least another decade, and perhaps for two or three decades, it is likely that wilderness legislation will continue to be a major natural-resource issue. *We expect a continuation of the trend in wilderness laws toward omnibus legislation covering more than one area, often in individual states, and the inclusion of more language to affirm and clarify management direction and address local concerns.* Congress will increasingly have to address smaller areas, those previously modified to some degree, and areas with strongly competing alternative uses. To resolve some of the very difficult wilderness designation and management issues in the future, a variety of special provisions, exceptions, and compromise provisions are likely to be proposed and included in wilderness designations. *However, the evidence suggests that Congress will generally hold the line on proposals for major exceptions and unique provisions in wilderness laws and/or amending the Wilderness Act.*

Part III: Management Implications from Litigation Involving Wilderness

Congress has the authority to regulate federal lands based on the U.S. Constitution, and the Wilderness Act of 1964 (U.S. Public Law 88-577) awarded the authority for designating wilderness areas to Congress as the means to create a National Wilderness Preserve System. This approach to an NWPS means that the legislative branch of government designates the areas, the executive branch carries out the management as directed by the act, and the judicial branch resolves conflicts in application of the act.

Like litigation on any contentious issue, wilderness litigation has been varied in content and reason for the action as well as in outcomes and decisions. It is difficult to obtain information on some cases that do not have written decisions, resulted in an out-of-court settlement, or were withdrawn. Similarly, some cases have been through multiple levels of the court system and have been overturned for various reasons. Litigation has tested the Wilderness Act and wilderness designation laws, and generally the courts have upheld the intent of the Wilderness Act (Ruckel 1999).

Litigation and the Wilderness Legislation

Citizens can bring suit against federal land-management agencies using the Administrative Procedures Act

(APA) of June 11, 1946, to challenge decisions that are unlawful or unsupported by the legislation and to ensure that the laws are executed in a responsible and legally appropriate manner. Dettmann (2008, 4) has expressed an opinion on why the Wilderness Act lends itself to litigation more than some other laws:

> The Wilderness Act…is a remarkable federal lands statute because it provides specific and clear directives on what wilderness is and how a system of wilderness is to be created and maintained. For established areas, the Act imposes a statutory structure with detailed requirements—such as no vehicles, no structures, and no commerce—along with a strict, overarching mandate to preserve wilderness character. These directives effectively remove the discretion that land management agencies might normally enjoy.

The idea that litigation is an affirmation of the Wilderness Act and not a punishment of the federal land-managing agencies is firmly outlined by Dettmann (2008, 5). His perspective is that the judicial oversight is a necessary and equally important third branch of government that has its place in executing the intent of the Wilderness Act.

> Our system of government seeks to resolve conflict by creating a forum for intelligent, civilized debate moderated by a judge vested with the power of the sovereign. The whole idea of litigation is to allow two disagreeing parties to come to a neutral authority that reviews the evidence and decides the case. The whole point of the judicial branch is to resolve conflict, not to create it. What better opportunity is there to advocate for the values of wilderness than through the judicial process? What more reasoned approach exists by which wilderness advocates can air their disagreements with agencies to an authority that is obligated to follow and enforce the law as it is written? Such a forum lends itself particularly well to a statute such as the Wilderness Act. It is hard to find a clearer definitive statutory command than, for example, there shall be no use of motor vehicles in wilderness….Litigation enables a broader public accounting of the issues as well as a broader understanding.

Whether litigation can be brought regarding wilderness under the Administrative Procedures Act (APA) has been narrowly construed by the U.S. Supreme Court in 2004 to relate to specific agency actions that are legally required (Konrad 2006). The issue was that an environmental group contended (*Norton v. Southern Utah Wilderness Alliance*) that the BLM had failed to adequately protect wilderness study areas from degradation by off-road vehicles as required under the Federal Land Policy and Management Act, NEPA, and the BLM Land Use Plan (Konrad 2006); the court found that the APA could not be used to comply with general statutory mandates. The Wilderness Act, as Dettmann points out, is much more directive and mandates agency action through statutory requirements.

Boundary Waters Canoe Area

The Boundary Waters Canoe Area (BWCA) is an example of a wilderness area that was founded in controversy and litigation that helped shape two wilderness laws related to wilderness designation of the BWCA. The area has one of the longest histories of legal battles between wilderness protection advocates and those who want to access and use the resources of the area for motorized transportation, timber harvest, resort and tourism development, mining development, and other uses. Protection of the wild character of the BWCA through legislation, administration, and litigation, starting in 1909 with the establishment of the Superior National Forest, did not reduce the conflicts and legal issues (Duncan and Proescholdt 1999). Multiple controversies, wilderness advocacy, and litigation eventually resulted in designating the BWCA as an original unit of the NWPS in the 1964 Wilderness Act. However, continued controversy and legal actions following the political compromises to permit nonconforming uses in the BWCA under the act resulted in the need for additional legislative clarification and the passage of the Boundary Waters Canoe Area Wilderness Act of 1978 (U.S. Public Law 95-495).

The BWCAW Act expanded the original boundaries of the BWCA by 57,000 acres to more than 1 million acres and renamed the area the Boundary Waters Canoe Area Wilderness. The BWCAW Act further restricted certain permitted and nonconforming activities, such as it "ended logging entirely, restricted mining and the use of snowmobiles, and cut motorboat use from sixty-four percent to twenty-four percent of the water surface" (Duncan and Proescholdt 1999, 626), but numerous permitted and nonconforming activities remained. The manner in which the BWCAW Act achieved these changes was through repeal of "section 4(d)(5) of the 1964 Wilderness Act, the only instance where the 1964 law was amended" (Duncan and Proescholdt 1999, 626).

The political compromises in the 1978 BWCAW Act brought the area into better compliance with other wildernesses in the NWPS, but the controversies remain and the lawsuits continue. A long history of litigation has shaped the BWCAW and will continue to affect the management of the area as the judicial process continues.

The Extent of Wilderness Litigation

Considering the 171 wilderness designation laws following the Wilderness Act and the more than 700 units and 107.4 million acres in the NWPS, it is surprising that the number of cases of litigation is small after more than forty-four years since the act was passed into law. There are very few compilations of litigation regarding federal land-management cases; however, one recent published study reported that litigation brought against the USFS during the 16-year period from 1989 through 2004 found that of the 888 documented cases with published decisions, only 26 cases (2.9 percent) were based on the Wilderness Act and another 29 were based on the Wild and Scenic Rivers Act (U.S. Public Law 90-542), and 3 of those cases were based on both the Wilderness and Wild and Scenic Rivers Acts (Malmsheimer et al. 2008).

No studies have yet been published that compile the entire time period since 1964, or compile litigation for the other three federal wilderness management agencies, the National Park Service, Bureau of Land Management, and U.S. Fish and Wildlife Service.

Between 1989 and 2004, the USFS won ten (38.5 percent) of the twenty-six lawsuits brought against it regarding the Wilderness Act, lost eleven cases (42.3 percent), and settled five cases (19.3 percent) (Malmsheimer et al. 2008). The statutory basis for these cases is not solely the Wilderness Act and often involved NEPA and other statutes (Table 5.2). The purposes of these cases were complex and Malmsheimer et al. (2008) break them down into only two categories: 77 percent of the cases were for less resource use (e.g., to restrict use or uses) and 23 percent were for greater resource use (e.g., to overturn restrictions on amount or type of use). Malmsheimer et al. (2008) note that twelve of the twenty-six cases involved some aspect of motorized vehicle use or access.

Litigation Themes and Management Implications

Three types of themes have emerged from litigation around the Wilderness Act: the definition and characteristics of wilderness, the scope of protection under the Wilderness Act, and issues involving wilderness management (Ruckel 1999).

Cases focusing on the definition of wilderness and the suitability of certain areas are exemplified by the *Parker v. United States* suit in 1971, which questioned whether a road leading into the proposed East Meadow Creek area disqualified the area to be designated as wilderness. This was in the Gore Range–Eagles Nest Primitive Area on the White River National Forest in north-central Colorado. Courts found that the area met the wilderness definition, such as the definitional phrase that an area "generally appears to have been affected primarily by the forces of nature, with the imprint of man's work substantially unnoticeable." Ruckel (1999, 613) concludes from this and other cases:

> The courts have adopted a practical definition, recognizing the primitive, untrammeled nature of wilderness,

Table 5.2. Number and Outcome of Cases Litigated Against the USFS Based on the Wilderness Act and Other Statutes from 1989 to 2004
(Malmsheimer et al. 2008)

Statutes Used in Case	Number of Cases Based on the Statute	Number of Outcomes on This Statute Won by USFS	Number of Outcomes on This Statute Lost by USFS	Number of Outcomes on This Statute Settled by USFS
The Wilderness Act	26	13	8	5
National Environmental Policy Act	18	12	2	4
National Forest Management Act	9	5	1	3
Wild and Scenic Rivers Act	3	2	1	0
Clean Water Act	3	2	0	1

but acknowledging the often unavoidable presence of conflicting circumstances or uses. Purist notions of completely untrammeled, unblemished land have not been sympathetically received. An important effect of this approach has been to maximize the availability of lands, some of which may be minimally impacted by a conflicting circumstance or use, for eventual wilderness classification.

Two cases illustrate that the scope of protection under the Wilderness Act is broad. In the 2003 case of *Wilderness Society v. U.S. Fish and Wildlife Service*, the courts decided that a commercial sockeye salmon fish-stocking project in the Kenai Wilderness of Alaska was expressly prohibited under the Wilderness Act because the project's purpose and effect were to support the commercial fishing industry (Rasband et al. 2004; Ryan 2005; Miller 2005). In the second case in 2000, *Potlatch Corp. v. United States* a decision by the courts ruled that Congress had intentionally and directly reserved the unappropriated water rights upstream of three Idaho wilderness areas in their designation under the 1964 Wilderness Act, which by extension affirms that federal water rights reservation applies to lands designated under the Wilderness Act of 1964 (Latimer 2000; Miller 2000).

Not all cases have been decided in favor of wilderness protection especially related to valid preexisting mining claims, road access across wilderness study areas to private inholdings, road construction and timber harvesting on lands adjoining a wilderness study area, and other situations on adjoining lands (Ruckel 1999).

Controversies and issues about the management of wilderness areas have been the source of some of the most well-known court cases regarding the Wilderness Act. *Wilderness Watch v. Mainella, Sierra Club v. Lyng, High Sierra Hikers Assn, et a v Bernie Weingardt, et al.,* and a USFS case banning fixed rock-climbing anchors in the Sawtooth Wilderness of Idaho illustrate the implications for managers who have some discretion to make management decisions; however, their decisions and management actions can be challenged by special interest and environmental groups.

The 2004 case of *Wilderness Watch v. Mainella* was about the NPS practice of transporting visitors across a portion of Cumberland Island Wilderness Area to provide access to historic sites, which was decided by the courts as violating the Wilderness Act and NEPA (Powers 2005).

The *Sierra Club v. Lyng* case in 1987 was over the USFS practice of extensive tree cutting and chemical spraying to control beetle infestations in southern USFS wilderness areas; the courts ruled against the program because it was justified based on protecting the economic interests and resources of adjacent private landowners (Glicksman and Coggins 1999). The courts remanded the USFS to file an EIS and redesign the program to limit the program's impacts on wilderness conditions—which the USFS carried out and received approval for implementation.

A seven-year legal battle in *High Sierra Hikers Assn, et al. v. Bernie Weingardt, et al.* was about commercial livestock use in the John Muir and Ansel Adams Wildernesses in California. The court decided on October 30, 2007, that the USFS violated the Wilderness Act by creating a plan that did not meet the requirement to preserve wilderness character and to properly regulate and limit commercial services for pack-stock use.

The fourth case involved the USFS decision to ban permanent fixed-anchor installations for rock climbing in Idaho's Sawtooth Wilderness. An administrative appeals hearing resulted in the USFS decision to initiate a negotiated rulemaking process that was to develop a national policy on permanent fixed-anchor installations for rock climbing in the wilderness areas (Cheever 1999; Dolan 2000).

Part III Summary

The judiciary has had a substantial and formative impact on the NWPS by hearing cases brought regarding suitability for wilderness designation, the breadth and scope of wilderness protection, and the complex and conflicting uses that Congress permits in wilderness as part of political compromises.

Ruckel (1999, 619) concludes after reviewing the published litigation on the Wilderness Act and designation legislation:

> It has taken strong action by the judiciary to strike balances and make determinations that have ensured that qualifying lands were all reviewed, and then, when formally dedicated to wilderness, protected. The courts have succeeded by carefully scrutinizing the facts of the cases brought before them and vigorously safeguarding wilderness principles and lands. They have maintained balance and objectivity by recognizing individual conflicting activities sanctioned by the Wilderness Act itself, or by specific legislation establishing individual wilderness areas.

In the final analysis, the attempt to litigate a disagreement over wilderness policies or management actions indicates much about the convictions of the parties involved regarding the issue in question. Litigation is very expensive and invariably follows a great deal of discussion and effort to resolve the matter without going to court. Thus, the threat of litigation brings a greater sense of urgency and importance to attempts to resolve wilderness issues than if the courts were never involved.

Study Questions

Part I. Evolution of the Designation Process

1. Discuss the factors that led to the RARE I process. What were some of the deficiencies of RARE I and how might they have been corrected? What were some of the benefits of RARE I?
2. In your judgment, how well did RARE II correct the deficiencies of RARE I? What problems still remain?
3. What were some of the factors that led to the increasing use of "omnibus" state wilderness legislation?
4. Describe the concepts of sufficiency, hard release, and soft release as they apply to the wilderness designation process. How have they been employed in the state where you live?
5. Discuss the problems in obtaining wilderness designation in national parks like Yellowstone and Yosemite. What are some of the reasons underlying the lack of progress?
6. Describe the various opportunities and problems facing the wilderness designation of lands under the administration of the BLM.
7. Describe the relative roles and importance of environmental groups versus the agencies' wilderness study processes in securing wilderness designation.
8. How did the RARE II lawsuit (*California v. Bergland*) influence the wilderness designation process, and thus the size and nature of the NWPS?
9. Why was designating wilderness in Alaska such a challenge? How did Congress resolve the unique conditions in making wilderness possible?

Part II. Management Implications in Wilderness Designation Acts

1. How do wilderness designation laws affect wilderness management?
2. How can legislative history influence the interpretation of a wilderness designation law? Where is the legislative history found?
3. Why are seven laws identified as "benchmark laws" in this chapter?
4. How were grazing provisions of the Wilderness Act clarified?
5. What are some management provisions of the Wilderness Act that have been clarified by subsequent wilderness designation, and what general direction have the clarifications taken?
6. When and where can you mine in wilderness?

Part III. Management Implications of Litigation

1. What place does the judicial branch of government have in executing the intent of the Wilderness Act?
2. Should the amount of litigation involving the Wilderness Act be described as small or large? Has that litigation challenged the intent of the Wilderness Act? Has that litigation helped clarify the issues surrounding compromises for permitted uses in wilderness?
3. What are the three types of themes that have emerged from litigation around the Wilderness Act? Give an example of a case for each theme.

Acknowledgments

John C. Hendee and Chad P. Dawson revised this chapter for the fourth edition, integrating and updating the material from previous editions (Part I: wilderness designation process, including agency roadless area reviews; and Part II: wilderness designation laws) and adding a new third part (management implications of litigation involving wilderness). The chapter builds on contributions to earlier editions from our colleagues Jay Watson, Joseph W. Roggenbuck, George H. Stankey, Dennis M. Roth, Jim Browning, and Ed Krumpe.

> **Case Discussion Questions: Olympic National Park Wilderness**
>
> Read the section in Part I of this chapter entitled "Olympic National Park Wilderness Review—A Case Study" on pages 132 to 134. The case outlines the entire process by which that wilderness study area eventually became a designated wilderness area during the period from 1972 to 1988.
>
> 1. Did the NPS meet its requirement to conduct a wilderness study review of four areas in the Olympic National Park? Did the NPS make a recommendation to Congress to designate those four areas as wilderness?
> 2. Why was the NPS wilderness designation recommendation for Olympic National Park not acted upon until 1988? What activities and influences caused the passage of the Washington Park Wilderness Act of 1988?
> 3. In the case of the Olympic National Park Wilderness, how much did professional expertise contribute to wilderness designation, and how much did political involvement influence passage of legislation to designate wilderness?

References

Administrative Procedure Act of June 11, 1946 (60 Stat. 237), and amended in 1966.

Allin, Craig W. 1982. *The Politics of Wilderness Preservation*. Westport, CT: Greenwood Press.

Bailey, Robert G. 1976. Ecoregions of the United States. Ogden, UT: U.S. Department of Agriculture, Forest Service, Intermountain Region. Map.

Baldwin, Pamela; Gorte, Ross W. 1984. National forest wilderness bill release alternatives: A legal analysis and policy issues. Congressional Research Service Report No. 84-542 ENR, February 21, 1984. Washington, DC: U.S. Library of Congress.

Brown, Perry. 2000. Issues in the quality of U.S. wilderness management. *International Journal of Wilderness*. 6(2): 3–4.

———. 2002. A summary of the report: Ensuring the Stewardship of the National Wilderness Preservation System. *International Journal of Wilderness*. 8(1): 10–12, 32.

Browning, James A.; Hendee, John C.; Roggenbuck, Joe W. 1988. *103 Wilderness Laws: Milestones and Management Direction in Wilderness Legislation, 1964-1987*. Bulletin 51. Moscow: University of Idaho, Idaho Forest, Wildlife and Range Experiment Station.

Cheever, F. 1999. Talking about wilderness. *Denver University Law Review*. 76(2): 335–345.

Church, Frank. 1977. Wilderness in a balanced land use framework. *First Annual Wilderness Resource Distinguished Lectureship*. Moscow: University of Idaho, Wilderness Research Center. 18p.

Coggins, George C.; Wilkinson, Charles F. 1987. *Federal Public Land and Resources Law*, 2nd ed. Mineola, NY: The Foundation Press.

Costley, Richard J. 1972. An enduring resource. *American Forests*. 78(6): 8–11.

Crumbo, Kim. 1996. Wilderness management at Grand Canyon—"Waiting for Godot?" *International Journal of Wilderness*. 2(1): 19–24.

Czech, Brian; Krausman, Paul. 1999. Controversial wildlife management issues in southwestern U.S. wilderness. *International Journal of Wilderness*. 5(3): 22–28.

Davis, George D. 1989. Preservation of natural diversity: The role of ecosystem representation within wilderness. In: Freilich, Helen R., comp. *Wilderness Benchmark 1988: Proceedings of the National Wilderness Colloquium*; January 13–14, 1988; Tampa, FL. General Technical Report SE-51. Athens, GA: U.S. Department of Agriculture, Forest Service, Southeastern Forest Experiment Station, pp. 76–82.

Dettmann, J. 2008. The need for wilderness litigation. *International Journal of Wilderness*. 14(1): 4–6.

Dolan, T. 2000. Fixed anchors and the Wilderness Act: Is the adventure over? *University of San Francisco Law Review*. 34: 355–378.

Dombeck, Mike. 1999. A wilderness agenda and legacy for the U.S. Forest Service. *International Journal of Wilderness*. 5(3): 4–6.

Duncan, R. A.; Proescholdt, K. 1999. Protecting the Boundary Waters Canoe Area Wilderness: Litigation and legislation. *Denver University Law Review*. 76(2): 621–658.

Evans, Kent E. 1986. Indian Mounds Wilderness Area: Perceived wilderness qualities and impacts of oil and gas development. In: Kulhavy, David L.; Conner, Richard, N., eds. *Wilderness and Natural Areas in the Eastern United States: A Management Challenge*. Nacogdoches, TX: Stephen F. Austin University, pp. 156–165.

Glicksman, R. L.; Coggins, G. C. 1999. Wilderness in context. *Denver University Law Review*. 76(2): 383–411.

Gorte, Ross W. 1987. History of release language in wilderness legislation, 1979–1984. CRS Report No. 87-559 ENR, June 12, 1987. Washington, DC: U.S. Library of Congress, Congressional Research Service.

———. 1988. Wilderness: Overview and statistics. Congressional Research Report No. 8846, ENR, January 5, 1988. Washington, DC: U.S. Library of Congress.

Gorte, Ross W.; Baldwin, Pamela. 1987. Wilderness release language history and effectiveness. Unpublished Congressional Research Service Memorandum, June 25, 1987. Washington, DC: U.S. Library of Congress.

Hendee, John C. 1977. Public involvement in the United States Forest Service roadless area review: Lessons from a case study. In: Coppock, Terrance; Sewell, Derrick, eds. *Public Participation in Planning*. New York: Wiley, pp. 89–103.

———. 1986. Wilderness: Important legal, social, philosophical and management perspectives. In: Kulhavy, David L.; Conner, Richard N., eds. *Wilderness and Natural Areas in the Eastern United States: A Management Challenge;* May 13–15, 1985; Nacogdoches, TX. Nacogdoches, TX: Stephen F. Austin University, School of Forestry, pp. 5–11.

Hendee, John C.; Smith, Zane G.; Lake, Robert. 1980. Public involvement in resource decisions: RARE I and II and implications for the future. *Proceedings: Multiple Use Symposium* September 1979; Clemson, SC. Clemson, SC: Clemson University.

Hendee, John C.; Stankey, George H.; Lucas, Robert C. 1991. *Wilderness Management.* 2nd ed., rev. Golden, CO: North American Press/Fulcrum.

Hummel, M. 2007. The Stikine-LeConte Wilderness in Alaska: Twenty-six years of management under ANILCA. *International Journal of Wilderness.* 13(2): 8–13, 17.

Konrad, J. C. 2006. The shrinking scope of judicial review in *Norton v. Southern Utah Wilderness Alliance. University of Colorado Law Review.* 77: 515–547.

Küchler, A. W. 1966. Potential natural vegetation (map). Natural Atlas of the United States. Washington, DC: U.S. Department of the Interior, Geological Survey, pp. 89–92.

Landres, Peter; Meyer, Shannon. 2000. National Wilderness Preservation System database. *Key attributes and trends 1964–1999,* 2nd ed., rev. RMRS-GTR-18. Fort Collins, CO: U.S. Department of Agriculture, Forest Service, Rocky Mountain Research Station, 97pp.

Latimer, K. J. 2000. Federal reserved water rights doctrine under the Wilderness Act: Is it finally here to stay? *Journal of Land Resources and Environmental Law.* 20: 335–356.

Leshy, J. D. 2005. Contemporary politics of wilderness preservation. *Journal of Land Resources and Environmental Law.* 25(1): 1–13.

Loop, Donna J. 1986. Claiming the Cabinets: The right to mine in wilderness areas. *The Public Land Law Review.* 7: 45–77.

MacCracken, John G.; O'Laughlin, J.; Merrill, T. 1993. Idaho roadless areas and wilderness proposals. Report 10. Moscow: University of Idaho, Idaho Forest, Wildlife, and Range Policy Analysis Group.

Malmsheimer, R. W.; Falco, C.; Anderson, A. M.; Keele, D. M.; Floyd, D. W. 2008. U.S. Forest Service litigation—The Wilderness Act and Wild and Scenic Rivers Act: 1989–2004. *International Journal of Wilderness.* 14(1): 7–14.

Matthews, Olen P.; Haak, Amy; Toffenetti, Kathryn. 1985. Mining and wilderness: Incompatible uses or justifiable compromises? *Environment.* 7(3): 12–36.

McClure, James. 1985. Congressional directives and expectations. In: Frome, Michael, ed. *Issues in Wilderness Management: Proceedings National Wilderness Management Conference;* October 1983; Moscow, ID. Boulder, CO: Westview Press.

Meyer, Shannon. 1999. The role of legislative history in agency decision making: A case study of wilderness airstrip management in the United States. *International Journal of Wilderness.* 5(2): 9–12.

Miller, J. 2005. Six million sockeye salmon and the Kenai Refuge: Analysis of the Ninth Circuit Court of Appeals' two decisions in *The Wilderness Society v. United States Fish and Wildlife Service. Journal of Land Resources and Environmental Law.* 25(1): 85–97.

Miller, R. C. 2000. Water Law—Wilderness areas and federal reserved water rights: Unlimited appropriations of unallocated water endorsed by Idaho Court, *Potlatch Corp. v. United States (in re SRBA)* No. 39576, 1999 WL 778325 (Idaho 1999). *Land and Water Law Review.* 35(2): 375–394.

Milton, William John, Jr. 1975. National forest roadless and undeveloped areas: Develop or preserve? *Land Economics.* 51(2): 139–143.

Mohai, Paul; Verbyla, David L. 1987. The RARE II wilderness decisions. *Journal of Forestry.* 85(1): 17–23.

Nickas, G.; Proescholdt, K. 2005. Keeping the wild in wilderness: Minimizing nonconforming uses in the National Wilderness Preservation System. *International Journal of Wilderness.* 11(3): 13–18.

O'Laughlin, Jay; Freemuth, J. 2000. Roadless area policy, politics and wilderness potential. *International Journal of Wilderness.* 6(1): 9–12.

Outdoor Recreation Resources Review Commission. 1962. *Wilderness and Recreation: A Report on Resources, Values and Problems.* ORRRC Study Report 3. Washington, DC: U.S. Government Printing Office.

Powers, S. 2005. *Wilderness Watch v. Mainella:* The Wilderness Act and National Environmental Policy Act are given meaning by the Eleventh Circuit. *Journal of Land Resources and Environmental Law.* 25(1): 99–108.

Rasband, J.; Salzman, J.; Squillace, M. 2004. *Natural Resources Law and Policy.* New York: Foundation Press.

Roth, Dennis M. 1984. *The Wilderness Movement and the National Forest: 1964–1980.* FS-391. Washington, DC: U.S. Department of Agriculture, Forest Service.

Ruckel, H. A. 1999. The Wilderness Act and the courts. *Denver University Law Review.* 76(2): 611–619.

Ryan, K. D. 2005. Preservation prevails over commercial interests in the Wilderness Act: *Wilderness Society v. United States Fish and Wildlife Service. Ecology Law Quarterly.* 32: 539–573.

Turner, R. F., Jr. 2004. The complexity of wilderness stewardship in Alaska. *International Journal of Wilderness.* 10(2): 18–22.

U.S. Department of Agriculture, Forest Service. 1973. Final environmental statement: Roadless and undeveloped areas (RARE). Washington, DC: United States Department of Agriculture, Forest Service.

———. 1978. Roadless area review and evaluation (RARE II). Draft environmental impact statement 78-04. Washington, DC: U.S. Department of Agriculture, Forest Service.

———. 1979. Final RARE II environmental impact statement. Washington, DC: U.S. Department of Agriculture, Forest Service.

———. 2005. What can I do in wilderness? Alaska National Interest Lands Conservation Act (ANILCA) and wilderness in national forests in Alaska. Ketchikan, AK: United States Department of Agriculture, Forest Service.

U.S. Department of the Interior, Bureau of Land Management. 1982. Wilderness study policy: Policies, criteria and guidelines for conducting wilderness studies on public lands. *Federal Register.* 47(23): 5098–5122.

U.S. Department of the Interior, U.S. Fish and Wildlife Service. [n.d.]. Red Rock Lakes Wilderness proposal.

———. 1970. Laguna Atascosa Wilderness study area.

U.S. Government Accounting Office. 1984. Private mineral rights complicate the management of eastern wilderness areas. GAO/RCED-84-101. July 26, 1984. Washington, DC: U.S. Government Accounting Office.

U.S. House of Representatives. 1973. *Ninth Annual Wilderness Report.* House Document 93-194. Washington, DC: U.S. Government Printing Office.

U.S. House of Representatives. 1974. House Report 93-872. 93rd Congress, 2nd Session. Washington, DC: U.S. Government Printing Office.

———. 1976. House Report 94-1562. 94th Congress, 2nd Session. Washington, DC: U.S. Government Printing Office.

———. 1977. House Report 95-540. 95th Congress, 1st Session. Washington, DC: U.S. Government Printing Office.

———. 1978. House Report 95-1460. 95th Congress, 2nd Session. Washington, DC: U.S. Government Printing Office.

———. 1979a. House Report 96-97, Part 1. 96th Congress, 1st Session. Washington, DC: U.S. Government Printing Office.

———. 1979b. House Report 96-617. 96th Congress, 1st Session. Washington, DC: U.S. Government Printing Office.

———. 1980. House Report 96-1126. 96th Congress, 2nd Session. Washington, DC: U.S. Government Printing Office.

———. 1983. House Report 98-40. 98th Congress, 1st Session. Washington, DC: U.S. Government Printing Office.

———. 1984. House Report 98-1019, Part 1. 98th Congress, 2nd Session. Washington, DC: U.S. Government Printing Office.

U.S. Public Law 86-517. Multiple-Use Sustained-Yield Act of June 12, 1960. 74 Stat. 215.

U.S. Public Law 88-577. The Wilderness Act of September 3, 1964. 78 Stat. 890.

U.S. Public Law 90-271. (San Rafael Wilderness) Act of March 21, 1968. 82 Stat. 51.

U.S. Public Law 90-542. Wild and Scenic Rivers Act of October 2, 1968. 82 Stat. 906.

U.S. Public Law 91-82. (Desolation Wilderness) Act of October 10, 1969. 83 Stat. 131.

U.S. Public Law 91-190. National Environmental Policy Act (NEPA) of January 1, 1970. 83 Stat. 852.

U.S. Public Law 91-504. (Omnibus Wildlife Refuge, National Park, and National Forest Wilderness) Act of October 23, 1970. 84 Stat. 1104.

U.S. Public Law 92-203. Alaska Native Claims Settlement Act of 1971 (ANCSA) of December 18, 1971. 92 Stat. 203.

U.S. Public Law 92-395. (Scapegoat Wilderness) Act of August 20, 1972. 86 Stat. 578.

U.S. Public Law 92-400. (Sawtooth NRA Wilderness Designation and Wilderness Study) Act of August 22, 1972. 86 Stat. 612.

U.S. Public Law 93-378. The Forest and Rangeland Renewable Resources Planning Act (RPA) of 1974. 88 Stat. 476.

U.S. Public Law 93-429. (Okefenokee Wilderness) Act of October 1, 1974. 88 Stat. 1179.

U.S. Public Law 93-622. (Eastern Wilderness) Act of January 3, 1975. 88 Stat. 2096.

U.S. Public Law 94-199. (Hells Canyon NRA and Wilderness Designation and Wilderness Study) Act of December 31, 1975. 89 Stat. 1117.

U.S. Public Law 94-357. Alpine Lakes Area Management (and Wilderness Designation) Act of July 12, 1976. 90 Stat. 905.

U.S. Public Law 94-544. (Point Reyes Wilderness) Act of October 18, 1976. 90 Stat. 2515.

U.S. Public Law 94-567. (National Park and Monument Omnibus Wilderness Designation and Coronado NF Wilderness Study) Act of October 20, 1976. 90 Stat. 2693.

U.S. Public Law 94-579. (Wilderness Study) Federal Land Policy and Management Act (FLPMA) of 1976. 90 Stat. 2743.

U.S. Public Law 94-579. Federal Land Policy and Management Act (FLPMA) of October 21, 1976. 90 Stat. 2743.

U.S. Public Law 94-588. National Forest Management Act (NFMA) of October 22, 1976. 90 Stat. 2949.

U.S. Public Law 95-237. Endangered American Wilderness Act of February 24, 1978. 92 Stat. 40.

U.S. Public Law 95-450. The Indian Peaks Wilderness Area, the Arapaho National Recreation Area and the Oregon Islands Wilderness Area Act of October 11, 1978. 92 Stat. 1095.

U.S. Public Law 95-495. (Boundary Waters Canoe Area Wilderness) Act of October 21, 1978. 92 Stat. 1649.

U.S. Public Law 96-312. Central Idaho Wilderness Act of July 23, 1980. 94 Stat. 948.

U.S. Public Law 96-487. Alaska National Interest Lands Conservation Act (ANILCA) of December 20, 1980. 94 Stat. 2371.

U.S. Public Law 96-550. (New Mexico Wilderness) Act of December 19, 1980. 94 Stat. 3221.

U.S. Public Law 96-560. (Colorado Wilderness) Act of December 22, 1980. 94 Stat. 3265.

U.S. Public Law 97-384. (Charles C. Dean Wilderness) Act of December 22, 1982. 96 Stat. 1942.

U.S. Public Law 97-407. Paddy Creek Wilderness Act of January 3, 1983. 96 Stat. 2033.

U.S. Public Law 97-466. (West Virginia Wilderness) Act of January 13, 1983. 96 Stat. 2538.

U.S. Public Law 98-322. Vermont Wilderness Act of June 19, 1984. 98 Stat. 253.

U.S. Public Law 98-328. Oregon Wilderness Act of June 26, 1984. 98 Stat. 272.

U.S. Public Law 98-339. Washington State Wilderness Act of July 3, 1984. 98 Stat. 299.

U.S. Public Law 98-406. Arizona Wilderness Act of August 28, 1984. 98 Stat. 1485.

U.S. Public Law 98-425. California Wilderness Act of September 28, 1984. 98 Stat. 1619.

U.S. Public Law 98-428. Utah Wilderness Act of September 28, 1984. 98 Stat. 1657.

U.S. Public Law 98-430. Florida Wilderness Act of September 28, 1984. 98 Stat. 1665.

U.S. Public Law 98-508. Arkansas Wilderness Act of October 19, 1984. 98 Stat. 2349.

U.S. Public Law 98-550. Wyoming Wilderness Act of October 30, 1984. 98 Stat. 2807.

U.S. Public Law 98-574. Texas Wilderness Act of October 30, 1984. 98 Stat. 3051.

U.S. Public Law 98-585. Pennsylvania Wilderness Act of October 30, 1984. 98 Stat. 3100.

U.S. Public Law 98-586. Virginia Wilderness Act of October 30, 1984. 98 Stat. 3105.

U.S. Public Law 99-504. Nebraska Wilderness Act of October 20, 1986. 100 Stat. 1802.

U.S. Public Law 100-184. Michigan Wilderness Act of December 8, 1987. 101 Stat. 1275.

U.S. Public Law 100-225. (El Malpais and Cebolla Wilderness) Act of December 31, 1987. 101 Stat. 1539.

U.S. Public Law 100-668. Washington Park Wilderness Act of November 16, 1988. 102 Stat. 3961.

U.S. Public Law 101-195. Nevada Wilderness Protection Act of December 5, 1989. 103 Stat. 1784.

U.S. Public Law 101-628. Arizona Desert Wilderness Act of November 28, 1990. 104 Stat. 4471.

U.S. Public Law 103-77. Colorado Wilderness Act of August 13, 1993. 107 Stat. 756.

U.S. Public Law 103-433. California Desert Protection Act of October 31, 1994. 108 Stat. 4473.

U.S. Public Law 106-399. Steens Mountain Cooperative Management and Protection Act of 2000 of October 30, 2000. 114 Stat. 1655.

U.S. Senate. 1969. Senate Report 91-97. 91st Congress, 1st Session. Washington, DC: U.S. Government Printing Office.

———. 1975. Senate Report 93-803. 93rd Congress, 2nd Session. Washington, DC: U.S. Government Printing Office. 47p.

———. 1980. Senate Report 96-914. 96th Congress, 2nd Session. Washington, DC: U.S. Government Printing Office. 64p.

———. 1984a. Senate Report 98-416. 98th Congress, 2nd Session. Washington, DC: U.S. Government Printing Office. 28p.

———. 1984b. Senate Report 98-463. 98th Congress, 2nd Session. Washington, DC: U.S. Government Printing Office. 37p.

———. 1984c. Senate Report 98-465. 98th Congress, 2nd Session. Washington, DC: U.S. Government Printing Office. 37p.

———. 1984d. Senate Report 98-580. 98th Congress, 2nd Session. Washington, DC: U.S. Government Printing Office. 16p.

———. 1984e. Senate Report 98-581. 98th Congress, 2nd Session. Washington, DC: U.S. Government Printing Office. 26p.

———. 1984f. Senate Report 98-614. 98th Congress, 2nd Session. Washington, DC: U.S. Government Printing Office. 27p.

Watson, A. E.; Patterson, M.; Christensen, N.; Puttkammer, A.; Meyer, S. 2004. Legislative intent, science and special provisions in wilderness: A process for navigating statutory compromises. *International Journal of Wilderness*. 10(1): 22–26.

Watson, Jay. 1996. Wilderness in the National Parks—now more than ever. *International Journal of Wilderness*. 2(1): 24.

Watson, Jay; Brink, Paul. 1996. The California Desert Protection Act: A time for desert parks and wilderness. *International Journal of Wilderness*. 2(2): 14–17.

Weaver, James; Cutler, Rupert. 1977. A colloquy between Congressman Weaver and Assistant Secretary Cutler. *Journal of Forestry*. 75(7): 392–394.

The Wilderness Society. 1974–75. The wilderness system: A report covering every existing or proposed wilderness. *The Living Wilderness*. 38(128): 38–47.

———. 2000. *The Wilderness Act Handbook*, 4th rev. ed. Washington, DC: The Wilderness Society.

Wilkinson, Charles F.; Anderson, H. Michael. 1987. *Land and Resource Planning in the National Forests*. Washington, DC: Island Press.

Worf, William A. 1980. Two faces of wilderness—a time for choice. *Idaho Law Review*. 16: 423–437.

Chapter 6

The National Wilderness Preservation System and Complementary Conservation Areas

Introduction 163
The National Wilderness Preservation System 164
    Ecological and Landscape Diversity Within the NWPS 166
    Federal Agency Management in the NWPS 167
        National Forests 167
        National Parks 167
        National Wildlife Refuge System 167
        Bureau of Land Management—Public Domain Lands 167
        Roadless Areas 167
Complementary Conservation Areas 168
    The National Trails System 168
    The Wild and Scenic Rivers System 169
    State Wilderness Systems 171
    Research Natural Areas 174
    Biosphere Reserves and World Heritage Sites 174
    Private Land Protection Efforts 175
Summary 176
Study Questions 176
Acknowledgments 177
Case Discussion: Recreational Use on the Colorado River in
        Grand Canyon National Park 177
Case Discussion Questions 178
References 178

Introduction

The Wilderness Act in 1964 (U.S. Public Law 88-577) created a National Wilderness Preservation System (NWPS) of areas protected in their natural condition at the primeval end of the environmental modification spectrum. The EMS concept, described in Chapter 1, is a continuum of environmental conditions from "the paved to the primeval." At one end of the continuum, cities and urbanized landscapes dominate; at the other end, more primeval conditions such as naturalness, solitude, and wildness characterize the setting. However, protection of areas with outstanding natural quality is not limited to the NWPS. Through a series of other federal and state laws and administrative actions, additional natural and wild areas also have been protected and managed to retain their unique natural qualities. For example, rivers designated as "wild" within the federal Wild and Scenic Rivers System (U.S. Public Law 90-542) are assured protection against dams and other developments along their free-flowing river segments.

Several states have also established state wilderness systems to protect state-owned wilderness areas that complement the federal program. Numerous but smaller areas of outstanding scientific value have been preserved as research natural areas (RNAs) under various administrative authorities. Protection of landscapes and biodiversity has increased with the support of private, nonprofit organizations such as The Nature Conservancy that own and manage reserves and often purchase sensitive flora and fauna land areas to hold for

Fig. 6.1. Oregon's Three Sisters Wilderness was one of the fifty-four U.S. Forest Service–managed "instant wildernesses" designated upon passage of the Wilderness Act of 1964. Photo courtesy U.S. Forest Service.

subsequent government purchase. Because these other areas may be near the primeval end of the environmental modification spectrum, it is important to understand how they complement and relate to the NWPS.

The National Wilderness Preservation System

Upon passage of the Wilderness Act, a core of fifty-four areas, totaling 9.1 million acres, was brought into the NWPS as "instant wilderness." The NWPS grew tremendously from 1964 to the turn of the century. Figure 6.1 shows the growth of the NWPS from 1964 to 2007 in total. The wilderness system grew slowly during the first few years following passage of the act. As discussed in Chapter 5, this was largely because it took the agencies time to develop procedures to carry out reviews of potentially suitable areas. During the first decade after passage of the Wilderness Act, the number of areas grew from 54 to 89 and the acreage from 9.1 million to slightly more than 11 million acres. By the end of the 1970s, the number of areas had doubled, to 118, as had the acreage, to 18.5 million.

In 1980, the NWPS experienced tremendous growth. Several omnibus bills designated a number of wilderness areas at once in individual states, but the system was expanded primarily by passage of the Alaska National Interest Lands Conservation Act (ANILCA; U.S. Public Law 96-487); that single act designated 35 units in Alaska and encompassed more than 56 million acres. In 1980 alone, 83 units were designated, covering more than 61 million acres, and the number of areas in the NWPS grew more than fourfold while acreage grew to nearly nine times that of the system created instantly in 1964.

The second major growth was 1984, when 223 units totaling 8.3 million acres were added. Many of these additions were in the eastern and southern United States, and the average size was small, about 37,000 acres. Only one area was added in 1985—the 13,300 acre Clifty Wilderness in Kentucky. With these additions, the NWPS almost reached a total of 89 million acres.

The third growth period was in 1990–1994 when numerous additions were designated with an emphasis on the southwestern U.S. The Arizona Desert Wilderness Act of 1990 (U.S. Public Law 101-628) added thirty-nine units and over 1 million acres to the NWPS. The California Desert Protection Act of 1994 (U.S. Public Law 103-433) added sixty-nine units and nearly 8 million acres to the system. The addition of these Arizona and California areas greatly increased the amount of low- and high-elevation desert environments in the NWPS, a component previously underrepresented.

The most recent growth period was in 2000 when fifteen areas, totaling more than 1 million acres, were added in five new wilderness designation laws. The single largest addition was ten areas in northwestern Nevada totaling 757,500 acres and managed by the BLM (U.S. Public Law 106-554). *By 2007, the NWPS included more than 703 units (Table 6.1 shows 735 units because some areas are managed by one or more of the four federal agencies) and covering more than 107.4 million acres.*

The distribution of the NWPS among the four federal land-managing agencies is uneven. *More than*

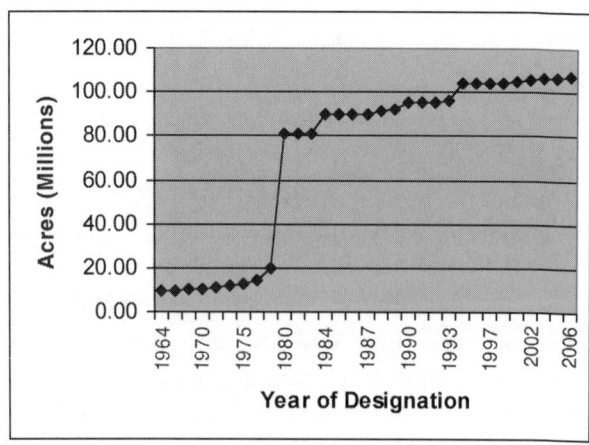

Fig. 6.2. Cumulative annual growth of the acres within the NWPS, 1964–2007 (www.wilderness.net accessed 3/29/08).

Table 6.1. Wilderness Units and Acres in the U.S. National Wilderness Preservation System (NWPS) by Federal Managing Agency in 2007

(www.wilderness.net, accessed March 29, 2008).

Agency	Units[1]	Federal Acres	Percentage of NWPS Acres
NWPS excluding Alaska			
Bureau of Land Management	190	7,796,842	15.6
U.S. Forest Service	399	29,618,974	59.3
U.S. Fish and Wildlife Service	50	2,009,735	4.0
National Park Service	48	10,557,241	21.1
Subtotal	687	49,982,792	100.0
NWPS in Alaska			
Bureau of Land Management	0	0	0.0
U.S. Forest Service	19	5,753,548	10.0
U.S. Fish and Wildlife Service	21	18,692,615	32.6
National Park Service	8	32,979,406	57.4
Subtotal	48	57,425,569	100.0
Entire NWPS			
Bureau of Land Management	190	7,796,842	7.3
U.S. Forest Service	418	35,372,522	32.9
U.S. Fish and Wildlife Service	71	20,702,350	19.3
National Park Service	56	43,536,647	40.5
Total	735	107,408,361	100.0

[1] The number of units managed by each agency does not equal the total because some areas are managed by more than one agency.

half of the wilderness areas are managed by the USFS, but the NPS manages more wilderness acreage. The BLM manages only a very small percentage of the NWPS (7.3 percent) acreage, but its role expanded during the 1990s and more growth is possible.

The average wilderness-area size has increased greatly since 1964; this is a reflection of the addition of some very large areas in Alaska, such as the Wrangell–Saint Elias Wilderness, totaling 9.0 million acres, and the Noatak and Gates of the Arctic Wilderness Complex, which together total 12.7 million acres. In all, 53 percent of the total NWPS acreage was in Alaska by 2007.

The largest wilderness areas in the lower forty-eight states are Death Valley Wilderness in California and Nevada, managed by the NPS (3.2 million acres); the Frank Church–River of No Return Wilderness in Idaho, managed by the USFS and BLM (2.4 million acres); and the Selway-Bitterroot Wilderness in Montana and Idaho, managed by the USFS (1.3 million acres).

The total area of the NWPS (107,408,361 acres) comprises approximately 4.5 percent of the entire U.S. land area. The bulk of the acreage is located in Alaska and the western United States; however, by 2007 some wilderness areas were located in all but six states: Connecticut, Delaware, Iowa, Kansas, Maryland, and Rhode Island.

The smallest wilderness is the five-acre Rocks and Islands Wilderness in California, administered by the BLM. As discussed in Chapter 4, the Wilderness Act defines a wilderness as containing at least 5,000 acres, but it qualifies this by stating "or is of sufficient size as to make practicable its preservation and use in an unimpaired condition." In all, more than 70 areas with fewer than 5,000 acres have been added to the NWPS and many, like 5.5-acre Pelican Island in Florida, are islands where protection of their wilderness character is possible, despite their size (Landres and Meyer 2000). Most wilderness areas (42 percent) are from 10,000 to 50,000 acres in size (Fig. 6.3), even though the areas range from less than 100 acres to millions of acres (Landres and Meyer 2000).

The distribution of the NWPS is not relative to the percentage of the total U.S. land area and federal land area found in each of the four census regions of the U.S. (Table 6.2). The NWPS is located predominantly in

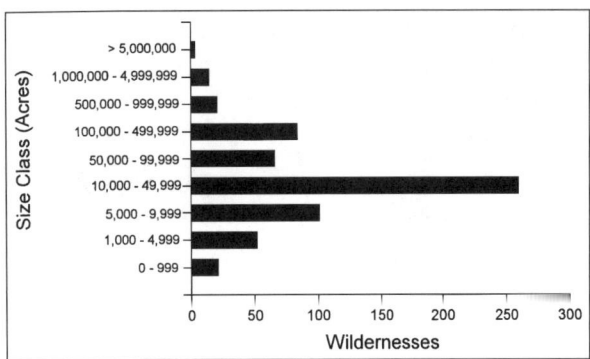

Fig. 6.3. The distribution of wilderness areas (number of areas) by size categories within the NWPS in 1999 (Landres and Meyer 2000, 9).

the western United States and Alaska, and away from most of the nation's population. The western United States has only about 22 percent of the population, but more than 95 percent of the NWPS. The disparity between population and wilderness is most dramatic in Alaska; less than 1 percent of the U.S. population lives there, but the state contains 53 percent of the entire area of the NWPS. Wilderness makes up 16 percent of all land in Alaska and, of the federal lands in Alaska, 26 percent are in the NWPS (Table 6.2), reflecting the relatively limited development and settlement in Alaska. Conversely, *less than 0.5 percent of the total land area in the south, midwest, and northeast regions is in the NWPS and less than 10 percent of the federal lands in those three regions is in the NWPS.*

What does the future hold for further growth in the NWPS? Although the future growth is not clear, some general observations can be made. First, *there is potential for growth in the number of areas and acreage in the NWPS.* Large areas and numerous wilderness proposals and study areas remain to be acted on, especially in national parks and on BLM-administered land. Furthermore, many millions of acres more will likely undergo careful study for wilderness designation in Alaska. Second, although the growth of the NWPS has slowed, the newly added and potential additions may be especially valuable from various perspectives, such as ecological diversity. Finally, *growth of the NWPS will be affected by decisions regarding the management of other lands. It appears likely that many additions will be made to the NWPS, perhaps 25 million acres* (Hendee and Dawson 2004).

Ecological and Landscape Diversity Within the NWPS

The NWPS should be examined in terms of the kinds of lands that it contains, because a major goal, albeit implied, of the Wilderness Act was to help ensure ecological and landscape diversity within the nation's wildlands. Many supporters of wilderness have cited preserving a diverse, representative sample of ecological variety in the United States as a major reason for having the NWPS. However, ecological diversity has never been an explicit criterion driving the wilderness designation process, and as a result, the diversity within the system is largely accidental; there has been little systematic decision-making to broaden the NWPS's ecological coverage. One exception to this general neglect was in the USFS's second Roadless Area Review and Evaluation (RARE II), which used ecosystem representation as a criterion for evaluating potential wilderness additions.

Nevertheless, some efforts have been made to assess the extent to which the NWPS includes representatives of the nation's ecological diversity; the most complete analysis revealed major shortcomings in the 1980s. By merging Bailey's ecoregion concept and Küchler's mapping of potential natural vegetation, the United States was divided into 261 distinct ecosystems (Davis 1989). *In 1989, 157 of the nation's 261 ecosystems (60 percent) were adequately represented within the NWPS, but representation was poor within the grasslands, deserts, eastern hardwoods, and coastal lowland ecosystems.* The addition in the 1990s of extensive desert areas in Arizona, California, and Nevada improved the representation to some extent.

Cordell et al. (2005b) reported that by 2003, forty-three of the fifty-two different provinces in *Bailey's Province-level Ecosystem* classification system were represented to some degree in the NWPS. The extent or area protected in some provinces is small and the least

Table 6.2. Percentage of Total U.S. Land Base and Federal Land Area Designated as Wilderness by U.S. Census Region and Alaska, 2004
(Cordell et al. 2005a, 64)

Census Regions	Percentage of Total U.S. Land Area	Percentage of Federal Land Area
West (less Alaska)	5.8	12.1
Alaska	15.9	26.3
South	0.5	9.6
Midwest	0.3	5.7
Northeast	0.2	8.3

well represented are in the eastern U.S., the prairies of the Midwest, the tropical forests of Puerto Rico, and the Hawaiian Islands (Cordell et al. 2005b). Overall, the fifty-two provinces in *Bailey's Province-level Ecosystem* classification system are organized into four domains; the percentage of total land area of the U.S. within each domain that is protected within the NWPS includes humid tropical domain (19.67 percent), polar domain (13.07 percent), dry domain (3.36 percent), and humid temperate domain (2.82 percent) (Cordell et al. 2005b). Even with further designation, recommendations by the four federal wilderness management agencies to include more examples of each basic ecosystem under their jurisdiction, *many ecosystems and provinces are still not adequately represented in the NWPS or in any equivalent state wilderness system.*

Federal Agency Management in the NWPS

Since passage of the Wilderness Act, the role of the four federal agencies managing wilderness in the United States have changed with the growth of the NWPS.

National Forests

The number of wilderness areas and acreages designated in the national forests and, thus, under USFS administration has grown considerably since 1964. The USFS managed more than 35.3 million acres of wilderness by 2007 (Table 6.1), with wilderness representing an important management category on the national forests. Among their 418 wilderness units, the USFS shares jurisdiction on only a few units with the BLM, FWS, or NPS.

National Parks

Although wilderness was not an administrative designation used in the national parks before the Wilderness Act, it was an important concept underlying the establishment of many parks. It was informally defined as land beyond 0.5 mile of roads; many parks were predominantly de facto wilderness despite the lack of any formal administrative designation. Because no formal system of wilderness reservation like the USFS's primitive-area program existed in the national parks, no NPS lands were included in the "instant" wilderness system created in 1964. Over recent decades, national park representation in the NWPS has expanded. By 2007, there were 56 national park units in the NWPS covering more than 43.5 million acres (Table 6.1). The large NPS wilderness units of 1 million acres or more are in Alaska, California, or Florida. Several parks have large roadless areas recommended for wilderness designation and managed as backcountry, but they still await congressional action to add them to the NWPS (Crumbo 1996).

National Wildlife Refuge System

Passage of ANILCA (U.S. Public Law 96-487) in 1980 resulted in a vast expansion of the U.S. Fish and Wildlife Service component of the NWPS. Under that legislation, more than 18 million acres of FWS land were added to the NWPS. By 2007, the FWS agency managed 71 units in the NWPS totaling more than 20.7 million acres, the majority of it in Alaska (Table 6.1).

Bureau of Land Management— Public Domain Lands

The BLM was not included in the Wilderness Act when it was passed in 1964. At that time, the BLM was primarily a custodial agency, charged with disposal of the public domain, and had no basic authority to manage public lands. However, a national review of public-land laws led to developing organic legislation giving BLM a mandate to manage the public lands under its jurisdiction, and it required the secretary of the interior to review areas of the public lands to determine if they had wilderness characteristics and to recommend to the president their suitability or nonsuitability for preservation as wilderness by 1991.

Until final congressional decisions were made, such lands were to be managed so as not to impair their suitability for wilderness designation. By 2007, the wilderness areas legislatively designated and administered by BLM had grown to 190 units with a total of 7.7 million acres (Table 6.1).

Roadless Areas

Many people seek recreation in areas that are roadless, but that are not wilderness. *Unless we provide for dispersed recreation in roadless areas, wilderness will eventually receive much of this activity, as it will be perceived as the only place such recreation can be accommodated.* This could lead to increased impact on wilderness and to more intensive wilderness management. For example, illegal wilderness use may increase as those pursuing activities such as driving off-road vehicles and motorized trail bikes, mountain bikes, or snowmobiles attempt to find places to enjoy these activities. Thus, in addition to wilderness, roadless recreation areas are needed to relieve demands for recreation use. A variety of names have been suggested for such areas, including backcountry and roadless areas. These relatively unmodified settings would enhance a variety of opportunities such as camping, fishing, and hiking. They would be distinguished from wilderness by the greater latitude provided management to increase recreational capacity, the kinds of permitted recreational activities, and the relatively lower priority assigned to the maintenance of natural ecological processes. Such areas could reduce the level of impact on designated wilderness while meeting the needs of many recreation users seeking roadless recreation environments.

New units and acreage are continual being added to the NWPS, and this reduces the acreage of roadless lands. A variety of means exist to provide a system of roadless areas. One alternative is creating a formal system, protected by law, analogous to the NWPS. Many favor this option because designation without a legal basis does not guarantee long-term protection. Another alternative would be to develop a formal administrative system of roadless recreation areas. The concept of the recreation opportunity spectrum (ROS) is now well established throughout the federal land-managing agencies. Application of the ROS concept suggests the importance of a system of roadless recreation areas to complement designated wilderness.

The concept of a roadless areas system on the national forests by President Clinton in 2000 attracted much controversy, including a legal challenge by President Bush in 2001. Only the latest in a seventy-five-year battle over how to use the remaining roadless areas on the national forests, the Clinton directive would have protected 40 million acres on the national forests from road building and timber harvest (O'Laughlin and Freemuth 2000). Commodity interests see it as only another lockup of resources on the public lands, whereas some wilderness proponents see it being used as a substitute for wilderness protection rather than as a complement to it.

Complementary Conservation Areas

In addition to formally designated wilderness, other kinds of conservation areas complement or supplement the purposes of the NWPS. Although some areas serve purposes similar to wilderness, others are quite different. Nevertheless, at least a portion of these other systems is located at the primitive end of the environmental modification spectrum, and their relationship to designated wilderness merits attention.

The National Trails System

In 1968, Congress established a National Trails System (U.S. Public Law 90-543) to provide for the ever-increasing outdoor recreation needs of an expanding population and to promote public access to travel within, and enjoyment and appreciation of, the open-air, outdoor areas of the nation. *The system now includes three different types of trails: (1) national recreation trails, (2) national scenic trails, and (3) national historic trails.* Some of these trails are within wilderness or areas proposed for wilderness designation and pose a potential conflict to the extent that they may attract use inappropriate to wilderness.

National recreation trails are intended to provide a variety of outdoor recreation uses near urban areas. The secretary of the interior or the secretary of agriculture may establish a national recreation trail with the consent of any other jurisdiction whose lands would also be involved, such as a state or local government. Thus, trails on state lands may also be designated as national recreation trails by the secretary of the interior, with consent of the state. Criteria for designating national recreation trails have been adopted by the secretaries of the interior and agriculture. Such trails may be relatively short (perhaps one-half mile), but must be continuous. National recreation trails are to be available to large numbers of people; consequently, locations such as stream valleys, utility rights-of-way, abandoned railroad rights-of-way, and levees or dikes are likely candidates. They may be designed solely for one use, such as hiking or recreational vehicle use, but opportunities for multiple use are to be explored. Before designation of a trail, the administering agency must provide the appropriate secretary (interior or agriculture) assurance that the trail will be available to the public for at least ten consecutive years.

National scenic trails differ from recreation trails in several respects. They can be designated only by an act of Congress. U.S. Public Law 90-543 established two such trails: the 2,170-mile Appalachian Trail and the 2,650-mile Pacific Crest Trail. *National scenic trails*

Fig. 6.4. Long-distance hikers take a break while backpacking on the Appalachian Trail (upper). The AT runs, in part, along the crest of the Presidential Range and along the upper boundary of the Great Gulf Wilderness (lower) on the White Mountain National Forest in Maine and New Hampshire. Photos courtesy Chad P. Dawson.

must possess superior scenic, historic, natural, or cultural qualities in combination with maximum outdoor recreation potential. They should avoid contact with developments such as transmission lines, highways, and industrial facilities; have adequate public access; and follow principal historic routes. Generally, they will be several hundred miles in length. Use of motorized vehicles on these trails is prohibited. By 2005, there were eight trails with more than 15,600 miles of existing, authorized, or proposed national scenic trails; however, only 10,386 miles of trail actually existed at that time (BLM et al. 2005).

Portions of the National Trail System pass through some units of the NWPS. For instance, the Pacific Crest Trail runs through several wildernesses in the Cascade Range and Sierra Nevada, and the Continental Divide National Scenic Trail follows the crest of the Rocky Mountains and runs through wilderness areas in several states.

Managers of these areas are faced with conflicting objectives when national scenic trails pass through wilderness because wilderness is managed for naturalness and solitude; national scenic trails can attract many users whose primary interest is not in wilderness-related values, and whose impacts may dilute wilderness values. Generally, the more restrictive management standards for wilderness will prevail, but the conflicting objectives and the heavier visitor use attracted to national trails does create management problems. For example, where the Pacific Crest Trail passes high mountain lakes, these scenic alpine locations can become overused by hikers. The Appalachian National Scenic Trail traverses twenty-one designated wilderness areas in the eastern U.S. and the presence of huts along the Appalachian Trail poses management problems when the trail crosses wilderness, such as in the Great Gulf Wilderness on the White Mountain National Forest.

National historic trails differ from scenic trails in that many partially follow roads or are "highway" trails for motorized traffic, with only some segments for hiking use. By 2005, there were twelve trails with more than 26,430 miles of existing, authorized, or proposed trails or roadways, but only 16,169 miles of trail or roadway actually existed as a marked and signed route (BLM et al. 2005). Most of these trails do not cross into NWPS areas although some small historic segments do enter wilderness or are near wilderness, such as the Mormon Pioneer National Historic Trail, which passes near the North Maricopa Mountains Wilderness (BLM) in Arizona.

The Wild and Scenic Rivers System

The Wild and Scenic Rivers Act of 1968 (U.S. Public Law 90-542) protected certain rivers throughout the nation to ensure their free and unimpaired flow and to preserve their outstanding scenic, recreational, geological, fish and wildlife, historic, and cultural values. Originally promoted as a means of countering federal dam-construction programs, the act evolved into an effort to limit development in general along rivers and their banks in the name of recreation (Tarlock and Tippy 1970).

The act designated portions of eight rivers and adjacent lands as immediate members of the National Wild and Scenic Rivers System. Another twenty-seven rivers were identified for study within ten years for possible inclusion in the system eighteen to be studied by the Department of the Interior and nine by the Department of Agriculture.

Fig. 6.5. Rafters on Idaho's Selway River, a wild and scenic river, ride heavy white water. The Wild and Scenic Rivers Act was originally intended to prevent dams on some of the nation's free-flowing streams; it became an instrument to limit development along river corridors and to protect the quality of water-oriented recreation. Photo courtesy U.S. Forest Service.

Designation of rivers under the Wild and Scenic Rivers Act can occur in two ways. First, specific congressional legislation can be enacted to protect a river or river segment. Second, a river can be added to the system by the secretary of the interior, provided certain nomination procedures are followed by the state in which the river is located, and provided there is no administrative cost to be borne by the federal government (Peters 1975).

The Klamath River in California and Oregon is an example of a river added under secretarial designations in 1981 and 1994. Six agencies, including the BLM, NPS, USFS, States of California and Oregon, and Hoopa Valley Indian Reservation, are responsible for management along 297 miles of the Klamath River designated as mainly recreational river, but with four river segments of wild and scenic river designation. The Allagash River in Maine was the first river designated in the national river system at the request of a state and with the state government then assuming primary management responsibility.

The Wild and Scenic Rivers Act recognizes three designations of rivers and provides guidelines for management (NPS and USFS 1982):

1. Wild Rivers—rivers or sections of rivers that are free of impoundments and generally inaccessible except by trail, with watersheds or shorelines essentially primitive and waters unpolluted. Many of these may border or be contained within wilderness areas.
2. Scenic Rivers—rivers or sections of rivers that are free of impoundments, with shorelines or watersheds still largely primitive and shorelines largely undeveloped, but accessible in places by roads.
3. Recreational Rivers—rivers or sections of rivers that are readily accessible by road or railroad that may have some development and agriculture along their shorelines, and that may have undergone some impoundment or diversion in the past.

Designating a river under the Wild and Scenic Rivers Act, combined with wilderness designation of the river and adjacent lands, could have very important complementary benefits. Wilderness designation generally affects broad reaches of land and would protect the broader watershed of which the river is a part. *Wild and scenic river designation, on the other hand, provides some important protection not afforded by the Wilderness Act.* First, designation of a river segment as *wild* provides complete protection against dam construction and other water-development projects. Such facilities can be developed in designated wilderness if judged by the president to be in the public interest. Second, the Wild and Scenic Rivers Act prohibits constructing power transmission lines. Third, although the Wilderness Act does not allow condemnation of private land inholdings, the Wild and Scenic Rivers Act permits administering agencies to condemn private land, if less than 50 percent of the entire river is owned by federal, state, or local government. Land within a city, village, or borough cannot be condemned if valid zoning ordinances protecting the river areas are in effect. Finally, although lands in the NWPS were open to mineral entry until 1984, there was complete withdrawal from mineral entry of lands within 0.25 mile of the bank of any river designated for management under the *wild* category of the Wild and Scenic Rivers System (Tarlock and Tippy 1970). *In general, the Wild and Scenic Rivers Act is more restrictive than the Wilderness Act, and in cases where the two laws overlap the more restrictive provisions prevail.*

By June of 2007, river segments on 165 rivers had been added to the National Wild and Scenic Rivers System (Table 6.3), covering more than 11,408.9

miles of river. The two states with the greatest mileage in the system are Alaska with 3,210 miles on twenty-five rivers and creeks, and Oregon with 1,778 miles on forty-eight rivers and creeks.

Designating rivers as wild, scenic, recreational, or a varying combination of the three designation types plays an important role in preserving diverse wild river communities. Continued support and interest in protecting free-flowing wild rivers will ensure their preservation. Congress continually adds new river sections to the National Wild and Scenic Rivers System. For example, in 2000, six congressional acts designated 363 miles of river, with some designations in all three categories. Those designations added eight rivers or creeks to the National Wild and Scenic Rivers System: (1) Lamprey River, New Hampshire (NPS and local government to manage 12 miles); (2) Wilson Creek, North Carolina (USFS to manage 23.3 miles); (3) Wekiva River, Florida (NPS and State of Florida to manage 41.6 miles); (4) White Clay Creek, Delaware and Pennsylvania (NPS and local government to manage 190 miles); (5) Wildhorse and Kiger Creeks, Oregon (BLM to manage 13.9 miles); (6) Donner und Blitzen River, Oregon (BLM to manage 14.8 miles); and (7) Lower Delaware River, New Jersey and Pennsylvania (NPS and local government to manage 67.3 miles) (www.rivers.gov/publications/rivers-table.pdf).

In addition to the National Wild and Scenic Rivers System, under authority of the federal government, twenty-three state river-protection systems were established by 1984. *Nearly 200 rivers were protected under state legislation involving more than 5,800 miles and with additional river segments under study for possible future state designation.* In 1984, nearly one river in five had some form of protective management under state legislation (Anderson and Morck 1984). Most state programs are fairly restrictive; they typically prohibit instream modification, establish land-use controls, and provide for the management of river use and users (Leatherberry et al. 1980).

State Wilderness Systems

The 1964 Wilderness Act pertains only to lands in federal ownership. Many states have also undertaken actions to preserve lands possessing wilderness qualities. This is an important step, for it represents another way to increase the geographic and ecological diversity of areas given wilderness protection. Many types of ecosystems simply are not found on federal lands; thus, even the most lenient approach to wilderness designation would fail to protect them. Where the states are important landowners, especially in the Midwest and in the East, state wilderness protection could help expand ecosystem representation in protected areas and extend wilderness protection and stewardship to more lands. Moreover, by preserving areas as wilderness, state programs also help ensure provision of a kind of primitive-recreation opportunity that might otherwise not be available.

A national survey undertaken in 1983 examined state-level protection of wilderness (Stankey 1984). Eighty-nine agencies in the fifty states plus Puerto Rico, Guam, and the Virgin Islands were queried. To qualify for recognition as state wilderness, programs had to satisfy five criteria: (1) statutory or administrative recognition of the program; (2) provision for preserving natural qualities and for providing primitive-recreational opportunities; (3) prohibition of resource-development activities; (4) establishment of area size, as specific acreage qualitative description (how large an area must be to meet objectives); and (5) recognition of other values, such as features of historic or scientific interest, considered consistent with management as wilderness.

Nine states were found to have wilderness preservation programs meeting these criteria (Stankey 1984). In addition, three states had designated areas for wilderness protection and purposes, but did not meet all the criteria. Most states have modeled their wilderness programs on the federal legislation. However, some important differences exist. For example, in some states, such as Alaska, wilderness was a zoning designation applied in 1984 to units of the state park system. In California, the state legislature can designate wilderness

Table 6.3. River Mileage Designations for the U.S. National Wild and Scenic Rivers System in 2007
(NPS, June 2007)

River Designation	Classified River Mileage	Percentage by Miles
Wild	5,384.7	47.2
Scenic	2,505.3	22.0
Recreational	3,518.9	30.8
Total	**11,408.9**	**100.0**

but, in lieu of action by that body, proposals can be brought before the California Park and Recreation Commission (Trumbly and Gray 1984).

By 1994, there were eight state wilderness programs still operating from the original nine studied by Stankey in 1984. Florida had its wilderness legislation repealed when it came up for reauthorization in 1989 and the ten wilderness areas were transferred to other state land-management programs (Peterson 1996).

In 2007, Propst and Dawson (2008) used the five criteria developed by Stankey (1984) and added a sixth criterion: a state had a wilderness program provided it had developed management plans to formally define wilderness area objectives that guided managers in activities and decision making. The 2007 study reported that the number of states had declined to seven state wilderness programs from 1984 (Florida had withdrawn its program and Minnesota did not qualify as having a wilderness program due to the sixth criterion); however, the number of designated state wilderness areas had increased from forty-eight to eighty-four areas and the total acreage had increased from 1.7 million to 2.9 million acres (Table 6.4), with important progress made in all seven states. About 96 percent of the acreage was in New York, California, and Alaska. Additionally, Propst and Dawson (2008) reported that they located five states (including Minnesota) that had one or two wilderness areas, but no formally defined wilderness program, and they included seven areas with 275,677 acres.

State activity in wilderness preservation represents an important complementary activity to federal efforts. Although federal legislation has set the general direction in defining wilderness and setting management guidelines, the states have adopted and modified these notions to apply to their situation. In addition, many states are actively engaged in other programs designed to protect the quality of their natural heritage through a variety of protected area designations other than wilderness designation. To see how state wilderness programs function, the California and New York systems are described in more detail (Dawson and Thorndike 2002).

California. In 1974, the California legislature established the California Wilderness Preservation System. Three basic criteria govern admission to the system: (1) the land must be state-owned; (2) the area must remain in, or have been returned to, or have

Table 6.4. States with Wilderness Programs or States with One or Two Wilderness Areas, but No program, in 2007
(Propst and Dawson 2008, 24)

State	Year Established	Number of Areas	Total Acreage
States with Wilderness Programs			
Alaska	1972	5	1,133,400
California	1974	11	475,725
Maryland	1971	30	43,733
Michigan	1972	4	57,733
Missouri	1977	11	22,993
New York	1972	22	1,214,217
Wisconsin	1973	1	5,939
Subtotal		84	2,953,740
States with Wilderness Areas, but No Program			
Minnesota	1975	1	18,000
Hawaii	1981	2	30,857
Maine	1966	2	204,733
Ohio	1988	1	8,000
Oklahoma	1918	1	14,087
Subtotal		7	275,677
TOTAL		91	3,229,417

substantially reestablished its principal, natural character and influence; and (3) the area must be of sufficient size to make its preservation practicable.

Although the system is legislatively founded, new areas can be added either administratively or legislatively. Legislatively established areas are called wilderness areas and are fully protected by law. State wildernesses also can be administratively designated on lands in the state park system by the California Park and Recreation Commission, a body appointed by the governor and overseeing management of the state park system. Although both types of areas are subject to protective management requirements of the act, those established by administrative designation can be reclassified and removed from the state wilderness system without approval of the legislature (Trumbly and Gray 1984).

The 1974 legislation created two wilderness areas: the Santa Rosa Mountains Wilderness Area, of about 87,000 acres, and the 10,000-acre Mount San Jacinto Wilderness Area, abutting the federally designated San Jacinto Wilderness. By 2007, the California

State Wilderness System contained more than 475,725 acres (Propst and Dawson 2008).

Although the California Wilderness Act closely parallels the federal legislation, there are some important differences. The California legislation explicitly states that wilderness areas need not be pristine but only "substantially restored to a near natural appearance" (California Public Resources Code 5093.30). It also contains language prohibiting aircraft flights below an altitude of 2,000 feet; no such prohibition is contained in the federal law. Special provisions permit the unobtrusive use of equipment for collecting hydrometeorological data and conducting weather modification activities; this latter activity in particular is one that has raised concerns with regard to its potential impact on wilderness. Finally, the California law prohibits any vehicle use within wilderness except for emergencies involving the health and safety of individuals. The Wilderness Act of 1964 is more permissive in this regard, permitting permanent roads and access if they are determined to be necessary for the minimum requirements for administration of the area as wilderness.

The California Wilderness System is small compared to the federal wilderness acreage in the state (475,725 acres in state ownership by 2007 as compared to 14.3 million acres administered by federal agencies in 2007), but the types of areas complement those in the NWPS. Much of the land in the state system is in the Anza-Borrego Desert State Park; other areas are along the coastline and nearby coast ranges. The state wilderness system adds to the broad diversity of ecosystems under wilderness protection.

New York. The wilderness preservation movement in New York state began in 1885 with legislation to create forest preserve lands. Numerous areas within the New York system had been logged, and there was concern about creating watershed protection and maintaining a forested landscape. The citizens of the state passed a referendum in 1894 to add constitutional protection to the forest preserve lands set aside within the Adirondack and Catskill Mountains. The most often-quoted portion of the legislation is Article XIV, which, in part, states: "The lands of the state, now owned or hereafter acquired, constituting the forest preserve as now fixed by law, shall be forever kept as wild forest lands. They shall not be leased, sold or exchanged, or be taken by any corporation, public or private, nor shall the timber thereon be sold, removed or destroyed." The state-owned lands within the Adirondack and Catskill forests were subsequently termed the "Forest Preserve." These lands, in combination with extensive private landholdings, were established as regional planning and management areas labeled the Adirondack and Catskill Parks.

Fig. 6.6. A palm oasis in the Borrego Palm Canyon wilderness unit benefits from protection as part of the California State Wilderness Act of 1974. The unit is one of twelve in the 386,000 acres of designated wilderness in Anza-Borrego Desert State Park. The act uses a very similar definition of wilderness to the U.S. Wilderness Act of 1964. Photo courtesy Marilyn Hendee.

The specific designation of some of the Adirondack Forest Preserve lands as "wilderness" was first proposed by the state legislature in 1960 and finally adopted in 1972. By 2007, there were eighteen wilderness management units (including the St. Regis Canoe Area) in the Adirondack Forest Preserve, totaling more than 1 million acres. In 1985, four wilderness units in the Catskill Forest Preserve were created by state agency action and total more than 100,000 acres.

The New York state definition of wilderness is nearly identical to the federal wilderness definition, except New York state requires a minimum size of 10,000 acres. Some of the guiding management phrases, in both the New York state and national wilderness definitions, relate to the resources and environment being of "primeval character" and manifesting only "natural conditions" within the wilderness. The guiding management phrases for recreation management are providing "outstanding opportunities for solitude" and providing for a "primitive and unconfined type of recreation." Management approaches for recreation visitors and resources are similar in New York state to those of federal agencies and include many parallel planning and management activities for wilderness areas. New York state allows few nonconforming uses or facilities compared to the federal system.

Fig. 6.7. Lake Colden and Avalanche Pass, in the High Peaks Wilderness of New York's Adirondack Mountains, are heavily used. The High Peaks Wilderness is the largest wilderness in the New York state system, at more than 226,400 acres, and it has the greatest recreation use per year, with 140,000 visits. Photo courtesy Chad P. Dawson.

Research Natural Areas

In 1966, the Department of the Interior, as part of the United States' participation in the International Biological Program (IBP), established the Federal Committee on Research Natural Areas. The committee was composed of representatives of the major federal land-management agencies along with liaison representatives from the Department of Defense, Atomic Energy Commission, and Tennessee Valley Authority. Its purpose was to inventory and prepare a directory of natural areas on federal lands.

Research natural areas (RNAs) are related to wilderness because one of their key objectives is to maintain natural processes and because they serve an important research and education role. Their specific objectives are:

1. To assist in preserving examples of all significant natural ecosystems for comparison with those influenced by humans.
2. To provide educational and research areas for scientists to study the ecology, successional trends, and other aspects of the natural environment.
3. To serve as gene pools and preserves for rare and endangered species of plants and animals (Federal Committee on Research Natural Areas 1968).

Moir (1972) has identified six basic characteristics of research natural areas:

1. They are examples of the natural environment.
2. Their natural features have been disturbed as little as possible by humans.
3. They are defined by ecological criteria.
4. They are assured the greatest possible degree of preservation and permanency.
5. Their withdrawal is for scientific and educational purposes.
6. They harbor genetic stock of possible value to society.

In 1990, approximately 440 RNAs in the United States were administered by eight federal land-managing agencies. They range in size from less than 1 acre to more than 100,000 acres and total nearly 5 million acres. However, given that one purpose of the RNA system is to preserve a representative array of the nation's ecosystems, much remains to be done (Burns 1984). For example, within the USFS component of the system there were more than 430 areas, totaling more than 500,000 acres of land, by 2006 but not all forest types were represented.

Most RNAs are surrounded or buffered by federal land. Research conducted on these areas must be essentially nondestructive and consistent with the purpose and character of the surrounding land. Studies involving manipulating the environment are generally not permitted.

Recreational use of RNAs is, by definition, limited and subordinate to the scientific and educational objectives for these areas. They nevertheless do serve as important recreational settings for activities that focus on learning and environmental awareness. More important, they supplement the scientific and educational objectives of the NWPS.

Biosphere Reserves and World Heritage Sites

In Chapter 3, international progress in protecting wilderness values was discussed and the authors noted that biosphere reserves and world heritage sites were two important sources of such protection. In the United States, both designations have been used in conjunction with other conservation classifications, such as national parks, to further recognize and protect important natural values. Like the other types of designations we have discussed, *biosphere reserves and world heritage sites represent another mechanism through which wilderness values can be protected. Because biosphere reserves have no legal standing, the protection they confer is through the international recognition that such designation conveys.*

The Biosphere Reserve Project was established in 1973 under the auspices of the United Nations Educational, Scientific, and Cultural Organization (UNESCO). *The basic objective of the program is to foster developing an international network of representative ecosystems* for use in research and education and to promote improved land-management practices. *Biosphere reserves* typically include portions where little or no human impact has occurred, a core zone, as well as areas of human use and impact, multiple-functions zone, and cultural zone. *Worldwide, 531 biosphere reserves had been established in 105 countries by February 2008.*

In the United States, the Biosphere Reserve Project is jointly coordinated by the NPS and the USFS. The first areas established were in 1976: Great Smoky Mountains National Park in North Carolina, Everglades National Park in Florida, Virgin Islands National Park, Coweeta Experimental Forest in North Carolina, Hubbard Brook Experimental Forest in New Hampshire, and Luquillo Experimental Forest in Puerto Rico. *By 2007, forty-seven areas in the United States had received recognition as biosphere reserves.* Portions of several reserves are also designated as wilderness, as, for example, in the Noatak National Arctic Range in Alaska where about one-half of the nearly 6-million-acre wilderness is recognized as a biosphere reserve. Nearly 81,000 acres of the 285,000-acre Three Sisters Wilderness in Oregon is similarly classified. The Admiralty Island Wilderness and Glacier Bay Wilderness have been classified as part of the Glacier Bay–Admiralty Island Biosphere Reserve.

World heritage sites, on the other hand, are products of the International Convention for the Protection of the World Cultural and Natural Heritage (World Heritage Convention) adopted by the UNESCO Geneva Convention in 1972. Rather than focusing on "representative ecosystems" as the Biosphere Reserve Project does, *areas qualifying for world heritage site listing must contain outstanding or superlative qualities of the natural environment.* Management of world heritage sites focuses on protection of these outstanding qualities. In some cases, areas may be recognized by both designations; however, both of these international-level conservation classifications are typically applied in conjunction with an existing designation in the country where the area is located (e.g., national forest, national park).

By 2007, a total of eighteen areas in the United States had been recognized as world heritage sites, plus two areas jointly administered by Canada and the United States. Several world heritage sites overlap areas where wilderness designations are also in effect, such as Yosemite National Park in California. Some areas in the United States are both biosphere reserves and world heritage sites, such as Everglades National Park, Great Smoky Mountains National Park, and Yellowstone National Park.

Selection of these areas as biosphere reserves, world heritage sites, or both reflects international recognition of their value as repositories of natural values, as a source of knowledge and understanding about natural processes, and their contribution to society as a source of pleasure and inspiration. Such designations represent another complementary form of recognition and, thus, protection of wilderness values.

Private Land Protection Efforts

The Nature Conservancy (TNC) is one of the largest and best-known private organizations that protect ecosystems and areas that would otherwise be lost to development. Its origins trace from 1917 when a group of scientists, under the auspices of the newly established Ecological Society of America, formed a committee for the preservation of natural environments. The early work of this group led to publication of *The Naturalists Guide to the Americas*, detailing the known outstanding and unprotected natural areas of North and South America. Later, in 1950, the Ecologists Union, one of the organizations arising from the Ecological Society, took the name "The Nature Conservancy" and launched a program to achieve better protection of wildlife and plants, including an effort to attain nonprofessional participation (Blair 1986).

Using a program of revolving funds—using grants and other monies to acquire land followed by fund-raising efforts to replenish the fund—TNC has become an important force in protecting natural environments; *it manages the largest privately owned group of nature sanctuaries in the world with professional managers who are assisted by volunteers.*

Because a major goal of TNC is to preserve natural diversity, the organization has developed four objectives: (1) *identifying* which endangered natural systems and species have not been preserved; (2) *protecting* the best areas where they are found; (3)

stewardship, managing the protected sites; and (4) *funding*, repaying the amount spent from the revolving fund.

Since the early 1950s, TNC has been involved in protecting more than 12 million acres of habitat, either individually or with partnerships of government, private landowners, and corporations in the United States (Blair 1986). In most cases, TNC has acquired critical habitats through purchases, gifts, and public-private partnerships. In Florida, TNC has acquired more than 1,700 acres for addition to the federally managed Crocodile Lake National Wildlife Refuge to assist in protecting the endangered crocodile (Blair 1986). The organization also relies on volunteers who assist professional managers. In Iowa, for example, the Fern Ridge Preserve provides habitat for balsam firs and an endangered species of snail; the area is managed by the local TNC chapter.

Areas under TNC management are managed primarily to protect natural values. Nevertheless, certain uses are allowed, including recreation. However, great care is taken to ensure that such use and supporting facilities like trails do not conflict with critical species or natural communities.

Summary

After a period of relatively slow growth following passage of the Wilderness Act, the National Wilderness Preservation System has grown rapidly. As the discussion in this chapter indicates, many key ecosystems are still not represented in the NWPS—but working toward representation of all ecosystems is nevertheless a worthwhile goal.

It remains clear, however, that wilderness alone cannot supply all the environmental values and recreational opportunities the public seeks at the primitive end of the environmental modification spectrum. Other roadless, nonwilderness settings must also be provided to more adequately meet public desires. Expansion of the National Trails System, the Wild and Scenic Rivers System and state wilderness systems, and providing roadless recreation areas managed intensively for recreation are needed to adequately cope with growing public interest and demand. The quality of tomorrow's wilderness will depend as much on our success in fully developing these alternative opportunities as on our achievements in developing and implementing innovative wilderness management programs.

Study Questions

1. In what important ways do state wilderness systems complement the federal wilderness system?
2. In what ways do programs such as the National Trails System and the Wild and Scenic Rivers System complement the NWPS? In what ways do they conflict with the NWPS? How might some of these conflicts be resolved?
3. What additional protection can wild and scenic river designation provide for a river in wilderness that is not provided by wilderness designation? Which law is more restrictive and which provisions prevail?
4. In what ways are the goals of the research natural area (RNA) program similar to those of the NWPS? In what ways do they differ?
5. Compare and contrast the biosphere reserve and world heritage site programs. What similarities do they share; how do they differ? Discuss the relationship between both programs and the goals of the NWPS.
6. Explain how biosphere reserve and/or world heritage recognition can help protect a site or area.

Acknowledgments

This chapter was extensively revised in the third and fourth editions by Chad P. Dawson, and John C. Hendee revised it for the second edition. George H. Stankey wrote the chapter for the first edition.

Case Discussion:
Recreational Use on the Colorado River in Grand Canyon National Park

The U.S. Congress established the Grand Canyon National Park (GCNP) in 1919 in northwestern Arizona and then enlarged it to 1.2 million acres in 1975. The GCNP was created for resource protection and enjoyment of the public. The GCNP includes 277 miles along the Colorado River. The most notable features of the GCNP are the spectacular landscape vistas that include the extensive and spectacular geological formations of the inner and outer Colorado River canyon and the numerous side canyons and waterfalls. While the rim areas are vegetated by conifer forests, the animal and plant communities of the canyon form a desert ecosystem intersected by a narrow riparian habitat along the river in the narrow inner canyon. The canyon system includes numerous historical and archeological features and sites.

While the GCNP is most often visited by those who come to look at the geological formations from the canyon rim or to hike down in the canyon, the river has had documented recreational boating use since the 1930s. Prior to 1960, recreational use included regularly scheduled motorized river trips and professionally outfitted trips as well as private, noncommercial trips on the river. The primary motorized and nonmotorized boating use occurs along a 226-mile segment from Lees Ferry to Diamond Creek.

The Wilderness Act of 1964 required a survey of all federal lands to evaluate and identify lands that could be recommended for possible designation as wilderness by Congress. The GCNP was included in the NPS review of these lands and was subsequently recommended for wilderness designation. Thus, the GCNP was required under the Wilderness Act to be managed as wilderness to preserve unimpaired its wilderness character and values in anticipation of its potential designation. The act in section 4(c) prohibits certain uses, including motorized use and mechanical transport. However, section 4(d) allows some exceptions that include nonconforming uses, such as previous existing motorboat use and commercial outfitting and guiding activities related to wilderness travel.

The issue of whether the inner canyon and the Colorado River should be designated by Congress as wilderness has been a pivotal debate that has kept any of the land area in the GCNP from yet being designated wilderness. A variety of wilderness designation proposals and plans have been developed by the National Park Service since 1970; some have recommended phasing out motorized use and others have not. Some have recommended that the river corridor be designated wilderness and others have recommended that it only be included as potential wilderness (due to motorized use). A plan to manage the Colorado River for recreational use has been developed, revised, and challenged in court by river user groups and preservationists. The various proposals to designate the majority (more than 90 percent) of the GCNP, about 1.1 million acres, as wilderness have stalled due to controversy over the recreational motorized use on the Colorado River through the GCNP. The GCNP remains in the category of potential wilderness.

The current version of the "Colorado River Management Plan" (CRMP) was signed in 2006 and allows motorized boat use on the Colorado River with regulations in four zones for the amount and type of use permitted (U.S. Department of the Interior 2006). The largest zone is along the popular 226-mile segment from Lees Ferry to Diamond Creek and it permits motorized and nonmotorized use by both commercial and private (noncommercial) user groups. Each of the four zones has different regulations for these user groups.

The main season in Zone 1 (the longest and most heavily used zone) is 5.5 months, with another 6.5 months for nonmotorized use. Maximum group size allowed varies between eight and thirty-two participants per group and the maximum number of groups allowed to launch per day varies from one to six every other day, with no more than sixty groups (985 passengers) on the river at one time. Maximum trip length varies from ten to twenty-five days depending on type of group and season of the year. An estimated 24,657 users spend 228,986 user days on the river each year. The numbers of allowable total annual recreational users, launches, and user days have increased substantially since the previous CRMP plan in 1989 (for example, 169,950 user days vs. 228,986 user days).

> **Case Discussion Questions:**
>
> 1. What are the pros and cons of designating the GCNP as wilderness and *prohibiting motorized boat use* on the Colorado River? Which wilderness values are protected and which are not protected?
> 2. What are the pros and cons of designating the GCNP as wilderness and *allowing motorized boat use* on the Colorado River? Which wilderness values are protected and which are not protected?
> 3. Discuss which types of wilderness visitor or user are affected by the two different approaches to wilderness designation—such as visitors looking into the canyon from the drive up sites on the rim, backpackers hiking into the inner canyon to camp, river runners, and hikers along the canyon rim. Compare how they would each react to what you reported for the previous two questions.
> 4. Should the majority of the GCNP be designated wilderness and the river corridor excluded from wilderness designation so that motorized use can continue? How would that proposal affect the wilderness experience of wilderness users in the rest of the GCNP? Are there adverse conditions created for wilderness experiences or not?

References

Anderson, Dorothy H.; Morck, Victoria L. 1984. The state of federal river recreation management. In: Popadic, Joseph S.; Butterfield, Dorothy I.; Anderson, Dorothy H.; Popadic, Mary R., eds. *Proceedings of the 1984 National River Recreation Symposium*; October 31–November 3, 1984; Baton Rouge, LA. Baton Rouge: Louisiana State University, School of Landscape Architecture, pp. 466–473.

Blair, William D., Jr. 1986. The Nature Conservancy: Conservation through cooperation. *Journal of Forest History.* 30: 37–41.

Bureau of Land Management, Federal Highway Administration, National Endowments for the Arts, National Park Service, USDA Forest Service. 2005. *National Historic and Scenic Trails: Accomplishments 2001–2005.* Washington, DC: Bureau of Land Management.

Burns, Russell M. 1984. Importance of baseline information to the research natural area program. In: Johnson, Janet L.; Franklin, Jerry F.; Krebill, Richard G., coords. *Research Natural Areas: Baseline Monitoring and Management; Proceedings of a symposium*; March 21, 1984; Missoula, MT. General Technical Report INT-173. Ogden, UT: U.S. Department of Agriculture, Forest Service, Intermountain Forest and Range Experiment Station, pp. 50–52.

Cordell, H. K.; Murphy, D.; Riitters, K.; Harvard, J. E., III. 2005a. The human context and natural character of wilderness lands. In: Cordell, H. K.; Bergstrom, J. C.; Bowker, J. M. *The multiple values of wilderness.* State College, PA: Venture Publishing, pp. 57–89.

———. 2005b. The natural ecological value of wilderness. In: Cordell, H. K.; Bergstrom, J. C.; and Bowker, J. M. *The multiple values of wilderness.* State College, PA: Venture Publishing, pp. 205–249.

Crumbo, Kim. 1996. Wilderness management of Grand Canyon—"Waiting for Godot." *International Journal of Wilderness.* 2(1): 19–23.

Davis, George D. 1989. Preservation of natural diversity: The role of ecosystem representation within wilderness. In: Freilich, Helen R., comp. *Wilderness Benchmark 1988: Proceedings of the National Wilderness Colloquium*; January 13–14, 1988; Tampa, FL. General Technical Report SE-51. Asheville, NC: U.S. Department of Agriculture, Forest Service, Southeastern Forest Experiment Station, pp. 76–82.

Dawson, C. P.; and Thorndike, P. 2002. State-designated wilderness programs in the United States. *International Journal of Wilderness.* 8(3): 21–26.

Federal Committee on Research Natural Areas. 1968. *A Directory of Research Natural Areas.* Washington, DC: U.S. Government Printing Office.

Hendee, J. C.; Dawson, C. P. 2004. Wilderness progress after forty years under the U.S. Wilderness Act. *International Journal of Wilderness.* 10(1): 4–7.

Landres, Peter B.; Meyer, S. 2000. *National wilderness preservation system database: Key attributes and trends, 1964 through 1999.* RMRS-GTR-18. Rev. ed. Ogden, UT: U.S. Department of Agriculture, Forest Service, Rocky Mountain Research Station.

Leatherberry, Earl C.; Lime, David W.; Thompson, Jerrilyn LaVarre. 1980. Trends in river recreation. In: *The 1980 National Outdoor Recreation Trends Symposium*; April 20–23, 1980; Durham, NH. General Technical Report NE-57. Broomall, PA: U.S. Department of Agriculture, Forest Service, Northeastern Forest Experiment Station, pp. 147–164.

Moir, William H. 1972. Natural areas. *Science.* 177(4047): 396–400.

National Park Service; U.S. Forest Service. 1982. Wild and Scenic Rivers Guidelines; National Wild and Scenic Rivers System; Final Revised Guidelines for Eligibility, Classification and Management of River Areas. *Federal Register.* 47(173): 39,454–39,461.

O'Laughlin, Jay; Freemuth, John. 2000. Roadless area policy, politics and wilderness potential: Toward understanding President Clinton's directive to the U.S. Forest Service. *International Journal of Wilderness.* 6(1): 9–12.

Peters, Clay E. 1975. A national system of wild and scenic rivers. *Naturalist.* 26(1): 28–31.

Peterson, M. R. 1996. Wilderness by state mandate: A survey of state-designated wilderness areas. *Natural Areas Journal.* 16(3): 192–197.

Propst, B. M.; Dawson, C. P. 2008. State-designated wilderness in the United States: A national review. *International Journal of Wilderness.* 14(1): 19–24.

Stankey, George H. 1984. Wilderness preservation activity at the state level: A national review. *Natural Areas Journal.* 4(4): 20–28.

Tarlock, A. Dan; Tippy, Roger. 1970. The Wild and Scenic Rivers Act of 1968. *Cornell Law Review.* 55(5): 707–739.

Trumbly, James M.; Gray, Kenneth L. 1984. The California Wilderness Preservation System. *Natural Areas Journal.* 4(4): 29–35.

U.S. Department of the Interior, National Park Service. 2006. Colorado River Management Plan. Grand Canyon, AZ: Grand Canyon National Park.

U.S. Public Law 88-577. The Wilderness Act of September 3, 1964. 78 Stat. 890.

U.S. Public Law 90-542. Wild and Scenic Rivers Act of October 2, 1968. 82 Stat. 906.

U.S. Public Law 90-543. National Trails Act. Act of October 2, 1968. 82 Stat. 919.

U.S. Public Law 96-487. Alaska National Interest Lands Conservation Act (ANILCA) of December 20, 1980. 94 Stat. 2371.

U.S. Public Law 101-628. Arizona Desert Wilderness Act of November 28, 1990. 104 Stat. 4471.

U.S. Public Law 103-433. California Desert Protection Act of October 31, 1994. 108 Stat. 4473.

U.S. Public Law 106-554. Black Rock Desert–High Rock Canyon Emigrant Trails National Conservation Area Act of December 21, 2000. 114 Stat. 2763.

III. Wilderness Management and Planning Concepts

Chapter 7
Principles of Wilderness Management

Introduction 180
Principle 1: Manage Wilderness as the Most Pristine Extreme on the
   Environmental Modification Spectrum 180
Principle 2: Manage Wilderness Comprehensively, Not as
   Separate Parts 181
Principle 3: Manage Wilderness, and Sites Within, Following a
   Concept of Nondegradation 182
Principle 4: Manage Human Influences, a Key to Wilderness
   Protection 183
Principle 5: Manage Wilderness Biocentrically to Produce Human
   Values and Benefits 184
Principle 6: Favor Wilderness-Dependent Activities 185
Principle 7: Guide Wilderness Management Using Written Plans with
   Specific Area Objectives 186
Principle 8: Set Carrying Capacities as Necessary to Prevent
   Unnatural Change 187
Principle 9: Focus Management on Threatened Sites and
   Damaging Activities 188
Principle 10: Apply Only the Minimum Tools, Regulations, or Force
   to Achieve Wilderness-Area Objectives 189
Principle 11: Involve the Public as a Key to the Success of
   Wilderness Management 189
Principle 12: Monitor Wilderness Conditions and Experience
   Opportunities to Guide Long-Term Wilderness Stewardship 190
Principle 13: Manage Wilderness in Relation to Its Adjacent Lands 191
Summary 193
Study Questions 193
Case Discussion: Assisted Trout Migration into Wilderness? 193
Case Discussion Questions 194
References 194

Introduction

Wilderness management is complex. When a problem arises, many solutions are possible; but seldom is there a single, unequivocal answer at hand. Instead, managers must devise and choose from alternative solutions, but it is important that the decision-making rationale produce solutions that are compatible with the wilderness idea. *This chapter offers a set of principles—fundamental assumptions and directions—that will help managers make consistent wilderness stewardship decisions.*

These thirteen principles can guide managers in solving day-to-day problems and in creating long-term policies. The principles offer perspectives on the nature of the wilderness resource, its uses, and its place in the spectrum of land uses. The list does not cover every situation, and managers may want to add a principle or two of their own. Nevertheless, applying these principles will usually result in appropriate action. The principles will often be cited and/or applied to specific management situations covered in subsequent chapters.

These thirteen principles have been developed with the U.S. wilderness system in mind, but many of them may also apply to wilderness in other countries.

Principle 1: Manage Wilderness as the Most Pristine Extreme on the Environmental Modification Spectrum

The concept of the environmental modification spectrum describes a continuum of settings that range from the "paved to the primeval" (Nash 1982); that is, from the totally modified landscape of modern cities to the most remote, pristine, and wild rural locations. *U.S. society encouraged Congress, through the Wilderness Act of 1964 (U.S. Public Law 88-577), to preserve selected areas at the undeveloped and most pristine end of the environmental spectrum as part of the National Wilderness Preservation System (NWPS):* Wording in the act describes its intent as "to assure that an increasing population, accompanied by expanding settlement and growing mechanization, does not occupy and modify all areas within the United States…leaving no lands designated for preservation and protection in their natural condition."

Several other countries (see Chapter 3) and U.S. state legislatures (see Chapter 6) have also decided by law or administrative policy that their most remote

Fig. 7.1. Two mountaineers enjoy solitude and the primeval character of Wolf Peak, in Montana's Absaroka-Beartooth Wilderness. One of thirteen wilderness management principles recognizes that wilderness is one extreme of the land-use spectrum, where naturalness and outstanding opportunities for solitude and primitive forms of recreation are legally mandated. Photo courtesy Richard Behan.

and pristine remaining areas should be protected as wilderness.

Principle 1 recognizes that lands range from the most pristine wilderness to those that are successively more modified than wilderness—from roadless but nonwilderness lands to wildlands with roads set aside for park, recreation, or scenic purposes; multiple-use lands serving resource production goals as well as public recreation, cultivated rural lands; and so on. A variation of this continuum is known as the recreation opportunity spectrum (ROS) and provides a guiding framework for land-use planning for recreation.

In the environmental modification spectrum, the wildest lands are in a relatively undisturbed condition, and the attributes of naturalness and solitude are indicative of wilderness. Uses that alter naturalness and solitude reduce wildness and thereby reduce the total spectrum of environmental conditions, opportunities, and ecosystem services available to meet public interests, needs, and desires. Such uses also threaten to erode the threshold along this spectrum that separates wilderness from other lands. *Thus, this principle mandates that a fundamental objective of wilderness management is to maintain the distinctive qualities that define and separate wilderness from other land uses— the relative naturalness and solitude (the wildness) of designated wilderness compared to adjacent lands.*

To achieve this fundamental objective, we need to resist pressures to modify environmental conditions in wilderness. Wilderness simply cannot, and should not, meet all of the demands that might be placed on it. To do so would directly violate provisions of the Wilderness Act and lead to a loss of those environmental qualities that prompted passage of the act in the first place; that is, the naturalness and solitude— the wilderness conditions—that such areas offer.

Yet it is impossible to protect these distinguishing attributes in the long run unless other opportunities along the environmental modification spectrum are also provided as alternatives so that commodity and recreation demands not requiring wilderness qualities can be met. For example, some nonwilderness but very wild areas may have some wilderness-like qualities that can meet some recreation demands and ecosystem services otherwise destined for wilderness, but they also allow for limited environmental changes and more intensive use and management. Such areas could thus accommodate many primitive-recreation uses outside of wilderness—thereby relieving pressures on wilderness.

A very controversial alternative proposed in 2000 by the U.S. Forest Service, under the title of "New Wilderness Recreation Strategy," for the national forests, would relax limiting use in some heavily used wilderness zones but maintain higher standards of naturalness and solitude in more remote and less visited zones (Oye 2001). Opponents of this view claim that it violates the Wilderness Act and the nondegradation principle of maintaining each area's wilderness standards existing at the time they were designated, and improving naturalness and solitude in areas below a minimum standard (Worf 2001). Others supported the new strategy, claiming that it recognizes use levels already existing in popular locations and politically impossible to roll back, and that physical impacts, not solitude, should govern use levels. "If we lock people out who will fight to save the wilderness" is an expression of this view (Spring 2001, 17).

We support a careful, targeted implementation of such a wilderness recreation strategy, while recognizing the merits of both pro and con arguments. However, some special response is needed in areas where there is unusually high demand for wilderness visitation and there are few alternatives (Cole 2001). The challenge will be in holding the line against physical impacts on naturalness, even while relaxing standards of solitude in popular, high-use areas. The strategy will only be worthwhile if it allows protection of high standards of naturalness and solitude in the surrounding wilderness—thereby maintaining overall the threshold that distinguishes wilderness as the most pristine extreme on the environmental modification spectrum.

Principle 2: Manage Wilderness Comprehensively, Not as Separate Parts

To label something as a *resource* is to say that it normally is useful to society, that it meets a vital need, and usually that it has economic value. Most of our wildland resources—timber, forage, water, and minerals—can be appraised in this fashion, but a few noncommodity resources, like air and especially wilderness, cannot.

Wilderness has only recently acquired value in our culture, and increasingly, we are recognizing economic values derived from protected and natural areas such as clean air, water, sequestered carbon, and other benefits.

Wilderness has clearly achieved the status of a resource in our society by virtue of its cultural values and ecosystem services, and the Wilderness Act refers to "an enduring resource of wilderness."

From a management standpoint, *one important attribute of the wilderness resource is the natural relationship among all its ecosystem parts: vegetation, water, forage, wildlife, and geology. Because wilderness is a composite resource with interrelated parts, its management must be focused on the whole and not on its parts. For wilderness, therefore, one should not develop separate management plans for vegetation, wildlife, or recreation. Rather, wilderness planning must deal comprehensively with the interrelationships between and among these and all other components of the wilderness resource.*

Criteria for managing and controlling the use of wilderness should be based on maintaining natural relationships. For instance, fishing regulations might be guided by effects of anglers on shoreline vegetation and soil and the solitude afforded visitors, as well as by

Fig. 7.2. Installing, maintaining, and monitoring wilderness boundary protects wilderness thresholds. A "No Vehicles" sign marks road's end in the Kofa Wildlife Refuge, Arizona. Photo courtesy Marilyn Hendee.

the impact on the fish populations. Likewise, wilderness recreation might be managed or controlled not only for site capacity, but also for the capacity of vegetation, wildlife distribution and behavior, and water quality to maintain their natural relationships. *Recognizing the need to manage wilderness comprehensively is the foundation of ecosystem management that is now being developed and applied to most wildlands. The idea is that if one part of the ecosystem is allowed to go out of balance, the rest will quickly follow. By strictly maintaining naturalness and solitude in wilderness through truly comprehensive, ecosystem-based management, it becomes possible to identify* ecosystem *benchmarks*.

If wilderness is viewed as a resource, what can be said about its renewability? Clearly, commercial exploitation—logging, mining (both mineral and hydrocarbon development)—would temporarily destroy what we define as the wilderness resource. However, just as clearly, the passage of time, coupled with restrictions on further disturbance, can eventually lead to restoring primitive, naturally appearing conditions—although perhaps without some plants or animals that were a part of the original ecosystem. In this sense, then, wilderness is renewable only to a degree.

Depending on the type of environment, renewing or restoring wilderness conditions could require many generations or even centuries. More interesting, Americans are now beginning to believe that wildlands, especially in the eastern United States, can once again become wilderness. Such a belief is not as widely held in the West, where wilderness is usually designated on lands not yet impacted by settlement. In the East, numerous areas have been designated as wilderness that were once farmed, had roads, and had been logged but are now reverting to wildness. Thus, for some of these lands under today's conditions, wilderness is being judged as their highest and best use (Hendee 1986). One might ask, which are more unnatural or permanently changed—wildernesses in the West that have had fire protection for eighty years or wildernesses in the East that were once extensively farmed and logged, but have now been returned to the influences of natural processes?

Renewability of the wilderness resource was a major theme and debate in the Wilderness Act. Many proponents have pointed to the importance of the act as a mechanism that might allow the wilderness characteristics of disturbed sites to recover. *One of the great promises of the Wilderness Act is that, within reasonable limits, we can designate as wilderness formerly disturbed areas because natural forces and processes can restore the primeval scene.* Thus, it is not easy to describe a precise status of wilderness on the renewable-to-nonrenewable continuum. The criteria for judging the degree of naturalness and solitude required for wilderness designation broadened during the roadless area reviews of the 1970s and 1980s and additional debate and designation in the 1990s. The idea that lands can revert to a wilder state and once again become wilderness is now more widely held. However, *even when management seeks to restore wilderness conditions, the focus must be on wilderness, allowing natural forces and processes to freely operate, thus protecting naturalness as a defining quality of wilderness.*

Principle 3: Manage Wilderness, and Sites Within, Following a Concept of Nondegradation

Not all areas designated as wilderness are identical in their primeval qualities; *wilderness areas vary in the degree to which naturalness has remained unspoiled or to which opportunities for solitude remain undiminished by current, established uses.* Such variations may occur within each wilderness, *thus providing a "wilderness opportunity spectrum (WOS)."* Expectations and definitions of wilderness may change with the condition of the areas surrounding it. The relativity of wilderness qualities and conditions was reflected in debates over the so-called Eastern Wilderness Act of 1975 (U.S. Public Law 93-622) and the subsequent designation of wilderness areas in the East that perhaps did not meet the more rigid, early criteria of the Wilderness Act of 1964 applied in the West (U.S. Public Law 88-577).

How are managers to work for a reasonably uniform standard for the NWPS when qualities such as naturalness and solitude vary widely, not only by region but also among and within individual wilderness areas? Wilderness managers have found the answer in the concept of nondegradation, which has been generally applied to management of air and water quality for decades. Basically, *the nondegradation concept calls for maintaining existing environmental conditions, if they equal or exceed minimum standards, and for restoring conditions that are below minimum levels.* For example, where air quality is currently higher than required by legislation, the higher level of quality should be preserved and not be allowed to deteriorate to minimum legal standards. Thus, the minimum legal standards of air quality do not constitute an acceptable level to which air quality everywhere will be allowed to deteriorate; rather, *the nondegradation objective is to maintain currently high standards, to prevent further degradation, and to restore below-minimum conditions to acceptable levels* (Meyers and Tarlock 1971; Mihaley 1972).

As applied to wilderness, the nondegradation principle recognizes that naturalness and solitude vary among individual wildernesses. The objective is to prevent degradation of current naturalness and solitude in each wilderness and to restore substandard naturalness and solitude to minimum levels, rather than letting all areas in the NWPS deteriorate to a common minimum standard. For example, certain wildernesses having minimum levels of naturalness and solitude do not set a lower standard to which higher-quality areas will be allowed to descend. The more pristine areas in Alaska or the northern Rocky Mountains should not be allowed to decline to the less wild conditions and greater levels of impact found in some urban-proximate wilderness in the southwestern or eastern U.S. Likewise, wilderness designation of heavily impacted areas in the eastern United States does not mean that the level of naturalness and solitude found in those areas should constitute an acceptable level for other areas in the East or in the West.

Generally, under the nondegradation principle, the conditions prevailing in areas when they were designated wilderness, or identified as wilderness study areas for potential future designation, establish the benchmark of naturalness to be sought by management—unless conditions are deemed below standard, and then the objective is to restore naturalness and solitude to some minimum, acceptable standard.

The nondegradation concept also provides for the opportunity to upgrade or to restore wilderness quality. Where existing conditions are judged to be below minimum acceptable levels of naturalness and solitude, an appropriate priority of management is to promote restoration of wilderness to minimum quality levels. This does not automatically imply that wilderness restoration will involve activities such as planting grass and shrubs, fertilizing, and watering (unless that is the minimum necessary approach). Numerous management activities and policies can promote natural restoration by methods such as controlling visitor numbers, timing of use, excluding use in some areas, and other measures. How purely to manage each wilderness has been a long-standing issue; the principle of managing wilderness and sites within, under a nondegradation concept, helps resolve this issue.

See Principle 1 for a discussion of a current initiative by the U.S. Forest Service called the New Wilderness Recreation Strategy (Oye 2001), which would relax standards of solitude in certain high-use wilderness locations. This strategy is to respond to high demand for wilderness visitation to such sites and not just divert such use to surrounding, low-use wilderness, thereby also diluting its naturalness and solitude.

Principle 4: Manage Human Influences, a Key to Wilderness Protection

The need for managing wilderness users hardly has to be argued (see Chapters 1, 15, and 16). Wilderness visitation has grown steadily, and many new areas that already sustain substantial primitive recreation have been added to the NWPS. Specific examples of site impacts and deterioration can be found in virtually any area. Human influence extends even to the most remote wilderness environments. Indirect influence

Fig. 7.3. Threats to wilderness resource conditions such as solitude and quality visitor experiences can be marred by heavy recreation use and congestion on mountain summits and overlooks. High Peaks Wilderness in New York State Adirondack Park. Photo courtesy Chad P. Dawson.

Fig. 7.4. An important principle of management is to favor wilderness-dependent recreational activities—those that require naturalness and solitude. A rider (top) in the Sycamore Canyon Wilderness, Arizona, and a solitary angler (bottom) in the Absaroka-Beartooth Wilderness, Montana, experience naturalness and solitude. Photo on top courtesy Starr Jenkins; Photo above courtesy Richard Behan.

of humans is evident, for example, in the unnatural vegetation patterns resulting from fire prevention and suppression, which in turn contribute to unnatural distributions of wildlife. The fragile, sometimes irreplaceable, qualities of these areas are easily lost unless thoughtful, deliberate stewardship protects against the direct impacts of use and the indirect but pervasive influences of civilization and global change.

Increasing the size of the NWPS offers only short-term protection. *Ultimately preserving wilderness resources and the values and benefits they offer people will depend on the stewardship and management of wilderness areas after they have been legally designated as wilderness or identified as wilderness study or candidate areas.*

The principal goals of wilderness preservation are to maintain long-term ecological processes and to provide opportunities for solitude and primitive forms of recreation. Where these processes and opportunities have been lost, people are generally the cause. *Thus, wilderness management is basically concerned with managing human use and influences to preserve naturalness and solitude.*

Recreation and permitted uses, like grazing, cause impacts that are currently among the most critical, unnatural influences in wilderness. Therefore, managers are challenged to guide, modify, and, if necessary as a last resort, directly control wilderness uses and influences to minimize their impact. This has become an accepted tenet of wilderness management, although specific controls on visitation and uses (who and where) are usually controversial.

To date, wilderness management has emphasized management of people. However, ecological problems are also becoming more important; wilderness managers are increasingly challenged to monitor the naturalness of wilderness ecosystems and to provide counterinfluences to human impacts. Fire, for example, needs to be restored to a closer approximation of its historical role and status in many areas (see Chapter 11). Noxious weeds like spotted knapweed, yellow star thistle, and leafy spurge have gained a foothold in many areas. Endangered wildlife may depend on natural conditions and solitude of wilderness (see Chapter 12). The growing number of impacted sites in wilderness, our increasing knowledge about how they occur, and our skill in controlling or correcting them are also extending wilderness management activities to deal directly with these ecological and physical impacts (see Chapters 15 and 16). Furthermore, the NWPS has grown and now includes some previously impacted areas that are in need of protection so that their natural wilderness qualities can return.

Principle 5: Manage Wilderness Biocentrically to Produce Human Values and Benefits

Wilderness is designated not just to protect its flora and fauna, but also for its enjoyment, values, and benefits (as wilderness) for people. The preservation goals for wilderness areas are clear in the Wilderness Act (U.S. Public Law 88-577): "it is…the policy of the Congress to secure for the American people of present and future generations the benefits of an enduring resource of wilderness." Thus, wilderness managers are legally mandated to produce the important and—in the judgment of some—necessary benefits that wilderness provides to people (see Chapter 16 for more discussion).

While benefits to wilderness users such as psychological restoration, physical challenges, or spiritual renewal are not easily measured, there also is no evidence that such benefits are produced only in wilderness. However, realizing that benefits are derived from wilderness preservation and use is implicit in the wilderness ideology and philosophy that led to passage of the wilderness acts and the subsequent designation of more than 107 million acres of land as U.S. wilderness (Scott 2004). These anticipated benefits reflect one of the central beliefs of the founding fathers of the U.S. wilderness system: that the character-building values of wilderness are vital to society (D. W. Scott 1984). The continued protection of wilderness naturalness and solitude as the source of these human values and benefits, and access to wilderness to experience them, are fundamental goals of wilderness stewardship.

There are two widely differing wilderness management philosophies; both are aimed at enhancing human benefits from wilderness. *The biocentric perspective emphasizes environmental integrity as the basis for human benefits, a hands-off approach. The anthropocentric perspective promotes active management of natural processes and provides facilities to increase aesthetic pleasure and facilitate wilderness use* (see discussion in Chapter 1).

The federal wilderness management agencies—the National Park Service (NPS), U.S. Forest Service (USFS), U.S. Fish and Wildlife Service (FWS), and Bureau of Land Management (BLM)—have established wilderness management policy manuals and/or guidelines reflecting a commitment to maintain high standards of naturalness and opportunities for solitude as mandated by the Wilderness Act. These agency policy manuals and handbooks provide direction for handling specific wilderness issues to preserve its contrasts to environments already modified by human activity. The agencies, in response to the Wilderness Act, might be considered generally biocentric, but there are differences in emphasis among them. With plenty of discretion remaining for agency wilderness managers in making day-to-day decisions, their personal philosophy is an important influence. We favor a biocentric philosophy for reasons outlined next and in Chapter 1.

We believe *it is from the primeval qualities of wilderness that human benefits are derived* (White and Hendee 2000). The anthropocentric approach *attempts to facilitate wilderness enjoyment by improving access to make visitation easier, more convenient, or simultaneously accessible to a larger number of people and can ultimately diminish wilderness's unique values and benefits. To us, this reality calls for a biocentric emphasis in wilderness management but applied with common sense.* In addition, strict standards for wilderness management ("purity in the extreme") can and have triggered a backlash from persons and groups whose access to wilderness has thus been unduly limited or controlled and from those who resent government regulations.

The biocentric-anthropocentric concepts are abstractions that facilitate discussion of management alternatives. The biocentric approach emphasizes keeping the wild in wilderness but allowing as much use as is consistent with that goal so people can directly benefit from wilderness experience. *How do your personal wilderness values rate on this biocentric-to-anthropocentric continuum? To find out, you may want to take the test at the end of Chapter 1* (Clark and Kozacek 1997).

Principle 6: Favor Wilderness-Dependent Activities

Wilderness is the setting for activities ranging from scientific study to recreational pursuits such as fishing, backpacking, day hiking, hunting, mountain climbing, rafting, and picnicking, some of which depend on a primeval setting. Some types of scientific study depend on access to a substantially unaltered ecosystem, perhaps covering a large area, and such conditions may only be found in wilderness. Thus, such use is wilderness-dependent (see Chapter 14 for more discussion of wilderness dependency). Conversely, other activities, such as certain kinds of fishing, are not dependent on a wilderness situation at all, although such a setting can enhance them, for example catching a native trout at a high mountain lake and preparing it for dinner over an open fire. *Whenever one or more uses conflict or compete for access, the principle of wilderness dependency, which calls for favoring activities most dependent on wilderness conditions, is used to guide visitor management toward preventing overuse.*

This principle is intended to ensure optimum use of wilderness resources, because activities that are not wilderness-dependent can, by definition, be enjoyed in other settings. Thus, what conflicts may arise should be resolved in favor of wilderness-dependent uses; that

Fig. 7.5. More than a dozen trail riders with pack stock enjoy Washington's Glacier Peak Wilderness. Wilderness managers must assess visitor use and consider their area's carrying capacity. Too many visitors can depreciate the quality of the wilderness environment and solitude. Photo courtesy Robert DeWitz.

policy avoids displacing those seeking true wilderness experiences, whose opportunities are in short supply compared to those satisfied with nonwilderness experiences. This point of view is not new and was championed long ago by Robert Marshall (1930, 1937), a founding father of the wilderness idea and system. Thus, providing an adequate range of nonwilderness recreation opportunities is important, as discussed under Principle 1, so persons not dependent on wilderness can use alternative areas for their activities.

Defining an activity as wilderness-dependent can be difficult. Often, it is not the activity itself that is dependent, but the particular style in which it is pursued. For example, hunting is not necessarily wilderness-dependent. However, certain styles of hunting, such as pursuing game under the most natural conditions away from roads, or stalking a bighorn sheep among high peaks, are highly dependent on wilderness settings. The importance of naturalness and solitude to the experience, not the mere quest for game, defines certain kinds of hunting as wilderness-dependent (see Chapter 12).

Likewise, although people can fish in many settings, certain styles of fishing may be wilderness-dependent. Those who desire remote, difficult-to-reach lakes, where one can fish under natural conditions without meeting other people, may rely more and more on wilderness for such opportunities. Many visitors report that fishing is an important part of their wilderness experience and complements satisfactions from other activities, such as observing aquatic life, photography, and so on.

Some higher-risk and adventure-recreation activities are not wilderness-dependent because their focus may be on the physical challenge and excitement and not the wilderness conditions and setting. In contrast, some commercial wilderness-experience programs use the wilderness environment as part of their method for education, therapy, or spiritual renewal.

Thus, favoring wilderness-dependent activities might call for reducing or discouraging—rather than eliminating—certain *forms* of some activities and encouraging other forms that are more wilderness-dependent. One is reminded again of the interdependency of wilderness with the rest of the recreation opportunity spectrum (ROS). One key to being able to emphasize wilderness-dependent activities in designated wilderness is to provide alternative nonwilderness lands to which nonwilderness-dependent uses can be diverted.

Principle 7: Guide Wilderness Management Using Written Plans with Specific Area Objectives

Wilderness management actions must be guided by formal plans that state specific area objectives and explain in detail how they will be achieved. Without such clear statements, management can become incremental, uncoordinated, and even counterproductive to the goals of the Wilderness Act. Local managers and the public need wilderness management plans to consider whether activities and strategies are appropriate for specific areas and are consistent with legislative goals and national policy. Public involvement is an important part of wilderness management planning and can be an educational process for both managers and citizens.

Wilderness management plans focus increasingly refined policy and action guidelines on stewardship of individual areas (see Chapter 8). There is a hierarchy of guidance, from the general outline of goals from the Wilderness Act, to narrower departmental regulations in the Code of Federal Regulations, to more explicit agency guidelines expressed in policy manuals, and to plans that translate into specific objectives for given areas and the policies and management actions to be used to achieve them. The planning process helps local agency officials and citizens review the varying situations of individual areas and develop local management strategies consistent with broad legislative goals. The legislative history and law designating a wilderness provide additional direction; for an example of legislative history influencing wilderness airstrip management see Meyer (1999).

Wilderness management plans must include specific objectives—clear statements of desired wilderness conditions, so proposed management actions can be evaluated for their potential contribution toward a specific objective. Clear objectives and the commitment to be guided by them are important because management actions can have enduring—even irreversible—results. Philosophies, perceptions, and definitions of wilderness can vary widely among managers. Plans with clear, formally stated objectives are needed to guide judgments about what management actions are necessary; to provide continuity when managers change; and to prevent potential damage from ill-conceived plans, no matter how well intended.

Excessive or poorly conceived management actions can be as damaging to wilderness values as the absence of management. For example, a series of relatively minor decisions outside an overall plan might result in too many trails being built to unnecessarily high standards, an excessive number of signs, or unnecessary restrictions on user activities. *The combined impact of such a tyranny of small decisions can depreciate wilderness values. Only the minimum actions necessary to achieve objectives should be planned.*

Articulating objectives gives benchmarks for measuring progress toward goals. If an objective is not met, an evaluator needs to ask why: Lack of feasibility? The need for different policies or implementing actions? The need for different kinds of administration or enforcement? Because the goals of the Wilderness Act are so broad, it is difficult to write clear objectives for all the various aspects of wilderness management. However, it is crucial to develop, through an orderly planning process, the clearest and most specific objectives possible and to use them as constant guides to management actions.

Principle 8: Set Carrying Capacities as Necessary to Prevent Unnatural Change

Wilderness has limited capacity to absorb the impacts of use and still retain its wilderness qualities. As use increases, or as damaging patterns of use develop at specific places or during particular times, wilderness qualities may disappear, either over time or rapidly. *Carrying capacity is the amount of use an area can tolerate without unacceptable change in conditions. The concept of carrying capacity offers a framework for managing use to preserve wilderness qualities.* (See Chapter 9 on carrying capacity and the limits of acceptable change—LAC.)

Change due to natural ecological processes inevitably occurs in wilderness. The purpose of wilderness management is not to freeze ecosystems at a point in their natural succession—such as a vignette of earlier times—but to allow natural processes and change to occur with an absolute minimum of human manipulation and influence. Ecosystems are in constant change and the goal of wilderness management is to see that such change is natural—staying within its historical range of variation. Both ecological conditions and available wilderness experiences may thus also change. Standards of naturalness (the ecological integrity of processes and conditions) and human solitude that are established for an area—and the specific area-management objectives that express these standards—help define the carrying capacity of an individual wilderness.

Applied to wilderness, the concept of carrying capacity has two important parameters: (1) physical-biological and (2) social-psychological. Physical and biological dimensions describe the amount and kind of use an ecosystem can sustain without undue evidence of unnatural impacts. Unnatural physical and biological impacts include such things as campsite deterioration and soil compaction, trampling vegetation, proliferation of paths to and from locations of concentrated human use, exposed and protruding tree roots, and the unnatural behavior and distribution of wildlife (see Chapter 15).

Social or psychological dimensions refer to the levels and concentrations of human use an area can accommodate before the solitude of wilderness experiences is unacceptably diminished. "Outstanding opportunities for solitude" is a wilderness condition specified in the Wilderness Act. Several studies document that solitude—privacy from persons in other parties, particularly from large parties and other users camping near one's campsite—is an important attribute of the wilderness experience. Concentrations of visitors at popular campsites and other attractions such as hot springs, alpine lakes, or shelters might indicate that the social carrying capacity is being approached or exceeded.

Four major points are summarized from Stankey et al. (1985) on carrying capacity (see Chapter 9 for more detail):

1. Carrying capacity is a relative term, not an absolute number to be discovered by managers and researchers. Its range depends on specific objectives and the standards established for an area.
2. Carrying capacity must be established and identified in the field by managerial judgments. No magic yardstick or formula can tell when it will be or when it has been exceeded.
3. Carrying capacity is tied to (a) the conditions of the physical-biological environment and (b) wilderness experience—the qualities of the human experience and values available in wilderness. Both physical-biological and social-psychological dimensions of carrying capacity are important.
4. The development of carrying-capacity limits is a necessary part of the planning process for those areas and locations where unacceptable change has occurred or may occur. To achieve long-term wilderness preservation goals, wilderness use must be managed over time and space to maintain its impacts within limits of acceptable change.

Principle 9: Focus Management on Threatened Sites and Damaging Activities

Wilderness management is mainly concerned with maintaining naturalness and solitude, and impacts on these qualities are usually concentrated in time and space, that is, physical impacts of use at popular locations, crowding impacts on group or individual visitor solitude, and the social-psychological experience of visitors. *This calls for a site-specific orientation of management, rather than an across-the-board approach that would impose restrictions everywhere in a wilderness to solve problems that might be concentrated at popular locations, along high-use corridors, and/or at certain times or seasons.*

Restrictions on use have been implemented for decades in some more heavily used wilderness areas and at certain popular locations in many others. The examples of intense demand for wilderness use (and thus the need for use limits) at Mount Whitney and Grand Canyon, cited in Chapter 2, are extreme but not uncommon across the NWPS. Whenever such restrictions are contemplated, several difficult questions must be addressed: Are use restrictions the minimum tool to solve the problem? Where and when should use be restricted? Who (what kind of users) should be restricted? Under what conditions and criteria? How should restrictions be implemented?

Obviously, not all uses produce equal impacts. *All types of wilderness use and activities can generally be ranked according to their relative physical, biological, and social-psychological impacts. When restrictions are necessary, those activities having the greatest long-term impact can be the first ones controlled.* For example, various types of wilderness use might be ranked, even in the absence of an actual study, in the following order of decreasing environmental and social-psychological impact: large parties of horse users, small parties of overnight campers building fires, small parties of overnight campers not building fires, and small parties of day hikers. Additional criteria—such as visitors' knowledge or skill levels—might also be used to establish or modify priorities among users as well as the degree of wilderness dependency of their planned activities (see Chapters 14 and 16).

The unacceptable impact from various users might require regulation in only a few locations in a wilderness. Furthermore, the vulnerability of the resource is greater at different times, such as in early spring when vegetation is emerging and is easily damaged or on peak weekends when use is heavier. *Thus, to minimize excessive environmental and social-psychological impacts, restrictions should be selective—to times, places, users, and activities having the greatest potential for damage.*

Selective restriction strives to promote equity by specifying those conditions under which uses will be regulated, thus minimizing overall restrictions. Discrimination against certain types of use, such as pack stock or large, organized groups of hikers, is based on their respective impacts on the wilderness environment. It relieves wilderness managers of deciding arbitrarily on a certain mix of horse users, organized groups, hikers, and so forth; or alternatively, it keeps managers from deciding that because they have no criteria for making such choices, everyone or no one must be restricted. Many judgments must be made in wilderness management and, even with principles and criteria to guide decision logic, it is not an exact science. Sensitive managers willing to make decisions are required.

Principle 10: Apply Only the Minimum Tools, Regulations, or Force to Achieve Wilderness-Area Objectives

Freedom, spontaneity, and escape are recognized as important qualities of the wilderness experience. However, as discussed elsewhere in this book, restrictions must sometimes be imposed to prevent these fragile attributes from being destroyed by overuse or misuse. Ironically, regulation itself can diminish the quality of the wilderness experience unless it is carefully implemented, and managers must always balance visitor freedom and resource protection (Cole 2000, 2001). *The guiding principle is that only the minimum tools, regulation, or force necessary to achieve established wilderness-area objectives are justified. This principle is sometimes called the minimum tool rule—apply only the* minimum *tools, equipment, device, force, regulation, action, or practice that will bring the desired result.* A "Minimum Regulation Decision Guide" is available online at www.wilderness.net.

Wilderness management actions fall on a continuum, ranging from subtle, light-handed, and indirect options (education and information) to direct and authoritarian options—such as telling visitors where they can travel and camp each day and how long they can stay. However, a key goal of wilderness management is to use indirect methods wherever and whenever possible to delay and minimize the need for direct controls (see Chapter 16). Which approach is appropriate depends on a manager's judgment about the degree of regulation necessary to achieve objectives and the likely effectiveness of various regulatory and nonregulatory actions in a certain situation.

Subtle, indirect controls include visitor education, appeals for self-regulated and low-impact camping practices, and locating trailheads and trails to minimize impacts on fragile areas. *Direct methods* might begin with managing specific, overused sites, where efforts would be made to educate, or sometimes to disperse users (keeping in mind that dispersing use may spread the impacts). Only as a last resort, when an array of specific and successively more restrictive measures have been exhausted, would direct control of visitation be considered.

If, for example, managers wish to more evenly distribute visitors, they might first use an indirect, educational approach—provide information about current visitor distribution, alternative trailheads and routes, times when use is low, and so on. However, if this approach appears to be inadequate or fails to redistribute use as desired, then a more direct, restrictive approach might be in order. A manager might then limit camping at heavily impacted sites, set entry quotas at each trailhead, or even assign campsites.

Fig. 7.6. A U.S. Forest Service ranger contacts wilderness visitors in the Three Sisters Wilderness in Oregon. Educating the public is one of a wilderness manager's most important tools. Direct visitor contact and guidance can help prevent undesirable impacts and postpone initiating stringent regulations. This follows the principle of using only the minimum level of regulation necessary to achieve wilderness management objectives. Photo courtesy U.S. Forest Service by Tom Iraci.

Ultimately, for every wilderness area, there is a point beyond which the wilderness experience becomes so diluted by restrictions that the feelings of freedom, spontaneity, and escape are in danger of being lost. It is at this point—which must be subjectively determined by managers, though perhaps guided by research findings—that rationing of entry, routes, and campsites must begin.

The U.S. Forest Service's so-called new wilderness recreation strategy was based, in part, on the fact that dispersing visitors to popular wilderness sites might spread impacts sufficient to dilute wilderness conditions in surrounding pristine areas that are lightly visited (Oye 2001). Therefore, their proposed strategy is to increase management at the high-density sites to contain the impacts, even though standards for solitude have to be relaxed. A positive trade-off here is being able to maintain high standards of naturalness and solitude in the surrounding lightly used wilderness.

Principle 11: Involve the Public as a Key to the Success of Wilderness Management

The Wilderness Act was one of the first major laws mandating *public involvement* in natural-resource management—in selecting and designating areas to be set aside as wilderness. Nevertheless, during the first decade or so following passage of the Wilderness Act, many wilderness managers thought that the public did not have enough expertise to participate in wilderness

Fig 7.7. Involving the public invests their interest in aspects of managing wilderness. Volunteers clean up trash at an old mining and homestead site in California's Old Woman Mountains Wilderness managed by the Bureau of Land Management. Photo courtesy Marilyn Hendee.

management decisions. They thought that only professional resource managers could understand the legal, ecological, and social complexities involved. However, subsequently, with the passage of other laws requiring public involvement, such as the National Environmental Policy Act (NEPA) in 1969, the public rapidly learned how to participate in wilderness and public-land management (U.S. Public Law 91-190). Wilderness managers also became more adept at working with people and providing for public involvement as an essential step in planning for wilderness designation and wilderness management. *Today, public involvement is recognized as perhaps the most important tool for successfully developing and implementing wilderness management plans and actions*—and all other management of public lands. Laws increasingly require public involvement for more and more decisions. Clearly, *any proposed wilderness management action needs public involvement as a source of practical information and essential public support, without which its implementation will fail.*

Public involvement is also extending beyond planning and decision making to wilderness management work. Volunteerism is a key to federal land management everywhere, and on state land too. Wilderness duty has been a favorite with volunteers for decades. *Managing volunteers and group service projects is a key wilderness management skill and task.* For example, managing volunteer rangers, wilderness information specialists, wilderness "cleanup" and "adopt-a-trail" projects, private contracting for projects such as trail construction and maintenance, and wilderness field-management projects with volunteers and partnerships with organizations all help increase citizen involvement in wilderness work. These volunteer efforts supplement diminishing budgets, but their greatest value goes beyond that: *they increase recognition and appreciation of wilderness values—and that is one of the ultimate purposes of the NWPS* (Hendee 1986).

Public involvement in wilderness management is firmly established and growing in importance. This has serious implications, because it is turning wilderness managers into public facilitators. No longer will most wilderness managers apprentice as seasonal employees and develop field skills for subsequent career positions. Field apprenticeships now are more likely to be filled by volunteers. Increasingly, career positions will deal with managing volunteer projects, coordinating cooperative work with conservation and service organizations, contracting with private firms, and educating the public.

Principle 12: Monitor Wilderness Conditions and Experience Opportunities to Guide Long-Term Wilderness Stewardship

Any management plan or program needs a monitoring system to evaluate progress toward stated objectives and to guide its long-term revisions, adjustments, and refinements. Devising monitoring systems remains one of the major challenges for advancing wilderness management. *A good plan describes what objectives are to be achieved. Only through monitoring—the systematic gathering, comparing, and evaluation of data—can one tell whether those objectives are being realized.* For example, under the limits of acceptable change (LAC) planning approach, data

Fig. 7.8. The National Wilderness Preservation System gained 516,200 acres with the designation of the Kofa Wildlife Refuge Wilderness in western Arizona, administered by the U.S. Fish and Wildlife Service. Monitoring the numbers of visitors, how they use wilderness, and wildlife behavior and interaction is an important task in wilderness management responsibility. Photo courtesy Marilyn Hendee.

are needed to establish baseline conditions and then a system to monitor indicators to determine the degree of change occurring in various opportunity zones (Stankey et al. 1985; also see Chapter 9).

Just what data should be collected, by what methods, and to what precision are important concerns for managers and scientists studying how to measure success. The challenge is to measure and evaluate specific indicators that can be compared to certain standards for *biological*, *physical*, and *social* conditions in wilderness (Merigliano and Krumpe 1986). *Biological conditions* included categorizing vegetation, mammals, and fish have to consider a total eighty-six potential indicators, which might singly or in combination reveal change. *Physical conditions* were categorized under soil, water, and air, with seventy-five potential indicators. *Social conditions* included describing visitors (e.g., day hikers, anglers, backpackers) and qualities of wilderness experiences, with fifty indicators considered. However, fieldwork has yet to test most of these proposed indicators in wilderness, or to use them to evaluate wilderness management plans. More recent work has focused on monitoring wilderness character, a hard-to-measure but important approach to check on the quality of the wilderness in the NWPS (Landres et al. 2005). This illustrates how important new monitoring is to wilderness management, yet how much work remains to integrate monitoring into it.

Wilderness monitoring is also important for realizing the ultimate values of the NWPS for science and environmental assessment. Critical questions facing humanity concern how human activity has modified natural processes globally as well as regionally and locally. Acid rain and related air pollution impacts, the effects of human activity on water quality and volume, climatic variation, heating and cooling of the earth, and other subtle global changes are all topics of concern. Such impacts may be reflected first in changes in vegetation and fish and wildlife habitats, and, subsequently, in loss of species. For example, in Alaska where climate change is pronounced, species migration and changes in relative abundance attributable to global climate change are occurring with species such as the polar bear (Regehr et al. 2007).

Wilderness provides enclaves of the earth's most natural remaining areas and can be a benchmark source of information on the degree of distortion of natural processes elsewhere—but only if sufficiently detailed information is collected and made available for use as

Fig. 7.9. The majority of visitor use in the Stikine-LeConte Wilderness in southeast Alaska is along the Stikine River corridor or its saltwater area. The wilderness's interior is pristine and dominated by glaciers and rugged topography. Photo courtesy John Hendee.

environmental baselines. Wilderness management and monitoring could partially provide such data, supplementing more detailed scientific studies of wilderness conditions elsewhere.

Principle 13: Manage Wilderness in Relation to Its Adjacent Lands

The principle—that wilderness managers must consider how adjacent lands are managed—is related to the environmental modification spectrum. Here, however, the concern broadens from outdoor recreation opportunities to a variety of other resource uses and management practices. Simply put, *wilderness does not exist in a vacuum—what goes on outside of, and adjacent to, a wilderness can have substantial impacts inside its boundary. Conversely, designating a tract of land as a wilderness can substantially affect the management of adjacent areas* (Hendee and Dawson 2001; see Chapter 13).

These interrelationships are illustrated by the impacts resulting from timber harvesting to building roads to the edge of a wilderness boundary that can dramatically affect the amount and character of its recreational use by providing easier access. Use of trail systems and target destination areas may then change, and thus the location of wilderness management emphasis to maintain natural conditions may need to change or be expanded.

Impacts resulting from increased use can compromise the goal of maintaining natural conditions. Reducing the number of roads, lowering road standards, and closing roads after timber harvesting might attenuate such impacts. Recent efforts to reduce road building and close many roads on the national forests may ease this problem, although road closures and roadless-area-protection initiatives by the Forest Service are very controversial (O'Laughlin and Freemuth 2000) and politically volatile.

Similarly, developing high-density recreational facilities (youth camps, picnic areas, paved roads, and parking lots) immediately adjacent to wilderness or undeveloped backcountry can bring serious management problems as users of these facilities access the wilderness for hiking, fishing, or other activities. For example, one of the most heavily used entrances into the San Gorgonio Wilderness in southern California is the South Fork Meadows Trail. Formerly, access to the wilderness boundary was over a route called "Poopout Hill," an appropriate name for a climb that required considerable effort. Ultimately, the hill was topped by a paved road and lined by numerous summer camps. Because of this improved access and the added visitation from the youth camps, use impacts in the area reached such a level that entry at that trailhead had to be rationed beginning in 1973 (Hay 1974). This rationing program, an extraordinary management step at the time, was necessitated by the improved road and youth camp development, but it has been successful in that it has allowed numerous youth to experience wilderness, while maintaining a wilderness threshold. This San Gorgonio Wilderness "use management" example was one of the first and is now replicated in many places across the NWPS.

Impacts can also move from wilderness to nearby nonwilderness areas. For example, programs to reestablish natural fire regimes through "let burn" or "prescribed fire" policies in wilderness can result in smoke pollution and a risk of fire losses in adjacent areas managed for commodity production or even summer-home development. Wildlife that thrives on summer range in high wilderness meadows may compete with domestic stock for forage in lower valleys during winter. In the Southeast, where the southern pine beetle has been a constant threat to the region's pine forests, there is long-standing concern that wilderness areas may harbor infestations that will spread to adjacent lands (Warren 1985).

Relating the management of wilderness to that of adjacent lands is complex and controversial. One common suggestion early in the history of the U.S. wilderness system was to create *buffer zones*—a band of land around the periphery of wilderness that would absorb impacts and help avert conflicts. However,

Congress spoke against buffer zones in the Endangered American Wilderness Act of 1978 (U.S. Public Law 95-237), so any buffer zone must now fall within the wilderness, which could dilute wilderness conditions (see Chapter 5 for more discussion). However, the best protection for wilderness from impacts originating on surrounding lands is through comprehensive land-use planning that anticipates potential conflicts and addresses the complementary and competitive relationships between wilderness and adjacent lands.

Managers planning to allocate various land uses should carefully define activities that are compatible with wilderness and take steps to reduce incompatibilities through their location. The recreation opportunity spectrum (ROS) is such an approach. A concentric circle application of the ROS could have high-density, facility-oriented sites located around the periphery of a planning region, and increasingly environment-oriented and primitive-style opportunities (wilderness) located at the core. This gradation of uses can help protect the environmental quality of the interior wilderness while keeping development-oriented opportunities accessible, but outside the wilderness. To do otherwise invites problems, particularly when the most intensive uses impinge on—and then ultimately displace—the least developed areas.

Explicitly defined use zones containing specific kinds of uses in particular locations also help protect against the invasion, succession, and displacement of primitive-site recreation users by new and more numerous, crowding-tolerant, and developed-site recreationists. The displaced users then move on to seek more primitive sites and can start a new wave of invasion, succession, and displacement. If this recreation-use displacement progression continues, the most pristine areas will have increased use, less solitude, and decreased natural conditions—the wilderness opportunity spectrum will disappear.

Managers may inadvertently aggravate undesirable impacts from this recreational displacement by responding to every increase in use with development to accommodate it—for example, by expanding campgrounds or parking lots, or allowing resort developments at the edge of wilderness. Furthermore, naturalness and solitude in the most remote sections of a wilderness can be impaired if wilderness managers are not aware of use interrelationships. Actions such as building trails to areas that are currently visited only by cross-country travelers seeking the greatest possible naturalness and solitude may ultimately eliminate the most remote and pristine wilderness experiences.

Summary

These thirteen principles provide a foundation for much of the detailed discussion about wilderness management in the following chapters. By themselves, the principles do not ensure quality wilderness stewardship, but they do provide basic concepts to guide that management. At the very least, for those in search of consistent policies and actions, the principles provide a framework for reviewing and evaluating approaches to wilderness management situations, issues, and problems.

Study Questions

1. How does the availability of recreational opportunities on nonwilderness land affect nearby wilderness?
2. Explain the environmental modification spectrum, the recreation opportunity spectrum (ROS), and the wilderness opportunity spectrum (WOS).
3. In what sense is wilderness a renewable resource? What are the implications for wilderness management?
4. What are the two major parameters or dimensions of carrying capacity that wilderness managers must consider? Explain each briefly.
5. What are wilderness-dependent uses? How can favoring them help protect wilderness from excessive impacts and at the same time protect users from unnecessary restrictions?
6. What is the minimum tool rule? Give an example of a light-handed or minimum tool approach to reducing damage to vegetation and crowding at a popular wilderness campsite.
7. What does the concept of nondegradation suggest for management of wilderness in the eastern United States, compared to more wild and pristine wilderness areas in the West?
8. How has increased public involvement affected the job of professional wilderness managers? Overall, do you think the effects have been or will be positive or negative? Why?
9. Explain the recreation-use displacement idea, how it can dilute the most pristine wilderness, and how managers can guard against it.

Case Discussion:
Assisted Trout Migration into Wilderness?

As environmental changes at the global and regional levels affect wilderness natural processes and conditions, managers will increasingly be faced with deciding when and how to intervene and to what extent interventions should follow the historic range of variation and natural processes and conditions. The spectrum of management options ranges from nonintervention (i.e., allow ecosystems to adapt and change as they will) to active intervention (i.e., restore historic natural conditions, resist change by maintaining current system, conserve some aspect of biodiversity, or transform to more resilient systems). Cole et al. (2008, 42) provide an example of a difficult decision to conserve a species over maintaining natural conditions:

Recently, in the Bob Marshall Wilderness, Montana, stewardship decisions have been made that some might consider inconsistent with wilderness. These decisions condone assisted migration—helping species relocate to places where they are more likely to persist—and they place more importance on species conservation than on naturalness. The decisions pertain to management of fish populations in about 20 lakes in the Bob Marshall Wilderness that historically were fishless but that have been stocked with nonnative trout for many decades.

The plan—approved but not yet implemented—is to remove all nonnative trout from these lakes. Then, rather than leave the lakes fishless as they originally were, they will be stocked with genetically pure westslope cutthroat trout. These lakes, which fish are unable to migrate to themselves, offer a refuge from other fish that hybridize with westslope cutthroat and pollute them genetically. Wilderness provides the most inviolate refuge and, therefore, is considered necessary to preserve the species, even though the requisite action compromises naturalness.

This situation is complicated by states' rights issues. This intervention, pushed by the Montana Fish, Wildlife and Parks Department, would almost certainly not have been proposed if the species at risk were not a game fish. Nevertheless, it illustrates the potential to give precedence to a conservation goal other than preserving natural conditions, even in wilderness.

> **Case Discussion Questions:**
> 1. Which of the thirteen principles in this chapter *support* this intervention of conserving the westslope cutthroat trout population? Why?
> 2. Which of the thirteen principles in this chapter *do not support* this intervention of conserving the westslope cutthroat trout population? Why?
> 3. How do you decide between conserving (in this case) species biodiversity in a regional context and maintaining the natural conditions present in a wilderness prior to European settlement in the region?

References

Clark, Kendall; Kozasek, Susan. 1997. How do your wilderness values rate? *International Journal of Wilderness*. 3(1): 12–13.

Cole, David. 2000. Natural, wild, uncrowded or free: Which of these should our wilderness be? *International Journal of Wilderness*. 6(2): 5–8.

———. 2001. Wilderness recreation management strategies: Balancing freedom and protection. *International Journal of Wilderness*. 7(1): 12.

Cole, David; Yung, L.; Zavaleta, E. S.; Aplet, G. H. et al. 2008. Naturalness and beyond: Protected area stewardship in an era of global environmental change. *The George Wright Forum*. 25(1): 36–56.

Hay, Edward. 1974. Wilderness experiment: It's working. *American Forests*. 80(12): 26–29.

Hendee, John C. 1986. Wilderness: Important legal, social, philosophical, and management perspectives. In: Kulhavy, David L.; Connor, Richard N., eds. *Wilderness and Natural Areas in the Eastern United States: A Management Challenge*. Symposium; May 13–15, 1985; Nacogdoches, TX. Nacogdoches, TX: Stephen F. Austin State University, School of Forestry, pp. 5–11.

Hendee, John C.; Dawson, Chad P. 2001. Stewardship to address the threats to wilderness resources and values. *International Journal of Wilderness*. 7(3): 5–10.

Landres, P., Boutcher, S.; Merigliano, L.; Barns, C.; Davis, D.; Hall, T.; Henry, S.; Hunter, B.; Janiga, P.; Laker, M.; McPherson, A.; Powell, D.; Rowan, M.; Sater, S. 2005. Monitoring selected conditions related to wilderness character: A national framework. General Technical Report RMRS-GTR-151. Fort Collins, CO: U.S. Department of Agriculture, Forest Service, Rocky Mountain Research Station. 38p.

Marshall, Robert. 1930. The problem of the wilderness. *Scientific Monthly*. 30(2): 141–148.

———. 1937. The universe of the wilderness is vanishing. *Nature*. 29(4): 235–240.

Merigliano, Linda; Krumpe, Ed. 1986. Scientists identify, evaluate indicators to monitor wilderness conditions. *Park Science*. 6(3): 19–20.

Meyer, Shannon. 1999. The role of legislative history in agency decision making: A case study of wilderness airstrip management in the United States. *International Journal of Wilderness*. 5(2): 9–12.

Meyers, Charles J.; Tarlock, A. Dan. 1971. Water pollution. In: Meyers, Charles J.; Tarlock, A. Dan, eds. *Water Resource Management*. Mineola, NY: Foundation Press, pp. 677–708.

Mihaley, Marc B. 1972. The Clean Air Act and the concept of nondegradation: *Sierra Club v. Rukelhaus. Ecological Law Review*. 2(4): 801–836.

Nash, Roderick. 1982. *Wilderness and the American Mind*, rev. ed. New Haven: Yale University Press. 425p.

O'Laughlin, Jay; Freemuth, John. 2000. Roadless area policy politics and wilderness potential. *International Journal of Wilderness*. 6(1): 9–12.

Oye, Garry. 2001. A new recreation strategy for national forest wilderness. *International Journal of Wilderness*. 7(1): 13–14.

Regehr, E. V.; Lunn, N. J.; Amstrup; Stirling, I. 2007. Effects of earlier sea ice breakup on survival and population size of polar bears in western Hudson Bay. *Journal of Wildlife Management*. 71(8), pp. 2673–2683.

Scott, D. W. 1984. The visionary role of Howard Zahniser. *Sierra*. 69(13): 40.

Scott, Douglas. 2004. *The enduring wilderness: Protecting Our Natural Heritage Through the Wilderness Act*. Golden, CO: Fulcrum.

Spring, Ira. 2001. If we lock people out who will fight to save the wilderness. *International Journal of Wilderness*. 7(1): 17–19.

Stankey, George H.; Cole, David N.; Lucas, Robert C. et al. 1985. The limits of acceptable change (LAC) system for wilderness planning. General Technical Report INT-176. Ogden, UT: U.S. Department of Agriculture, Forest Service, Intermountain Forest and Range Experiment Station.

U.S. Public Law 88-577. The Wilderness Act of September 3, 1964. 78 Stat. 890.

U.S. Public Law 91-190. The National Environmental Policy Act (NEPA) of January 1, 1970. 83 Stat. 852.

U.S. Public Law 93-622. (Eastern Wilderness) Act of January 3, 1975. 88 Stat. 2096.

U.S. Public Law 95-237. The Endangered American Wilderness Act of February 24, 1978. 92 Stat. 40.

Warren, B. Jack. 1985. Why we need to control pine beetles in wilderness areas. *Forest Farmer*. 44(4): 6–8.

White, David; Hendee, John C. 2000. Primal hypotheses: The relationships between naturalness, solitude and the wilderness experience benefits of development of self, development of community, and spiritual development. In: McCool, S.; Cole, D.; Borrie, W. T.; O'Loughlin, J., comps. *2000 proceedings: Wilderness Science in a Time of Change*, Vol. 3; *Wilderness as a Place for Scientific Inquiry*; May 23–27, 1999; Missoula, MT. RMRS-P-15. Fort Collins, CO: U.S. Department of Agriculture, Forest Service Proceedings, pp. 223–228.

Worf, William. 2001. The new Forest Service recreation strategy spells doom to the National Wilderness Preservation System. *International Journal of Wilderness*. 7(1): 15–16.

Chapter 8
Wilderness Management Planning

Introduction	196
The Need for Planning	196
Trends and Evolution of Wilderness Planning Direction	197
Plan Flexibility and Effectiveness	198
Wilderness Planning Under the National Environmental Policy Act of 1970	198
Wilderness Management Planning in U.S. Federal Agencies	199
National Park Service	199
U.S. Forest Service	200
U.S. Fish and Wildlife Service	202
Bureau of Land Management	202
Planning Tools and Techniques	203
Terminology and Logic	203
A Goal-Achievement for Writing Wilderness Management Plans	204
The Limits of Acceptable Change for Wilderness Planning	206
Preparing Wilderness Management Plans: Problems and Suggestions	207
Writing the Plan	207
Zoning—A Wilderness Opportunity Spectrum	209
Public Involvement and Plan-Review Process	210
Bear Trap Canyon (BLM)	211
Denali National Park and Preserve (NPS)	211
Maricopa Complex Wilderness (BLM)	212
Criteria for Evaluating Wilderness Management Plans	212
Summary	214
Study Questions	214
Acknowledgments	214
Case Discussion: Comprehensive Conservation Planning for the Kenai National Wildlife Refuge, Alaska	215
Case Discussion Questions	215
References	216

Introduction

Wilderness management needs to be guided by formal plans that clearly state goals, objectives, and standards, and prescribe specific actions necessary to meet the objectives and standards. *Wilderness management translates the general directions of legislation, departmental regulations, and agency policy into guides for management in the field.*

In this chapter we discuss the background and need for planning; review the wilderness management planning processes of the National Park Service (NPS), U.S. Forest Service (USFS), U.S. Fish and Wildlife Service (FWS), and Bureau of Land Management (BLM); outline tools and techniques for developing plans, including a framework and format for preparing wilderness management plans; provide excerpts from actual plans to show how the suggested framework is applied; and discuss criteria for evaluating management plans.

Because wilderness management plans are developed under the larger land-use planning organization of the different management agencies, there are differences in agency format and process for wilderness management planning. *Regardless of the particular agency process, wilderness management needs to be guided by clear statements of goals and objectives, by selecting management actions necessary to achieve the goals and objectives, and by monitoring to ensure that the goals and objectives are being met.*

The Need for Planning

Since the passage of the Wilderness Act of 1964 (U.S. Public Law 88-577) the United States has embraced the concept of identifying, protecting, and managing vast amounts of land in the National Wilderness Preservation System (NWPS). Presently, there are more than 107 million acres protected in over 700 wilderness areas nationwide, or about 4.5 percent of the United States. Many of the political battles fought to protect and preserve these landscapes have been lengthy and intensely debated, and the resulting legislation often includes compromises that pose problems for future management (see Chapter 5). It was apparent in the beginning that wilderness designation alone is not sufficient to protect and perpetuate the human and ecological values for which these areas have been set aside. An ongoing stewardship program is necessary to deal with the human influences from both outside and within wilderness, and their accompanying undesirable impacts. Along with recognizing that management was necessary, even in wilderness, came realization of the need to develop plans that would direct wilderness management toward long-range goals. In fact, it is argued in Chapter 2 that the drive for passage of the Wilderness Act was spurred by the very lack of management and long-term planning afforded the original wild, wilderness, and primitive areas designated under the L-20 Regulation and the U regulations of the Forest Service. Thus, the need for wilderness planning has been an integral part of wilderness management since the inception of the NWPS (Krumpe 2000).

Three realities underlie the need for wilderness management planning: (1) human-caused change is inevitable; (2) recreation use is not self-limiting; and (3) good planning anticipates problems.

Fig. 8.1. Wilderness, such as the Selway-Bitterroot on the Idaho-Montana border, (above) is managed under constraints of the Wilderness Act and agency direction. Management direction is contained in specific wilderness area management plans, or in resource management plans that include both wilderness and nonwilderness land. Wilderness management, to preserve an area's distinctive qualities of naturalness and solitude, can be no better than the underlying planning process. Photo courtesy U.S. Forest Service.

1. *Human-caused change is inevitable: we cannot prevent conditions from changing in wilderness.* Wilderness resources are not static but are constantly evolving naturally. All human use within and nearby wilderness areas causes some impact to natural wilderness conditions. Given our growing global population, threats to wilderness conditions (see Chapter 13), and increasing participation in a variety of wilderness activities, resource conditions

and experiences will continue to change. Planning is the logical process to anticipate change and to consider how much human-caused change is acceptable in the areas where it occurs.

2. *Recreation use is not self-limiting.* As wilderness use increases, people who seek quieter, less impacted conditions will typically modify their behavior to visit during less busy times or may decide to visit other, less crowded areas (hence, be displaced). These visitors are replaced by people willing to accept more crowded conditions and more evidence of use. As use continues to increase, this cycle repeats itself again and again. Thus, many wilderness places gradually become more crowded with more impacts and conflicts, unless a strategy is developed to stop this cycle with actions to protect and provide the desired wilderness settings and experiences. Thus, the question is not whether limits and management actions are needed, but rather which wilderness settings an area should provide and which management actions are needed to achieve them.

3. *Good planning anticipates problems: it is far better to begin planning early.* The best plans are those developed before unacceptable wilderness conditions are widespread. It is better to define when "corrective" management actions will be needed than to try to restore wilderness conditions or change use patterns or amounts of use when the situation is perceived to be causing undesirable impacts.

Without management plans developed by an orderly process, wilderness management may be no more than a series of uncoordinated reactions to immediate problems. This can lead to practically irreversible impacts on wilderness conditions. Managers need to look ahead to identify undesirable trends and implement actions to achieve desired conditions and to prevent, reverse, or restore undesirable conditions. For example, complete suppression of fires, trail building in sensitive wildlife habitat, overuse of popular locations, or introduction of exotic fish, wildlife, and plant species can lead to impacts, which are cumulative and could take decades to reverse.

Management-by-reaction works against wilderness preservation goals because management direction can easily be shaped by a succession of minor decisions. The cumulative results of such incremental decisions may be undesirable and hard to reverse. Unplanned management can be recognized by a shifting of focus from problem to problem; inconsistent, conflicting actions; and a loss of overall direction toward wilderness preservation goals.

Through planning, managers can reconcile differences in management philosophy and ideas before taking actions that have long-range effects on the wilderness resource. Good plans have a stabilizing influence on management, despite changes in personnel or the influence of multiple administrative units in a wilderness where several managers may have different philosophies, perceptions, and definitions of wilderness. *Plans that establish clear, attainable, measurable, and acceptable objectives and standards for an area, and the actions by which such objectives will be pursued, are essential for guiding wilderness management toward desired outcomes.*

Successful wilderness management depends on the quality of plans that guide management actions for individual areas. In an era of heightened concern about cost effectiveness and competition for scarce federal dollars, the potential for obtaining the personnel and money essential for wilderness management, or optimizing effectiveness without them, will be increased by plans that identify clear objectives and the management actions necessary to achieve them.

Furthermore, involving the public in the planning process will attract and focus volunteer efforts to help accomplish essential tasks. More important now than ever before, the planning process gives the interested public an opportunity to learn about, evaluate, provide input, and become involved in a wilderness area's management. Effectiveness and consistency of wilderness management, as well as the public's acceptance of that management and its implementation, are highly dependent on involving them in plans and the planning process.

Trends and Evolution of Wilderness Planning Direction

Wilderness management planning among the four federal wilderness management agencies has changed and evolved to reflect concurrent changes in regulations, policies, legislation, and society that affect planning. In addition, wilderness managers and planners have gained experience and expertise and have embraced evolving directions in natural-resource planning.

There are four trends in resource planning:

1. *Use allocation versus land allocation.* The first wilderness plans typically dealt with allocating land areas to various uses, primarily focusing on designating land as wilderness or nonwilderness. Planning attention has now shifted to directing stewardship of lands allocated to various uses, including wilderness. The current trend focuses on allocating resource use among various compatible human activities and excluding noncompatible uses from wilderness.
2. *An ecosystem approach to planning.* Planning in the 1960s tended to focus on separate aspects of the wilderness, such as recreation, vegetation, water, forage, wildlife, or fire within wilderness boundaries. More recently planners recognize that the greatest value of wilderness is its naturalness, and plans must address wilderness as a composite resource (see Chapter 7). This, of course, is the key value of all ecosystem planning. Furthermore, the conditions of naturalness and solitude in wilderness can provide benefits to the larger landscape. *The ultimate goal of planning today is to protect and perpetuate interrelated natural ecosystem processes that often function on a landscape scale and that extend beyond wilderness boundaries.*
3. *A focus on desired resource and social conditions rather than outputs.* Early plans focused on quantifying "outputs" to be produced, such as recreation visitor days, animal-unit months of grazing, miles of trail developed, etc. *Plans now present desired future resource and social conditions as goals for which specific objectives are developed.* Outcomes are now portrayed as protecting or restoring resource conditions, which produces desired services or benefits such as recreation, personal growth, improved health, species diversity, naturally functioning aquatic systems, natural fire regimes, endangered species recovery, and so forth.
4. *A focus on collaborative decision-making processes.* Traditional planning models relied heavily on experts and data and used rational and comprehensive approaches. Such planning was often paralyzed by the lack of "good" data, and implementation was often hampered by lack of public understanding and acceptance. Today, planning is a collaborative effort with the public and users of the wilderness. In wilderness planning, the question has shifted from "how much use is too much" to a focus on the desired resource and social conditions and "how much change in conditions is acceptable." This raises the question "acceptable to whom?" which is a value-based rather than technical question. Science and data play important roles in understanding how an ecosystem works and identifying the consequences of various options, but the decision about what is desired or acceptable is based on society's values and beliefs (Krumpe and McCool 1997). Recent wilderness plans have placed greater emphasis on utilizing processes that involve the public in discussion throughout the planning process.

Plan Flexibility and Effectiveness

Wilderness management planning processes are constantly evolving, and several approaches and styles of plans have been developed during more than four decades of designated wilderness management. One approach, called a "goal-achievement framework," is described later in the chapter. Increasingly, managers face conflicting goals, such as providing freedom from restrictions versus protecting resource conditions and solitude. In these situations, a limits of acceptable change (LAC) approach is useful, and it is discussed in Chapter 9. Other planning approaches and frameworks may also be useful (McCool et al. 2007).

There is room for flexibility in format and level of detail in wilderness plans as long as they effectively translate overall management direction—laws, regulations, and policy—into specific objectives and necessary management actions to achieve them. Whatever the format, well-reasoned plans are necessary to explain management intentions, to guide public discussion, to achieve public understanding of necessary management actions, to justify budget requests and meet legal requirements, and to ensure protection and stewardship of an enduring wilderness resource.

Wilderness Planning Under the National Environmental Policy Act of 1970

Following passage of the National Environmental Policy Act (NEPA; U.S. Public Law 91-190) in 1970, the federal wilderness managing agencies incorporated environmental analysis procedures into their planning processes. *NEPA requires that environmental impacts be considered through an analysis of a proposed action and its alternatives, and that the public be allowed to comment on the actions under consideration. Under the guidelines for implementing NEPA, agencies must assess the environmental and social impacts of alternative management actions before implementing them.* Thus, whatever wilderness management planning approach is used, the plan must conform to NEPA's requirements that alternatives be considered and analyzed, the effects of alternatives be disclosed, and the public have an opportunity to provide input.

These alternatives must include a range of actions, including taking no action at all. All alternatives must meet the need for taking action and respond to the major issues. During this NEPA analysis process, managers have an opportunity to consider prospects for the type of wilderness environment described in section 2(c) of the Wilderness Act of 1964 (U.S Public Law 88-577)—one featuring naturalness and outstanding opportunities for solitude and primitive recreation—in light of conditions in the area under consideration and the impacts of actions proposed in the plan or proposed project. A significant requirement under NEPA is that all federal plans are required to develop an environmental impact statement (EIS) or environmental assessment (EA).

Wilderness Management Planning in U.S. Federal Agencies

Wilderness management planning is an integral part of the overall management-planning approaches for the natural-resource units managed by the four federal wilderness managing agencies. Thus, plans for wilderness, wilderness study areas, or proposed wilderness are subordinate to—meaning they fall within the larger planning framework of the agency jurisdiction—a national park or national forest plan, BLM resource-management plan, or FWS refuge plan. For example, the White Mountain National Forest Land and Resource Management Plan (U.S. Department of Agriculture, Forest Service 2005) incorporated broad, strategic consideration of the five wilderness areas within the forest throughout the forest plan and addressed the specific management direction for those areas within a separate appendix (Wilderness Management Plan). The forest plan included recommendations for additional wilderness areas and acreage as some of its specific recommendations. Congress subsequently passed the New England Wilderness Act of 2006 (U.S. Public Law 109-382) and added a new area, wild river wilderness, to the NWPS and expanded the existing Sandwich Range Wilderness based on recommendations in the 2005 forest plan.

The following discussion of planning procedures used by the federal agencies assumes the completion of the environmental analysis requirements mandated by NEPA. Most agencies have now combined NEPA's environmental analysis of environmental impact statements (EIS) and environmental assessment (EA) requirements and their land-use planning procedures, so that all NEPA requirements are met when a plan is completed.

National Park Service

Management direction for wilderness in the National Park System comes from the Wilderness Act, the National Park Service Act of 1916 (U.S. Public Law 235), legislation establishing individual national park units, and wilderness designation laws designating wilderness areas within particular parks. Interpretation of this legislation for wilderness preservation and management is found in chapter 6 of the revised Management Policies of the NPS (U.S. Department of the Interior, National Park Service 2006) and is supplemented by the NPS Director's Order #41 on Wilderness Preservation and the Management and Reference Manual 41, which was included with that order. The congressional intent reflected in these laws and policies is implemented predominantly through multiple levels of planning applicable to all national park units, including national recreation areas and national monuments administered by the NPS. The overall plan is the General Management Plan (GMP), with an EIS for each NPS unit.

1. *General Management Plans* (GMPs) are required for each national park under the National Parks and Recreation Act of 1978 (U.S. Public Law 95-625). The NPS Program Standards explain (U.S. Department of the Interior, National Park Service 2004, 6):

   General management planning results in a shared understanding among NPS managers and the public about the kinds of resource conditions and visitor experiences that will best fulfill the purpose of the park. General management plans zone the park for some variety of resource conditions and experiences (consistent with the discretion allowed under the NPS Management Policies) based on the intrinsic qualities of particular locations and considering the range of stakeholder interests and concerns. The plan looks at the park as a whole and as a part of larger ecological, cultural, and socioeconomic systems. This comprehensive approach helps ensure that the decisions made through general management planning are widely supported and sustainable over time. General management plans direct park managers to focus on achieving the conditions and experiences prescribed by the plan, but they do not provide direction for specific actions, recognizing that managers may have to continuously adapt their approaches to current situations based on changing information or conditions, including changing staffing, budgeting, and scheduling opportunities and constraints.

2. The GMP is supported by three other levels of planning as needed: (a) a program plan to identify and recommend the best strategies for achieving the long-term desired conditions and visitor experiences in that park unit; (b) a strategic plan to make decisions about how to achieve the desired conditions in the next three to five years of those identified in the park program plan; and (c) an implementation plan to describe in detail how to achieve the elements of the strategic plan over the next several years (U.S. Department of the Interior, National Park Service 2004, 2005c).

National parks with a potential or actual designation as wilderness address these opportunities and requirements in the GMP. A park with a potential wilderness must go through a formal wilderness study process prior to proposing an area or park for wilderness designation by Congress. NPS managers consider the extent to which the requirements of wilderness stewardship planning could be met when developing a park GMP. *The Park Planning Source Book* (U.S. Department of the Interior, National Park Service 2005c, 39) suggests:

> A wilderness stewardship plan done along with the GMP should address zoning and desired conditions and establish indicators and standards for achieving the desired conditions. These are GMP requirements already and would not result in extra work for the GMP team. Decisions about trails and other public facilities, campfires, user capacity, etc., can be addressed through zoning and desired conditions without mentioning each trail or cabin. For parks with few issues or little wilderness use, this level of wilderness planning may meet most of their needs. This plan would then provide the broad framework for more detailed implementation plans, such as a fire management plan or trail plan. Parks with complex wilderness issues, such as heavy overnight or day use, commercial pack trips, or climbing, will likely need a separate wilderness stewardship plan to provide management guidance for these issues.

NPS policy requires each administrative unit of the National Park System to address user capacity and impact issues and identify and offer mitigating actions for problems, such as overcrowding, stock damage to trails and tundra, disposal of human waste, threats to visitor safety, or danger from or to grizzly bears. The Park Service developed a planning process called Visitor Experience and Resource Protection (VERP) in response to the mandate to develop a carrying-capacity framework for each park.

According to an NPS report on National Park Service wilderness (U.S. Department of the Interior, National Park Service, 2005b), wilderness areas had been designated in forty-five units of the National Park System by 2004 and thirty two wilderness study areas were under NPS management. Of those areas, seventeen were proposed through the formal study process for congressional action during the Nixon, Ford, and Carter administrations in the 1970s. Although environmentalists take heart from the protection these areas receive from being in the NPS system, completion of the proposed wilderness designation, and subsequent completion of wilderness management plans for each designated area, are a goal for many citizens and NPS managers (U.S. Department of the Interior, National Park Service 1993).

U.S. Forest Service

Management direction for wilderness in the National Forest System comes from several laws: the Multiple-Use, Sustained-Yield Act of 1960 (U.S. Public Law 86-517); the Wilderness Act of 1964 (U.S. Public Law 88-577); the Forest and Rangeland Renewable Resources Planning Act of 1974 (U.S. Public Law 93-378); the National Forest Management Act (NFMA) of 1976 (U.S. Public Law 94-588), the Clean Air Act and amendments of 1977 (U.S. Public Law 95-95), and sometimes specific wilderness designation laws establishing areas on national forests. The U.S. Department of Agriculture Regulations 36 *Code of Federal Regulations*, parts 219, 228, 261, and 293, and USFS policy guidelines for wilderness management and for land-use planning (chapter 2320 of the USFS manual) contain further direction (U.S. Department of Agriculture, Forest Service 2007).

The USFS has two primary land and resource plans that relate to wilderness: (1) Renewable Resources Assessment and Program, a nationwide plan that guides the overall management of the National Forest System; and (2) Forest Land and Resource Management Plan usually called Land Management Plan or Forest Plan, which includes management direction for wilderness areas of the national forest.

1. *The Renewable Resources Assessment and Program* is required by the 1974 Forest and Rangeland Renewable Resources Planning Act (RPA), as amended by the 1976 National Forest Management Act (NFMA). Together, these laws provide a comprehensive framework for planning management of the National Forest System. The RPA calls for (a) preparation of an assessment of the supply and demand for the nation's forest and rangeland resources and (b) development of a management program for national forests that considers alternative management directions and the role of the national forests. The RPA program proposes (subject to annual appropriations by Congress) national direction and output levels for the National Forest System. These proposals and associated management implications are based on an assessment of supply and demand and fiscal and political considerations. Wilderness is one of the resources analyzed.

2. *The Forest Land and Resource Management Plan,* commonly known as the Forest Plan, defines the management emphasis for different areas of the forest and sets forest-wide and area-specific goals and objectives. This is the primary document providing long-term management direction for a particular wilderness. Six types of decisions are made in Forest Plans: (a) forest-wide goals and objectives, (b) forest-wide management requirements, (c) management-area direction, (d) suitability of land for resource use and production, (e) monitoring and evaluation requirements, and (f) recommendations for wilderness and wild and scenic river designation.

Each wilderness is a separate management area for which the Forest Plan specifies desired conditions, standards, and guidelines for management activities and monitoring requirements. The types of decisions made in a Forest Plan are programmatic in nature, meaning that the plan establishes what is to be achieved and parameters to guide the management and human uses that may occur, but it does not prescribe specific actions or commit to how goals and objectives will be achieved.

To implement the Forest Plan, a number of field-level documents are prepared. The Forest Service Manual (U.S. Department of Agriculture, Forest Service 2007) sets forth the guidelines for planning and managing wilderness areas with the National Forest System. The main objectives for wilderness planning are to ensure that wilderness resource considerations are integrated in the Forest Plan and that all resources and activities within a designated wilderness are in line with USFS wilderness policies and direction. A fish and wildlife management policy provides guidance for USFS planning activities in wilderness (U.S. Department of Agriculture, Forest Service and U.S. Department of the Interior, Bureau of Land Management 2006).

In February 2005 the chief of the USFS, Tom L. Thompson, issued a 10-Year Wilderness Stewardship Challenge to raise wilderness area stewardship to a higher level of stewardship because a report in 2003 found only "18 percent of the 406 wildernesses under our care were managed to a minimum stewardship level" (Thompson 2005, n.p.). The 10-Year Wilderness Stewardship Challenge includes consideration of and management of ten elements (Thompson 2005, n.p.):

1. Wilderness is covered by a fire plan that allows for the full range of management responses.
2. Wilderness is successfully treated for noxious/invasive plants.
3. Air quality monitoring is conducted and baseline is established.
4. Wilderness education plans are implemented.
5. Wilderness has adequate standards, in which monitored conditions are within Forest Plan standards, and opportunities for solitude or primitive and unconfined recreation are stable or increasing.
6. Wilderness has completed recreation site inventory.
7. Outfitter and guide permit operating plans are in place that direct outfitters to model appropriate wilderness practices and incorporate appreciation for wilderness values in their interaction with clients.
8. Wilderness has a full range of adequate standards that prevent degradation of the wilderness resource.

9. Wilderness managers have their priority information needs addressed through data collection and corporate applications.
10. Wilderness has a baseline workforce (from workforce assessment) in place for each wilderness.

U.S. Fish and Wildlife Service

Management direction for wilderness in the National Wildlife Refuge System comes from the Wilderness Act; the National Wildlife Refuge System Administration Act of 1966 (U.S. Public Law 89-669), as amended; the Alaska National Interest Lands Conservation Act (ANILCA) of 1980 (U.S. Public Law 96-487); specific authorities (including executive orders) establishing individual units of the refuge system; and specific legislation designating wilderness-areas within particular refuges. Further national direction comes from FWS regulations for wilderness preservation and management (50 *Code of Federal Regulations,* part 35), and the agency's wilderness area management policy (U.S. Department of the Interior, U.S. Fish and Wildlife Service 2000). The wilderness-area management policy provides guidelines for administrative and public use and preparation of wilderness management plans.

Wilderness management direction is developed and implemented through the refuge planning process. Two types of planning documents guide management of refuge wilderness areas: comprehensive conservation plans (CCPs) and step-down management plans.

All refuges are to be managed in accordance with an approved CCP. The CCP is a land-use plan that describes the desired future conditions of a refuge or planning unit and provides long-range guidance and management direction to achieve the purpose of the refuge; helps fulfill the mission of the refuge system; maintains and, where appropriate, restores the biological integrity, diversity, and environmental health of each refuge and the refuge system; helps achieve the goals of the NWPS; and meets other mandates. The CCP sets forth goals, objectives, and strategies to guide management decisions and accomplish these ends. Public involvement and NEPA compliance are integrated into the CCP process. CCPs are prepared by the refuge manager and refuge staff and approved by the regional director.

Step-down management plans are prepared for a specific management program or resource; they provide detailed strategies and implementation schedules for meeting the broader goals and objectives identified in the CCP. Wilderness Management Plans (WMPs) are a form of step-down management plan that guide the preservation, management, and use of a particular designated wilderness in a refuge or unit. The WMP describes specific management goals, objectives, and strategies based on the refuge or unit's purposes; the principal wildlife conservation mission of the refuge system; and wilderness management principles. WMPs are developed under a 1986 policy (U.S. Department of the Interior, U.S. Fish and Wildlife Service 2000) that was revised in November 2008 (73 Federal Register 222: 67876–67882).

Bureau of Land Management

The Federal Land Policy and Management Act (FLPMA) of 1976 (U.S. Public Law 94-579) called for the review of all roadless areas and roadless islands on BLM-administered land by 1991 to determine their suitability for wilderness designation (see Chapter 5). At the end of the review, the secretary of the interior was to make recommendations to the president as to which wilderness study areas (WSAs) were suitable or unsuitable for wilderness designation. The timetable for wilderness designation of USDI Bureau of Land Management (BLM)–administered land has not progressed as fast or as far as FLPMA anticipated, but by 2006 the BLM managed 7.7 million acres of wilderness in 190 areas, and managed twice as many acres in WSAs as potential wilderness.

Wilderness areas administered by the BLM are managed under direction contained in BLM's wilderness management regulations and policy (43 *Code of Federal Regulations,* part 6300) (U.S. Department of the Interior, Bureau of Land Management 2000), which replaced the BLM's 1983 *Management of Designated Wilderness Areas* manual (section 8560). WSAs are managed to protect their wilderness values until Congress either drops them from further wilderness consideration or designates them as wilderness. The wilderness management policy addresses a wide range of potential uses and management actions within wilderness areas and provides guidance for preparing wilderness management plans. A fish and wildlife management policy provides guidance for BLM planning activities in wilderness (U.S. Department of Agriculture, Forest Service and U.S. Department of the Interior, Bureau of Land Management 2006). The BLM's approach to writing wilderness management plans is based on concepts of the goal-achievement planning framework described later in this chapter.

BLM wilderness management plans are developed within the context of the bureau's overall planning system. *The BLM has two types of planning documents for wilderness areas: Resource Management Plans (RMPs) and activity plans. Wilderness Management Plans are one kind of activity plan.* Activity plans are also prepared for other BLM-managed resources, such as livestock grazing allotments, wildlife habitats, and recreation.

1. The Resource Management Plan (RMP) is a type of land-use plan designed to guide the management of BLM lands administered by a field office and may include several million acres of public land. RMPs may address both wilderness management and wilderness designation, that is, making wilderness suitability recommendations. Each alternative considered in the RMP contains a description of how the various resources and uses of WSAs would be managed. Planning decisions inconsistent with WSA protection requirements may not be implemented unless Congress determines that an area should not be designated a unit of the NWPS. In the case of areas that are eventually designated as wilderness, RMP decisions would be included in a wilderness management plan where they are consistent with the Wilderness Act.

2. An activity plan—specifically a wilderness management plan—is prepared for a specific wilderness area or two or more closely related areas. The plan may also apply in some cases to areas adjacent to a wilderness, such as trailheads. Wilderness Management Plans help to implement the BLM's wilderness management policy. They show the actions that will be taken to preserve the wilderness resource and the connection between these actions and wilderness management objectives. However, because of the interrelationships between wilderness and other lands, a coordinated activity plan that addresses many issues may be prepared in place of a wilderness-specific activity plan.

3. BLM Wilderness Management Plans have three functions: (a) to explain the goals, objectives, policies, and specific actions for management of the wilderness resource and all associated resources; (b) to establish the general sequence for implementing necessary management actions; and (c) to establish time frames and procedures for monitoring and revising plans.

As the preceding descriptions of agency planning processes illustrate, each agency has integrated wilderness planning into its overall planning procedures for the larger land-management system of which it is a part. These all use basic planning concepts but differ in their details and terminology.

Planning Tools and Techniques
Terminology and Logic

Regardless of the particular organization or format used in a Wilderness Management Plan, it should include certain basic planning concepts. When labeled, these concepts provide a terminology for discussing management direction, ranging from general goals and objectives to specific actions. Goals and objectives are hierarchical, with one or more objectives under each goal to provide more specific direction to the plan. *The relationship among the components of the framework—the planning logic—is important, that is, goals and objectives are pursued through planned actions.*

Goals are general portraits of ideal ends or effects. They limit the range of potential objectives by providing direction and purpose. Goals are often lofty statements of intent. One example from the Wilderness Act is "to secure for the American people the benefits of an enduring resource of wilderness" (section 2). Sometimes there is confusion about whether a particular statement is a goal or an objective. Distinction should rest on specificity and attainability. *Objectives are attainable in the short term and are more specific than goals.* Broad goals for the NWPS are found in the Wilderness Act, although, as explained earlier, other legislation can also shape goals and influence direction for specific agencies and particular areas.

Objectives are statements of specific conditions to be achieved—reference points that if attained, will ensure progress in the direction of established goals. In the suggested planning framework that follows, objectives are used to describe wilderness conditions to be achieved and/or maintained through management. Objectives are shaped by the goals they serve. They are descriptions of the field conditions sought through management and serve as criteria for identifying necessary management policies and actions. *Clearly stated objectives are the key to effective management plans.* Managers may need to develop objectives for all-important aspects of the wilderness resource and its use, such as visitor access, wildlife, fire management, and recreation.

Situation and assumption statements define local conditions and expectations about particular aspects of the wilderness area covered by the plan. Important information regarding the current situation and assumptions about how things will change in the future should be identified because wilderness areas have different physical characteristics, attractions, and types and levels of use. This information can be helpful in setting feasible objectives and in identifying measures necessary to achieve them. For example, current levels of rafting and kayaking in a portion of a wilderness, combined with expectations about future use, might influence objectives as well as the policies and management actions needed to achieve them.

Policies are explicit expressions of intent describing what will be done to attain objectives. Sometimes a policy describes what will not be done or otherwise prescribes constraints on management activity.

Programs are sets of related actions that are combined to help achieve particular objectives under established policy, such as the Leave No Trace program (LNT).

Actions are specific management practices applied to achieve objectives within the constraints of agency policy and regulations.

Standards are measurable statements—based on the objectives—that define minimally acceptable conditions. They serve as reference points that can trigger corrective management actions if conditions are not met.

The internal logic reflected in the planning framework is important. Managers develop feasible *objectives* consistent with wilderness goals suited to local conditions in specific areas and acceptable under agency management guidelines that interpret the Wilderness Act. The *objectives* and standards of quality established in the plan, considering current *situations* and *assumptions* about the future, lead to *actions*, which are designed to achieve desired conditions described under *standards* of quality.

A Goal-Achievement Framework for Writing Wilderness Management Plans

The planning framework described in this chapter adapts basic planning principles to wilderness management. It emphasizes planning as a decision-making process that seeks to attain clearly stated management goals and objectives. Goals and objectives stated in plans serve two purposes: (1) they are intentions and aims for deciding which management policies and actions are necessary and (2) they are the targets against which the effectiveness of wilderness management is judged.

Table 8.1 shows a *goal-achievement framework* for organizing and writing wilderness management plans using the previously described concepts and terminology. The approach is a type of management-by-objectives framework. It features straightforward statements of goals and objectives followed by the management actions needed to achieve them.

A variety of plan formats can adapt the logic of this framework. Table 8.2 is a suggested outline showing how the framework could be applied to a Wilderness Management Plan. Plan organization and format are

Table 8.1. Outline of the Goal-Achievement Framework Applied to Wilderness Management Planning

Goals	Objectives	Current Situation and Assumptions	Management Actions	Standards
Broad statements of intent, direction, and purpose, found in (1) legislation, for example, the Wilderness Act; (2) departmental regulations; and (3) agency national policy and philosophy that interpret legislation and establish management direction.	Statements that describe specific conditions sought in a particular wilderness, serve as criteria for deciding what management actions are needed, and used as the basis for later evaluation of the effectiveness of management.	Statements of local conditions and situations, and predictions about changes, that help determine the need for specific management mechanisms.	Specific policies, programs, and actions through which objectives for a given wilderness are achieved, and standards by which their attainment can be measured through monitoring.	Measurable statements of minimally acceptable conditions.

normally dictated by agency guidelines. This outline is merely one format for focusing wilderness management direction through a logical planning framework.

The following examples are from existing management plans that use concepts described in the goal-achievement framework and the integration of wilderness management and stewardship into the overall planning document for the forest, park, refuge, or planning area. The overall goals (see Table 8.3) are from a plan for the BLM's Bear Trap Canyon unit of the Lee Metcalf Wilderness in Montana (U.S. Department of the Interior, Bureau of Land Management 1985). This plan was prepared as a prototype for future BLM wilderness management plans using the goal-achievement planning framework.

Another example is the integrated approach to the Land and Resource Management Plan (LRMP) in the White Mountain National Forest (WMNF) of New Hampshire and Maine (U.S. Department of Agriculture, Forest Service 2005). Wilderness is addressed throughout the LRMP for the WMNF and more specifically in a separate appendix of the LRMP for the wilderness areas in the WMNF.

The Wilderness Management Plan specifically addresses all five wilderness areas within the WMNF: Great Gulf (5,500 acres), Presidential Range–Dry River (29,000 acres), Pemigewasset (45,000 acres), Sandwich Range (25,000 acres), and Caribou-Speckled Mountain (14,000 acres). The Wilderness Management Plan (U.S. Department of Agriculture, Forest Service 2005, E-3) begins by setting the stage for implementing the LRMP within wilderness areas:

> The many components of the 1964 Wilderness Act created numerous challenges for land management. In addition to recognizing Wilderness as "an area where the earth and its community of life are untrammeled by man," the act provides for recreational access as well as consideration of ecological, geological, scientific, educational, scenic, and historic values. These different values can lead to contradictory management objectives. This plan is aimed at managing the White Mountain National Forest Wildernesses in such a way that these somewhat incongruous values all receive proper attention. Thus, the plan sets forth an agenda and a program of work for WMNF Wilderness management that aims to assure we maintain a balance among primitive recreation, ecological integrity, and other values of a heavily used urban national forest.

Table 8.2. A Goal Achievement Framework for Wilderness Management Planning— Expanded Outline for Organizing and Writing a Plan

Plan Framework	Section of Plan	Content
	Introduction	Brief description of the area, purpose, and organization of the plan.
	Summary and overview of overall situation and management strategy	An overview or summary of current conditions affecting management, such as use levels and patterns, special situations, personnel, and general management strategy.
Goals	National direction	Concise summary of legislative requirements, departmental guidelines, and national agency policy and philosophy.
	Overall area goals	A statement of goals for the management of the particular wilderness.
Objectives	Objectives for all important aspects of wilderness management	Specific wilderness conditions sought for all important aspects of the wilderness, such as vegetation, recreation, wildlife, fire, trails, and travel. (Topics may vary by agency or area.)
Current situation and assumptions	Current situation	Summary of trends, conditions, and assumptions pertinent to each objective.
	Assumptions	Judgments about future trends, pressures, and problems pertinent to each objective.
Management direction mechanisms	Management policies	Guiding policies that—considering current situation and assumptions about the future—are necessary to guide actions toward established objectives.
	Management actions	Specific practices, programs, and actions that are judged necessary to achieve established objectives.
	Standards	Minimally acceptable standards that trigger corrective action if not met.

These lands are managed to allow natural processes to continue with minimal impediment, to minimize the effects and impacts of human use, to provide primitive and unconfined recreation opportunities, to foster appreciation of the qualities of wilderness landscapes, to continue use for educational and scientific purposes, and to recognize their evolving roles in the history of the landscape.

This management plan describes processes and actions aimed toward further realizing these goals. Our intent is to provide strong, clear management, in order to maintain Wilderness character. These values include a balance of use and preservation, an understanding of and support for protection of these lands, and a perpetuation of Wildernesses' roles as representatives of landscapes minimally affected by the impacts of human use.

The WMNF Wilderness Management Plan uses the components of the goal-achievement framework as a second level of planning detail within the LRMP to explicitly state the standards, monitoring methods, and management actions necessary to achieve the overall goals and objectives. This WMNF Wilderness Management Plan example is particularly useful because it also uses the framework of the limits of acceptable change and area zoning to implement the plan goals.

The Limits of Acceptable Change for Wilderness Planning

The goal-achievement framework for wilderness management planning presented in this chapter appeared in previous editions of *Wilderness Management*. Subsequently, recreation researchers and planners have continued to refine this approach and to combine it with a related framework to wilderness management planning known as the limits of acceptable change (LAC). The LAC method has become established as a useful wilderness management and planning tool, especially where there are conflicts between uses, trade-offs between goals, and concerns about the control of unacceptable change (Cole and Stankey 1997; McCoy et al. 1995; Stankey et al. 1985). The LAC approach is discussed in detail in Chapter 9. The short discussion of LAC here is only to relate it to the goal-achievement framework.

The goal-achievement framework for wilderness management planning and the LAC method focus on developing wilderness management actions to achieve desired conditions. However, the two planning frameworks differ mainly in that the LAC approach uses very specific standards rather than more general objectives. Some of the differences between the objectives in the goal-achievement framework and LAC standards are:

1. Objectives in the goal-achievement framework define a target to work toward—directional statement: based on what is considered desirable (what you are striving to achieve), do not clearly identify when there is a serious problem, may be written to reflect minimally acceptable management performance, and do not identify an "end point"; thus, once an objective is reached, there is no decision on whether to continue current management actions or not.
2. LAC standards define an absolute limit and an end point (a "state"): based on what are considered minimally acceptable conditions (maximum amount of deviation from "ideal" conditions considered acceptable), explicitly define the compromise between

Table 8.3. Overall Goals—Bear Trap Canyon Unit, Lee Metcalf Wilderness, BLM
(U.S. Department of the Interior, Bureau of Land Management, 1985)

First goal	To provide for the long-term protection and preservation of the area's wilderness character under a principle of nondegradation. The area's natural condition; opportunities for solitude; opportunities for primitive and unconfined types of recreation; and any ecological, geological, or other features of scientific, educational, scenic, or historical value will be managed so that they will remain unimpaired.
Second goal	To manage the wilderness area for the use and enjoyment of visitors in a manner that will leave the area unimpaired for future use and enjoyment as wilderness. The wilderness resource will be dominant in all management decisions where a choice must be made between preservation of wilderness character and visitor use.
Third goal	To manage the area using a minimum of tools, equipment, or structures to successfully, safely, and economically accomplish the objective. The chosen tool, equipment, or structure should be the one that least degrades wilderness values temporarily or permanently. Management will seek to preserve spontaneity of use and as much freedom from regulation as possible.
Fourth goal	To manage nonconforming but accepted uses permitted by the Wilderness Act and subsequent laws so as to prevent unnecessary or undue degradation of the area's wilderness character. Nonconforming uses are the exception rather than the rule; therefore, emphasis is placed on maintaining wilderness character.

conflicting goals, clearly identify when there is a serious problem (conditions are no longer acceptable), and provide assurance to the public that "restrictive" actions will not be implemented unless the standard is not being met.

LAC standards are on different tiers or levels in planning time with long-term versus short-term frames and different scales of area being planned (large forest and park size units compared to specific management areas). Use the goal-achievement framework objectives for general management planning and LAC standards for more specific implementation plans. Especially consider using the LAC approach when there is an actual or potential conflict between two or more goals, and when there is a need to balance between those goals to some extent. For example, when there is an opportunity to obtain solitude versus providing freedom from access restrictions, it may be appropriate to use a standard of solitude with a 90 percent probability of encountering no more than ten groups per day and a maximum of twenty people observed at one time (at a particular attraction area).

The LAC concept is applicable to any management problem in wilderness in which there are conflicting goals and concerns about the control of unacceptable change, for example conflicts between managing for recreation use and preserving natural conditions. The LAC concept provides a framework within which the acceptable amount and extent of change in wilderness can be identified. It is designed to alert managers to the need for action when changes exceed standards.

The LAC process consists of four major components: (1) specifying acceptable and achievable resource and social conditions, defined by a series of measurable parameters; (2) analyzing the relationship between existing conditions and those judged acceptable; (3) identifying management actions judged to best achieve these desired conditions; and (4) monitoring and evaluating management effectiveness. These four components, in turn, are broken down into ten planning steps to facilitate application (see Chapter 9). For example, managers (with help from the public) identify the issues and concerns that need to be resolved through the planning process and use indicators of resource and social conditions to determine standards for comparison (Merigliano 1989; Merigliano and Krumpe 1986).

Both the goal-achievement and LAC frameworks incorporate monitoring and management-action revision procedures. Management actions are not fixed but can be revised if monitoring shows that desired wilderness conditions are not being achieved. The LAC framework adds more specificity in the monitoring and evaluation of whether the use impacts have exceeded the limits defined as the minimum acceptable conditions for a given wilderness area. The LAC process can be used independently for a specific management situation, such as conflicts in use between wilderness hikers and pack-stock users, or as a component in an overall goal-achievement planning process (that is, the LAC framework for specific use issues or conditions is subordinate to the overall wilderness goal-achievement plan).

The wide array of wilderness environments, with their differing types and intensities of wilderness management concerns, lead us to conclude that multiple management planning approaches and frameworks (for example, LAC and ROS) are necessary to cover all the situations encountered in planning processes (McCool et al. 2007). Both the goal-achievement and LAC frameworks are flexible enough to adapt to various situations—and we refer to them in plural in recognition of the fact that variations in both approaches are possible.

Preparing Wilderness Management Plans: Problems and Suggestions

Regardless of the planning approach selected, we have found the following advice to be relevant in improving plans.

Writing the Plan

The actual writing of a Wilderness Management Plan has some pitfalls, one of which is the *potential for investing too much effort in stating current situations and assumptions about the future.* Some of each are needed, but if they are too detailed, they can become the focus of the plan. In extreme situations, a plan can attain a problem-solution focus, with current situation statements and assumptions describing problems in detail and presenting dire forecasts of worsening situations. This orientation can lead to heavy-handed management prescriptions to fulfill assumptions, whether or not the assumptions are well founded. In many cases, accurate data are lacking, so that assumptions about the future are no better than guesses. *A good plan needs to balance clear descriptions of current situations against explicit assumptions, both supported with field experience and data.*

Situation and assumption statements are easy to dwell on because they are easier to write than management objectives. However, a plan that focuses too much on situation and assumption statements will lack clear objectives as criteria for prescribing management policies and actions, and standards for judging their attainment. For example, a current situation statement describing heavy use, combined with assumptions predicting substantial increases in use, could lead indirectly to very restrictive policies and actions. Although these might ultimately be necessary, a recreational-use objective that describes the specific experience opportunities to be provided and standards for acceptable conditions would provide a better basis for identifying necessary actions. This approach is also easier to explain and to justify to critics, because debate and disagreement over wilderness management may be focused on the stated objectives and actions necessary to achieve them, rather than on the accuracy of situation and assumption statements or on individual management actions that, considered in isolation, might not appear necessary.

Clearly stated objectives are the key to good Wilderness Management Plans. Objectives are particularly difficult to write for wilderness because of the subjective nature of the resource and the experiences it offers. Nevertheless, it is important to have clear objectives so that they can guide policies and actions and set standards for measuring attainment. *It is helpful to think of objectives as positive statements of wilderness condition or as experience opportunities that management seeks to preserve or provide.*

Fig. 8.2. Wilderness Management Plans may identify particular locations where special management is needed, such as travel corridors, heavy-use locations, unusual attractions, or vegetation zones where special restrictions apply. A no-campfire regulation in high alpine zones may help preserve aesthetic features such as this ancient whitebark pine snag in the John Muir Wilderness, California; the tree might otherwise be used as firewood. Photo courtesy U.S. Forest Service.

Not all objectives can be described with equal specificity. The precision of individual objectives varies with the aspect of wilderness under consideration—but usually the more quantifiable the objective, the better. An objective for interpretive signs, for example, might be more specific than an objective for wildlife, vegetation, or water. Objectives ranging from *general* to *specific* can still suggest management direction. However, *vague objectives give few clues to necessary management direction. Compare the following examples that focus on planning:*

- *A Specific Objective*—To maintain lakes and watercourses in their current natural condition, with a high degree of water clarity and water quality suitable for primary human contact, subject to natural forces and free of human-caused contaminants.
- *A General Objective*—To maintain lakes and watercourses free of human-caused contaminants.
- *A Vague Objective*—To protect lakes, watercourses, and water quality.

Understanding the distinction between general and vague objectives is important. Objectives can be stated in broad, general terms and still retain fairly definite implications about the kinds of actions needed to achieve them. Vague objectives, on the other hand, are abstract and lack clear implications for management actions. *Although an objective might be general, its management implications should be clear.*

It is also important to recognize that the specificity of goals and objectives is directly related to their location in the planning hierarchy or framework. *The closer an objective is to the field-action level, the more specific it should be.* Compared to the lofty goals of the Wilderness Act ("to preserve, unimpaired, a wilderness resource"), management objectives for one particular aspect of wilderness will seem quite specific—for example, "to provide for primitive recreation only to the extent that naturalness, outstanding opportunities for solitude, and physical and mental challenge are preserved." Farther down the hierarchy, a field manager's objectives may be to implement very specific actions derived from goals, objectives, and policies formulated higher up in the planning process. For example, one field manager indicated that the objectives he was interested in were the number of signs to be posted that summer, the number of campsites to be relocated, and the miles of trail to be cleared or maintained—these are annual operational-level objectives that fit within higher-level goals. Perspective varies in the framework. At lower levels in the planning and management process, field objectives *are* statements of actions to be carried out—but actions derived from goals and objectives set forth in the planning process. *If a plan has internal logic, it should be possible to trace field actions to higher levels of the planning process (goals, objectives, and policies) by asking why the activity is being carried out.*

The level of detail that wilderness management plans should contain is an important consideration. A plan containing 200 pages of single-spaced typing might be so overwhelming that managers and the public may never read it. On the other hand, except for the simplest circumstances, a plan of only ten pages will not contain enough detail, even though it might give general direction. Again, balance of generality and detail is needed, depending on the size, complexity, levels of visitation, degree of public interest, and problem situations in the particular wilderness. *The plan should contain sufficient detail to describe all objectives, policies, and the what and where of particular actions; but the when, how, and by whom level of detail is more appropriately included in action, operating, or annual work plans.*

Management plans are most useful if they are straightforward, well organized, and readable. They should clearly inform managers and users of the management direction for the wilderness, what this entails in policies that govern field actions, and the kind of major actions that will be carried out at particular locations. The plan should not be so long and detailed that only its authors or the affected managers are willing to read it. Some recent Wilderness Management Plans are quite complex and reflect the growing sophistication of wilderness management, the diversity of management approaches, and differences in views of users competing for access to the wilderness. However, if plans are so complex that their internal logic is not apparent to the public, the advantages they gain through complexity will be lost.

Zoning—A Wilderness Opportunity Spectrum

Recreation managers and researchers have made a concerted effort to classify the recreation opportunity spectrum on public lands and the kinds of facilities, management practices, and visitor behavior appropriate to each type of opportunity (Buist and Hoots 1982; Clark and Stankey 1979). This *recreation opportunity spectrum* (ROS) idea is, in effect, zoning on a macro scale. It classifies *opportunity* from primitive, semiprimitive nonmotorized, semiprimitive motorized, roaded and natural, rural, and urban zones. *Not surprisingly, as managers began to address increasingly complex wilderness management situations, the idea of a wilderness opportunity spectrum (WOS) emerged as part of the management strategy to address finer gradations of primitive and semiprimitive classes. The WOS includes, for example, pristine, primitive, and portal (entry) designations, indicating decreasing degrees of naturalness and solitude. The WOS, like the ROS, is a kind of zoning—delineating particular areas where different management prescriptions or restrictions on visitor behavior apply.* Other examples include no-camping zones; trailless zones, where only cross-country travel and special minimum-impact camping practices are allowed; and special management zones having particular problems, high-use sites, impacted locations, or perhaps sensitive wildlife areas. Areas characterized by relatively dense concentrations of visitors and use impacts are also a kind of zoning.

The WOS and other zoning approaches are appropriate in certain situations. *We offer three criteria for judging the utility and appropriateness of zoning.* First, remember the cardinal rule of wilderness management—*do only what is necessary.* Implement zoning restrictions only when required to protect the wilderness resource and consider the impact that the zoning would have on visitor perceptions and use. Second, *make zoning clear to users*, along with its rules and expectations. To be effective, any zone that requires different visitor behavior must be indicated by some reliable means, such as a use permit or map. There is some concern over advertising the most pristine zones because of attracting use. Third, *zoning should not be used to justify nonwilderness conditions that presently exist*; when such conditions exist, they should be corrected, *rather than* assigned to a "special management zone." Congress established wilderness as a land designation, and agencies should not try to change its status through zoning that allows for standards lower than those intended under the Wilderness Act.

Identifying zones where special management will be needed may be necessary. However, minimum wilderness standards, at least, need to be maintained. The invasion of a wild area by too many users, the displacement of less tolerant vistors by more crowd-tolerant visitors, and declining wild conditions are well-established phenomena in recreation use and management. In fact, they are factors that triggered the establishment of the NWPS. A major goal of wilderness managers is to ensure that wilderness remains wild to meet the intent of the Wilderness Act.

Public Involvement and Plan-Review Process

Public participation is especially important for managing wilderness. The public must have an opportunity to help formulate Wilderness Management Plans if these documents are to have credibility and acceptance. Plans not supported by the public will be difficult to implement—in fact they may be legally appealed and their implementation blocked under NEPA regulations requiring public involvement in a significant environmental action such as a Wilderness Management Plan.

Involving the public is an accepted part of the decision-making process in the federal agencies with wilderness management responsibilities. The 1964 Wilderness Act was one of the first resource-management laws that required public involvement in decision making (U.S. Public Law 88-577, Sec. 3[d][1][d] and was followed in 1970 by NEPA (U.S. Public Law 91-190), which mandated public involvement in all natural-resource decisions having potential for significant environmental impact.

The public is a source of ideas, a sounding board for the acceptability of proposed management direction and policy, and a potential partner in its formation and implementation. Furthermore, wilderness planning now takes place in a political arena, in which consensus and negotiation are every bit as important as scientific data and logic (Krumpe and Stokes 1993). Planners and managers recognize that dual conditions are required for effective planning. First, a technically sound planning process is required for explicitness and to facilitate the search for reasonable alternatives by systematically working through a logical sequence. This is a necessary but not sufficient condition for effective planning. Managers understand that they need informed consent among those affected by the plan about the proposed course of action. In the politicized settings in which wilderness planning take place, the values in conflict are often well articulated, expressed, and pursued by the various contending interest groups. The arena of conflict may shift over time, but it still encompasses the agency and its perceived mission. Indeed, one or several groups may in reality hold the power of implementation rather than the planning agency. This power, held in the political realm, is in practice the power of veto over new plans (Krumpe and McCool 1997).

Planners and wilderness managers can become frustrated when politics gets in the way of rational planning. They become frustrated when decisions seem to be motivated more by political considerations than by purely scientific or philosophical considerations of fairness, equity, or other idealized values they hope would guide the management of publicly held natural resources (Krumpe 2000).

The public, on the other hand, experiences equal frustration at the extensive time and effort going into planning that often result in no change, or in plans that do not address the needs of a particular interest or may be so legalistic and bureaucratic as to confuse and alienate readers. As a result, both managers and the public have become disillusioned that science does not, or often cannot, give them the facts they need to answer the thorny questions raised in wilderness planning. In fact, it is a common delaying tactic for one user group or another to simply question the legitimacy of (and thus dismiss) any science that runs counter to their values or expectations.

Most managers who have successfully used public involvement are impressed with its helpfulness in making better and more acceptable decisions. Many user groups are affected by wilderness management: hiking and climbing clubs, conservation groups, hunters and anglers, horse riders, photographers, youth organizations, outfitters, and others. They *deserve* to be involved in planning. Without the understanding and support of the involved public, wilderness management will fall short of its goals. Public involvement in wilderness matters extends beyond wilderness planning and management to include involvement in wilderness work as well. Volunteers, conservation group work parties, adopt-a-trail efforts, and even private contracting of wilderness work projects all increase the involvement of the public in wilderness management.

As mandated by the NEPA, the four federal wilderness management agencies have adopted procedures that integrate public involvement with the planning process. Agency staff now routinely prepare environmental assessments and environmental impact statements during the wilderness planning process and before making land-use decisions. Such documents, with their sections detailing potential alternatives and environmental consequences, provide multiple opportunities for public involvement, from the initial identification of issues (called scoping) to lengthy public comment periods on the draft and final publications of EIS and wilderness plans.

Many different strategies may be successful for securing public involvement, but few have widespread applicability. Beginning in the late 1990s, the Internet has been widely used to facilitate public involvement, with plans posted for downloading from World Wide Web sites and comments received over electronic mail. Face-to-face methods are still the heart of good public involvement. Three public-involvement approaches used in connection with the BLM's Bear Trap Canyon plan, the Denali National Park and Preserve plan, and the Maricopa Complex Wilderness Management Plan are described later. Public-involvement procedures are flexible to fit different circumstances. Both managers and the public must be prepared for the time required to get public input, incorporate it into new drafts, and then get agency staff and the public response to the revisions. One to two years, or more, may be required to adequately involve the public in a wilderness management and EIS planning process.

Bear Trap Canyon (BLM)

The public-involvement process used during development of the Bear Trap Canyon unit's Wilderness Management Plan in 1984 is an early example of a successful program that began by establishing a five-member advisory group. This Bear Trap Canyon Wilderness Committee, representing a variety of interests both private and commercial, met as a group in January. BLM staff provided the group with a list of management issues to be addressed in the management plan and requested comments.

In March, members of the committee, key individuals, and agency officials were sent a copy of an in-house draft of the plan to review. Committee members were asked to disseminate pertinent portions of the document for comment to their constituencies. Next, individuals on the BLM's district office wilderness mailing list were sent postcards and asked to return them if they wanted copies of the draft Bear Trap plan for review.

A printed draft, revised in response to concerns of reviewers, was sent to members of the public who had indicated an interest in reviewing the plan, to the advisory group, and to agencies. Public notices in local newspapers informed the public that copies of the draft were available. The document sent to the public consisted of a draft plan and draft EA.

Due to public interest, the BLM extended the closing date of the original forty-five-day comment period from August 31 to October 12, 1984. During the comment period, the agency hosted two open houses in local communities. Twenty-seven written comments were received; in general, they supported continuing existing management of the area. Most respondents supported long-term protection and preservation of wilderness values, and they believed that existing use levels did not warrant restrictive measures.

What has become a standard procedure is for managers to print the public comments on a draft environmental analysis document and their response to such comments as part of the final version of the environmental analysis. In the Bear Trap plan, the planning team organized its response to public comment issue by issue.

The following issues illustrate the variety of concerns that can emerge from public comment: management for long-term protection of wilderness values versus maximizing visitor enjoyment, prohibition of overnight camping for rafting parties, trail closures and maintenance, prohibition of horse use, application of a let-fires-burn policy, construction of backcountry toilets, the number of commercial outfitters, use of paddleboats, and Air Force overflights.

In response to public comment and additional in-agency review, the BLM revised portions of the draft document before printing the final draft of the plan. The final plan and EA were published and distributed to the public in January 1985, one year after the first public-involvement efforts began.

Denali National Park and Preserve (NPS)

The 6-million-acre Denali National Park and Preserve includes 2.1 million acres that were designated by Congress as wilderness in 1980. The designated wilderness and much of the park are protected and managed collectively as backcountry. Development of the Backcountry Management Plan for Denali National Park and Preserve (U.S. Department of the Interior, National Park Service 2005a) included an amendment to the General Management Plan for Denali National Park and an EIS and used an extensive public-involvement and consultation process with federal, state, and local governments and native tribal governments throughout the scoping and planning processes.

NPS planners began this planning process by publishing their Notice of Intent to prepare an EIS and then held four public scoping meetings that were attended by 150 people. The planners received sixty-five written comments during fall 1998 through the spring of 1999 regarding issues and impacts that should be addressed in the Backcountry Management Plan. As a follow-up to the scoping meetings, a summary document describing the primary range of activities and alternatives to be evaluated in the plan and EIS was sent to more than 2,000 individuals, agencies, and organizations in January 2001. The alternatives for management and impact topics were further refined as part of five open house meetings in February 2001.

The NPS consulted with federal agencies, state and local governments, native tribal governments, and numerous user groups and organizations. During the winter and spring of 2002, collaborative planning workshops were held in two communities to discuss alternatives and issues associated with climbing and mountaineering, airplane use, subsistence, and other uses, especially those uses allowed under ANILCA.

When NPS planners and managers released the Draft Backcountry Management Plan and EIS in February 2003, they held informational workshops in six communities. The public comment period on the draft plan was initially open for seventy-five days, from February 25 to May 7, 2003, and then was extended to May 30, 2003. During that time, public hearings were held in seven communities and 9,370 written comments were received from two agencies, 27 organizations, and 9,341 individuals. As a result of the extensive public comments, the NPS staff revised the draft plan and solicited some additional public comment before releasing the final plan in April 2005.

Maricopa Complex Wilderness (BLM)

The Maricopa Complex Wilderness management area—Sierra Estrella, North Maricopa Mountains, South Maricopa Mountains, Table Top—includes 172,000 acres of desert under BLM administration as designated under the Arizona Desert Wilderness Act of 1990 (U.S. Public Law 101-628). The planning process, begun in 1992, used several types of public involvement: (1) four public scoping meetings in different local communities with a period of time for written comments on the issues; (2) formation of a working group of eight interested public representatives and agency personnel who met five times during 1992–1993 to review planning progress and to provide input; (3) a public briefing held in 1993 to discuss planning issues and input; and (4) a public review of the draft Maricopa Complex Wilderness plan held in 1994 at two community meetings and through written comments. A final plan was completed in 1995 to be implemented over a ten-year period (U.S. Department of the Interior, Bureau of Land Management 1995).

These examples are only three of many strategies that could be followed to involve the public in preparing Wilderness Management Plans. Each of the federal land-management agencies has developed its own public-involvement procedures based on their experience. *To summarize: (1) public involvement in wilderness management is essential and valuable, (2) many approaches are possible,* and *(3) adequate time (one to two years or more) is required for public involvement.*

Although this discussion focuses on how the public may assist with the review of draft plans, considerable technical and policy review of plans takes place at different levels within the agency and within other public agencies. It is essential that any other governmental units, such as fish and wildlife, forestry, environmental, or other resource-protection and management agencies, have a chance to review and provide input to plans. The challenge is to secure the optimum review and input from the public and agencies to afford managers the benefit of diverse opinions and to better represent the public and wilderness users in plan development. All input is advisory to the wilderness planners who must ultimately develop a plan within the legal requirements set by the Wilderness Act, the legislation designating the wilderness area, and agency policies and regulations.

Criteria for Evaluating Wilderness Management Plans

Growing public involvement ensures that Wilderness Management Plans will come under increasing public scrutiny; moreover, the growth of wilderness management as a resource specialty means that other managers will review plans. What criteria are appropriate for evaluating plans regardless of their format, specific agency requirements, or type of public involvement? We suggest the following questions to guide the review and evaluation of Wilderness Management Plans:

1. Does the plan summarize and explain the relationship of management goals to the Wilderness Act, legislation designating the area, departmental guidelines, national agency policy that guides and directs management of the area, and any higher-order plan of which it is a part, such as a Forest Plan in the USFS or a General Management Plan in the NPS? Including such goals will help relate management of the individual wilderness to the larger agency land-management system and the NWPS, of which it is a part.
2. Are local conditions, user activities, and internal and external threats relevant to management of the wilderness concisely described and explained?
3. Is the general management strategy concisely explained? This strategy might include (a) differing administrative methods by various units of the agency, the numbers of managers, and their responsibilities; (b) a description of user requirements such as a permit system; (c) use of the WOS or other zoning scheme; and (d) other details essential to implementing the management strategy.
4. Does the plan have an internal logic that links objectives to prescribed management policies and actions? It is essential that some kind of framework guide prescription of management policies and actions based on their *intention* to achieve objectives or some desired condition in the area.
5. How well does the plan consider alternative actions for meeting management objectives? The NEPA requires managers to analyze alternatives and how each alternative would attain goals or desired conditions.
6. Does the plan address the need for coordination of its resource-management activities and non-conforming uses, such as a national scenic trail crossing a wilderness, grazing activities, access to private land inholdings, or endangered species protection? Coordinating with other affected parties, such as adjacent landowners or state wildlife departments responsible for managing wildlife populations, is essential.
7. Were the managers who administered the wilderness area directly involved in the preparation of the plan? Involvement builds commitment to implementing the plan, because the best plan is of little value if it is not fully understood by managers and used to guide everyday stewardship.

Fig. 8.3. An early spring visitor in the Selway-Bitterroot Wilderness on the Idaho-Montana border plunge-steps down a snow slope. Wilderness area management plans need to consider all types of visitor use throughout its entire area and through the year, not just seasonal heavy use. Photo courtesy Chad P. Dawson.

8. Does the wilderness management direction contained in the plan help to resolve issues and management concerns facing field managers? Does the plan respond to important issues, concerns, and opportunities raised during public involvement efforts? To be effective and useful, plans need to provide site-specific application of wilderness management policy to resolve the everyday challenges of field managers seeking to achieve the plan's objectives.
9. Does the plan provide for a monitoring system, using field measures of indicators of wilderness conditions—biological, physical, and social—to determine if objectives are being achieved to the desired level of the standards? How often is monitoring done, and how will the data be used? The plan is not complete unless it includes provisions for objective data collection to see if the plan's objectives and standards are being achieved.
10. Key measures of a plan's success are found in answers to these four questions:
- Does the plan address and resolve key problems/conflicts/potential threats?

Fig. 8.4. Wilderness planning in Alaska and some other wilderness areas in the western U.S. needs to consider planning related to such activities as sight seeing by aircraft. Photo courtesy Chad P. Dawson.

Fig. 8.5. Monitoring wilderness conditions and experiences involves studying the resources and its visitors; here visitors are asked to fill out surveys about their overnight camping experiences on a canoe trip. Photo courtesy Chad P. Dawson.

- Does the plan contain an explicit rationale for decisions?
- Is there an agency commitment to move forward with plan implementation?
- Will implementing the plan's direction improve the current wilderness situation and condition?

11. Does the plan specify when and by whom the plan will be reviewed and updated? Does it identify what conditions or situations might prompt an earlier review? This information will help keep a plan from becoming obsolete.

Summary

Good planning is essential to good management. Planning is a formal process of thinking ahead about what conditions are desired and how to achieve them, the management problems likely to be encountered, and alternative methods for resolving them. This chapter described how Wilderness Management Plans fit into the planning processes of federal agencies, and suggested some elements of good planning, a framework for developing plans, and several examples of plans. A planning process should prescribe site-specific policies and actions implementing the framework of principles of wilderness management outlined in Chapter 7.

Study Questions

1. Why is it important to have a management plan for a wilderness area?
2. How does the NEPA affect Wilderness Management Plans?
3. Each of the four federal wilderness managing agencies (BLM, FWS, USFS, and NPS) has two or more levels of planning applicable to its wilderness areas. Briefly describe them.
4. Define the following basic planning concepts: goals, objectives, situation and assumption statements, management actions, and standards.
5. How can too strong a focus on describing current situations and assumptions about the future cause problems in preparing and implementing a Wilderness Management Plan?
6. What are three criteria for deciding whether zoning restrictions are useful and appropriate in a particular wilderness area?
7. Why is it important to involve the public in the development of a Wilderness Management Plan?
8. What are some questions to consider in evaluating Wilderness Management Plans?
9. How are the goal-achievement framework and the LAC process alike? How do they differ?

Acknowledgments

This chapter was written for the first and second editions by John C. Hendee and Russell von Koch, recreation planner, Bureau of Land Management, Moab, Utah. John C. Hendee and Chad P. Dawson, with help from Ed Krumpe, professor of resource recreation and tourism at the University of Idaho, and Linda Merigliano, wilderness coordinator, Bridger-Teton National Forest, Jackson, Wyoming, revised and updated this chapter for the third edition. For review of the third edition we thanked Greg Hansen, wilderness coordinator, Mesa Ranger District, Tonto National Forest, Mesa, Arizona; Jim Mahoney and Rick Hansen, BLM Phoenix District Office, Phoenix, Arizona; Russ von Koch, recreation planner, BLM, Moab, Utah; Jeff Jarvis, national wilderness program leader, BLM, Washington, DC; Donita Cotter, wilderness program analyst, U.S. Fish and Wildlife Service, Washington, DC; and Nancy Roeper, national wilderness program leader, U.S. Fish and Wildlife Service, Washington, DC.

Chad P. Dawson and John C. Hendee updated this chapter for the fourth edition.

Case Discussion: Comprehensive Conservation Planning for the Kenai National Wildlife Refuge, Alaska

The Kenai National Wildlife Refuge (KNWR) extends over 1.98 million acres on the Kenai Peninsula of Alaska and includes 1.3 million acres of designated wilderness. The refuge was originally established as the Kenai National Moose Range to protect the Alaska-Yukon moose, and it has brown and black bears, caribou, Dall sheep, mountain goats, wolves, lynx, wolverines, eagles, and thousands of shorebirds and waterfowl. The waters of the refuge include habitat for chinook, sockeye, coho, and pink salmon; Dolly Varden char; rainbow trout; and arctic grayling. Landscapes vary from the Harding Ice Field to mature spruce forests and glacial lakes. The KNWR's scenic landscape and wildlife include what has been described as a miniature version of many of the habitats in Alaska and, because of its proximity to Anchorage, draws half a million visitors a year, more than any other wildlife refuge in Alaska.

All refuges are required to be managed in accordance with an approved comprehensive conservation plan. The KNWR is involved in the planning process of revising the refuge's comprehensive conservation plan (CCP), which had been in place since 1985 (U.S. Department of the Interior, U.S. Fish and Wildlife Service 1985). Developing a CCP requires following guidance found in the National Wildlife Refuge System Improvement Act of 1997 (U.S. Public Law 105-57), ANILCA, the Wilderness Act, and NEPA. Revising the KNWR CCP will provide direction for management of the refuge for fifteen years after being approved. During 2003 to 2007, the KNWR staff progressed in developing the revised CCP and EIS through the leadership of the refuge's *interdisciplinary planning team*, consisting of U.S. Fish and Wildlife Service, Alaska Department of Fish and Game, and Alaska Department of Natural Resources staff. Distribution of the public review draft was published in May 2008, with a final plan expected by 2009. As part of the public input process in 2005, the *interdisciplinary planning team* requested public input on a variety of issues and concerns and requested public feedback (U.S. Department of the Interior, U.S. Fish and Wildlife Service 2005).

One of the preliminary management issues in 2005 was how the KNWR could enhance wildlife-oriented recreation such as wildlife viewing opportunities. The public was asked to comment on four general alternatives (not specific to wilderness or nonwilderness areas of the KNWR), which were summarized as:

1. Current management situation is that opportunities and facilities are limited and habitat projects with associated wildlife viewing facilities are not provided.
2. Improve existing wildlife viewing facilities (e.g., pullouts on roads and waysides) only.
3. Improve existing wildlife viewing facilities and develop new additional wildlife viewing facilities (e.g., overlooks, towers, blinds, boardwalks).
4. Improve existing wildlife viewing facilities and develop additional wildlife habitat improvement projects with new additional wildlife viewing facilities (e.g., overlooks, towers, blinds, boardwalks).

Wildlife viewing is challenging on the KNWR for several reasons: the vegetation is dense and the line of sight is limited except on roads, water bodies, and glacial areas; wildlife generally do not concentrate for easy viewing; wildlife are not habituated to humans in their environment and generally move away from humans; and some large animals can be a threat to human safety if startled or approached too closely.

Case Discussion Questions:

1. Is the range of the four preliminary management alternatives presented to enhance wildlife viewing opportunities acceptable or unacceptable to you as a local visitor, as a tourist from outside Alaska, as a wilderness advocate, and as a wilderness manager? Why might the responses of different people vary between being acceptable and unacceptable?
2. To what degree should the KNWR enhance wildlife-oriented recreation such as wildlife viewing opportunities in the wilderness areas of the refuge and the nonwilderness areas of the refuge?
3. Balancing between human enjoyment and protecting the resource is a complex process. What planning framework may be helpful to the interdisciplinary planning team? Why?

References

Buist, Leon; Hoots, Thomas A. 1982. Recreation Opportunity Spectrum approach to resource planning. *Journal of Forestry.* 80: 84–86.

Clark, Roger N.; Stankey, George H. 1979. The Recreation Opportunity Spectrum: A framework for planning, management, and research. General Technical Report PNW-98. Portland, OR: U.S. Department of Agriculture, Forest Service, Pacific Northwest Forest and Range Experiment Station.

Cole, David N.; Stankey, George H. 1997. Historical development of limits of acceptable change: Conceptual clarifications and possible extensions. In: McCool, Stephen F.; Cole, David, comps. *1997 Proceedings: Limits of Acceptable Change and Related Planning Processes: Progress and Future Directions;* May 20–22, 1997; Missoula, MT. INT-GTR-371. Ogden, UT: U.S. Department of Agriculture, Forest Service, Rocky Mountain Research Station, pp. 5–9.

Krumpe, Edward E. 2000. The role of science in wilderness planning—a state-of-knowledge review. In: Cole, D. N.; McCool, S. F.; Borrie, W. T.; O'Loughlin, J., comps. *Wilderness science in a time of change conference,* Vol. 4; *Wilderness Visitors, Experiences, and Visitor Management;* May 23–27, 1999; Missoula, MT. RMRS-P-15-Vol-4. Ogden, UT: U.S. Department of Agriculture, Forest Service, Rocky Mountain Research Station, pp. 5–12.

Krumpe, Edward E.; McCool, Stephen F. 1997. Role of public involvement in the limits of acceptable change wilderness planning system. In: McCool, Stephen F.; Cole, David, comps. *1997 Proceedings: Limits of Acceptable Change and Related Planning Processes: Progress and Future Directions.* May 20–22, 1997; Missoula, MT. INT-GTR-371. Ogden, UT: U.S. Department of Agriculture, Forest Service, Rocky Mountain Research Station, pp. 16–20.

Krumpe, Edward E.; Stokes, Jerry. 1993. Application of the limits of acceptable change planning process in United States Forest Service wilderness management. In: Hendee, J. C.; Martin, V. G., comps. *Proceedings of the 5th World Wilderness Congress Symposium on international wilderness allocation, management and research;* September 24–October 2, 1993; Tromso, Norway. Fort Collins, CO: International Wilderness Leadership Foundation, pp. 186–191.

McCool, Stephen F.; Clark, R. N.; and Stankey, G. H. 2007. An assessment of frameworks useful for public land recreation planning. General Technical Report PNW-GTR-705. Portland, OR: U.S. Department of Agriculture, Forest Service, Pacific Northwest Research Station.

McCoy, K. Lynn; Krumpe, Edwin E.; Allen, Stuart. 1995. Limits of acceptable change planning evaluating implementation by the U.S. Forest Service. *International Journal of Wilderness.* 1(2): 18–22.

Merigliano, Linda L. 1989. Indicators to monitor the wilderness recreation experience. In: Lime, D. W., ed. *Managing America's Enduring Wilderness Resource.* Conference Proceedings; September 11–17, 1989; Minneapolis, MN. St. Paul: University of MN. Extension Service, pp. 156–162.

Merigliano, Linda L.; Krumpe, Edwin E. 1986. Scientists identify, evaluate indicators to monitor wilderness conditions. *Park Science.* 6(3): 18–19.

Stankey, George H.; Cole, David N.; Lucas, Robert C.; et al. 1985. The limits of acceptable change (LAC) system for wilderness planning. General Technical Report INT-176. Ogden, UT: U.S. Department of Agriculture, Forest Service, Intermountain Forest and Range Experiment Station. 37p.

Thompson, T. L. 2005. 10-year wilderness stewardship challenge. Memo from U.S. Forest Service Chief T. L. Thompson to staff; dated February 7, 2005; file code 2320.

U.S. Department of Agriculture, Forest Service; U.S. Department of the Interior, Bureau of Land Management. 2006. Policies and Guidelines for Fish and Wildlife Management in National Forest and Bureau of Land Management Wilderness. Washington, DC: U.S. Department of Agriculture, Forest Service; U.S. Department of the Interior, Bureau of Land Management.

U.S. Department of Agriculture, Forest Service. 2005. White Mountain National Forest Land and Resource Management Plan. Laconia, NH: U.S. Department of Agriculture, Forest Service, Eastern Region.

———. 2007. Wildernesses, primitive areas, and wilderness study areas. In: *Forest Service Manual,* chap. 2320. Washington, DC: U.S. Department of Agriculture, Forest Service.

U.S. Department of the Interior, Bureau of Land Management. 1985. Wilderness management plan for the Bear Trap Canyon Unit of the Lee Metcalf Wilderness. BLM-MT-ES PS-003-4332. Billings, MT: U.S. Department of the Interior, Bureau of Land Management.

———. 1995. Maricopa Complex Wilderness Management Plan, Environmental Assessment and Decision Record. EA AZ-026-94-20. Phoenix, AZ: U.S. Department of the Interior, Bureau of Land Management. Phoenix District, Lower Gila Resource Area.

———. 2000. *Wilderness Management.* 43 *Code of Federal Regulations,* Parts 6300 and 8560. *Federal Register* 65 (241): 78,358–78,376. December 14, 2000. Washington, DC.

U.S. Department of Interior, National Park Service. 1993. Wilderness Task Force: Report on improving wilderness management in the National Park Service. Washington, DC: U.S. Department of the Interior, National Park Service.

———. 2004. Program Standards: Park Planning. Washington, DC: U.S. Department of the Interior, National Park Service.

———. 2005a. Denali National Park and Preserve Revised Draft Backcountry Management Plan: General Management Plan Amendment and Environmental Impact Statement. Denali Park, AK: U.S. Department of the Interior, National Park Service.

———. 2005b. The National Parks: Index 2005–2007 Revised to include the Actions of the 108th Congress ending December 31, 2004. Washington, DC: U.S. Department of the Interior, National Park Service.

———. 2005c. *Park Planning Program Sourcebook: General Management Planning.* Washington, DC: U.S. Department of the Interior, National Park Service.

———. 2006. Wilderness Preservation and Management. In: *NPS Management Policies Handbook.* Washington, DC: U.S. Department of the Interior, National Park Service.

U.S. Department of the Interior, U.S. Fish and Wildlife Service (USDI, FWS). 1985. Kenai National Wildlife Refuge Final Comprehensive Conservation Plan, Environmental Impact Statement and Wilderness Review. Anchorage, AK: U.S. Department of the Interior, U.S. Fish and Wildlife Service.

———. 2000. Refuge Planning Policy. In: *Refuge Manual.* Part 602, Chap. 1, 3, 4. Washington, DC: U.S. Department of the Interior, U.S. Fish and Wildlife Service.

———. 2005. Kenai National Wildlife Refuge: Planning Update Winter–Spring 2005. Anchorage, AK: U.S. Department of the Interior, U.S. Fish and Wildlife Service.

U.S. Public Law 235 (codified at 16 U.S.C. sec. 1-18f). National Park Service Act of August 25, 1916. 39 Stat. 535.

U.S. Public Law 86-517. Multiple Use Sustained-Yield Act of June 12, 1960. 74 Stat. 215.

U.S. Public Law 88-577. The Wilderness Act of September 3, 1964. 78 Stat. 890.

U.S. Public Law 89-669. National Wildlife Refuge System Administration Act of October 15, 1966. 80 Stat. 927.

U.S. Public Law 91-190. National Environmental Policy Act (NEPA) of January 1, 1970. 83 Stat. 852.

U.S. Public Law 93-378. Forest and Rangeland Renewable Resources Planning Act of August 17, 1974. 88 Stat. 476.

U.S. Public Law 94-579. (Wilderness Study) Federal Land Policy and Management Act (FLPMA) of 1976. 90 Stat. 2743.

U.S. Public Law 94-588. National Forest Management Act of October 22, 1976. 90 Stat. 2949.

U.S. Public Law 95-95. Clean Air Act Amended as of August 7, 1977. 42 Stat. 7401.

U.S. Public Law 95-625. National Parks and Recreation Act of November 10, 1978. 92 Stat. 3467.

U.S. Public Law 96-487. Alaska National Interest Lands Conservation Act (ANILCA) of December 12, 1980. 78 Stat. 890.

U.S. Public Law 103-433. California Desert Protection Act of October 31, 1994. 108 Stat. 4473.

U.S. Public Law 105-57. National Wildlife Refuge System Improvement Act of October 9, 1997. 111 Stat. 1252.

U.S. Public Law 109-382. New England Wilderness Act of December 1, 2006. 120 Stat. 2673.

Chapter 9
Managing for Appropriate Wilderness Conditions: The Limits of Acceptable Change Process

Introduction	218
Considerations in Managing Visitor Use and Impact	219
The Carrying-Capacity Concept	219
The Limits of Acceptable Change (LAC) Concept	222
When Should the LAC Process Be Used?	223
The LAC Process	223
Step 1: Define Goals and Desired Conditions	224
Step 2: Identify Area Issues, Concerns, and Threats	225
Step 3: Define and Describe Acceptable Conditions	226
Step 4: Select Indicators of Resource and Social Conditions	228
Step 5: Inventory Existing Resource and Social Conditions	229
Step 6: Specify Standards for Resource and Social Indicators for Each Opportunity Class	230
Step 7: Identify Alternative Opportunity Class Allocations	231
Step 8: Identify Management Actions for Each Alternative	231
Step 9: Evaluate and Select a Preferred Alternative	232
Step 10: Implement Actions and Monitor Conditions	232
Potential Barriers to the LAC Process	233
Positive Benefits to Wilderness from the LAC Process	233
Implementing the LAC Process	234
The Bob Marshall Wilderness Complex	234
The San Juan–Rio Grande National Forests Wilderness Management Direction	238
The White Mountain National Forest Wilderness Management Plan	240
The Maricopa Complex Wilderness Management Plan	242
The Superstition Wilderness Implementation Plan	244
Direct Visitor Management May Be Necessary When Standards Are Exceeded	246
Summary	247
Study Questions	247
Acknowledgments	247
Case Discussion: Campsite Indicators on the White Mountain National Forest	248
Case Discussion Questions	249
References	249

Introduction

Although many qualities and conditions are associated with wilderness, naturalness and solitude are the focus of the Wilderness Act and this book. Most wilderness qualities and conditions are sensitive to the type and amount of use an area receives, and excessive use can impact the quality of the natural setting as well as the sense of solitude that visitors experience. Management concern with overuse by visitors is not a recent phenomenon, and there is strong legislative intent to provide for public use while at the same time preventing degradation from overuse and inappropriate use. This intent is stated clearly in the Wilderness Act of 1964 (U.S. Public Law 88-577, sect. 2[a]): that wilderness areas "shall be administered for the use and enjoyment of the American people in such a manner as will leave them unimpaired for future use and enjoyment as wilderness, and so as to provide for the protection of these areas, the preservation of their wilderness character."

As related in Chapter 2, there was concern with uncontrolled overuse and its associated impacts in the high country of California's Sierra as early as the mid-1930s. Lowell Sumner's call in 1942 for the restriction of use within an area's carrying capacity or recreational saturation point reflects both concern and recognition that continued increases in use could destroy the area's wild qualities (Sumner 1942).

Many overuse concerns still remain. Most wilderness visitors and managers can relate personal experiences with undesirable conditions: heavy visitor

Fig. 9.1. Carrying capacity of wilderness is the amount, kind, and distribution of use that can occur without leading to unacceptable impacts on either the physical-biological resource or the available wilderness experience. The limits of acceptable change (LAC) planning process was used by the Bureau of Land Management in Arizona's North Maricopa Mountains Wilderness to ensure that hikers would be able to experience solitude. Photo courtesy Chad P. Dawson.

Fig. 9.2. Use limits can protect quality wilderness environment such as the Three Sisters Wilderness, Oregon. Excessive use, however, can lead to resource damage and to a degraded wilderness experience. Photo courtesy U.S. Forest Service.

use on some trails, campsites devoid of vegetation, or mud and horse manure on trails. Wilderness has been lost when conditions like these prevail. Although such conditions do exist in some areas, pristine settings and outstanding opportunities for solitude are found in much of the NWPS. Clearly, there is concern among users and managers alike about overuse and resource damage, but in the face of substantial variations in use intensity within and between areas, how does one approach the matter of deciding what is an appropriate level and type of use in wilderness?

Establishing appropriate use levels in wilderness typically has been addressed through the concept of carrying capacity. In Chapter 7, we argued that preserving qualities essential to wilderness required a carrying-capacity constraint as a fundamental principle of wilderness management. Our concern is to determine at what point environmental and social conditions within wilderness become inconsistent with the qualities required by the Wilderness Act. Each set of wilderness conditions and visitor experiences sought as management objectives implicitly carries it some limit in the kinds and amounts of recreational use that can be considered acceptable; it also implies the need for various policies and actions to see that these acceptable limits are not exceeded. This approach to monitoring of visitor impacts is termed limits of acceptable change (LAC).

In this chapter, we explore the origins of the carrying-capacity concept, consider limitations in how the concept has been applied in wilderness settings, and propose LAC as a reformulated conception of the carrying-capacity model designed to help citizens and managers establish guidelines for acceptable wilderness conditions. Brief case studies examining how such guidelines have been developed for several wilderness areas are also described.

Considerations in Managing Visitor Use and Impact

Several basic considerations must be taken into account in managing recreational use and its impact. First, to determine the kinds of impacts that will occur and their possible implications, three components must be considered:

Natural resources—The physical and biological characteristics of the natural-resource base greatly influence the amount of change in the environment resulting from recreation activities during visitor use. Although recreation use inevitably causes change in the environment, some resources are inherently more fragile than others.

Social—The needs and wants of people are important in determining appropriate uses of natural resources. User perceptions and opinions of what types and level of use are preferred are an essential element in developing prescriptions of appropriate visitor use.

Managerial—Legal directives and agency missions play a major role in determining appropriate and feasible resource, social-political, and management conditions. Managerial factors help identify what conditions should be maintained, what recreation-related impacts need to be managed to maintain desired conditions, and what management actions might be used to achieve those conditions (Manning 1999).

Second, any recreation use of an area leads to change in the character of these three components. *The inevitability of change is a critical consideration* that managers must take into account in prescribing carrying capacities related to visitor management. For example, many studies of ecological impact related to recreation activities point to the fact that most of the total impact recorded occurs under fairly low use levels (see Chapter 15). It may be necessary to establish lower use limitations than what would be acceptable to some public groups; this would be accomplished by restricting the number of visitors in an area. Rather than attempting to define how much visitor use is too much, the limits of acceptable change (LAC) focus attention on deciding the amount of change that will be allowed to occur.

Third, determining the amount of change that will be considered acceptable involves a *subjective value judgment,* and such judgments are derived from the management objectives prescribed for the area (see Chapter 8). These objectives provide the framework within which judgments regarding desired conditions for an area—resource, social, and managerial—can be defined. These desired conditions need to be expressed explicitly and quantitatively to reduce disputes over interpretation and to allow for determination if objectives have been met or not. Area-management objectives contain explicit standards that define the extent to which conditions will be allowed to change, if at all. The Wilderness Act provides general guidelines as to what management objectives are to be achieved. For instance, environmental conditions must be highly natural whereas visitor-use conditions provide opportunities for primitive recreation. The exact meaning of these conditions is open to interpretation; people will have different ideas about what they mean and their implications for management. Specifying standards helps give objective meaning to them.

Finally, although the historic focus of concern in visitor use and impact management was on the number of users, the impacts of use—on resources and other users—are often related to other aspects, such as the type of use, timing and location of use, and visitor behavior. This means the management approach of regulating visitor-use numbers may not sufficiently control impacts, which is really what we are most concerned about in wilderness. More attention needs to focus on other management strategies that more directly prevent or mitigate visitor impacts defined as unacceptable in wilderness.

These basic considerations underlie much of the discussion in this chapter. The goal of management is to identify the desired resource, social, and managerial conditions to be maintained or restored in wilderness, with these desired conditions expressed as explicit, measurable standards. Thus, the focus of management attention shifts from defining maximum recreation use to identifying desired conditions and managing use levels and visitor impacts to not exceed the standards (Shelby and Heberlein 1986).

The Carrying-Capacity Concept

Carrying capacity is a fundamental concept in natural-resource and environmental management (Dasmann

1964; Godschalk and Parker 1975). It can be defined as the maximum level of use an area can sustain as set by natural factors of environmental resistance such as food, shelter, or water. Beyond this natural limit, no major increases in the dependent population can occur (Odum 1959). In the field of wildlife management, for instance, carrying capacity describes the number of animals of a particular species that can use the range on a sustained basis, given available food, shelter, and water. If the balance between animals and the range's capacity is upset, either by an increase in the number of animals or through loss of the range's productive capacity (e.g., drought), problems will occur. Food resources become depleted as animals search for nutrition. Even though animal numbers might increase in the short term, loss of range productivity means that, in the long term, fewer animals can be supported; in extreme cases, irreversible environmental impacts can occur.

This description may seem to imply that carrying capacity is an uncomplicated, straightforward notion. In fact, it is not. The capacity of the range would be quite different if populated by elk or sheep rather than deer; were it a mix of species, it would be even more complicated. The dynamic nature of ecosystems makes a static determination of wildlife carrying capacity difficult to calculate. On any given area of land, calculations of available forage and cover may vary widely within a single decade based upon perturbations such as wildfire, the stage of plant succession, presence or absence of drought conditions, and abundance or absence of predators, as well as changes in hunting seasons and regulations. Furthermore, the natural factors of environmental resistance can be influenced by resource managers food crops can be planted, water storage reservoirs created, and so forth. Thus, wildlife carrying capacity can be increased or decreased by management actions; it is not an inherent, fixed value of the land.

Practitioners and researchers in recreation management in general and wilderness management in particular early on adopted carrying capacity as a guide to making management decisions in order to identify how much use a resource could sustain. They sought to determine how many people could use a given recreational setting before unacceptable impact occurred, critical information for managers. In 1964, Wagar published one of the first substantive discussions of the carrying-capacity concept in recreational management. A major contribution of this analysis was the inclusion of the impacts of use on social or experiential considerations in addition to the typical environmental concerns (Manning 1999). As Wagar (1964) noted, "The study...was initiated with the view that the carrying capacity of recreation lands could be determined primarily in terms of ecology and the deterioration of areas. It soon became obvious that the resource-oriented point of view must be augmented by consideration of human values."

Under this broadened concept of carrying capacity, recreational areas had not only an *ecological or biological capacity but a social capacity* as well. Use could impact not only an area's biophysical resources, such as soils and vegetation, but also the character of the recreational experience. The recognition of a *social dimension* of carrying capacity implies that determination of carrying capacity is a sociopolitical process as well as a biophysical one (Burch 1984).

Wagar's monograph marked the beginning of a major effort by researchers and managers to define the carrying capacity of recreational areas. This effort was driven, in part, by the great surge in recreational use following the end of World War II. Wilderness in the United States, once remote and little used, began to receive increasing amounts of use as what Clawson (1985) called the *four fueling factors*—leisure time, income, access, and population—grew substantially. Between 1955 and 1964, for instance, the average annual growth of recreational use on national forest wilderness exceeded 10 percent (Stankey and Lucas 1986). Since passage of the Wilderness Act, wilderness recreation use is still increasing and probably will continue unless it is controlled (Cole 1996). Increasing use has generated concerns about ecological and social impacts that might jeopardize the essential wilderness qualities of naturalness and solitude.

Such concern over visitor impacts is the impetus for continuing interest in the use of carrying capacity as a framework within which decisions could be made to control impacts that would otherwise erode the qualities and conditions of wilderness. Literature on carrying capacity and the associated management techniques for applying the concept of carrying capacity has grown considerably since 1970 (Graefe et al. 1984; Manning 1999; Manning and Lime 2000).

Historically, there was limited progress in developing a well-understood procedure for applying the carrying-capacity concept in the field. Some authors even argued that the concept should be abandoned (Bury 1976; Wagar 1974). Many reasons underlie reservations about the utility of the carrying-capacity concept. For instance, recreation lands are used by many different people seeking many different and potentially conflicting experiences. Some want solitude, others look for companionship; what is an appropriate visitor-visitor encounter level for one person represents congestion or loneliness for another. Brown and Haas (1980) reported five different types of visitors in Colorado's Rawah Wilderness and noted that the psychological domain of "meeting/observing other people" showed the most difference among the five types. When there is agreement as to the type of experience that an area is to provide, there is generally agreement about what constitutes appropriate use of an area. Shelby (1981), for instance, found remarkable similarity in the definition of appropriate encounter levels in three study areas when the general management direction (wilderness, semiwilderness, undeveloped recreation) was specified.

Just monitoring impacts on biophysical resources does not help establish obvious capacity limits. Any recreational use of an area produces some change; typically, much of the total impact found in an area occurs with only light recreational use (Cole 1985). Thus, if a manager elects to allow a level of use producing little or no change, it will be necessary to restrict visitor use in many wilderness areas (Wagar 1968).

Carrying capacity implies a strong cause-and-effect relationship between the amount of visitor use an area receives and subsequent impact. However, many studies of this relationship point out that use intensity alone is a poor predictor of total impact. The season and type of use involved, for instance, frequently are more important in explaining impact than the amount of use (Cole 1985; Kuss 1986).

One the most serious problems in using carrying-capacity analysis in wilderness management was that some people considered the carrying capacity of an area as an inherent value of the resource base, one that could be determined through careful measurement and research. As Wagar (1968) notes, this idea seems to have been borrowed from range management, where resource capacity is defined largely by natural factors such as soils and precipitation. The concept of carrying capacity implied to some people that complex, controversial decisions about recreation visitor use could be easily formulated. However, more important, such a naïve understanding of the concept obscured the important distinction between carrying capacity as the product of a technical assessment and its establishment through value judgments that weighed resource and social impacts, along with human needs and values. Carrying capacity was seen as a scientific concept whose measurement was only constrained by the level of research effort used by managers and researchers, rather than the result of a judgment process.

In practice, attempting to apply a carrying-capacity concept to wilderness management produced two serious kinds of backlash in the public. First, the carrying-capacity concept focused the public's attention on the wrong question—how many users are too many? Even though the concept's originators never meant it to be, the public invariably linked the term *carrying capacity* with controlling visitor numbers rather than managing for desired wilderness resource and social conditions. Second, the public often did not accept the carrying-capacity decisions made by managers. People who wanted more access and use and those who wanted use and impacts to be lessened protested any attempt to establish recreation carrying capacity because research could not demonstrate a clear cause-and-effect relationship between levels of use and impacts. This forced researchers and managers to reexamine the carrying-capacity concept and its application in wilderness management.

Perhaps one of the most important developments in understanding the carrying-capacity concept over the past fifty years is that *carrying capacities are the product of value judgments as well as science*. It follows that if carrying capacity is a product of technical assessment (a scientific question) and value judgments, then the question becomes "Whose value judgments?" This means that determining carrying capacity is a scientific assessment and a political process through public involvement. Carrying-capacity determination becomes a process by which biophysical and social research is integrated with agency policies and the values of managers and users to reach a collective judgment.

In developing a wilderness management program, managers and the public must recognize that value judgments reflect the philosophical, emotional, spiritual, experiential, and economic responses of

Fig. 9.3. Horses typically have more impact on wilderness ecosystems than an equal number of hikers, especially on fragile sites such as lakeshores. To prevent irreversible damage, horses are now prohibited at Reflection Pond in the Glacier Peak Wilderness, Washington. Photo courtesy U.S. Forest Service.

those making the judgments. Obviously, few people will have identical responses, and therefore few will make identical value judgments. The task facing wilderness managers is to determine some level of informed consent or consensus in value judgments as one input to what constitutes desired wilderness conditions and how those conditions should be maintained within the constraints imposed by the Wilderness Act and other wilderness-related legislation. Because the management of wilderness is actually management of wilderness *users* and their impacts, decisions that reflect value judgments must be made with the support, consent, and/or agreement of those managed (the users). This implies that wilderness planning and management are a political process incorporating biophysical and social data and framed by enabling legislative mandates and agency policy.

Research information is one important input into the wilderness carrying-capacity determination process (Burch 1984; Hammitt and Cole 1998). This input involves describing the social and ecological consequences of alternative visitor-use levels, thus providing the opportunity for managers to judge whether these consequences are consistent with area-management objectives (Cole and Stankey 1997; Stankey 1979). Research can help wilderness managers who are concerned with carrying capacity, but it cannot supply definitive answers about the carrying capacity of a site or area.

The evolving recognition of carrying capacity as a *conceptual approach, derived from social and ecological judgments about appropriate wilderness conditions,* has led to increasing attention to the factors that shape and influence these judgments. The concern for applying the carrying-capacity concept (Graefe et al. 1986; Shelby and Heberlein 1986) has led to a series of similar planning efforts to develop more effective frameworks for managing visitor recreation use and associated impacts: the process for Visitor Impact Management (VIM), the Visitor Experience and Resource Protection (VERP) approach, the Management Process for Visitor Activities (known as VAMP), and the concept of the limits of acceptable change (LAC) all represent efforts to correct limitations in applying the carrying-capacity concept to recreation management (Nilsen and Tayler 1997; McCool et al. 2007). These frameworks share a common focus on identifying measurable objectives regarding desired conditions and on the distinction between steps involving objective description and analysis and those involving judgmental evaluations.

In this chapter, we describe how the LAC concept can be used as a framework for managing wilderness within a carrying-capacity concept (McCool and Cole 1997b; Stankey et al. 1985). The LAC approach is similar in structure to other frameworks that use goals, objectives, indicators, and standards to evaluate visitor impacts on biophysical and social conditions (Nilsen and Tayler 1997; McCool et al. 2007). We recognize that federal agencies use different frameworks, but the LAC framework serves here as an example of how to apply a framework—within the carrying-capacity concept—in a wilderness planning process.

The Limits of Acceptable Change (LAC) Concept

Establishing appropriate visitor-use levels and acceptable recreation-activity impacts in wilderness is a major concern for managers (Washburne and Cole 1983). The concept of the *limits of acceptable change* (LAC) represents an alternative approach for resolving the carrying-capacity issue. The general idea underlying the LAC is not new, nor does it represent a radical change in how wilderness planning is conducted. The LAC process recognizes that change in response to visitor use is inevitable and that decisions have to be made with regard to how much change will be permitted to occur. Lime (1970), for instance, called for establishing standards in the Boundary Waters Canoe Area (MN) that defined acceptable limits to the impacts that would be permitted to occur on both the various resource elements and visitors. Frissell and Stankey (1972) outlined the basic framework for the LAC concept. They focused attention on the control of human-induced change on conditions. The goal of management, they argued, is to halt the character and ... of change that would lead to conditions judged as unacceptable in legally defined wilderness.

The LAC concept recognizes and seeks to enhance and protect a variety of wilderness conditions (Clark and Stankey 1979; Driver and Brown 1978; U.S. Department of Agriculture, Forest Service 1982). Some of this variety is related to physical-biological differences. Other variability is related to visitor-use patterns that are partially the result of an area's trail system and location of lakes, streams, attractions, and entry points (Stankey et al. 1976). This inherent variability produces differences in the conditions found within the area. In some areas, pristine physical-biological conditions and outstanding opportunities for solitude are found. Elsewhere, the evidence of visitor use is more apparent and contact with others is more common.

Variability in wilderness conditions is inevitable and desirable, to some extent, because then the most pristine areas can have more stringent restrictions to keep them in the most natural condition. No single condition constitutes wilderness—a range exists from the absolutely pristine to slightly modified. A point of debate is just how modified a situation can be before it no longer represents wilderness. Because there are few explicit standards describing unacceptable wilderness conditions, it is not surprising that conditions in some areas have deteriorated to where they represent unacceptable conditions to the majority of visitors—ankle-deep mud on miles of trails, crowded campsites around popular lakes, and so on. However, nonwilderness conditions should not be tolerated or accepted in wilderness. Establishing a clear measure of what constitutes acceptable wilderness conditions in the form of explicit, measurable standards is the intent of the LAC process.

The focus of our discussion about the LAC concept in this chapter is on the impacts associated with recreation use. Recreation is a major concern in many wildernesses and the impacts it causes can adversely affect both the integrity of natural ecosystems and the quality of the experience for other visitors (Hammitt and Cole 1998; Washburne and Cole 1983).

When Should the LAC Process Be Used?

The LAC concept is meant to support the overall wilderness planning process and is not a replacement for it, nor is LAC necessary in every wilderness planning project. The LAC concept is applicable to any management problem in wilderness where there are actual or potential conflicting goals and concerns about the control of unacceptable change, for example conflicts between managing for recreation use and preserving natural conditions. The LAC concept provides a framework within which the acceptable amount and extent of change in wilderness can be identified. It is designed to alert managers to the need for action when changes exceed standards, for example excessive impacts on vegetation in wilderness from pack stock, used by recreation visitors and outfitters, who are foraging for preferred plant species while in wilderness.

The LAC process is not useful when there are not actual or potential conflicts or trade-offs in implementing multiple management goals (Cole and McCool 1997a). Some trade-offs or compromises between the competing goals must be possible, such as comparing the consequences and outcomes between alternative management approaches or actions. It is possible that some goals constrain others, such as the need for preserving wilderness conditions being considered a constraint on recreation use. Furthermore, LAC is only useful as a framework if measurable and attainable standards can be developed to determine if the current management practices are maintaining or improving the wilderness resource and social conditions, or if the conditions are degrading and require management action.

The LAC Process

The historic model of carrying capacity focused principally on identifying how much recreation use an area could tolerate before unacceptable impacts occurred. For such an approach to be effective, it was necessary to know the relationship between recreation use and impact: if so much use occurs, what type of impact will result? Because recreation use was not a constant or predictable measure, it was a difficult issue to evaluate. The level of impact resulting from a given number of visitors can vary considerably, depending on the characteristics of those visitors and their use and the nature of the wilderness, whether desert, alpine, or other.

Consider a hypothetical situation where it is estimated that an area can tolerate a hundred visitors before unacceptable environmental impact occurs. What happens if fifty of these visitors are backpackers and fifty are using pack stock? Or if the visitation occurs in the early spring, when soils are wet and soft, as opposed to midsummer, when soils are dry and hard? Or if some of the visitors follow minimum-impact camping procedures carefully while the remainder do not? As the characteristics of visitors or their specific recreation behaviors change, the associated impacts will change as well, perhaps substantially so, even though the number of visitors remains constant. To set a carrying capacity based on the number of visitors would require a different capacity for each situation, and that is not realistic for most management situations.

The real concern in the preceding example is not the number of visitors involved, but the impacts on the conditions of the area that result from recreation use. It is these conditions on which the LAC process focuses. Given that any use produces at least some impact, the LAC process requires managers to identify where, and to what extent, varying degrees of change are appropriate and acceptable. Once the appropriate and acceptable degree of change has been identified, managers can select from an array of management techniques that target causes of impacts to ensure that desired conditions are maintained or restored. These techniques range from light-handed measures, such as providing information to visitors on appropriate camping methods, to restricting the number of visitors to the area. The conditions that characterize a particular area of wilderness and distinguish it from others are specified by measurable objectives.

The LAC process consists of four major components: (1) specifying acceptable and achievable resource and social conditions, defined by a series of measurable parameters; (2) analyzing the relationship between existing conditions and those judged acceptable; (3) identifying management actions judged to best achieve these desired conditions; and (4) monitoring and evaluating management effectiveness. These four components, in turn, are broken down into planning steps to facilitate application.

The original LAC process was a nine-step planning process (Stankey et al. 1985), with each step designed to achieve a particular task and provide the basis for later activities. By 1992, the LAC process had become widely

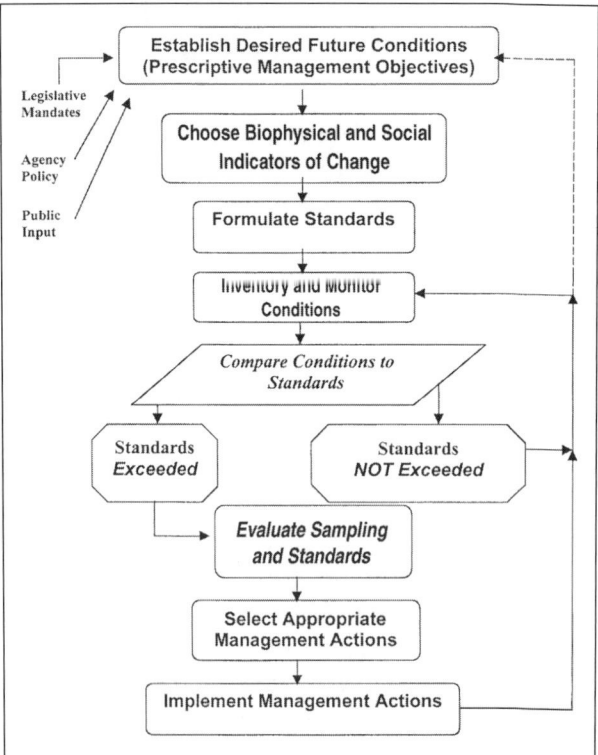

Fig. 9.4 Essential elements of the limits of acceptable change process.

applied, reportedly being used by 92 percent of 50 western national forests that contained 116 wilderness areas (McCoy et al. 1995). A majority of these LAC plans did not strictly adhere to the original nine-step process because some of the steps were cumbersome for public work groups participating in the planning processes, and some elements seemed to be missing, such as defining goals and desired conditions. Figure 9.4 portrays the essential elements in the LAC process, including several important additions to the original process, "establish desired future conditions" and "evaluate sampling and standards" when standards are exceeded.

The authors of the original LAC process clearly stated that they intended that the process be adapted to meet the planning needs for each particular area. In the following discussion, we present a ten-step process (Cole and McCool 1997b) and review the rationale for each step and the specific activities involved in completing it. *We expect that these ten steps can be modified as necessary to accommodate ongoing agency planning efforts that are interrelated, such as preparing an EIS under NEPA.*

Step 1: Define Goals and Desired Conditions

Start the planning process by reviewing the Wilderness Act, area-specific enabling legislation, and agency policies related to the wilderness area in question. The idea is to develop a perspective on relative position of this wilderness to others on the wilderness opportunity spectrum (WOS) and to highlight the unique significance of the area to the region and the NWPS. This information will generally describe the overall goals for any wilderness: preserving or restoring natural conditions and processes, maintaining or restoring outstanding opportunities for solitude, and minimizing restrictions on recreation behavior to allow for primitive recreation experiences. Articulate these goals and desired conditions as soon as they have been identified, because they will provide the guidance needed to direct the entire planning process. Specific goals will vary among areas and relate to the unique attributes or type of visitor use that will, in turn, lead to being able to state the desired conditions needing to be maintained or restored. For example, the North Maricopa Mountains Wilderness in Arizona is within twenty miles of the Phoenix metropolitan area, and yet it had only a few hundred visitors per year in the late 1990s. It is a lower Sonoran Desert environment with desert bighorn sheep; this fragile and unique ecosystem needs to be preserved and protected from heavy visitor use.

Step 2: Identify Area Issues, Concerns, and Threats

Identify public issues and managerial concerns that relate to distinctive features and characteristics of the wilderness area and the relationship of the individual area to other units of the wilderness system and to nonwilderness areas offering primitive recreation opportunities. This step builds on Step 1 by refining management direction to deal with the specific situation in each area. Managers could consider matters such as:

1. Does the area contain outstanding ecological, scientific, recreational, educational, historic, spiritual, or conservation values that warrant special attention?
2. Does the area provide critical habitat for threatened or endangered species?
3. Do land uses on contiguous areas represent situations requiring special management attention? For example, are timber harvests planned, or are changes in access likely?
4. Are there existing or potential nonconforming uses in the area that will require special attention?
5. Are there regional and/or national issues that need consideration? For example:
   a. What is the availability of and demand for wilderness and dispersed recreation opportunities in the planning region?
   b. Are the physical-biological features of the area found elsewhere in the region, or does it possess unique features?
   c. Are the types of recreation opportunities offered by the area available in other wildernesses, or does the area offer unique opportunities? For example, are opportunities for long-distance backcountry horse rides available in many other areas or just this one?
6. Are there sociopolitical factors specific to the area that will influence the planning process and its possible outcomes? For example, are there established outfitter use and historical patterns of stock use?
7. Have managers or the public identified threats or issues that merit special attention?

Answers to such questions help managers identify the values of the area, the desired future conditions, and the area's role in the region and in the NWPS. For example, in some areas, primary management attention might focus on preserving or restoring a high level of environmental protection. This might be based on the presence of an endangered species such as the grizzly bear or on the basis that the area contains ecosystems otherwise nonexistent in the NWPS. In such a case, primary management emphasis would be on environmental protection, with wilderness recreation given relatively less attention. In another area, more attention might be directed at maintaining or restoring outstanding opportunities for solitude or for specific forms of primitive recreation, such as extended cross-country travel. In either example, only relative emphases are involved, and the need to preserve environmental conditions and to provide opportunities for solitude and primitive recreation must be accommodated, to some extent, in all areas.

Some issues and concerns identified in this step might conflict. For example, managers might identify solitude as a major value in the area, while the public supports increased access. *The LAC process exists for planning how to proceed with such conflicts or trade-offs in achieving goals; otherwise, the LAC process may not be necessary for wilderness planning in that area*

(Cole and McCool 1997b). The inevitable diversity of situations highlights the importance of examining individual wildernesses within a regional framework. In Step 7, managers can partially accommodate diverse concerns because they allocate the area to different wilderness opportunity classes.

Step 3: Define and Describe Acceptable Conditions

Develop a series of acceptable condition zones or opportunity classes for the wilderness. An *opportunity class* defines the resource, social, and managerial conditions considered desirable and appropriate within the wilderness. The designation of opportunity classes follows the basic recreation opportunity spectrum (ROS) system and represents the range of wilderness recreation settings for which to manage within any given area. Cole and McCool (1997b) suggest that the intent of this step may be better understood as developing "acceptable condition" or "prescriptive management zones" because the emphasis should be on the acceptable, nondegradation conditions rather than present conditions that may be degraded from the wilderness ideal. The range or diversity of acceptable conditions may be divided into two or more classes, which implies that opportunity classes represent a continuum of social, resource, or managerial variables. The underlying variables or dimensions used should be carefully identified and developed before determining the number and description of opportunity classes. For example, a wilderness may vary in the level of contact among visitors and the amount of visible human-induced impact. These two dimensions would constitute the underlying continuum that would be divided into two or more opportunity classes. Each dimension included should be addressed in each opportunity class description. Often, the dimensions selected will directly relate to the issues identified in Step 2. Thus, each opportunity class is defined relative to the other; more (or less) impact on the environment is considered acceptable; more (or less) contact with others is acceptable; and so on.

The opportunity class description represents the varied wilderness setting conditions that are considered to be acceptable and appropriate. Actual geographic mapping of opportunity class designations occurs in Step 7. The outcome of this step, the opportunity class descriptions, provides a basis for identifying indicators (Step 4), developing standards (Step 6), and suggesting management actions (Step 8).

Historically the ROS *defines six categories of classes*: primitive, semiprimitive nonmotorized, semiprimitive motorized, roaded natural, rural, and urban. Typically, within wilderness areas, the primitive and semiprimitive nonmotorized categories of classes would apply. In general terms, these two categories of classes can be characterized as follows:

Primitive

An area characterized by an essentially unmodified natural environment
- Fairly large in size
- Interaction between users is very low
- Evidence of other users is minimal
- Area is managed to use off-site restrictions and controls on recreation use (they are not evident in the wilderness)
- Motorized use within the area is not permitted

Semiprimitive Nonmotorized

An area is characterized by a predominantly natural or natural-appearing environment
- Moderate to large size
- Interaction between users is low
- Often evidence of other users is present
- Area is managed in such a way that minimum on-site controls and restrictions for recreation use may be present, but are subtle
- Motorized use is not permitted

These setting descriptions are broad and qualitative, and within each it is possible to describe several classes. For example, at major entry points, use levels may be relatively high, with fairly frequent contact among visitor parties. Similarly, resource impacts can be moderately visible in these areas. Elsewhere in the same wilderness, there are areas where few visit and where ecological conditions are almost undisturbed. Between these extremes, a continuum of conditions exist, all within the same wilderness. Eliminating this internal variability would be difficult and could only be attained with a highly regulated system of entry and use dispersal. Managers may explicitly consider maintaining this variability and it would be reflected in the opportunity class descriptions.

Managers decide not only how each opportunity class is defined, but also how many are appropriate for each wilderness. For example, smaller wildernesses may have only one or two classes, whereas larger areas may have up to six classes. The question of how many classes to designate can be answered only after analysis of the issues, the current range of conditions, the manageability and demands for wilderness recreation, and the regional supply of different wilderness settings.

Descriptions of resource conditions are influenced by the issues, concerns, and threats identified in Step 1, but typically they include the type and extent of recreation visitor impacts. In writing statements regarding acceptable resource conditions for each opportunity class, managers should consider:

- Type of impact
- Severity of impact
- Prevalence and extent of impact
- Apparentness of impact and extent to which impact is noticeable to visitors

To contrast the kinds of resource conditions appropriate for different opportunity classes, consider the following statements written for an opportunity class preserving the most pristine condition in an area, *Class I*, and one representing a situation where more resource and social impacts are judged acceptable, *Class IV*. These examples of two opportunity classes represent the extremes of a hypothetical four-class spectrum.

Class I
- Resource impacts are minimal; restricted to minor temporary loss of vegetation where camping occurs and along some travel routes.
- Impacts typically recover on an annual basis and are subtle in nature.
- Impacts generally not apparent to most visitors.

Class IV
- Resource impacts found in many high-use locations and some can be substantial in a few places, such as near major entry points.
- Impacts may persist from year to year; possibly substantial loss of vegetation and soil at some sites.
- Impacts are readily apparent to most visitors.

Social conditions must also be covered in the description, including the characteristics of visitors, type of visitor use, levels and types of encounters occurring among different types of users, and other factors. For example, managers could consider the extent and location of interparty contacts to compare conditions in a Class I opportunity class with those in a Class IV opportunity class.

Class I
- Few, if any, contacts with other groups.
- Contact limited to trails.
- Camping out of sight and sound of others almost always possible.

Class IV
- Contact with others moderately frequent.
- Fairly high level of interparty contact can occur on the trail.
- Camping has fairly high level of interparty contact.

Such descriptions outline very different kinds of social settings for these two opportunity classes. They indicate that Class I will provide high levels of solitude whereas Class IV is an area of visitor-use concentration and relatively frequent contact.

Managers need to develop a descriptive statement of managerial conditions because it establishes a framework for what will be done to achieve the acceptable resource and social conditions. A clear description of appropriate management conditions is important because management standards will not be prescribed in Step 6, because they are for desired resource and social conditions and because management conditions deal primarily with the means by which the resource and social conditions will be achieved. A carefully developed descriptive statement of managerial conditions addresses the following kinds of management issues:

- Presence of management personnel
- On-site versus off-site management strategies
- Site modification
- Rules and regulations on behavior
- Facilities and trail construction standards

Comparing the managerial settings in Class I with those in Class IV, the philosophy is to preserve class

I areas and restore wilderness conditions in other class areas. Therefore, descriptors might read as follows:

Class I
- Direct on-site management of visitors not practiced (unless required to reduce area degradation); little or no evidence of site management.
- Rules and regulations communicated to visitors outside the area; little evidence of management personnel.

Class IV
- Use of on-site management and site modification is evident.
- Rules and regulations enforced with signs and management personnel in the area; substantial use of regulations to influence visitor behavior.

Although the descriptive statements of managerial conditions indicate the prescribed intensity and intrusiveness of action, some situations will exist where these management actions differ considerably from what is prescribed. For example, an area may be designated as Class I, but current conditions may be far worse than those defined as acceptable by the standards adopted for this area. To achieve these standards, managers may have to adopt fairly intrusive actions—such as designating campsites and consequent enforcement—to restore wilderness conditions to an acceptable level. Once such conditions are achieved, management direction may then revert to what was prescribed in the opportunity class description.

Collectively, these prescriptions of the resource, social, and managerial conditions for each opportunity class constitute the *management objectives* for the area. They prescribe the conditions sought in the wilderness and serve as criteria for identifying what and where specific management actions are needed. These objectives serve throughout the process to determine what types of information are needed, what standards need to be developed, the appropriateness of various activities, and what management actions are needed.

Step 4: Select Indicators of Resource and Social Conditions

Managers need to identify indicators—specific variables—that, singly or in combination, are taken as indications of the acceptable condition of the overall opportunity class. Such measures enable managers to define acceptable conditions and to assess the effectiveness of various management practices. In practice, the selection of a limited number of indicators—five to ten indicators—has proven to be one of the most critical and difficult steps in the LAC process.

Managers need to first review the broadly defined issues, concerns, and threats in Step 2 that require attention. For example, there might be concern with issues such as excessive use levels along trails in the area or with the amount of biophysical impact at campsites. The following examples suggest some resource and social topics:

Resource examples:
- Trail conditions
- Campsite conditions
- Water quality
- Exotic species
- Wildlife populations
- Threatened and endangered species
- Range condition

Social examples:
- Solitude while traveling
- Campsite solitude
- Conflicts among visitors with different travel methods
- Conflicts regarding party size
- Noise

Within these broad categories, however, managers need to identify one or more indicators that reflect the overall condition. Overall criteria that can help guide selecting indicators include:

- Early-warning ability. Does the indicator act as an early warning, alerting managers to deteriorating conditions before unacceptable levels of change have occurred?
- Significant. Does the indicator detect a change in conditions that persists for a long time, disrupts ecosystem functioning, or reduces the future desirability of the area?
- Indicative. Does the indicator reflect the condition of more than itself?
- Discriminative. Does the indicator discriminate between a change in conditions caused by human activities and natural causes?

- Sensitive. Can the indicator detect a change in conditions that occurs within one year?
- Responsive. Does the indicator detect a change in conditions that is responsive to management control?
- Quantitative. Can the indicator be measured quantitatively?
- Reliable. Can the indicator be measured reliably (with training, will different observers collect the same information)?
- Feasible. Can the indicator be measured by field personnel using simple equipment and sampling techniques?

Not every indicator will meet all of these criteria, but indicators should be selected that meet as many criteria as possible.

For example, campsite condition encompasses a number of concerns. What specific indicators should be selected for measurement? Indicators that could be used to measure campsite condition include total area of bare ground, number of damaged trees in the campsite area, soil compaction, presence of exotic vegetation, or a composite index reflecting overall campsite condition. For campsite encounters, indicators might include the number of other persons camped within sight or sound or the total number of sites located within some unit area.

No single indicator constitutes a comprehensive measure; it will reflect only a portion of what the objective seeks to achieve. For example, if provision of outstanding opportunities for solitude is the objective, managers might use indicators such as the number of interparty contacts while on the trail or while at the campsite. If interparty contacts can be held to two or fewer per day while traveling, the objective of providing outstanding opportunities for solitude may be considered attained. Other factors, such as whether contact is with a horse or hiker party, also influence whether or not the objective is achieved. Thus, two or more indicators can be used as a way of comprehensively measuring performance in terms of the objectives.

It is important to select indicators that relate as directly as possible to the objective and are developed for outputs (e.g., social conditions) and not for inputs (e.g., total daily visitor use). For example, managers might select interparty contact levels as the indicator for a solitude objective, instead of visitor density levels.

Managers should seek indicators that will specifically enable them to evaluate acceptable conditions in the wilderness area they manage, rather than select indicators because they have been used elsewhere. Several studies have reviewed what makes a good indicator and developed a list of indicators that have been used in wilderness plans (Merigliano 1987; Watson and Cole 1992). Indicators are driven largely by the issues, concerns, and threats that are identified in the planning effort; therefore, it is important that managers select only those that relate to their wilderness. In addition, definitions as to what are the most important or useful indicators will often change over time; thus, the list of selected indicators for a wilderness should be reviewed periodically for its appropriateness.

Finally, when selecting indicators managers need to consider how the data will be collected and used to make a decision about whether acceptable conditions were achieved or not.

Step 5: Inventory Existing Resource and Social Conditions

The indicators in Step 4 specify the variable(s) to be inventoried and identify the unit of analysis. For example, managers might be concerned about water quality. In selecting indicators that will define water-quality standards, they might select coliform counts in lakes or streams adjacent to campsites. Thus, the resulting water-quality inventory has a specific focus that defines what data are to be collected and where. During the inventory, data need to be collected that provide information on the coliform counts throughout the area. The inventory must be conducted in an objective and systematic fashion to be of value to management.

Inventory data provide managers with the range of conditions for each indicator. Such information must be recorded for easy analysis of its spatial patterns, because it facilitates comparison between existing conditions and those defined as acceptable for an opportunity class. Managers might also be concerned with the distribution of opportunity classes over different landscape types within the wilderness.

Resource inventories can be conducted at different levels of detail. Often, managers will have inventory data from previous fieldwork, or they might have partially completed inventory data. Although it is most desirable to have an up-to-date comprehensive inventory of the condition of the indicators, managers often have to work with incomplete or non-current data. Where this is the case, the limits of the data should be carefully documented and an improved database should be a priority in the monitoring process in Step 10.

Step 6: Specify Standards for Resource and Social Indicators for Each Opportunity Class

The task is to assign quantitative or highly specific measures to the indicators. This greater specificity is obtained by establishing *standards*—measurable aspects of the indicators defined in Step 4. These standards provide a basis for judging whether a particular condition is acceptable or not.

By using data collected in Step 5, it is possible to specify standards that describe the acceptable and appropriate conditions for each indicator in each opportunity class. Setting standards is a judgmental process; however, the process is explicit, traceable, and subject to public involvement and review. Standards relative to appropriate use conditions can be derived with the input of the visitors themselves (Shelby and Heberlein 1986).

Standards are conditions that managers think can be achieved over a reasonable time. In some cases, standards might be merely statements of current pristine conditions. In other cases, standards can be written to direct modification of degraded wilderness conditions toward a more natural state. Standards should be stringent enough to be meaningful, such as restrictions to obtain acceptable resource and experience conditions. Three general guidelines apply to the process of establishing standards:

A. *Standards Follow Descriptors.* The qualitative descriptions developed in Step 3 provide clues to the kinds of conditions characterizing each opportunity class. For example, if a description written for a Class IV area suggests that "contacts are fairly frequent while traveling," managers could use the inventory data to help specify how "fairly frequent" might be quantitatively defined, such as contact levels on trails near major entry points may average ten to fifteen parties per day. These data could be used to help set the standard for the "average contacts with others per day" indicator to define the Class IV area. Normally, it is important that the standards are not established for existing conditions. For example, there are places where conditions have deteriorated to the point that they no longer meet acceptable wilderness standards. In such cases, managers are legally bound to restore these areas to a condition that is, *at the minimum*, acceptable in wilderness, or to seek opportunities to improve them by establishing more stringent standards. In no way is the LAC process designed to condone maintenance of unacceptable wilderness conditions. In formulating standards, there needs to be a consideration of existing conditions to lend realism to the specific standards and professional judgment along with public input to set the standards at levels that can lead to improvements in conditions.

B. *Standards Describe a Range of Conditions.* When looking at the opportunity classes for any given indicator, the standards should describe a logical progression or gradation of conditions. For example, managers might select "other parties camped within sight or sound at night" as an indicator for solitude. In a Class I opportunity class, a standard of "no other parties camped within sight or sound" might be prescribed. Then, remembering that the intent is to provide a logical progression or gradation of conditions relative to this particular indicator, managers might set standards of "no more than one," "no more than two," and so on for the remaining opportunity classes. On occasion, the standards set for an indicator might be shared by two or more opportunity classes, but the opportunity classes would be distinguished by the standards set for other indicators. In some instances, an indicator might apply only to a single opportunity class, for example, an indicator related to a threatened and endangered species whose range is limited. Although a progression of standards across opportunity classes will be typical, there might be certain conditions that apply area-wide and that do not discriminate between classes, such as air quality and water quality. Moreover, baseline standards might prescribe conditions that must be met in all areas; namely, under no situation could a condition in a wilderness fall below this baseline standard. Baseline standards do not preclude more stringent standards within individual areas.

C. *Standards Express the Typical Situation.* Standards are often best expressed in terms of probabilities. For example, a standard for daily contacts while traveling in the primitive opportunity class might be expressed as "Interparty contact levels on the trail will not exceed two per day on at least 90 percent of the days during the summer-use period." This recognizes the fact that the high degree of resource and social variability in a complex wilderness system often makes specific, absolute standards unrealistic. Choosing indicators and writing standards are crucial steps because they determine, to a great extent, the future character of the wilderness. Public input, research information, and managerial experience will be helpful guides. Because monitoring and evaluating are an integral part of this procedure, managers will be able to revise indicators and standards in response to improved information. Moreover, the judgments about selecting indicators and standards are documented so that they can be reviewed by others.

Step 7: Identify Alternative Opportunity Class Allocations

The task is to decide what resource and social conditions (in the form of specific standards) are to be maintained or achieved in specific areas of the wilderness while reflecting area issues and concerns and existing resource and social conditions. This is a *prescriptive step*, concerned with establishing what *should be*, and input from both managers and the public should be used to make these decisions. This step involves analyzing the inventory data collected in Step 5, along with the area issues and concerns identified in Step 2. These issues and concerns, however, do not prescribe what should be done. They have to be checked against the realities of what exists, as revealed by the spatial analysis of existing condition for each indicator, as well as what is possible in terms of agency resources.

Maps of alternative opportunity classes, reflecting area issues and concerns and existing resource and social conditions, result from this step. Some issues will be contradictory, such as "increase opportunities for easier access into most portions of the wilderness" and "provide greater opportunities for solitude." Managers could respond in a variety of ways. They might attempt to provide the full range of opportunity classes. Or, they might elect to manage primarily for only a couple of the opportunity classes, because the other classes are adequately represented elsewhere in the region. Finally, they might propose a variety of management alternatives that reflect a range of opportunity class mixes.

Step 8: Identify Management Actions for Each Alternative

After alternative opportunity classes have been formulated, managers need to identify the differences, if any, that exist between current conditions (inventoried in Step 5) and the standards (identified in Step 6). This will help identify problems and management actions that are needed. Then managers need to consider what actions can achieve the conditions specified by each alternative and to evaluate the costs and appropriateness of implementing these actions. If an alternative calls for a set of opportunity areas that closely match the current situation, the management actions needed to achieve this may not be costly. On the other hand, if a major change is proposed, considerable costs may be incurred.

Where existing resource and social conditions are better than standards, we generally assume there is little need for change in management, although there might be a need to evaluate whether existing management actions should be changed or eliminated. For example, if existing conditions are better than the standard, but there is evidence that the trend in conditions is degrading, managers should initiate some preliminary actions to prevent this trend from continuing and eventually violating the standard. Where current conditions are close to or substantially worse than standards, managers must consider new management actions.

For any given alternative, a number of possible management actions could be undertaken to achieve the standards. The qualitative descriptions for each opportunity class developed in Step 3 serve as guidelines as to whether or not a particular management action is appropriate. However, these descriptions are only guidelines, not standards. As a general rule, apply the principle of minimum regulation (see Chapter 7); use only that level of control necessary to achieve a specific objective. However, if the existing resource and social conditions are degraded from those desired, the management actions needed to achieve those standards should be employed, even if they are not consistent with the management condition descriptor written in Step 3.

For example, if a heavily used and impacted wilderness area were to be transformed to a pristine condition, intensive management would be needed. Such a program might include restrictions on where and how long visitors could camp, restriction of recreational stock, and closures of certain areas. Normally, direct on-site management of visitors and obvious site management actions would be considered inappropriate in a pristine opportunity class, but without such measures, achieving management objectives in a reasonable time span would be difficult. Hence, more restrictive management is imposed until progress is made toward achieving the standards.

Managers should remember that *standards define minimally acceptable conditions sought in an area*. Nevertheless, such standards do not preclude providing protection in part of an opportunity class above that specified by the standards. For example, some Class IV wilderness areas consist of frequently visited valley-bottom trail corridors bounded by trailless, relatively pristine valley walls—thus, some pristine areas exist within the geographic area classified for a Class IV wilderness experience and should be maintained as pristine as possible (not allowed to degrade).

Step 9: Evaluate and Select a Preferred Alternative

The selection of a preferred alternative will reflect the evaluation of both managers and involved publics. No simple approach exists for making such a decision. Some questions to guide this selection include the following:

- Which wilderness user groups are affected and how (does it facilitate or restrict use by certain groups)? Which uses are wilderness-dependent?
- Which wilderness values are promoted and which are diminished?
- How does a particular alternative fit into the regional or national wilderness opportunity spectrum? Does the alternative contribute a unique kind of wilderness setting to the system?
- What is the feasibility of managing the wilderness areas as prescribed, given constraints of personnel, budgets, etc?

In analyzing the alternatives, a variety of costs need to be qualitatively and quantitatively considered. These would include the financial costs, such as for personnel and materials; information costs, such as for acquiring information needed to implement actions; opportunity costs for not carrying out a proposed action; and other resource and social costs. These latter costs are difficult to quantify, particularly in monetary terms, but they are extremely important (Lucas 1982).

Although it is difficult to quantify the costs and benefits of the various alternatives, their qualitative presence or absence usually can be identified, such as the kinds of costs (e.g., increased impacts on vegetation) and benefits (e.g., increased opportunities for solitude) associated with a management action. Recognizing their existence will improve the ability of managers and citizens to evaluate the alternative.

Deciding what constitutes the "best" alternative is obviously not easy. Information on the issues identified previously should clarify the costs and benefits associated with each alternative. The information also enables different groups to better understand how different alternatives affect their own interests. In addition, the LAC process allows public groups to focus their comments on specific assumptions, actions, or areas among the alternatives and through their participation play an important role in selecting a final alternative.

Step 10: Implement Actions and Monitor Conditions

With selection of an alternative and its associated management actions, the management program must be implemented and its performance assessed. Monitoring provides systematic feedback on how well management actions are working and identifies trends in conditions that may require changes in management actions.

Priorities for monitoring should consider situations in which (a) conditions were very close to standards at the time of the last assessment; (b) rates of resource or social change are judged to be the highest; (c) the quality of the database is poorest; (d) the understanding of management action effects is poorest; or (e) unanticipated changes in factors have occurred, such as increased visitor access or increases in commercial uses of adjacent lands.

Monitoring could reveal a number of possible outcomes relative to the standards—both the condition existing at the time of monitoring and the trend in that condition are important. For instance, existing conditions might be better than the standards call for but are nevertheless degrading over time, or existing conditions might be worse than those called for by the standards but their trend could be improving. *Monitoring alerts managers to the current conditions and trends in relation to the standards.* Depending on the specific circumstances, managers might want to implement actions to prevent changes in conditions, even when standards have not been violated. For example, when a trend clearly indicates that conditions are degrading, even though within the standard, managers might begin implementing management actions to halt this trend.

One concern is the appropriate frequency of monitoring. Ideally, all indicators addressed by standards should be monitored throughout an area to provide timely information in relation to the issues and concerns that caused the need for management actions. Given budgetary and personnel constraints, however, certain indicators will be monitored less frequently than others, and certain areas will be monitored less closely than others.

The results of monitoring help evaluate program effectiveness and improve future programs. If monitoring shows that previously acceptable wilderness resource or social conditions have deteriorated and are now worse than the standards, new or additional management actions are called necessary. If resource or social conditions had previously violated standards, and monitoring shows they still do and the trend is not improving, the management actions can be judged ineffective, at least within the time since being initiated. A management action might prove ineffective for various reasons. Perhaps the action was appropriate, but its implementation was not effective or the programs have not had enough time to work. Trends reflected in the monitoring data should indicate where the problem lies. *Monitoring provides feedback regarding the effectiveness of certain management actions in solving particular kinds of problems.*

Managers need to be alert to changes in external circumstances that could affect the resource and social conditions within the wilderness. These could include external visitor-access systems, adjacent-land uses and management activities, population growth, or the relative availability of alternative types of recreational opportunities. In some cases, impacts stemming from such alterations can be coped with through different wilderness management actions. In the case of major changes, fundamental alterations in area wilderness management objectives might need to be considered.

Potential Barriers to the LAC Process

Five potential barriers to implementing an LAC process were discussed by participants at a 1997 workshop on the LAC planning approach (McCool and Cole 1997b). First, wilderness and protected-area planners and managers expressed concern that government agencies did not have adequate staff and funding to conduct good planning and management; they lacked funding to collect baseline information on resource and social conditions in wilderness. Second, there were difficulties integrating various and sometimes complex planning and management functions necessary to develop and conduct an LAC process from research inventories and public involvement through monitoring and enforcement. For example, they had difficulty combining a prescribed natural-fire policy and the policy's implications for visitor management. Third, the legal framework for involving public individuals and organizations in the LAC process is complex and confusing for federal agencies even though wilderness planners and managers recognize the value of such input into the wilderness planning process. Fourth, scientifically gathered information is valuable to the planning process, but less rigorously gathered information collected by agency personnel and wilderness visitors and organizations during wilderness experiences can be just as valuable for the planning process. Finally, wilderness planning and management are a political process involving competing and sometimes polarized groups; the groups can influence federal agencies involved in wilderness planning and management to be unwilling to make needed but controversial wilderness management decisions.

Positive Benefits to Wilderness from the LAC Process

Over the time that the LAC process has been used in wilderness planning, the potential benefits of using the LAC process have been outlined and found to outweigh the potential barriers to conducting such a process (McCool and Cole 1997a). The LAC planning process is a pragmatic and effective approach to understanding the carrying-capacity concept and then applying it in a direct and positive way—placing attention on achieving appropriate management of the acceptable resource and social conditions in wilderness. The management actions are then monitored to measure their effectiveness, and this provides feedback on how to adjust management actions and provides more general information about the potential utility of that management action in similar situations in other wilderness areas.

The LAC step to delineate opportunity classes, or prescriptive management zones, allows for more pristine areas within wilderness to be maintained and human-caused impacts limited—although some critics of zoning in wilderness express the opposite concern that zoning could allow degradation of more heavily used and impacted areas, and they argue that all wilderness should be managed for nondegradation. The critics are correct in one aspect—the LAC process is meant to find some compromise between the competing goals inherent in wilderness management under current wilderness legislation: wilderness resource preservation and wilderness recreation use. However, the LAC process makes explicit what the acceptable conditions are, how to manage for those conditions, how to measure the relative success toward those standards, and, if standards are exceeded, the management direction to improve wilderness conditions.

Finally, the LAC process involves public participation throughout the planning process. This has a number of beneficial outcomes for wilderness and wilderness management. By making the planning and management processes more evident, wilderness managers make interest groups and the public more aware of the costs and benefits of wilderness management. Such awareness extends from the diverse issues and concerns of various users, to the consequences of different management alternatives, and to the effectiveness and equity of management actions to achieve the desired conditions. Furthermore, public involvement can lead to more informed discourse on how wilderness values can be protected and, potentially, to more public support for agency wilderness programs.

Implementing the LAC Process

The LAC process provides a general framework within which acceptable wilderness conditions can be prescribed and the management actions to maintain or restore these conditions are identified. However, like any planning framework, the value of the LAC lies in its ability to help shape wilderness planning and management in the real world. In the following discussion, we examine how the LAC has been applied in the development of plans to guide wilderness management for entire wilderness areas and specific management issues.

The Bob Marshall Wilderness Complex

The LAC process was first applied and developed in the Bob Marshall Wilderness Complex (BMWC) planning and management process. The BMWC is composed of three wildernesses lying astride the Continental Divide in Montana: the Bob Marshall Wilderness, the Scapegoat Wilderness, and the Great Bear Wilderness. Together they comprise more than 1.5 million acres, one of the largest unroaded areas in the lower forty-eight states. Four national forests (Flathead, Helena, Lewis and Clark, and Lolo) within the complex are administered through five ranger districts.

The BMWC is large enough to contain complete ecosystems and contains watersheds of three major drainages: the Middle and South Forks of the Flathead River, and the Sun River. The BMWC contains nearly all large game and nongame wildlife residents in the area when the Europeans arrived, including grizzly bear, elk, black bear, deer, and gray wolf. Vegetation has been significantly influenced by fire, with the major fire type being low-frequency, high-intensity stand-replacement fires.

Fig. 9.5. Woodward Lake, in the Bob Marshall Wilderness, is one of three separate but contiguous wildernesses comprising the Bob Marshall Wilderness Complex, which covers approximately 1.5 million acres in western Montana. Conditions range from pristine to heavily impacted, presenting a difficult challenge to managers. Photo courtesy U.S. Forest Service.

As with many other wildernesses, use is not dispersed evenly over the complex. More than seventy trailheads and one active airstrip provide access to about 1,500 miles of trails; however, only seven trailheads account for about 50 percent of the total use. From 1970 to 1982, the average number of parties encountered increased from 1.3 per day to 1.6, yet this was still relatively small when compared to other wildernesses (Lucas 1985).

Recreation use of the complex totaled approximately 200,000 visitor days annually in the early and mid-1980s. Most of the visits to the complex were hikers, comprising approximately 57 percent of the people, with horseback riders totaling about 36 percent of the visitors (Lucas 1985). This is a change from 1970 when horseback riders outnumbered hikers about two to one. Horseback riders tend to come in larger group sizes and stay longer than hikers, however, so the majority of total use in visitor days comes from horseback riders. Between 1970 and 1982, party sizes and lengths of stay tended to drop for both groups.

More than fifty outfitters, with about fifty-five base camps, provide hunting and summer-oriented horseback and rafting recreation opportunities. More hunters hire outfitters than do nonhunters. About 14 percent of all visitors in the summer and 24 percent of the fall visitors use outfitters. Compared to 1970, the number of outfitted guests has remained about the same or declined slightly (Lucas 1985). In 1970, about 32 percent of all visitors hunted, contrasted to 1982 when 16 percent hunted.

A Wilderness Management Plan for the Bob Marshall Wilderness (designated in 1964) was developed in 1972, but much of it was never implemented because it lacked public support. The limited implementation of the 1972 Bob Marshall Wilderness Management Plan, the designation of the Scapegoat Wilderness in 1972 and Great Bear Wilderness in 1978, and the need for updated comprehensive management direction for the entire BMWC prompted the four national forests to develop a coordinated approach to all three wildernesses through their respective Forest Land and Resource Management Plans beginning in 1982.

LAC was selected to provide the technical framework for developing common wilderness management direction for the entire wilderness complex. A comprehensive management document for the BMWC would then be appended to each forest's respective Forest Land and Resource Management Plan. A major concern to managers was the lack of vested interest in and support for the 1972 Bob Marshall Wilderness Plan by wilderness users and interest groups. Thus, the public-involvement process was considered a critical element to ensure success of the LAC.

Fig. 9.6. Public input is critical to formulating Wilderness Management Plans. U.S. Forest Service planners included citizens as members of the LAC task force for the Bob Marshall Wilderness Complex in Montana. Because the number and type of visitors change over the years, more recent LAC planning efforts may need to consider changing users, such as these hikers using llamas as pack animals in the BMWC. Photo courtesy U.S. Forest Service.

One major question facing managers of the BMWC was how to solicit public comment and review through the planning process. They selected a working group approach in the early 1980s, in which participants would share their wilderness knowledge and use experiences with the planners, who, in turn, would share information about the wilderness planning model approach and systematic data analysis with the citizens. The dialogue that would develop between citizen and planner leads to mutual learning. Through the dialogue and learning processes, the working group, using its accumulated knowledge, makes an informed decision about a course of action (McCool et al. 1986). These working groups, if adequately represented by a cross-section of involved individuals and interest groups, had the potential to form a viable political coalition that can ensure the plan has adequate political support to achieve implementation.

Issues confronting the BMWC are similar to those in other wildernesses: visitors were concerned about campsite impact and opportunities for solitude; trail conditions were frequently identified by both managers and visitors as unacceptable; conflict between hikers and horseback riders had increased; flights into the Schafer Meadows airstrip, an area included within the Great Bear Wilderness with its designation in 1978, produced noise that disturbed nearby wilderness visitors; forage for horses was relatively scarce, and recreational stock and wildlife competed for it; the entire complex was occupied grizzly bear habitat; and a natural-fire-management program, while accepted by most visitors (Lucas 1985), reportedly had negative impacts on specific trails and campsites. In response to these issues as well as other legislation, the U.S. Forest Service, with the Flathead National Forest serving as the lead forest, initiated a planning effort in 1982 to develop more specific and additional direction than that established in the 1972 plan.

The approach to public involvement chosen was to develop a task force for the BMWC planning project—there are other ways to involve public participation in an LAC process. The task force public-participation format was based on nine assumptions that reflected the underlying working group planning approach used in the BMWC situation (Stokes 1986). For example, the citizen component included a sufficiently broad spectrum of interest groups to constitute a microcosm of local, regional, and national interests in the BMWC; however, the task force was not representative of all wilderness interest groups because the formal public-review process provided the opportunity for groups or individuals not included in the task force to make their views known.

The task force was composed of managers, researchers, and citizens. This composition allowed the opportunity to share technical/scientific knowledge and personal knowledge among participants. Most citizens' representatives had personal knowledge of the BMWC based on their experience as users. Many of them also had technical knowledge to share with others. The managers had personal knowledge of the area and scientific/technical background and knowledge; the researchers provided concepts such as LAC and the best scientific data and analysis that were available. Through discussions and dialogues at meetings, the personal knowledge of all representatives became integrated with the collective scientific/technical knowledge of the group.

The LAC Task Force for the BMWC consisted of up to fifty persons. Over a period of nearly four years, with an independent facilitator helping in the last two years, the Recreation Management Direction for the BMWC was established. Nine meetings of the full task force were held, along with numerous other meetings of geographical- or problem-based subgroups. A major public information and involvement program was also initiated. In total, more than eighty formal and informal meetings and information and working sessions occurred in the four-year period. A seven-member Agency Core Team, composed of ranger district representatives, the Flathead National Forest recreation staff officer, and the independent facilitator, provided the task force with information and tentative proposals from which to work.

The LAC steps were used to focus task force input. For example, the task force developed the preliminary list of issues and indicators, identified alternative standards, mapped alternative opportunity-class allocations, and identified potential management actions. Much of the core team and researcher effort was focused on educating the task force about the LAC process, interpreting research data, and resolving other administrative issues. In addition, because the LAC process was new and still being developed, the core team was continuously being educated. Task force members contributed by validating or pointing out weaknesses in data and in the LAC process itself.

The LAC process resulted in four wilderness recreation-opportunity classes. Class I, the most pristine, permits almost no evidence of human occupancy and use, whereas Class IV allows somewhat more change to occur. For example, signs indicating trail names or numbers and administrative signs and signing for use dispersal and resource protection are permitted in Class IV areas, but in Class I, no signs are permitted. Mapping the opportunity classes centered on identifying heavily used travel corridors that were mapped as either Class III or IV. These travel corridors were generally mapped at about one mile in width under the assumption that most visitor impacts would be found that close to the trail. The remaining area was then mapped as Class I or II, depending on existing resource and social conditions. A portion of the map is shown in Fig. 9.7.

The final opportunity class allocation for the Bob Marshall, Great Bear, and Scapegoat Wildernesses, in 1987, for the 1,535,352 acres was: Class I, 60 percent; Class II, 18 percent; Class III, 16 percent; and Class IV, 6 percent. The corridors were mapped with a one-mile width, but because the trails are very narrow, only 6 percent of the complex was in Class IV, the least pristine opportunity class (U.S. Department of Agriculture, Forest Service 1987).

Indicators, and their standards, were also developed through task force participation. Various small-group techniques were used to develop and validate these. The indicators and standards are shown in Table 9.1. The standards were developed, with the exception of the encounter standards, by examining the inventory data available. Encounter standards were developed with the help of visitor preference data generated by Lucas (1985). Using these data allowed realistic and attainable standards and does not legitimize existing unacceptable conditions in the wilderness. The task force thoroughly discussed the standards, their meaning, and application. This discussion was helpful to managers because it extended their knowledge of LAC and its implications, and it also helped in developing standards that were consistent with preserving the character of the BMWC.

Management actions were developed to address potential or existing unacceptable resource and social conditions. Task force members appeared to have the greatest difficulty with this step, primarily because it placed the citizen and the researcher in the role of manager. A general list of actions identified by type of problem and by opportunity class was developed. Actions are listed in order of their preference by task force members, generally from the least restrictive to the most restrictive. For example, information and education were favored as the most acceptable action to control campsite impacts in Classes I and IV (the most and least pristine classes). This was based on a concern to follow the principle of minimum regimentation. However, the acceptability or appropriateness of the other management actions was tied to the kinds of conditions defined as consistent with the opportunity class definition outlined in Step 3 of the LAC process.

The BMWC was the first place where LAC was attempted as a full process. Applying the LAC process to planning of the BMWC focused on managing the area's recreational use because much of the impact on the area's quality stemmed from such use. However, managers must be concerned about many issues other than recreation. For example, the BMWC contains both diverse wildlife populations and key threatened and endangered species such as the grizzly bear and gray wolf. Although the LAC process did not directly address wildlife concerns, nonetheless it was possible to examine how the alternative opportunity class allocations would affect the wildlife values contained within them.

The 1987 BMWC Recreation Management Direction was accepted with four inventory and monitoring requirements: (1) determine overall use patterns, activities, and levels; (2) conduct an extensive social survey; (3) inventory trail conditions; and (4) determine range trend and condition. In the ten years following implementation of the BMWC Recreation Management Direction, Warren (1997) reported that the information collected was very incomplete in achieving these four requirements. The 1987 BMWC Recreation Management Direction also included seven resource condition standards: (1) trail, campsite, and river encounters with other parties; (2) number of human impact sites; (3) occurrences of litter on wild and scenic river banks; (4) wild and scenic river recreation user experience quality; (5) encounters with other float parties at Schaefer Meadows; (6) forage utilization; and (7) aircraft landings at Schaefer Meadows airstrip. Only four of these standards were measured in 1987–1997 as planned—two standards were mostly attained, one was partially attained, and one was not attained—the other three standards were incompletely monitored and of limited use to managers (Warren 1997). The available information has allowed managers to take some adaptive management actions; however, if the BMWC Recreation Management Direction is to be fully realized, then complete monitoring needs to be conducted and the necessary management must be implemented.

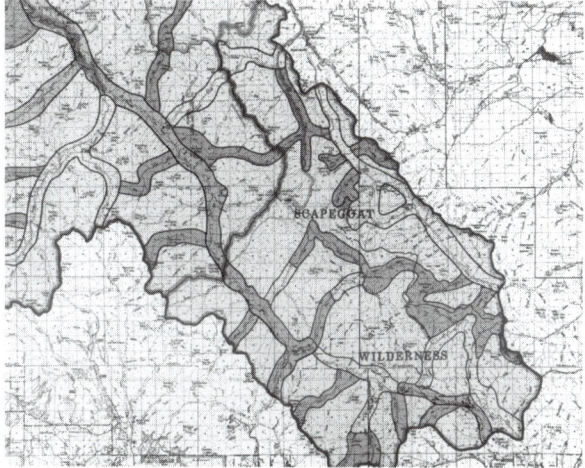

Fig. 9.7. Portion of a map displaying recreational opportunity classes in the Scapegoat Wilderness, part of the BMWC, Montana. Shadings indicate levels of naturalness ranging from pristine to heavily impacted.

The San Juan–Rio Grande National Forests Wilderness Management Direction

The Forest Plan Direction for wilderness was originally developed in 1985 for the San Juan and in 1993 for the Rio Grande National Forest in Colorado. A 1996 revision was limited to changes in their Wilderness Management Direction. The LAC process was used in the 1996 development of the San Juan–Rio Grande National Forests Wilderness Management Direction, which includes five designated wilderness areas—Weminuche, South San Juan, Lizard Head, Sangre de Cristo, and La Garita—and the Piedra Area, a congressionally designated study area to be managed to retain its existing wilderness character and for potential inclusion in the NWPS.

The public participation for the planning process began with a Weminuche Wilderness Study Group in 1993 and provided input that was used to develop issues and alternatives for all six areas in 1996. Subsequently, a formal scoping document with proposed amendments to the Wilderness Management Direction was sent to more than 800 people and organizations, plus public open houses were held in five communities. The initial environmental assessment was sent to 150 people and organizations that had requested copies or provided input to the document. The final Wilderness Management Direction

Table 9.1. Proposed Standards for Resource and Social Indicators for Each Opportunity Class in the BMWC
(U.S. Department of Agriculture, Forest Service 1987)

Indicators	Opportunity Class I	Opportunity Class II	Opportunity Class III	Opportunity Class IV
Social				
1. Number of trail encounters with other parties	80 percent probability of 0 encounters per day	80 percent probability of 1 or fewer encounters per day	80 percent probability of 3 or fewer encounters per day	80 percent probability of 5 or fewer encounters per day
2. Number of other parties camped within sight of parties or continuous sound	80 percent probability of 0 parties per per day	80 percent probability of 0 parties per day	80 percent probability of 1 or 0 parties per day	80 percent probability of 3 or fewer parties per day
Human-Impacted Sites				
3. Area of barren core (sq ft)[1]	100	500	1,000	2,000
4. Number of human-impacted sites per 640-acre area[2]	1 permitted	2 permitted	3 permitted	6 permitted
5. Number of human-impacted sites above a particular conditioned class index per 640 acres	No moderately or highly impacted campsites/section	No more than 1 moderately impacted site and 0 highly impacted sites/section	No more than 2 moderately impacted sites and 0 highly impacted sites/section	No more than 3 moderately impacted sites and 1 highly impacted site/section
Range				
6. Degree of forage used	No more than 20 percent forage used	No more than 20 percent forage used	No more than 40 percent forage used	No more than 40 percent forage used
7. General range trend	Static or improving	Static or improving	Static or improving	Improving
8. Overall range condition	Excellent	Excellent	Generally good or better	Generally good

1 Excludes authorized horse-handling facilities. A variance will be given to outfitter base camps not currently in compliance and a timetable for compliance will be developed and administered through the outfitter operation plans.
2 Human-impacted sites defined as any site with evidence of human impact, normally centered on a fire ring, regardless of its prior use for camping.

was accepted in August 1998 (U.S. Department of Agriculture, Forest Service 1998a).

Through Step 1 in the LAC process, nine goals were developed to guide the Wilderness Management Direction in the San Juan–Rio Grande National Forests:

- Manage wilderness so that changes in the ecosystem are primarily a consequence of natural forces, or within a range of natural variability and succession.
- Maintain wilderness in a natural and untrammeled condition while accommodating human uses.
- Provide outstanding opportunities for solitude and a primitive and unconfined recreation experience.
- Sustain wilderness as a place of peace, solitude, and sanctuary.
- Preserve natural resources for their inherent ecosystem and biological diversity values and for scientific research purposes.
- Provide the opportunity for challenge and risk.
- Minimize long-term impacts caused by human uses.
- Sustain natural and indigenous life-forms.
- Protect and preserve historic and cultural resources found in wilderness.

These goals were determined to comply with national wilderness legislation and other laws and regulations related to national forest management.

Three major categories of issues and concerns were identified in the public scoping process and were used in Step 2 of the LAC process:

- Human effects on wilderness ecosystems, such as increasing wilderness visits, habituation of black bears and mountain goats to the presence of humans, stocking nonindigenous fish species in high mountain lakes, and the introduction of nonnative plants and noxious weeds.
- Human effects on wilderness experience opportunities, such as crowding, visitor conflicts, group size, visitor impacts on trails and campsites, and domestic dog behavior; however, visitors were also concerned about restrictions on the number, type, and size of visitor groups.
- Recreation pack-stock effects on wilderness resources, such as grazing affecting native plants, impacts on trails and campsites, and introducing nonnative plants and noxious weeds.

Based on wilderness planning and LAC processes, the Wilderness Management Direction chosen was "to improve wilderness qualities while maintaining recreation opportunities." Some of the management approaches and actions chosen to support this direction (Steps 8 and 9 of the LAC process) are listed here according to the three major categories of issues and concerns:

- *Human effects on wilderness ecosystems:*
  1. Prohibit camping in the Twin Lakes area where the mountain goat population has the most contact with visitors.
  2. Reduce visitor crowding and campsite densities to reduce black bear habituation to the presence of humans.
  3. Evaluate fish stocking on a case-by-case basis and use native cutthroat trout species where stocking is used to provide fishing opportunities in high mountain lakes.
  4. Monitor populations of noxious weeds and develop recommendations for treatment alternatives.

- *Human effects on wilderness experience opportunities:*
  1. Set standards for visitor-visitor encounters on trails and at campsites for each opportunity class.
  2. Limit visitor-group size in all areas to fifteen people, with a maximum combination of twenty-five people and stock in groups using pack stock, except for the Piedra River where group size is twenty people for river-running activities.
  3. Require domestic dogs to be on a leash or respond to voice commands.
  4. A quota permit system on recreation visitors will be used only if less restrictive management measures are shown through monitoring to not meet standards and guidelines in the plan.

- *Recreation pack-stock effects on wilderness resources:*

1. Establish standards and guidelines for riparian areas and meadows that result in the reduction or prohibition of recreation pack stock grazing or restraining of stock within these areas.
2. Limit group size for pack-stock users (see above).

One of the generally stated guidelines in the Wilderness Management Direction is to use the minimum regulation necessary to achieve the objectives. The implementation and monitoring step of the LAC process began in 1998, and progress will be evaluated in the future to ensure that standards are not exceeded by the management prescriptions set by this Wilderness Management Direction. The plan states that should the standards indicate a need for more restrictive management, then it will be implemented in an effort to achieve the goals and objectives stated in the Wilderness Management Direction.

The White Mountain National Forest Wilderness Management Plan

The Land and Resource Management Plan (LRMP) in the White Mountain National Forest (WMNF) of New Hampshire and Maine (U.S. Department of Agriculture, Forest Service 2005) addressed wilderness issues and management direction throughout the document. The LRMP specifically outlines the LAC process in appendix E for all five wilderness areas within the WMNF: Great Gulf (5,500 acres), Presidential Range–Dry River (29,000 acres), Pemigewasset (45,000 acres), Sandwich Range (25,000 acres), and Caribou-Speckled Mountain (14,000 acres). This WMNF Wilderness Management Plan uses the LAC and four opportunity class zones to monitor progress toward the plan goals.

The plan delineated four different wilderness zones and mapped the zones across the five wilderness areas. The opportunity class zones are labeled A, B, C, and D and are based on ecological characteristics, social conditions, and management needs. Although visitor use levels were not the determining factor in applying the zoning scheme, use estimates were helpful in understanding the distinctions among zones; the zones generally run from least (Zone A) to most heavily used (Zone D). The following is from the U.S. Department of Agriculture, Forest Service plan (2005, E-5 to E-9):

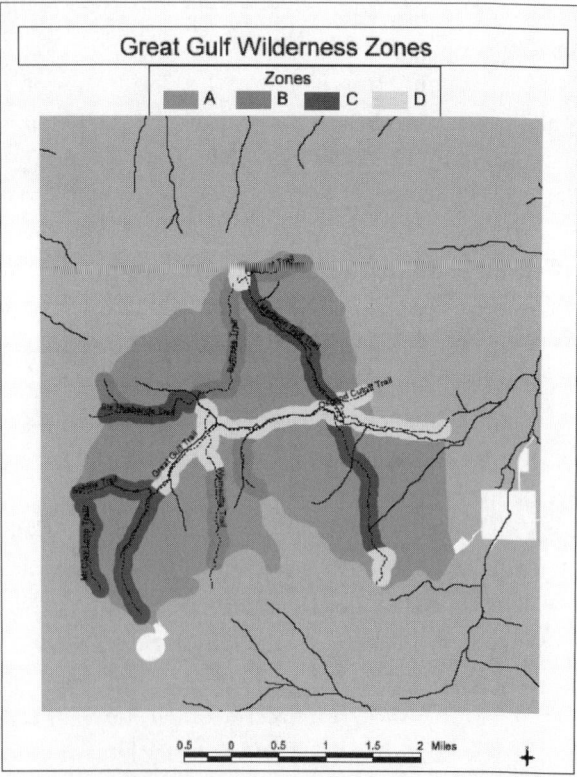

Fig. 9.8 Four opportunity class zones in the Great Gulf Wilderness, part of the White Mountain National Forest, New Hampshire. U.S. Department of Agriculture, Forest Service (2005, E-35).

Zone A—Areas 500 feet or more from all trails
This zone includes the trailless areas of WMNF Wilderness, and represents the largest area of WMNF Wilderness. The landscape appears largely unmodified, supports no maintained trails or facilities, has few restrictions, has low managerial regulation, has little direct management activity, and has exceptional opportunities for visitors to experience both solitude and a very primitive and unconfined recreation.

Social Conditions: Encounters with other visitors or with management are nonexistent to infrequent. The environment offers the highest degree of challenge, self reliance, and risk. There is an outstanding opportunity for solitude, and visitors will experience primitive, unconfined recreation within this area.

Facilities/Infrastructure: No maintained or constructed facilities present. Very little or no obvious on-the-ground evidence of human presence or activity, except for occasional historical artifacts.

Campsites: Very low density of campsites. Campsite impacts are not visible from year to year; sites are difficult to discern and generally are rehabilitating naturally. Designated sites are not established.

Vegetation/Soils: The Forest vegetative composition may have been affected by predesignation activities such as timber harvesting. There is very little or no vegetation loss, soil compaction, or lasting alteration of the duff and litter layer resulting from human use. Areas do not receive regular, recurring use. Any existing impacts in these areas are generally rehabilitating.

Managerial Presence: Management focuses on sustaining and protecting the natural ecosystem, allowing natural events and processes to occur with minimal or no management. Agency patrols are rare, primarily to monitor ecological conditions. Efforts will be made to minimize regulations, but they may be utilized in specific areas for protection of Wilderness character. Signs will not be present except in rare instances for resource protection.

Zone B—Areas within 500 feet of low-use trails

This zone includes the lowest-use, least developed trails within WMNF Wilderness. It offers the greatest opportunity for solitude and/or an unconfined recreation experience along a maintained trail system. With the exception of the developed trail system, the landscape appears largely unmodified, supports only these minimally maintained trails but no other facilities, and has regular opportunities for visitors to experience both solitude and a primitive recreation confined only by the presence of the trail system.

Social Conditions: Encounters with other visitors or with management are infrequent. The environment offers a high degree of challenge, self-reliance, and risk. There is a great opportunity for solitude, and visitors will generally experience primitive and unconfined recreation within this area.

Facilities/Infrastructure: The trail system is the primary infrastructure. Primitive trails and trail structures consistent with WMNF Level 1 trail specifications (FSH 2309.18) may be present. No other facilities will be constructed or maintained. Historical artifacts may be present and are sometimes concentrated and may be obvious. Other impacts will not be readily apparent.

Campsites: Very low density of campsites. Campsites may be discernable, but are generally rehabilitating and not receiving regular, recurring use. Designated sites are not established.

Vegetation/Soils: The Forest vegetative composition may have been affected by predesignation activities such as timber harvesting. There is very little or no vegetation loss, soil compaction, or lasting alteration of the duff and litter layer resulting from human use except on trails. These trails are more primitive and receive less maintenance. Areas do not receive regular, recurring use outside the trail corridor. Any existing impacts in these areas are generally rehabilitating.

Managerial Presence: Management focuses on sustaining and protecting the natural ecosystem and providing primitive access for visitors. Agency patrol will be on a regular basis, primarily for monitoring and education. Efforts will be made to minimize a regulatory approach; however, regulations will be utilized for protection of Wilderness character. Signs may be present at trail junctions and in rare cases for resource protection.

Zone C—Areas within 500 feet of moderate-use trails

This zone includes the moderate-use, moderately developed trails within WMNF Wilderness. As outlined below, Zone C is in general more highly used and more highly developed than Zone B. Despite this, Zone C offers visitors an opportunity to experience escape from more highly developed landscapes while still being able to access a maintained trail system. In most places, the landscape appears largely unmodified. Exceptions include the trail system and associated structures and lasting campsites, including some designated sites. Facilities such as bridges may exist, but shelters and toilets do not. The area is likely to have site-specific as well as blanket regulations, with generally frequent managerial presence. Direct management activity including enforcement of regulations occurs.

Social Conditions: Encounters with other visitors or with management are likely, especially along trails and at established campsites. There is a high degree of challenge and risk, and a lower degree of self-reliance than in Zones A and B. There is a generally moderate opportunity for solitude.

Facilities/Infrastructure: The trail system and associated structures are the primary evidence of past human presence and activity. Trails and associated structures are consistent with WMNF Level 2 trail specifications (FSH 2309.18). Bridges may exist for public safety or resource protection only. No other facilities will be maintained or constructed. Historical artifacts may be present and are sometimes concentrated and may be obvious. Other impacts will not be readily apparent.

Campsites: Campsite density is low to moderate. Within standards, there are sufficient sites to accommodate peak use without the creation of new sites. Bare mineral soil may exist on sites, and most sites will persist from year to year. Designated campsites may be present and exist for resource protection.

Vegetation/Soils: The Forest vegetative composition may have been affected by predesignation activities such as timber harvesting. Moderate soil compaction and loss of vegetation, litter and duff are expected on many trails and campsites. User-created trails may be present, especially in destinations and camping areas. Minimal erosion may occur on a small percentage of the disturbed sites and may be mitigated to ensure resource protection. Riparian and lakeshore conditions may show signs of human impacts in localized areas, and these are expected to persist from year to year.

Managerial Presence: Management emphasizes sustaining and protecting natural conditions, while providing access for and accommodating a moderate level of human recreation use. Agency patrol will be on a regular basis, for monitoring, education, and enforcement purposes.

Management actions will be necessary to protect Wilderness character, and may be indirect or direct. Overall management presence will be more noticeable to visitors. Site-specific or blanket area regulations may be implemented, especially related to camping or campfires. Signs will be present at trail junctions and at designated campsites and will be used for resource protection.

Zone D—Areas within 0.25 mile of developed facilities or 500 feet of high-use trails

This zone includes the most heavily used and most highly developed trails and areas within WMNF Wilderness. It represents the smallest area of WMNF Wilderness. The landscape within this zone is modified by the developed trail system and associated structures, and may include bridges, primitive shelters and/or toilets, designated campsites, and impacts resulting from recurring recreation use. However, in most places the landscape still appears largely unmodified. To manage use and protect resource conditions the area likely has site-specific as well as blanket regulations, with frequent managerial presence. Direct management activity including enforcement of regulations occurs. This area has occasional opportunities for visitors to experience solitude as well as primitive and unconfined recreation bounded by the presence of the trail system, existing regulations, shelters, toilets, and campsites.

Social Conditions: Depending on the season, encounters with other visitors or with management are very likely, especially along trails and at established campsites. There is a moderate degree of challenge and risk, and a lower degree of self-reliance than in other zones. There is a moderate opportunity for solitude.

Facilities/Infrastructure: Bridges may exist for public safety or resource protection. Shelters and toilets may exist where identified in Wilderness enabling legislation or where consistent with standards described in this plan. The trail system and associated trail improvements are the primary evidence of past human presence and activity. Trails are managed consistent with WMNF Level 2 trail specifications (FSH 2309.18). Other evidence may include shelters and toilet structures. Historic artifacts may be present and are sometimes concentrated and may be obvious. Other impacts may be apparent.

Campsites: A moderate to high density of established sites may exist. Bare mineral soil may exist on sites, and impacts are recurring and will persist from year to year. Designated campsites may be present and exist for resource protection and to accommodate visitor use.

Vegetation/Soils: The Forest vegetative composition may have been affected by predesignation activities such as timber harvesting. Moderate to high soil compaction and loss of vegetation, litter and duff are expected in localized areas on many trails and campsites. User-created trails may be present, especially in destinations and camping areas. Minimal erosion occurs on the disturbed sites and may be mitigated to ensure resource protection. Riparian and lakeshore conditions may show signs of human impacts in localized areas, and are expected to persist from year to year.

Managerial Presence: Management emphasizes sustaining and protecting natural conditions, while providing access for and accommodating a moderate to high level of human recreation use. Agency patrol occurs frequently for monitoring, education, and enforcement purposes. Management actions are necessary to protect Wilderness character, and may be indirect or direct. Overall management presence is noticeable to visitors. Site-specific or blanket area regulations may be implemented, especially related to camping or campfires. Signs are frequently present at trail junctions and at designated campsites and are used for resource protection.

The WMNF Wilderness Management Plan used the LAC process to develop four categories of indicators and standards: biophysical, social, aesthetic, and ecosystem process indicators. These indicators also include some monitoring efforts for wilderness conditions that are not part of the LAC process and address broader concerns about wilderness conditions such as suggested in a Forest Service framework for monitoring conditions related to wilderness character (Landres et al. 2005). The Wilderness Management Plan concludes with an education plan to address concerns and issues present in each of the four opportunity class zones and outlines educational program actions and message content.

The Maricopa Complex Wilderness Management Plan

The Arizona Desert Wilderness Act of 1990 (U.S. Public Law 101-628) designated four areas near Phoenix—Sierra Estrella, North Maricopa Mountains, South Maricopa Mountains, Table Top—to be managed as wilderness by the BLM. The Maricopa Complex Wilderness management area includes 172,000 acres of lower Sonoran Desert. These four wilderness areas contain little or no evidence of human disturbance. There are former vehicle ways and cattle-grazing fences or water devices, and the four areas provide good to outstanding opportunities for solitude and naturalness because they are generally inaccessible due to rugged terrain and rough access roads (U.S. Department of the Interior, Bureau of Land Management 1995). Visitation by recreation users is relatively light, reportedly fewer than 2,000 annual visitor-use days, with dispersed hiking over a few marked trails and along former vehicle ways, and some access by visitors on horseback. The demand for recreation access into these wilderness areas is expected to increase due to its proximity to Phoenix, with its population of more than 2 million and only a two-hour drive away.

The wilderness planning and LAC processes for the Maricopa Complex Wilderness were used to address five categories of issues: protecting and enhancing the natural character of wilderness, providing opportunities for solitude and primitive recreation, managing other land uses and activities provided for by the Wilderness Act, managing wildlife, and managing vegetation. We will only highlight some of the planning outcomes by focusing on two of the more notable problems for these areas: (1) removing evidence of former vehicle access, and (2) maintaining wildlife water-development structures built in the 1960s prior to designation.

The Maricopa Complex Wilderness planning process used several types of public involvement, all of which was less formal and at a smaller scale than what was used in the Bob Marshall Wilderness Complex LAC process. Public involvement included (1) four public scoping meetings in different local communities in 1992 and a period of time for written comments on the issues; (2) formation of a working group of eight interested public representatives and agency personnel who met five times during 1992–1993 to review planning progress and provide input; (3) a public briefing held in 1993 to discuss planning issues and input; and (4) a public review of the draft Maricopa Complex Wilderness plan held in 1994 at two community meetings and through written comments. A final plan was completed in 1995 and designed to be implemented over a ten-year period.

The final plan document stated that the overall management strategy for the Maricopa Complex Wilderness Management Plan (U.S. Department of the Interior, Bureau of Land Management 1995, 25) was to

> maintain or improve the natural character of each of the four wildernesses by protecting many of the near-pristine conditions while rehabilitating existing human impacts. The strategy recognizes that as metropolitan Phoenix and the surrounding communities expand their borders and populations over the next 10 years, the wildernesses will be subject to dramatic increases in visitor use and its associated consequences.

Some of the issues surrounding mechanical vehicle access arose due to past and present situations. There were approximately ninety-five miles of former vehicle ways in these four wilderness areas and the boundary designations allowed for some primitive dirt roads to continue to exist as corridors, or so-called cherrystems, into the wilderness area. Due to the lack of ground cover, illegal vehicle trespass into the wilderness can occur and is a concern because evidence of cross-country travel by mechanical vehicles can be seen for as long as twenty years after the intrusion. One type of proposed management action was to install vehicle barriers and to reduce the evidence of the ninety-five miles of former vehicle ways by rehabilitating vehicle ways (sixteen miles) to trail standards, actively rehabilitate eighteen former vehicle ways (nineteen miles) to natural conditions, and allow forty-one vehicle ways (sixty miles) to rehabilitate naturally. Notable progress had been achieved by 2000. The measurement standards of acceptable change were to monitor barricaded and rehabilitated vehicle ways twice a year to (1) determine if any unauthorized vehicle use had occurred and to respond, if necessary, with management modifications to stop such activities, and (2) document progress toward vehicle way rehabilitation, although natural processes are relatively slow in the Sonoran Desert environment and natural reclamation and revegetation are expected to take fifty to a hundred years.

Livestock grazing is permitted in these wilderness areas, and there is evidence of livestock fences and water developments in the grazing allotment areas, such as diked water diversions into earthen livestock tanks. However, the issues surrounding nine wildlife water-development structures—eight rainwater catchments and one well, used exclusively for wildlife, such as mule deer and desert bighorn sheep—were more complex and challenging to plan for in the LAC process. The need for these wildlife water-development structures occurs because of the lack of surface water in the wilderness areas, and the surface water located adjacent to the wilderness areas requires wildlife to approach or cross roads and developed agricultural or residential lands. In an effort to protect a population of 200 to 300 desert bighorn sheep, wildlife water-development structures were allowed to be maintained and, at times, refilled by tank trucks driving into the wilderness areas or by helicopter. One type of proposed management action was to allow mechanized and motorized equipment to enter the wilderness areas to permanently modify six wildlife water catchments, so that further mechanized transport of water would be reduced or eliminated. The measurement standards of acceptable change were to monitor the water-level reliability and availability annually in each catchment with the Arizona Game and Fish Commission, which conducts overflights to monitor water levels in catchments and wildlife populations and conduct field observations throughout the year. Final decisions to end mechanized transport of water to supplement catchment water levels will be a joint decision of the BLM and the Arizona Game and Fish Commission.

The Superstition Wilderness Implementation Plan

Originally established as a primitive area in 1939, the Superstition Wilderness in Arizona was designated under the 1964 Wilderness Act. The Superstition Wilderness now totals more than 159,000 acres after more area was added in 1984. An implementation plan was written in 1998 for the wilderness, as part of the Tonto National Forest, to further specify and act on the prescriptions of the Tonto Land and Resource Plan of 1985 (U.S. Department of Agriculture, Forest Service 1998b). The four implementation plan objectives were to preserve the integrity of the wilderness resource; to provide uniform and consistent administration by the three ranger districts; to conduct necessary administrative activities in a manner most protective of the wilderness resource and visitor experience; and to initiate law enforcement action, when necessary, to ensure wilderness resource protection and public safety.

The LAC process was used extensively in the Superstition Wilderness Implementation Plan (U.S. Department of Agriculture, Forest Service 1998b) and is most evident in the monitoring standards established for each WOS opportunity class. For example, visitor management was an important issue and use ranged from light to very heavy, particularly visitor day use due to close proximity to metropolitan Phoenix. Visitor group size is limited to fifteen persons throughout the wilderness. There is a limit of fifteen livestock that can be used by visitor groups. A very diverse use by hikers, backpackers, and horseback riders is spread over 180 miles of trails, but the majority of use is via two trailheads.

The monitoring of visitor use has been institutionalized into management, with LAC indicators monitored every day by wilderness personnel who are in the field (U.S. Department of Agriculture, Forest Service 1998b). Survey techniques and forms were developed for LAC daily traveling surveys and monthly summaries for each WOS opportunity class. A total of sixteen indicators were selected and ten were measured during the wilderness travel by personnel (Table 9.2). In addition, a backcountry campsite inventory form was developed and used to inventory all campsites, with subsequent inventories to be conducted again within five years. A total of thirteen indicators were selected and measured during the campsite inventory by personnel; a selected list of six indicators is shown in Table 9.3.

The Superstition Wilderness experienced rapid development and use by rock climbers between 1987 and 1992. More climbers registered at trailheads in 1987 than the total number of climbers from 1979 through 1986. The controversy that surrounded this activity was begun when new climbing guides were published and when interested sport-climbing individuals and groups in Arizona rapidly escalated the number of routes and placed hundreds of fixed bolts. Because of the soft nature of the exposed rock, climbers were using steel bolts and hangers or web slings as anchor systems to protect them in climbing routes that would otherwise be unsafe without the fixed anchors. The issues revolved around damage being done to rocks while drilling to place the bolts, the visibility of the permanent bolt anchors, the concern that this type of sport climbing was not a wilderness-dependent activity, and that there were other areas outside wilderness for sport climbing.

The LAC process related to bolting in the Superstition Wilderness resulted in developing a rock-climbing-route inventory methodology and monitoring standards (U.S. Department of Agriculture, Forest Service 1992). However, following the LAC process related to the bolting issue for the Superstition Wilderness and a series of recommendations made by a National Task Force on Fixed Anchors in Wilderness, the Tonto National Forest supervisor issued a moratorium on installation and/or removal of bolts pending a national policy decision from the USFS (U.S. Department of Agriculture, Forest Service 1992). This is a good example of the use of an LAC process for a specific recreation issue like placing fixed anchors in wilderness, or so-called bolting.

Table 9.2. Selected Limits of Acceptable Change Indicators and Standards for Visitor Experiences in the Superstition Wilderness, Arizona
(U.S. Department of Agriculture, Forest Service 1998b)

Indicator	WOS Class I	II	III	IV
1. Horseback groups encountered per day	0	3	4	4
2. Backpack groups encountered per day	1	3	4	4
3. Day-hiking groups encountered per day	1	3	6	40
4. Maximum size of group	6	15	15	15
5. Maximum number of horses/mules in a group	0	15	15	15
6. Number of occupied backcountry camps within 100 yards of selected camp(s)	0	1	3	3
7. Number of low-flying aircraft seen or heard nearby (2,000 feet or less above ground level) per day	0	0	1	1
8. Number of uncontrolled dogs encountered per day	0	0	1	1
9. Number of nonindigenous animals encountered per day (excluding dogs, horses, mules, burros, cattle)	0	0	0	0
10. Number of gunshots heard, not including legal hunting activities	0	0	0	0

Table 9.3. Selected Limits of Acceptable Change Indicators and Standards for Backcountry Camps in the Superstition Wilderness, Arizona
(U.S. Department of Agriculture, Forest Service 1998b)

Indicator	Level of Severity 0	1	3	5	WOS Class I	II	III	IV
1. Litter	None	Few small pieces of litter and/or 30 sec. cleanup	Moderate amount of litter and/or up to 3 min. cleanup	Widespread litter and/or over 10 min. cleanup	0	1	3	4
2. Rock fire rings	None	1 small fire ring and/or up to 2 min. cleanup	1 large or 2 medium fire rings and/or up to 10 min. cleanup	More than 3 fire rings and/or more than 20 min. cleanup	1	1	2	3
3. Charcoal/fire scars outside of fire ring	None	Up to 10 sq ft of area	Up to 50 sq ft of area	Up to 75 sq ft of area	0	1	2	3
4. Camp modifications	None	Slight evidence of linear or radial rock/log arrangements	Some construction: rock/log arrangements nails in trees, trenching; up to 10 min. cleanup	Well-developed construction: benches, shelves, tables, leveled tent pad; up to 20 min. cleanup	0	1	2	2
5. Nearby camps	None	1 camp within 100 yards	3 camps within 100 yards	5 or more camps within 100 yards	0	1	3	3
6. Barren core area	None	Up to 10 sq ft	31 to 60 sq ft	More than 100 sq ft	0	2	3	4

Fig. 9.9. Whether or not to allow permanent bolts for rock climbing is a hotly contested wilderness management issue in the Superstition Wilderness, Arizona. The cliffs (left and center) are popular sport-climbing areas. A bolt (above), drilled into rock with a hanger attached, is a permanent point of protection on a climbing route. Some wilderness users argue that sport climbing is not a wilderness-dependent activity. Photos courtesy John Hendee, left; and Greg Hansen, U.S. Forest Service, center and above.

Direct Visitor Management May Be Necessary When Standards Are Exceeded

The LAC approach, as part of the wilderness planning process, determines the standards—limiting factors—that are most sensitive to increasing recreation use and related use impacts in wilderness. Monitoring these standards serves as feedback to the wilderness managers, indicating when previously implemented management actions have or have not succeeded in maintaining acceptable conditions. When previous management actions have proven ineffective, a visitor use limit may remain the only alternative available—minimum tool or regulation—and is not an action to avoid; rather, a use limit is viewed as one of a number of management approaches that can be taken in response to exceeding the standards that are the acceptable limits of change. Because of projected increases in use levels or negative changes in conditions, managers may decide to implement a use limit before standards are violated, rather than waiting until the standard has been reached.

Instead of focusing on identifying how many people can use an area without generating unacceptable impacts, the LAC process seeks to define the acceptable conditions for an area. As long as these acceptable conditions are present, the logic of the LAC process leads us to conclude that the visitor use levels occurring within the wilderness area are acceptable or within the area's carrying capacity, because those visitor use levels and activities are not producing impacts judged unacceptable according to the standards of resource and social conditions defined for that area.

When wilderness conditions begin to degrade toward the standard that was defined as acceptable, managers must consider how an area's conditions can be protected. In Chapters 15 and 16, specific techniques to accomplish this are discussed. In general, *the concept of minimum regulation should guide management*. As the level of use increases, it will become necessary at some point to use direct visitor management techniques, such as limiting the number of recreation users and types of activities, so that unacceptable impacts and conditions do not occur or can be improved.

The LAC process is a powerful planning tool that can be used as a framework for managing wilderness within a carrying-capacity concept. The goal of management is to identify the desired resource and social and managerial conditions to be maintained or restored in wilderness, with these desired conditions expressed as explicit, measurable standards. Thus, the focus of management attention shifts from defining maximum visitor use to identifying desired conditions and managing recreation use levels and impacts so that resource, social, and managerial conditions are what was intended under the Wilderness Act.

Summary

Three key concepts form the theme of this chapter. First, the challenge facing wilderness managers in dealing with the issue of *carrying capacity* is not a matter of developing maximum numbers that describe how much use is too much. Rather, it is a matter of prescribing what kinds of social and resource conditions are desired, comparing these desired states against existing conditions, and identifying the kinds of policies and actions needed to maintain or restore the desired conditions. *Managerial judgment* is one key element in this process. Judgment is a product of experience, research data, basic inventory information, public input, careful analysis, and common sense.

The second concept is that establishing appropriate conditions depends on formulating explicit *management objectives* that (1) provide specific, measurable indicators related to desired conditions and (2) define the standards for the resource and social elements identified as critical indicators. One key factor in the success or failure of a Wilderness Management Plan is the process by which the *public participates* in the development of informed consent and helps shape the value judgments on which the plan is based; that is, in the formation of what are or are not appropriate wilderness conditions.

Third, either *resource or social indicators* can determine if an area has reached or exceeded its recreation visitor capacity. As area conditions degrade and the standards for an indicator are approached, managers may need to employ different management actions to control impacts. Some of these management actions will be more effective than others, and it may become necessary to implement direct visitor management. At some level of visitor use, it will become necessary to restrict further increases in the number of users to protect desired conditions, but it is the condition of the area, not the number of visitors, that is the focus of management attention.

Study Questions

1. Discuss some of the factors that will influence determining a recreation carrying capacity for a wilderness area. Identify ways in which managers might alter the carrying capacity of that area.
2. Identify four considerations in managing visitor recreation use and impacts in wilderness.
3. What information would you consider useful in selecting indicators to measure resource and social condition impacts in wilderness?
4. What information and criteria would you use in specifying standards to identify acceptable levels of resource or social impact for a wilderness area?
5. The idea of a recreation carrying capacity implies that at some point no further increases in recreation use should be allowed. Some argue that visitor restrictions are necessary to protect wilderness-area resources and the nature of the wilderness experience. Others, however, argue that restricting visitors' use of wilderness areas is inappropriate and inequitable. Discuss some of the pros and cons underlying these respective positions.

Acknowledgments

This chapter is a major revision in the third and fourth editions by Chad P. Dawson and John C. Hendee of Chapter 9 of the second edition; we recognize and thank our colleagues, George H. Stankey, Stephen F. McCool, and Gerald L. Stokes, for their seminal efforts on these topics in the previous editions. For contributions and review comments on the third edition of this chapter, we thank the following colleagues. We have benefited from their review comments, but we alone bear responsibility for any errors, mistakes, and opinions reported herein. Ed Krumpe, professor of Resource Recreation and Tourism, University of Idaho, Moscow, Idaho; Liese Dean, wilderness program coordinator, Stanley Ranger Station, Sawtooth National Recreation Area and National Forest, USDA, Forest Service; Gregory Hansen, wilderness staff officer, Mesa Ranger District, Tonto National Forest, USDA, Forest Service, Mesa, Arizona; Richard Hanson, supervisory outdoor recreation planner, USDI, BLM Field Office, Phoenix, Arizona; James

Mahoney, wilderness manager, USDI, BLM Field Office, Phoenix, Arizona; Linda Merigliano, wilderness, trails, and recreation planner, Bridger-Teton National Forest, USDA, Forest Service; R. Gregory Thompson, Rio Grande National Forest, USDA, Forest Service.

Case Discussion:
Campsite Indicators on the White Mountain National Forest

The Land and Resource Management Plan (LRMP) in the White Mountain National Forest (WMNF) of New Hampshire and Maine (U.S. Department of Agriculture, Forest Service 2005) outlined the LAC process in appendix E for all five wilderness areas within the WMNF: Great Gulf (5,500 acres), Presidential Range–Dry River (29,000 acres), Pemigewasset (45,000 acres), Sandwich Range (25,000 acres), and Caribou-Speckled Mountain (14,000 acres).

The LAC process resulted in developing four opportunity class zones ranging from (A) least used and developed to (D) most used and developed. The relatively small wilderness areas and heavy visitor use were a consideration when the impacted condition of designated campsites and proliferation of off-trail campsites was reviewed. The LAC standards, monitoring methods, and management actions for maintaining or improving campsite density were developed for off-trail campsites (U.S. Department of Agriculture, Forest Service 2005, E-17) as follows:

	Zone A	Zone B	Zone C	Zone D
Standard	0 lasting campsites with no visible impacts lasting more than 1 year.	0 sites within 500 feet of each other. 0 sites within 200 feet of trail.	0 sites within 200 feet of each other, maximum total of 2 sites within 500 feet of each other.	3 sites within 200 feet of each other, maximum total of 5 sites within 500 feet of each other.
Method of Measure, Frequency	Survey along 1 selected stream drainage within each wilderness each year. Survey 1 trailless peak above 2,999 feet within each wilderness each year, as appropriate.	Complete inventory once during the life of the plan.		
Management Action	1. Active site revegetation. Written reminder to all VIS centers reinforcing the established education message for this zone. Examine management that may contribute to a change in use patterns. 2. Increase focused patrols in the affected area. 3. If initial actions do not resolve issue, conduct focused management assessment to consider: • Enact closure order for affected area. • Implementation of limited overnight-use system.		1. Post revegetation signs. Written reminder to all VIS centers reinforcing the established education messages for this zone. Examine management that may contribute to a change in use patterns. Analyze group-use policies and act accordingly. 2. Increase focused patrols in the affected area. 3. If initial actions do not resolve issue, conduct focused management assessment to consider: • Enact or expand closure order for affected area. • Implementation of limited overnight-use system.	

Case Discussion Questions:

1. Does this WMNF indicator for campsite density meet the nine criteria for selecting a good indicator: early-warning ability, significance, indicative, discriminative, sensitive, responsive, quantitative, reliable, and feasible?
2. Does the campsite density standard appropriately describe a range of conditions across the four opportunity classes?
3. Is the monitoring method appropriate for the indicator and frequent enough for management responsiveness?
4. Are the management actions clearly identified if the standard (minimum acceptable condition) is exceeded in each opportunity class?

References

Brown, Perry J.; Haas, Glenn E. 1980. Wilderness recreation experiences: The Rawah case. *Journal of Leisure Research.* 12(3): 229–241.

Burch, William R., Jr. 1984. Much ado about nothing—some reflections on the wider and wilder implications of social carrying capacity. *Leisure Sciences.* 6(4): 487–496.

Bury, Richard L. 1976. Recreation carrying capacity—hypothesis or reality? *Parks and Recreation.* 11(1): 22–25, 56–57.

Clark, Roger N.; Stankey, George H. 1979. The recreation opportunity spectrum: A framework for planning, management, and research. General Technical Report. PNW-98. Portland, OR: U.S. Department of Agriculture Forest Service, Pacific Northwest Forest and Range Experiment Station. 32p.

Clawson, Marion. 1985. Outdoor recreation: Twenty-five years of history, twenty-five years of projection. *Leisure Sciences.* 7(1): 73–100.

Cole, David N. 1985. Management of ecological impacts in wilderness areas in the United States. In: Bayfield, N. G.; Barrow, G. C., eds. The ecological impacts of outdoor recreation on mountain areas in Europe and North America. Recreation Ecology Research Group Report No. 9. Wye, England: Recreation Ecology Research Group, pp. 138–154.

———. 1996. Wilderness recreation use trends, 1965 through 1994. Research Paper INT-RP-488. Ogden, UT: U.S. Department of Agriculture, Forest Service, Intermountain Research Station.

Cole, David N.; McCool, S. F. 1997a. Limits of acceptable change and natural resources planning: When is LAC useful, when is it not? In: McCool, Stephen F.; Cole, David N., comps. 1997. *Proceedings: Limits of Acceptable Change and Related Planning Processes: Progress and Future Directions;* May 20–22, 1997; Missoula, MT. INT-GTR-371. Ogden, UT: United States Department of Agriculture, Forest Service, Rocky Mountain Research Station, pp. 69–71.

———. 1997b. The limits of acceptable change process: Modifications and clarifications. In: McCool, Stephen F.; Cole, David N., comps. *Proceedings: Limits of Acceptable Change and Related Planning Processes: Progress and Future Directions;* May 20–22, 1997; Missoula, MT. INT-GTR-371. Ogden, UT: U.S. Department of Agriculture, Forest Service, Rocky Mountain Research Station, pp. 61–68.

Cole, David N.; Stankey, G. H. 1997. Historical development of limits of acceptable change: Conceptual clarifications and possible extensions. In: McCool, Stephen F.; Cole, David N., comps. *Proceedings: Limits of Acceptable Change and Related Planning Processes: Progress and Future Directions;* May 20–22, 1997; Missoula, MT. INT-GTR-371. Ogden, UT: U.S. Department of Agriculture, Forest Service, Rocky Mountain Research Station, pp. 5–9.

Dasmann, Raymond F. 1964. *Wildlife Biology.* New York: John Wiley and Sons.

Driver, Beverly L.; Brown, Perry J. 1978. The opportunity spectrum concept and behavioral information in outdoor recreation resource supply inventories: A rationale. In: Lund, G. H.; LaBau, V. J.; Folliott, P. F.; Robinson, D. W., tech. coords. Integrated inventories of renewable natural resources. General Technical Report RM-55. Fort Collins, CO: U.S. Department of Agriculture, Forest Service, Rocky Mountain Forest and Range Experiment Station, pp. 24–31.

Frissell, Sidney S., Jr.; Stankey, George H. 1972. Wilderness environmental quality: Search for social and ecological harmony. In: *Proceedings of the 1972 National Convention;* October 1–5, 1972; Hot Springs, AR. Washington, DC: Society of American Foresters, pp. 170–183.

Godschalk, David R.; Parker, Francis H. 1975. Carrying capacity: A key to environmental planning? *Journal of Soil and Water Conservation.* 30(4): 160–165.

Graefe, Alan R.; Kuss, Fred R.; Loomis, Laura. 1986. Visitor impact management in wildland settings. In: Lucas, Robert C., comp. *Proceedings: National Wilderness Research Conference: Current research;* July 23–26, 1985; Fort Collins, CO. INT-GTR-212. Ogden, UT: U.S. Department of Agriculture, Forest Service, Intermountain Research Station, pp. 432–439.

Graefe, Alan R.; Vaske, J. J.; Kuss, R. R. 1984. Social carrying capacity: An integration and synthesis of twenty years of research. *Leisure Sciences.* 8(3): 395–431.

Hammitt, W. E.; Cole, D. N. 1998. *Wildland Recreation: Ecology and Management,* 2nd ed. New York: John Wiley.

Kuss, Fred R. 1986. Impact ecology knowledge is basic. In: Lucas, Robert C., comp. *Proceedings: National Wilderness Research Conference: Current research;* July 23–26, 1985; Fort Collins, CO. INT-GTR-212. Ogden, UT: U.S. Department of the Interior Forest Service, Intermountain Research Station, pp. 92–93.

Landres, P.; Boutcher, S.; Merigliano, L.; Barns, C.; Davis, D.; Hall, T.; Henry, S.; Hunter, B.; Janiga, P.; Laker, M.; McPherson, A.; Powell, D.; Rowan, M.; Sater, S. 2005. Monitoring selected conditions related to wilderness character: A national framework. General Technical Report RMRS-GTR-151. Fort Collins, CO: U.S. Department of Agriculture, Forest Service, Rocky Mountain Research Station.

Lime, David W. 1970. Research for determining use capacities of the Boundary Waters Canoe Area. *Naturalist.* 21: 8–13.

Lucas, Robert C. 1982. Recreation regulations—when are they needed? *Journal of Forestry.* 80(3): 148–151.

———. 1985. Visitor characteristics, attitudes, and use patterns in the Bob Marshall Wilderness Complex, 1970–82. Research Paper INT-345. Ogden, UT: U.S. Department of Agriculture, Forest Service, Intermountain Research Station.

Manning, Robert E. 1999. *Studies in Outdoor Recreation: Search and Research for Satisfaction,* 2nd ed. Corvallis: Oregon State University. 374p.

Manning, Robert E.; Lime, D. W. 2000. Defining and managing the quality of wilderness recreation experiences. In: Cole, D. N.; McCool, S. F; Borrie W. T.; O'Loughlin, J., comps. *Proceedings: Wilderness science in a time of change conference,* Vol. 4: *Wilderness Visitors, Experiences, and Visitor Management;* May 23–27, 1999; Missoula, MT. RMRS-P-15-Vol-4. Ogden, UT: U.S. Department of Agriculture, Forest Service, Rocky Mountain Research Station, pp. 13–52.

McCool, Stephen F.; Ashor, Joseph L.; Stokes, Gerald L. 1986. An alternative to rational-comprehensive planning: Transactive planning. In: Lucas, Robert C., comp. *Proceedings: National Wilderness Research Conference: Current research;* July 23–26, 1985; Fort Collins, CO. INT-GTR-212. Ogden, UT: U.S. Department of Agriculture, Forest Service, Intermountain Research Station, pp. 544–545.

McCool, Stephen F.; Clark, R. N.; Stankey, G. H. 2007. An assessment of frameworks useful for public land recreation planning. General Technical Report PNW-GTR-705. Portland, OR: U.S. Department of Agriculture, Forest Service, Pacific Northwest Research Station.

McCool, Stephen F.; Cole, D. N. 1997a. Experiencing limits of acceptable change: Some thoughts after a decade of implementation. In: McCool, S. F.; Cole, D. N., comps. *Proceedings: Limits of Acceptable Change and Related Planning Processes: Progress and Future Directions;* May 20–22, 1997; Missoula, MT. INT-GTR-371. Ogden, UT: U.S. Department of Agriculture, Forest Service, Rocky Mountain Research Station, pp. 72–78.

———. 1997b. *Proceedings: Limits of Acceptable Change and Related Planning Processes: Progress and Future Directions;* May 20–22, 1997; Missoula, MT. INT-GTR-371. Ogden, UT: U.S. Department of Agriculture, Forest Service, Rocky Mountain Research Station.

McCoy, L.; Krumpe, E. E.; Allen S. 1995. Limits of acceptable change planning—evaluating implementation by the U.S. Forest Service. *International Journal of Wilderness.* 1(2): 18–22.

Merigliano, Linda L. 1987. *The identification and evaluation of indicators to monitor wilderness conditions.* Thesis. Moscow: University of Idaho, College of Forestry, Wildlife and Range Sciences.

Nilsen, P.; Tayler, G. 1997. A comparative analysis of protected area planning and management frameworks. In: McCool, S. F.; Cole, D. N., comps. *Proceedings: Limits of Acceptable Change and Related Planning Processes: Progress and Future Directions;* May 20–22, 1997; Missoula, MT. INT-GTR-371. Ogden, UT: U.S. Department of Agriculture, Forest Service, Rocky Mountain Research Station, pp. 49–57.

Odum, Eugene P. 1959. *Fundamentals of Biology.* Philadelphia: W. B. Saunders.

Shelby, Bo. 1981. Encounter norms in backcountry settings. *Journal of Leisure Research.* 13(2): 129–138.

Shelby, Bo; Heberlein, Thomas A. 1986. *Carrying Capacity in Recreation Settings.* Corvallis: Oregon State University Press.

Stankey, George H. 1979. A framework for social-behavioral research: Applied issues. In: Burch, William R., Jr., ed. *Long-Distance Trails: The Appalachian Trail as a Guide to Future Research and Management Needs.* New Haven, CT: Yale University, School of Forestry and Environmental Studies, pp. 43–53.

Stankey, George H.; Cole, David N.; Lucas, Robert C.; Petersen, Margaret E.; Frissell, Sidney S. 1985. The limits of acceptable change (LAC) system for wilderness planning. INT-GTR-176. Ogden, UT: U.S. Department of Agriculture, Forest Service, Intermountain Forest and Range Experiment Station. 37p.

Stankey, George H.; Lucas, Robert C. 1986. Shifting trends in backcountry and wilderness use. Unpublished paper on file at U.S. Department of Agriculture, Forest Service, Intermountain Research Station, Forestry Sciences Laboratory, Missoula, MT.

Stankey, George H.; Lucas, Robert C.; Lime, David W. 1976. Crowding in parks and wilderness. *Design and Environment.* 7(3): 38–41.

Stokes, Gerald L. 1986. LAC task force role. In: Lucas, Robert C., comp. *Proceedings: National Wilderness Research Conference: Current Research;* July 23–26, 1985; Fort Collins, Colorado. INT-GTR-212. Ogden, UT: U.S. Department of Agriculture, Forest Service, Intermountain Research Station, pp. 546–547.

Sumner, E. Lowell. 1942. The biology of wilderness protection. *Sierra Club Bulletin.* 27(8): 14–22.

U.S. Department of Agriculture, Forest Service. 1982. ROS users' guide. Washington, DC: U.S. Department of Agriculture, Forest Service. 38p.

———. 1987. Opportunity class allocation for the Bob Marshall, Great Bear, and Scapegoat Wildernesses and map. Forest Service, Flathead, Lolo, Helena, Lewis and Clark National Forests.

———. 1992. Superstition Wilderness: Rock Climbing/LAC Operations and Management. Booklet. Tonto National Forest, Mesa Ranger District.

———. 1996. Environmental Assessment: Amendment to Sawtooth National Forest Land and Resource Management Plan; Sawtooth Wilderness Management Direction. Sawtooth National Forest, Idaho. Ketchum, ID.

———. 1997. Decision notice and finding of no significant impact: Sawtooth Wilderness management direction. Sawtooth National Forest, Idaho. Ketchum, ID.

———. 1998a. San Juan–Rio Grande National Forests wilderness management direction. San Juan–Rio Grande National Forests.

———. 1998b. Superstition Wilderness Implementation Plan. Tonto National Forest. Phoenix, AZ.

———. 2005. White Mountain National Forest Land and Resource Management Plan. Laconia, NH: U.S. Department of Agriculture, Forest Service, Eastern Region.

U.S. Department of Interior, Bureau of Land Management. 1995. Maricopa Complex Wilderness Management Plan, Environmental Assessment and Decision Record. Arizona State Office, Phoenix District Office. Phoenix, AZ.

U.S. Public Law 101-628. Arizona Desert Wilderness Act of November 28, 1990. 104 Stat. 4471.

Wagar, J. Alan. 1964. *The Carrying Capacity of Wild Lands for Recreation.* Forest Science Monograph 7. Washington, DC: Society of American Foresters.

———. 1968. The place of carrying capacity in the management of recreation lands. *Rocky Mountain–High Plains Parks and Recreation Journal.* 3(1): 37–45.

———. 1974. Recreational carrying capacity reconsidered. *Journal of Forestry.* 72: 274–278.

Warren, G. A. 1997. Recreation management in the Bob Marshall, Great Bear, and Scapegoat Wildernesses: 1987 to 1997. In: McCool, S. F.; Cole, D. N., comps. *Proceedings: Limits of Acceptable Change and Related Planning Processes: Progress and Future Directions;* May 20–22, 1997; Missoula, MT. INT-GTR-371. Ogden, UT: U.S. Department of Agriculture, Forest Service, Rocky Mountain Research Station, pp. 21–24.

Washburne, Randel F.; Cole, David N. 1983. Problems and practices in wilderness management: A survey of managers. Research Paper INT-304. Ogden, UT: U.S. Department of Agriculture, Forest Service, Intermountain Forest and Range Experiment Station.

Watson, Alan E.; Cole, David N. 1992. LAC indicators: An evaluation of progress and list of proposed indicators. In: Merigliano, Linda, ed. *Ideas for Limits of Acceptable Change Process,* Book II: *Selected Papers on Wilderness Management Planning Efforts and the LAC Process.* Washington, DC: U.S. Department of Agriculture, Forest Service, Recreation, Cultural Resources and Wilderness Management Staff.

IV. Wilderness Resources, Values, and Threats to Them

Chapter 10
Wilderness Ecosystems

by Jerry F. Franklin and Gregory H. Aplet

Introduction	252
An Ecosystems Primer	252
The Three Components of Ecosystems: Composition, Structure, and Function	252
Dynamics of Ecosystems	254
Disturbance and Succession	254
Historical Range of Variability (HRV)	255
Human Disturbances Affect the Historical Range of Variability	255
The Importance of Scale	256
The Nature of Wilderness Ecosystems	257
What Is a "Wilderness Ecosystem"?	257
Humankind's Historical Role in Wilderness	258
Changes in Wilderness Ecosystems	259
The Pristine Myth	260
Fire Control	260
Grazing	260
Loss of Animal Species	261
Loss of Plant Species	261
Exotic Weeds	261
Introduced Fish and Animals	262
Direct Human Impact	262
Outside Influences	262
Profound Versus Cosmetic Change	262
Managing Wilderness Ecosystems	264
Managing Wilderness as Part of the Larger Landscape	264
Boundary Effects	264
The "Greater Ecosystem" Concept	264
Ecological Restoration in Wilderness	265
Assessing Naturalness for Wilderness Planning	267
Monitoring and Adaptive Management	268
Summary	270
Study Questions	271
Acknowledgments	271
Case Discussion: Protecting Hemlocks in the Southeastern U.S.	271
Case Discussion Questions	272
References	273

Introduction

Wilderness ecosystems have long been revered as places where nature is allowed to run free, unhindered by human activity. Their often large size, relatively intact biota, and high water and air quality have made them important places for ecosystem study. Wilderness areas are among the last places where large predators still roam and may be among the last places where evolutionary forces still operate without significant human impacts. In general, wilderness ecosystems have been found to be among the healthiest systems left in the United States and world, befitting their role, in Aldo Leopold's words, as a "base datum of normality" for a "science of land health" (Leopold 1949, 274). However, despite this reputation, not all wilderness areas are in a healthy ecological condition. Many suffer threats from both inside and outside their boundaries (see Chapter 13). On close inspection, the special qualities that make a place a wilderness ecosystem seem quite elusive. Nevertheless, all wilderness areas contain ecosystems or representative portions of ecosystems, and their ecological characteristics are increasingly becoming issues of management concern.

This chapter considers major features of all wilderness ecosystems and describes the qualities of wilderness ecosystems. The emphasis is on the dynamic nature of ecosystems, their strong internal linkages, and the ways various human activities have affected and continue to affect the "naturalness" of wilderness ecosystems. It examines the concept of "historical range of variability," or the bounded behavior of ecosystems over time, and its relevance to wilderness management, especially in view of emerging information on global climate change. The chapter closes with a brief treatment of several current ecosystem issues in wilderness management. Only when ecosystem dynamics, including interrelationships with humans, are fully understood, can the consequences of alternative management strategies be assessed.

Some readers may think that ecosystem concepts are only marginally related to wilderness management because so much of the book's material concerns people management. However, a large body of literature reminds us that *ignorance of ecosystem concepts—internal linkages among environment, plants, and animals, and successional dynamics—lies at the root of many wilderness and park stewardship problems.* Many concerns about fire-management policies

Fig. 10.1. Clusters of subalpine fir invade a meadow near Big Agnes Mountain, Mount Zirkel Wilderness, Colorado. Wilderness ecosystems are constantly changing as a result of normal successional processes and patterns of periodic disruption. Wilderness management should ensure that natural processes proceed with minimal disruption. Photo courtesy Jay Higgins.

are based on ecosystem considerations (see Chapter 11). Management of large ungulates, such as bighorn sheep, elk, caribou, and carnivores, such as grizzly bears and wolves, inevitably must be based on ecosystem concepts (see Chapter 12). Even limits on human use under the naturalness constraint of the Wilderness Act—biophysical carrying capacity—rest on ecosystem concepts (see Chapters 7, 8, and 9). The idea of regulated human use and management of wilderness so as not to distort naturalness and the idea of humans as an integral but not dominant part of wilderness are important ecosystem concepts.

An Ecosystems Primer

The Three Components of Ecosystems: Composition, Structure, and Function

An ecosystem includes all the organisms of an area, their environment, and a series of linkages or interactions among them. As Odum (1971, n.p.) summarizes, "The ecosystem is the basic fundamental unit in ecology, because it includes both *organisms* and *abiotic environments*, each influencing the properties of the other and both necessary for the maintenance of life." The *abiotic environment* includes climatic conditions, such as temperature and moisture regimes, and inorganic substances supplied by mineral soil. The *biotic community* contains all living organisms within an ecosystem. Green plants provide the ecosystem's entire energy base by fixing solar energy and using simple inorganic substances to build up complex organic substances. Animals, microbes, and fungi use the complex substances produced by plants as a food base and rearrange or decompose them.

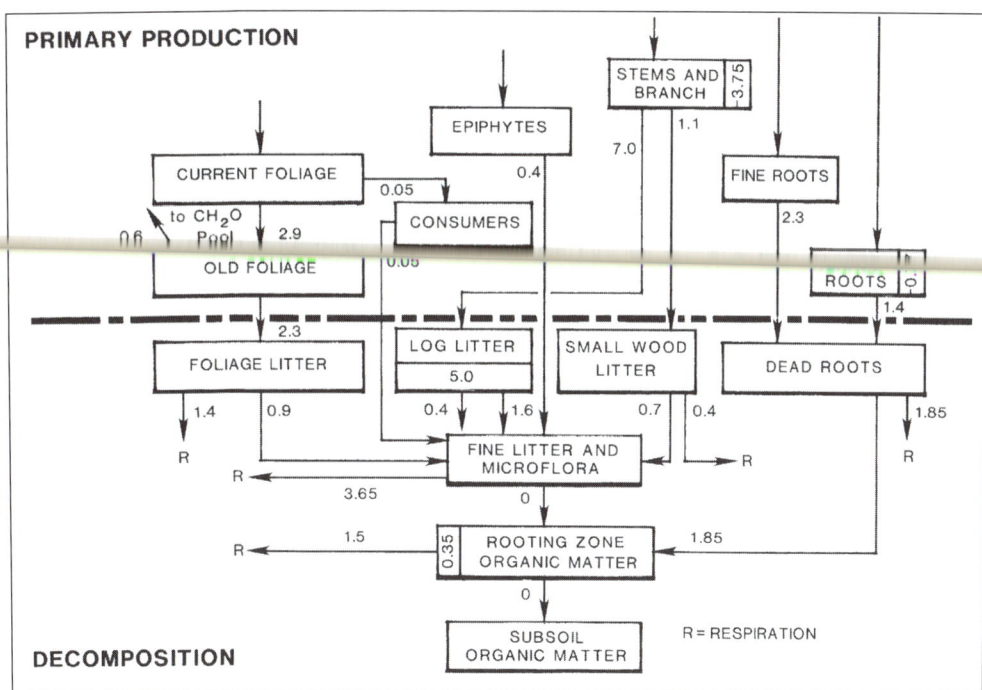

Fig. 10.2. Any ecosystem, such as forest or meadow, contains many linked components, each possessing given amounts of biomass, energy, nutrients, or other material.

All ecosystems can be characterized in terms of their composition, structure, and function. Composition refers to the array of abiotic factors and plant and animal species in various proportions. Each biological component contains given amounts of biomass or energy, nutrients, and other materials—the compartments or states recognized in computer models of ecosystems. In any ecosystem, these elements have a spatial arrangement, or *structure,* that produces ecosystems we recognize as forest, meadow, or savanna.

Ecosystem function refers collectively to the various processes—photosynthesis, transpiration, consumption (by ingestion), and decomposition—going on within an ecosystem. These processes are characterized by *linkages* or *energy and material flows* between parts of an ecosystem. Processes like photosynthesis are carried out exclusively by organisms, but other processes, such as erosion, weathering of rocks, and evaporation, are essentially physical processes. *Environmental factors, such as light, temperature, and moisture, are important in controlling the rates at which the flows take place; biological composition and spatial arrangement largely determine the flow paths.* In addition, not all environmental or abiotic factors are the classical climatic influences of sunlight and precipitation: periodic driving forces, such as fires and floods, also influence ecosystems. These fundamental ecosystem processes are extremely important, for *it is by altering paths and rates of flows that humans most profoundly influence ecosystems.*

Composition, structure, and function are all tightly linked and sustain each other. For example, the species composition of the plant community determines the kind of fuels that may burn in a fire, and their structure contributes to the flammability of the fuel present on a site. Weather and flammability determine the behavior of fire in the ecosystem. Low-intensity fire may have little effect on live vegetation, but high-intensity fire may convert much of the living biomass to dead organic matter as well as consume modest amounts (10 to 20 percent) of the material. Here, composition affects structure, structure affects function, and function affects composition and structure. All ecosystems are the product of these complex interactions.

To illustrate these ecosystem characteristics in the real world, consider a lodgepole pine stand. The stand has a complement of plant and animal species of varying abundance that define its biological composition: lodgepole pine, dwarf huckleberry, fungus, squirrel, and Steller's jay. The spatial arrangement of these organisms and abiotic elements defines the ecosystem's structure: an overstory canopy, shrub layer, litter layer, rooting zone, and nesting territories. By combining elements of composition and structure, we can define the major compartments of the ecosystem: leaves, stems, roots, primary or plant consumers, and secondary and tertiary consumers or predators. Each compartment contains a certain amount of biomass energy, nutrients, and water. Flows or transfers of energy or materials take

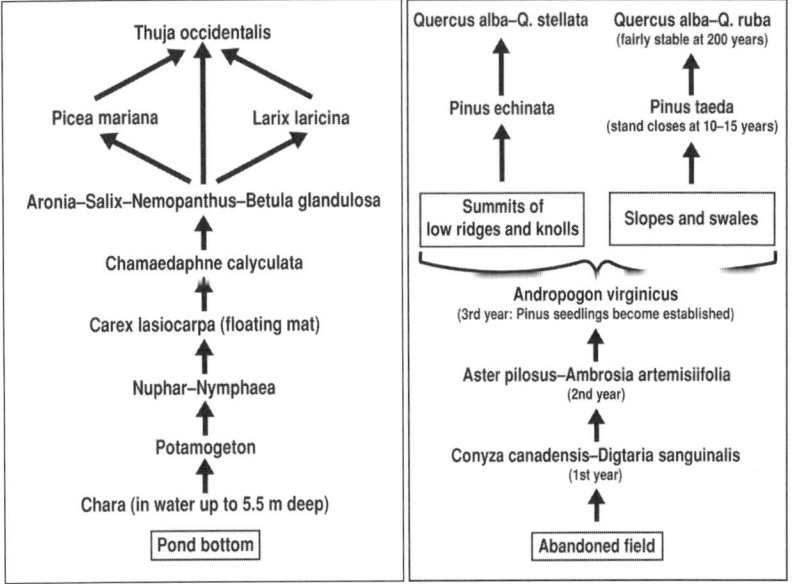

Fig. 10.3. Classical successional sequences are either primary or secondary, and end in a stable climax community; succession of dominants in a pond in Michigan (left); and succession in the Piedmont Plateau (right) (from Daubenmire 1968).

Disturbance and Succession

Strictly speaking, the idea that ecosystems change is not a new one. The description of succession, or the orderly process of development of a plant community, was the focus of some of the earliest work in the field of ecology (Clements 1916; Cowles 1899). It has long been recognized that following the creation of new earth (lava flows, sandbars, mudslides) or the clearing of vegetation (through fire, avalanche, wind throw), plant communities developed in a generally predictable direction from grasses and forbs to more "mature" vegetation, such as forest (Daubenmire 1968; Odum 1971). In these early characterizations of succession, vegetation was believed to develop in a single direction toward a stable "climax," where it would remain indefinitely (Fig. 10.3). This belief was predominant among ecologists and managers at the time of passage of the Wilderness Act. As a result, it was commonly believed that wilderness ecosystems would remain relatively unchanged over time and would be maintained in approximately the condition in which explorers first found them.

Although succession remains an important concept in ecology, scientific studies over recent decades have led to the realization that periodic disturbances—fire, flood, storm, avalanche, pathogens, or geomorphologic processes (such as mass soil movements)—are always at work, allowing very few ecosystems ever to reach a stable state. Periodic disturbances of some type almost inevitably intercede before a climax state is reached. In the Great Lakes states, in most of the ecosystems of the western United States, and in interior Alaska, fire is the primary disrupter, and it occurs at fairly frequent intervals, thereby initiating new successional sequences (explained more fully in Chapter 11). In the redwood region, both flood and fire have been intrinsic environmental elements that periodically alter the ecosystem, although redwood is quite capable of perpetuating itself without them. Hurricanes function as periodic disrupters or rejuvenators of forest ecosystems in much of the eastern temperate forest, as do typhoons in Japan and strong winter winds along the northern Pacific coast. Insects and pathogens

place from one compartment to another—for example, from the leaves and branches to animals by grazing, or from live green plants to the decomposing organisms by litter-fall (Fig. 10.2). Periodically, disturbances such as fire or insect outbreaks kill the live trees, driving large changes in the system. Processes and transfer rates are largely controlled or driven by environmental or abiotic factors such as temperature and moisture availability. *All of the elements of the lodgepole pine stand are connected by the paths of energy and material flow—tree, bird, deer, nematode, and bacterium. Through these flow pathways, ecosystem composition and structure are intimately linked to ecosystem function.*

Dynamics of Ecosystems

The discussion so far has explained the *first of two fundamental ecosystem principles: all parts of an ecosystem are interrelated.* This has been a tenet of ecology since before John Muir issued his famous observation that "when we try to pick out anything by itself, we find it hitched to everything else in the universe" (Muir 1911 n.p.). *A second ecosystem principle that has achieved widespread acceptance is that ecosystems constantly change.* This principle modifies an earlier idea that all ecosystems move toward a stable end point, or "climax" state. This new emphasis on "constant change" has been called a paradigm shift in ecology, or one of the most profound intellectual developments in science (Fiedler et al. 1997).

(e.g., pine beetles, defoliators, and fungal diseases) are the periodic disturbers of some wildland ecosystems, such as some lodgepole pine forests that escape fire.

Historical Range of Variability (HRV)

Although the weight of scientific evidence shows that ecosystems are dynamic, it also shows that they are not infinitely so (Pickett et al. 1992; White et al. 2000). *As long as the major factors controlling ecosystems (i.e., climate, the organism pool, soil parent material, topography, disturbances, etc.) remain relatively constant, the behavior of the ecosystem will likewise remain bounded in its dynamic behavior* (Aplet and Keeton 1999; Landres et al. 1999; Morgan et al. 1994; Swanson et al. 1994). The bounded behavior of an ecosystem over time may be referred to as its historical range of variability (HRV) (Morgan et al. 1994).

For example, the composition and age structure of a forest within a given area are controlled by the regional climate, the species available to inhabit the area, the composition and topography of the soil, and the frequency and intensity of disturbances, such as fires and insect outbreaks. When all these factors remain relatively consistent over time, the composition and structure of the forest will remain relatively bounded. Disturbances may kill some older forest and allow the establishment of younger successional stages, but overall, the forest retains a consistent character (Fig. 10.4).

This historical behavior changes when the factors controlling ecosystems change. The introduction of a new species or the elimination of another, the alteration of fire regimes, and changes in climate drive major changes in the composition, structure, and function of ecosystems—and, thus, deviations from the historical range of variability. For example, tree-ring research in the southwestern United States has shown that fire prevention and control in the past century have eliminated fire from ecosystems that experienced frequent fires for hundreds, if not thousands, of years, resulting in forests of altered structure and composition in many places (Swetnam et al. 1999). In the Sierra Nevada, a warming climate over the past several centuries has shifted the elevation of tree line measurably, changing the character of the vegetation from tundra to forest (Millar 1996). Ecologists are increasingly realizing that *it is impossible to understand the conditions of ecosystems existing today without understanding the dynamic histories that produced them.*

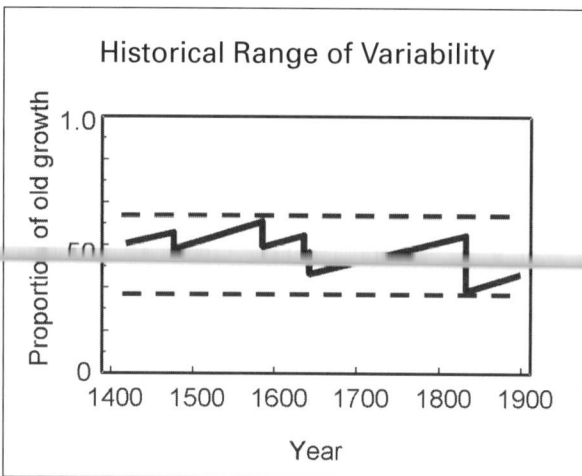

Fig. 10.4. Historical range of variability (HRV) represents the bounded behavior of ecosystems over time. HRV includes not just the range in values, but also the duration and rate of change of ecosystem conditions (Aplet and Keeton 1999).

Accompanying the general appreciation for the dynamics of ecosystems has been a growing sense among scientists and land managers that the best way to sustain the composition, structure, and function of an ecosystem is to try to prevent its behavior from departing dramatically from its HRV (Manley et al. 1995; Quigley and Arbelbide 1997). The emphasis is on sustaining into the future the dynamics that characterized ecosystems before the dramatic, human-induced modifications of the recent past. The challenge for managers is to sustain the natural variability while minimizing the disruptive influence of *technology-driven and human-caused* change, such as exotic species introduction, fire prevention, pollution, and global climate change and consumptive practices, such as intensive timber harvest.

Human Disturbances Affect the Historical Range of Variability

A Millennium Ecosystem Assessment (2005, 1) was compiled from extensive published research information and an international team of scientists who conclude that "over the past 50 years, humans have changed the ecosystems more rapidly and extensively than in any comparable period in human history, largely to meet rapidly growing demands for food, fresh water, timber, fiber, and fuel. This has resulted in a substantial and largely irreversible loss of diversity of life on Earth."

The change agents that affect and threaten wilderness areas are numerous and growing (see Chapter 13). The future range of variability depends on both the HRV and the human-caused changes that affect ecosystems now. The future range of variability is modified by the complex interaction of the ecosystem components with an increasing number and growing extent of human-caused disturbances. Wilderness areas can serve as benchmarks against which estimates of the degree and direction of some changes in ecosystems can be measured and compared.

The Importance of Scale

An ecosystem can be large or small, based on how we define its boundaries. It can be an aquarium, forest stand, watershed, entire park or wilderness, biome, or the entire world. It is necessary only that the area have the characteristics of an ecosystem—constituent organisms interacting with their environment, energy flow, and material cycles. *Scientists and managers often find a watershed to be a useful unit for ecosystem study because it is relatively easy to define the physical boundaries and to measure many of the flows into and out of the ecosystem*; it also incorporates both terrestrial and aquatic elements and allows study of their interaction. Nevertheless, ecosystems can be of widely varying size or scale, depending on the objectives of the work.

In wilderness management, the size of an ecosystem unit of interest will vary. In some cases, it might be an individual stand of trees or community of plants, such as a meadow. When locating camping areas or determining carrying capacity, it might be a small lake basin or watershed—or the "heavy use areas," "trail corridors," or "subalpine zone" described for planning purposes. For many planning and management activities, the entire wilderness can be the ecosystem unit. With respect to habitat for large mammals, such as elk and grizzly bears, the relevant ecosystem might need to transcend wilderness boundaries to take in a large river basin or more. Determining the size of an ecosystem unit or the number of units to be recognized in a wilderness will depend on management objectives and the kinds of problems, both biological and social, facing managers.

It is also important to understand that *ecosystems are nested inside of larger ecosystems in a hierarchy*. One of the most obvious and convenient hierarchies to recognize is a watershed, in which the watersheds of small streams are nested inside those of larger creeks, which are nested inside of the watershed of a large river. Its scale affects the behavior of each of these ecosystems. For example, a localized rainstorm may cause a small creek to flood, but that flood may have no noticeable

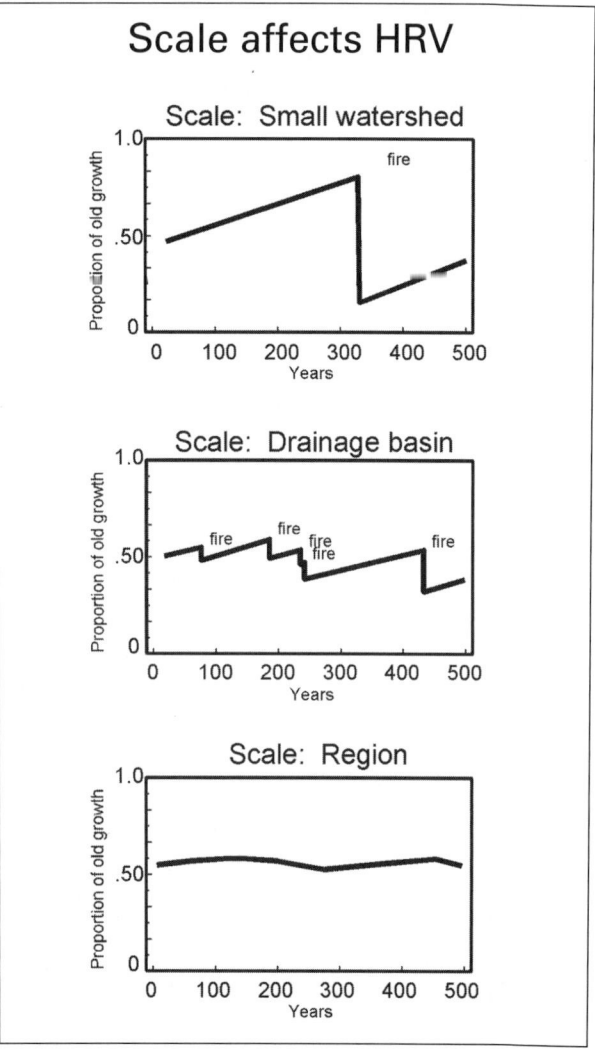

Fig. 10.5. The range of variability in composition, structure, or function changes with the scale of observation. Here, variability in the proportion of an ecosystem in old-growth forest declines (top, small watershed) as the size of the ecosystem increases (middle, drainage basin; bottom, region) (Aplet and Keeton 1999).

effect on the big river downstream. Conversely, heavy snows over the region may cause no flooding in headwater streams, but the cumulative effect of all that snowmelt may spell disaster downstream.

The geographical and temporal scale over which an ecosystem is described is critical to the description of HRV (Aplet and Keeton 1999; Landres et al. 1999). For example, fluctuations in the amount of old-growth forest may be quite large within a small area, as the forest grows following disturbance to a great age, while at the scale of a region, the fluctuation is dampened—old growth comes and goes on an annual basis as fires burn and winds blow (Fig. 10.5). The perceived dynamics of ecosystems are directly affected by the scale of observation.

The Nature of Wilderness Ecosystems

There are a number of ways of conceiving of wilderness, ranging from the references to remote or undeveloped places to specific descriptions of areas in the NWPS, lands legally designated as wilderness. Areas now included within the NWPS range from large and apparently "pristine" areas in Alaska to small and historically altered places, such as may be found throughout the Appalachian Mountains of the eastern United States. Thus, even within the United States, it is hard to describe in general terms the kind of ecosystems that we call wilderness.

One way to look at the nature of wilderness ecosystems is to examine the content of the lands we recognize as wilderness. The first attempt to do this in a formal way was during the RARE II process (see Chapter 5), in which the Forest Service determined that 131 "ecosystems" were "adequately represented" in the NWPS in 1978, accounting for 50 percent of the 261 ecosystems eventually recognized by Bailey (1980). Davis (1989) repeated the analysis almost a decade later, and found that the NWPS, despite having more than tripled in size, included only 157 ecosystems, or 60 percent of the total. Using a more simplified system involving only thirty-five ecosystem "provinces" (Bailey 1995), Loomis and Echohawk (1999) determined that, despite the growth of the NWPS to more than 100 million acres, barely one-third of the ecosystems of the lower forty-eight states had more than 1 percent of their area protected as wilderness. Studies such as these help to describe the ecological contents of the lands we have designated as wilderness, but they leave unanswered the fundamental question of what defines a wilderness ecosystem.

What Is a "Wilderness Ecosystem"?

It is tempting, given the U.S. Wilderness Act's (U.S. Public Law 88-577) description of wilderness as areas "where man himself is a visitor who does not remain," to think of wilderness ecosystems as areas of earth that are unaffected by human activity. However, ecosystem science over the past several decades has shown that no such place is left on earth. Changes in atmospheric carbon dioxide, global nitrogen cycles, and climate ensure that even the most remote places on the planet have been touched by human-caused environmental change (Vitousek et al. 2000). To be sure, some places are more affected than others, but as the history of wilderness designation in the eastern United States shows (Chapter 5), there is no threshold level of human alteration that defines a wilderness ecosystem. In the eastern U.S., there was a great debate as to whether its apparently natural remaining lands qualified for wilderness designation or belonged in a separate "wild-areas" system (see Chapter 5). Subsequently, even farmed and otherwise altered lands were included in the Wilderness System under provisions of the so-called Eastern Wilderness Act (U.S. Public Law 93-622).

Despite the variety of ecosystems in the NWPS, one thing unites them all. According to section 4(b) of the Wilderness Act, they must all be administered "to preserve [their] *wilderness character*" (emphasis added). State wilderness laws may define the standard somewhat differently (Chapter 6), and, as Chapter 3 makes clear, conceptions of wilderness vary around the world, but generally, it can be said that once land is designated as wilderness, it must be managed to sustain its special qualities as wilderness. *For the purposes of this chapter, we define wilderness ecosystems as those*

Fig. 10.6. Ecosystems represented in the NWPS: semiarid Gila Wilderness, New Mexico (left); the sere landscape of Craters of the Moon National Monument, Idaho (center); and deciduous forest of the Presidential Range–Dry River Wilderness, New Hampshire (right). Left photo courtesy U.S. Forest Service; center courtesy National Park Service; right courtesy U.S. Forest Service.

managed to sustain their wilderness character, regardless of their current or historical condition.

Although this definition helps bring focus to wilderness ecosystems, it does not help resolve what is meant by "wilderness character." Fortunately, section 2 of the Wilderness Act provides some clues. According to section 2, wilderness is "untrammeled" by humans, or free from tight human control. Wilderness also "appears to have been affected primarily by the forces of nature" and is free of permanent developments or habitation. Thus, wilderness ecosystems must be managed to be free from human controls and preserved and protected in their natural condition. *Perhaps no language from the Wilderness Act better captures these twin goals than the description of wilderness as "retaining its primeval character and influence." Wilderness is land that retains not only the composition and structure of its historical ecosystems, but also the influences or processes that shaped those ecosystems over time.*

Elsewhere in the book, we have described wilderness character in terms of naturalness and solitude. The ability of the land to provide opportunities for solitude is clearly an important determinant of primeval influence. With increased occupation of the land comes increased control and use and, inevitably, change. Similarly, the remoteness of wilderness is important to minimize the influence of "mechanical sites and sounds and smells" (Marshall 1933). *However, for the purposes of this chapter, the critical determinant of primeval influence is the degree to which ecological processes are free to operate according to historical rhythms. A place is more wild where fires continue to burn, floods erode and create, and migratory animals still roam. Only where ecological processes remain outside of direct human control can the land be called "untrammeled."*

However, ecosystems do not exist in isolation. All are nested within larger ecosystems and are affected by what goes on beyond their boundaries. Although they may be managed to preserve their wilderness character, wilderness ecosystems sit within a larger matrix, a spectrum or gradient of lands that may not be managed for the same goal. On these surrounding nonwilderness lands, habitat manipulation, road construction, timber harvest, fire suppression, water development, pest management, and other activities are likely to affect the composition, structure, and function of entire ecosystems. Under U.S. law—the Colorado Wilderness Act of 1980 (U.S. Public Law 96-560)—Congress has decided that lands adjacent to wilderness cannot be managed as buffers to protect wilderness character. Therefore, *only land within wilderness boundaries may be defined as wilderness ecosystems: that is, those parts of larger ecosystems that are managed to preserve their wilderness character.*

Wilderness character, or the "wildness" of an ecosystem, increases as both its composition and structure tend toward primeval character and as its function tends toward the "untrammeled" (i.e., primeval influence). Any ecosystem can be described in terms of these two qualities, with wilderness representing the highest expression of both qualities (Fig. 10.8). Nonwilderness lands may be tightly controlled to produce natural conditions, such as the highly managed "natural" habitat exhibits at some zoos, while other ecosystems may decline in naturalness in an untrammeled condition, such as where exotic species alter ecosystem composition. The greatest wilderness character is achieved where natural conditions are attained in an untrammeled state. Thus, *ecosystems simultaneously expressing both primeval character and influence are the best expressions of wilderness ecosystems.*

Humankind's Historical Role in Wilderness

The attention to HRV and natural processes that is so important to wilderness management necessarily raises the question of the human role in determining primeval character and influence. *Are humans an integral component of the wilderness and are their influences natural or unnatural?* The answer is relative rather than absolute. Native Americans unquestionably played a role in shaping the wilderness landscapes of North America (Cronon 1983; Denevan 1992; Krech 1999; Landres et al. 1999; McCann 1999a, 1999b; Whitney 1994). Humans burned and hunted, raised crops, and built settlements. In some cases, large population centers, agriculture, and elaborate irrigation systems transformed the land. In other cases, aboriginal use of fire and hunting practices clearly influenced the composition and structure of vegetation and fauna. Current estimates place the population of North America at the time of European contact at about 20 million, compared to the 2007 population level of more than 300 million. Just as today where populations are dense, the effect of human technology on ecosystems was substantial; elsewhere it was not, or it joined a background of lightning, floods, tree falls, and other influences on ecosystems. There, humankind was an influencing factor, but only one of many (Krech 1999; Vale 1998).

This situation changed with the arrival of European settlement and technology. Plows, saws, guns, and livestock provided the means for a wholesale transformation of large parts of North America even before the industrial revolution (Cronon 1983; Whitney 1994). Since then, modern machinery and chemicals have added to the scope and scale of change. Current human technology is extremely powerful and in the short term can buffer humans from the ecosystems in which they live. *"Natural" human influences may be considered those that have been elements in the long-term evolution of precolonial ecosystems—present for hundreds or thousands of years, such as hunting, gathering, burning, and clearing of floodplains for agriculture—influences contributing to the HRV in ecosystems.* The impacts of modern humankind are not this type. The wild ecosystems we are concerned with here—those in designated wilderness—are intended to be managed to protect them from the influence of modern, transformative technologies.

In summary, humans are a natural part of wilderness, but modern technology is not. Modern technological forces are so novel and so dramatic in their effects on ecosystems that they cannot be accepted as a natural component of wilderness. For this reason, the Wilderness Act legally restricts exposure to modern technology.

Changes in Wilderness Ecosystems

Realistically, we must recognize that no completely unaltered ecosystems are left on this planet. The effects of modern humans and their products are pervasive. This is as true in the Antarctic, where DDT is found in the tissues of penguins, as it is in Central Park in New York City. None of the world remains unaltered. Even in areas we perceive as pristine, modern humans have already had a significant impact. Nevertheless, human impacts vary in their effect on ecosystems such that in every landscape some parts are less affected than others. In a few places, such as Alaska's Arctic National Wildlife Refuge, wilderness ecosystems are large enough and whole enough that they still function largely according to natural influences and reflect historical patterns—the primary threats to their wilderness character are pervasive, global climatic changes and the invasion of modern technology. *In most places in the lower forty-eight states, though, wilderness ecosystems have suffered exposure to some degree of unnatural influence that has forced, or threatens to force, a departure from their historical range of variability.*

Fig. 10.7. Wilderness ecosystems exist at a variety of scales: from small islands in Washington's San Juan Islands Wilderness (top left), to vast landscapes in the Absaroka-Beartooth Wilderness, Wyoming (top right); to an arid desert game range in Nevada (bottom left), and a remnant roadless area in Monongahela National Forest, West Virginia (bottom right). Top left photo courtesy Geroge Devan, FWS; top right courtesy L. J. Prater, U.S. Forest Service; bottom left courtesy David B. Marshall, FWS; and bottom right courtesy U.S. Forest Service.

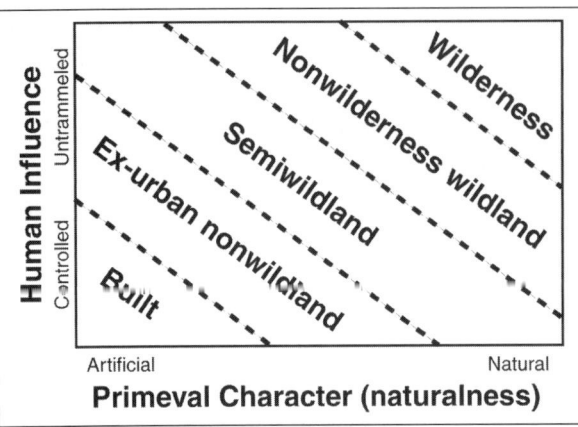

Fig. 10.8. Wilderness character (wildness) can be said to increase in two dimensions: along a gradient from artificial to natural and along a gradient of human influence from controlled to untrammeled.

The Pristine Myth

William Denevan (1992) wrote a controversial article in the *Annals of the Association of American Geographers* titled "The Pristine Myth: The Landscape of the Americas in 1492" in which he argued that the Native American influence in pre-Columbian North America was so pervasive that it cannot be said to have ever been (untrammeled) wilderness, at least since the arrival of humans 10,000 to 20,000 years ago. This article did much to illuminate misconceptions of the precolonial landscape as "untouched," but it failed to address the fact that human influence since 1492 has been of a dramatically different sort than that which preceded it for the previous hundreds to thousands of years. *Modern influences now threaten the ecosystem composition, structure, and function that characterized the continent for most of the last 10,000 years.*

Ironically, perhaps because of the extent of human-induced change that has attended the conquest of North America, we tend to think of our remaining natural areas as remnants of pristine America, and wilderness most of all. The language of the Wilderness Act, in defining wilderness as "land retaining its primeval character and influence," indicates that the act's framers saw wilderness as remnants of pristine, historical conditions. *However, in many places, even designated wilderness does not exist in a pristine state. It has been modified directly or indirectly by modern human activities.*

What are some of the changes in ecological influences that have been wrought by modern humans? Following is a brief review of some of these influences.

Fire Control

A visible and well-known ecological influence on wilderness is fire control by managers (see Chapter 11). Fires at some intervals are a natural feature of almost all of our forest and range ecosystems. It has been argued that western coniferous forests have actually evolved in such a way as to increase flammability and ensure periodic burning. Certainly many tree species of great interest to us depend on fire for their perpetuation—giant sequoia (Kilgore 1973), Douglas-fir and ponderosa pine on some sites (Cooper 1960; Habeck and Mutch 1973), red pine and white pine in the Great Lakes states (Frissell 1973), the "closed-cone" pines (Vogl 1973), and several pines of the southeastern United States, notably longleaf pine (see Chapter 11).

Elimination of fire, when periodic low-intensity fires are naturally part of an ecosystem's environment, can have catastrophic consequences. First, a successful fire-suppression program will change the composition of ecosystems in many cases. Generally, with fire exclusion, forests become denser with less herbaceous undergrowth. Not only will the plant composition of the forest change, but the habitat for many animal species also will be altered with consequent changes in the animal composition of an area. Organic matter and nutrients accumulate in slowly decomposing woody plant material on the forest floor. Continuity develops between the crown and surface fuels as the density of saplings and poles increases. One result is likely to be a loss of early-stage successional communities and the growth of shade-tolerant tree species, ultimately leading to a catastrophic fire to which the ecosystems in the area are not adapted.

Grazing

Grazing by domestic animals is another unnatural influence in wilderness ecosystems introduced by modern humans. Concentrated grazing by sheep or cattle, or even by large wild ungulates such as elk or mountain goats, introduced to regions where they are not native or maintained at artificially high populations, is not part of the natural regimen under which many of our wilderness meadows and savannas have evolved. Thus these ecosystems may be poorly adapted to high rates of grazing. Changes in the composition of meadows may occur due to preferential feedings on, and/or sensitivity of, some native plant species. The site may be physically altered due to soil compaction, accelerated erosion, and resulting changes in water table levels. Exotic plant species, introduced through livestock, horse feed, and stock feces, may become established due to altered conditions. Animal pathogens may be introduced into native ungulate populations with catastrophic results—such as when bacterial disease of domestic sheep, was introduced into bighorn sheep populations (see Chapter 12 for other wildlife examples).

Some changes, such as addition of exotic species and site degradation, can be particularly important because they can permanently alter the ecological potential of the site—its ability to return to or stay within its HRV. An outstanding example of the effects of grazing on the ecosystem is found in the sagebrush-bunchgrass and Palouse prairie types of eastern Washington and Oregon (Daubenmire 1970). Heavy grazing by large herbivores reduces the vigor of, and can eliminate, the bunchgrass dominants such as blue bunch wheatgrass and Idaho fescue. Exotic cheatgrass or Kentucky bluegrass invades, and, once established, native species can never recolonize the sites. Exacerbating the situation, the exotic grasses facilitate the spread of fire, which causes further mortality of the native species.

Loss of Animal Species

Eliminating animal species is another way modern humans have altered, and continue to alter, the naturalness of wilderness ecosystems. The best-known examples are the elimination of predatory animals—for example, wolves, wolverines, grizzly bears, cougars, and various raptors. Elimination or reduction of these species directly affects species composition of wilderness ecosystems. Furthermore, loss of large predators, such as wolves, can have significant ecological impacts when these natural controls on large herbivore populations and medium-sized predators are removed. Irruptions of herbivores can result in overgrazing, and "mesopredator release" (of raccoons, opossums) can have devastating impacts on forest-nesting birds. Because hunting and trapping also reduce animal population levels, these activities can alter the dynamics and states of wilderness ecosystems. For this reason, national park wildernesses (where hunting and trapping are prohibited) may be of greater value for research on natural ecosystems compared to wildernesses where hunting and trapping are allowed.

Loss of Plant Species

Plant species have also been eliminated or drastically reduced in some areas. Chronic pollution by such gases as sulfur dioxide and ozone can affect compositional changes in plants or animals, or even rates of productivity or decomposition. These losses significantly alter the ecosystem's structure and its basic functions of energy fixation and nutrient conservation. Disregarding grazing and fire, losses in plant species are most frequently caused by human introduction of exotic pathogens. Examples that affect designated wildernesses include the nearly complete loss of American chestnut to the nonnative chestnut blight, and substantial losses of five-needle pines to exotic white pine blister rust. The loss of white bark pine to blister rust is of great concern, because grizzly bears in the Yellowstone area and elsewhere depend on white bark nuts as a major seasonal food supplement. A current catastrophic example is the destruction of eastern hemlock forest by the introduced wooly adelgid, which is advancing farther north due to warmer winter weather patterns. Exotic pathogens of this type can have extremely profound and often poorly perceived effects on otherwise pristine landscapes. Once introduced, exotic pathogens are virtually uncontrollable, because they lack natural enemies to keep their populations in check. Thus, if the host species lack significant genetic or other mechanisms for resistance, effects (alterations) are essentially permanent.

Many forest types in North America have been dramatically impacted by the arrival of exotic diseases and insects, altering their composition and natural functioning. Yet, these impacts typically go unnoticed by most visitors who are not trained in the natural sciences.

Exotic Weeds

Invasive, exotic weeds are also degrading the naturalness of wilderness ecosystems (Asher and Harmon 1995). In a survey published in 2000, 15 percent of wilderness managers identified invasive plants as among their top ten management concerns, particularly in California, the Rocky Mountains, and eastern forests—although the true extent of the problem is unknown, because 70 percent of respondents reported that they do not monitor plant invasions (Marler 2000). Invasive plants can cause a host of problems for wilderness ecosystems, including displacement of native species, promotion of nonnative animals, increased erosion, and alteration of nutrient cycles and fire regimes

(Randall 2000). Particularly vexing are the speed and permanence of some of these invasions. For example, in northern California, yellow star thistle is estimated to have spread from 1 million acres in 1977 to 10 million acres in 1995, and in the Selway-Bitterroot Wilderness in Idaho, spotted knapweed continues to spread, although grazing, which is often implicated in the spread of exotic weeds, has not occurred for more than fifty years (Asher and Harmon 1995).

Introduced Fish and Animals

Fish and animals introduced into pristine landscapes, often to make the areas more attractive recreationally, have affected wilderness ecosystems. The introduction of sport fish to originally "barren" lakes and streams is an outstanding example. There can be no question that when fish introductions of this type are successful, the affected aquatic ecosystem is markedly changed. An entirely new component may be added if a fish at the same trophic level was not previously present, and native fish may be displaced or eliminated entirely. The native brook trout in the Great Smoky Mountains is now found only in high, isolated streams. An extensive study of stocked and nonstocked lakes in the Sierra Nevada Mountains of California documented important impacts on the rare mountain yellow-legged frog after introduction of exotic sport fish (Matthews and Knapp 1999). Elsewhere, mountain goats live in wilderness where they are not native. Alterations in the composition of other organisms, and paths and rates of energy and nutrient flows, are to be expected from the introduction of nonnative fish and animals.

Direct Human Impact

Direct impacts from human use tend to be concentrated in relatively small areas but may be acute, including soil compaction, destruction or alteration of vegetation at campsites, and impacts on wildlife—dispersing some and attracting others, such as jays, mice, and sometimes bears. Although heavily impacted locations may be small relative to the total area, they may be among the most popular, aesthetic, and frequently visited. Clearly, in local areas, input of human wastes can reach levels sufficient to cause health concerns about pollution. In such cases, the threat to human health probably becomes a management concern before the threat of eutrophication (see Chapter 15 on ecological impacts of wilderness recreation use). In wilderness in the eastern United States, past human activities have sometimes included logging, road building, and clearing land for agriculture. Modern humans are introducing a variety of substances, such as plastic and other refuse to many wilderness areas; petroleum products in the Boundary Waters Canoe Area Wilderness, where motors are allowed; and exhaust in the wilderness areas of Alaska, where airplanes are allowed.

Outside Influences

Introducing unnatural substances to wilderness ecosystems occurs on a large scale by activities outside of the wilderness. Pollutants present in the atmosphere, such as sulfurous gases and pesticides, may be deposited in rain or as dust or brought in by migrant organisms. Atmospheric pollution, including acid rain and ozone, is of increasing concern, even in western wilderness, and is the subject of major research and monitoring programs. In some cases, pollution is clearly affecting the ecosystem and producing pathological effects on organisms, such as damage to trees in the Los Angeles basin and several southern California wildernesses, and acidification of lakes and streams in the Appalachian and Adirondack Mountains in the eastern United States. For example, Miller (1973) describes accelerated mortality of ponderosa pine in the San Bernardino Mountains of California, and acid precipitation has been blamed for accelerated mortality in spruce and fir stands in the northeastern and southeastern United States.

Profound Versus Cosmetic Change

The ecological importance of human activity is a function of the magnitude and permanence of its effects. In assessing its importance, one needs to know if the change in ecosystem composition, structure, or function is large or small, and if it is transient or essentially permanent. *In general, managers and visitors most easily perceive changes in composition and structure, but changes in function may be more important in the long run, though they are often very difficult to identify until they have progressed beyond correction.*

Introduction of the chestnut blight resulted in permanent elimination of the American chestnut (Shugart and West 1977). The hardwood ecosystems of which it was a part underwent a permanent change in composition and population structure, not only as a result of the loss of a dominant tree but also because of adjustments in animal species dependent on chestnuts for food. Rates of energy and nutrient cycling, forest structure, and successional sequences were altered as the ecosystems adjusted and other species filled the gaps. Despite these changes, visitors to forests from which chestnuts have been eliminated do not typically perceive these as unnatural or human-altered ecosystems—the space they once filled has been taken over by other hardwood tree species. Likewise, presence of planted sport fish is rarely perceived by visitors as an unnatural influence, although the largely unseen effects on the structure and function of the lake, pond, or stream ecosystem may be significant and permanent.

On the other hand, visitors are more likely to be aware of the elimination of grizzly bears and wolves from many wilderness ecosystems. Without reintroduction or natural recovery, the composition of these ecosystems has been changed permanently, and where these large predators no longer keep prey populations in check, ecosystem processes, such as nutrient cycling and erosion, may be altered substantially. Despite these important effects, the visitor is probably more likely to notice their absence as a missing sound or an enhanced sense of safety—not because of their direct effects on the ecosystem.

The effect of adding materials to an ecosystem may be either profound or cosmetic. For example, burning fossil fuels along the eastern seaboard has added unprecedented amounts of nitrogen to the forests of New England, potentially driving a cascade of ecosystem-level responses, including soil and stream acidification, nutrient imbalances, and reduced growth rates (Fenn et al. 1998). In contrast, the main concern over air pollution in the wilderness ecosystems of the Southwest has been its effect on regional haze, diminishing views and the quality of visitors' experiences (Tonnessen 2000). Ecosystem response is affected by the amount and kind of material added, the environmental regime (e.g., precipitation, temperature), nutrient status, and the susceptibility of the constituent organisms.

Logging historically removed a considerable amount of material in some wildernesses, particularly in the eastern United States and Upper Midwest and Great Lakes states. Ecosystem composition, structure, and function were drastically altered; however, many aspects of system function, such as conservation of nutrients, quickly recovered to near prelogging levels. The species composition and structure of the forest will be measurably different for decades to come, but the forest has returned, and to most visitors, eastern wilderness areas possess wild character. It is probable that the rapid vegetative regrowth of the summer-wet and humid forest environments of the Appalachian Mountains shortens the duration of the logging impacts on structure and composition. Elsewhere, such as in arid western coniferous forests, slower growth and recovery ensure that ecosystem-level impacts and visitor perceptions will be altered for decades or centuries.

Physical alteration of a site may have profound ecological effects, such as where a roadcut permanently alters ecosystem hydrology, or it may be relatively benign, such as constructing a bridge across a stream. In wilderness, physical alteration tends to be relatively localized, such as the compaction of campsites, or low-impact, such as trail maintenance. However, physical alterations are easily noticed by the public and degrade visitors' experiences (see Chapter 15).

Alterations of natural disturbance regimes, through fire control programs (Chapter 11), water impoundments, and so forth, are some of the most pervasive and important changes humans have wrought on wilderness ecosystems. Yet, effects are gradual and, initially, are not as drastic and noticeable as logging impacts, for example. Some research (Chang 1996; Quigley and Arbelbide 1997) suggests that higher elevations in the western United States, where most large wilderness areas are located, have been less affected by fire-suppression programs than lower elevations, but Hessburg et al. (1999) found that fire suppression has altered forest spatial patterns throughout the interior Columbia River basin.

Finally, human activity can modify the basic environmental regime of light, temperature, moisture, sound, and so forth. In some cases, such as where humans alter regional climate, the effects on ecosystems are likely to be profound, although their effects will take time to be expressed and may not be noticed by visitors. Other factors, such as light pollution from nearby developments or noise from aircraft, may have less of an effect on the ecosystem but are likely to be readily noticed by visitors. In some cases, noise, particularly from aircraft, may cause stress in wildlife populations and is suspected of disrupting processes like reproduction.

Ecologically, the most important human alterations of natural ecosystems are not necessarily the most obvious. What a visitor perceives as natural may have been profoundly and permanently altered. The degree of naturalness in ecological terms is a function of ecosystem factors and, if quantified, will often differ from the visitors' perception of naturalness. Here, *both managers and users need to broaden their perspectives so they can distinguish between cosmetic and profound ecological impacts* (see Franklin 1987).

Managing Wilderness Ecosystems

What does the foregoing tell us about managing wilderness ecosystems? First, that wilderness ecosystems exist at a number of scales, with linkages across boundaries to nonwilderness ecosystems. They are complex, consisting of many components, arranged in certain ways, according to historical influences. *It tells us that we expect a lot from our wilderness ecosystems, because we call on them to remain as insulated as possible from unnatural influences, both free of the domination of modern technological human society and retaining patterns and processes reflecting their HRV.*

Ultimately, maintaining healthy wilderness ecosystems will require following the essential steps of adaptive management: ecosystem assessment, setting goals, developing and implementing a management strategy, monitoring the ecosystem to determine if goals are being met, and modifying management appropriately. *For wilderness managers, key steps will be assessing ecosystem naturalness and monitoring to determine trends relative to the HRV. Management actions must focus on seeing that processes, such as disturbance regimes, remain within historical ranges* (Parsons 1999). Most importantly, follow the wilderness management principle of doing only what is necessary and using minimum tools to maintain ecosystem processes and disturbances within the area's HRV.

Managing Wilderness as Part of the Larger Landscape

One of the most important lessons we have learned from studying wilderness ecosystems is that they do not exist in isolation. *Wilderness ecosystems are embedded in larger ecosystems and interact with them through material and energy flows.*

Boundary Effects

For most of the history of wilderness and primitive-area designation, decision makers and managers relied on a belief that ecosystem protection required only that we designate a protected-area boundary and stop activities inside that threatened its integrity. With time, though, we have come to understand that *boundaries do not necessarily protect ecosystems.* Many threats to wilderness are oblivious to boundaries, and, in some cases, boundaries can create problems (see Chapter 13; Knight and Landres 1998; White et al. 2000).

Fire is an excellent example. Managers may deem that inside of wilderness, fire should be allowed to play its natural role in the ecosystem, but outside, it should be suppressed to protect nonwilderness resources and private property. This policy ignores the fact that many fires start at low elevations, outside of wilderness, and burn upslope. Fire suppression outside of wilderness may prevent some natural fires from ever entering wilderness, thus disrupting a historically natural and basic ecosystem function. The problem also works in the opposite direction, where fires inside of wilderness are suppressed for fear they will spread outside (Parsons 2000).

Another example is wildlife movement. Wilderness may be established to protect certain places, but if it does not also protect migratory corridors, wildlife will continue to leave the protected area and thereby be subject to unnatural influences (see Chapter 12). Such is the case near Yellowstone National Park, where hunters may shoot native wildlife as they migrate from the protected park area in search of winter range. In Arizona, desert bighorn sheep are regularly killed in automobile accidents when they leave wilderness to reach traditional watering holes.

The "Greater Ecosystem" Concept

Realizing that existing protected areas are rarely large enough to protect all of the components of the ecosystem has given rise to *the "greater ecosystem" concept,* which accounts for all of the resources outside a protected area that are necessary to sustain the protected area. One of the first applications of this was in the region of Yellowstone National Park. There, after studying the movement patterns of grizzlies, bison, and elk, ecologists realized that the park, even at 2 million acres, was insufficient to sustain viable populations of many species. They found that animals were using resources beyond even the wilderness areas surrounding the park, including considerable private property, where they were vulnerable to many threats, ranging from habitat destruction to direct mortality. In response, managers and citizen activists began to promote the

"Greater Yellowstone Ecosystem," which is really more a way of thinking about the interconnections across the region than a fixed line on a map. Today, the Greater Yellowstone Ecosystem (GYE) has spawned a number of programs and management reforms aimed at protecting all of the resources in the surrounding region necessary to sustain the park's ecosystem.

Another example of "thinking outside the lines" is the Northwest Forest Planning process, crafted in 1994 (see www.reo.gov/general/aboutNWFP.htm) to protect the old-growth ecosystems of the Pacific Northwest, including the habitat of the northern spotted owl. This plan, although not specifically designed to protect a wilderness ecosystem, identified and protected critical habitat in a series of reserves distributed across the region. It also explicitly addressed the question of links between reserves by placing rules on what activities could be conducted on the intervening lands. As such, it was designed with the intent of protecting ecosystem composition, structure, and function across a region.

In the future, if wilderness ecosystems are to be sustained as largely retaining their HRV and pre–western settlement (natural) influences, they must be managed as part of a "greater ecosystem." They cannot be managed in isolation. Rather, policies on surrounding lands must complement and integrate with wilderness management goals. Fire policies outside wilderness should be supportive of natural processes inside wilderness. Important landscape linkages and corridors should be maintained. If policies such as these can be established, wilderness can play a more vital role in the health of the entire ecosystem, not left to struggle against overwhelming external threats. *However, all this must happen through the integration of resource management programs, since law prohibits establishing buffer zones around wilderness.*

Ecological Restoration in Wilderness

For ecologists, one of the most compelling arguments for wilderness has long been its value to science. Aldo Leopold (1949, 196) wrote of this value and ultimately concluded that wilderness is an essential reference point, providing "a base datum of normality, a picture of how healthy land maintains itself." An ecological baseline against which to compare the effects of management has long been one of the great values of wilderness.

Historically, it was thought that if we just left wilderness alone, it would continue to provide a control against which to judge management and a model of healthy land. Now, recognition of changes to wilderness ecosystems blurs the value of wilderness as a baseline. *In many places, wilderness remains the least altered part of the landscape and still holds value as a comparison to the managed landscape, but extinctions, biological invasions, and altered ecological processes ensure that the baseline itself is changing.* Anthropogenic ecosystem changes, such as global climate change, threaten to drive wilderness ecosystems around the world even further from historical conditions (Millennium Ecosystem Assessment 2005).

Under such conditions, "retaining primeval character and influence" is an enormous and sometimes impossible challenge that is directed by the Wilderness Act of 1964. In some places, conditions are so altered that ecosystem function no longer retains its "primeval influence." Restoring processes and conditions more representative of pre–western settlement conditions may require intervention, "trammeling" the land to some extent, such as the intentional introduction of fire or extirpated wildlife species. Alternatively, decisions not to intervene may cause further degradation, such as where exotic species have invaded.

Reconciling conflict between current conditions and HRV is among the most difficult decisions faced by wilderness managers today. So difficult is this task that Cole (1996, 2000) and Landres et al. (2000) have called it "the dilemma of wilderness management." Should we emphasize the value of wilderness as

Fig. 10.9. Dense stands of small-diameter ponderosa pines encroach on the open, old-growth forest, Mount Trumbull Wilderness, in northern Arizona. BLM managers propose restoring the old-growth forest structure and disturbance regime of frequent low-intensity fires. Photo courtesy Gregory H. Aplet.

a control and leave it alone, accepting what we perceive to be ecosystem change or "degradation" as a consequence? Or should we manage for land health, thereby losing the control for our management experiments? To intervene or not intervene is often a question for wilderness stewardship.

Keeping in mind fundamental versus cosmetic change, there are times when it is obvious that intervention is appropriate in wilderness. These cases will often involve small areas without obvious, far-reaching ecological impacts, or sites where an incipient problem is developing. *Campsite revegetation, trail maintenance, and eradication of new weed infestations will be needed in certain places* (see Chapters 13 and 15). *Localized intervening actions such as these may be thought of as rehabilitation and not the trammeling of wilderness.*

Landscape-scale intervention, or *restoration*, however, presents new challenges to wilderness managers. *Restoration involves large areas and results in intentionally large ecological effects.* Actions such as reintroducing extirpated species, prescribed fire and other vegetation treatments, liming acidic rivers, and so forth are intended to alter the composition, structure, or function of ecosystems to restore naturalness and primeval influences, especially where time alone is not expected to improve wilderness character. For example, Sydoriak et al. (2000) believe that past grazing and fire suppression have pushed the Bandelier Wilderness in New Mexico across a threshold where recovery of herbaceous vegetation and the reintroduction of fire cannot be achieved without intervention. Mechanically thinning the pinyon-juniper woodland and scattering the sawn branches are needed to protect the soil and allow establishment of native grasses. Such treatment is recognized as at least a temporary trammeling of the ecosystem to restore wilderness character.

The choices faced by managers of wilderness ecosystems may be summarized using the same dimensions used to describe wilderness character (Aplet 1999) (see Fig. 10.10). In this concept, the term *neglect* describes a situation in which an ecosystem is allowed to undergo continued ecological decline in order that it is natural or "untrammeled." *Development* then describes the act of bringing such an ecosystem under control to *diminish* its naturalness (i.e., putting it to use). *Restoration*, then, is the act of intervening or "trammeling" an ecosystem to restore it to its HRV. Finally, *release* describes the situation in which the naturalness of an ecosystem increases as it is set free from human control.

For wilderness managers, release is the ideal choice, because it retains or increases both natural and untrammeled conditions. In many cases, even severely degraded ecosystems exhibit enormous capacity for resilience if simply left alone (McDonald 2000; Aplet 1999). Often, however, as is apparently the case at Bandelier, a return to naturalness cannot be achieved without intervening (i.e., mechanical thinning). In this case, getting the ecosystem to the point that it can recover will require temporary trammeling (Fig. 10.11).

The concept of natural recovery or "release" is central to the decision to engage in restoration in wilderness. *Allowing an ecosystem to completely degrade is inconsistent with the goals of wilderness management, as is keeping tight control on the ecosystem for the purpose of increasing naturalness. The decision whether to intervene will hinge on whether the potential for natural restoration outweighs the ecological uncertainties and the magnitude and duration of the required trammeling.* Wilderness managers must always keep in mind that they are guardians, not gardeners—continuing intervention is not acceptable.

As Cole (2000) notes, it does not make sense ecologically or financially to try to re-create specific, historical landscapes except in restricted, culturally important areas, such as historic sites and battlefields. Likewise, it does not make sense to apply a description

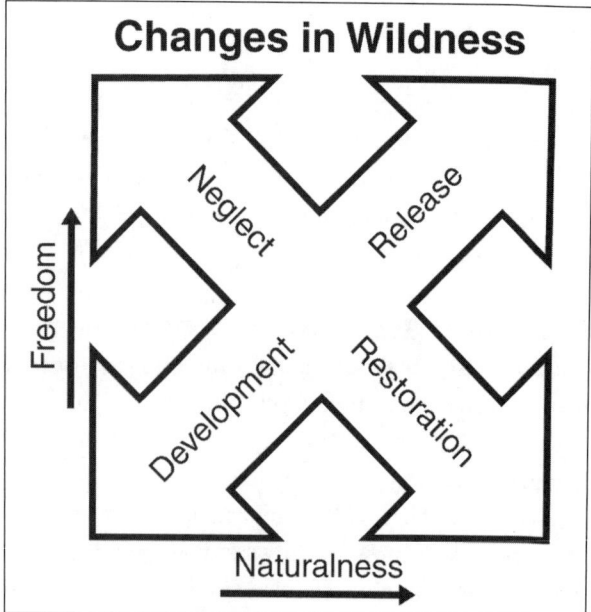

Fig. 10.10. Management can move a landscape in any direction with regard to primeval character and influence.

of historical conditions, even dynamic historical conditions, to an area that has been substantially altered in its driving factors. For example, the wilderness character of Mount St. Helens was not diminished by its 1980 eruption, despite the fact that it is substantially altered from its HRV. As a description of natural conditions, HRV must account for historical dynamics and recent nonanthropogenic change. *The goal of restoration should be to steer the ecosystem toward conditions that would have existed today in the absence of recent, human-induced modification.*

Global climate change presents an especially vexing challenge for wilderness managers (Vitousek et al. 2000). *As global climate changes, so does one of the primary drivers of ecosystem behavior, and we can expect increasing departures from HRV.* One could argue that such change, if anthropogenic, is unnatural, and wilderness managers should, therefore, manage wilderness to keep it within HRV in the face of climate change. However, as Keeley and Stephenson (2000) point out, such an approach presents "intractable" management challenges. Indeed, our current problem with wildland fire is the result of a century of fighting against the force of climate. *Rather, climate change requires that we better understand the historical interactions between climate and ecosystems and manage wilderness to allow processes to play out as naturally as possible. This will require alleviating, to the extent possible, the disruptive, modern influences, such as roads, water diversions, and other traditional land uses, and in some cases subsidizing processes, such as through the use of prescribed fire* (Keeley and Stephenson 2000; White et al. 2000).

Cole (2007) proposed that the Minimum Requirements Decision Guide (MRDG), developed by the four federal land-management agencies under the Wilderness Act, be used to address big wilderness issues such as restoration and rehabilitation. While MRDG was designed to address small projects and manage site-specific prohibited uses in wilderness (e.g., use of mechanical access and equipment by managers in wilderness), Cole explains how the MRDG process could be modified for use on large spatial and temporal scales to address restoration programs in wilderness. His main example is the USFS decision to maintain hemlock stands in the southeastern U.S. that have increasing rates of mortality due to the invasion of hemlock woolly adelgid from Asia (see case discussion at the end of this chapter).

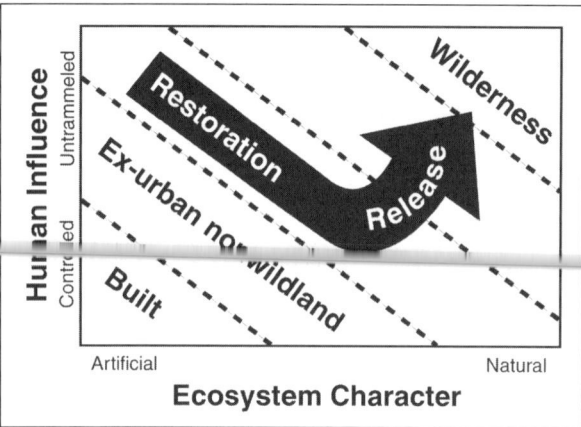

Fig. 10.11. For wilderness managers, nonintervention is the preferred ecosystem trajectory. If restoration is undertaken, it should lead to future nonintervention.

It is important to note that some changes are irreversible and must simply be accepted as fundamental changes in the condition of wilderness ecosystems. Examples are loss of keystone plant species to exotic pathogens (such as chestnut in eastern hardwood forests) and establishment of exotic plants and animals that alter fundamental processes, including disturbance regimes (such as cheatgrass, star thistle, and spotted knapweed in western rangelands).

Assessing Naturalness for Wilderness Planning

Knowledge of ecological conditions in wilderness is more important now than it was in the first three decades following passage of the Wilderness Act, when wilderness management was essentially people management. Today, knowledge of ecosystems and ecosystem dynamics is critical to determining naturalness, setting wilderness management goals, and developing management approaches to achieving those goals. *In this section, we present some general concepts to guide the development and use of ecological information in planning for management to protect wilderness ecosystems.*

Planning for wilderness management must begin with a basic description of the ecosystem, although the required detail may vary with area management objectives and problems. Characterizing the naturalness of an ecosystem requires answers to a few key questions:

1. What kinds of ecosystems are present? (classification)
2. Where are they located? (distribution, mapping)
3. What are their biological and physical characteristics and conditions? (characterizations of species

composition, population status, ecosystem structure, climate, geology, etc.)
4. What are their dynamic processes? (rates and directions of historical change)
5. How have those processes been affected by human activity?

Recent advances in technology make it much easier for wilderness managers to characterize ecosystem conditions. Foremost are remote-sensing technology, geographic information systems (GIS), and portable microcomputers. These developments dramatically expand information gathering, storage, and data analysis capabilities. These tools can supply stored information on demand about the state of wilderness ecosystems, information not available in the past.

Inventory requires classifying the landscape into useful units. Landscape ecology is a major subdiscipline of ecology (see Forman and Godron 1986), *and concepts such as landscape hierarchies, patch characteristics (size, shape, context), and corridors can be helpful in evaluating ecosystems.* Although it is a principle of wilderness management to "manage for the whole, rather than for the parts," there will be times when sustaining ecosystem composition requires focusing on an ecosystem's constituent parts. For example, sustaining elk in the ecosystem may require landscape-level attention to the distribution of habitat patches and their connectivity. Very large areas of diverse habitat may be required for top carnivores such as the grizzly bear, larger than available in all but the largest wilderness complexes, to provide for viable populations of this wide-ranging species. Sustaining aquatic habitat and organisms, such as anadromous fish (e.g., salmon), also requires a landscape perspective. The movement of materials along the continuum from headwater stream to river, and considerations of terrestrial and aquatic ecosystem interactions, force such a perspective (Swanson et al. 1982). A study of aquatic habitats in the Hoh River in Olympic National Park (WA) illustrates some of the interactions that occur at larger scales, which are essential to aquatic productivity, as well as the critical importance of some spatially limited "biological hotspots" (Franklin et al. 1982). Building an inventory program according to landscape hierarchies can help managers account for phenomena at multiple ecological scales.

Determining historical dynamics (HRV) is critical to assessing naturalness and will often require in-depth analysis. Methods of historical analysis are improving and yielding more complete pictures of the past (Sisk 1998; Swetnam et al. 1999). Moreover, several initiatives have shown how these descriptions of historical change may be converted into useful guidelines for management (Hessburg et al. 1999; Manley et al. 1995).

Monitoring and Adaptive Management

Determining whether an ecosystem is on a trajectory of sustained naturalness or is departing from HRV requires monitoring the ecosystem components and the threats to its integrity (see Chapter 13). A number of ecological monitoring methods are available, and Merigliano (1987) outlines a wide range of indicators to be measured to assess wilderness conditions. The USFS recently began developing a wilderness character monitoring framework that uses the definition of wilderness from section 2(c) of the Wilderness Act to identify four qualities of wilderness related to wilderness character: untrammeled, natural, undeveloped, and outstanding opportunities for solitude or a primitive and unconfined type of recreation (Landres et al. 2005). The natural qualities are the wilderness concerns that relate most directly to ecosystem management, and monitoring increases the capacity of managers to manage adaptively with new information on changes in natural processes and conditions.

Each ecosystem must be considered in terms of the specific objectives for management of that wilderness and the current conditions and assumptions about the future, which combine to identify problems (see Chapters 8 and 9). Features requiring intensive inventory and monitoring may include sensitive and/or intensively used areas, such as meadows, lakes, and streams, and some animal components. These are ecosystems or elements that can change relatively rapidly and are focal points of visitor interest and use. In addition to these sites, monitoring should track changes across the broader landscape, although this may require less frequent and intensive sampling.

Subalpine meadows are the classic example of ecosystems for which detailed information might be needed, particularly in heavily used areas. Documented instances of compositional change and rapid tree invasion of meadows are abundant. Some of these have been ascribed to natural causes such as tree invasion in the subalpine meadows of the Pacific Northwest (Franklin et al. 1971) resulting from climatic change.

More often, humans and domestic livestock are responsible. Lodgepole pine has invaded Sierra Nevada and Rocky Mountain meadows because of reduced wildfire, livestock grazing, changes in soil moisture (due to erosion and/or trail trenching), and climatic change. Grazing, trampling, and compaction have altered meadow composition in many wildernesses.

Meadows, margins of lakes and ponds, and timberline forests are examples of fragile ecosystems particularly likely to be heavily impacted by visitors. They should receive special attention in inventory and, especially, in monitoring programs. The initial state (composition) of these ecosystems needs to be known in greater detail. Greater numbers of permanent sample plots and photo points are appropriate and they should be remeasured more frequently than in heavily forested and lightly used ecosystems.

An example of a pioneering effort in monitoring such sensitive areas (and the value of such an effort) is Thornburgh's (1962) study of the Image Lake area in the Glacier Peak Wilderness, Washington. Substantial deterioration of the vegetation was apparent at that time. Thornburgh provided lists of use-susceptible and use-resistant species and recommended exclusion of livestock from the area. Remeasurements of twenty-five permanent transects made in 1966 and 1971 have served as a basis for management programs to rehabilitate damaged sites and to control types and intensity of use in the lake basin. Shifts in impacted areas from the vicinity of the lakeshore to higher benches (use of the immediate lakeshore area was restricted) were detected by continuous monitoring. Similar studies at other locations and research on techniques for rehabilitating damaged sites (such as seeding native plants or using plug transplants) have been a major emphasis in recent decades (see Chapter 15).

This discussion has focused on monitoring biological changes, such as changes in vegetation composition and structure on permanent plots or transects. Inferences drawn from the initial inventory on rates and directions of change are useful, but they are no substitute for observing actual changes on permanent, long-term samples. *Some general principles for setting up a series of permanent plots, points, and photo points for monitoring are the following:*

1. Sample more intensively and remeasure plots more frequently in areas where unnatural changes are likely, such as where people congregate.
2. When sampling areas are likely to be heavily impacted, set up control plots in comparable areas where you expect little change in use.
3. Ecotones between different types of vegetation, such as forest and meadow, are sensitive locations for monitoring biological changes.
4. Permanently locate or reference monitoring sites on the ground; document in detail their location and the techniques used.

With the emphasis on lakes and meadows as areas of special visitor attraction, changes in forest ecosystems are often overlooked. It is important that at least some monitoring be carried out in forested areas over long periods to identify actual, not inferred, trends in forest structure and composition. Careful documentation and referencing of plot location are critical because of the long time spans and difficulty of relocating plots in forested landscapes. Frequency of remeasurement will depend on objectives and likely rate of change. Seral forests of short-lived species (aspen and alder), and fire types where fire intervals are (or were) frequent, require more frequent attention than the conifer forests in wetter locations such as in the Pacific Northwest, which are composed of long-lived species and have long return intervals for fire. When dealing with forest (or even shrub-dominated) ecosystems, it is extremely important to pay close attention to entries or "births" of new individuals.

Special-interest species, such as those that are known to be rare or endangered, may be another focus of wilderness monitoring programs. Such programs are especially important for species that are threatened or endangered. Because wilderness, as large natural tracts, sometimes provides major reservoirs of such species, wilderness managers can expect increased emphasis on monitoring and management of particular threatened or endangered flora and fauna in the future.

Sound data-management programs are absolutely critical to the success of any long-term research and monitoring program, and guidelines are available on methods of managing such data sets (Michener 1986). Past failures of such efforts are often attributable to failures in data documentation and archiving.

Currently, monitoring and collecting of baseline data in wilderness are woefully inadequate. In most areas, they are essentially absent. What little monitoring is being done is generally not part of a systematic, comprehensive plan. Documentation is poor. Most work is focused on "high-impact areas" or immediate problem areas. Such sites merit high priority but tend to overshadow needs for monitoring and gathering baseline data over the wilderness as a whole. Other natural-resource managers do not tolerate such an inadequate inventory base in their programs. In timber-management programs, for example, there are extensive systems of continuous inventory plots, comprehensive stand examinations, simulation models, and complex data storage and retrieval systems. Certainly this type and intensity of inventory and monitoring are not advocated for wilderness. However, it should make wilderness managers aware of the total inadequacy of past efforts and the imperative for an improving database in the future. Even providing for large-scale GIS databases of wilderness tracts, something sorely lacking for many areas, would be a major improvement.

Unfortunately, there is no simple formula to follow to determine the proper course of wilderness management once ecosystem conditions are well understood. *The decision as to what action to take will require evaluating a set of management alternatives that cover a range of possible methods and outcomes. Reaching the right choice will require weighing the effects of taking or not taking a potential action, the probability of success and risk of unforeseen outcomes, and public opinion.* Because of the uncertainty that attends all such decisions, it is essential that the results of monitoring feed directly back to an evaluation of the soundness of past decisions. Such a process, in which management is undertaken with the intent of learning and adjusting to new information, is called "adaptive management," and is recognized as a cornerstone of sound ecosystem management. In wilderness ecosystems, this process should be guided always by the goal of maintaining natural influences to produce both untrammeled and natural conditions.

Summary

This chapter discussed the major features of all wilderness ecosystems and described the components of

Fig. 10.12. A campsite on Grassy Lake, John Muir Wilderness, California, shows impacts of human use, including pronounced trampling. Fragile vegetation renders alpine and subalpine meadows and lake shorelines sensitive to heavy use, and a short growing season hampers recovery of damaged areas. Photo courtesy U.S. Forest Service.

wilderness ecosystems—composition, structure, and function. We emphasized the dynamic nature of ecosystems, their strong internal linkages, and the ways various human activities have affected and continue to affect the "naturalness" of wilderness ecosystems. The concept of "historical range of variability (HRV)," or the bounded behavior of ecosystems over time, was discussed, as was their modification due to global climate change. Current ecosystem issues in wilderness management include fire control, grazing, loss of plant and animal species, exotic weeds, introduction of fish and animals to wilderness, direct human impacts, global climate change, and influences from outside wilderness. Only when ecosystem dynamics, including interrelationships with humans, are fully understood can the consequences of alternative management strategies be assessed. The decision as to what action to take will require evaluating a set of management alternatives that cover a range of possible methods and outcomes. Reaching the right choice will require weighing the effects of taking or not taking a potential action, the likely modifications of HRV due to anthropogenic disturbances (e.g., global climate change, land-use change, population growth), the probability of success and risk of unforeseen outcomes, and public opinion.

> **Study Questions**
> 1. Define the term *ecosystem*. What are the three main components of an ecosystem?
> 2. What are two qualities of ecosystems that wilderness managers must continually bear in mind?
> 3. Describe the classic concept of forest succession. Is it relevant to wilderness ecology?
> 4. Describe what is meant by "historical range of variability." Why is it important to wilderness management, and why do managers need to be aware that these patterns are being modified by anthropogenic disturbances?
> 5. To what extent are humans and their influences affecting wilderness ecosystems?
> 6. What is the main focus of wilderness ecosystem management?
> 7. How does "adaptive management" apply to wilderness management?
> 8. Where should the most data plots be located to monitor wilderness ecosystems, and why?
> 9. Why is the invasion of exotics a threat to the naturalness of wilderness ecosystems? Give some examples.
> 10. What is the difference between rehabilitation and restoration as discussed in this chapter?

Acknowledgments

Dr. Jerry F. Franklin, Bloedel Professor of Ecosystems Analysis, College of Forest Resources, University of Washington, Seattle, wrote this chapter for the first two editions of the book and coauthored a major revision for the third edition with Dr. Gregory H. Aplet, forest ecologist and director, Center for Landscape Analysis, The Wilderness Society, Denver. Minor revisions and updating were added by Chad Dawson and John Hendee for the fourth edition.

We are indebted to the following colleagues who provided review and/or input to earlier drafts of this chapter: Peter Landres, research ecologist, Aldo Leopold Wilderness Research Institute, Missoula, Montana; Pete Morton, resource economist, The Wilderness Society.

Case Discussion:
Protecting Hemlocks in the Southeastern U.S.

Ecological restoration in wilderness raises questions about how to develop compromises between the naturalness and untrammeled qualities of wilderness character. This concern becomes even more acute as large spatial scales are considered for management activities to restore vegetation and wildlife species or other wilderness conditions to their historic pattern. Cole (2007) suggests that careful consideration of the management actions could benefit by applying the same concepts that are central to the Minimum Requirements Decision Guide (MRDG) and ensure that managers are taking the minimum action necessary to optimize wilderness character. Cole, in his 2007 paper (pages 9–11), outlines the example of how the USFS went about using the MRDG to decide how, when, and where to protect hemlocks in wilderness areas in the southeastern U.S.:

> An excellent example is provided by planning efforts of the National Forests in North Carolina devoted to preservation of hemlocks (USDA Forest Service 2005). Two species of hemlock (eastern and Carolina) are experiencing high rates of mortality due to a small aphid-like insect, the hemlock woolly adelgid, native to Asia and first detected in the eastern United States in 1951. Eastern hemlock is the second most common conifer in these forests, is a significant component of old growth forests, notably in some wilderness areas, is often important in riparian communities and often lends a distinctive, scenic component to landscapes.
>
> By 2001, the adelgid had spread to the forests of North Carolina and, by 2004, mortality of hemlocks was occurring. Research conducted elsewhere suggests that tree mortality can occur in as few as three years and that more than 90% mortality of hemlocks can be expected within 10 to 12 years of a stand becoming heavily infested (Mayer et al. 2002). Without intervention, it is likely that most hemlocks—among the oldest-lived trees (600 plus years) in the east—would be lost from eastern forests. Carolina hemlock might go extinct, since its range is primarily in western North Carolina. Extinction of the more widely distributed eastern hemlock is also possible. Even with intervention, the result of the adelgid infestation will be a loss of biodiversity, degradation of aquatic habitat and scenic values, and a reduction in wilderness character, through a loss of naturalness. Many of the finest hemlock stands, in terms of condition, age and character, are in wilderness. Moreover, many of the most intact ecosystems in the east are in wilderness.

Intervention options exist that appear capable of protecting hemlocks. Injection of the insecticide imidacloprid into the soil close to trees kills the adelgid, resulting in dramatic recovery (Steward and Horner 1994). In close proximity to water and where soil is highly permeable, tree stems must be injected, a technique that can damage trees and is less long lasting. In addition, introduction of nonnative beetles (from China, Japan and the northwestern United States) can reduce adelgid populations sufficiently to allow infested trees to recover (Cheah and McClure 2002).

The choice facing the Forest Service, both inside and outside wilderness, was whether to let hemlocks disappear from these forests or to use insecticides and introduction of another nonnative species to protect these trees. Wilderness character was doomed to decline as soon as the first adelgid arrived in the United States. The choice facing planners was which aspects of wilderness character to protect, where and how. As noted before, wilderness system values are optimized when different compromises are reached in different places because outstanding examples of all components of wilderness character are preserved at least somewhere in the system.

In this case, the Forest Service decided to compromise both the untrammeled and naturalness components of wilderness character, by intervening in some but not all stands. They adopted an objective of maintaining "reproducing populations of eastern and Carolina hemlock throughout their historical and elevational range." This objective is quite different from such possible objectives as protecting all hemlock stands or protecting stands wherever resources can be mustered to protect them. Their decision for the first step of a minimum requirements analysis was that administrative action is necessary because the desired outcome in wilderness is maintenance of some hemlock stands in wilderness. This decision could not have been made without a decision about desired outcomes in wilderness.

The planners used the concepts of the metapopulation and minimum viable population size to decide how many trees and conservation areas to protect, as well as the minimum intervention needed to protect the trees in each conservation area. The outcome of the second step in the minimum requirements analysis, then, was a decision about which specific actions in which specific places collectively constitute the minimum necessary. Ultimately, from nearly 400 hemlock stands, they decided to release predatory beetles in 159 hemlock areas (typically 125 acres [50.6 ha] in size) across the Forests. To ensure maintenance of an adequate gene pool until effective biocontrol is established, trees will be treated with insecticide in as many as half of these areas. The minimum activity is not the least obtrusive single action. Rather it is the combination of actions, varying in obtrusiveness and applied in the minimum number of stands, that minimizes loss of the untrammeled quality of wilderness character while meeting the desired outcome.

Since the objective of maintaining hemlock in some of these forests applies equally inside wilderness and outside wilderness, many of the treated stands will be in wilderness. It might have been possible to meet the overall objective of maintaining "reproducing populations of eastern and Carolina hemlock throughout their historical and elevational range" by only intervening in stands outside wilderness, but this would have impacted wilderness character unacceptably. The keys to deciding what to do in wilderness, then, came from deciding about desired future conditions and how to compromise between the components of wilderness character, not from attempting to apply interventions outside wilderness.

While one might disagree with this decision, the process is true to the spirit of the minimum requirements analysis. Primary attention was given to optimizing wilderness character, in this case crafting a desired future condition that represented a compromise between the conflicting components of naturalness and untrammeled (Landres et al. 2005). This compromise was codified in a management objective that defined the desired future condition. A management prescription was developed that was a combination of different treatments being conducted in a carefully specified number of stands. The "minimum" activity designation comes as much from intervening in the minimum number of places as from the minimum obtrusiveness of the intervention.

Case Discussion Questions

1. What ecological outcome do you think is desirable in southeastern U.S. wilderness areas that historically had viable hemlock forests?
2. Do you agree or disagree that the proposed USFS management plans and activities to maintain hemlock stands in southeastern U.S. wilderness areas are appropriate as the minimum action under the Wilderness Act? Why or why not?
3. How would you decide when to attempt restoration to a historic condition and when to allow human-caused disturbances to an ecosystem (that originate outside that wilderness area) to run their course and arrive at a new condition? How much trammeling do you think is acceptable to restore historic naturalness?

References

Aplet, Gregory H. 1999. On the nature of wildness: Exploring what wilderness really protects. *Denver University Law Review.* 76: 347–367.

Aplet, Gregory H.; Keeton, William S. 1999. Application of historical range of variability concepts to the conservation of biodiversity. In: Baydack, R. K.; Campa, H., III; Haufler, J. B., eds. *Practical Approaches to the Conservation of Biological Diversity.* Covelo, CA: Island Press, pp. 71–86.

Asher, Jerry E.; Harmon, David W. 1995. Invasive exotic plants are destroying the naturalness of U.S. wilderness areas. *International Journal of Wilderness.* 1(2): 35–37.

Bailey, Robert G. 1980. Description of the eco-regions of the United States. Miscellaneous Publication 1391. Washington, DC: U.S. Department of Agriculture, Forest Service.

———. 1995. Description of the eco-regions of the United States, 2nd ed. Miscellaneous Publication 1391. Washington, DC: U.S. Department of Agriculture, Forest Service.

Chang, Chi-ru. 1996. Ecosystem responses to fire and variations in fire regimes. In: *Sierra Nevada Ecosystem Project: Final Report to Congress*, Vol. II, Assessments and scientific basis for management options. Davis: University of California, Centers for Water and Wild Land Resources, pp. 1071–1099.

Cheah, C. A. S.; McClure, M. S. 2002. *Pseudoscymnus tsugae* in Connecticut forests: The first five years. In: Onken, B.; Readeon, R.; Lashomb, J. eds. *Proceedings, Hemlock woolly adelgid in the eastern United States symposium.* Rutgers University, East Brunswick, NJ, pp. 150–165.

Clements, Frederic. 1916. *Plant Succession: An Analysis of the Development of Vegetation.* Washington, DC: Carnegie Institution of Washington.

Cole, David N. 1996. Ecological manipulation in wilderness: An emerging management dilemma. *International Journal of Wilderness.* 4(3): 28–31.

———. 2000. Paradox of the primeval: Ecological restoration in wilderness. *Ecological Restoration.* 18(2): 77–86.

———. 2007. Scaling-up the minimum requirements analysis for wilderness issues. *International Journal of Wilderness.* 13(1): 8–12.

Cooper, Charles F. 1960. Changes in vegetation, structure, and growth of southwestern pine forests since white settlement. *Ecological Monographs.* 30(2): 120–164.

Cowles, Henry C. 1899. The ecological relations of the vegetation on the sand dunes of Lake Michigan. *Botanical Gazette.* 27.

Cronon, William. 1983. *Changes in the Land: Indians, Colonists, and the Ecology of New England.* New York: Hill and Wang.

Daubenmire, Rexford. 1968. *Plant Communities.* Evanston, NY: Harper and Row. 300p.

———. 1970. Steppe vegetation of Washington. Technical Bulletin 62. Pullman: Washington State University, College of Agriculture, Washington Agricultural Experiment Station.

Davis, George D. 1989. Preservation of natural diversity: The role of ecosystem representation within wilderness. In: Freilich, Helen R., comp. *Wilderness Benchmark 1988: Proceedings of the National Wilderness Colloquium*; January 13–14, 1988; Tampa, FL. Service General Technical Report SE-51. Asheville, NC: U.S. Department Agriculture, Forest Service, Southeast Forest Experiment Station, pp. 76–82.

Denevan, William M. 1992. The pristine myth: The landscape of the Americas in 1492. *Annals of the Association of American Geographers.* 82: 369–385.

Fenn, M. E.; Poth, M. A.; Aber, J. D.; Baron, J. S.; Bormann, B. T.; Johnson, D. W.; Lemly, A. D.; McNulty, S. G.; Ryan, D. F.; Stottlemyer, R. 1998. Nitrogen excess in North American ecosystems: Predisposing factors, ecosystem responses, and management strategies. *Ecological Applications.* 8: 706–733.

Fiedler, Peggy; White, Peter S.; Leidy, Robert A. 1997. The paradigm shift in ecology and its implications for conservation. In: Pickett, S. T. A.; Ostfeld, R. S.; Shachack, M.; Likens, G. E., eds. *The Ecological Basis of Conservation: Heterogeneity, Ecosystems, and Biodiversity.* New York: Chapman and Hall, pp. 83–92.

Forman, Richard T.; Godron, Michel. 1986. *Landscape Ecology.* New York: John Wiley.

Franklin, Jerry F. 1987. Scientific use of wilderness. In: Lucas, Robert C., comp. *Proceedings–National Wilderness Research Conference: Issues, State-of-Knowledge, Future Directions*; July 23–26, 1985; Fort Collins, CO. INT-GTR-220. Ogden, UT: U.S. Department of Agriculture, Forest Service, Intermountain Research Station, pp. 42–46.

Franklin, Jerry F.; Moir, William H.; Douglas, George W.; Wiberg, Curt. 1971. Invasion of subalpine meadows by trees in the Cascade Range, Washington and Oregon. *Arctic and Alpine Research.* 3(3): 215–224.

Franklin, Jerry F.; Swanson, Frederick J.; Sedell, J. R. 1982. Relationships within the valley floor ecosystems in western Olympic National Park: A summary. In: Starkey, Edward E.; Franklin, Jerry F.; Matthews, Jean W., eds. *Ecological Research in National Parks of the Pacific Northwest.* Corvallis: Oregon State University, Forest Research Laboratory, pp. 43–45.

Frissell, Sidney S., Jr. 1973. The importance of fire as a natural ecological factor in Itasca State Park, Minnesota. *Quaternary Research.* 3(3): 397–407.

Habeck, James R.; Mutch, Robert W. 1973. Fire-dependent forests in the northern Rocky Mountains. *Quaternary Research.* 3(3): 408–424.

Heinselman, Miron L. 1973. Fire in the virgin forests of the Boundary Waters Canoe Area, Minnesota. *Quaternary Research.* 3(3): 329–383.

Hessburg, Paul F.; Smith, Bradley G.; Salter, R. Brion. 1999. Detecting change in forest spatial patterns from reference conditions. *Ecological Applications.* 9: 1232–1252.

Keeley, Jon; Stephenson, Nathan L. 2000. Restoring natural fire regimes to the Sierra Nevada in an era of global change. In: Cole, D. N.; McCool, S. F.; Borrie, W. T.; O'Loughlin, J., comps. *Wilderness science in a time of change conference*, Vol. 5: *Wilderness Ecosystems, Threats, and Management*; May 23–27, 1999; Missoula, MT. RMRS-P-15. Ogden, UT: U.S. Department of Agriculture, Forest Service, pp. 255–265.

Kilgore, Bruce M. 1973. The ecological role of fire in Sierran conifer forests. *Quaternary Research.* 3(3): 496–513.

Knight, Richard L.; Landres, Peter B., eds. 1998. *Stewardship Across Boundaries.* Washington, DC: Island Press.

Krech, Shepard, III. 1999. *The Ecological Indian: Myth and History.* New York: W.W. Norton.

Landres, P.; Boutcher, S.; et al. 2005. Monitoring selected conditions related to wilderness character: A national framework. General Technical Report RMRS-GTR-151. U.S. Department of Agriculture, Forest Service, Rocky Mountain Research Station, Fort Collins, CO.

Landres, Peter B.; Brunson, Mark W.; Merigliano, Linda; Sydoriak, Charisse; Morton, Steve. 2000. Naturalness and wildness: The dilemma and irony of managing wilderness. In: Cole, D. N.; McCool, S. F.; Borrie, W. T.; O'Loughlin, J., comps. *Wilderness science in a time of change conference*, Vol. 5: *Wilderness Ecosystems, Threats, and Management*; May 23–27, 1999; Missoula, MT. RMRS-P-15. Ogden, UT: U.S. Department of Agriculture, Forest Service, pp. 377–381.

Landres, Peter B.; Morgan, Penelope; Swanson, Frederick J. 1999. Overview of the use of natural variability concepts in managing ecological systems. *Ecological Applications.* 9: 1179–1188.

Leopold, Aldo. 1949. *A Sand County Almanac.* New York: Oxford University Press.

Loomis, John; Echohawk, Chris. 1999. Using GIS to identify under-represented ecosystems in the National Wilderness Preservation System in the USA. *Environmental Conservation* 26: 53–58.

Manley, Patricia, et al. 1995. *Sustaining Ecosystems: A Conceptual Framework.* R-5-EM-TP-001. San Francisco: U.S. Department of Agriculture, Forest Service, Pacific Southwest Region.

Marler, Marilyn. 2000. A survey of exotic plants in federal wilderness areas. In: Cole, D. N.; McCool, S. F.; Borrie, W. T.; O'Loughlin, J., comps. *Wilderness science in a time of change conference*, Vol. 5: *Wilderness Ecosystems, Threats, and Management*; May 23–27, 1999; Missoula, MT. RMRS-P-15. Ogden, UT: U.S. Department of Agriculture, Forest Service, pp. 319–327.

Marshall, Robert. 1933. *The People's Forest.* New York: Harrison Smith and Robert Haas.

Matthews, Kathleen R.; Knapp, Roland A. 1999. A study of high mountain lake fish stocking effects in the U.S. Sierra Nevada wilderness. *International Journal of Wilderness.* 5(1): 24–26.

Mayer, M.; Chianese, R.; Scudder, T.; White, J.; Vongpaseuth, K.; Ward, R. 2002. Thirteen years of monitoring hemlock woolly adelgid in New Jersey forests. In: Onken, B.; Readeon, R.; Lashomb, J. eds. *Proceedings, Hemlock woolly adelgid in the eastern United States symposium.* Rutgers University, East Brunswick, NJ, pp. 50–60.

McCann, Joseph M. 1999a. Before 1492: The making of the pre-Columbian landscape, Part I: The environment. *Ecological Restoration.* 17: 15–30.

———. 1999b. Before 1492: The making of the pre-Columbian landscape, Part II: The vegetation, and implications for restoration for 2000 and beyond. *Ecological Restoration.* 17: 107–119.

McDonald, Tein. 2000. Resilience, recovery and the practice of restoration. *Ecological Restoration.* 18: 10–20.

Merigliano, Linda L. 1987. *The identification and evaluation of indicators to monitor wilderness conditions.* Thesis. Moscow: University of Idaho, College of Forestry, Wildlife and Range Sciences.

Michener, William K., ed. 1986. *Research Data Management in the Ecological Sciences.* No. 16. Columbia: University of South Carolina Press, Belle W. Baruch Library in Marine Science.

Millar, Constance I. 1996. The Mammoth-June Ecosystem Management Project, Inyo National Forest. In: *Sierra Nevada Ecosystem Project, final report to Congress,* vol. II, *Assessments and scientific basis for management options.* Davis: University of California, Centers for Water and Wild Land Resources, pp. 1273–1346.

Millennium Ecosystem Assessment. 2005. *Ecosystems and Human Well-Being: Synthesis.* Washington, DC: Island Press.

Miller, Paul L. 1973. Exidant-introduced community change in a mixed conifer forest. *American Chemistry Society Advances in Chemistry.* No. 122. Washington, DC: American Chemistry Society, pp. 101–117.

Morgan, P.; Aplet, G. H.; Haufler, J. B.; Humphries, H. C.; Moore, M. M.; Wilson, W. D. 1994. Historical range of variability: A useful tool for evaluating ecosystem change. *Journal of Sustainable Forestry.* 2: 87–111.

Muir, John. 1911. *My First Summer in the Sierra.* Boston: Houghton Mifflin.

Odum, Eugene P. 1971. *Fundamentals of Ecology,* 3rd ed. Philadelphia: W. B. Saunders.

Parsons, D. J.; Swetnam, T. W.; Christensen, N. L., eds. 1999. Uses and limitations of historical variability concepts in managing ecosystems. *Ecological Applications.* 9: 1177–1277.

Parsons, David S. 2000. Restoration of natural fire to United States wilderness areas. In: Watson, A. E.; Aplet, G. H.; Hendee, J. C., comps. *Personal, Societal, and Ecological Values of Wilderness: 6th World Wilderness Congress Proceedings on Research, Management and Allocation,* Vol. II; October 24–29, 1998; Bangalore, India. RMRS-P-14. Ogden, UT: U.S. Department of Agriculture, Forest Service, Rocky Mountain Research Station, pp. 42–47.

Pickett, S. T. A.; Parker, V. T.; Fiedler, P. L. 1992. The new paradigm in ecology: Implications for conservation biology above the species level. In: Fiedler, P. L.; Jain, S. K., eds. *Conservation Biology: The Theory and Practice of Nature Conservation, Preservation, and Management.* New York: Chapman Hall, pp. 65–88.

Quigley, T. M; Arbelbide, S. J., tech. eds. 1997. *An Assessment of Ecosystem Components in the Interior Columbia Basin and Portions of the Klamath and Great Basins*; Vol. 2. PNW-GTR-405. Portland, OR: U.S. Department of Agriculture, Forest Service, Pacific Northwest Station, pp. 337–1055.

Randall, John M. 2000. Improving management of nonnative invasive plants in wilderness and other natural areas. In: Cole, D. N.; McCool, S. F.; Borrie, W. T.; O'Loughlin, J., comps. *Wilderness science in a time of change conference,* Vol. 5: *Wilderness Ecosystems, Threats, and Management*; May 23–27, 1999; Missoula, MT. RMRS-P-15. Ogden, UT: U.S. Department of Agriculture, Forest Service, pp. 64–73.

Shugart, Herman H., Jr.; West, D. C. 1977. Development of an Appalachian deciduous forest succession model and its application to assessment of the impact of chestnut blight. *Journal of Environmental Management.* 5(2): 161–179.

Sisk, Thomas, ed. 1998. Perspectives on the land use history of North America: A context for understanding our changing environment. USGS/BRD/BSR-1998-0003. Washington, DC: U.S. Department of the Interior, U.S. Geological Survey.

Steward, V. B.; Horner, T. A. 1994. Control of hemlock woolly adelgid using soil injections of systemic insecticides. *Journal of Arboriculture.* 20: 287–288.

Swanson, Frederick J.; Gregory, Stanley G.; Sedell, James R.; Campbell, A. G. 1982. Land-water interactions: The riparian zone. In: Edmonds, Robert L., ed. *Analysis of Coniferous Forest Ecosystems in the Western United States.* Stroudsburg, PA: Hutchinson Ross Publishing Co., pp. 292–332.

Swanson, Frederick J.; Jones, Julia A.; Wallin, D. O.; Cissel, John H. 1994. Natural variability—implications for ecosystem management. In: Everett, R. L., team leader. *Ecosystem Management: Principles and Application.* Vol. II of the *Eastside Forest Ecosystem Health Assessment.* PNW-GTR-318. Portland, OR: U.S. Department of Agriculture, Forest Service, Pacific Northwest Experiment Station, pp. 80–94.

Swetnam, Thomas W.; Allen, Craig D.; Betancourt, Julio L. 1999. Applied historical ecology: Using the past to manage for the future. *Ecological Applications.* 9: 1189–1206.

Sydoriak, Charisse A.; Allen, Craig D.; Jacobs, Brian F. 2000. Would ecological landscape restoration make the Bandelier Wilderness more or less of a wilderness? In: Cole, D. N.; McCool, S. F.; Borrie, W. T.; O'Loughlin, J., comps. *Wilderness science in a time of change conference,* Vol. 5: *Wilderness Ecosystems, Threats, and Management*; May 23–27, 1999; Missoula, MT. RMRS-P-15. Ogden, UT: U.S. Department of Agriculture, Forest Service, pp. 209–215.

Thornburgh, Dale. 1962. *Image Lake report.* Thesis. Berkeley: University of California, Berkeley.

Tonnessen, K. 2000. Protecting wilderness air quality in the United States. In: Cole, D. N.; McCool, S. F.; Borrie, W. T.; O'Loughlin, J., comps. *Wilderness science in a time of change conference,* Vol. 5: *Wilderness Ecosystems, Threats, and Management*; May 23–27, 1999; Missoula, MT. RMRS-P-15. Ogden, UT: U.S. Department of Agriculture, Forest Service, pp. 74–96.

U.S. Department of Agriculture, Forest Service. 2005. *Environmental Assessment: Suppression of Hemlock Woolly Adelgid Infestations.* National Forests in North Carolina, Asheville, NC.

U.S. Public Law 88–577. The Wilderness Act of September 3, 1964. 78 Stat. 890.

U.S. Public Law 93–622. (Eastern Wilderness) Act of January 3, 1975. 88 Stat. 2096.

U.S. Public Law 96–560. (Colorado Wilderness) Act of December 22, 1980. 94 Stat. 3265.

Vale, Thomas R. 1998. The myth of the humanized landscape: An example from Yosemite National Park. *Natural Areas Journal.* 18(3): 231–236.

Vitousek, Peter M.; Aber, John D.; Goodale, Christine L.; Aplet, Gregory H. 2000. Global change and wilderness science. In: Cole, D. N.; McCool, S. F.; Freimund, W.; O'Loughlin, J., comps. *Wilderness science in a time of change conference,* Vol. 1: *Changing Perspectives and Future Directions*; May 23–27, 1999; Missoula, MT. RMRS-P-15-Vol-1. Ogden, UT: U.S. Department of Agriculture, Forest Service, pp. 5–9.

Vogl, Richard J. 1973. Ecology of knobcone pine in the Santa Anna Mountains of California. *Ecological Monographs.* 43(2): 125–143.

White, Peter S.; Harrod, Jonathan; Walker, Joan L.; Jentsch, Anke. 2000. Disturbance, scale, and boundary in wilderness management. In: McCool, S. F.; Cole, D. N.; Borrie, W. T.; O'Loughlin, J., comps. *Wilderness science in a time of change conference,* Vol. 2: *Wilderness Within the Context of Larger Systems*; May 23–27, 1999; Missoula, Montana. RMRS-P-15. Ogden, UT: U.S. Department of Agriculture, Forest Service, pp. 27–42.

Whitney, G. G. 1994. *From Coastal Wilderness to Fruited Plain: A History of Environmental Change in Temperate North America, 1500 to the Present.* New York: Cambridge University Press.

Wright, H. E., Jr. 1974. Landscape development, forest fires, and wilderness management. *Science.* 186(4163): 487–495.

Chapter 11
Fire in Wilderness Ecosystems

Introduction 276
Fire Occurrence and Behavior 277
    Ignition and Spread Determinants 277
        Fuels 277
        Weather 277
        Ignition Source 277
        Topography/Landscape Features 277
    Seasonal and Climatic Factors 278
    Major Drought Episodes 279
    Fire Size and Intensity 279
    Fire Regimes 279
The Natural (Historical) Role of Fire in Wilderness Ecosystems 280
How Has the Historical Role of Fire Been Modified by Fire Suppression? .. 283
Objectives of Wilderness Fire Management 285
    Agency Wilderness Fire Policy/Objectives 285
    What Is "Natural" in Wilderness Fire Management? 287
Wilderness Fire Policy Alternatives and Their Consequences 289
    Alternative 1: Attempt to Suppress All Fires 289
    Alternative 2: Allow All Fires to Burn 289
    Alternative 3: Manage Lightning-Caused Fires 290
    Alternative 4: Ignite Prescribed Fires 290
    Alternative 5: Manipulate Vegetation and Fuels Without Fire 290
    Alternative 6: Combined Alternatives 290
Wilderness Fire-Management Planning: Considerations and Constraints .. 291
    Economic Considerations 292
    Interagency Planning 292
    Other Constraints on Planning 292
    Fire Behavior Prediction 294
    Limiting Fire-Suppression Impacts 294
    Monitoring and Evaluating Wilderness Fires 295
    Learning from Past Mistakes 295
History and Evolution of Wilderness Fire-Management Programs 296
    Sequoia–Kings Canyon and Yosemite National Parks, California 296
    Selway-Bitterroot Wilderness, Idaho and Montana 297
    Grand Teton and Yellowstone National Parks and Surrounding
        National Forest Wildernesses, Wyoming and Montana 297
    Controversy and Support for Wilderness Fire-Management Programs .. 297
Fire-Dependent Ecosystems in American Wilderness Regions 298
    The Sierra Nevada Region of California 298
    The California-Oregon Coast Ranges Region 298
    The Coast Redwoods Region, California 298
    The Cascades Range of Oregon and Washington 299
    The Intermountain Region and Southwest 299
    The Rocky Mountain Region 299
    The Lake Superior Region 301

The Boreal Forest Region	301
The Eastern Deciduous Forest, Appalachian, and Gulf Region	302
Trends and Future Needs	302
Wilderness Fire Programs and Ecosystem Management	303
Study Questions	303
Acknowledgments	303
Case Discussion: Ignition Is One of the Challenges to Managing Fires in Wilderness	304
Case Discussion Questions	304
References	305

Introduction

Periodic fires are a natural part of most wilderness ecosystems. In designated wilderness areas where the earth and its community of life are to be kept as untrammeled by humans as possible, fire is as natural and vital a process as rain, snow, or wind. From jack pine and aspen in the Boundary Waters Canoe Area Wilderness (BWCAW) of Minnesota; through lodgepole pine in Yellowstone National Park, Wyoming; to ponderosa pine in Grand Canyon National Park, Arizona; and to the giant sequoia of California's Sequoia–Kings Canyon and Yosemite National Parks, fire plays an essential role in the structure and functioning of the ecosystems in these areas. *Many plant communities and ecosystems actually depend on fire for their well-being, and are thus called fire-dependent plant communities and ecosystems.*

Fire has long been an important focus of forest and park management (i.e., to prevent and control fires) and research beginning in the late 1970s through the present. Many important studies on fire frequency, effects of fires, and characterization of fire regimes took place in the 1980s through the present and form the foundation for wilderness fire management and for this chapter.

Evidence for the past role of fire is found in charred wood and cones in glacial deposits, in charcoal stratigraphy of laminated lake-bottom sediments, and in fire-scarred cross-sections, increment borings, or wedges from both recent and ancient trees (McBride 1983; Swain 1973). Such fires were ignited by lightning or by Indians. To a great extent, the ignition source was unimportant; when fuels and weather conditions were right, the vegetation would burn, and with typical patterns of frequency and intensity, region by region. *The kind of fire activity that characterizes a specific region is known as its "fire regime."* Such regimes vary from ponderosa-pine forests where frequent low-intensity surface fires were the pattern to true fir forests where at very long intervals (more than 300 years) high-intensity crown or surface fires occurred that led to the replacement of most of the forest stand. Such fires are sometimes called "stand-replacement" fires.

The goal of wilderness fire management is to restore fire as nearly as possible to its natural role. Before 1970 (and still to some extent today), our urban-based society perceived fire as a destroyer and attempted to ban it from wilderness. The past four decades have brought a more enlightened, biocentric approach to wilderness management that acknowledges fire must play its natural role to the greatest possible extent—consistent with the safety of people and adjacent nonwilderness resources. Of course, this recognition is still more an ideal than a practical reality. *This chapter explores our current knowledge of fire ecology and alternatives for restoring fire as an important natural force to wilderness.*

Fig. 11.1. Elk on the East Fork of the Bitterroot River during a fire on the Sula Complex, August 6, 2000. Fire has been a historic force shaping the character of the wilderness. Restoring fire to its natural role after years of a policy of suppression is a major challenge to wilderness managers. Photo courtesy John McColgan, Bureau of Land Management.

Wildland fire management is becoming a more prominent natural-resource management specialty. Wilderness fire management is a part of that. The NPS, USFS, and state agencies managing wilderness fire programs recognize that well-trained professionals are essential to direct and implement the complex fire-management programs involved in wilderness fire management. These agencies have in-house and intra-agency training programs for managers, professionals, and technicians in fire behavior, fire effects, fire monitoring, and fire-suppression concepts and techniques. The organizational training programs supplement college and university fire-ecology and fire-management curricula and short courses, and periodic interagency conferences on park and wilderness fire management. Wilderness fire is a prominent topic in natural-resource wildland and fire-management journals and magazines, building an evolving information base on fire in ecosystem management and for training fire managers.

Fire Occurrence and Behavior

Over the past three decades, we have developed an improved capability to predict the behavior of wildland fire (Keown 1985a; Rothermel 1983). However, fire models are not enough; personal knowledge of fire behavior and actual fire experience are still required to ensure valid input of data and interpretation of the output.

Ignition and Spread Determinants

Key fire behavior factors that determine whether a fire will be ignited and spread are fuels, weather, topography, and ignition source.

Fuels

The accumulation of some minimum level of smaller fuels, sufficiently dried, and properly arranged is critical to ignition and spread of fire. Even heavy fuel loads may not burn when exposed to a moving, self-perpetuating fire if they are discontinuous, poorly arranged in terms of heat transfer, or too moist. The chemistry of certain fuels such as chaparral, however, renders them more flammable when living than when dead and dry. In forests, unlike chaparral, crown fires usually require adequate surface fuels beneath the stand and some ladder fuels to carry fire up into the tree crowns.

Fig. 11.2. Lightning is the primary source of ignition in wilderness ecosystems. Photo courtesy John Cossett.

Weather

Suitable burning weather is characterized by one or more factors that promote drying of available fuels: (1) a precipitation-free period long enough to reduce the moisture content of fine- to medium-sized fuels to the critical level, (2) sufficiently high temperatures, and (3) low relative humidity. In level terrain, some wind is generally required to make wildfires spread rapidly. In mountainous terrain, fires create their own drafts uphill. In certain chaparral areas, like southern California, dry, warm Santa Ana or Foehn winds are important contributors to fire weather.

Ignition Source

The major natural ignition source is lightning. If thunderstorms are not accompanied by heavy rains, fires can result from strikes in snags, trees with dry rot, flammable crowns, or ignition of surface fuels. On many wilderness travel routes or boundary roads and nearby highways, humans are an ignition source. In historic times, Native Americans commonly burned much of what is wilderness today (Arno 1985; Gruell 1985). These indigenous people burned for many reasons, including signaling, hunting, managing forage and animal populations, managing vegetation, maintaining habitat diversity, protecting villages from fire, and waging war (Dennis and Wauer 1985).

Topography/Landscape Features

Fire movements are strongly related to elevation, aspect, slope, and geology of local landforms. Fire tends to move upslope. Its movements are particularly influenced by aspect and steepness of slopes. The steeper south- and southwest-facing slopes and infertile or dry landforms are particularly fire-prone. Dense forests on moist and fertile sites will burn only after long drying periods. Other landscape features that limit fire spread include nonflammable areas, such as lakes, rivers, swamps, barren rock areas, snowfields, timber lines, and fresh burns where fuels are lacking.

Seasonal and Climatic Factors

Wildfires spread in the dry season when fuels are sufficiently dry to burn. The annual climatic regime has much to do with the expected occurrence of wildland fires and, therefore, the recommended timing of prescribed burning or wildfires that might be allowed to burn. Many North American wildernesses at higher elevations have snow cover from early fall through late spring or early summer. Except for this common element, seasonal patterns in North America vary significantly from one region to another, and fall into one of three major types: (1) wet winter and dry summers, (2) dry winters and wet summers, and (3) more complex patterns.

1. *Wet winters and dry summers* characterize the maritime climates of the Pacific coast and the Cascade Range and Sierra Nevada from Washington to Baja California. Much of the total annual precipitation in the mountains is snow. Winters are not very cold; summers are clear, warm, and dry. Occasional summer thunderstorms occur and they carry little rainfall; hence mid to late summer is the fire season here.

2. *Dry winters and wet summers* are typical of the Northeast, the Great Lakes, the Midwest, most of the Canadian boreal forest region, and the eastern area of the Rocky Mountains from Alberta to New Mexico. Summers have frequent frontal rainstorms, and thunderstorms are common from spring through fall, but particularly in midsummer. Enough rain to extinguish lightning ignitions accompanies most storms. Some dry storms, however, occur in most years, and prolonged summer droughts occur at intervals of five to thirty years.

3. *More complex patterns* are found in the Intermountain West, the Southeast, and Gulf states. Characteristics of both the Pacific maritime and continental climates are found from interior British Columbia south through eastern Washington and Oregon, all of Idaho, and extreme western Montana to Utah, Nevada, and Arizona. Winter snowfall and snow packs are moderate to heavy, and occasional periods of frontal summer rainfall occur. Extended summer dry periods are frequent, however, and dry thunderstorms are common. Lightning occurrence decreases northward.

The southeastern states and the Gulf Coast have considerable maritime precipitation derived from the Gulf of Mexico and the Atlantic Ocean. Annual precipitation is heavy and well distributed over the year, but temperatures are high, and even short droughts create conditions conducive to burning. Winter fires

Fig. 11.3. Two photos, taken eighty years apart, in the Mariposa Grove, Yosemite National Park, California, illustrate plant succession in the absence of fire. The 1890 photo (left) shows few understory trees. By 1970 (right), a dense thicket of white fir has grown up. Such thickets provide a fuel ladder that could result in a fire intense enough to kill mature giant sequoias. Left photo courtesy Mrs. Dorothy Whitener and Mary and Bill Hood; right, courtesy Dan Taylor, National Park Service.

are possible in most areas because vegetation is cured and snow cover is usually lacking. Although thunderstorms are frequent, most (but not all) lightning ignitions are extinguished by rainfall. The vegetation of many areas is not very flammable except for the southern pine regions of the Coastal Plain and Piedmont.

The effect of global climate change on annual climatic regimes—and fire regimes—are not presently well documented and will take several decades to understand. Climatic changes may best be characterized as more unpredictable and with more variability (e.g., less snow or more drought events) than the historic climatic regimes would suggest. Some authors predict longer fire seasons and more area burned each year (McKenzie et al. 2004).

Major Drought Episodes

In the northern United States and Canada, "fire weather" means much drier than average weather. In primeval times, stand-replacing fires evidently occurred during infrequent major regional or subcontinental droughts. The extensive Greater Yellowstone Area fires of 1988 seem to fall in this category (Christensen et al. 1989). Elsewhere, it is clear from studies of fire-scarred pines, Douglas-fir, and giant sequoia that surface fires burned at closely spaced intervals (Baisan and Swetman 1995; Frissell 1973; Heinselman 1973; Houston 1973; Kilgore and Taylor 1979). Certainly not all these fires occurred in only major drought years, but studies generally indicate that more large and intense fires in these areas may have coincided with such droughts.

Fire Size and Intensity

Most lightning-caused fires are small and soon go out. However, some smolder or ignite and develop into surface fires that burn over hundreds or even thousands of acres—scarring trees, significantly changing the understory vegetation, killing occasional trees or groups of trees, recycling nutrients, and reducing surface fuels. Between 2001 and 2006, only 51 to 86 percent of the acres of wildland fires each year were started by lightning (National Interagency Fire Center 2007).

An occasional blaze develops into a major high-intensity fire, killing forests over large areas, such as occurred in the Greater Yellowstone Area in 1988. Large, intense wildfires have occurred in recent years in remote areas as well as urban proximate areas. In 2004, wildfires in Alaska burned more than 6.38 million acres, with the largest fire, known as the Taylor Complex Fire, burning more than 1.3 million acres. The largest fire ever recorded in the Sonoran Desert occurred in June 2005 in Arizona, where the Cave Creek Complex Fire burned 248,310 acres (National Interagency Fire Center 2007).

Even in intense crown fires usually some areas remain lightly burned or unburned—protected by the vagaries of wind, topography, and fuel. Such sites are commonly found on north slopes, near lakes or streams, and in draws, canyons, or swamps. Bypassed areas are important as sources of plant propagate for recolonizing the burn. It is commonly supposed that small, relatively undamaging fires are best for the wilderness. However, in most forested ecosystems, including Yellowstone's lodgepole pine forests, infrequent but large crown fires covering thousands of acres are responsible for an area's distinctive character and typical vegetation mosaic (Romme and Despain 1989). Wilderness managers need to try to provide—safely—for some burns of this character, or such ecosystems cannot be maintained in an approximation of their natural state, that is, within their historical range of variability.

Fire Regimes

Although chance plays a role, how a given ignition subsequently develops is mainly determined by the relevant *fire regime*, that is, *the kind of fire activity that characterizes a specific region* (Heinselman 1985). We classify ecosystems into varying fire regimes made up of factors such as (1) fire type and intensity (distinguishing crown fires or severe surface fires from low-intensity surface fires), (2) frequency or return intervals typical for the vegetation type or geographic unit, and (3) size of area burned in a typical ecologically significant fire. For this discussion, we have established the following regimes:

1. Frequent low-intensity surface fires—one- to twenty-five-year return intervals.
2. Infrequent low-intensity surface fires—more than twenty-five-year return intervals.
3. Infrequent high-intensity surface fires—more than twenty-five-year return intervals.
4. Short-interval, stand-replacement crown fires—twenty-five- to one-hundred-year return intervals.
5. Variable regime: frequent low-intensity surface fires and long-interval, stand-replacement fires—one hundred- to three-hundred-year return intervals.

6. Very long-interval stand-replacement fires—more than three hundred-year return intervals.

Combinations of regimes are typical of many regions; we have included one mixed regime to illustrate such variation. Although the concept of fire regime may be an "intuitive rather than rigorous concept," it is nevertheless useful in that it lends a pattern and semblance of order to an otherwise confusing, contradictory, and voluminous literature of fire ecology that tends to be highly specific and descriptive only of a particular fire at a particular time and site (Pyne 1984).

The Natural (Historical) Role of Fire in Wilderness Ecosystems

One of the most fundamental concepts on which all wilderness fire programs are based is that fire plays a natural role with important effects on many native-plant communities, and that many natural ecosystems are fire-dependent (see the descriptions of fire-dependent ecosystems in wilderness regions later in this chapter). An ecosystem can be called fire-dependent if periodic perturbations by fire significantly influence the functioning of the system. Many of the forest, grassland, and savanna ecosystems of the primeval American wilderness were fire-dependent (Kozlowski and Ahlgren 1974; Lotan et al. 1985).

Fire-dependent plant communities burn more readily than non-fire-dependent communities because natural selection has favored development of characteristics that make them more flammable. For example, because they drop highly flammable dry needles annually, ponderosa-pine stands are frequently swept by fire and thus the pine gains a competitive advantage over other species that occur in mixed conifer communities.

Fire performs many roles in most ecosystems or plant communities. Some of these roles, and especially the visual effects, are vividly illustrated in "Fire and Vegetative Trends in the Northern Rockies: Interpretations from 1871–1982 Photographs" (Gruell 1983). Fire plays its roles in concert with other important environmental factors such as topography, elevation, soils, and climate. The most important roles of fire are:

1. Influences species composition of plant communities.
2. Interrupts or alters succession.
3. Influences scale of the vegetation mosaic.
4. Regulates fuel accumulations.
5. Influences nutrient cycles and energy flows.
6. Affects wildlife habitat.
7. Interacts with insects and diseases.
8. Influences ecosystem productivity, diversity, and stability.
9. Influences succession, diversity, and ecosystem stability.
10. Affects water yields, water quality, and sediment.

1. *Fire influences species composition of plant communities in many ways including* (a) triggering release of seeds (jack pine, lodgepole pine); (b) stimulating flowering and fruiting; (c) altering seedbeds by reducing litter and humus and exposing bare soil, ash, or thin humus, thus favoring germination and survival (pines, Douglas-fir, giant sequoia, larch, and many other trees, shrubs, and herbs—including some nonnative species in certain systems); (d) stimulating vegetative reproduction of many species when the overstory is killed; (e) reducing competition for moisture, nutrients, heat, and light; (f) selectively eliminating parts of a plant community; and (g) controlling species and age composition for types reproduced by crown fires only (jack pine, lodgepole pine, and chaparral).

2. *Fire interrupts and alters succession.* More than any other factor, fire interrupts succession in brush fields, forests, grasslands, prairies, and deserts, thereby altering the course of succession. The nature of

Fig. 11.4. The scale of a fire shapes the resulting forest communities. Here a large fire burns in the Scapegoat Wilderness, part of the Bob Marshall Wilderness Complex, Montana, in summer 1988. Photo courtesy Bert Linder, *Great Falls Tribune*.

such disturbances depends on the fire regime typical of the vegetation, topography, and region. For example, the giant sequoia–mixed conifer forests of the Sierra Nevada are thought to have typically experienced light- or moderate-intensity surface fires at short intervals (five to eighteen years). These fires suppressed invading shrubs, true firs and incense cedar. They scarred, but seldom killed, the giant sequoias. At long intervals—perhaps every five hundred to a thousand years—fuel and weather would permit more intense burning, perhaps including crowning out or torching of individual giant sequoias, and created openings that favored sequoia reproduction (Harvey et al. 1980; Kilgore 1973; Parsons 1995). This pattern contrasts sharply with that of jack pine and lodgepole pine in the boreal and Rocky Mountain forests, where the primeval regime was primarily crown fire or severe surface fires covering thousands of acres at intervals of fifty to two hundred years. Such fires killed most or all of the trees and opened the closed cones of the pines to reseed the area (Heinselman 1973; Rowe and Scotter 1973).

Postfire changes in vegetation may follow various pathways, depending on species-specific attributes related to reproduction and survival. For example, a Rocky Mountain community made up of aspen, lodgepole pine, and western larch may develop in several directions, depending on the intervals between fires and the life-history traits of the three species. Aspen has a 130-year lifespan and is capable of vegetative reproduction; lodgepole pine has a 200-year lifespan and needs twenty years to produce mature cones; and larch has a 300- to 400-year lifespan and can become established from seeds dispersed at a considerable distance. If a stand is reburned in less than twenty years, lodgepole pine may be lost if it is not also present in surrounding communities. If the community burns between 20 and 130 years after the last fire, all three species will be present. If the community burns after 200 years, only larch may remain. Chance conditions soon after disturbance, as well as long-term climate changes, play an important role in successional change and contribute to heterogeneity in wilderness (Christensen et al. 1989).

3. *Fire influences the scale of the vegetation mosaic.* Periodic fire occurrence results in a vegetation

Fig. 11.5. Between 1968 and 1986, about 1,200 lightning-caused fires were allowed to burn approximately 190,000 acres on specially designated zones in national parks and national forest wildernesses. Most of these fires were small, low-intensity fires, which burned less than one-fourth of an acre (top), but a few covered thousands of acres over several months, usually closely monitored by fire specialists (above). Photos courtesy Bruce Kilgore, National Park Service.

mosaic, a patchwork, on the landscape, the scale of the patches of contrasting age classes, species compositions, successional stages, or vegetation types reflecting past fire size and the kind of fire regime involved. The physiographic base on which fire works also heavily influences scale. Steep and broken terrain often shows more complex patterns than level, gently rolling, or uniformly graded terrain, because fire and other important environmental factors are more varied in behavior and effects on complex slopes. In regions such as the boreal forest, where large crown fires or severe surface fires are common, individual patches of the mosaic may cover many thousands of acres in dynamic patterns. Although there are basic principles at work, the patterns of disturbance and recovery from fire are uncertain and—like the pieces in a kaleidoscope—are periodically rearranged by fire and succession, and a variety of configurations are possible (Christensen 1995; Wright 1974).

4. *Fire regulates fuel accumulations.* In most coniferous forest ecosystems, the production of plant biomass exceeds decomposition for many years following initiation of a new plant-community succession. Although production of all plant biomass increases with temperature, precipitation, and fertility, so does the rate of decomposition. The overall result is that *plant materials accumulate throughout an ecosystem in both living and dead trees, shrubs, and grasses, and in rotting organic debris comprising the ground litter and humus, fuel that is periodically reduced by fire.*

All of these materials are potential fuel, but it is important to recognize that at any one time, some biomass, like living tree boles, simply will not burn. Even buildup of downed dead biomass occurs in an irregular manner as branches and tree boles accumulate on the ground in response to natural causes of mortality and downfall. Although the process is complex—neither circular nor linear—the general trend is toward an increase in fuels with elapsed time following fire (Brown 1985; Christensen 1995). Then when fire does occur again, all materials combustible under the prevailing weather and fuel moisture conditions may be reduced to ash or volatilized, and the fuel cycle starts again. Some living trees and shrubs are killed but not burned by a given fire; such snags (standing dead tree trunks) and dead shrub skeletons gradually fall and become part of a new ground fuel buildup for a subsequent fire.

5. *Fire influences nutrient cycles and energy flows.* In fire-dependent ecosystems, fire is a major factor influencing nutrient cycling and energy flow. The immediate effect of fire is to convert organic matter stored in litter, twigs, leaves, branches, and dead and down tree trunks to ash and charred materials, with some loss to the atmosphere by volatilization and some material remaining incompletely burned. Available phosphorus, potassium, calcium, and magnesium levels generally increase following burning; while total nitrogen decreases, available nitrogen often increases (Wells et al. 1979). Rates and pathways are influenced by fire frequency and intensity. Overall, low-intensity surface fires facilitate cycling of nutrients and generally do not increase soil erosion, while *high-intensity fires volatilize large amounts of nitrogen, disrupt soil structure, and may induce water repellency and erosion* (Wells et al. 1979).

6. *Fire affects wildlife habitat.* With the great differences in fire frequencies and intensities found in various fire regimes, one would expect fire to have diverse impacts on the broad array of wildlife species and habitats found in different ecosystems. *The influence of fires on wildlife habitat can be summarized with several broad principles*: (a) quantity and quality of browse are often increased immediately after fires; (b) fires increase quantity or availability of berries and seeds; (c) in boreal forests, fires tend to eliminate some forage plants associated with older stands of timber, such as arboreal lichens; (d) fires may increase populations of surface and wood-boring insects that are important to quail, woodpeckers, and other insectivorous birds, but may decrease populations of other insects and animal parasites; (e) fires impact "cover" by changing the scale and pattern of vegetation mosaics, such as "edge" and diversity of related wildlife habitat through frequency, intensity, and size of burn, and may temporarily destroy habitat for many small mammals; and (f) fires interrupt succession and alter plant-species composition and vegetation structure in ways that favor some wildlife species and may not adversely affect others.

Considerable variation in wildlife population response to habitat changes caused by fire may be related to differences in (a) intensity, severity, and duration of the fire; (b) the season of burning; (c) the vegetation type and animal species involved; and (d) whether we are considering short-term effects or longer-term effects. In a state-of-knowledge review of the effects of fire on fauna, Lyon et al. (1978) concluded that although fire may temporarily displace species dependent on late stages of plant-community development, such as caribou, marten, wolverine, and fisher, there is a remarkable stability of species numbers and populations of smaller birds and animals following fire; and in general, grouse and larger animals such as moose, deer, elk, and black bear increase in numbers after fire. One study (Hutto 1995) documented an increase in abundance of several species of birds in areas burned with intense crown fires compared to unburned forest types.

7. *Fire interacts with insects and diseases.* Broadly, fire, or its absence, regulates the total vegetative mosaic and the age structure of individual forest stands that, in turn, influence insect populations. For example, when extensive stands of balsam fir or lodgepole pine reach maturity, outbreaks of the spruce budworm or mountain pine beetle can kill trees and create fuel concentrations that make large-scale fires possible. Such fires then terminate the outbreak by eliminating the host trees until new stands attain susceptible ages. Stand-replacing fires may thus temporarily remove such pests as spruce budworm or mountain beetle, or plant parasites such as dwarf mistletoe on black spruce, lodgepole pine, and other species. Thus, attempts to reduce the frequency of disturbance by fire or insects may lead to situations where large areas of otherwise unmanaged forests are more susceptible than normal to catastrophic disturbances by insects, fire, or other factors. Additionally, global climate change has allowed some insect species to spread beyond historic ranges; for example, warmer early winter temperatures allow beetles that infest and kill spruce trees to survive in the Pacific Northwest, thereby increasing the fuel load available for fires.

8. *Fire influences ecosystem productivity, diversity, and stability.* Fire speeds up the recycling of nutrients that might otherwise move very slowly, and production of vegetation is heavily linked to complete nutrient cycling. Periodic light surface fires or severe fires at long intervals prevent the accumulation of nutrients, dry matter, and energy in organic soil layers; reduce peat formation; and prevent permafrost encroachment (Heinselman 1974). Recurrent burns may be necessary to maintain long-term system productivity in many ecosystems.

   Because there is considerable disagreement about how to measure diversity and stability, it is difficult to compare the effects of fire on these factors in wilderness ecosystems, but vegetative cycles maintained and driven in long-term patterns by fire must be considered to be stable. For example, grasslands, chaparral shrub fields, and lodgepole pine communities would be considered very stable, because fires in such seral communities (intermediate stage of ecological succession) result in replacement vegetation similar to that originally found there. Even in vegetation where large crown fires are typical, individual patches of a mosaic may cover thousands of acres in dynamic patterns, yet the mosaic as a whole changes little over time, thereby establishing and maintaining a historical range of variation.

9. *Fire influences succession, diversity, and ecosystem stability.* Early ecological concepts held that climax communities were inherently more diverse and stable than pioneer communities that follow fire. However, "undisturbed" natural systems apparently did not exist in the real world. *Fire interrupts successional trends at intervals related to the operational fire regime*, thus creating an ever-changing, fire-created mix of successional stages, communities, and stand ages in the vegetation mosaic of most wilderness ecosystems. This pattern creates diversity and stabilizes the system as a whole (Heinselman 1981).

10. *Fire affects water yields, water quality, and sediment.* Because fire may change vegetation cover over large areas, it affects the rate and extent of surface water runoff and thereby the amount of minerals, contaminants, and sediment deposited in watercourses. This impacts water quality.

How Has the Historical Role of Fire Been Modified by Fire Suppression?

Since the early part of the twentieth century, fire has been systematically suppressed in national forests, national parks, and other public and private lands. Early suppression efforts were sometimes feeble, but by the 1930s and 1940s it had dramatically improved. Thus, most season-long fires that had burned forests, shrublands, and grasslands periodically for thousands of years were suppressed. This disrupted the mosaics of vegetation described in the discussion of fire regimes. As such, many lightning fires that may have naturally covered large acreages in the Sierra Nevada, or the forests or grasslands of Florida or Washington, were instead suppressed at less than the size they would have attained. This has been particularly true as aerial detection systems, smoke jumpers, helitack crews, and aerially delivered fire retardants have been used to suppress lightning ignitions while they are still small, even in remote areas.

In addition, European settlers gradually put an end to the burning done by Native Americans in many parts of North America. Such burning had been virtually eliminated by the 1700s in eastern America and as late as the 1870s in the Sierra Nevada of California. Removal of such fires, together with the advent of fire suppression, resulted in major changes in frequency of fires in certain areas, such as the giant sequoia–mixed conifer forests of Kings Canyon National Park (Kilgore and Taylor 1979; Parsons 1995). Native American–set fires were fairly widespread in the West (Arno 1985; Gruell 1985; Lewis 1985) but at the same time were often localized in distribution, many occurring at lower-elevation areas. Hence, removal of this ignition source had varied impacts on different fire regimes in present park and wilderness lands (Phillips 1985).

However, despite fire prevention and suppression, and the loss of Indian ignitions, national data suggest a trend of increasing area burned over the past ten to twenty years. The trend is clear, though we might argue over the causes, such as global warming and associated weather fluctuations, increased use of prescribed fire, or the effects of fuel buildups from fire prevention and control. The following description from a USDA Forest Service Assessment (U.S. Department of Agriculture, Forest Service 2001, 10) of the 2000 fire season in its Northern and Intermountain regions is indicative of additional serious fire seasons in the new millennium: "Although it is difficult to rank the significance of individual large fire years, the extent and effects of the 2000 fire season in the Northern Rockies and Great Basin arguably rank the 2000 fire season in the top five most serious since 1870, and perhaps the second most serious fire season in the last 100 years."

More than 3 million acres burned during 2000 in the portions of six states making up these regions, four hundred homes or other structures were destroyed, thousands of people were evacuated, sixty communities were besieged with smoke for extended periods, and tourism was impacted. All lands are included here, including wilderness.

The role that fire now plays in near-natural zones, such as parks and wilderness, is affected not only by suppression of Native American and lightning fires, but by developments adjacent to and downslope from such areas. Commercial forests, farms, home sites and towns, or even paved parking lots and condominiums now surround many wilderness units and parks. Vegetation often has been subjected to unnatural forces—pinyon-juniper stands removed by chaining, grasslands invaded by chaparral, and plant densities reduced by grazing livestock. Fires that historically would have been ignited in lower-elevation zones and moved up into the present wilderness and park units will no longer play a role in maintaining the historical fire frequency. Moreover, any management plans to allow natural lightning ignitions to burn within such parks and wilderness units are seriously compromised by having high-value developments and private lands adjacent to the wilderness boundaries. So other alternatives for restoring fire to wilderness and park ecosystems are now seriously considered, if not commonly applied, including use of agency-ignited prescribed burns under conditions subject to complete management control. Of course, because fires ignited or allowed to burn in wilderness are restricted to "safe" burning periods, they often fall short of mimicking natural fire effects (Christensen 1988).

The reduced presence of fire for the past fifty to eighty years in certain western wildernesses may have resulted in two primary changes of concern to wilderness managers: (1) an increased incidence of unnatural fuel accumulations (perhaps reflected in recent increases in acreage burned and fire severity) and (2) some modifications of vegetation structure beyond the range found in presettlement ecosystems (Bonnicksen and Stone 1985; Parsons et al. 1986).

For example, in short-fire-interval ecosystems, where fires occur every ten years or so on the average, lack of fire for a fifty- to eighty-year suppression period could allow accumulation of fuels to an unnatural level (van Wagtendonk 1985). By contrast, in long-fire-interval ecosystems, where natural fire cycles range from one hundred to five hundred years or more, fire suppression for the past fifty to eighty years has probably not affected these ecosystems greatly (Habeck 1985).

Forest structure is often divided into four conceptual aspects: *age structure, species composition, horizontal structure* or *mosaic pattern,* and *vertical*

Fig. 11.6. Prescribed burns (left) are used in certain wildernesses and parks when lightning fires cannot be allowed to burn because of hazardous accumulations of fuel and changes in forest structure induced by fire suppression. Such prescribed burning is also appropriate in small wildernesses and near wilderness boundaries, where natural fire might threaten commercial developments and private lands (right). Photos courtesy Bruce Kilgore and William Jones, National Park Service.

structure or *fuel ladders*. Each of these aspects can be modified by fire exclusion. In the frequent, low-intensity fire regime, lack of fire has allowed shifts in species composition from sun-loving pine to more shade-tolerant true fir, and related changes in mosaic patterns and vertical fuel ladders as well as changes in age composition of the various mixed conifer species (Kilgore and Taylor 1979; Parsons and DeBenedetti 1979; van Wagtendonk 1985).

As specific examples, fire suppression has brought the following structural changes to ponderosa-pine and sequoia–mixed conifer forests: (1) a large increase in the younger age classes of pine (in nature ponderosa-pine types) and shade-tolerant white fir (in sequoia–mixed conifer forests); (2) survival of saplings beneath mature trees (in ponderosa-pine forests) and one or more vertical layers of white fir beneath the overstory canopy of sequoia and pine (in sequoia–mixed conifer forests) providing ladder fuels that can lead to high-intensity crown fires; (3) many more trees per acre, particularly of young shade-tolerant saplings; and (4) a blending of what had been discrete patchy units into a more uniform forest, with more uniform burning intensities, gradually destroying the identity of individual even-aged groups or aggregations found in presettlement times (Kilgore and Taylor 1979; Parsons and DeBenedetti 1979; Weaver 1974).

Objectives of Wilderness Fire Management

Considering the important natural roles played by fire in various wilderness ecosystems and the impacts of fire suppression and exclusion on short-fire-interval systems during the past fifty to eighty years, it is not surprising that restoring a more natural role for fire is an objective of most wilderness fire-management programs. *The basic objective can usually be stated as follows: "To restore fire to its natural role in the ecosystem to the maximum extent consistent with safety of persons, property, and other resources."* Note that the goal is *not* to produce any specific mix of vegetation types, to create desirable wildlife habitat, to reduce fuels, to improve aesthetics, or to attain related specific benefits. Some of these benefits might accrue from a successful program, but the *objective is to "restore the naturalness of the environment and let natural processes take over"* to the maximum extent possible.

In large national parks and wildernesses, the objectives are also to try to perpetuate landscapes and landscape processes. Said another way, managers want to aim for a moving-picture vignette made up of a range of spatial patches that would have existed in our time had we not interfered with the processes responsible for creating and maintaining such patches. The range of patches and relative importance of different disturbance processes should be regarded as variables likely to change in the future. Similarly, in national forest wilderness, the objective is to use fire to create desired future conditions in wilderness. These ideas are embraced in the concept of historical range of variability (HRV) that has emerged in importance in the last decade (see Chapter 10): the notion that *ecosystems may vary in structure and appearance over time, but they are still considered natural if they remain within their HRV. Fire is a key determinant of the HRV.*

Agency Wilderness Fire Policy/Objectives

Despite four decades of effort to restore fire to a more natural role in wilderness, the accomplishments lag far behind the acreage that would have been burned historically with no intervention by modern humans.

Although some good fire-management plans have been implemented in the larger wilderness, most wildernesses are small and do not have fire plans. Every bad fire year or incident, such as the Yellowstone fires of 1988, tightens fire policies in the direction of suppression (Mutch 1995; Parsons and Landres 1998).

Aplet (2006) observes that while there are many excellent managers throughout the federal land management agencies who support wilderness fire, some managers still struggle against three types of barriers: risk-averse agency administrators and a system that has historically suppressed fires; institutional complex planning and management requirements as well as the liability and responsibility when operating as the fire manager; and political and public concerns about the nuisance of smoke and fire, the fear for human health and welfare, and concern about the loss of structures and other economic losses.

The largest wildernesses are managed by the USFS and the NPS. Wilderness fire management in these two agencies has changed tremendously in the past four decades toward allowing natural fire a more natural, historic role in wilderness ecosystems—when conditions are right (prescribed conditions).

In 1968, largely based on ecological research findings, the NPS modified its policy of immediate suppression of all fires, which had been standard practice since the NPS was created. In the early 1970s, the USFS also began allowing lightning-caused fires to play a more natural role in wilderness, an exception to its policy of suppressing all fires by 10 AM the following day were authorized as approved by the chief of the USFS. This gradual modification of policy (see Kilgore 1976, 1982) culminated in early 1985 in a policy revision whereby the USFS may even ignite fires in wilderness under certain conditions. It is important here to note that all fires allowed to burn in wilderness—whether naturally or deliberately ignited—are prescribed fires, because fires are only allowed to burn under specified conditions. In 1997, the National Wildfire Coordinating Group (an interagency group of fire managers) adopted new terminology that recognizes only two kinds of fire: (1) wildland fire (nonprescribed) and (2) prescribed fire—any fire ignited by management actions to meet specific objectives (Parsons and Landres 1998). Fires that are allowed to burn will henceforth be called wildland fires.

One of the important aspects of the 1985 USFS policy was the use of prescribed fire to reduce unnatural buildups of fuels when necessary to (1) permit lightning-caused fires to play a more natural ecological role within wilderness and (2) reduce the risks and consequences of wildfire within wilderness, or escaping from wilderness (U.S. Department of Agriculture, Forest Service 1985). Subsequently, desired future conditions, in wilderness and elsewhere, have become a focus of USFS fire policy (Stokes 1995). USFS personnel would ignite fires only where—because of past fire suppression—"acceptable" lightning-caused fires alone cannot achieve wilderness management objectives. Thus, before using prescribed ignitions, wilderness managers need to be sure the fire danger in a given area is greater than would have existed had fire been allowed to occur naturally during the past fifty to eighty years (Kilgore 1987).

In the NPS, the objectives for wilderness fire management, as well as most other resource-management objectives for the natural areas of the national parks, are grounded in principles laid down by the Leopold report of 1963 (Leopold et al. 1963). Despite the fact that the original document can be interpreted as seeing park ecosystems as static rather than dynamic entities, the Leopold report was the first document to explicitly call for an ecologically based management philosophy: "Above all other policies, the maintenance of naturalness should prevail" (Leopold et al. 1963, n.p.)

How to achieve the goals suggested by the Leopold Committee has been the focus of much debate since then. Several NPS researchers proposed that the principal aim of National Park Service resource management in natural areas should be the unimpeded interaction of native ecosystem processes and structural elements (Parsons et al. 1986). Any need for intervention in such natural ecosystem operations would be limited, such as (1) to reverse or mitigate anthropogenic factors where knowledge and tools exist to do so, (2) to protect a featured resource, and (3) to protect human life and property. *Included in the limited situations requiring intervention—meaning the use of prescribed fire—would be the situation where a fire would now burn with greater intensity than would have occurred under natural conditions,* for example, an unnaturally intense crown fire caused by unnatural fuel buildup that drastically changes a community type adapted only to frequent low-intensity surface fires.

Further refining this idea, Agee and Huff (1986) believed that neither structure nor process alone could be defended as the appropriate vegetation management goal for wilderness. Both can be interpreted as being mandated by the Wilderness Act; conversely, both can be criticized for not producing truly natural vegetation. The issue becomes which is most appropriate in a certain situation. After analysis of a wilderness situation having frequent fire (Crater Lake, OR), they concluded that the natural-fire regime, size of the area, and degree of past human intervention all influence setting goals. They indicated that a more reasonable approach at Crater Lake may be to integrate both types of goals into a hybrid structure/process goal in which fire is reintroduced in a low-intensity mosaic pattern as the restorative tool, but with attention paid to maintaining structural elements (mature pine) that are not easily replaced if removed by too intense fire (Agee and Huff 1986). Clearly some situations, such as Crater Lake forests where logging had reduced pine/white fir ratios, require special care in restoring fire to reproduce natural conditions (Thomas and Agee 1986).

Managers wrestling with these concepts tend toward a conservative management strategy of "minimum intervention." That goal is to let natural processes operate as freely as possible while minimizing the impacts of human actions—past and present—on wilderness ecosystems (Kilgore 1985).

Following the high-intensity fires in Greater Yellowstone in 1988, an Interagency Fire Management Policy Review Team appointed by the secretaries of agriculture and interior recommended modifications to agencies' fire-management policies for national parks and federally designated wilderness areas (Christensen et al. 1989). Although the review team confirmed that the objectives of prescribed natural-fire programs in national parks and wildernesses are sound, it suggested that implementing these policies needed to be refined, strengthened, and reaffirmed. The report, and *political reactions to the Yellowstone fires and others since then, led to policy changes to even more carefully screen prescribed natural fires* and to encourage use of management-ignited prescribed burns to complement prescribed natural-fire programs. All fire-management plans are now carefully reviewed to ensure that current policy requirements are met and that they include interagency planning, stronger weather and fuel prescriptions, and daily certification that adequate resources are available

Fig. 11.7. Smoke is natural in many forest ecosystems and cannot be eliminated without consequences. Yet when fires are deliberately set as part of a wilderness management program, the impacts of smoke on visitors and nearby communities must be carefully considered and controlled. Photo courtesy National Park Service.

to ensure that prescribed fires will remain within prescription, given reasonably foreseeable weather conditions and fire behavior. Of course, even the highest level of planning and the strongest level of policy do not totally eliminate the risk from prescribed fire. The goal should be to understand and manage the risk.

The evolution of NPS fire policy still favors the most liberal and untouched natural fires possible in parks but with policies and practices carried out at the park level (Norum 1995). The idea is that parks are parks because they are unique, special places, and managers in any of them must consider a natural fire in light of a park's vegetation types, topography, fuels, cultural resources, neighbors, typical fire behavior, size and shape, founding legislation, previous management, and other conditions (Norum 1995).

What Is "Natural" in Wilderness Fire Management?

The term *natural* does not yet have a common meaning in wilderness fire management or fire ecology. It can mean (1) fire occurring during the pre-European, pretechnological, or presettlement periods; (2) fire occurring without human past or present intervention or influence; (3) the attributes of the fire process known or presumed to be an intrinsic part of a given vegetative type; or (4) the role fire played in the evolution of an ecosystem. Managers of national parks and wilderness units are still seeking to clarify whether Native Americans should be considered as sources of "natural" or "primeval" fire.

Fig. 11.8. Efficient fire control, using helicopters (top) and fire-retardant bombing (middle), has contributed to less than natural incidence of fire in some wildernesses. Photos courtesy USFS. Light-handed tactics are being used in areas such as the Eagle Cap Wilderness, Wallowa-Whitman National Forest, Oregon (bottom), whereby fire is allowed to burn in low-fuel areas and to either go out or be extinguished with minimal effort and disturbance of the natural setting. Photo courtesy Francis Mohr, U.S. Forest Service.

A working definition of natural fire, first developed for the 1983 Wilderness Fire Symposium (Lotan et al. 1985) and evolving since then, considers natural fire for any given ecosystem (1) burns within the range (and frequency distribution) of fire intensities, frequencies, seasons, and sizes found in that ecosystem before arrival of Western technological humans; and (2) yields the historic range of fire effect results found in that ecosystem before the arrival of technological humans. Thus, natural fire and its effects stay within the HRV for the ecosystem involved.

Even with this definition in mind, philosophical and policy questions remain about the appropriateness of fire-restoration efforts. Should wilderness and park managers (1) simply allow natural fires to burn, (2) reduce obvious fuel accumulations in certain zones with prescribed fires and then allow natural fires to burn, or (3) carefully restore natural stand structure to estimated presettlement conditions before allowing natural fires to burn?

Allowing natural fires to burn has a broad appeal, but it has many drawbacks, too. In presettlement times, many lightning fires originated outside and spread into areas now designated as wilderness. In addition, some lightning fires currently ignited in wilderness will be suppressed. Hence, an array of options need to be carefully evaluated (Kilgore 1987). These options include allowing natural fires to burn where possible and using prescribed ignition fire where natural fires cannot be allowed to burn, where ignitions from outside wilderness are no longer possible, and where unnatural fuel accumulations must be reduced.

As we learn more about how fire operates, the definition of *naturalness* becomes more complex. Johnson and Agee (1988) point out that national park and wilderness ecosystems cannot be defined at a particular level that will unequivocally be perceived as "natural," because the word *natural* involves individual value judgments. Park and wilderness preservation goals will have to be stated in more precise system-component terms, depending on the values represented by the individual area.

Although it may be difficult to define such concepts as "wilderness," "wildness," and "natural," the management objectives for parks and wilderness require that we try. If not, we will have lost contact with the philosophy behind the original establishment of national parks and wilderness. In our approach to such objectives, we need to realize that whatever we thought we knew from the research of the past may be supplanted by newer knowledge. We need to be continually open to the best and latest thinking about the more complex role fire and fire suppression have had on given ecosystems.

Wilderness Fire Policy Alternatives and Their Consequences

Five broad policy alternatives are available as wilderness fire alternatives: (1) attempt to suppress all fires; (2) allow all fires to burn; (3) manage lightning-caused fires; (4) ignite prescribed fires; and (5) manipulate vegetation and fuels without fire. Although these options can be combined, it is useful to describe each separately. Failure to pursue a specific policy will still result in the unintended or haphazard implementation of one or more of these options.

Alternative 1: Attempt to Suppress All Fires

A fire-exclusion policy requires an immediate, aggressive attempt to suppress all fires, regardless of cause, location, or expected damage. It might fail in application, but if the policy requires prompt suppression, without exceptions, then the objective is to exclude fire from the ecosystem. Until the early 1970s, this was the standard policy in most wildernesses and national parks, and it remains the policy in many. At the very least, it is often defensible as a holding action until the expertise and equipment to implement a more desirable option are available, or until a rational judgment concerning the best alternative can be made. However, if it is known that an ecosystem was strongly fire-dependent, managers must recognize that fire exclusion is really a powerful form of vegetation manipulation. If the policy succeeds, there will probably be major changes in both vegetation and wildlife populations and perhaps also major changes in the productivity, diversity, and stability of the ecosystem. Often, excluding fire is only temporary, and thus, ultimately the policy fails and the deleterious effects become obvious and long-lasting.

When fire is excluded, accumulation of fuels and changes in vegetation structure may occur. In the kinds of ecosystems we are discussing here, fire was one of nature's ways of reducing fuels and recycling biomass and nutrients. With both lightning and humans as sources of ignition, one wonders if fire exclusion is really a viable option in such ecosystems. *By following a policy of exclusion, we may be setting the stage for major high-intensity fires that could not only endanger human life and property but also disrupt the very ecosystems we are purportedly protecting.*

Fire exclusion often requires the use of bulldozers or other heavy line-building equipment, aircraft, smoke jumpers, and retardants. When major fires in wilderness threaten lives or property outside the wilderness, managers are tempted to use all available techniques, regardless of the consequences to natural resources. Fire exclusion, if practiced too long, can force managers to inflict major damage on the landscape in the name of saving the wilderness from fire. For example, tractor-built or hand-constructed fire lines can erode, leading to lasting environmental damage. Old trails and temporary roads opened for access to the fire can also erode and can change subsequent access patterns to and in the area. Fire camps can leave impacts, even after the temporary mess is cleaned up. Aerially applied retardant can pollute streams. Unless it is not a significant factor in the ecosystem, fire should not be excluded except as an interim measure.

More reasonable variations of an aggressive attack policy in wilderness are included in the USFS's "confine, contain, and control" tactics. *Confine* means to restrict a fire within predetermined boundaries; *contain* means to surround the fire with a fire line to check its spread; whereas *control* means to put it out, involving fire-line construction, burning out, and otherwise removing any threat of subsequent escape. *Some fairly creative efforts at minimizing impacts on wilderness resources can be developed under the "confine" tactic,* including careful observation and monitoring of the fire's progress to ensure that it stops at reasonable natural boundaries.

It is also becoming very clear that aggressive fire suppression or exclusion strategies are very expensive and that significant amounts of money can be saved with less aggressive strategies if they are feasible.

Alternative 2: Allow All Fires to Burn

Allowing all fires to burn is the opposite of fire exclusion, and such a program may endanger human life and property—including lives and property beyond wilderness boundaries. Responsible management must provide for the safety of persons and property—both outside the wilderness and for visitors and agency personnel within the area. Therefore, this option usually must be rejected, except in the most remote, large wilderness areas, such as in interior Alaska.

Alternative 3: Manage Lightning-Caused Fires

A policy of managing lightning-caused fires rests on the belief that they are natural and desirable in wilderness ecosystems. This policy attempts to restore lightning fires in wilderness to their natural status while protecting lives and property and coping with unnatural fuel accumulations that, if ignited, might even damage wilderness resources. *This policy approach avoids direct manipulation as much as possible by allowing nature to select the time, place, vegetation, and fuels for fires through lightning ignitions.* The management involves selective fire control based on both safety and ecological considerations. For many large areas, this is a viable approach to the fire problem, and we will be discussing it further.

Alternative 4: Ignite Prescribed Fires

The goal of an "ignited prescribed fire policy" is to *restore the natural fire regime by substituting deliberate ignitions for lightning-caused or Native American–caused fires.* The commonly held assumption that the ecological effects of fire will be the same, whether human- or lightning-caused, is not totally valid. Although it might be possible through skillful and well-planned ignitions to create significant burns that closely resemble lightning-caused fires, it might also be possible to burn in seasons when lightning never occurs or at closer than natural intervals. So these circumstances should be considered too in trying to stimulate naturalness when igniting prescribed fires. The main advantage of this option is that fires can be managed best if the time and place of ignition are selected in advance, thus allowing time to ready personnel and equipment, monitor weather forecasts, assess fuels, and prepare control lines. Many situations exist—such as small management unit size and proximity to wilderness boundaries—where lightning fires are not allowed to burn. Prescribed fire then becomes a reasonable substitute. This option will also be explored later in greater detail.

Fig. 11.9. In the summer of 1988—a period of extreme drought—a series of lightning-caused fires and human-caused fires burned some 793,880 acres in Yellowstone National Park. The Clover Fire in the background joined several other fires that eventually covered about 1.4 million acres in Yellowstone National Park and adjacent national forests. Photo courtesy Jim Peaco, National Park Service.

Alternative 5: Manipulate Vegetation and Fuels Without Fire

This policy substitutes mechanical manipulations such as harvest of the forest, soil disturbance, or planting for the periodic natural perturbations caused by fire. The policy rejects fire, both lightning-caused and prescribed, as an unacceptable or unsafe agent of change. Safety is usually given as the reason for favoring this option. Vegetation removal need not be commercially motivated, and no product must necessarily be moved from the site, and in fact would usually be illegal in a wilderness or park.

There are many ecological problems with this approach, not the least of which is that it is unnatural, and that the scientific values associated with natural vegetation would be lost when virgin forests or natural vegetation of any type are heavily manipulated or removed. As a method of restoring vegetation already altered by commercial logging, the method might have some merit. However, once a little manipulation is begun, it is easy to conclude that more would be better.

Alternative 6: Combined Alternatives

Various combinations of the five wilderness fire policy alternatives would be possible, but the most practical combination—one now being used in various parks and wildernesses—involves (a) allowing lightning-caused fires to assume their natural role as much as possible; (b) using ignited prescribed burns where it is impractical to allow lightning fires to burn or where past suppression requires a reduction in unnatural fuel accumulations and (c) suppressing fires near wilderness boundaries where valuable resources outside the area are at risk, a confine tactic.

The panel of scientists assessing the ecological consequences of the 1988 fires in the Greater Yellowstone Area pointed out that it was relatively easy to set broad objectives to protect a particular national park or wilderness scene; however, it is far more difficult to develop objectives to preserve and protect particular *natural processes*. The next generation of managers will need to rewrite ideals into management objectives and answer the questions: What kind of manipulation is acceptable? By what means? For what purposes? On what scale? According to what social and political processes? (Christensen et al. 1989).

Four policies related to manipulating wilderness fire must be faced: (1) Under what conditions are scheduled prescribed burns or human-ignited burns appropriate in parks and wildernesses? (2) Do we need to simulate the historic role of Native American burning in certain wilderness units? (3) How do we deal with the natural role of high-intensity, stand-replacing fires characteristic of many northern wilderness ecosystems? (4) If natural-ignition fires are part of a given unit's fire plan, what scale of fire (patch size) is acceptable?

Lacking evidence on the importance of Native American burning and recognizing the difficulties of simulating it, the NPS has minimized such ignitions in management plans for parks in the Sierra Nevada (Parsons et al. 1986).

A major problem in trying to simulate Native American burning and in restoring fire in general is that fuel conditions have changed as a result of decades of fire suppression. Fuel has built up, and this is evident in the evolving trend of increasing acres burned in large, severe fires in recent years, a trend increasing into the new millennium. High-intensity fires in wilderness now pose one of the most difficult policy and program dilemmas for the future. Yet the question remains of whether the ecological effects of historic, stand-replacing crown fires can be duplicated with gentle and manageable prescribed surface burns. It seems important, therefore, that we either develop the capability to allow intense fires to burn in wilderness or substitute higher-intensity prescribed burns that produce the ecological effects of historic stand-replacing fires. Although a number of such fires were allowed to burn in large wilderness units in Idaho and Montana between 1979 and 2000, few covered the acreage or reached the intensity of the Greater Yellowstone fires of 1988 (Jeffery 1989) and others in California and Alaska since 1988, and those that did burn large patches at high intensity did not necessarily remain within the wilderness boundaries. Hence, the issue of how to handle large-scale, high-intensity fires will continue to be a major challenge for fire managers.

Wilderness Fire-Management Planning: Considerations and Constraints

Wilderness fire-management planning aims to provide a guide for all fire-management actions within the wilderness portion of a park, wilderness, or other natural area. Such planning involves determining the appropriate response to lightning fires and use of manager-ignited fires to accomplish wilderness management objectives. Important considerations include fire and ecosystem interactions, including fire history, potential, and effects; and special resource and use considerations, such as archeology, rare and endangered species, administrative sites, grazing allotments, and so forth (Fischer 1985).

Managers need a clear understanding of the natural-fire regime in the area of concern, such as the historical role of fire in the ecosystem in question for developing and evaluating a park and wilderness fire-management plan. But, there is some question about whether a given "natural" regime from the past would be appropriate in the future without allowance for the considerable yet difficult-to-predict variations that would occur under long-term climate changes in response to global warming.

In places like Sequoia and Kings Canyon National Parks, fire management is now focused primarily on restoring fire as an ecological process rather than restoring precise vegetation structure. Given our limited and imperfect knowledge about vegetation structure, fire regimes, the role of Native American burning, and possible shifts in long-term climate, the NPS has decided it can best re-create and maintain an ecosystem in which fire can function fairly naturally, by using a strategy of minimum intervention.

At the same time, every effort is being made to further define *natural*, to carefully monitor behavior and impact of both lightning fires and prescribed burns carried out in these areas, and to continually reevaluate both objectives and methods as these programs continue (Parsons 1995; Parsons et al. 1985). Modern management tools can help computer mapping of fuels and vegetation with the use of GIS and remote-sensing tools to help managers understand the landscape dimension of the potential fire regimes. Computer simulation of historical fire patterns, intensities, and frequencies can be used along with additional fire history studies to obtain a better understanding of the historical range of predictable fire intensities found in natural fires.

Economic Considerations

Wilderness fire-management programs have been planned and evaluated largely on an ecological basis rather than on cost effectiveness (Agee 1985). Yet one of the arguments for allowing lightning fires to burn in parks and wildernesses is that society will save costs of suppressing fires as well as gain positive ecological values by perpetuating wilderness character. However, identifying dollar value returned for dollar value spent in various wilderness fire programs is difficult (Condon 1985; Towle 1985). In a survey of twelve national parks with significant fire-management programs, Agee (1985) found that few generalizations could be made about the cost effectiveness of complex fire-management plans. None of the plans compared the costs of the current plan with costs of total suppression. Average costs of lightning-caused or agency-ignited prescribed fires were from $7 to $19 per acre, whereas suppression costs averaged $1,830 per acre (Agee 1985). Today the costs of suppression are usually much higher.

Interagency Planning

One of the most important advances in wilderness fire planning has been the movement toward joint efforts across agency boundaries. This began in 1976 with the Interagency Wilderness Fire Management Program, involving one-half million acres of the Teton Wilderness in Wyoming (Bridger-Teton National Forest) adjacent to Yellowstone National Park. For the first time on a planned basis, fire on national parklands would be allowed to cross the boundary onto national forestlands, and fires from the Teton Wilderness would be allowed to cross into Yellowstone National Park (Kilgore 1982). Cooperative agreements were in effect in 1982 that involved more than 4 million acres on five national forests and two national parks in the Greater Yellowstone region.

The 1988 Greater Yellowstone Area fires were the first test of this concept, and postfire reviews by the Interagency Fire Management Policy Review Team urged that such fire-management planning be further strengthened. Additional early interagency fire planning efforts involved Lassen National Forest and Lassen Volcanic National Park in California (Swanson and Denniston 1985) and extensive efforts in Alaska (Taylor et al. 1985).

These experiences all indicated the importance of the informal as well as the formal aspects of the planning process, involving all levels of the organization to promote feelings of ownership, and getting the public involved early in the planning process. Integrating the planning team with interagency members was important too so that differences in procedures and terminology did not interfere with the overall objective of preparing a plan to allow fire to burn in the combined park and forest ecosystems in a prescribed, rational pattern. Today, interagency park and wilderness fire planning is a widespread, accepted practice wherever park and wilderness jurisdictions adjoin.

In 1999 and 2000, the U.S. Government Accounting Office (USGAO) recommended that the U.S. Forest Service and National Park Service develop more cohesive strategies for agency response to wildland fire threats and events and then evaluated their program five years later (U.S. Government Accounting Office 2005). The 2005 USGAO report acknowledged that progress had been made by both agencies; however, the report highlighted the important shortcomings and challenges to complete a more cohesive strategy and plan of action to address wildland fires in a timely and cost-effective manner.

Other Constraints on Planning

Other constraints that need to be considered in the wilderness fire-management planning process include (1) air-quality impacts; (2) unnatural buildup of fuels and changes in vegetation structure following extensive and extended suppression activities; (3) wilderness-user safety; (4) impacts of fire management on recreation and cultural resources, and particularly on archeological resources; (5) limitations imposed by relatively small wilderness units and wilderness boundary issues, such as fires starting near boundaries in any wilderness unit; and (6) wilderness and wildland fires escaping into adjoining developed lands and residence areas.

1. *Air Quality—All planning for use of prescribed fire must take into consideration the Clean Air Act and public tolerance for smoke. Prescribed fires—including those ignited by lightning and allowed to burn under prescribed conditions—produce varying quantities of smoke. Such smoke is an integral part of many ecosystems, yet when* manager-ignited fires contribute to naturally, already occurring smoke as part of a wilderness fire-management program, public reaction may be severe. Smoke-management objectives and techniques must be carefully applied to keep local and regional air quality within acceptable limits.

   Although prescribed fire appears essential to meet wilderness management objectives, the short-term effects of fire on air quality may violate certain air-quality standards, either inside or outside the wilderness. The Environmental Protection Agency (EPA) has developed National Ambient Air Quality Standards (NAAQS) for six air pollutants, including particulate matter. The Clean Air Act requires all federal agencies to comply with all federal, state, and local air-quality regulations (Haddow 1985).

   Proposed standards for particulates will include both inhalable and respirable particulates, much smaller particles than previous standards have addressed. Most smoke particulates emitted from prescribed burning are in these small size classes. Land managers need to work closely with EPA and state air regulatory personnel to explain the importance of uses and control methods available for wilderness fire. Tools have been developed to predict and monitor air-quality impacts and to coordinate policy making with federal and state regulatory agencies (Haddow 1985). In this way, both wilderness fire and air-quality objectives can be achieved.

2. *Unnatural Fuels and Forest Structure—We have* already discussed *the possible impacts that fire exclusion had in creating an unnatural increase in fuels or a change in the vegetation structure. The danger is in the possibility that such conditions could foster high-intensity crown fires, risking damage to resources beyond the wilderness, or to local wilderness resources from too intense heat.* Thus, where unnaturally high fuel loads exist, managers may favor prescribed burning to reduce fuel before allowing lightning fires to burn.

3. *Wilderness-User Safety—As numbers of visitors increase in parks and wildernesses where fires are allowed to burn for long periods of time (months in some cases), the likelihood of human contact with such fires also increases.* Although public safety has not been directly threatened, visitors have been inconvenienced when routes are closed because of fire and anyone present would be irritated by smoke. The potential exists for life-threatening events involving wilderness visitors and wildfire. *Managers need to prepare visitors and agency personnel to avoid accidents and disasters. Yet, in wilderness, safety from risk is not guaranteed, and this is part of the appeal for wilderness experiences.* There is more risk in wilderness recreation than elsewhere because of the emphasis on natural processes and the chance to experience one's place in a system where natural forces are free to operate (Silverman 1995).

   Safety procedures that can help prevent prescribed fire accidents include interpretative contacts made with visitors near ongoing fires; consistent, accurate monitoring and evaluation of fire behavior as a basis for effective briefings and contingency plans; informing nearby residents and visitors about fire occurrences, status, and actions; caution signs on roads and trails to warn travelers that a fire is in progress; and maps and brochures that instruct/inform about safety hazards and precautions.

   Mutch and Davis (1985) describe the possibility of a disaster resulting from prescribed fire programs in wilderness. Their scenario is that ten to twenty small lightning fires have gradually been allowed to burn under prescription during June and July in a large wilderness where numerous visitors, traveling on foot and on horseback, are scattered at unknown locations. Over a period of weeks, the weather turns hot and dry until one day in mid-August, "red flag" (dangerous conditions) weather is forecast: strong afternoon winds with gusts to fifty miles per hour, temperatures of 95° to 100°F, and humidities below 15 percent. The "prescribed conditions" under which the twenty small lightning fires had been allowed to burn have changed. The extreme, red flag conditions could not be forecast in June or July, when the prescribed conditions for letting the fires burn were met.

Wilderness managers must recognize subtle warning signals during the early stages and must constantly monitor conditions and adjust their wilderness fire program's information and access policies to ensure the safety of both recreationists and agency personnel. Thus, the possibility for disaster is a constant "warning" influence on agency plans for natural and prescribed ignited fires in wilderness, and calls for contingency plans to curtail or constrain prescribed fires under extreme conditions.

4. *Impacts on Cultural Resources*—Wilderness fires may damage recreation and cultural resources, and suppression activities may do even greater damage (Anderson 1985; Switzer 1974). Although fire intensity and duration of heat are potentially damaging to cultural resources, heavy equipment used during suppression is the most important threat to archeological resources. During the 1977 wildfire at Bandelier National Monument, New Mexico, archeologists guided fire-line construction to prevent needless destruction (Anderson 1985). Managers of park and forest wildernesses with important recreation and cultural resources should consider their protection in fire-management planning and involve recreation and cultural resource specialists in fire-suppression implementation when such resources may be damaged.

5. *Small Wildernesses and Boundary Issues*—Allowing lightning fires to burn is usually not practical in smaller wilderness units, such as where distance from the ignition point to the boundary may be less than a mile or where topography would allow the fire to move quickly to lands beyond the boundary. This same limitation, of course, can apply to ignitions near boundaries of even larger wilderness units. In such situations, fuel management is especially important. Particularly where developments and private lands adjoin parks or wilderness boundaries, agency-ignited prescribed burns can be used to create a buffer zone of reduced fuels. Near many small wildernesses and large wilderness boundaries, neighboring landowners suppress all fires. Thus, a major source of natural ignitions outside the park or wilderness boundary is lost. Agency-ignited prescribed burns must then be substituted for natural ignitions to maintain near-natural fuel levels and forest structure.

6. *Wilderness and wildland fires escaping into adjoining developed lands and residence areas*—As human habitation and second home developments increase around wilderness areas, so does concern for the loss of structures and human life, and economic costs (Miller 2006). Since 1988 several large fires have caused extensive damage to residences and commercial buildings, such as the June 1996 Millers Reach Fire in Alaska, which destroyed 344 structures and burned 37,336 acres, and the June 2002 Hayman Fire in Colorado, which destroyed 600 structures and burned 136,000 acres (National Interagency Fire Center 2007). More urban-proximate fires have increased public concern over wildfires because of extensive structure damage and loss of life, such as two fires in California: the October 1991 Oakland Hills Fire (1,500 acres), which destroyed 2,900 structures and caused the loss of 25 lives, and the October 2003 Cedar Fire (275,000 acres), which destroyed 2,400 structures and caused the loss of 15 lives (National Interagency Fire Center 2007).

Fire Behavior Prediction

In the 1980s and 1990s, fire-behavior-prediction technology advanced rapidly and models were developed to predict fire behavior and intensities (Keown 1985b; Rothermel 1983). The potential for intense crown fires, effects on air quality, and visitor safety are important outputs from those techniques. Fire-behavior predictions are also useful for understanding historical fire behavior in a wilderness, and "gaming" techniques can be used to compare predictions of fire size, intensity, and rates of spread with the actual results. Knowledge of long-term climatic cycles and fluctuations (e.g., see Finklin 1983) is essential to adequately make such comparisons. Fire-simulation models are useful for training and for estimates based on quantified prior experience. All the research tools available to land managers, however, can only supplement—not replace—the planner/manager's basic fire-behavior knowledge and experience. Judgments based on available data, but grounded in manager judgments, are fundamental components of wilderness fire-management plans.

Limiting Fire-Suppression Impacts

The special characteristics of parks and wildernesses require that routine fire-suppression actions be adjusted

Fig. 11.10. The 1988 fire season in Yellowstone National Park was characterized by unusually dry fuels and periods of high winds. Once a fire was ignited, burning embers might be lofted across several miles to start numerous spot fires. Some spot fires grew to thousands of acres; others burned out at comparatively small size. The landscape of Yellowstone became a mosaic of burned and unburned forest, as shown in an aerial view of the Madison River near Madison Junction (left) and of the Old Faithful Complex (right), which was overrun by the North Fork Fire, September 7. Photos courtesy Jim Peaco, National Park Service.

to minimize signs of human activities. Tractor fire lines, felled snags, helispots, and areas clear-cut of standing trees and snags can cause longer-lasting adverse impacts on the wilderness resource than the wildfire itself. Decisions must be made about what equipment will be allowed in wilderness and under what circumstances. Special concern is needed for impacts on wilderness resources during suppression actions, including suppression priorities beyond protection of human life, such as property, threatened and endangered species, recreation, historical or archeological sites, and Native American religious sites.

Fire-suppression strategies and tactics to minimize damage to the wilderness are now common. Both prefire training in special suppression techniques and postfire monitoring and evaluation of the results can help improve efforts to be easy on the land with wilderness fire-suppression efforts. However, it all starts in the wilderness fire-management plan.

Monitoring and Evaluating Wilderness Fires

As in most planning systems, wilderness fire planning calls for monitoring and evaluating as a final step. Operational monitoring of prescribed fire calls for collecting and recording data on fire behavior, weather, fuels, topography, air quality, and fire effects to provide a basis for immediate evaluation of whether the fire is still in prescription, or later evaluation of whether objectives were achieved. The primary purposes for monitoring fire behavior and weather on wilderness fires are to ensure the fire remains in prescription and inside park or wilderness boundaries, and that human life and property are not threatened. The main reasons for gathering data on fuels, fire effects, and air quality are to ensure that resource-management objectives are met and to minimize impacts on resources—such as air and water—outside the wilderness or park. The resource objective in wilderness and parks usually involves the restoration of natural processes, including intensities and frequencies of fire, to maintain/restore a natural range of conditions. Techniques for monitoring and evaluating wilderness fires are outlined by Ewell and Nichols (1985), Ryan and Noste (1985), and Reeburg (1995).

Learning from Past Mistakes

Although the use of prescribed fire will always incur some risk, we can and must learn from ongoing programs, particularly examples where fires have not gone as planned. Such situations may be the result of mistakes in judgment or lack of knowledge and training. Reviewing case studies of fires that escaped their plans can be instructive and the fire literature provides many of them (e.g., Christensen et al. 1989).

Following are a few overviews. Several agencies had prescribed burns that went awry, some in wilderness and some in nonwilderness: the Mack Lake Burn (USFS), the Ouzel Fire (NPS), and the Bandelier National Monument fire (NPS).

The Mack Lake Burn in Michigan was ignited on May 5, 1980, by foresters of the Huron-Manistee

National Forest in what was intended to be a 210-acre prescribed burn in a cutover area of jack pine. The objective was to provide critical nesting habitat for an endangered species, the Kirtland's warbler. Yet, by 7 PM, one firefighter was dead and 20,000 acres of jack pine had burned along with forty-one homes and summer cottages (Simard 1981).

The Ouzel Fire, by contrast, was a wilderness fire ignited by lightning, and the managing agency personnel (NPS) decided to allow the fire to burn. Ouzel began on August 9, 1978, in a fire-management unit above 10,000-feet elevation in spruce-fir forest of the Rocky Mountain National Park in Colorado. Although it initially smoldered and crept along, by early September high winds caused crowning and spotting (Butts 1985). On September 15 and 16, it made substantial runs outside the prescribed fire zone in the direction of Allenspark, a small community near the park's boundary. The fire had to be suppressed.

The Cerro Grande Fire was started as a prescribed burn of 900 acres by 19 staff and crew in Bandelier National Monument (NPS) on May 4, 2000. The fire spread, due to weather changes, into the Santa Fe National Forest and threatened adjoining structures. By May 19, the fire had been substantially contained by more than a thousand firefighters but not before the fire had burned 48,000 acres, caused the evacuation of 18,000 residents, destroyed or damaged 280 residences and 40 structures at the Las Alamos National Laboratory, and caused over $1 billion in damages (U.S. Government Accounting Office 2000).

History and Evolution of Wilderness Fire-Management Programs

Wilderness fire-management programs were initiated years ago in a number of national parks, wildernesses, and related reserves. During the late 1980s and 1990s, such programs and their supporting policies were extended to virtually all parks, wildernesses, and comparable reserves that are in fire-dependent ecosystems.

The beginnings of park and wilderness fire-management programs can be traced in part to the so-called Leopold Committee's 1963 report on "Wildlife Management in the National Parks" (Leopold et al. 1963). Although not aimed directly at fire problems, it identified the ecosystem changes resulting from eliminating natural fire as a key element influencing national park wildlife habitat. In 1965, an NPS directive called for implementing the Leopold report, particularly for areas in the natural category in several of the large parks, and encouraged the use of fire, including prescribed fire, as an appropriate natural agent in ecosystem-restoration programs. Fire-management programs were then not long in appearing.

Sequoia–Kings Canyon and Yosemite National Parks, California

The first designation of a natural-fire-management zone in forested western wilderness, and the first breakthrough in prescribed burning in the West, came in Sequoia and Kings Canyon National Parks in 1968. Beginning then, lightning-caused fires were allowed to burn under surveillance unless safety problems or resource damage was anticipated, in a special management zone above 8,000 feet in the Middle Fork Kings River drainage. This zone was gradually expanded and now encompasses about 740,000 acres, or roughly 86 percent of these parks. The forests are generally open and subalpine, and much of the terrain is extremely rocky or above timberline. Some 305 fires were allowed to burn within this zone up to 1986, burning about 29,000 acres. Partial suppression was needed for only a few fires, and no serious problems developed (Bancroft et al. 1985).

In 1969, starting in the Redwood Mountain Area of Kings Canyon National Park, prescribed burning to reduce unnatural fuels and invading shade-tolerant understory trees began in the lower-elevation mixed-conifer forests, including some giant sequoia groves (Kilgore and Sando 1975). Initially, some cutting and piling of small trees and litter were used to avoid control problems and unnatural overstory damage. Similar work was done later on a twenty-acre unit of the Mariposa Grove of giant sequoias in nearby Yosemite National Park. Later, more extensive understory prescribed burning in mixed conifers was carried out without significant modification of fuels in both the Yosemite and Sequoia–Kings Canyon areas. This type of burning became part of a large-scale operational program (Bancroft et al. 1985).

In 1972, a natural-fire zone was also established in Yosemite and, by 1982, it included 594,000 acres, or about 78 percent of the park. The judgment was that essentially any fire within this zone could safely be allowed to burn because of prevailing fuel, vegetation, and physiographic factors *if* specified weather and fuel prescriptions were met.

An additional 57,000 acres were in a *conditional fire zone*, an area of restricted burning because of the fuel accumulations generated by previous fire exclusion. As these fuels were reduced, parts of the conditional zone became areas in the natural-fire zone. In the remainder of the park, which included most of the park's developments, a full-scale fire-suppression program was in effect, but prescribed burning was used to reduce fuels in critical areas (van Wagtendonk 1978). Prescribed fire-management planning, and its evaluation with research and monitoring, have continued at Sequoia–Kings Canyon and Yosemite National Parks, thereby providing valuable knowledge in support of fire, management programs elsewhere (e.g., Parsons 1995).

Selway-Bitterroot Wilderness, Idaho and Montana

The Selway-Bitterroot Wilderness, managed by the USFS, is one of the largest units (1,337,910 acres) of the NWPS in the contiguous forty-eight states, and it is the first USFS unit having operational experience with a sizable wilderness fire-management program. The White Cap fire-management area in the Selway-Bitterroot in 1972 became the first approved exception to the USFS 10 AM (total suppression) policy. In a pioneering joint research and management effort, fire-management prescriptions were written for each vegetation-management zone of the hundred-square-mile area (Habeck 1976).

The first major test of the White Cap plan was the 1,200-acre Fitz Creek Fire in 1973. Although the fire had to be suppressed in one area where it had escaped the approved fire-management area, the experiment was successful and presaged the incorporation of fire-management considerations into wilderness planning throughout the USFS. Between 1974 and 1979, additional fire-management plans were developed for other units of the Selway-Bitterroot Wilderness. The Independence Fire of 1979 burned more than 16,300 acres during a three-month period, the largest pre-1988 fire allowed to burn under a natural-fire program by any agency (Keown 1985a). By 1982, more than 1 million acres in the Selway-Bitterroot Wilderness were covered by plans that allow lightning-caused fires to play a more natural role. During the first ten years of the program, seventy-six lightning fires were allowed to burn nearly 39,000 acres (Kilgore 1982).

Grand Teton and Yellowstone National Parks and Surrounding National Forest Wildernesses, Wyoming and Montana

Natural-fire-management programs in Grand Teton and Yellowstone National Parks in Wyoming began in 1972. During the first fourteen years, to 1986, of this program, more than 177 lightning fires were allowed to burn more than 33,000 acres. Among these were thirteen larger fires, ranging from 160 to 7,400 acres (Kilgore 1982). This early history, however, was dwarfed by the 1988 fires, which burned some 1.4 million acres in the Greater Yellowstone Area, including about 793,880 acres in Yellowstone National Park. These included both prescribed natural fires and wildfires (Christensen et al. 1989). A series of four fires that started as prescribed natural fires covered more than 450,000 acres in Yellowstone National Park, including two fires (or fire complexes) of nearly 200,000 acres each.

One of the largest natural-fire areas and programs in the United States was created in 1982 under the revised Teton Wilderness Plan. Wilderness segments of Yellowstone and Grand Teton National Parks were coordinated in a more than 4-million-acre prescribed natural-fire program that included designated natural-fire areas in the adjacent USFS Teton, Washakie, North Absaroka, and Absaroka-Beartooth Wildernesses (Kilgore 1987). Fire on national parklands would be allowed to cross the boundary onto national forestlands, and fires in any of the national forest wildernesses would be allowed to cross into Yellowstone and Grand Teton National Parks.

Controversy and Support for Wilderness Fire-Management Programs

Studies of fire history and fire effects on various aspects of different ecosystems provided a site-specific database to support early fire-management plans and continue to do so today. This basic information is important every time a fire escapes its prescription and attracts public attention to the concept of allowing lightning or ignited prescribed fires to burn in parks and wildernesses. *However, it is still difficult to gain support for natural fires, particularly where they are visible in residential, commercial, or aesthetically important areas,* or where annoying smoke and bad air quality result. Public reluctance to accept fire is a legacy of seventy years of fire-prevention campaigns. Studies, however, document growing public knowledge about the natural effects of fire and growing public acceptance of fire management in wilderness (McCool and Stankey 1986). In particular, recent studies document that public acceptance of wildland fire often is influenced both by their understanding about the ecological process of fire and their trust in the federal land-management agencies to carefully manage wildland fires that are in line with ecological stewardship and protection of public welfare and safety (Knotek 2006; Liljeblad and Borrie 2006). *Public information and education are the key, for the research documents increased acceptance of prescribed fire as knowledge about the natural role of fire increases.*

Fire-Dependent Ecosystems in American Wilderness Regions

The following brief descriptions of the role of fire in ecosystems of various wilderness regions—including differences in fire history and fire effects—will help explain the need for active wilderness fire-management programs.

The Sierra Nevada Region of California

This high mountain country contains some of the world's best-known national parks and wilderness units and some of the most magnificent conifer forests on earth, including Sequoia–Kings Canyon and Yosemite National Parks and the surrounding national forest wildernesses. The climate is winter-wet/summer-dry, with snow remaining into June in higher elevations. Dry lightning storms occur infrequently, but ignitions are common because of the dry summers (twenty-one to forty lightning-caused fires per million acres per year). Episodes of dry thunderstorms combined with unusual dryness cause widespread fires at intervals of several years; low-intensity surface fires occur at five- to twenty-year intervals in giant sequoia and ponderosa pine–sugar pine stands (Biswell 1967; Kilgore and Taylor 1979).

Many giant sequoias bear impressive multiple-fire scars, some of them more than a thousand years old. Most individual fires were probably small because the terrain is broken and the vegetation varied. Fire intensity has traditionally been thought to have been generally low, with local hotspots, depending on fuel and topography. Lightning-caused fires are also common in the high country, but seldom become large or intense because of the open character of the forest and fuel discontinuities caused by exposed bedrock, boulder fields, wet meadows, lakes, and snowfields. The giant sequoia and mid-elevation pine forests were kept open and structured by frequent periodic surface fires in primeval times, with some evidence of patchy, higher-intensity burning. Fire suppression over many decades has caused unnatural fuel accumulations and invasions of true firs and incense cedar beneath the pines and sequoias. These conditions set the stage for unnatural conflagrations that could kill even the fire-resistant pines and sequoias (Biswell 1967; Kilgore 1973; Kilgore and Sando 1975; Weaver 1974).

The California-Oregon Coast Ranges Region

The string of wildernesses in the coastal mountains, from southern Oregon to southern California, contain chaparral or oak-madrone at lower elevations and ponderosa and lodgepole pine and other species at higher elevations. The climate is similar to that of the Sierra Nevada, but with fewer lightning ignitions (five to twenty fires per million acres per year). The vegetation of the chaparral–oak–madrone–digger pine zones is extremely flammable, and in nature was probably subject to periodic high-intensity fires that killed most of it. Attempted fire exclusion has created very difficult fire-control problems in the chaparral and related vegetation zones due to resulting fuel buildups (Biswell 1974; Dodge 1972; Weaver 1974). The natural fire regime in the ponderosa–Jeffrey pine zones was probably similar to that in the Sierra Nevada.

The Coast Redwoods Region, California

Several small de facto wildernesses and protected reserves occur within Redwoods National Park and in some of the larger California state (redwoods) parks. This region is within the coastal fog belt of northern California—a winter-wet/summer-dry climate, mitigated by fog drip. Elevations are slight, and snow, when it does fall, soon melts. Lightning-caused ignitions occur, although summer thunderstorms are not common. The redwoods were clearly subject to intermittent fires—many of the largest veterans bear deep fire scars hundreds of years old. The natural-fire regime was probably one of long-interval, severe surface fires (Veirs 1980). Redwood sprouts from the root crown, and some groups of trees are sprouts from fire-killed individuals.

The Cascades Range of Oregon and Washington

The Cascades are dominated by a series of spectacular and geologically recent volcanic peaks, many now included in wildernesses or national parks, for example, Glacier Park Wilderness and Mount Rainier National Park. Climate and fire history vary greatly from west to east; whereas summer weather is mostly clear and dry, thunderstorms do occur and lightning ignitions result in eleven to forty lightning fires per million acres per year.

On well-watered west slopes, the natural-fire regime seems to have been mostly one of large-scale but very long-interval crown fires or severe ground fires. Return intervals were perhaps one hundred fifty to five hundred years or more for various sites. Extensive even-aged stands of Douglas-fir attest to past fires, and many fire boundaries are still evident (Franklin and Hemstrom 1981). The lower-elevation, east side, ponderosa-pine forests probably had a regime of frequent light surface fires. East of the Cascade Crest, Fahnestock (1976) documented a lightning-caused fire occurrence rate of thirteen fires per million acres per year in the Pasayten Wilderness (WA) from 1910 to 1969. Two fires, each exceeding 20,000 acres, accounted for 59 percent of the lightning-caused burn area. This history suggests a long-interval crown fire or severe surface fire regime for the Pasayten. Fire control has not yet significantly affected fuels or disrupted the age structure of the forest.

The Intermountain Region and Southwest

Between the Cascade and Sierra mountain systems and the main Rocky Mountain system, there is a discontinuous series of more isolated ranges from Washington and western Montana south to Arizona and New Mexico having a vegetation and fire history somewhat different from either system. The climate retains some Pacific maritime influence, with heavy snow in the winters and dry summers in the north, but with late summer rains in Arizona and New Mexico. Summer thunderstorms are frequent, with twenty to sixty lightning fires per million acres per year, especially during periods of prolonged drought (seventy per million acres in the White Cap drainage of the Selway-Bitterroot Wilderness). June through September is the fire season in the north; May and June are the fire months in New Mexico and Arizona.

Throughout the region, light to moderate small fires occurred every six to fifteen years in ponderosa-pine and Douglas-fir stands. These fires, mainly burning on the forest floor, were severe enough to kill back most of the Douglas-fir regeneration, thin out overdense ponderosa-pine saplings, and occasionally kill off individuals, clumps, or small groves of aged ponderosa pine or Douglas-fir. Such a history maintained the open ponderosa-pine stands originally characteristic of lower elevations throughout the region (Habeck and Mutch 1973; Weaver 1974). Frequent fires kept juniper woodland restricted to shallow, rocky soils and rough topography in many parts of the West. Minimum fire frequency in sagebrush-scrub communities has been reported at thirty-two to seventy years (Houston 1973).

At higher elevations, on some north slopes, and farther north on slopes with more maritime climates, the fire regime has often been one of long-interval (one hundred fifty to three hundred years), severe surface fires or crown fires that killed whole stands on individual slopes or drainages and regenerated relatively even-aged stands of western white pine, western larch, and lodgepole pine. In some areas, these stands also contain mixtures of western red cedar, western hemlock, and grand fir. Fire suppression in the last sixty years has probably altered fuels and vegetation most in areas that were subject to periodic light surface fires. The heavy-fuel, long-interval, crown fire regime areas and higher-elevation forests have probably not been affected so much because many areas would not have burned in the period in any case (Habeck 1985; van Wagtendonk 1985).

The Rocky Mountain Region

The Rocky Mountain system, as discussed here, includes all of the eastern front ranges and the secondary western ranges extending from Jasper National Park in the Canadian Rockies south to Colorado. It includes such widely known park and wilderness units as Glacier and Yellowstone National Parks and surrounding national forest wilderness units, the Bob Marshall Wilderness of Montana, and many lesser-known wilderness units in Montana, Wyoming, Utah, and Colorado.

This region holds a broad cross-section of vegetation, ranging from white bark pine, alpine larch, and extensive meadows and alpine tundra in the north and at timberline to local areas of western hemlock, western red cedar, western white pine, western larch, Douglas-fir, and ponderosa pine; but the primary species at middle to upper elevations are lodgepole pine, Engelmann spruce, subalpine fir, and quaking aspen. The climate is more continental than in the preceding regions, with long, very cold winters at high elevations and warm summers with considerable rainfall, much of it as thunderstorms. Most lightning fires are extinguished by rains, as such occurrence is only two to fifteen fires per million acres per year; summer droughts increase northward, with highest fire occurrences in Montana and parts of Colorado.

Fire regimes in the Rockies are extremely complex, reflecting the great variation in climate, topography, vegetation, and productivity of mountainous regions (Heinselman 1985). One aspect of the vegetative complexity is the wide ecological amplitude of lodgepole pine. Lodgepole is an extremely adaptable species. In some areas it occurs in even-aged stands resulting from periodic stand-replacing fires, whereas in other sites it occurs with multiple ages, sizes, densities, and height classes, interspersed with small even-aged stands.

The major vegetation pattern found in lodgepole pine today was caused by stand-replacement fires, although many uneven-aged lodgepole-pine stands result from lower-intensity surface fires. Most individual fires were low-intensity, creeping, surface fires, but most acreage was burned by the occasional high-intensity crown fires that occurred during severely dry and windy weather (Lotan et al. 1985).

Before 1988 in Yellowstone National Park and adjacent wilderness, it was believed that most stands would not sustain crown fires until they developed a significant understory component of Engelmann spruce and subalpine fir 300 years or more after the previous fire. Until 1988, such fires had ranged from 1,000 to 8,000 acres (Heinselman 1985). The extensive fires of 1988 in the Greater Yellowstone Area increased our understanding of the variable role of large high-intensity fires in lodgepole pine in the Rocky Mountain region (Romme and Despain 1989). The extremely dry fuel conditions combined with repeated episodes of high winds resulted in 1.4 million acres burned in the Greater Yellowstone Area, 68 percent of this in Yellowstone National Park. Of this acreage, 60 percent was burned by canopy fire and 33 percent by surface fire (Christensen et al. 1989). Because of the dryness of fuels and severity of winds, a number of younger forests adjacent to older forests did in fact burn. However, the heterogeneous behavior of the fires led to a complex mosaic of burned and unburned areas that will be a dominant feature of the Yellowstone landscape (Christensen et al. 1989; Romme and Despain 1989).

In the Rocky Mountain region, Engelmann spruce and subalpine fir stands often occur in valleys, coves, and around lakes and streams where they escape most fires, but many stands are also clearly of fire origin (Day 1972; Loope and Gruell 1973). Most quaking-aspen stands were fire-maintained through root suckering, but fire exclusion has prevented their renewal for sixty years, and many stands that might have burned are now in decline.

Fire has also been a factor in maintaining meadow and grassland communities in the parks of river valleys and flats. Ancient, fire-scarred Douglas-firs occurring in groves or as scattered individuals along the margins of these local grasslands tell of their fire history from Jasper Park in the north to the Yellowstone–Grand Teton region in the south (Houston 1973; Loope and Gruell 1973; Tande 1979). Fires apparently burned these grassland, meadow, and sagebrush areas and crept into the Douglas-fir groves around their margins, at intervals of six to sixty years over at least the past four hundred years. Both lightning and Native American ignitions were probably involved, but lightning alone is clearly a major ignition factor.

In summary, the two dominant fire regimes in most presettlement Rocky Mountain wildernesses were (1) long-interval crown fires (perhaps one hundred to three hundred years) in the continuous forests of lodgepole pine mixed with spruce and fir, *and (2) short-interval* (five to sixty years), low- to moderate-intensity surface fires in the lower-elevation Douglas-fir, aspen, and ponderosa-pine stands, in grassy parklands, and in adjacent open lodgepole-pine stands.

The Lake Superior Region

The ancient Laurentian Highlands surrounding Lake Superior contain several reserves that include the last major remnants of the old "Northwoods": Isle Royale and Voyageurs National Parks, Quetico Provincial Park in Ontario, the adjacent Boundary Waters Canoe Wilderness Area, and Seney Wildlife Refuge and Porcupine Mountains State Park in Michigan. This is spectacular lake country, and together these areas contain several thousand small to medium-sized lakes, largely in glacially dammed bedrock basins. All areas have had some logging, but more than 1 million acres of virgin country still remain, largely in the BWCAW-Quetico area and on Isle Royale.

The climate is continental, with long, cold winters and short, warm summers. More than one-half of the annual twenty-eight inches of precipitation falls during May through September, when thunderstorms are common. The frequent lightning ignitions are usually extinguished by rains but some dry storms occur. Lightning fires occur at the rate of about one to five per million acres per year. Fast-moving crown fires tend to occur in spring and fall.

Heinselman (1973, 1981, 1985) has noted that the presettlement Great Lakes forests had three distinct fire regimes: (1) jack pine and spruce-fir forests with very large stand-replacement crown fires or severe surface fires every fifty to one hundred years in the west and one hundred fifty to two hundred years in the east (such fires in the BWCAW sometimes exceeded 250,000 acres in size); (2) red pine and white pine forests with combinations of moderate-intensity surface fires at twenty- to forty-year intervals, and more intense crown fires at one-hundred-fifty- to three hundred-year intervals; and (3) mixed aspen-birch-conifer forests with high-intensity surface or crown fires. (Although intervals are less sure here, spruce budworm outbreaks occurred every forty to seventy years, creating tremendous fuel loads at those intervals.)

In summary, the natural-fire regime of most coniferous forests in this region was one of long-interval crown fires or severe surface fires. As an example, in the well-studied BWCAW, most of the area burned and most stand origins in the last three hundred years can be accounted for in about a dozen fire years. This history suggests that most of the ecologically significant fires occurred during infrequent periods of severe drought. There were many other fires, but they account for only small areas. The natural-fire rotation for the BWCAW as a whole is about one hundred years. This is the only region where studies of charcoal stratigraphy in annually laminated lake sediments have been combined with pollen analysis to document the pre-Columbian fire regime (Swain 1973; Wright 1974). This work shows that periodic forest fires have occurred in the region for at least 9,300 years.

The Boreal Forest Region

This vast region stretches from Quebec and Labrador across northern Canada to interior Alaska (Helmers 1974). Only a few forested nature reserves having wilderness qualities have yet been designated, but great areas are still de facto wilderness, and more reserves will be established. Present reserves include Riding Mountain, Prince Albert, Wood Buffalo, and South Nahanni River National Parks in Canada and Denali National Park and several newer national parks and wildernesses in Alaska.

The forests are relatively simple for so vast a region. Jack pine and balsam fir cover the land from Quebec to Alberta; lodgepole pine and subalpine fir reach from Alberta to the Yukon. White spruce and black spruce, tamarack, plus quaking aspen, balsam poplar, and paper birch cross the continent. The climate is characterized by long, very cold winters and a winter-dry/summer-wet pattern, with occasional extended droughts when periods of warm, clear weather occur. The long days at northern latitudes permit severe drying during June and July; thunderstorms are infrequent, with lightning-fire occurrence less than one per million acres per year.

During presettlement times, the dominant fire regime in the main boreal forest regions of Canada and interior Alaska was apparently one of high-intensity, short- to long-interval crown fires (or severe surface fires). These were large to very large in size (Heinselman 1981), often covering more than 25,000 acres and sometimes more than 1 million acres (Heinselman 1985). In the drier regions of northwestern Canada and interior Alaska, fire cycles probably averaged fifty to one hundred years; by contrast, cycles of one hundred to three hundred years were found in eastern Canada, with its wetter climate, and near tree line in the open subarctic spruce-lichen woodlands. The fire regimes of some jack-pine and lodgepole-pine forests in western

Canada include medium-intensity surface fires that do not kill whole stands at twenty-five-year intervals.

In summary, the typical natural-fire regime in this region is one of long-interval crown fires or severe surface fires, killing most of the stands in given areas. Most of the area burned in given subregions probably burned during major droughts at intervals of ten to forty years (Rowe and Scotter 1973).

The Eastern Deciduous Forest, Appalachian, and Gulf Region

Many small to medium-sized wilderness units have been designated or are under study in this region, including Baxter State Park in Maine, Adirondack Park in New York, and the Great Smoky Mountains National Park in North Carolina and Tennessee, Shining Rock, Joyce Kilmer/Slickrock, and other wildernesses in the national forests. The vegetation of this large region is too diverse and complex to treat here, but in general, it is largely a mixture of broadleaf hardwoods, several pines, and some fir, spruce, and northern white cedar. The climate varies from north to south, but in general, precipitation is abundant throughout the seasons, temperatures are relatively high, and severe and short-term droughts do occur. Thunderstorms are frequent but include rain, and forest fuels decompose rapidly. These climatic patterns result in *a fire regime of infrequent surface fires during the dormant season in the hardwood forests and slightly more frequent but long-interval fires in conifer forests.* The southern coastal plain pine region and the Everglades were clearly the exception; here frequent light surface fires were the rule (Komarek 1974).

Trends and Future Needs

Wilderness fire management is complex and demanding work, requiring professional competence and intimate knowledge of the specific ecosystem and land unit to be managed. More people trained in ecology, botany, and forestry with understanding of fire behavior and an interest in preservation management are needed to manage these programs.

The need for natural-fire programs in wilderness is generally understood by informed supporters of parks and wilderness, but not among people who have been indoctrinated since childhood with the negative aspects of fire. To many such people *all* fire is bad, all smoke is pollution, and most fires kill wildlife. Elected officials can be expected to follow public opinion and to take a conservative view of prescribed fire, because it will usually come to their attention only in response to problems. This legacy from fire-prevention campaigns can be overcome only by new and innovative public information and education efforts with the media and schools.

The more people understand about the function of fire in maintaining the naturalness of wilderness ecosystems, the more likely that they will support the prescribed use of fire. But the task of facing wilderness fire programs is daunting because of the legacy of fire prevention and control, and because past suppression efforts have allowed fuels to accumulate, thus setting the stage for very intense fires in many areas. Fire, including natural fire in wilderness ecosystems, is an important forest research topic, as the numerous references in this chapter attest. The research is continuing (see the publication list by the Aldo Leopold Wilderness Research Institute, Missoula, Montana, on numerous wilderness fire topics, http://leopold.wilderness.net/pubs.cfm), further documenting fire histories, frequencies, ecosystem effects, and emerging concerns such as climate change (McKenzie et al.

Fig. 11.11. Scientists need whole, functioning, natural ecosystems to study fire as a natural force in fire-dependent communities. Photo courtesy Bruce Kilgore, National Park Service.

2004), fire impacts on local populations, and wilderness and recreation use (Borrie et al. 2006).

Wilderness Fire Programs and Ecosystem Management

Wilderness fire management is important not only because it can maintain the natural landscapes and biota of wilderness as a cultural and recreational resource, but also because it can maintain large-scale functioning ecosystems that will contribute to basic scientific knowledge. This is especially important given the current focus on ecosystem management for all wildlands. Wilderness ecosystems can be an important benchmark, classroom, and laboratory for ecosystem knowledge. Our understanding of natural ecosystems—as systems—is still in its infancy. Future scientists will require whole, functioning, natural ecosystems as research laboratories to probe questions vital to the future of humans, as fire is a crucial ecological process to understand. Many of the most widespread natural ecosystems of the earth—including most coniferous forests, savannas, glades, and grasslands—are fire-dependent. In addition, many of our most important domestic plants and animals and useful forest trees evolved in such ecosystems. It is chiefly in our larger wildernesses and national parks that natural ecosystems can be studied because people are rapidly converting the rest of our planet to farms, pastures, commercial forests, mines, highways, and urban areas. As this book has emphasized throughout, the value of natural wilderness processes and conditions will surely increase in the next few decades. And we know that it holds many secrets important to our future on this planet.

Study Questions

1. What factors determine fire regimes? Name some typical regimes.
2. List some roles that fire plays in most ecosystems or plant communities. Explain the importance of each.
3. What impact has the attempted suppression of fire had on fuels and forest structure of wilderness ecosystems?
4. What is the main objective of wilderness fire management in national parks and forests?
5. How do scientists determine the fire history of an area?
6. List some wilderness fire policy alternatives and the advantages and disadvantages of each. Which alternatives can or should be combined in a wilderness fire program?
7. Name some constraints on planning for wilderness fire management. Indicate how each impacts development of wilderness fire plans.
8. What are things to consider in addressing visitor safety with respect to wilderness programs?
9. How has research supported development of wilderness fire-management programs? What are some of the questions and issues research needs to address in the future?
10. Describe and contrast the fire history and fire effects found in two very different wilderness regions of North America, such as the Sierra Nevada and Rocky Mountains, or other regions.

Acknowledgments

This chapter was revised by John Hendee and Chad Dawson for the fourth edition and was substantially revised for the third edition by John C. Hendee with help from Leon Neuenschwander and review and suggestions by Dr. David Parsons, director, Aldo Leopold Wilderness Research Institute, Missoula, Montana. It is largely an updated summary of the more detailed and referenced fire chapter in the first and second editions by Bruce Kilgore of the National Park Service and the late Myron "Bud" Heinselman, U.S. Forest Service and the University of Minnesota.

Case Discussion: Ignition Is One of the Challenges to Managing Fires in Wilderness

Forest and land management in the twentieth century often focused on fire suppression to protect resources, communities, and human life. In fire-dependent ecosystems there has been an accumulation of dead and live vegetation that is greater than historic fuel conditions and this has reportedly contributed to larger and more intense wildland fires (U.S. Government Accounting Office 2005). The average annual acreage of area burned by wildland fires (National Interagency Fire Center 2007) has increased in recent years:

Decade	Average Annual Acreage Burned
1970–1979	3.2 million
1980–1989	3.0 million
1990–1999	3.3 million
2000–2006	7.0 million

A report by the U.S. Government Accounting Office (2005) noted that fuel reduction was one approach for wildland fire management and planning. Although harvesting and thinning operations in disconnected patterns or mechanical removal of vegetation that comprises fuel loads may be appropriate and effective in nonwilderness wildland areas, such mechanical techniques cannot be legally used in wilderness. Equally important, some amount of fire is necessary in fire-dependent ecological communities to sustain the community structure, form, and functions. The degree to which historic fire intensity, extent, frequency, and annual timing is maintained is one of the challenges of managing fire in wilderness. To date, wilderness managers have relied on lightning-caused fires to provide the natural ignition to cause fires to burn either with or without management, such as lightning-ignited fires in prescribed burn areas under approved wilderness fire plans. Between 2001 and 2006, only 51 to 86 percent of the acres of wildland fires each year were started by lightning (National Interagency Fire Center 2007).

The question of using alternatives to natural ignitions along with prescribed fire plans within wilderness has repeatedly been raised as one approach to reducing fuel loads while maintaining fire as a process in wilderness (Miller 2006).

Morton et al. (2006, 17) report that during a U.S. Fish and Wildlife Service workshop in Alaska, "five situations were identified in which prescribed fire might be an appropriate tool in Alaskan wilderness:

1. To restore or enhance habitats of federally listed threatened and endangered species
2. To control or eradicate invasive flora
3. To increase the likelihood of naturally ignited fire to burn unimpeded (by reducing hazardous fuel loads around structures and the urban interface)
4. To restore a natural fire regime that has been temporarily altered (e.g., extreme fuel loads due to blow-down from a hurricane)
5. To mimic a natural fire regime that has been altered and is not expected to be restored due to constraints on wildfire management."

The idea of management-ignited fire that is carefully managed in wilderness by following prescriptive fire-management plans is not philosophically acceptable to many managers and raises serious questions about to what extent we manage natural processes like fire.

Case Discussion Questions:

1. Outline the role of fire in fire-dependent communities. What will happen to that ecological community if fire is suppressed?
2. Does the Wilderness Act permit the use of management-ignited fires along with prescribed fire plans within wilderness?
3. What are some philosophical objections to management-ignited fires?
4. What are the ecological benefits to management-ignited fires, along with using prescribed fire plans, within wilderness when otherwise fires would be suppressed?
5. What are the pros and cons of using management-ignited fires in the five situations listed above where "prescribed fire might be an appropriate tool in Alaskan wilderness"?

References

Agee, James K. 1985. Cost-effective fire management in National Parks. In: Lotan, James E., et al., tech. coords. *Proceedings: Symposium and Workshop on Wilderness Fire,* November 15–18, 1983; Missoula, MT. General Technical Report INT-182. Ogden, UT: U.S. Department of Agriculture, Forest Service, Intermountain Forest and Range Experiment Station, pp. 193–198.

Agee, James K.; Huff, Mark H. 1986. Structure and process goals for vegetation in wilderness areas. In: Lucas, Robert C., comp. *Proceedings: National Wilderness Research Conference: Current Research;* July 23–26, 1985; Fort Collins, CO. General Technical Report INT-212. Ogden, UT: U.S. Department of Agriculture, Forest Service, Intermountain Research Station, pp. 17–25.

Anderson, Bruce A. 1985. Archeological considerations for park and wilderness fire management planning. In: Lotan, James E., et al., tech. coords. *Proceedings: Symposium and Workshop on Wilderness Fire*; November 15–18, 1983; Missoula, MT. General Technical Report INT-182. Ogden, UT: U.S. Department of Agriculture, Forest Service, Intermountain Forest and Range Experiment Station, pp. 145–148.

Aplet, G. H. 2006. Evolution of wilderness fire policy. *International Journal of Wilderness*. 12(1): 9–13.

Arno, Stephen F. 1985. Ecological effects and management implications of Indian fires. In: Lotan, James E., et al., tech. coords. *Proceedings: Symposium and Workshop on Wilderness Fire*; November 15–18, 1983; Missoula, MT. General Technical Report INT-182. Ogden, UT: U.S. Department of Agriculture, Forest Service, Intermountain Forest and Range Experiment Station, pp. 81–86.

Baisan, Christopher H.; Swetman, Thomas. 1995. Historical fire occurrence in remote mountains of southwestern New Mexico and northern Mexico. *Proceedings: Symposium on Fire in Wilderness and Park Management*; March 30–April 1, 1993; Missoula, MT. General Technical Report INT-GTR-320. Ogden, UT: U.S. Department of Agriculture, Forest Service, pp. 153–158.

Bancroft, Larry; Nichols, Thomas; Parsons, David; Graber, David; Evison, Boyd; van Wagtendonk, Jan. 1985. Evolution of the natural fire management program at Sequoia and Kings Canyon National Parks. In: Lotan, James E. et al., tech. coords. *Proceedings: Symposium and Workshop on Wilderness Fire*; November 15–18, 1983; Missoula, MT. General Technical Report INT-182. Ogden, UT: U.S. Department of Agriculture, Forest Service, Intermountain Forest and Range Experiment Station, pp. 174–180.

Biswell, H. H. 1967. The use of fire in wildland management in California. In: *Natural Resources: Quality and Quantity.* Berkeley: University of California Press, pp. 71–87.

———. 1974. Effects of fire on chaparral. In: Kozlowski, T. T.; Ahlgren, C. E., eds. *Fire and Ecosystems.* New York: Academic Press, pp. 321–364.

Bonnicksen, Thomas M.; Stone, Edward C. 1985. Restoring naturalness to national parks. *Environmental Management*. 9(6): 479–486.

Borrie, W. T.; McCool, S. F.; Whitmore, J. G. 2006. Wildland fire effects on visits and visitors to the Bob Marshall Wilderness Complex. *International Journal of Wilderness*. 12(1): 32–35, 38.

Brown, James K. 1985. The "unnatural fuel buildup" issue. In: Lotan, James E., et al., tech. coords. *Proceedings: Symposium and Workshop on Wilderness Fire*; November 15–18, 1983; Missoula, MT. General Technical Report INT-182. Ogden, UT: U.S. Department of Agriculture, Forest Service, Intermountain Forest and Range Experiment Station, pp. 127–128.

Butts, David B. 1985. Case study: The Ouzel Fire, Rocky Mountain National Park. In: Lotan, James E., et al., tech. coords. *Proceedings: Symposium and Workshop on Wilderness Fire*; November 15–18, 1983; Missoula, MT. General Technical Report INT-182. Ogden, UT: U.S. Department of Agriculture, Forest Service, Intermountain Forest and Range Experiment Station, pp. 248–251.

Christensen, Norman L. 1988. Succession and natural disturbance: Paradigm, problems, and preservation of natural ecosystems. In: Agee, James K.; Johnson, Darryll B., eds. *Ecosystem Management for Parks and Wilderness.* Seattle: University of Washington Press, pp. 62–86.

———. 1995. Fire and wilderness. *International Journal of Wilderness*. 1(1): 30–34.

Christensen, Norman L.; Agee, James K.; Brussard, Peter F.; Hughes, Jay; Knight, Dennis H.; Minshall, G. Wayne; Peek, James M.; Pyne, Stephen J.; Swanson, Frederick J.; Wells, Stephen; Thomas, Jack Ward; Williams, Stephen E.; Wright, Henry A. 1989. Interpreting the Yellowstone fires. *Bioscience*. 39: 677–685.

Condon, Michael K. 1985. Economic analysis for wilderness fire management: A case study. In: Lotan, James E., et al., tech. coords. *Proceedings: Symposium and Workshop on Wilderness Fire*; November 15–18, 1983; Missoula, MT. General Technical Report INT-182. Ogden, UT: U.S. Department of Agriculture, Forest Service, Intermountain Forest and Range Experiment Station, pp. 199–205.

Day, R. J. 1972. Stand structure, succession, and use of southern Alberta's Rocky Mountain forest. *Ecology*. 53(3): 472–478.

Dennis, John G.; Wauer, Roland H. 1985. Role of Indian burning in wilderness fire planning. In: Lotan, James E., et al., tech. coords. *Proceedings: Symposium and Workshop on Wilderness Fire*; November 15–18, 1983; Missoula, MT. General Technical Report INT-182. Ogden, UT: U.S. Department of Agriculture, Forest Service, Intermountain Forest and Range Experiment Station, pp. 296–298.

Dodge, M. 1972. Forest fuel accumulation—A growing problem. *Science*. 177(4044): 139–142.

Ewell, Diane M.; Nichols, H. Thomas. 1985. Prescribed fire monitoring in Sequoia and Kings Canyon National Parks. In: Lotan, James E., et al., tech. coords. *Proceedings: Symposium and Workshop on Wilderness Fire*; November 15–18, 1983; Missoula, MT. General Technical Report INT-182. Ogden, UT: U.S. Department of Agriculture, Forest Service, Intermountain Forest and Range Experiment Station, pp. 327–330.

Fahnestock, G. R. 1976. Fires, fuels, and flora as factors in wilderness management: The Pasayten case. In: *Proceedings Annual Tall Timbers Fire Ecology Conference No. 15*; October 16–17, 1974; Portland, OR. Tallahassee, FL: Tall Timbers Research Station, pp. 33–69.

Finklin, A. I. 1983. *Weather and Climate of the Selway-Bitterroot Wilderness.* Moscow: University of Idaho Press.

Fischer, William C. 1985. Elements of wilderness fire management planning. In: Lotan, James E., et al., tech. coords. *Proceedings: Symposium and Workshop on Wilderness Fire*; November 15–18, 1983; Missoula, MT. General Technical Report INT-182. Ogden, UT: U.S. Department of Agriculture, Forest Service, Intermountain Forest and Range Experiment Station, pp. 138–144.

Franklin, J. F.; Hemstrom, M. A. 1981. Aspects of succession in the coniferous forests of the Pacific Northwest. In: West, D. C.; Shugart, H. H.; Botkin, D. B., eds. *Forest Succession, Concepts, and Application.* New York: Springer-Verlag, pp. 212–229.

Frissell, S. S., Jr. 1973. The importance of fire as a natural ecological factor in Itasca State Park, Minnesota. *Quaternary Research*. 3(3): 397–407.

Gruell, George E. 1983. *Fire and vegetative trends in the northern Rockies: Interpretations from 1871–1982 photographs.* General Technical Report INT-158. Odgen, UT: U.S. Department of Agriculture, Forest Service.

———. 1985. Indian fires in the interior West: A widespread influence. In: Lotan, James E., et al., tech. coords. *Proceedings: Symposium and Workshop on Wilderness Fire*; November 15–18, 1983; Missoula, MT. General Technical Report INT-GTR-182. Ogden, UT: U.S. Department of Agriculture, Forest Service, Intermountain Forest and Range Experiment Station: 68–74.

Habeck, J. R. 1976. Forests, fuels and fire in the Selway-Bitterroot Wilderness, Idaho. In: *Proceedings, Tall Timbers Fire Ecology Conference,* No. 14; October 8–10, 1974; Missoula, MT. Tallahassee, FL: Tall Timbers Research Station, pp. 305–353.

———. 1985. Impact of fire suppression on forest succession and fuel accumulations in long-fire-interval wilderness habitat types. In: Lotan, James E., et al., tech. coords. *Proceedings: Symposium and Workshop on Wilderness Fire*; November 15–18, 1983; Missoula, MT. General Technical Report INT-182. Ogden, UT: U.S. Department of Agriculture, Forest Service, Intermountain Forest and Range Experiment Station, pp. 110–118.

Habeck, J. R.; Mutch, R. W. 1973. Fire-dependent forests in the northern Rocky Mountains. *Quaternary Research*. 3: 408–424.

Haddow, Dennis V. 1985. Wilderness fire management and air quality. In: Lotan, James E., et al., tech. coords. *Proceedings: Symposium and Workshop on Wilderness Fire*; November 15–18, 1983; Missoula, MT. General Technical Report INT-182. Ogden, UT: U.S. Department of Agriculture, Forest Service, Intermountain Forest and Range Experiment Station, pp. 129–131.

Harvey, H. Thomas; Shellhammer, Howard S.; Stecker, Ronald E. 1980. Giant sequoia ecology: Fire and reproduction. *Science Monograph Series*, No. 12. Washington, DC: U.S. Department of the Interior, National Park Service.

Heinselman, M. L. 1973. Fire in the virgin forests of the Boundary Waters Canoe Area, Minnesota. *Quaternary Research*. 3(3): 329–382.

———. 1974. Restoring fire to the canoe country. *Naturalist*. 24(4): 21–31.

———. 1981. Fire intensity and frequency as factors in the distribution and structure of northern ecosystems. In: Mooney, H. A., et al., eds. *Proceedings of the Conference Fire Regimes and Ecosystem Properties*; December 11–15, 1978; Honolulu, HI. General Technical Report W0-26. Washington, DC: U.S. Department of Agriculture, Forest Service, pp. 7–57.

———. 1985. Fire regimes and management options in ecosystems with large high-intensity fires. In: Lotan, James E., et al., tech. coords. *Proceedings: Symposium and Workshop on Wilderness Fire*; November 15–18, 1983; Missoula, MT. General Technical Report INT-182. Ogden, UT: U.S. Department of Agriculture, Forest Service, Intermountain Forest and Range Experiment Station, pp. 101–109.

Helmers, A. E. 1974. Interior Alaska (includes reprint of map "Major Ecosystems of Alaska"). *Naturalist*. 25(1): 16–23.

Houston, D. B. 1973. Wildfire in northern Yellowstone National Park. *Ecology*. 54(5): 1111–1117.

Hutto, Richard L. 1995. The importance of intense crown fires to some bird species in Rocky Mountain coniferous forests. In: Brown, James; Mutch, Robert W.; Spoon, Charles W.; Wakimoto, Ronald H., tech coords. *Proceedings: Symposium on Fire in Wilderness and Park Management*. General Technical Report INT-GTR-320. Ogden, UT: U.S. Department of Agriculture, Forest Service.

Jeffery, David. 1989. Yellowstone: The great fires of 1988. *National Geographic*. 175(2): 255–273.

Johnson, Darryll R.; Agee, James K. 1988. Introduction to ecosystem management. In: Agee, James K.; Johnson, Darryll R., eds. *Ecosystem Management for Parks and Wilderness*. Seattle: University of Washington Press, pp. 3–14.

Keown, Larry D. 1985a. Case study: The Independence Fire, Selway-Bitterroot Wilderness. In: Lotan, James E., et al., tech coords. *Proceedings: Symposium and Workshop on Wilderness Fire*; November 15–18, 1983; Missoula, MT. General Technical Report INT-182. Ogden, UT: U.S. Department of Agriculture, Forest Service, pp. 239–247.

———. 1985b. Fire behavior prediction techniques for park and wilderness fire planning. In: Lotan, James E. et al., tech. coords. *Proceedings: Symposium and Workshop on Wilderness Fire*; November 15–18, 1983; Missoula, MT. General Technical Report INT-182. Ogden, UT: U.S. Department of Agriculture, Forest Service, Intermountain Forest and Range Experiment Station, pp. 162–167.

Kilgore, B. M. 1973. The ecological role of fire in Sierran conifer forests: Its application to national park management. *Quaternary Research*. 3(3): 496–513.

———. 1976. From fire control to fire management: An ecological basis for policies. In: *Transactions of 41st North American Wildlife and Natural Resources Conference*; March 1976. Washington, DC: Wildlife Management Institute, pp. 477–493.

———. 1982. Fire management programs in parks and wilderness. In: Lotan, James E., ed. *Fire—Its Field Effects: Proceedings of the Symposium*; October 19–21, 1982; Jackson, WY. Missoula, MT: Intermountain Fire Council; Pierre, SD: Rocky Mountain Fire Council, pp. 61–91.

———. 1985. Human-ignited prescribed fires in wilderness: A response to Bill Worf. In: Lotan, James E., et al., tech. coords. *Proceedings: Symposium and Workshop on Wilderness Fire*; November 15–18, 1983; Missoula, MT. General Technical Report INT-182. Ogden, UT: U.S. Department of Agriculture, Forest Service, Intermountain Forest and Range Experiment Station, pp. 283–285.

———. 1987. The role of fire in wilderness: A state-of-knowledge review. In: Lucas, Robert C., comp. *Proceedings: National Wilderness Research Conference*: Issues, state-of-knowledge, future directions; July 23–26, 1985; Fort Collins, CO. General Technical Report INT-220. Ogden, UT: U.S. Department of Agriculture, Forest Service, Intermountain Research Station, pp. 70–103.

Kilgore, B. M.; Sando, R. W. 1975. Crown-fire potential in a sequoia forest after prescribed burning. *Forest Science*. 21(1): 83–87.

Kilgore, B. M.; Taylor, D. 1979. Fire history of a sequoia–mixed conifer forest. *Ecology*. 60: 129–142.

Knotek, K. 2006. Understanding social influences on wilderness fire stewardship decisions. *International Journal of Wilderness*. 12(1): 22–253.

Komarek, E. V. 1974. Effects of fire on temperate forests and related ecosystems: Southeastern United States. In: Kozlowski, T. T.; Ahlgren, C. E., eds. *Fire and Ecosystems*. New York: Academic Press, pp. 251–277.

Kozlowski, T. T.; Ahlgren, C. E., eds. 1974. *Fire and Ecosystems*. New York: Academic Press.

Leopold, A. S.; Cain, S. A.; Cottam, C. M.; Gabrielson, I. N.; Kimball, T. 1963. Study of wildlife problems in national parks: Wildlife management in national parks. In: *Transactions of the North American Wildlife and Natural Resources Conference*; Washington, DC: Wildlife Management Institute (28): 28–45.

Lewis, Henry T. 1985. Why Indians burned: Specific versus general reasons. In: Lotan, James E., et al., tech. coords. *Proceedings: Symposium and Workshop on Wilderness Fire*; November 15–18, 1983; Missoula, MT. General Technical Report INT-182. Ogden, UT: U.S. Department of the Interior, Forest Service, Intermountain Forest and Range Experiment Station, pp. 75–80.

Liljeblad, A.; Borrie, W. T. 2006. Trust in wildland fire and fuel management decisions. *International Journal of Wilderness*. 12(1): 39–43.

Loope, L. L.; Gruell, G. E. 1973. The ecological role of fire in the Jackson Hole area, northwestern Wyoming. *Quaternary Research*. 3(3): 425–443.

Lotan, James E.; Brown, James K.; Neuenschwander, Leon F. 1985. Role of fire in lodgepole pine forests. In: *Lodgepole Pine: The Species and Its Management: Proceedings of a symposium*; May 8–10, 1984; Spokane, WA. Pullman: Washington State University, Cooperative Extension, pp. 133–152.

Lyon, L. Jack; Crawford, Hewlette S.; Czuhai, Eugene; et al. 1978. Effects of *Fire on Fauna: A State-of-Knowledge Review*. General Technical Report W0-6. Washington, DC: U.S. Department of Agriculture, Forest Service.

McBride, Joe R. 1983. Analysis of tree rings and fire scars to establish fire history. *Tree-Ring Bulletin*. 43: 51–67.

McCool, Stephen F.; Stankey, George H. 1986. Visitor Attitudes Toward Wilderness Fire Management Policy—1971–84. Research Paper INT-357. Ogden, UT: U.S. Department of Agriculture, Forest Service, Intermountain Research Station.

McKenzie, D.; Gedalof, Z.; Peterson, D. L.; Mote, P. 2004. Climatic change, wildlfire, and conservation. *Conservation Biology*. 18: 890–902.

Miller, C. 2006. Wilderness fire management in a changing world. *International Journal of Wilderness*. 12(1): 18–21, 13.

Morton, J. M.; Berg, E.; Newbould, D.; MacLean, D.; O'Brien, L. 2006. Wilderness fire stewardship on the Kenai National Wildlife Refuge, Alaska. *International Journal of Wilderness*. 12(1): 14–17.

Mutch, R. W. 1995. Prescribe fires in wilderness: How successful? In: Brown, James K.; Mutch, Robert W.; Spoon, Charles W.; Wakimoto, Ronald H., tech. coords. *Proceedings: Symposium on Fire in Wilderness and Park Management*; March 30–April 1, 1993; Missoula, MT. General Technical Report INT-GTRP-320. Ogden, UT: U.S. Department of Agriculture, Forest Service, pp. 38–41.

Mutch, Robert W.; Davis, Kathleen M. 1985. Visitor protection in parks and wildernesses: Preventing fire-related accidents and disasters. In: Lotan, James E.; et al., tech. coords. *Proceedings: Symposium and Workshop on Wilderness Fire*; November 15–18, 1983; Missoula, MT. General Technical Report INT-182. Ogden, UT: U.S. Department of Agriculture, Forest Service, Intermountain Forest and Range Experiment Station, pp. 149–158.

National Interagency Fire Center. 2007. www.mifc.gov; accessed January 1, 2007.

Norum, Rob 1995. The National Park Service (fire) program. In: Brown, James K.; Mutch, Robert W.; Spoon, Charles W.; Wakimoto, Ronald H., tech coords. *Proceedings: Symposium on Fire in Wilderness and Park Management*; March 30–April 1, 1993; Missoula, MT. General Technical Report INT-GTRP-320. Ogden, UT: U.S. Department of Agriculture, Forest Service. 122p.

Parsons, David J. 1995. Restoring fire to giant sequoia groves: What have we learned in 25 years? In: Brown, James K.; Mutch, Robert W.; Spoon, Charles W.; Wakimoto, Ronald H., tech coords. *Proceedings: Symposium on Fire in Park and Wilderness Management*;

March 30–April 1, 1993; Missoula, MT. General Technical Report INT-GTR-320. Ogden, UT: U.S. Department of Agriculture, Forest Service, pp. 256–258.

Parsons, D. J.; Bancroft, Larry; Nichols, Thomas; Stohlgren, Thomas. 1985. Information needs for natural fire management planning. In: Lotan, James E., et al., tech. coords. *Proceedings: Symposium and Workshop on Wilderness Fire;* November 15–18, 1983; Missoula, MT. General Technical Report INT-182. Ogden, UT: U.S. Department of Agriculture, Forest Service, Intermountain Forest and Range Experiment Station, pp. 356–359.

Parsons, D. J.; DeBenedetti, S. H. 1979. Impact of fire suppression on mixed-conifer forest. *Forest Ecology and Management.* 2: 21–33.

Parsons, D. J.; Graber, D. M.; Agee, J. K.; van Wagtendonk, J. W. 1986. Natural fire management in national parks. *Environmental Management.* 10(1): 21–24.

Parsons, David J.; Landres, Peter. 1998. Restoring natural fire to wilderness: How are we doing? In Pruden, Teresa; Brennan, Leonard, eds. Fire in ecosystem management: Sifting the paradigm from suppression to prescription. *Tall Timbers Fire Ecology Conference Proceedings No. 20.* Tallahassee, FL: Tall Timbers Research Station, pp. 366–373.

Phillips, Clinton B. 1985. The relevance of past Indian fires to current fire management programs. In: Lotan, James E., et al., tech. coords. *Proceedings: Symposium and Workshop on Wilderness Fire;* November 15–18, 1983; Missoula, MT. General Technical Report INT-182. Ogden, UT: U.S. Department of Agriculture, Forest Service, Intermountain Forest and Range Experiment Station, pp. 87–92.

Pyne, Stephen J. 1984. *Introduction to Wild Land Fire: Fire Management in the United States.* New York: John Wiley.

Reeburg, Paul. 1995. The western region fire monitoring handbook. In: Brown, James K.; Mutch, Robert W.; Spoon, Charles W.; Wakimoto, Ronald H., tech coords. *Proceedings: Symposium on Fire in Wilderness and Park Management;* March 30–April 1, 1993; Missoula, MT. General Technical Report INT-GTR-320. Ogden, UT: U.S. Department of the Interior, Forest Service, pp. 259–261.

Romme, W. H.; Despain, D. G. 1989. Historical perspective on the Yellowstone fires of 1988. *BioScience.* 39(10): 695–699.

Rothermel, R. C. 1983. How to predict the behavior of forest and range fires. General Technical Report INT-143. Ogden, UT: U.S. Department of Agriculture, Forest Service, Intermountain Forest and Range Experiment Station.

Rowe, J. S.; Scotter, G. W. 1973. Fire in the boreal forest. *Quaternary Research.* 3(3): 444–464.

Ryan, Kevin C.; Noste, Nonan V. 1985. Evaluating prescribed fires. In: Lotan, James E., et al., tech. coords. *Proceedings: Symposium and Workshop on Wilderness Fire;* November 15–18, 1983; Missoula, MT. General Technical Report INT-182. Ogden, UT: U.S. Department of Agriculture, Forest Service, Intermountain Forest and Range Experiment Station, pp. 230–238.

Silverman, Arnold 1995. Appropriate risks for recreation in wilderness. In: Brown, James K.; Mutch, Robert W.; Spoon, Charles W.; Wakimoto, Ronald H., tech coords. *Proceedings: Symposium on Fire in Wilderness and Park Management;* March 30–April 1, 1993; Missoula, MT. General Technical Report INT-GTR-320. Ogden, UT: U.S. Department of Agriculture, Forest Service, pp. 89–90.

Simard, A. J. 1981. The Mack Lake fire. *Fire Management Notes.* 42(2): 5–6.

Stokes, Jerry. 1995. Planning for desired future conditions in wilderness. In: Brown, James K.; Mutch, Robert W.; Spoon, Charles W.; Wakimoto, Ronald H., tech coords. *Proceedings: Symposium on Fire in Wilderness and Park Management;* March 30–April 1, 1993; Missoula, MT. General Technical Report INT-GTR-320. Ogden, UT: U.S. Department of Agriculture, Forest Service.

Swain, A. M. 1973. A history of fire and vegetation in northeastern Minnesota as recorded in lake sediments. *Quaternary Research.* 3(3): 383–396.

Swanson, John R.; Denniston, Alan E. 1985. The Park-Caribou Plan: An example of integrated planning. In: Lotan, James E., et al., tech. coords. *Proceedings: Symposium and Workshop on Wilderness Fire;* November 15–18, 1983; Missoula, MT. General Technical Report INT-182. Ogden, UT: U.S. Department of Agriculture, Forest Service, Intermountain Forest and Range Experiment Station, pp. 215–219.

Switzer, Ronald R. 1974. The effects of forest fire on archeological sites in Mesa Verde National Park, Colorado. *Artifact.* 12(3): 1–8.

Tande, G. F. 1979. Fire history and vegetation pattern of coniferous forest in Jasper National Park, Alberta. *Canadian Journal of Botany.* 57: 1912–1931.

Taylor, Dale L.; Malotte, Frenchie; Erskine, Douglas. 1985. Cooperative fire planning for large areas: A federal, private, and State of Alaska example. In: Lotan, James E., et al., tech. coords. *Proceedings: Symposium and Workshop on Wilderness Fire;* November 15–18, 1983; Missoula, MT. General Technical Report INT-182. Ogden, UT: U.S. Department of Agriculture, Forest Service, Intermountain Forest and Range Experiment Station, pp. 206–214.

Thomas, Terri L.; Agee, James K. 1986. Prescribed fire effects on mixed conifer forest structure at Crater Lake, Oregon. *Canadian Journal of Forest Research.* 16(5): 1082–1087.

Towle, Everett L. 1985. Management considerations for a cost-effective fire management program in national forest wilderness. In: Lotan, James E., et al., tech. coords. *Proceedings: Symposium and Workshop on Wilderness Fire;* November 15–18, 1983; Missoula, MT. General Technical Report INT-182. Ogden, UT: U.S. Department of Agriculture, Forest Service, Intermountain Forest and Range Experiment Station, pp. 191–192.

U.S. Department of Agriculture, Forest Service. 1985. Chapter 2320. Wilderness and primitive areas. Forest Service Manual, Amendment 93. Washington, DC: U.S. Department of Agriculture, Forest Service.

———. 2001. Toward restoration and recovery—Overview assessment of the 2000 fire season in the Northern and Intermountain Regions. Washington, DC: U.S. Department of Agriculture, Forest Service.

U.S. Government Accounting Office. 2000. Fire management: Lessons learned from the Cerro Grande (Los Alamos) fire. GAO/T-RCED-00-257. Washington, DC.

———. 2005. Wildland fire management: Important progress has been made, but challenges remain to completing a cohesive strategy. GAO-05-147. Washington, DC.

van Wagtendonk, Jan W. 1978. Wilderness fire management in Yosemite National Park. In: *Proceedings: 14th Biennial Wilderness Conference.* Sponsored by the Sierra Club and National Audubon Society. Boulder, CO: Westview Press.

———. 1985. Fire suppression effects on fuels and succession in short-fire-interval wilderness ecosystems. In: Lotan, James E., et al., tech. coords. *Proceedings: Symposium and Workshop on Wilderness Fire;* November 15–18, 1983; Missoula, MT. General Technical Report INT-182. Ogden, UT: U.S. Department of Agriculture, Forest Service, Intermountain Forest and Range Experiment Station, pp. 119–126.

Veirs, Stephen D., Jr. 1980. The influence of fire in coast redwood forests. In: *Proceedings of the Fire History Workshop;* October 20–24, 1980; Tucson, AZ. General Technical Report RM-81. Fort Collins, CO: U.S. Department of Agriculture, Forest Service, Rocky Mountain Forest and Range Experiment Station, pp. 93–95.

Weaver, Harold. 1974. Effects of fire on temperate forests: Western United States. In: Kozlowski, T. T.; Ahlgren, C. E., eds. *Fire and Ecosystems.* New York: Academic Press, pp. 279–319.

Wells, C. G.; Campbell, R. E.; DeBano, L. F.; et al. 1979. *Effects of Fire on Soil: A State-of-Knowledge Review.* General Technical Report W0-7. Washington, DC: U.S. Department of Agriculture, Forest Service.

Wright, H. E. 1974. Landscape development, forest fires, and wilderness management. *Science.* 186(4163): 487–495.

Chapter 12

Wildlife in Wilderness: A North American and International Perspective

by John C. Hendee and David J. Mattson

Introduction	309
The Wilderness Wildlife Resource	309
Categories of Wilderness Wildlife	310
Wilderness-Dependent Wildlife	310
Wilderness-Associated Wildlife	310
Common Native Wildlife Found in Wilderness	311
Nonnative Species Found in Wilderness	311
Wilderness Wildlife Relationships	311
Wildlife as a Measure of Wilderness Character	312
Wildlife's Role in Wilderness	312
A Wilderness Role in Wildlife Preservation	313
Wilderness Wildlife as an Environmental Baseline and Laboratory	315
Recreational, Aesthetic, Economic, and Political Values	316
Wildlife-Related Problems in Wilderness Stewardship	318
Multiple Agencies and Missions	318
Legislative and Administrative Direction	319
Disease	320
Depredation	320
Human Safety	322
Reintroductions	323
Livestock Grazing	323
Hunting and Poaching	324
Human-Caused Disturbance	325
Long-Term, Long-Distance Effects	325
Wilderness Management Objectives and Guidelines for Wildlife	327
Proposed Wilderness Management Objectives (Conditions to Be Sought) for Wildlife in Wilderness	327
Some Key Guidelines	327
Allow Natural Process to Shape Wilderness Habitat	327
Encourage Angling and Hunting Styles Compatible with Wilderness Experiences	328
Focus Research on Essential Priorities Using Appropriate Methods	329
Summary	330
Study Questions	331
Acknowledgments	331
Case Discussion: Wolf Reintroduction Controversy	331
Case Discussion Questions	332
References	332

Introduction

The presence of certain indigenous wildlife, and their natural distribution, abundance, and behavior, reflect wilderness conditions. Some wildlife are icons of certain wilderness areas.

Internationally, Tibetan antelope and Mongolian gazelle symbolize the vast Asian steppe; wildebeest and white-eared kob symbolize the trackless East African wilderness; the deep taiga of the Amur evokes images of the Siberian tiger; the Serengeti's sere landscape evokes images of leopards, cheetahs, and African lions; Nepal's Chitwan National Park is rhino and tiger country.

In North America, the Bob Marshall Wilderness in Montana is grizzly country; the canyon lands evoke images of the cougar; caribou symbolize the vastness of the Arctic wilderness; and the wolf is now successfully reestablishing its species as a symbol of wilderness in the Yellowstone ecosystem.

Without wilderness, many species of wildlife could not survive or behave as truly wild animals. Wilderness without wildlife, and wildlife without the freedom of wilderness, are virtually unthinkable. *However, wilderness stewardship must be comprehensive because wilderness is more than wildlife habitat.* Thus, important as it is, *wildlife is only one component of the wilderness resource and must be managed as such—one consideration in an overall wilderness stewardship scheme* (see Chapter 7).

This chapter categorizes and describes wilderness wildlife, some important interrelationships between wildlife and wilderness, and the many wildlife-related problems in wilderness stewardship, including legal, administrative, and cultural constraints that influence

Fig. 12.2. Wild horses roam adjacent to a BLM roadless area in Nevada; they compete for forage with native wildlife. Photo courtesy Bureau of Land Management.

managing wildlife in wilderness. The perspective is international, because some of the world's most dramatic wilderness wildlife are in developing regions. Reference to literature is abundant because wilderness wildlife worldwide has been the focus of many studies and books. Finally, with more of an eye to the United States and North America, *the chapter proposes some stewardship objectives for wildlife in wilderness and guidelines for attaining them. Basic principles of wilderness management (see Chapter 7) should prevail. That is, when management actions are necessary, only the minimum tools, methods, and force should be used to meet planned area objectives. In designated wilderness, natural forces should normally be allowed to shape wildlife habitat, populations, and behavior to the fullest possible extent.*

The Wilderness Wildlife Resource

A literal definition of "wilderness" would be "place of the wild beasts" (Nash 1970). But what wild beasts?

From ecological and economic perspectives, several definitions of wilderness wildlife have been suggested. Aldo Leopold (1933) considered wilderness wildlife as species harmful to or harmed by economic land uses specifically—wapiti (elk), caribou, bison, grizzly bear, moose, mountain (bighorn) sheep, and mountain goat. A. Starker Leopold (1966) identified five North American ungulates associated primarily with wilderness climax-forage types: caribou, bighorn sheep, mountain goat, musk ox, and bison. Dasmann (1966) defined wilderness species as those that are obligate members of a climax community or wilderness area, for example, the passenger pigeon, caribou, musk ox, bighorn sheep, and grizzly bear. Durward

Fig. 12.1. A bighorn watches a photographer. Wildlife is a part of all wilderness ecosystems; its distribution, abundance, and behavior reflect the naturalness of a wilderness. Photo courtesy Richard Behan.

Allen (1966) called the cougar, grizzly bear, and wolf true wilderness animals because they are capable, wide-ranging, and at odds with the livestock industry. Hochbaum (1970) includes many species of migratory waterfowl, like swans and geese, that seek undisturbed (wilderness) wetlands for nesting purposes.

Mattson (1997) considered wilderness species to be those vulnerable to contact with humans or, like Leopold (1933), fundamentally incompatible with traditional human economic activities. By these standards, megaherbivores such as elephants and rhinos and large carnivores such as bears, wolves, and large cats depend on wilderness naturalness and solitude for their survival and viability. Migratory ungulates such as caribou, wildebeest, and Mongolian gazelle also require wilderness conditions with unimpeded access to forage over vast areas.

Theoretically, salamanders and butterflies are as important in wilderness as grizzly bears, mountain lions, eagles, and even smaller and less familiar creatures—what Edward O. Wilson (1984, 12) calls "the miniature wilderness that can take almost forever to explore." *However, as a practical matter, the ungulates and carnivores at or near the top of their food chains have historically depended on or symbolized wilderness.* Thus, this chapter focuses on these large species.

Categories of Wilderness Wildlife

Wilderness wildlife may be categorized as (1) wilderness-dependent wildlife, (2) wilderness-associated wildlife, (3) common native wildlife found in wilderness, and (4) nonnative species found in wilderness. We recognize the difficulty of placing a particular species into any single category because of overlap, regional and local variation, and exceptions that do not fit neatly into compartments. However, these categories are useful for discussion because they suggest something about the relationship of the species to conditions found in wilderness.

Wilderness-Dependent Wildlife

Wilderness-dependent wildlife includes species that require wild habitat found in wilderness and that are vulnerable to contact with humans, including those species whose relationship with humans is intractable because they threaten human safety or chronically damage livestock or crops. Such species require refuge from humans or the absence of human development such as roads, railways, and fences.

Although a host of organisms, large and small, are a part of wilderness ecosystems, many of these species are not themselves dependent on wilderness, though they may benefit by wilderness conditions. Species that depend in some way on wilderness are a potential focus of management attention in wilderness settings, and thereby are a symbol of wilderness itself.

Polar bears, Siberian tigers, African elephants, and Pantanal jaguars are examples of wilderness-dependent wildlife, although even these species may, in places, lead a tenuous existence in modified environments outside wilderness. In any particular area, locally or regionally rare species that are dependent on the continued naturalness of their habitats can also be placed in the wilderness-dependent category. In the U.S. northern Rocky Mountains, for example, such locally or regionally rare species as grizzly bear, mountain caribou, native "west slope" cutthroat trout, Canada lynx, wolverine, mountain goat, Richardson's blue grouse, fisher, marten, peregrine falcon, bald and golden eagles, osprey, mountain sheep, and northern white-tailed ptarmigan are associated with wilderness. In the Southwest, desert tortoise and desert bighorn sheep would be on the list.

Internationally, megaherbivores like rhinos, elephants, African Cape buffalo, and large carnivores such as tigers, leopards, African lions, brown bears, pumas, jaguars, Komodo dragons, and large crocodiles are wilderness-dependent—meaning they also depend on naturalness and solitude. In the case of rhinos, elephants, bears, and tigers, problematic relations with humanity are exacerbated by poaching motivated by the high value placed on their body parts.

Wilderness-Associated Wildlife

Wilderness-associated wildlife includes species commonly associated in human perception and ecological reality with wilderness conditions and species displaced to wilderness by human activities. For example, this category includes wildlife common in the high-elevation habitat of the western United States (e.g. whistling marmots) and species associated with conditions characteristic of wilderness in the East, including southern swamps (alligators) and hardwood forests (black bears).

The relationship of wilderness-associated species with humans is not as intractable as that of wilderness-dependent wildlife. Human impacts on them are potentially subject to remedy. For example, construction of the Ulaanbataar-Beijing railroad severed traditional migratory routes of the Mongolian gazelle, resulting in massive declines in gazelle populations due to lost access to forage and reduced opportunities to escape harsh winter conditions (Lhagvasuren et al. 1999). By contrast, with judicious forethought and design modifications providing for migration, construction of the oil pipeline from the North Slope of Alaska to ports to the south did not interfere with migrations of caribou, which have since flourished (Cronin et al. 1998).

Wilderness-associated wildlife also includes species displaced to wilderness for protection by local demand for meat and by dependence on habitat conditions lost with the expansion of agriculture and domestic livestock. A host of ungulates, primates, and their interdependent mid-size or large carnivores in Africa, Asia, and South America, as well as prairie dogs and black-footed ferrets in North America, have been relegated to wilderness for these reasons. A solution? Alternate sources of protein or modest changes in husbandry and farming practices could alleviate some displacement of these kinds of wilderness-associated species, and thus their human-created dependence on wilderness conditions. Such is less often the case for wilderness-dependent wildlife. This is not to say that wilderness is unimportant to the long-term survival of animals such as the orangutan, spectacled bear, red deer (wapiti), or Tibetan antelope. Rather, compared to wilderness-dependent wildlife, humans are graced with more options for coexisting with wilderness-associated wildlife.

Common Native Wildlife Found in Wilderness
Common native wildlife found in wilderness includes species that happen to be found in wilderness but that also often live in more modified environments. Their relationship to wilderness is incidental. They are not necessarily associated in ecological fact or in human perception with especially wild places. Some examples include deer, coyotes, bobcats, raccoons, rabbits, muskrats, and minks; a host of rodents like squirrels, field mice, and rats; and many kinds of birds, such as some raptors, grouse, woodpeckers, sparrows, juncos, and thrushes.

Unlike the other two categories, common native wildlife species do not have specific relationships to wilderness. They neither are dependent on such areas ecologically nor have any real or perceived association with wilderness in the human mind. However, when found in wilderness, they are no less important in their natural roles. In fact, these common species, because they are adapted to more modified environments (and may even be benefited by them), may reveal through their natural place in the wilderness scheme an important comparison with nonwilderness conditions. Thus, keeping wilderness wild provides opportunity for keeping "wildness" in species otherwise adapted to civilization.

Nonnative Species Found in Wilderness
Nonnative species are, unfortunately, a fourth category that is increasingly a threat to wilderness character (see Chapter 13). This category would include *those introduced species that have adapted to wilderness habitat and compete with resident native wildlife, leading to changes in native ecosystems.* This would include such species as wild (feral) hogs, horses and burros, feral goats, nutria, chukkar partridge, and eastern brook trout in wilderness in the western United States. Whether or not a species is native to a particular wilderness can be controversial, such as with mountain goats in the High Uintas of Utah and in Olympic National Park in Washington. The ecological impacts of many nonnative species have been well documented in some proposed or designated wilderness areas, so wilderness wildlife managers are often involved in conceiving, implementing, and evaluating alternatives for the management, impact mitigation, or control of nonnative species. However, in places, nonnatives can benefit the conservation of wilderness species, as with the Patagonian puma, which consumes nonnatives in much of its range because native prey have been nearly extirpated (Novaro et al. 2000).

These four categories are useful guides to inventories of wilderness wildlife and the characteristics of natural habitat on which they depend. Such information is essential for management as well as useful in assessing some of the values dependent on an area's wilderness designation. In the following, some of the most dramatic examples to illustrate the wilderness wildlife categories come from outside the United States.

Wilderness Wildlife Relationships
Wildlife relates to wilderness, and vice versa, in many ways. The presence of certain wildlife can be a measure of wilderness character; wildlife also play a role in the wilderness ecosystems they inhabit and wilderness certainly plays a role in wildlife preservation; wilderness wildlife can serve as an environmental baseline and

laboratory; finally, wilderness wildlife has recreational, educational, aesthetic, economic, and political values.

Wildlife as a Measure of Wilderness Character
The distribution, numbers, diversity, and behavior of wildlife species can be a measure of the naturalness and solitude of a wilderness. Wildlife reflects ecological conditions and their changes over time, so wildlife can serve as indicators of wilderness character and quality—in fact as well as in human perception.

Culturally, for many people the concept of wilderness is linked to some form of wildlife. For example, when Margaret and Olaus Murie (1966) wrote of their American elk studies in Wyoming, they called their book *Wapiti Wilderness*. Andy Russell (1971) titled his book on Canadian wilderness *Grizzly Country*. The presence of particular kinds of wildlife suggests the relative absence of human influence and the existence of primitive harmonies, for example, grizzlies and wolves. For many people, simply knowing that such wildlife are present is important to the meaning of wilderness. If key wildlife are removed, although everything else remains visibly the same, the intensity of the sense of wilderness may be diminished (Nash 1970).

Leopold described the diminishment of wilderness after the demise of an old grizzly, the last of its kind roaming the Arizona high country. When a government hunter shot the bear for bounty, Leopold wrote a eulogy: "(Mount) Escudilla still hangs on the horizon but when you see it you no longer think of bear. It's only a mountain now" (Leopold 1949, 137). The extent to which wildlife is perceived as a wilderness criterion is striking. For example, John Milton (1972) described wildlife species, especially caribou, as carrying the soul of the Alaskan wilderness because they require freedom and vast space in which to range. To Milton, Alaska's wildlife symbolizes its wilderness—the space, the openness, and the silence of the North. Sigurd Olson (1963) wrote of the wild-laughing choruses of the common loon as the sound that more than any other typifies the rocks and waters and forests of wilderness. These examples suggest the importance of wildlife as a cultural reference to wilderness character. However, just as important—some would say more important—is wildlife as a biological indicator of wilderness conditions.

Wildlife that potentially indicate wilderness conditions include megaherbivores, large carnivores, and migratory ungulates, that is, certain wilderness-dependent wildlife. Species that survive only where there are few humans, minimal human activity, and little agriculture logically denote wilderness in a sense that is meaningful to most industrialized societies. Thus, animals such as grizzly bears, wolves, rhinos, lions, tigers, and saltwater crocodiles are prime indicators of wilderness conditions in their native habitats.

Wide-ranging wilderness-dependent species potentially serve as an indicator of wilderness conditions, and can have *umbrella effects,* that is, the protection of natural processes and nontarget organisms as a consequence of managing for a few "featured" species (Caro and O'Doherty 1999). For example, in the case of wide-ranging species vulnerable to contact with humans, protection of wilderness areas (especially large ones) may also preserve natural migratory movements and disturbance regimes and, thus, a full complement of biodiversity.

Organisms that combine wide-ranging movements with wilderness dependence are readily identified. In the Far North, these animals include large carnivores such as wolves, polar bears, and interior brown or grizzly bears. An individual male grizzly bear in the Northwest Territories of Canada can annually range over 2,600 square miles whereas an individual polar bear can wander in excess of almost 18,000 miles across the Arctic ice in search of aquatic prey such as seals, walruses, and beluga whales. Nearer the equator, wild dogs and elephants thrive in similarly large wilderness areas. A viable population of African elephants typically requires eight hundred square miles of high-quality safe habitat while a pack of wild dogs can range over six hundred to eight hundred square miles in search of prey during the dry season. Everywhere they still survive, ungulates that undertake long-range migrations are associated with the most extensive primeval conditions still to be found on earth. This holds for wildebeest and white-eared kob in East Africa, springbok and eland in southern Africa, Mongolian gazelle and Tibetan antelope on the steppes of east-central Asia, and caribou and wild reindeer in circumpolar Arctic and boreal regions. Vast herds of these herbivores can still move hundreds of miles between seasonal ranges only because humans have spared their traditional migratory routes.

Wildlife's Role in Wilderness
Wildlife is an inseparable part of the wilderness resource. It plays a vital role in developing, maintaining, and modifying

soil, vegetation, and ecosystems that cover wilderness topography; in dispersal, planting, and germination of seeds; in pollination; in fertilization; in distribution of nutrients; and in conversion of dead plants into organic matter that is more usable by living plants.

The alligator and its relentless search for water in the dry season is an example from the Everglades wilderness of Florida. Alligators seek low places where the water table lies just below the surface and work either to deepen the existing waterholes or to excavate new ones, breaking the caked earth with their powerful tails and shoveling away the debris with their broad snouts. During the worst droughts, alligators have been known to dig their way down through four feet of compacted mud and peat before coaxing water from the porous substrate. Such gator holes, found throughout the parched glades, attract many thirsty creatures ranging from otters to herons, and soon become biological microcosms of the whole region. The alligators, conserving energy and living on their own fat, largely ignore these boarders; the refugees sustain their lives on the gator holes' remaining fish, insect life, and vegetation, and live side by side in a relative state of truce. When the rains finally return, it is from these gator-made oases that the various species go forth to repopulate the Everglades (Carr 1973).

Alligators in Florida are a good example of a *"keystone" species—wildlife that have a disproportional effect on the structure and function of the ecosystems within which they live* (Simberloff 1998). Such species are consequently of great importance. Large carnivores have keystone effects in wilderness settings attributable to a cascade of influences down the trophic ladder, from top predator, to herbivores, to vegetation, to microorganisms. This has been shown for wolves and moose on Isle Royale, in Lake Superior in the United States (D. L. Allen 1979); for orcas and sea otters along Pacific coastal North America; and for jaguars, puma, and their prey in tropical Central and South America (Terborgh et al. 1999).

Elephants also have a well-documented ability to dramatically transform ecosystems, especially where their natural movements are curtailed by confinement to artificially small protected areas. Elephants have virtually deforested some of these areas and consequently changed fire regimes and ecosystem productivity (Dublin 1995). Prairie dogs and termites can have similarly transforming effects on ecosystems

Fig. 12.3. Elk browse on summer range on the Gallatin National Forest near Yellowstone National Park, Montana. Animals, such as elk and deer, may migrate seasonally in and out of designated wilderness areas, which makes their management a multiagency stewardship challenge. Photo courtesy W. E. Steuerwald, U.S. Forest Service.

where they exist in sufficient densities. Anadromous salmon are important vectors for transporting nutrients from productive oceanic systems to impoverished terrestrial ones; bears preying on the fish finish the job by depositing nutrients contained in fish upslope from streams either as carcasses or bear feces. Finally, some wilderness animals, such as bears, can have inordinate effects at a smaller scale. Grizzlies are well-known diggers and can dramatically alter soil structure, nutrient flows, and vegetation composition at sites that are habitually dug for roots and rodents (Tardiff and Stanford 1998). The rootings of wild pigs impact deciduous forests in the Great Smoky Mountains in the southeastern United States in much the same way (Singer 1984). Although all organisms affect the ecosystems within which they live, keystone species—those that have a disproportionate impact on their environment—are of special interest in managing wilderness ecosystems.

A Wilderness Role in Wildlife Preservation

By definition, wilderness is critical to the survival of wilderness-dependent wildlife and a major factor in conserving wilderness-associated species. Wilderness may directly enhance survival of species with certain specialized habitat needs and those that are vulnerable to contact with humans by providing them with security. Although many species with an affinity for wilderness conditions may in fact survive in less pristine habitats, they live and behave most naturally in wilderness. Certain animals injure or kill humans regularly enough to preclude peaceful coexistence near even sparse human settlement. These animals include grizzly bears, polar bears, lions, tigers, leopards, pumas, crocodilians, Komodo dragons, and elephants. Often, it is young, injured, or senescent animals that prey on or otherwise kill humans, but human antagonism is often generalized to the entire species. Among humans' cultures, there is widespread intolerance of animals that kill and potentially eat people. Perhaps more than any, it is these species that depend on wilderness sanctuaries—or areas where the intrusions of people are tightly controlled for their survival.

For many of these same animals, as well as for a host of others, wilderness conditions can be critical to the preservation of essential food and cover. For example, the 64,000-acre Badlands Wilderness in Badlands National Park, South Dakota, is the primary site for reintroducing endangered black-footed ferrets to the wild. This wilderness is one of the last locations with extensive intact prairie dog towns—home of the ferret's primary prey.

Over large parts of the world, humans often exacerbate conflicts with larger carnivores and herbivores by reducing or eliminating prey and forage. In many instances, this is accompanied by dramatic increases in numbers of domesticated livestock. Substituting domesticated animals for native prey leads to spiraling conflict with predators because they resort to preying on cattle or sheep as wild ungulates decline. The spread of intensive agriculture and pastoralism also typically escalates crop damage by large herbivores forced to seek out forage in agricultural areas as forage in adjacent or former rangelands is depleted by livestock or converted to crops. This scenario is pervasive in Africa, India, the Himalayas, Southeast Asia, and parts of South and Central America, involving predators such as jaguar, snow leopards, wild dogs, cheetahs, and lions, and herbivores as diverse as Cape buffalo, eland, red deer, kiang, Nilgiri tahr, guanaco, and elephants. For animals like these, wilderness conditions can be critical to their survival, especially where economic and cultural influences do not foster or allow for changes in human behavior or tolerance. Such is the case in many impoverished countries experiencing rapid population growth.

Wildlife that are prone to overharvest by humans also benefit from wilderness conditions. For some, wilderness is even critical to their survival. In most industrialized, developed countries, wildlife harvests are closely regulated with an eye to sustainability. Elsewhere, this is often not the case. Overharvest of vulnerable animals occurs in developing countries for two primary reasons—because body parts of the animal are highly valued or because the animal is an important source of protein. For some animals, such as the African elephant, these causes are combined. Most residents of Western countries do not think of wild game as a primary source of protein—yet in many tropical regions of Africa and South America, wildlife is an important source of food. In fact, protein from wild game is considered to be the factor most limiting to human population growth in African rain forests (Barnes and Lahm 1997). It is thus not surprising that tapirs and larger-bodied primates and antelopes are overharvested apace with the destruction of tropical forests and the local growth of human populations on several continents. Short of the unlikely event that alternate sources of protein will be found, the fates of these animals are inextricably tied to vast tracts of unexploited tropical forest, that is, wilderness conditions. The emphasis is on large areas, because, in the absence of enforced protected areas, harvests of larger-bodied primates or antelope are often unsustainable within about seven miles of tropical human settlements (Muchaal and Ngandju 1999; Wilkie et al. 1998).

The high monetary value placed on the body parts of some species can make the wilderness dependence of these animals even more extreme and can jeopardize their survival. Although the ranges of tigers, rhinos, elephants, and tropical bears have declined primarily because of competition with humans and related losses of habitat, poaching or unregulated harvest driven by demand for bones, horns, tusks, or gall bladders is a grave threat, even with current restrictions on trade under the Convention on International Trade in Endangered Species (CITES). To a lesser extent, markets, legal or otherwise, for their skins and pelts also threaten animals like crocodiles, caimans, and leopards. The potential wealth derived from an elephant tusk or rhino horn is so great that space often does not provide much protection from highly motivated and well-outfitted poachers. Antipoacher patrols and extreme regulations can give a false sense of security and are often disrupted by even a short period of civil unrest. The precariousness of civil order in many countries around the world argues for the importance of very large wilderness reserves for the long-term survival of elephants, rhinos, tigers, and tropical bears.

Extensive wilderness conditions can be important to maintaining natural evolutionary selection and related behaviors among wilderness-dependent or wilderness-associated species, because wilderness protects habitats that have been least modified from conditions under which their biotic communities evolved. For example, selective harvest of large-tusked African elephants has led to increasing numbers of tuskless or small-tusked animals, along with alteration of mating and foraging behaviors related to tusk size (Jachmann et al. 1995). Strategies to reduce poaching on African rhinos by the deliberate removal and sale of rhino horns also promise to change selective pressures affecting survival and breeding (Berger and Cunningham 1994). The evolution of pachyderms has come to increasingly reflect the influences of international human commerce rather than their natural habitat. Elephants and rhinos may persist, but without wilderness they will no longer be the products of a natural and wild environment. The same could be said of the remaining migratory ungulates. Although all of these species could survive in small populations in small protected areas, their numbers and behaviors that define both a presence on the landscape and effects on whole ecosystems would have been lost, as has already happened with American bison and the black wildebeest of South Africa.

Wilderness can protect conditions, such as wetlands, grasslands, or old-growth primary forest, that are potentially important to numerous species at some stage in their life cycle, but prone to be eliminated in nonwilderness settings. For example, large, hollow trees are highly preferred for natal dens by American and Asiatic bears and are more abundant in wilderness settings (Mattson 1990). Similarly, although the old-growth forests that spotted owls depend on in the U.S. Pacific Northwest can persist in settled areas, they are highly associated with wilderness conditions because of historical timber harvests. The list of species benefited in this way could go on, including flamingos, endangered cranes, and ungulates as obscure as the newly discovered Southeast Asian bovid, the saola, or snow geese, brant, eiders, scaups, tundra swans, and white-fronted geese that nest in Arctic wetlands. *The point is that, although humans can choose not to drain wetlands or cut down old trees anywhere, such protection of natural conditions is guaranteed in areas set aside as wilderness.* Along with such protection come many benefits to many species dependent on the natural patterns and cycles of nature, whether involving water, wind, or fire.

Finally, for all of those species that benefit from wilderness conditions, wilderness can serve as a wildlife bank, especially valuable following times of overharvest and depleted populations. For example, between 1892 and 1962 many of the elk ranges of the western United States were restocked, after several decades of massive elk declines, with animals live-trapped in the backcountry of Yellowstone National Park. Similarly, wildlands of northern Minnesota in the United States and of Spain, Italy, and Poland were the source of wolves that have established themselves widely in surrounding semiwild areas since protections were instituted in the 1950s and 1970s.

Having made the case for the benefits of wilderness to wildlife conservation, wilderness conditions are obviously not of equal benefit to all species or of any benefit to some. Many species, including crows, starlings, robins, Norway rats, certain species of cockroaches, and even white-tailed deer, have benefited from the conversion of wildlands to urban, industrial, or agricultural landscapes. Because the habitat requirements of various wildlife species are so diverse, there is no question that some species will benefit and some species will not by wilderness designation. This, of course, makes the case for identifying wilderness-dependent and wilderness-associated wildlife for purposes of planning wilderness management for wildlife.

In short, the key to diverse wildlife is diversity of habitat. The U.S. wilderness system offers the possibility of preserving most of the nation's 261 ecosystems, as defined by the Bailey-Küchler method, which reflects physical and biological factors. However, traditional emphasis on areas with dramatic high-country appeal at the expense of less scenic lowland areas will need to be changed if the wilderness system is to fulfill its potential in providing that diversity. By 1987, only 157 of the 261 ecosystems were represented in designated wilderness (Davis 1989).

Wilderness Wildlife as an Environmental Baseline and Laboratory

Wilderness areas have three crucial roles as permanent and natural laboratories: (1) as *invaluable reservoirs of genetic material* that preserve resource options and serve as storehouses of potential new products; (2) as *biological standards of comparison* with ecological communities more heavily affected by human activities; and (3) as sites for *integrated studies of the structure and function of natural ecosystems* (Cutler 1980).

Aldo Leopold promoted the theme in his writings that wilderness provides a standard against which the alteration of developed lands can be measured—

Fig. 12.4. Grizzly bears are one of several species of wildlife requiring large tracts of undisturbed habitat. Most wilderness areas are too small or boundaries do not coincide with the natural movements of such animals, thus complicating effective management. Grizzly bears are currently confined to a few isolated and comparatively small wilderness ranges in the contiguous United States, primarily because humans are intolerant of the risks they pose to human safety and to livestock. Photo courtesy U.S. Fish and Wildlife Service.

that wild places reveal what the land was, what it is, and what it ought to be. He emphasized the importance of areas where evolution operates without hindrance from humans, thereby providing standards against which to measure the effects of development. Each biotic province, said Leopold (1949, 196), "needs its own wilderness for comparative studies of used and unused land."

The value of wilderness to understanding ecosystem relationships, including the regulation of wildlife populations, predator-prey relationships, diseases and hosts, and relations between herbivores and their physical and vegetal environments, has been amply demonstrated in many classical studies. For example, the intact flora and fauna of the Serengeti and Yellowstone wildlands have been priceless to science (Clark et al. 1999; Sinclair and Arcese 1995). The wildlands of boreal and Arctic regions have been similarly important to describing the simpler web of relations among caribou, moose, wolves, bears, and their biophysical environment (e.g., Gasaway et al. 1983), or simpler even yet, the classic relations among wolves, moose, and moose browse on Isle Royale (D. L. Allen 1979; McLaren and Peterson 1994). Wilderness areas also have offered the opportunity for numerous in-depth studies of individual species in comparatively intact ecosystems, including cougar in central Idaho (Hornocker 1970), black bears in deciduous forests of Great Smoky Mountains National Park (Pelton and van Manen 1996), chimpanzees and mountain gorillas in central and eastern Africa (Goodall 1986; Schaller 1963), and leopards in Kruger National Park, South Africa (Bailey 1993). Clearly, many insights into biological systems have depended and will continue to depend on setting aside and fully protecting wilderness areas and continuing ongoing research or initiating additional long-term integrated studies in relatively undisturbed wilderness laboratories.

Recreational, Aesthetic, Economic, and Political Values

The recreational and aesthetic uses of wilderness have increased rapidly during the past four decades, and the presence of wildlife is surely part of this wilderness lure. Many people come to view or photograph native species in natural settings. In season, others come to hunt or fish under primitive conditions. The fact that many other people are not deliberately seeking contact with wildlife in wilderness makes little difference; the incidental contact—the chance observation under natural conditions—can immeasurably enrich a wilderness experience. For some, even *danger* can be a positive aspect of a wilderness experience—knowing that the possibility exists for confrontation with dangerous wildlife. Close encounters with grizzlies, lions, rhinos, or other dangerous animals, or even hearing or sighting them, make for impressive recollections.

Millions of people also enjoy wilderness wildlife vicariously by appreciating it through friends, stories, photos, and art; through television and movies; or simply by knowing that it is there. Even those who have never personally experienced wilderness have some vision of its enchantment due to the rich profusion of these media celebrating its beauty in images, words, and poetry that appeal to a strong human instinct for maintaining some connection to a more primitive life and time.

Quantifying the economic value of wilderness wildlife may be even more difficult than putting a dollar value on wilderness itself, but efforts to date show that both are extremely valuable. Wildlife is an important component for many visitors who use outfitters and guides to access wilderness—and supports a multi-million-dollar-a-year industry in western states such as Idaho and Montana where a guided wilderness elk hunt may cost $5,000 or more per hunter. Interestingly, in the Frank Church–River of No Return Wilderness where a recovering population of wolves has been reestablishing itself, at least part of the appeal is a chance of hearing wolves and seeing wolf sign. Internationally, wilderness wildlife is the appeal for a big business in both hunting and camera safaris, presenting wilderness wildlife in its native habitat, in films, videos, and television.

In developing countries, wilderness wildlife can have a more pragmatic (as well as economic) value to humans. The fish, birds, and mammals that live in wildlands are an important source of food and potentially a reason to value wilderness in areas where there are few other reasons to do so. For example, the Selous Game Reserve in Tanzania was valued by nearby residents because of the game meat that it provided, despite negative attitudes toward the state wildlife management authority (Gillingham and Lee 1999). The value of game meat is so high in developing countries that wildlife in most protected areas experience substantial harvest, mostly by market hunters. Whether legal or not, this activity provides a livelihood for many people living nearby and is important to local economies. In Kenya's Arabuko-Sokoke Forest the annual harvest of 2,000 pounds/square mile (350 sq km) was valued at $35,000—a veritable fortune in an impoverished region such as this one (Fitzgibbon et al. 1995). Unfortunately, harvests of larger species are rarely sustainable in developing countries, yet it is the value placed on such animals as food that potentially motivates humans to change their practices to preserve wildlands and sustain the wildlife within.

Kellert's studies (1985) of Americans' attitudes toward wildlife indicate strong public support—especially by younger, more highly educated, and urban-dwelling Americans—for protecting many wilderness wildlife species such as wolves, eagles, and grizzly bears. Kellert argues that wildlife makes direct and vicarious contributions to the American quality of life, especially where contact with and awareness of such wildlife are inextricably woven into the fabric of a region's heritage. For example, findings from his study of Minnesota residents' attitudes toward timber wolves revealed that some people had *seen* and *heard* a timber wolf in the wild, and a majority reported having read an article about wolves within the previous year. The study revealed a strong "existence value" for wolves, and a majority of all publics except farmers agreed that (1) it was important to them to "know that wolves exist in Minnesota"; (2) "timber wolves are essential to maintaining the balance of nature"; and (3) they "would very much like to see a timber wolf in the wild" and "hear a wolf howl." A majority of all respondents except farmers and trappers opposed the idea of "giving preference to people who derive a living from the land over protection of the timber wolf." However, relatively few residents were willing to protect timber wolves if it meant excluding people from northern areas of Minnesota (Kellert 1985).

Other findings complicate the wolf picture. A national survey (Kellert 1985) indicated that wolves were among the least liked animals. Only 42 percent nationwide said they did not dislike wolves, but that figure jumped to 74 percent among Alaskans. Again, younger age, higher education, and especially knowledge of animals were correlated with a favorable view of wolves. Kellert observes that affection for wolves has increased, and that this trend can be expected to continue, especially among residents of regions containing wolves that are not potentially impacted by them. It is interesting to consider Kellert's findings from the mid-1980s that project increasing acceptance of wolves in light of the substantial progress achieved since that time in wolf conservation and the animal's recovery in the wild.

Politically, wildlife is the most powerful focus for environmental protection in the world. Two of the largest citizen environmental organizations in the United States are wildlife oriented (Hendee and Pitstick 1993). *The largest, the National Wildlife Federation, with 5.8 million members (including its affiliates), increased nearly threefold between 1970 and the mid-1980s. The Audubon Society, with more than 500,000 members, grew even more during the same period. Although wildlife is a common thread of these organizations, their interests go beyond wildlife and certainly include wilderness protection. They are a strong political force.*

In the participatory society of the United States, wilderness preservation could not have come to pass without a viable political constituency. Although this constituency has been composed of many groups, for one hundred years, wildlife supporters, including hunters and anglers, have played an important role, spurred by their recognition of wilderness as a fish and wildlife refuge. For example, the major 132-million-acre expansion of U.S. national forests came during the 1901 to 1909 presidency of Teddy Roosevelt, a founder of the Boone and Crockett Club—a big-game hunting club—who believed forest conservation and game conservation to be synonymous. It was the Boone and Crockett Club that urged the passage of the Yellowstone Game Protection Act of May 7, 1894, which was designed in part to protect Yellowstone wildlife from poachers and thus established the precedent of the national parks as game preserves (Trefethen 1975).

Fig. 12.5. Supplementing natural water catchments for wildlife in wilderness areas raises management questions. In some national wildlife refuge wildernesses, legislation establishing the refuge or game range (and superseding wilderness designation within it) may call for management measures to ensure the survival of particular species. A rock tank catchment (left) in the Cabeza Prieta Wilderness, Arizona, helps ensure the survival of desert bighorn. A "guzzler" (center) and a water flue (right), both in the Desert National Wildlife Range, Nevada, provide vital water for wildlife. Photo courtesy U.S. Fish and Wildlife Service.

When a wilderness bill was first introduced in the U.S. Congress in 1956, among its charter supporters were the National Wildlife Federation and the Wildlife Management Institute (Mercure and Ross 1970).

To be sure, hunters and anglers are not always supportive of a particular wilderness. They exhibit considerable concern about the ultimate levels of hunting and fishing to be allowed in wilderness, and there is concern that designation of wilderness, and in certain wilderness areas restoring wolves and grizzly bears, may limit harvests of big game. These concerns demonstrate the need for a basic philosophy incorporating wildlife into a wilderness framework—a philosophy emphasizing wildlife uses including appreciation, hunting, and fishing carried out in a manner consistent with wilderness values.

Wildlife-Related Problems in Wilderness Stewardship

Wildlife issues are frequently a source of controversy in wilderness management. One survey of agency managers in the southwestern United States reported controversial wildlife-management practices in 53 of 273 wilderness areas (Czech and Krausman 1999). Water provision for wildlife, trout stocking, feral burro control, and big-game surveys using motorized vehicles accounted for over half the controversies. The other half spanned diverse issues, including research with motorized access, mechanical and chemical vegetation control, reintroductions, mechanical animal capture, and restrictions on hunting and human access. Wilderness management involving wildlife is loaded with potential conflict and difficult problems—not just in the United States but internationally. Following are ten categories of wildlife-related problems in wilderness management with examples drawn from the United States and around the world.

Multiple Agencies and Missions

A designated wilderness in the United States is almost always part of a larger federal agency jurisdiction, and the four federal land-managing agencies have wildlife-management missions and traditions predating the 1964 Wilderness Act. The situation is further complicated by the legal tradition dating back to the Magna Carta under which fish and wildlife devolved to state (not federal) authority, a status quo not changed by the Wilderness Act. For practical purposes, this means that the state manages the wildlife and the federal agency manages the habitat everywhere except in certain national parks. Obviously, careful coordination is needed, because populations and habitats are interdependent. There are differences of opinion about how far a state's legal authority extends to federal lands, and cutbacks in federal budgets can weaken the ability of federal agencies to maintain their traditional strength in some federal-state controversies over wilderness wildlife policies, for example, control of wild horse and burro populations to protect habitat on federal lands, reintroduction of wolves and grizzlies, and fish stocking in high mountain lakes.

Management of wilderness wildlife is affected by a variety of constraints that have the potential to conflict with the goals of naturalness and solitude set forth in the Wilderness Act. For example, hunting and fishing are established activities in all national forest and BLM wildernesses, and in some national wildlife refuges; and fishing is allowed in some places in national parks. The Endangered Species Act (ESA), passed in 1973 (U.S. Public Law 93-205), requires preserving certain threatened or endangered species and critical habitat for those species in wilderness. This requirement may even call for using artificial devices and alterations if necessary, thereby conflicting with wilderness goals. Furthermore, ecological trends precipitated by previous management, such as unnatural conditions resulting from excluding fire, may be reflected unnaturally in current wilderness wildlife populations.

Although we do not specifically review policies of U.S. agencies for wildlife in wilderness, we want to make the point that each agency responds to a different set of interest groups. Each has different philosophical traditions, a particular management legacy, and a different balance of disciplines among its professional personnel. These are important influences that directly and subtly shape management policy for wilderness (and its wildlife) within each agency. Funding for wilderness and wildlife management in wilderness is subordinate to other pressing concerns in all agencies. This holds true for agencies that manage wilderness in other countries too.

Legislative and Administrative Direction

Wilderness management operates under a number of legislative and administrative constraints, including some directly related to wildlife. In the United States, the hierarchy of wilderness management direction includes the Wilderness Act of 1964 (U.S. Public Law 88-577); the *Code of Federal Regulations* that interpret and clarify that act; national wilderness management policies of the administering agencies as set forth in their manuals and handbooks; management plans for jurisdictions (such as parks, refuges, national forests, or public land in a state or region) that include wilderness; and individual wilderness management operating plans that specify how national direction will be applied on the ground. Other laws, such as the Endangered Species Act of 1973 (U.S. Public Law 93-205) and the Clean Air Act Amendments of 1977 (U.S. Public Law 95-95), further constrain management. In some cases, wilderness designation acts—laws designating individual areas as wilderness—specify how particularly controversial wilderness management issues will be handled (see Chapters 4 and 5), and these issues often affect fish and wildlife; for example, in wilderness designated by the 1994 California Desert Protection Act (U.S. Public Law 103-433), motorized access for fish and wildlife management purposes is allowed.

Grazing by horses, goats, sheep, and cattle is allowed by the Wilderness Act in some national forest and BLM wilderness, with often substantial impacts on wildlife. The Colorado Wilderness Act of 1980 (U.S. Public Law 96-560) mandated guidelines for grazing of livestock in new wilderness areas in that state that have been adopted elsewhere (see Chapter 5). Grazing also brings pressures for predator control to the affected areas, additional human activity, pressure for developing fences and water sources, and consumption of forage, all with obvious local impacts on fish and wildlife.

Endangered species protection in the United States illustrates the problem of conflicting legislative direction. For example, the Wilderness Act limits managerial freedom to manipulate vegetative cover to perpetuate, improve, or alter an area's value for any particular purpose, including wildlife. On the other hand, the Endangered Species Act directs agencies to make sure no actions are taken that would "jeopardize the continued existence of any endangered species or threatened species or result in the destruction or adverse modification of critical habitat" (U.S. Public Law 93-205, 7a), including in wilderness. Thus, conforming to the Endangered Species Act may require management intervention (e.g., vegetative manipulation, predator control, and mechanized access) that would otherwise be restricted in wilderness. For example, in wilderness on the Cape Romain National Wildlife Refuge off the coast of South Carolina, raccoons prey on the eggs of the endangered sea turtle. A raccoon-trapping program to protect sea turtles, by itself an unnatural alteration of raccoon numbers, also requires periodic motorized access to tend traps.

Endangered species protection in wilderness may also require (under the Endangered Species Act) that managers intervene in natural processes. For example, taking condor eggs to be hatched and raised in captivity is not a natural solution, but it has been applied as a necessary approach to save the condor (Campbell 1984). In 1986, there were only six California condors remaining in the wild (many in wilderness) when adult condor number eight was captured. One condor produced twelve offspring in captivity before being released in April 2000 to join other condors flying free in California and Arizona.

The Historic Preservation Act of 1966 (U.S. Public Law 89-665) and related executive orders also result in activities that may be contrary to the intent of the NWPS—such as preserving or refurbishing historic structures that would otherwise be removed or allowed to deteriorate. Examples are cabins in some western wildernesses and lighthouses in wilderness on East Coast wildlife refuge islands and capes. Because such attractions draw excessive recreational use, wilderness wildlife may be locally impacted.

Wilderness in Alaska represents a special case because of provisions in ANILCA, the Alaska National Interest Lands Conservation Act of 1980 (U.S. Public Law 96-487). Following years of debate on ANILCA, Congress established wilderness areas in Alaska that are different in some important ways from those in the lower forty-eight states. In general, these units are open to most "customary" uses unless specifically closed following public hearings. Motorized access by plane, motorboat, or snowmobile is guaranteed to private inholdings such as homesteads, mining claims, trade centers, management cabins, and native lands. Subsistence hunting, trapping, and fishing may continue in wilderness, even in some designated national park units, and temporary facilities, equipment, and sometimes motorized access directly related to such activities are allowed. Subsistence taking of fish and wildlife provides important food, material, and livelihood for native and other communities (Muth 1990). However, subsistence uses continue to be controversial, with debate over who should qualify as a subsistence taker, and how the *need* for subsistence taking should be established. A 2005 publication clarified ANILCA provisions for wilderness on the national forests (U.S. Department of Agriculture, Forest Service 2005), and a 2007 case study by Hummel (2007) illustrates their application over twenty-six years to the Stikine-LeConte Wilderness.

Disease

In places, wilderness animals may be reservoirs of diseases that threaten humans or domesticated livestock. The dynamics of the brucellosis bacterium and the rinderpest virus, harbored by bison in Yellowstone and African antelope in the Serengeti, respectively, are among the best studied in this regard (Meagher and Meyer 1994; Dobson 1995). Brucellosis can induce abortions whereas rinderpest causes death outright in domesticated cattle. Concerns among livestock producers in areas as far apart as Wood Buffalo National Park, Canada; Yellowstone National Park, Wyoming; and Serengeti National Park, Tanzania, have led to heated debates about the merits and means of limiting potential for the exchange of these diseases between wild and domesticated bovids. Ironically, both brucellosis and rinderpest were introduced to native ungulates in these wild areas from infected cattle. More extreme yet in its effects is trypanosomiasis, otherwise known as sleeping sickness (Murray and Njogu 1989). This disease causing parasite is endemic among native African herbivores and is typically transmitted to humans and domesticated livestock by the tsetse fly. Not only are the tsetse fly and trypanosomiasis associated with wild areas, this lethal combination of organisms has been implicated in maintaining vast tracts of de facto wilderness, especially in central Africa.

Although wild animals can harbor diseases harmful to domesticated animals, the reverse also is true. Wild canids are probably most affected by diseases transmitted by domesticated animals—primarily dogs. Rabies threatens the endangered Ethiopian wolf and wild dogs in Africa; distemper destroys bat-eared foxes in the Serengeti; leishmaniasis affects crab-eating foxes in Brazil; and mange imperils the extremely rare Mednyi Island arctic fox (Fanshawe et al. 1991; Goltsman et al. 1996; Gottelli and Sillero-Zubiri 1992).

Control of disease transmittal in wildlands or in transition zones between wilderness and agricultural lands is difficult and complicated. The effects of mange on arctic foxes and the effects of rinderpest on cattle and wild bovids have been curbed in the short term by administering vaccines. However, the long-term efficacy of these vaccination programs remains in doubt (Dobson 1995); proposals to control brucellosis have ranged from eradicating infected wildlife, to vaccinating wild and domesticated animals, to creating buffer zones. In all cases, development of solutions has been hindered either by lack of information, lack of resources, or conflict among management agencies and affected groups. In most cases, resolution of wilderness-related disease problems requires broad-scale integrated programs involving many wilderness and nonwilderness stakeholders, including wildlife- and land-managing agencies and diverse political interests (Wobeser 1994).

Depredation

Wilderness areas are often a source of native animals that prey on domesticated animals or consume or trample crops on nearby agricultural lands. These problems range from depredation of oats and beehives by bears in northern regions, to predation on domesticated dogs in rural and urban areas by wolves and cougars, to trampling and depredation of grain by Asiatic elephants, to predation on cattle, goats, sheep, and yaks by large cats wherever they occur. In many instances, the depredating animals are injured, young, and inexperienced, or old and in decline. Quite often, they are dispersing juveniles, pushed out of prime wilderness habitats by older individuals. For example, most of the black bears involved in conflicts over beehives in the Peace River region of Alberta, Canada, were juvenile males, presumably dispersing from wildlands to establish their own adult ranges (Mattson 1990). In most cases, these sorts of problems are exacerbated by or are the direct outcome of human actions. Predation by cougars on dogs in California is associated with urban areas expanding into historical cougar range (Torres et al. 1996). Depredation of livestock is often the result of humans removing native prey and replacing it with cattle or other domesticated herbivores. This has been observed for wolves and bears in the southwestern United States (Brown 1983, 1985); for tigers in southeastern Tibet and snow leopards in adjacent southwest China (Qiu 1996; Schaller et al. 1988); for leopards, lions, and tigers in India (Oza 1974; Sekhar 1998); and for wolves in Italy (Meriggi and Lovari 1996). Similarly, increased depredation of crops by elephants in the Shivalik Ranges of India is attributed to the reduction of native forage in conservation areas by increasing numbers of livestock (Chowdhury 1995).

Wild herbivores can cause changes in the vegetation of wilderness areas that are incompatible with management goals or the values of some people. In the case of introduced nonnative ungulates, this can take the form of unprecedented change in plant communities not adapted to large-scale herbivores. Such was the case with red deer and Himalayan thar introduced into New Zealand by European settlers wanting hunting opportunities. These species caused widespread damage to native vegetation and were then subjected to massive, sustained animal-control programs, but only after damage was well advanced (Lever 1985). An interesting twist to this scenario is provided by the history of mountain goats in Olympic National Park, Washington. Here, a species native to the continent was introduced outside of its recent historical range with damaging effects to sensitive native vegetation (Houston et al. 1994) and accompanying concerns about maintaining the integrity of the wilderness ecosystem. Among native ungulates, there are well-documented effects by white-tailed deer on woody vegetation in the eastern United States where populations of this species have surged to historically unprecedented densities within the last thirty years (Porter 1996). More controversial are contentions that dense populations of wapiti in North America have caused accelerated erosion and the decline of aspen, willows, and other browse in wilderness parks ranging from Banff in Canada to Yellowstone and Rocky Mountain National Parks in the United States (Wagner et al. 1995). For both species, extirpation of native predators and declining harvest by humans are implicated as ultimate causes of "unnatural" densities of these native herbivores.

Nonnative fish have threatened the integrity of aquatic systems in wilderness areas in many parts of the world. In the western United States, lake trout introduced by humans, legally or otherwise, have decimated native cutthroat trout populations. Most recently, lake trout were discovered in Yellowstone Lake in Yellowstone National Park, where they threaten to reduce cutthroat trout populations by as much as 70 percent, with dramatic negative effects on the thirty species of vertebrates that prey on cutthroat, but not lake trout (Kaeding et al. 1996). Native and nonnative fish stocked in mountain lakes formerly without fish predators because of physical barriers have similarly devastated resident populations of native amphibians. For example, a study of 2,200 lakes in the John Muir Wilderness and Kings Canyon National Park backcountry in California concluded that fish stocking had an adverse effect on native biota, especially the mountain yellow-legged frog (Knapp and Matthews 2000; Matthews and Knapp 1999).

Although resolving depredation problems can be technically simple, it is often politically controversial. Where depredation on livestock or crops is a problem because native prey have been replaced by livestock, or because livestock have depleted native forage on which wilderness wildlife depend, removal of the livestock or restoration of native herbivores is an obvious solution. However, in most instances such actions are not acceptable because they threaten people's livelihoods and traditional lifeways. Monetary compensation for losses and selective removal of depredating animals have often been successful in both preventing future losses and recruiting the acceptance of native predators by livestock owners. Proponents of wolf reintroduction and recovery in U.S. wilderness have proposed such accommodation, such as the Defenders of Wildlife's Wolf Compensation Fund in Yellowstone National Park and the northern Rocky Mountains (Defenders of Wildlife 2008). Electric fences have been used to reduce crop damage by elephants and beehive damage by bears. Where there are perceived problems with overgrazing or overbrowsing by native ungulates, reintroduction of native predators and reinstitution of human harvest are obvious solutions. Ironically, a reduction in predator controls might solve the problem. However, violent prejudices against predators or hunting often preclude such responses and contribute to protracted debates about whether wild areas are overpopulated and the nature of the solution. Where nonnative species are destroying native species, the obvious wilderness solution is removal of the nonnatives. However, in most instances, there is no means for their complete removal; thus the often costly control programs are an uncertain, long-term solution. As with the solution of disease problems, resolving depredation problems usually requires the development of broad integrated programs; responsive to the outlooks and concerns of affected stakeholders and the interests and sensibilities of the involved management agencies.

Human Safety

A host of animals, ranging from invertebrates such as scorpions and tarantulas to megaherbivores such as bison and Cape buffalo, pose a threat to human safety in wilderness. Annually worldwide, up to three thousand people are seized, mutilated, and, in most cases, eaten by crocodiles (Alderton 1991). In Africa, 40 of 444 Nile crocodiles (nearly 10 percent) had human remains in their stomachs at the time of their death. Elephants in Southeast Asia regularly kill people and cause such damage in Sumatra that a "school" was set up in the 1980s to modify the behavior of individual elephants thought to be amenable to reform (McNeeley and Wachtel 1988). However, in most cases the perception of risk from wilderness wildlife is at variance with reality. Although cougars are thought by many to pose a grave threat to humans in the United States and Canada, only ten people were killed by cougars in these countries from 1890 to 1990 (Beier 1991). Even where an animal species causes significant harm to humans, most injury or death occurs at boundaries of settled and wild lands or is associated with incursions by humans involved with economic activities such as honey or firewood gathering in tiger reserves in India. In places, risk to humans within wilderness areas has been distinguished from risk incurred at boundaries. For example, in North America, about 1 out of 320,000 visitors is injured by black or grizzly bears in the wilderness settings of Kluane and Denali National Parks (Herrero and Fleck 1990). Wherever such rates have been compiled, rates of injury caused by animals are almost always trivial compared to injury by other causes.

Although some risk is incurred by the mere presence of large carnivores and herbivores or poisonous reptiles and invertebrates, in most cases there are identifiable factors, including human behavior, that strongly affect that risk. In the case of crocodilians, only those greater than nine feet in length pose much threat to humans. As with cases of depredation, the mammalian predators most prone to injuring humans are young, sick, or enfeebled. Among bears, risk of injury is aggravated where individual animals have either lost their fear of humans through constant exposure (i.e., habituated) or sought out foods made available by humans (Herrero and Fleck 1990). In general, habituation and unsanitary conditions around human facilities aggravate the risks posed by most large carnivores and herbivores. Ignorant behavior by humans greatly increases any risk. *Most injuries to humans caused by large herbivores, such as bison in U.S. national parks, are precipitated by humans unknowingly trespassing on an animal's personal space, or unwittingly posing a threat to a female's young.*

Given the factors that precondition risk of human injury by animals in wilderness settings, potential solutions are often straightforward. In virtually all cases, research and education that identify risks and preventative behavior are important (Sanyal et al. 2006). *Keeping wildlife wild (and thus shy of people) is a key. Sanitizing human facilities, such as campsites, can alleviate many conflicts with carnivores. Restriction of humans to certain places and times also can minimize volatile encounters and allow animals to avoid humans.* These kinds of restrictions are the basis for the outstanding safety record of the bear-viewing program at McNeil River Falls, Alaska (Aumiller and Matt 1994). *As a final recourse, removal of animals known to injure humans may be necessary, especially if this behavior has become habitual or was an obvious act of predation.* Enforcement of policies to reduce risks is a necessary adjunct to any of these measures.

Reintroductions

Wilderness areas are often prime sites for reintroducing native species extirpated or endangered by historical human practices. This general principle is well illustrated by efforts to reintroduce or otherwise restore wolves and grizzly bears in the Rocky Mountains of the United States (see "Case Discussion" at the end of the chapter). The essential feature of habitat suitable for both species is extensive areas free of roads and human activity, which largely corresponds with designated or candidate wilderness areas. Other features critical to recovery of these carnivores, such as abundant native prey, also are associated with wildlands. Wilderness is a similarly desirable setting for reintroducing and restoring a host of other species, including ungulates such as the Arabian oryx, North American desert bighorn sheep, and Przewalski's horse. For these species, wilderness conditions help ensure freedom from human disturbance, competition from livestock for forage, and transmittal of disease from domesticated animals. However, most restoration projects are fraught with controversy because of feared curtailment of traditional human prerogatives and economic activities in reintroduction areas. Ignorance of the risks or impositions posed by species contemplated for reintroduction typically aggravates fears. As in many other areas, disagreement over goals for reintroduction of grizzly bears into central Idaho was further exacerbated by disagreement over goals between state and federal governments (MacCracken et al. 1994).

Reintroduction programs have been subject to numerous appraisals designed to elicit general lessons and principles. These principles apply as much in wilderness settings as in any other. *In general, the release of more animals over a longer period of time in good to excellent habitat within the core of historical range greatly enhances the likelihood of success* (Griffith et al. 1989; Wolf et al. 1996). *Omnivores and mammals are more likely to successfully establish compared to animals with specialized dietary needs and birds.* However, successful programs are built on more than biological principles. Reintroduction success is more likely where local residents are heavily involved or where economic, political, and cultural factors are otherwise given great consideration, and where organizational arrangements are made to enhance learning, flexibility, and attention to local and regional power structures (Reading et al. 1997). As with the management of disease and

Fig. 12.6. Fishing can be an important part of an overall wilderness experience. An angler tries his luck at Johnson Lake in the Absaroka-Beartooth Wilderness, Montana. Photo courtesy W. E. Steuerwald, U.S. Forest Service.

depredation problems, successful reintroduction of animals into wilderness areas must be built on broad-based integrated programs that are attentive to biological, economic, political, social, and cultural perspectives.

Livestock Grazing

Worldwide, livestock use many of the areas that are designated wilderness by law, policy, or custom, and where they do not graze inside such areas they may use those that are adjacent. Livestock can create problems when they become the prey of native carnivores. However, in most instances, depredation becomes an issue only when wildlands have been overstocked with livestock and native prey have declined. Although poaching and hunting by the pastoralists who accompany livestock contribute to native prey declines, competition for forage is often the most important factor. Such has been the case for Tibetan antelope (chirus), wild asses (kiang), wild yaks, and Tibetan gazelle in competition with increasing numbers of domesticated sheep, goats, and yak on the Tibetan plateau (Schaller 1998); Mongolian and goitered gazelles in competition with 33 million head of livestock in Mongolia (Lhagvasuren et al. 1999); wild reindeer in competition with domesticated reindeer in Siberia (Heptner et al. 1989); and Nilgiri tahr in competition with cattle and goats in south India (Mishra and Johnsingh 1998).

In the United States, livestock grazing was provided for in wilderness on the national forests and BLM lands where it was already established (see Chapter 14). Concerns about effects on native animals, historic and ongoing, have been greatest in the more arid southwestern regions of the country. Livestock have affected native animals by altering hydrologic regimes, depleting forage, displacing other herbivores, and providing a rationale for predator "control." Historically, livestock aggravated conflicts with bears, wolves, and cougars in the United States, leading to aggressive control programs that resulted in the extirpation of these animals over large parts of their range (Brown 1983, 1985). The negative effects of livestock on wilderness animals are well documented and numerous.

The simplest solution to the negative effects of livestock on wilderness animals is to remove the livestock. However, this solution is rarely politically viable because such action can harm the economic well-being of livestock owners and threaten traditional lifeways. Livestock owners can be politically powerful. Moreover, livestock owners can cause harm if they resort to increased poaching and other illegal practices in retaliation for policies that they perceive as harmful to their interests. It is, in fact, livestock owners that often conduct their lives nearest to wilderness wildlife (see Chapter 14 for discussion of grazing in U.S. wilderness). In developing countries especially, alleviating grazing damage caused by domestic animals requires provision of alternate livelihoods or compensation to ranchers and increasing the levels of education and economic well-being in rural areas (Mattson 1997).

Hunting and Poaching

Humans kill wilderness animals for many reasons, including economic gain, prestige from trophies, acquiring food, resolution of conflict, and alleviating danger. Legal harvest of wildlife under established regulations distinguishes hunting from poaching. However, regardless of the legality, where the motives to harvest outstrip political will or resources for enforcement, wilderness animals subject to overharvest can decline in number, even to the point of extirpation. Problems of this nature are most acute in developing nations, where the acquisition of protein or cash or the preservation of property and other sources of livelihood can literally be a matter of life and death. In developed countries, it is easier to enforce sustainable harvests of wilderness animals.

Hunting to provide protein and to generate income for market hunters is commonplace in tropical areas of Africa, Asia, and South America, with often dire consequences for the affected wilderness animals. Larger species and those animals that live within nine miles of human settlements or protected area boundaries are often subject to unsustainable harvest, especially where there are gaps in enforcement of existing regulations (e.g., Campbell and Hofer 1995; Wilkie et al. 1998). Primates, antelopes, tapirs, pachyderms, and birds are all affected. For those animals that are unfortunate enough to sport highly valued body parts, such as pelts, horns, and medicinal bones and organs, space offers little protection. Poachers travel to the most remote regions because of often-substantial monetary rewards and this affects such animals as elephants, rhinos, bears, leopards, and tigers (e.g., Douglas-Hamilton and Douglas-Hamilton 1992, Peluso 1993). However, in addition to regulations and enforcement, enduring solutions lie in tackling root causes such as poverty, malnutrition, human population growth, and the demand for animal body parts fostered in affluent countries (Heinen 1996; Poffenberger 1990; Shaw 1989). Although not addressing demand, some of the greatest successes in conserving tigers and elephants have been achieved by banning international trade in body parts of these species (Chadwick 1992).

In many parts of Europe and North America, conservation programs have been so successful that the debate in places may be whether there are too many rather than too few animals. For example, the alteration of vegetation communities associated with extraordinarily high densities of elk and deer in many U.S. and Canadian national parks is a consequence of eliminating native predators and terminating hunting or other herd-reduction programs (Wagner et al. 1995). In places, low densities of moose (called elk in Eurasia) have been ascribed to high densities of predators such as wolves and bears and a change in predator-prey dynamics induced by past human harvests of both predators and prey (Ballard and van Ballenberghe 1997). Whether it is desirable or not to harvest elk and deer in U.S. and North American parks, or to reduce densities of predators to increase moose populations in boreal wilderness areas, is a matter of goals and philosophies. To date, most parks have opted to minimize direct intervention with natural predators whereas provincial and state game-management agencies have killed predators under a variety of programs designed to boost moose numbers, with varying success.

Hunting can cause changes in prey behavior and the selective pressures acting on prey species. Whether such changes are philosophically good or bad is a function of values, but in wilderness the ecological goal is to preserve "natural" behaviors and selective pressures. Even so, in the United States, hunting pressures forcing white-tailed deer into remote areas likely benefited one wilderness species, the highly endangered Florida panther (Kilgo et al. 1998).

A greater concern about hunting is how harvest biased toward trophy males might affect the vigor and evolutionary trajectory of hunted species. Trophy hunting by highly technologically advanced and well-equipped humans, supported by guides and trackers, does not resemble natural predation under primeval conditions, where predators, human or otherwise, usually took sick, young, or old animals and were relatively unsuccessful at capturing the most vigorous. However, in modern trophy hunting, males in their prime are the principal targets. Selective removal of such animals reduces the likelihood that their genes associated with vigor, strength, and breeding success will survive. Over the long term, this could produce species that are less resilient to environmental stressors, with less evolutionary potential. Although this issue is conceptually well founded and there is evidence for these effects, trophy hunting remains commonplace.

Human-Caused Disturbance

Human-caused disturbances or human-induced changes in wildlife behavior are of greatest concern when they affect the survival or reproduction of wildlife species or lead to changes in behavior, such as conditioning to humans or human-related foods, that increase chances of human injury or death. Although short-term responses by mammals to disturbances by humans have been observed for many species in wilderness settings, few effects on survival or reproduction have been documented (Anderson 1995). In fact, with frequent exposure, flight or other signs of alarm and stress abate for many mammals as they habituate to the presence of humans. Although this minimizes direct harm to responding animals, it reduces wildness and many habituated rodents and carnivores turn to exploiting human-related foods. When such food conditioning or habituation involves large mammals, chances of human injury escalate (Herrero and Fleck 1990), and habituated animals threatening humans are likely to be shot.

In contrast to mammals, there is conclusive evidence of widespread effects of human intrusion on the reproductive success and survival of birds, especially waterfowl (Anderson 1995; Anthony et al. 1995; Bowles 1995; Knight and Gutzwiller 1995). Humans can easily displace breeding birds from their nests, resulting in increased chances of nest failure. Moreover, when humans visit nests, they often attract predators that eat the eggs. Under extreme circumstances, these kinds of human-related effects have been blamed for 70 percent to 90 percent declines in hatching rates among ground-nesting waterfowl. Wildlife photographers cause great harm because they often closely approach birds and their nests. For the same reasons, photographers have been singled out for disturbing mammals and contributing to undesirable changes in their behavior. In concept, the means for minimizing harm to nesting birds are straightforward. If the location and timing of nesting are known, public education is key and such areas can be placed off-limits to humans for the duration of the nesting season, as with bald eagle fishing sites in Glacier National Park, Montana.

Fig. 12.7. University of Arizona wildlife researcher Paul Krausman studies desert bighorn sheep in the Pusch Ridge Wilderness near Tucson. Most wilderness wildlife is studied using simple means and methods, such as observation, that are compatible with wilderness conditions. Using high-technology instruments such as radiotelemetry requires agency approval. Photo courtesy John C. Hendee.

Whether wildlife flee humans or not in wilderness settings raises interesting questions about the presumed "natural" behavior of the wild animals. Many wilderness visitors cherish the opportunity to view wildlife and at close quarters. However, functionally, where frequent encounters with humans result in rewards for wildlife, such as food, most animals lose what fear they have and tolerate or even seek out the presence of humans. Conversely, where encounters with humans typically lead to distress, injury, or death, most animals come to fear humans and flee our presence. The latter response seems more closely related to the wilderness goal of keeping wildlife wild.

Long-Term, Long-Distance Effects

Increasingly, wilderness areas are islands in an ocean of humans and human-influenced environments. At the very least, most wilderness areas are small in comparison to surrounding lands managed for other goals. Moreover, wilderness boundaries are highly permeable—animals migrate in and out and humans and their effects intrude. As wilderness size declines, the proportional effects of "edge" increase to the point where no core area remains immune to outside influences (Noss et al. 1999). For example, most populations of northern birds and most populations of larger animals cross the boundaries of wilderness areas that often constitute the core of their ranges to use nearby agricultural or multiple-use lands outside. Whereas these animals might be protected from harassment and hunting inside wilderness boundaries, they are often disturbed, pursued, and killed by humans outside, with the effects on behavior, population size, and population structure transmitted back. Similarly, and especially in developing countries, humans often intrude from the periphery with undesirable effects that include poaching, grazing, deforestation, and other alterations of vegetation. These effects, from movements of wildlife or humans, typically operate at the scale of 6 to 24 miles, rendering smaller to medium-size wilderness areas of 74,000 to 1,240,000 acres vulnerable to humans and their practices on adjacent lands. This leaves only the largest wilderness areas with portions free of these smaller-scale transboundary effects, and such wilderness areas are now rare outside of harsh Arctic, boreal, or tropical environments.

Human impacts are also transmitted at broader scales. Oceanic harvest of anadromous fish such as salmonids or disruption by dams of their river travel routes impacts distant terrestrial spawning grounds. Loss of salmon in the upper Columbia River basin in the Pacific Northwest of the United States has been implicated in the declines and extirpations of several salmon predators, including grizzly bears (Merrill et al. 1999). In fact, grizzly bears occupying the Yellowstone ecosystem in Wyoming are a stellar example of a wilderness species affected by long-distance human impacts. In the eastern part of the ecosystem, grizzly bears obtain much of their energy from adult army cutworm moths that aggregate in alpine talus to feed on tundra nectar during crepuscular and nighttime hours (French et al. 1994). However, moth populations are heavily affected by distant agricultural practices where the moths complete their life cycle in croplands. Yellowstone's bears are similarly threatened by human introduction of threats to two of their critically important foods, such as spawning cutthroat trout and seeds of the whitebark pine tree (Reinhart et al. 2001). Introduction of the nonnative cutthroat predator, the lake trout, and the highly lethal fungal pathogen, white pine blister rust, threatens to substantially reduce numbers of cutthroat trout and white bark pine, respectively. Thus, even in a wilderness ecosystem of more than 10,000 square miles, wilderness animals are affected by human activities far outside the area.

At the global scale, no wilderness area is immune to changes in atmospheric chemistry due to industrial and agricultural activities that have changed and will continue to change the global climate. Prognoses for changes in regional precipitation are highly uncertain. However, climatologists are gaining confidence in their predictions for the warming of high latitudes, along with the melting of polar ice, the northern advance of warmer water in the Pacific, and possible retreat south of the Gulf Stream in the Atlantic. Animals of the most remote wilderness regions of the world—walruses, belugas, narwhals, ringed seals, and polar bears—stand to be devastated by such changes. The receding Arctic sea ice has reduced the hunting range of polar bears seeking seals as food among ice floes (Regehr et al. 2007) and caused polar bears to compete with brown bears along shorelines. Polar bears have been known to swim more than fifty miles out to sea in search of floes as platforms for hunting seals.

The northern advance of tree line stands to rob caribou and wild reindeer of much of their prime summertime barren ground forage. Salmon, already buffeted by loss of spawning grounds in inland streams, would lose much of their oceanic food in the Pacific. In small wilderness islands farther south, populations of wilderness-dependent and wilderness-associated species will likely be lost as suitable habitat declines in their current ranges (Dobson et al. 1999).

Transboundary effects yield fairly straightforward management implications. If wilderness managers want to achieve their wildlife-related goals, their efforts must be integrated with those of others managing adjacent lands, even across international boundaries, the snow leopard being an Asian example (Singh and Jackson 1999) and the mountain caribou a North American example. Whether the above are part of such integrated management or not, wilderness managers must at least be aware of broader-scale transboundary influences and, where there is little option to control such effects at their source, strive to manage to mitigate them within wilderness boundaries. This kind of broad-scale thinking and integration is arguably one of the greatest challenges to land managers, including those responsible for wildlife in wilderness.

Wilderness Management Objectives and Guidelines for Wildlife

What do all the foregoing examples and reasoning lead to in terms of what we should be trying to achieve (objectives) for wildlife in wilderness, and how we should go about doing that (guidelines)? This section recommends objectives for wildlife in the U.S. NWPS and suggests guidelines for wilderness management to meet those objectives.

The objectives for wilderness wildlife describe conditions to be sought with respect to wildlife in wilderness; they provide criteria against which to evaluate management policies and actions—that is, will such policies/actions move things toward such conditions? The objectives are deliberately broad to apply to all types of wilderness, regardless of agency jurisdiction. Again, although these objectives are aimed at the U.S. wilderness system, we believe there is considerable relevance to wildlife in wilderness systems around the world.

Proposed Wilderness Management Objectives (Conditions to Be Sought) for Wildlife in Wilderness

1. Seek natural distribution, numbers, population composition, and interaction of indigenous species of wildlife.
2. Allow natural processes, as far as possible, to control wilderness ecosystems and their wildlife.
3. Keep wildlife wild, with its behavior altered as little as possible by human influence.
4. Permit viewing, hunting, and fishing where such activities are (a) biologically sound, (b) legal, and (c) carried out in the spirit of a wilderness experience.
5. Favor the protection and restoration of threatened and endangered species and wildlife dependent on or associated with wilderness conditions.
6. Minimize degradation of wilderness qualities—naturalness, solitude, and absence of permanent, visible evidence of human activity—while managing wildlife in wilderness.

Some Key Guidelines

These objectives for wildlife do not displace the book's overriding theme that a wilderness philosophy and ethic, based on an appreciation of all affected values, should guide all wilderness management actions. *Only necessary actions to achieve objectives are justified, and they must employ the minimum methods (tools, regulations, and force) that are required.* Here are guidelines for three key elements of wilderness wildlife in the U.S. NWPS: the role of (1) natural processes, (2) fishing and hunting, and (3) wildlife research in wilderness.

Allow Natural Processes to Shape Wilderness Habitat

Designated wilderness is ideally a place where nature rolls the dice, and the resulting naturalness, whatever its characteristics, defines that wilderness. Artificial-habitat manipulations that may be desirable for fish and wildlife management on nonwilderness lands and waters are not consistent with either the Wilderness Act or the wilderness ethic that should guide its implementation. When there are legal exceptions to the natural-process criterion, *the most natural practices and tools should be used in a manner exerting the minimum impact on wilderness naturalness and solitude.*

Wilderness means *natural*. Vegetation should reflect natural conditions, substantially unaltered by people and their influence. Lakes and streams should reflect undisturbed watersheds and channels. Animal life should approach natural numbers of native species; at some times and places this will mean reduction or loss of some wildlife populations. The principal task facing wilderness managers is to promote conditions that will allow natural processes to operate as freely as possible. Where it is clear that human actions have compromised natural processes, managers may have to mimic or simulate natural processes, for example by eliminating nonnative species or introducing fire where it has been unnaturally suppressed. However, this should be done as briefly and naturally as possible. The idea that wilderness should be a "vignette of primitive America" should not suggest a static picture of a pre-European settlement landscape maintained in suspended animation by whatever techniques are needed. *Wilderness, rather, should be a vignette of natural dynamic forces characteristic of early America, operating as freely as possible.* Ecological change will be inevitable and constant, its velocity dependent on the ecosystem involved. This means that there may be times in its ecological succession when a wilderness or portions thereof will not be a particularly good habitat for some wildlife species or particularly appealing to visitors.

Sometimes it will be possible even to restore aspects of a wilderness that was previously impacted by human influence, and natural recovery may occur. Native animals that have disappeared may be reintroduced or recover naturally. The effects of minor grazing or forest cutting can be erased, over time, by plant succession. Fire and other natural disturbances can initiate new cycles of plant and animal life, as they did before the coming of modern humans. Consider the example of recovery of natural conditions in the BWCAW in Minnesota. In 1948, about 14 percent of this area was privately owned, and there were about forty-five resorts plus some one hundred individual cabins in what was to be designated in 1964 as wilderness (Lucas 1972). Now the private properties are gone, except for a few cabins whose owners have life estates, and canoeists paddling Basswood Lake today cannot discern the shores where the Peterson Lodge entertained fifty guests at a time in 1955. However, the loons know things have changed; they have returned to Hoist Bay in response to restored naturalness and solitude. In some ways at least, you *can* turn back the clock.

A natural-processes criterion argues for artificial control of native predatory animals, insects, or plant diseases only when resources are threatened outside the wilderness. We believe that in wilderness, outbreaks of insects or disease should simply run their course as a contribution to a constantly changing natural ecosystem. Of course, massive infestations of nonnative organisms originating outside wilderness can be a major threat, as with the introduced gypsy moth in Shenandoah National Park, Virginia; the balsam woolly aphid in declining stands of native Fraser fir in Great Smoky Mountains National Park; and the southern pine beetle and hemlock woolly adelgid in the southeastern United States. White bark pine is declining because of infestations by the nonnative white pine blister rust in the western U.S.

Where predators are present, they should be permitted their natural role. Where the predators are absent, however, prey populations may irrupt and unnaturally threaten the continued existence of a natural environment. In such a case, native predatory species might be reintroduced, their prey cropped, or both, the goal being to return the wilderness to natural conditions.

Moreover, again we must be cautious in our definition of "natural." In the United States, an Adirondack forest in the East, by virtue of natural plant succession, may have recovered from logging a century ago, and in some ways be more natural today than a western mountain forest never logged but held for the past seventy-five years in unnatural plant succession by efficient fire prevention and fire-control programs. Because wilderness fire may be the most effective and appropriate habitat-management tool, fire management and wilderness wildlife management must go hand in hand.

Encourage Angling and Hunting Styles Compatible with Wilderness Experiences

Fishing is a traditional recreation activity in many wilderness areas and is allowed by the U.S. Wilderness Act under the direction of state fish and wildlife agencies. Fishing can achieve its finest quality in wilderness, can be a scarce wilderness-dependent experience, and can be a means to realize wilderness values. Wilderness is a place where one can partially experience or imagine a primitive experience of living off the land. Catching a few fish and eating them in camp with family or friends can contribute to the overall quality of a wilderness experience for many users. Such fishing is best as *part* of a wilderness experience, rather than the *reason* for a wilderness visit. Thus, *fishing regulations for wilderness should encourage a focus on the overall wilderness experience rather than on the size of the catch*. Naturalness of the experience, not a bulging creel, should be emphasized.

Stocking of certain high lakes in many western states was begun years ago to transport to suitable waters those species prevented from entering by natural barriers. Some of these lakes are now in designated wilderness, and their stocking and the method used (aerial stocking) are among the most controversial wilderness management issues. There is no question that artificial stocking compromises the naturalness of wilderness, particularly the specific aquatic ecosystems that are affected (Knapp and Matthews 2000; Matthews and Knapp 1999). *Stocked fisheries are advisable in designated wilderness only where the practice is clearly established and carried out in a manner to minimize its impact on wilderness ecosystems, qualities, and the experiences of visitors.* The long-range policy question is whether any artificial stocking in designated wilderness is too harsh a conflict with naturalness. Although fishing is an attraction for some visitors to high mountain lakes, hikers and others were just as numerous in one thirty-year-old study, and they spent just as much time at the lakeshores (Hendee et al. 1977). We expect that this pattern holds today and may be even more pronounced.

Ideally, wilderness hunting and trapping should be carefully managed to ensure that the natural ecosystem is not impaired; just intensive and selective enough (with regard to age and sex) to protect the natural behavior and dynamics of game populations; mimicking as closely as possible the pattern of harvest by natural predators; with very little or no hunting of predators near the top of the food chain, such as wolves and grizzlies; and with no hunting where either a lack of knowledge or lack of regulatory staff makes the consequences on naturalness uncertain. In reality, *lack of control and knowledge about the effects of predation makes it impossible to predict with certainty how hunting affects wildlife naturalness. The need for a conservative approach is thus imperative.*

Where hunting is allowed in wilderness, it should feature those aspects of the sport that are uniquely wilderness-dependent—in keeping with wilderness criteria of naturalness, solitude, and contrast with civilization. It is not realistic to expect that all wilderness hunters would voluntarily restrict themselves to muzzle loaders or bow and arrows. However, conforming to a wilderness ethic implies restrictions on extravagant base camps or extensive communication or GPS equipment that alters the remoteness and solitude of the wilderness experience. Wilderness is the one place hunters can confront game on its terms, and this opportunity should be encouraged for those who can appreciate the unique values of such an experience. *In wilderness, hunting, like fishing, should be part of a wilderness experience and not the sole reason for it.*

Focus Research on Essential Priorities Using Appropriate Methods

In using wilderness as a wildlife research laboratory, scientists should be under the same constraints as everyone else not to degrade its naturalness or solitude. Although some instrumentation may be necessary to adequately characterize the biological systems of wilderness, physical structures as a general rule are not in keeping with wilderness character, nor are such visible techniques as painting large markings on live-trapped animals for later observation from an obtrusive helicopter.

We must recognize, however, that radiotelemetry and aircraft, used with discretion, are invaluable means—sometimes the only means—of obtaining data on the natural distributions and behaviors of large wilderness wildlife species. However, temperance is needed. For example, radio collars are an intrusion on wilderness and may constitute a form of harassment that threatened species such as grizzlies do not need and which possibly distorts natural behavior (Krausman and Hewert 1983). At the very least, radio collars can be color matched to the animal to minimize visual impact. Marking with paint or dye, colored tags, and streamers should be discouraged, and methods such as radioactive tagging and tracing with nucleotides should not be tolerated for fear of seriously compromising natural processes.

Many wildlife research opportunities are essentially wilderness-dependent; they require vast natural areas that may ultimately exist only in wilderness. Nonetheless, *the scientific community should be sure that investigations in wilderness seek essential data, that wilderness is not open to casual inquiry with intruding mechanization, and that the scholar above all is governed by a wilderness ethic. Wilderness wildlife research should, in short, be restricted by overriding principles of wilderness management: do only what is necessary and use the minimum methods, approaches, or tools required.* Wilderness wildlife research proposals need to be strongly justified as sources of information about natural processes that are necessary to wilderness wildlife protection, to improved understanding of global human impacts on nature, and thus to the continued well-being of humans on the planet. If sophisticated mechanization is allowed in wilderness for scientific purposes that are not so justified, it could set a precedent, opening the door to excessive development for other purposes, such as radar installations for monitoring air traffic, telecommunication installations, instrumentation of headwaters for water-yield information for commercial irrigation, climatological data gathering for weather forecasting, and so forth. Currently, agency policies permit sophisticated instrumentation and mechanization for wilderness wildlife research, usually on a case-by-case basis.

Long-standing questions about what factors control populations of ungulates and larger carnivores can only be studied in large, undisturbed areas (Peek 1989). Research to test competing hypotheses about how ungulate populations are naturally regulated (habitat vs. predation), in addition to other important hypotheses,

Fig. 12.8. Bison range in Yellowstone National Park, habitat for one of the largest remaining free-roaming bison herds in the world. Bison roaming from the park may be shot or relocated due to concern that they may transmit brucellosis to cattle. Photo courtesy David Mattson, USGS Forest and Rangeland Ecosystem Science Center.

will require relatively intact ecosystems found only in large wilderness areas. This work has important ramifications for management of native ungulates outside and inside wilderness. In the United States, several benchmark studies of predator-prey and habitat relationships have been conducted in relatively undisturbed wilderness settings. For example, studies such as those of moose-wolf interaction in Isle Royale National Park (Allen 1979; McLaren and Peterson 1994), cougar-ungulate interaction in the Frank Church–River of No Return Wilderness (Hornocker 1970), and desert bighorn sheep habitat relationships in Arizona (Etchberger et al. 1989; Krausman et al. 1989) should be extended and replicated in other systems and with other species. Such research requires large wilderness areas.

In summary, *the greatest scientific values of wilderness are based on the opportunity to study, over long periods of time, large ecosystems that are relatively undisturbed.* Wilderness wildlife is an integral part of these ecosystems, sometimes dependent on and reflective of their degree of naturalness and solitude. Study of wilderness wildlife ecosystems can reveal how natural processes work, thereby providing baseline information for assessing human influence on the area, and from which management guidelines can be derived. Such information is increasingly important as resource management becomes more complex and human influences expand. At the same time, wildlife research methods in wilderness must be balanced against the need to maintain wilderness values.

Summary

We recommend applying, as necessary, the foregoing objectives and guidelines for wilderness wildlife management. These guidelines are intended to supplement, not substitute for, wilderness management principles suggested in Chapter 7 and elsewhere in the book. However, it is unrealistic to seek one monolithic approach to wilderness wildlife management. Wilderness management needs to be tailored to the unavoidable constraints of the individual situations. For this reason, *a wilderness ethic is always needed as a guiding policy*. This chapter has sought to set the stage for such considerations. It has also sought to remind readers of the interdependency of wilderness and wildlife.

Wildlife is only one component of the wilderness resource and cannot be managed separately. Wildlife uses, such as recreational fishing and hunting, must be governed not just by the capacity to produce "surplus" populations of fish and game but by the impact of such uses on the naturalness of relationships among all components of the wilderness resource. A fish population in a particular lake may have the capacity to support a certain level of fishing pressure, but the acceptable level must also be measured against the aggregate impact of anglers and other users on the vegetation and soil around the lake or along the stream, and on the solitude of visitors' experiences.

Finally, we wish to emphasize the importance of wildlife in the wilderness web. Wilderness lets us view the natural processes by which the land and the living things on it have achieved their characteristic forms and by which they maintain their existence. It makes us aware of the incredible intricacies of plant and animal communities, their intrinsic beauty, and of contrasts with creeping degradation in many non-wilderness settings. These things add up to the great lesson of human-environment interdependency: that air, water, soils, trees, plants, insects, birds, fishes, mammals, and people are all part of the same scheme—an

Fig. 12.9. Sighting a bald eagle near Anchorage, Alaska, a national symbol of the U.S. and a charismatic bird, is a long-remembered event by visitors in and around wilderness. Photo courtesy Carol Dawson.

intricately woven landscape fabric. Snip one thread and the entire cloth begins to unravel; stitch up one tear and you begin to repair the whole.

Exposure to wilderness and wildlife, even indirectly, can be a doorway to the ecological understanding of the complete human interdependence with our environment and with life everywhere. The ultimate wilderness and wildlife value may be their contribution to the development of a culture that will secure the future of an environment fit for life and fit for living, and to an appreciation of all those amenities that are inexorably linked to the prosperity of the human spirit.

Study Questions

1. What are four categories of wilderness wildlife? How important is wilderness to each?
2. What are some ways in which wilderness contributes to the preservation of wildlife?
3. How should hunting and fishing in wilderness be different from those activities outside of wilderness?
4. What are some wilderness uses allowed under the Wilderness Act, and some mandates under other legislation, that have the potential to complicate wilderness wildlife management?
5. What are some effects on wildlife of human activities in wilderness, on adjacent nonwilderness lands, and elsewhere?
6. What are some broad objectives for wilderness wildlife management that may be used as criteria for choosing and judging management guidelines (policies and actions)?
7. What are some of the problems wilderness managers face as they attempt to allow "natural processes" to shape wilderness wildlife habitat?
8. How can wilderness wildlife research benefit wildlife and humankind? How can wilderness managers and researchers preserve other wilderness values while conducting necessary research?

Acknowledgments

This chapter was written for the first and second editions by John C. Hendee and Clay Schoenfeld (now deceased). David Mattson of the U.S. Geological Survey's Forest and Rangeland Ecosystem Science Center, Northern Arizona University, Flagstaff, joined John C. Hendee in coauthoring a substantial revision for the third edition. The fourth edition includes minor editing by John C. Hendee and Chad P. Dawson.

Case Discussion: Wolf Reintroduction Controversy

In 1995, fourteen wolves were released in Yellowstone National Park from a trap-and-transfer agreement with Alberta, Canada. In 1996, another twenty wolves from British Columbia were released in Yellowstone and Montana. Wolves were confirmed in various adjoining states within two years, and the population of wolves recovered in the greater Northern Rockies (Idaho, Montana and Wyoming) to an estimated seventy-one breeding pairs and more than a thousand wolves by 2005 (U.S. Department of the Interior 2008, Defenders of Wildlife 2008). The wolf recovery projects were strongly supported by numerous individuals during various hearings regarding proposed plans and management actions and by the Defenders of Wildlife, National Wildlife Federation, and Friends of Animals (Defenders of Wildlife 2008).

The gray wolf was protected under the Endangered Species Act (U.S. Public Law 93-205) and was listed as endangered and under the management authority of the U.S. Fish and Wildlife Service until February 27, 2008 (U.S. Department of the Interior 2008). In spite of this federal protection, several wolves were killed by hunters and ranchers in the area after the reintroduction, who sometimes claimed they thought they were shooting coyotes or dogs running in the wild.

In an effort to compensate ranchers and livestock owners for losses due to wolf predation on livestock, the Defenders of Wildlife established a Wolf Compensation Fund from donations to help offset any verified economic losses by ranchers. Between 1987 and 1999 the fund paid $80,000 to 85 ranchers. The number of compensation payments increased as the wolf recovery grew, until by 2005 and 2006, the fund paid out more than $100,000 each year to 70 or 80 ranchers (Defenders of Wildlife 2008).

During 1995 to 2007, several serious legal actions were taken to block the wolf reintroduction by the American Farm Bureau, Montana Stock Growers,

Fig. 12.10. Dall sheep, which have large home ranges, graze outside Chugach National Forest near Anchorage, Alaska. Photo courtesy Carol Dawson

Wyoming Farm Bureau, and others (Defenders of Wildlife 2008). In addition, Idaho, Montana, and Wyoming proposed to the FWS in 2003 through 2007 that they be allowed to manage wolves under state wolf management plans they developed, following the delisting of wolves in the region from protection under the ESA (U.S. Department of the Interior 2008). In 2006, the Idaho Department of Fish and Game announced plans to radio-collar one wolf in each pack, including those in the Frank Church–River–of–No–Return Wilderness and the Selway-Bitterroot Wilderness.

In 2007, the FWS proposed removing the gray wolf from the endangered listing under the ESA and allowing states to manage wolf populations according to their state management plans. In 2008, the FWS completed the process to remove the gray wolf from ESA protection and returned wolf management back to state agencies (U.S. Department of the Interior 2008). Returning wolf management back to the state fish and wildlife agencies means that public hunting and other wolf control programs would be implemented, such as allowing states to kill wolves in areas where deer and elk populations are declining and allowing landowners to kill wolves caught attacking dogs or livestock in Idaho, Montana, and Wyoming. However, the ESA decision is being challenged in federal courts and the final outcome is not yet known.

Case Discussion Questions:

1. Are gray wolves in the northern Rockies and Yellowstone National Park wilderness-dependent or wilderness–associated wildlife?
2. Which types of groups of people are for and against wolf reintroduction and why? What are their arguments for and against wolves?
3. How does this wolf reintroduction case match up with the six wildlife-management objectives proposed in this chapter?
4. Some recent research (U.S. Department of the Interior 2008) suggests that wolves are moving elk out of meadow areas that had been overgrazed by elk and allowing for recovery of the natural vegetation and other wildlife species, but which then means elk populations are reduced. Others claim that elk are declining due to direct wolf predation and that wolves have become too abundant. How do you focus research on these issues to decide what needs to be done for naturalness?

References

Alderton, David. 1991. *Crocodiles and Alligators of the World*. New York: Facts on File.

Allen, D. L. 1979. *Wolves of Minong*. Boston, MA: Houghton Mifflin.

Allen, Durward. 1966. The preservation of endangered habitats and vertebrates of North America. In: Darling, F. Fraser; Milton, John, eds. *Future environments of North America*. Garden City, NY: The Natural History Press, pp. 22–37.

Anderson, Stanley. 1995. Recreational disturbances and wildlife populations. In: Knight, Richard L.; Gutzwiller, Kevin J., eds. *Wildlife and Recreationists*. Washington, DC: Island Press, pp. 157–181.

Anthony, Robert G.; Steidl, Robert J.; McGarigal, Kevin. 1995. Recreation and bald eagles in the Pacific Northwest. In: Knight, Richard L.; Gutzwiller, Kevin J., eds. *Wildlife and Recreationists*. Washington, DC: Island Press, pp. 223–241.

Aumiller, Larry D.; Matt, Colleen A. 1994. Management of McNeil River State Game Sanctuary for viewing of brown bears. *International Conference on Bear Research and Management*. 9: 51–61.

Bailey, Theodore N. 1993. *The African Leopard: Ecology and Behavior of a Solitary Felid*. New York: Columbia University Press.

Ballard, Warren B.; van Ballenberghe, Victor. 1997. Predator/prey relationships. In: Franzmann, Albert W.; Schwartz, Charles C., eds. *Ecology and Management of the North American Moose*. Washington, DC: Smithsonian Institution Press, pp. 247–273.

Barnes, R. F. W.; Lahm, S. A. 1997. An ecological perspective on human densities in the central African forests. *Journal of Applied Ecology*. 34: 245–260.

Beier, Paul. 1991. Cougar attacks on humans in the United States and Canada. *Wildlife Society Bulletin*. 19: 403–412.

Berger, Joel; Cunningham, Carol. 1994. Active intervention and conservation: Africa's pachyderm problem. *Science*. 263: 1241–1242.

Bowles, Ann E. 1995. Responses of wildlife to noise. In: Knight, Richard L.; Gutzwiller, Kevin J., eds. *Wildlife and Recreationists*. Washington, DC: Island Press, pp. 109–156.

Brown, David E., ed. 1983. *The Wolf in the Southwest: The Making of an Endangered Species*. Tucson: University of Arizona Press.

———.1985. *The Grizzly in the Southwest*. Norman: University of Oklahoma Press.

Campbell, Ken; Hofer, Heribert. 1995. People and wildlife: Spatial dynamics and zones of interaction. In: Sinclair, A. R. E.; Arcese, Peter, eds. *Serengeti II: Dynamics, Management, and Conservation of an Ecosystem*. Chicago: University of Chicago Press, pp. 534–570.

Campbell, Sheldon. 1984. Restocking the wilderness with captive-bred animals: The California condor story. In: Martin, Vance; Inglis,

Mary, eds. *Wilderness: The Way Ahead*. Middleton, WI: Lorian Press, pp. 108–113.

Caro, Tim M.; O'Doherty, Gillian. 1999. On the use of surrogate species in conservation biology. *Conservation Biology*. 13: 805–814.

Carr, Archie. 1973. *The Everglades*. New York: Time-Life Books. 184p.

Chadwick, Douglas H. 1992. *The Fate of the Elephant*. San Francisco: Sierra Club Books.

Chowdhury, Keshab N. 1995. India's Rajaji National Park and threatened elephants. *Environmental Conservation*. 22: 79–80.

Clark, Tim W.; Curlee, A. Peyton; Minta, Steven C.; Kareiva, Peter M., eds. 1999. *Carnivores in Ecosystems: The Yellowstone Experience*. New Haven, CT: Yale University Press.

Cronin, Matthew A.; Ballard, Warren B.; Bryan, James D.; Pierson, Barbara J.; McKendrick, Jay D. 1998. Northern Alaska oil fields and caribou: A commentary. *Biological Conservation*. 83: 195–208.

Cutler, M. Rupert. 1980. Wilderness decisions: Values and challenges to science. *Journal of Forestry*. 78(2): 74–77.

Czech, Brian; Krausman, Paul R. 1999. Controversial wildlife management issues in southwestern U.S. wilderness. *International Journal of Wilderness*. 5(3): 22–28.

Dasmann, Raymond F. 1966. *Wildlife Biology*. New York: John Wiley.

Davis, George D. 1989. Preservation of natural diversity: The role of ecosystem representation within wilderness. In: Freilich, Helen R., comp. *Wilderness Benchmark 1988: Proceedings of the National Wilderness Colloquium*; January 13–14, 1988; Tampa, FL. Asheville, NC: U.S. Department of Agriculture, Forest Service, Southeastern Forest Experiment Station, pp. 76–82.

Defenders of Wildlife. 2008. A chronology of wolf recovery in the northern Rockies. www.defenders.org/programs_and_policy/wildlife_conservation/imperiled_species/wolves; accessed January 31, 2008.

Dobson, Andy. 1995. The ecology and epidemiology of rinderpest virus in Serengeti and Ngorongoro Conservation Area. In: Sinclair, A. R. E.; Arcese, Peter, eds. *Serengeti II: Dynamics, Management, and Conservation of an Ecosystem*. Chicago: University of Chicago Press, pp. 485–505.

Dobson, Andy; Ralls, Katherine; Foster, Mercedes; Soulé, Michael E.; Simberloff, Daniel; Doak, Dan; Estes, James A.; Mills, L. Scott; Mattson, David; Dirzo, Rodolfo; Arita, Héctor; Ryan, Sadie; Norse, Elliott A.; Noss, Reed F.; Johns, David. 1999. Corridors: reconnecting fragmented landscapes. In: Soulé, Michael E.; Terborgh, John, eds. *Continental Conservation*. Washington, DC: Island Press, pp. 129–170.

Douglas-Hamilton, Ian; Douglas-Hamilton, Oria. 1992. *Battle for the Elephants*. New York: Viking.

Dublin, Holly T. 1995. Vegetation dynamics in the Serengeti-Mara ecosystem: The role of elephants, fire, and other factors. In: Sinclair, A. R. E.; Arcese, Peter, eds. *Serengeti II: Dynamics, Management, and Conservation of an Ecosystem*. Chicago: University of Chicago Press, pp. 71–90.

Etchberger, Richard C.; Krausman, Paul R.; Mazaika, Rosemary. 1989. Mountain sheep habitat characteristics in the Pusch Ridge Wilderness, Arizona. *Journal of Wildlife Management*. 53(4): 902–907.

Fanshawe, John H.; Frame, Lory H.; Ginsburg, Joshua R. 1991. The wild dog—Africa's vanishing carnivore. *Oryx*. 25: 137–146.

Fitzgibbon, Clare D.; Mogaka, Hezron; Fanshawe, John H. 1994. Subsistence hunting in Arabuko-Sokoke Forest, Kenya, and its effects on mammal populations. *Conservation Biology*. 9: 1116–1126.

French, Steven P.; French, Marilynn G.; Knight, Richard R. 1994. Grizzly bear use of army cutworm moths in the Yellowstone ecosystem. *International Conference on Bear Research and Management*. 9: 389–399.

Gasaway, William C.; Stephenson, Robert O.; Davis, James L.; Sheperd, Peter E. K.; Burris, Oliver E. 1983. Interrelationships of wolves, prey, and man in interior Alaska. *Wildlife Monographs*. 84: 1–50.

Gillingham, Sarah; Lee, Phyllis C. 1999. The impact of wildlife-related benefits on the conservation attitudes of local people around the Selous Game Reserve, Tanzania. *Environmental Conservation*. 26: 218–228.

Goltsman, M.; Kruchenkova, E. P.; Macdonald, D. W. 1996. The Mednyi Arctic foxes: Treating a population imperilled by disease. *Oryx*. 30: 251–258.

Goodall, Jane. 1986. *The Chimpanzees of Gombee: Patterns of Behavior*. Cambridge, MA: Belknap Press of Harvard University Press.

Gottelli, Dada; Sillero-Zubiri, Claudio. 1992. The Ethiopian wolf—an endangered endemic canid. *Oryx*. 26: 205–214.

Griffith, Brad; Scott, J. Michael; Carpenter, John W.; Reed, Christine. 1989. Translocation as a species conservation tool: Status and strategy. *Science*. 245: 477–480.

Heinen, Joel T. 1996. Human behavior, incentives, and protected area management. *Conservation Biology*. 10: 681–684.

Hendee, John C.; Clark, Roger N.; Dailey, Thomas E. 1977. Fishing and other recreation behavior at roadless high lakes: Some management implications. Research Note PNW-304. Portland, OR: U.S. Department of Agriculture, Forest Service, Pacific Northwest Forest and Range Experiment Station.

Hendee, John C.; Pitstick, Randal. 1993. The growth of environmental and conservation-related organizations: 1980–1991: *Renewable Resources Journal*. 10(2): 6–11.

Heptner, H. O. N.; Nasimovich, A. A.; Bannikov, A. G. 1989. *Mammals of the Soviet Union*. Vol. I: *Ungulates*. New York: E. J. Brill.

Herrero, Stephen; Fleck, Susan. 1990. Injury to people inflicted by black, grizzly, or polar bears: Recent trends and new insights. *International Conference on Bear Research and Management*. 8: 25–32.

Hochbaum, H. Albert. 1970. Wilderness wildlife in Canada. In: McCloskey, Maxine E., ed. *Wilderness: The Edge of Knowledge*. San Francisco: Sierra Club, pp. 23–33.

Hornocker, Maurice. 1970. An analysis of mountain lion predation on mule deer and elk in the Idaho Primitive Area. *Wildlife Monographs*. 21.

Houston, Douglas B.; Schreiner, Edward G.; Moorhead, Bruce B. 1994. *Mountain Goats in Olympic National Park: Biology and Management of an Introduced Species*. NPS/NROLYM/NRSM-94/25. Washington, DC: National Park Service.

Hummel, M. 2007. The Stikine-LeConte Wilderness in Alaska: Twenty-six years of management under ANILCA. *International Journal of Wilderness*. 13(2): 8–13, 17.

Jachmann, H.; Berry, P. S. M.; Imae, H. 1995. Tusklessness in African elephants: A future trend. *African Journal of Ecology*. 33: 230–235.

Kaeding, Lynn R.; Boltz, Gary L.; Carty, Daniel G. 1996. Lake trout discovered in Yellowstone Lake threaten native cutthroat trout. *Fisheries*. 21: 16–20.

Kellert, Stephen R. 1985. Public perceptions of predators, particularly the wolf and coyote. *Biological Conservation*. 31: 167–189.

Kilgo, John C.; Labisky, Ronald F.; Fritzen, Duane E. 1998. Influences of hunting on the behavior of white-tailed deer: Implications for conservation of the Florida panther. *Conservation Biology*. 12: 1359–1364.

Knapp, Roland; Matthews, Kathleen. 2000. Nonnative fish introductions and the decline of mountain yellow-legged frog from within protected areas. *Conservation Biology*. 14: 428–438.

Knight, Richard L.; Gutzwiller, Kevin J., eds. 1995. *Wildlife and Recreationists*. Washington, DC: Island Press.

Krausman, Paul R.; Hervert, John J. 1983. Mountain sheep responses to aerial surveys. *Wildlife Society Bulletin*. 11(4): 372–375.

Krausman, Paul R.; Leopold, Bruce D.; Seegmiller, Rick F.; Torres, Steven G. 1989. Relationships between desert bighorn sheep and habitat in western Arizona. *Wildlife Monographs*. 102.

Leopold, A. Starker. 1966. Adaptability of animals to habitat change. In: Darling, F. Fraser; Milton, John, eds. *Future Environments of North America*. Garden City, NY: The Natural History Press, pp. 66–75.

Leopold, Aldo. 1933. *Wildlife Management*. New York: C. Scribner's Sons.

———. *A Sand County Almanac, and Sketches Here and There*. New York: Oxford University Press.

Lever, Christopher. 1985. *Naturalized Mammals of the World*. New York: Longman.

Lhagvasuren, B.; Dulamtseren, S.; Amgalan, L.; Mallon, D.; Schaller, G.; Reading, R.; Mix, H. 1999. Status and conservation of antelopes in Mongolia. *Scientific Research of the Institute of Biological Sciences*. 1: 96–108.

Lucas, Robert C. 1972. Wilderness perception and use. In: Thompson, Dennis L., ed. *Politics, Policy, and Natural Resources*. New York: Macmillan, pp. 309–323.

MacCracken, James G.; Goble, Dale; O'Laughlin, Jay. 1994. Grizzly bear recovery in Idaho. Idaho Forest, Wildlife and Range Policy Analysis Group, Report No. 12. Moscow: University of Idaho.

Matthews, Kathleen R.; Knapp, Roland A. 1999. A study of high mountain lake fish stocking in the U.S. Sierra Nevada Wilderness. *International Journal of Wilderness*. 5(1): 24–26.

Mattson, David J. 1990. Human impacts on bear habitat use. *International Conference on Bear Research and Management*. 8: 33–56.

———. 1997. Wilderness-dependent wildlife: The large and the carnivorous. *International Journal of Wilderness*. 3(4): 34–38.

McLaren, B. E.; Peterson, R. O. 1994. Wolves, moose, and tree rings on Isle Royale. *Science*. 266: 1555–1558.

McNeeley, Jeffrey A.; Wachtel, Paul S. 1988. *Soul of the Tiger.* New York: Doubleday.

Meagher, Mary; Meyer, M. E. 1994. On the origin of brucellosis in bison of Yellowstone National Park: A review. *Conservation Biology.* 8: 645–653.

Mercure, Delbert V., Jr.; Ross, William M. 1970. The Wilderness Act: A product of congressional compromise. In: Cooley, Richard; Wandesforde-Smith, Geoffry, eds. *Congress and the Environment.* Seattle: University of Washington Press, pp. 47–64.

Meriggi, Alberto; Lovari, Sandro. 1996. A review of wolf predation in southern Europe: Does the wolf prefer wild prey or livestock? *Journal of Applied Ecology.* 33: 1561–1571.

Merrill, Troy; Mattson, David J.; Wright, R. Gerald; Quigley, Howard B. 1999. Defining landscapes suitable for restoration of grizzly bears *Ursus arctos* in Idaho. *Biological Conservation.* 87: 231–248.

Milton, John P. 1972. The web of wilderness. *Living Wilderness.* 35(16): 14–19.

Mishra, Charudutt; Johnsingh, A. J. T. 1998. Population and conservation status of the Nilgiri tahr *Hemitragus hylocrius* in Anamalai Hills, south India. *Biological Conservation.* 86: 199–206.

Muchaal, Pia K.; Ngandjui, Germain. 1999. Impact of village hunting on wildlife populations in the western Dja Reserve, Cameroon. *Conservation Biology.* 13: 385–396.

Murie, Margaret E.; Murie, Olaus. 1966. *Wapiti Wilderness.* New York: Knopf.

Murray, Max; Njogu, A. R. 1989. African trypanosomiasis in wild and domestic ungulates: The problem and its control. *Symposium of the Zoological Society of London.* 61: 217–240.

Muth, Robert M. 1990. Community stability as social structure: The role of subsistence uses of natural resources in southeast Alaska. In: Lee, Robert G.; Birch, Donald R., Jr., eds. *Community and Forestry.* Boulder, CO: Westview Press.

Nash, Roderick. 1970. Wild-deer-ness. In: McCloskey, Maxine E., ed. *Wilderness: The Edge of Knowledge.* San Francisco: Sierra Club, p. 36.

Noss, Reed F.; Dinerstein, Eric; Gilbert, Barrie; Gilpin, Michael; Miller, Brian J.; Terborgh, John; Trombulak, Steve. 1999. Core areas: Where nature reigns. In: Soulé, Michael E.; Terborgh, John, eds. *Continental Conservation.* Washington, DC: Island Press, pp. 99–128.

Novaro, André J.; Funes, Martín C.; Walker, R. S. 2000. Ecological extinction of native prey of a carnivore assemblage in Argentine Patagonia. *Biological Conservation.* 92: 25–33.

Olson, Sigurd. 1963. *Listening Point.* New York: Knopf.

Oza, G. M. 1974. Conservation of the Asiatic lion: Now limited to Gujarat State, India. *Biological Conservation.* 6: 225–227.

Peek, James M. 1989. Natural regulation of ungulates: What constitutes a real wilderness? *Wildlife Society Bulletin.* 8(3): 217–227.

Pelton, Michael R.; van Manen, Frank T. 1996. Benefits and pitfalls of long-term research: A case study of black bears in Great Smoky Mountains National Park. *Wildlife Society Bulletin.* 24: 443–450.

Penny, Malcolm. 1988. *Rhinos: Endangered Species.* New York: Facts on File.

Poffenberger, Mark, ed. 1990. *Keepers of the Forest: Land Management Alternatives in Southeast Asia.* West Hartford, CT: Kumarian Press.

Porter, William F. 1996. Management of overabundant species in protected areas: The white-tailed deer as a case example. In: Wright, R. Gerald, ed. *National Parks and Protected Areas: Their Role in Environmental Protection.* Cambridge, MA: Blackwell Science, pp. 223–248.

Qui, Ming J. 1996. Tiger-human conflict in southeastern Tibet. *Oryx.* 30: 5–6.

Reading, Richard P.; Clark, Timothy W.; Griffith, Brad. 1997. The influence of valuational and organizational considerations on the success of rare species translocations. *Biological Conservation.* 79: 217–225.

Regehr, E. V.; Lunn, N. J.; Amstrup, S. C.; Stirling, I. 2007. Effects of earlier sea ice breakup on survival and population size of polar bears in western Hudson Bay. *Journal of Wildlife Management.* 71(8): 2673–2683.

Reinhart, Daniel P.; Haroldson, Mark; Mattson, David J.; Gunther, Kerry A. 2001. Effects of exotic species on Yellowstone's grizzly bears. *Western North American Naturalist.* 61(3): 277–288.

Russell, Andy. 1971. *Grizzly Country.* New York: Knopf.

Sanyal, N.; Krumpe, E. E.; Vanormer, C. 2006. Modeling encounters between backcountry recreationists and grizzly bears in Glacier National Park. *International Journal of Wilderness.* 12(3): 24–29, 48.

Schaller, George B. 1963. *The Mountain Gorilla: Ecology and Behavior.* Chicago: University of Chicago Press.

———. 1998. *Wildlife of the Tibetan Steppe.* Chicago: University of Chicago Press.

Schaller, George B.; Junrang, Ren; Mingjiang, Qiu. 1988. Status of the snow leopard *Panthera uncia* in Quinghai and Gansu Provinces, China. *Biological Conservation.* 45: 179–194.

Sekhar, Nagothu U. 1998. Crop and livestock depredation caused by wild animals in protected areas: The case of the Sariska Tiger Reserve, Rajasthan, India. *Environmental Conservation.* 25: 160–171.

Shaw, R. Paul. 1989. Rapid population growth and environmental degradation: Ultimate *versus* proximate factors. *Environmental Conservation.* 16: 199–200.

Simberloff, Daniel. 1998. Flagships, umbrellas, and keystones: Is single-species management passé in the landscape era? *Biological Conservation.* 83: 247–257.

Sinclair, A. R. E.; Arcese, Peter, eds. 1995. *Serengeti II: Dynamics, Management, and Conservation of an Ecosystem.* Chicago: University of Chicago Press.

Singer, Francis J. 1984. Effects of wild pig rooting in a deciduous forest. *Journal of Wildlife Management.* 48(2): 466–473.

Singh, Jaidev "Jay"; Jackson, Rodney. 1999. Transfrontier conservation areas: Creating opportunities for conservation, peace and the snow leopard in Central Asia. *International Journal of Wilderness.* 5(3): 7–12.

Tardiff, Sandy, E.; Stanford, Jack A. 1998. Grizzly bear digging: Effects on subalpine meadow plants in relation to mineral nitrogen availability. *Ecology.* 79: 2219–2228.

Terborgh, John; Estes, James A.; Paquet, P.; Ralls, K.; Boyd-Heger, Diane; Miller, Brian J.; Noss, Reed F. 1999. The role of top carnivores in regulating terrestrial ecosystems. In: Soulé, Michael E.; Terborgh, John, eds. *Continental Conservation.* Washington, DC: Island Press, pp. 39–64.

Torres, Steven G.; Mansfield, Terry M.; Foley, Janet E.; Lupo, Thomas; Brinkhaus, Amy. 1996. Mountain lion and human activity in California: Testing speculations. *Wildlife Society Bulletin.* 24: 451–460.

Trefethen, James B. 1975. *An American Crusade for Wildlife.* New York: Boone and Crockett Club.

U.S. Department of Agriculture, Forest Service. 2005. What can I do in wilderness? Alaska National Interest Lands Conservation Act (ANILCA) and wilderness in national forests. Ketchikan, AK: U.S. Department of Agriculture, Forest Service.

U.S. Department of the Interior. 2008. Endangered and threatened wildlife and plants. Final rule designating the northern Rocky Mountain population of gray wolf as a distinct population segment and removing this distinct population segment from the federal list of endangered and threatened wildlife. 50 *Code of Federal Regulations*, Part 17: February 27, 2008; *Federal Register.* 70(39): 10,514–10,560.

U.S. Public Law 88-577. The Wilderness Act of September 3, 1964. 78 Stat. 890.

U.S. Public Law 89-665. National Historic Preservation Act of October 15, 1966. 80 Stat. 915.

U.S. Public Law 93-205. Endangered Species Act of December 28, 1973. 87 Stat. 884.

U.S. Public Law 95-95. Clean Air Act of August 7, 1977, as amended. 42 Stat. 7401.

U.S. Public Law 96-487. Alaska National Interest Lands Conservation Act of December 2, 1980. 96 Stat. 487.

U.S. Public Law 96-560. Colorado Wilderness Act of December 22, 1980. 94 Stat. 3265.

U.S. Public Law 103-433. California Desert Protection Act of October 31, 1994. 108 Stat. 4473.

Wagner, Frederic H.; Foresta, Ronald; Gill, R. Bruce; McCullough, Dale R.; Pelton, Michael R.; Porter, William F.; Salwasser, Hal. 1995. *Wildlife Policies in the U.S. National Parks.* Washington DC: Island Press.

Wilkie, David S.; Curran, Bryan; Tshombe, Richard; Morelli, Gilda A. 1998. Modeling the sustainability of subsistence farming and hunting in the Ituri Forest of Zaire. *Conservation Biology.* 12: 137–147.

Wilson, Edward O. 1984. Million-year histories: Species diversity as an ethical goal. *Wilderness.* 48(165): 12–17.

Wobeser, Gary A. 1994. *Investigation and Management of Disease in Wild Animals.* New York: Plenum Press.

Wolf, C. Magdalena; Griffith, Brad; Reed, Christine; Temple, Stanley A. 1996. Avian and mammalian translocations: Update and reanalysis of 1987 survey data. *Conservation Biology.* 10: 1142–1154.

Chapter 13
Potential Threats to Wilderness Resources and Values

Introduction	336
Wilderness Conditions and Values Affected by Threats	337
Potential Threats to Wilderness	339
Fragmentation and Isolation of Wilderness as Ecological Islands	340
Threatened and Endangered Species	341
Increasing Commercial and Public Recreational Use	341
Livestock Grazing	343
Nonnative Species	343
Administrative Access, Facilities, and Intrusive Management	344
Adjacent-Land Management and Use	345
Inholdings	345
Mining	346
Wildland Fire Suppression	346
Air Quality	347
Water Projects and Water Quality	347
Advanced Technology	348
Motorized and Mechanical Equipment Trespass and Legal Use	348
Aircraft Noise and Airspace Reservations	349
Urbanization and Encroaching Development	349
Global Climate Change	350
Legislation Designating Areas with Compromised Wilderness Conditions	350
Lack of Political and Financial Support for Wilderness Protection and Management	351
Monitoring Threats to Wilderness	352
Summary	352
Study Questions	353
Acknowledgments	353
Case Discussion: Interactions Between Potential Threats to Wilderness	354
Case Discussion Questions	354
References	355

Introduction

Formal protection of wilderness conditions and natural processes starts with wilderness designation and then relies on good planning and management to achieve the purposes of the Wilderness Act of 1964 (U.S. Public Law 88-577), to "secure for the American people of present and future generations the benefits of an enduring resource of wilderness." These National Wilderness Preservation System areas, individually and collectively, exist in a global context and the nation's larger, natural landscape and public-land heritage. *Wilderness is affected both internally and externally by many environmental and social conditions, influences, and changes that may threaten wilderness resources and values, now or in the future.*

Since 1964, the NWPS has grown from 9 million acres to more than 107 million acres. These areas encompass more than 4 percent of the land area of the United States and more than 2 percent of the land area of the contiguous forty-eight states, with additional designations expected in the future. *The intended resources and values of the Wilderness Act—which we interpret in this book as naturalness, opportunities for solitude, and wildness—will become even more scarce in the future because the vast remainder of the land will have been urbanized, cultivated, roaded, harvested, mined, drilled, scraped, heavily managed, or otherwise developed for human uses.* Thus, the 4.5 percent of the landscape set aside to remain "untrammeled" as wilderness will also represent remnants of many ecosystems, wild conditions, and natural landscapes that have either disappeared or been altered. Given the growth and urbanization of our population in the United States and the world, wilderness may represent the remaining wild conditions and natural landscapes in the future, plus one of the only remaining opportunities for solitude and some forms of unconfined and primitive recreation. Wilderness, like parks—another wildland resource—is jeopardized worldwide by at least forty-eight specific threats to ecological and social values, as noted by Machlis and Tichnell (1985) in their survey of threats to the world's parks identified by park managers.

The loss of naturalness and wildness in wilderness may be incremental, but ultimately, in the long term, it is dramatic. The exponential development and use of our natural-resource base over the last one and one-half centuries are having profound direct and indirect impacts on all other resources of the planet. Consider the profound changes to the natural ecosystems and landscape in the brief human history on the planet. Drive the Lewis and Clark Trail today from St. Louis to the Pacific Northwest, and read their diaries and narratives of discovery from only two hundred years ago. Or ask anyone who is fifty to eighty years old or a lifelong wilderness traveler what changes they have seen in the landscape or wilderness in their lifetime. Even we remember drinking pure, untreated water in wilderness forty years ago, whereas the risk of giardia or other diseases makes water filters or purifiers mandatory in wilderness today.

Of course, population growth and economic development are driving the worldwide use of resources and contribute directly and indirectly to the loss of naturalness and wildness. Environmentally concerned individuals and wilderness visitors, while enjoying the resources, also contribute incrementally to losing naturalness and solitude. The dramatic changes of the past and present on our global environment and the unnatural influences manifested by them will have a profound effect on wild landscapes and wilderness areas of the world.

The incremental loss of naturalness is subtle and harder to detect than more dramatic changes, especially by the majority of people who live and work in urbanized environments and do not have the experience or training to observe and comprehend the longer-term and more gradual changes that are occurring within wilderness. We rely on scientists, managers, and wilderness advocates who study and monitor changes to report them; yet the information may be overlooked

Fig. 13.1. Visitors hike a popular trail on Mount Washington, New Hampshire. They have views of the Presidential Range–Dry River Wilderness on the White Mountain National Forest. High-use recreation access is one threat to wilderness resource conditions and to visitor wilderness experiences. Photo courtesy Chad P. Dawson.

by the public and be regarded as less relevant than more urgent day-to-day matters. Moreover, the ubiquitous focus on today's urgent issues makes it difficult to appreciate the important long-term impacts and changes that are occurring in wilderness areas over time. Only through long-term resource monitoring in wilderness and other natural, protected areas can we document and understand the global changes that are taking place. This is essential not just for wilderness protection today and in the future, but to safeguard the human condition.

Many public controversies about wilderness arise when a particular user group or special-interest group contends that their interests are not being served in some way by agency management practices in a particular wilderness. It is well documented that recreational use of wilderness areas is growing, and the public continues to support and be committed to the concept of wilderness and the NWPS. However, there is concern among the federal agencies, such as the U.S. Forest Service, that the resources committed to protect and manage the wilderness have not kept pace with their needs (Dombeck 1999). Some wilderness managers believe that we should be planning wilderness for a century or longer time frame and not the ten-year plans typical of most federal and state agencies. A longer planning horizon might help avoid short-term contentiousness between vested interests and managers who focuses primarily on today's needs and wants, often to the detriment of long-term naturalness and "untrammeled" conditions. *Managing the long-term threats to wilderness resources and values is an important challenge, if we are to achieve the purposes of the Wilderness Act of 1964—to preserve wilderness now and for future generations.*

Wilderness Conditions and Values Affected by Threats

We define threats to wilderness as a general concept and focus on change agents or processes that negatively or adversely impact wilderness resource conditions and values. Although we present some examples of impacts on wilderness resource conditions and values, our focus is mostly on the *change agents—what causes the impacts.* These change agents or causes of impacts on natural conditions and values come directly or indirectly from human activities, because natural disturbances (e.g., lightning-caused fires, volcanoes, and hurricanes) *are considered natural processes.* For example, increasing visitor use of wilderness areas (i.e., a change agent) can impact opportunities to experience wilderness values due to crowding, visitor conflicts, loss of solitude, and from impacts on wilderness resources such as loss of vegetative ground cover at campsites and trail erosion.

Our intent is to summarize threats (i.e., causes of impacts) to wilderness and the NWPS. We do not intend *to exhaustively list the impacts* to wilderness conditions (i.e., ecosystem effects) nor the impacts to wilderness values (i.e., human-dimension effects). In this chapter, we draw on the recent, excellent work by scientists at the Aldo Leopold Wilderness Research Center in Missoula, Montana (Cole 1994; Cole 2001b; Cole and Landres 1996; Landres, Marsh et al. 1998a; Landres, Morgan et al. 1999; Landres, Brunson et al. 2000; Cole, McCool et al. 2000).

Changes that affect wilderness can be considered threats when they are not part of the natural conditions and processes that shaped, or were historically part of, the wilderness. The Wilderness Act of 1964 (U.S. Public Law 88-577) describes wilderness as "an area where the earth and community of life are untrammeled by man…retaining its primeval character and influence…protected and managed so as to preserve its natural conditions…affected primarily by the forces of nature." The natural conditions and attributes of wilderness character that are of most concern have been previously summarized as including air quality, aquatic systems, landforms, soils, vegetation, fish and wildlife, ecosystems and landscapes, and wilderness experiences (Cole 1994; Cole and Landres 1996). These threats will be expanded in this chapter, based on available information and including more consideration of some of the social values of wilderness, such as recreation.

In defining natural conditions and naturalness, we think of the ecological concept of historical range of variation (HRV), which recognizes that natural conditions are not static, but in constant change (Aplet and Keeton 1999; see Chapter 10). For example, during the two-hundred- to three-hundred-year period since white settlement began in the United States, a given wilderness site might have appeared quite different from time to time, although remaining within its HRV, due to ecological succession and perturbations of fire, wind, and natural forces.

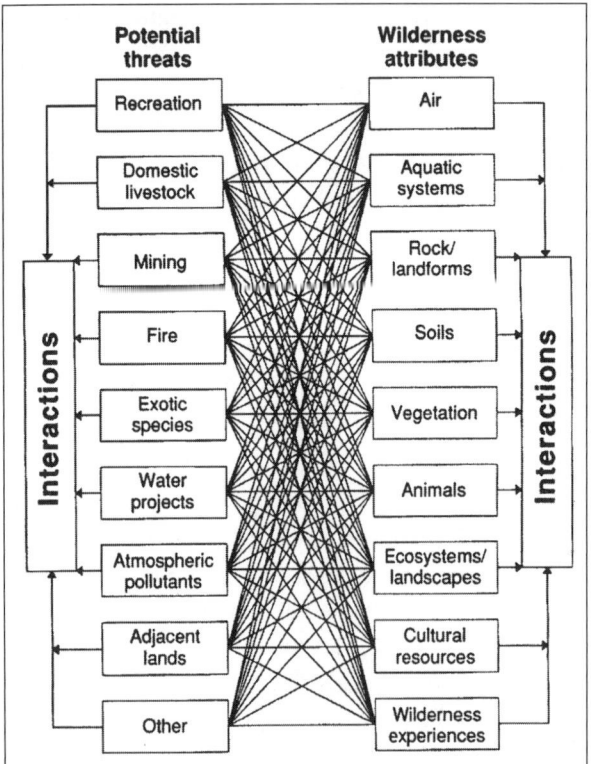

Fig. 13.2. Cole and Landres (1996, 169) charted the complex linkages and interactions among potential threats to wilderness and wilderness attributes.

Naturalness is a relative concept and is often applied to wilderness, but it is difficult to relate to threats or degradation to wilderness conditions and functions. Landres, White et al. (1998b) have defined *naturalness* as the compositions, structures, and functions within an area that are unaffected by contemporary (roughly since the time of European settlement) anthropogenic (human) influences. This definition of naturalness requires explicit spatial bounds, temporal bounds, and value judgments. Natural variability applies too because none of the wilderness conditions remains the same over time or at other locations. *Natural variability* is the natural change in the compositions, structures, and functions of an ecological system over time (temporal variability) and from one place to another (spatial variability), and this variability can be described statistically (Landres, White et al. 1998b).

In wilderness, the management guideline is to allow ecosystems to function within their HRV, without being manipulated. This does not mean returning ecosystems to any preexisting condition (e.g., presettlement conditions) or maintaining them in any particular set of conditions, but rather to allow natural processes and variability to continue (Landres, Morgan et al. 1999).

Wildness refers to the absence of human controls and manipulation on the naturalness and natural variability within an area (Landres, Brunson et al. 2000). While the biological aspects of *naturalness* are a requisite component of wilderness, the concept of *wildness* carries with it the connotation of social and biological values that benefit society because the area is unmodified by civilization and human manipulation.

Threats to wilderness conditions, attributes, and values are complex, interactive, and both internal and external to wilderness. *Internal threats to wilderness may come from remnants of previous activities, as well as the so-called permitted but nonconforming uses under the Wilderness Act,* like grazing, mineral exploration, and mining. For example, previous human activities, such as logging, homesteading, or road building, can affect natural conditions even in the present-day wilderness. Grazing is an obvious impact, even though it may be legal. These impacted conditions may recover to some degree once natural processes are allowed to operate freely, but such impacts are not likely to be completely removed. *External threats* may come from air pollution, acid rain, pesticide drift from agriculture, or forest-management practices on nearby lands. External threats may originate some distance from a particular wilderness and still alter its biota, like hunting migratory wildlife species, invasions of nonnative plant species, or some larger influences like global climate change.

Actual or potential impacts from changing conditions will vary depending on a myriad of factors

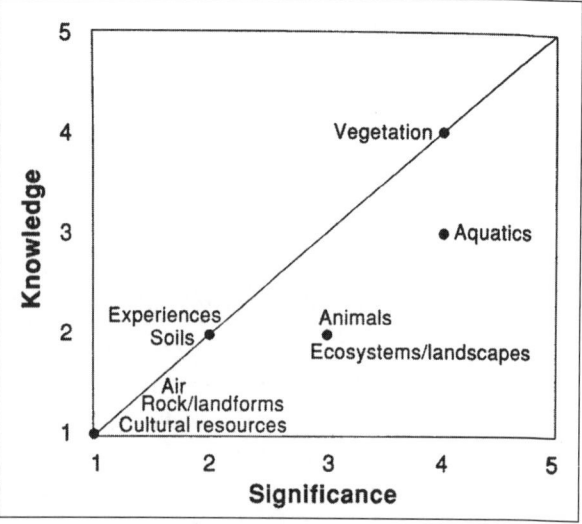

Fig. 13.3. Cole (1994, 9) charted the significance of impacts to wilderness attributes caused by nonnative species of flora and fauna and the state of knowledge about those impacts for the USFS's Northern Region.

ranging from social and political trends to global climate shifts. *Further complicating any prediction about the impacts is that threats to wilderness conditions can be multiple and interactive so that the cumulative changes and cascading effects are more extensive or far-reaching than any single threat.* Given the complexity of ecosystems and the natural environment, and our limited knowledge of historic conditions and processes, predicting impacts from changing conditions is problematic. Research and professional experience, however, suggest that there are some high-priority threats, and these are discussed in the following sections. More information on many of the impacts is in Chapters 10, 11, 12, and 15.

Potential Threats to Wilderness

Threats to wilderness and their impacts are so numerous that it would be too exhaustive to present all of them here (Cole, McCool et al. 2000). For example, grazing (a potential threat) can have different impacts on vegetation, soil, erosion, water quality, and native animal and fish populations. Because our knowledge about these threats and how to manage them in wilderness is limited, we have to organize and limit our discussion to them.

A study of threats to wilderness in the U.S. Forest Service's Northern Region (Cole 1994) provides a prototype matrix of how to organize potential threats and information about them. For example, comparing the importance of impacts to wilderness, such as those caused by nonnative species of flora and fauna (noxious weeds, nonnative fish and wildlife) and the state of knowledge about those impacts suggests that more knowledge is needed to fully understand ecosystem impacts. When the knowledge of potential threats is combined with the importance of each for wilderness management, a more comprehensive pattern can emerge for scientists and managers. Cole (1994) compiled such a table for the U.S. Forest Service's Northern Region and concluded that a better understanding of the effects of fire on the wilderness attributes of soils, vegetation, and ecosystems was one of the most important information needs for wilderness managers (Table 13.1).

The following information focuses on the categories of threats to wilderness that are mentioned in the relevant literature and that have come up in our discussions about wilderness threats with wilderness managers and researchers. From this information, we identified nineteen threats to wilderness conditions and values, including *fragmentation and isolation of wilderness as ecological islands; threatened and endangered species; increasing commercial and public recreation use; livestock grazing; nonnative species; administrative access, facilities, and intrusive management; adjacent-land management and use; inholdings; mining; wildland fire suppression; air quality; water projects and water quality; advanced technology; motorized and mechanical equipment trespass and legal use; aircraft noise and airspace reservations; urbanization and encroaching development; global climate change; legislation that designates new wilderness areas with compromised wilderness conditions; and lack of political and financial support for wilderness protection and management.*

Table 13.1. The Significance Ratings (1 = low to 5 = high) for the Impacts of Potential Threats to Wilderness Attributes for the USFS's Northern Region, as Reported by Cole (1994)

Attributes of Wilderness Character	Potential Threats								
	Recreation	Livestock	Mining	Fire	Exotic Species	Water Projects	Atmospheric Pollutants	Adjacent Lands	Other
Air	1	1	1	2	1	1	4	3	0
Aquatic systems	4	3	3	4	4	3	4	3	0
Rock/landforms	1	2	2	1	1	2	1	1	0
Soils	3	3	2	5	2	2	4	2	0
Vegetation	3	3	2	5	4	3	4	2	0
Animals	4	2	2	4	3	2	2	4	0
Ecosystems/landscapes	2	3	2	5	3	2	4	5	0
Cultural resources	3	2	2	2	1	1	1	1	0
Wilderness experiences	4	3	2	3	2	2	2	3	0

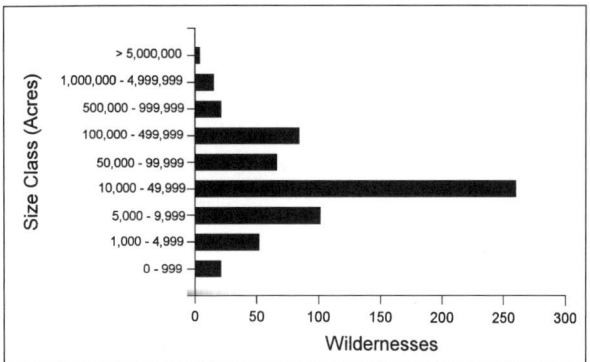

Fig. 13.4. The distribution of wilderness areas by size categories within the NWPS (Landres and Meyer 2000, 9).

These internal and external threats are acting on wilderness in the United States (and some other countries) at this time, with considerable variation among geographic regions and between remote and more urban-proximate wilderness. Some of these threats are the focus of management actions to minimize or mitigate them, and these threats may diminish over time even as new threats and controversies appear.

Fragmentation and Isolation of Wilderness as Ecological Islands

The threat here is that wholly functioning wilderness ecosystems cannot be adequately protected if they are not large enough and/or are disconnected from surrounding natural habitat. The more than 700 units within the NWPS tend toward smaller rather than larger units. The NWPS does include some very large areas, such as the four wildernesses in Alaska that include more than 5 million acres, but beyond those there are only nineteen wildernesses that are more than 1 million acres in size within the NWPS (Landres and Meyer 2000).

The most common (42 percent) wilderness size is from 10,000 to 50,000 acres (Landres and Meyer 2000). Thus, because of their relatively small size, most wilderness areas are ecological "islands," and more vulnerable to external forces than larger areas, especially those of 1 million acres or more. Due to the small size of most wilderness areas and the lack of connecting corridors of wildland between them, wildlife naturalness is strained because entire home ranges of some species are not provided. Furthermore, beyond the necessary movement for seasonal changes, a mixing of species populations is also essential to genetic diversity upon which their long-term health and survival depend. *One of the greatest impacts of fragmentation may be on the species that require large undisturbed home ranges, such as wolves and grizzly bears.* Further aggravating fragmentation in other countries may be transboundary effects where international borders may increase isolation of wilderness from other habitat, with major impacts on key wildlife species (Singh and Jackson 1999) (see Chapter 12 for international wildlife examples).

The lack of large or "corridor-connected" wilderness units is most pronounced in the eastern United States. Landres and Meyer (2000, 9) report that "western states have an average of 37 wildernesses per state, compared to an average of five wildernesses per state in the east. Western states have an average of 5.4 percent of their land area in wilderness, compared to the eastern states average of 0.5 percent land area in wilderness. The average size of a wilderness in the western states is 101,610 acres, compared to 23,657 acres in the eastern states." Thus, fragmentation threats to wilderness ecosystems and landscapes is greater in the East and further complicated because creating corridors to connect ecological islands is more difficult and expensive.

So how can the undesirable impacts of fragmentation be minimized (i.e., loss of genetic diversity and protection of wholly functioning ecosystems)? We think the following actions would help. *Pursue additional wilderness designations and protection of other park and wildland areas, with connecting corridors that create larger blocks of natural functioning ecosystems.* The campaign to establish a mosaic of protected lands from Yellowstone to the Yukon (Y2Y) with connecting corridors is an example (Posewitz 1998). Furthermore, *manage wildlands that are related to or adjacent to wilderness with policies that support or restore ecosystem integrity for the combined area* (Landres, White et al. 1998b). Proposals calling for restoring wild conditions on lands adjacent to wilderness, such as in the northeastern U.S. (Klyza 2001), require enhanced communications and coordination among different land-management agencies and units. The biosphere-reserve concept, which may have wilderness in a central core with gradations of decreasing wildness and coordinated management radiating from the central area, is a good strategy that has been applied in the United States (e.g., Yellowstone National Park) and many other countries. Internationally, fragmentation threats may only be lessened with diplomacy to subjugate hostilities between countries in favor of higher conservation goals that transcend national differences. It is worth pursuing, and positive examples do exist (Singh and Jackson 1999).

Threatened and Endangered Species

Losing threatened and endangered species, or sometimes intrusive actions to save them, can threaten wilderness naturalness and solitude. Some threatened and endangered species are wilderness-dependent, requiring naturalness and solitude, whereas others merely find favorable protected conditions in wilderness. The impacts of adjoining-land-management activities and global changes add to the pressures on threatened and endangered species in wilderness, and the smaller or more fragmented the area, the greater the pressure.

Protecting endangered species may require (and this is legally allowed in U.S. wilderness) special efforts, including manipulating the wilderness environment to favor the continuation of such species. At first, this may seem like the most logical course of events to "restore" that which is threatened and endangered. However, there is concern that such well-intentioned activities may cause a secondary and tertiary wave of perturbations within the ecosystem that we cannot predict due to our limited knowledge of interactions and interdependencies. For example, some people may question whether efforts to protect bighorn sheep in the Cabeza Prieta Wilderness, Arizona, by providing water actually are necessary or whether such action sustains bighorn sheep populations at unnaturally high levels (or, for that matter, other species benefiting from the water). Furthermore, is the invasion of naturalness and solitude by truck traffic to supply the water worth the impacts? The answers to such questions are not easy and pose a two-edged sword. Losing an endangered species would impact naturalness but so might intrusive efforts to save it.

The many external threats and global changes compound pressures on threatened and endangered species in wilderness such that their probability for survival may be reduced, in spite of what wilderness and wildlife managers might do to save them. In addition, with any intervention to solve one problem, there may be undesirable side effects. Therefore, managers should be cautious and humbled by the complexity of ecosystems and not assume that we can surgically or technologically solve all the problems that occur.

Increasing Commercial and Public Recreational Use

Wilderness has visitation levels, at least in places, that cause unacceptable biological and ecological impacts. *As commercial and public recreation use of wilderness increases, impacts accumulate, and managers then increasingly regulate and control use* (Cole 1996a; also see Chapters 14, 15, 16). Recreation use and visitor management are already intense in some high-use locations in the more popular wilderness areas and in some places, some times, and some seasons, even in many large and remote areas. *Therefore, the impacts of commercial and recreation use and efforts to control them may threaten wilderness resources and values.* The choice described by David Cole (2000) was natural, wild, uncrowded (meaning regulated), or free (unregulated): which of these should our wilderness be? Clearly, if we want to preserve wilderness resources it will take considerable regulation of visitors in many places. However, such regulation takes away the freedom and spontaneity that many visitors associate with wilderness experiences.

The social and ecological impacts of users and their activities on the wilderness flora and fauna, especially at heavily used sites, have been the subject of concern by managers for decades (see Chapter 15; Hammitt and Cole 1998; Liddle 1997; Manning 1997; Marion et al. 1993). For example, the biophysical impacts from recreational activities were studied in six wilderness areas in Oregon and Washington. The impacts were found to be considerable in specific sites (Cole, Watson et al. 1997), but may represent only a small percentage of the total wilderness area (Table 13.2). However, these impacts are highly noticeable

Fig. 13.5. Wilderness managers have to balance what some may perceive as intrusive water facilities and using mechanized vehicles against sustaining an endangered species, in this case desert bighorn. This four-part facility, a catchment, underground storage, water diverson, and walk-in drink tanks, in Arizona's North Maricopa Mountains Wilderness (BLM) eliminates the use of a tank truck and helps the sheep that are cut off from their natural water supply by agricultural and residential development. Photo courtesy Chad P. Dawson.

Table 13.2. Biophysical Impacts of Total Disturbed Area, Barren Area, and Type of Disturbance to Six Wilderness Areas in Washington and Oregon
(Cole, Watson et al. 1997)

	Snow Lake	Rachel Lake	Rampart Lakes	Marion Lake	Sunshine-Obsidian	Green Lakes
Total land area (ha)	160	52	50	293	309	179
System trails						
Disturbed (sq m)	11,400	1,200	200	10,200	15,600	6,000
Barren (sq m)	5,700	600	100	5,100	7,800	3,000
Social trails						
Disturbed (sq m)	3,768	1,674	2,545	2,000[a]	1,949	1,488
Barren (sq m)	2,600	1,222	1,909	1,500[a]	1,696	1,250
Campsites						
Disturbed (sq m)	4,392	4,382	1,205	20,093	19,902	6,016
Barren (sq m)	2,562	3,797	663	15,664	8,367	1,600
Total area						
Disturbed (sq m)	19,560	7,256	3,950	32,293	37,451	13,504
Barren (sq m)	10,862	5,619	2,672	22,264	17,863	5,850
Percent of land area						
Disturbed (sq m)	1.2	1.4	0.8	1.1	1.2	0.8
Barren (sq m)	0.7	1.1	0.5	0.8	0.6	0.3

[a]These are conservative estimates because only about one-third of the social trails at Marion Lake were inventoried.

to users, especially at heavily visited campsite areas or vistas along lakeshores (Table 13.3), and become a high priority for management restrictions or closure of such areas to allow for recovery.

The need to balance wilderness recreation use among user groups and against environmental protection standards is increasing because of rising demand and because commercial and public recreation users often have divergent motivations, expectations, and management preferences (Hunger et al. 1999). Although commercial outfitters provide access to members of the public who might not otherwise have a wilderness or wild river experience, they do charge fees and operate to make a profit. The cost of acquiring a commercial outfitter and guiding permit can be very high, tens or hundreds of thousands of dollars, which requires steep user fees to offset and still make a business profit. This profit motive and competition between outfitters often encourages a "catering to customer comforts and convenience" orientation, which may further aggravate impacts on naturalness and solitude. This also creates pressures on the wilderness managing agencies against tightening environmental standards for uses taking place under permits for which fees are charged, especially as they become more dependent on recreation revenues.

We believe the guiding principles for allocating wilderness uses should be to protect wilderness

Table 13.3. Extent of Biophysical Impacts Along Lakeshores in Five Wilderness Areas in Washington and Oregon
(Cole, Watson et al. 1997)

	Snow Lake	Rachel Lake	Rampart Lakes	Marion Lake	Green Lakes
Total shoreline (m)	5,024	1,776	2,150	7,200	4,800
Disturbed shoreline length (m)	403	59	80	120	84
% of total	8.0	3.3	3.7	1.7	1.8

conditions and processes, as well as opportunities for solitude and primitive recreation. Already allocation decisions between commercial and public recreation use have proven difficult administratively, and conflicts between these two categories of use have stalled completion of some Wilderness Management Plans. The Forest Service has not enforced social standards for wilderness visitation in some high-use locations and fears that limiting visits there would only displace use to pristine locations, with an overall loss of wilderness values (Oye 2001). However, others are critical of the Forest Service for *not* enforcing social and environmental standards in its plans and demanding a rollback in use, a strict non-degradation policy, and use limits wherever ecological conditions or solitude are threatened (Worf 2001).

Management response to the express needs of special-interest recreation groups will be difficult for a variety of reasons. Many special-interest groups in wilderness are also some of the strongest supporters of wilderness, and their continued support is essential. Some accommodation may be necessary because wilderness management is also political and becoming more so (Mackey 1998). Besides, the need to maintain and build partnerships with established users, just to get wilderness work done, makes them special stakeholders. Furthermore, special-interest groups can often rationalize their requests as appropriate in wilderness because they may have been doing that activity or something similar for years or decades, with no negative feedback (e.g., no specific regulations against the activity or no enforcement if there was such a regulation). For example, placing fixed anchors on rock-climbing routes in wilderness may seem reasonable to climbers (e.g., safe anchors for ascents, rappelling, or rescue), and where fixed anchors have been tolerated for years, enforcement now may be difficult.

Livestock Grazing

Livestock grazing by cattle, sheep, and recreational pack stock is allowable in wilderness where it existed before wilderness designation, but it is a definite threat to naturalness and social values. The use of public lands for grazing is a long-standing one, but increasingly controversial, and is more so in wilderness because of its impacts on naturalness of the ecosystem. Permitting grazing in wilderness was a compromise that was politically necessary to achieve passage of the Wilderness Act, but it is in conflict with the wilderness goals of maintaining primeval character, preserving natural conditions, and leaving wilderness affected primarily by the forces of nature (McLaran 1990). Thus, livestock grazing, like mining, is called a "nonconforming but allowed" use. Fortunately, because of the economic costs of operating in remote wilderness locations, many grazing permits are no longer actively used.

Besides the impacts on soil and water (Derlet et al. 2008), consuming and trampling vegetation for forage may be in direct competition with wildlife, or cause changes in the composition of the forage base, which may also affect wildlife (Murray 1997). Livestock grazing is selective of plant species and influences plant composition on wilderness meadows and range. Such influences do not necessarily favor the wildlife that must compete for the same forage. Moreover, where livestock graze there are accompanying pressures for predator control. A related concern is that livestock and associated management activities and facilities may negatively impact visitors' experiences—although one study indicated visitors are split in their perceptions of whether livestock grazing is acceptable on wilderness lands (Laura Johnson et al. 1997). Another negative impact on social values is the public perception of the loss of naturalness and wildness through the direct ecological impacts of grazing and associated management activities on wilderness conditions.

The legal basis and various issues surrounding grazing as a permitted use in wilderness are discussed further in Chapter 14. Grazing is an allowable (even though nonconforming) use of wilderness, and this was further confirmed in the Colorado Wilderness Act (U.S. Public Law 96-560) in 1980 and its accompanying committee report (see Chapter 5). In addition, grazing is strongly protected by the politically powerful livestock industry. *However, there is a continuing struggle between that industry and wilderness supporters to keep grazing and its accompanying facilities within limits that will sustain wilderness resources and values. Wilderness managers need public support for a strong, pro-wilderness stand on this issue.*

Nonnative Species

The invasion of wilderness ecosystems by nonnative species is an increasing, direct threat to wilderness naturalness. Nonnative and nonindigenous species are increasing in places in wilderness at an alarming rate and can trigger ecosystem changes and displace native species. For example, noxious weeds like knapweed, star thistle, cheatgrass, leafy spurge, and others have outcompeted native species and are rapidly spreading (Asher and Harmon 1995). Secondary impacts may then result from control efforts, which may include biological, chemical, and physical control mechanisms. Although the goal of such action is to maintain or restore natural conditions and natural functioning ecosystems, some management actions can cause other problems because often previous conditions cannot be restored without other perturbations to the ecosystem (Cole 1996a; Worf 1997). Manipulation for restoration may be appropriate if it is the minimum tool to restore natural conditions, but there is often a lack of baseline information about natural and wild conditions and processes to guide such intrusive management. The loss of wildness is one of the largest concerns because management actions and impacts may not be justified even if they can restore naturalness.

How do nonnative species get introduced to wilderness? Some have been introduced inadvertently and others deliberately. For example, noxious weeds may be introduced through pack-stock feces, even though "clean" feed may be packed for the duration of the pack trip in wilderness. Burros and wild horses may also migrate in and out of wilderness, importing seeds. Some noxious-weed seeds come in on the feet and clothes of hikers, or in the droppings of seed-eating birds. It is hard to prevent these inadvertent introductions. However, some nonnative species have been deliberately introduced, such as eastern brook trout stocked in western wilderness lakes. Even fish stocking beyond natural population levels, or where fish did not naturally occur, may impact other natural, native species like yellow-legged frogs displaced by fish stocking in wilderness of the High Sierra of California (Matthews and Knapp 1999). The translocation of mountain goats to areas where they were not indigenous is another example of deliberate introduction of a nonindigenous species (Carter 1997).

Furthermore, attempts to remove nonnative species can cause additional changes beyond restoring the preexisting "natural" conditions. Wilderness is not isolated from the global climate changes that are occurring and which may favor nonindigenous species, such as noxious weeds whose spread is encouraged in some places by drought and warming of the climate. Some nonnative pests are so firmly established that they have defied serious efforts to control them in nonwilderness for example, blister rust invading white bark pine in high-elevation wilderness sites in the West and chestnut blight in the East. In such cases, as with infestation of native pests, allowing natural ecological response and succession to the place is probably the best approach. *It seems wilderness managers, in trying to preserve naturalness, must often decide which is a more serious intrusion—the problem or the cure. The cure, in the end, may not work and could wreak havoc of its own in a natural ecosystem and diminish wildness.*

Administrative Access, Facilities, and Intrusive Management

Wilderness management can threaten naturalness with too-frequent mechanized access, excessive facilities, and methods that are too intrusive. Mechanized access to wilderness by managers is legal under the Wilderness Act when it is the minimum method to accomplish a legitimate wilderness or endangered species purpose, such as facility construction and maintenance (e.g., fencing in an approved grazing allotment), forest-fire suppression, search-and-rescue operations, accessing inholdings, and so forth. Such management activities may be in support of any legitimate wilderness purpose, such as visitor use, grazing, mining, commercial outfitting, maintaining historic structures, and trail construction. The test is whether such access is the minimum necessary method to accomplish the purposes of the Wilderness Act.

Mechanized access can be a threat to wilderness naturalness and solitude, in spite of a manager's best intentions. So, rigorous interpretation of what is the minimum method to accomplish the purposes of the Wilderness Act will go a long way toward protecting the wilderness from mechanized access, excessive facilities, and intrusive management. Such a minimum-tool requirement becomes even more important with passage of wilderness laws like the California Desert Protection Act (U.S. Public Law 103-433), which provides for mechanized access to support fish and wildlife management (not just endangered species) and law enforcement in the sixty-nine BLM wilderness areas it established.

Some examples of administrative management impacts on wilderness are evident in wildlife and fisheries management. For example, such management can include the removal of nonindigenous species, such as mechanical capture of feral burros, aerial application of chemicals to kill "undesirable" fish species before reintroducing native species, fish stocking, removal of nonindigenous species by physical or chemical applications, water developments (guzzlers) for wildlife and their servicing in arid environments, and mechanized wildlife surveys or research methods. Czech and Krausman (1999) reported that 53 of 273 wilderness areas in the southwestern United States acknowledged using controversial wildlife-management activities, a great number of which involved mechanized access. Often the purpose of the activity is questionable, such as stocking to maintain unnaturally productive fisheries (Duff 1995); translocation or maintenance of nonindigenous (but huntable) populations of mountain goats; or construction, reconstruction, maintenance, and servicing of water developments for desert bighorn sheep—maybe to support a few more hunting permits.

The entire issue of hunting in wilderness (legal on national forest and BLM wildernesses) is controversial because it so often has focused on harvesting "trophy" animals. The effects of removing the males with larger antlers or horns from the gene pool are not fully understood but are suspected of being unnatural. Therefore, when hunting (or fishing) is supported with mechanized intrusions in wilderness, it further inflames controversy.

For today's wilderness managers, who may not be familiar with primitive tools used in earlier days, such as crosscut saws, pack strings, hand tools, and pioneering skills, it may be especially tempting to use mechanized access and tools, or to establish and maintain supporting facilities. However, a lax attitude toward minimum tools by managers will threaten wilderness naturalness and solitude by making approval of mechanized intrusions too easy.

Adjacent-Land Management and Use

Wilderness can be impacted by management activities and use on adjacent lands. The influence of external factors on wilderness management is a concern for managers because they often have little or no control over what happens on lands adjacent to wilderness areas (Stokes 1996). A survey of U.S. wilderness managers in 1995 reported sixty different perceived impacts that adjacent land uses had on the wilderness (Kelson and Lilieholm 1997). The top five were fire management, military overflights, nonnative plant introduction, air pollution, and off-road vehicle use. Kelson and Lilieholm (1997, 25) concluded that "one of the most challenging issues facing U.S. wilderness managers was the impact of activities located beyond wilderness borders"—even on lands managed by the same agency for multiple use. For example, timber harvests and grazing disrupt naturalness in adjacent ecosystems with ripple effects on wilderness. These are probably best illustrated by big-game wildlife examples but may be as important to other species. Suppression of insects, disease, and fire, and predator control all impact adjacent ecosystems. Roads serving management and use on adjacent lands may make access to wilderness easier or harder, thereby affecting the amount or location of recreation use. Everything is connected within and adjacent to wilderness. *Wilderness planners and managers need to expand their awareness and communication beyond wilderness boundaries and seek to coordinate adjacent-land management activities to minimize their impacts on wilderness* (Landres, White et al. 1998b).

Inholdings

Inholdings of private or public lands within wilderness areas can create impacts because inholders have a right to reasonable access and use of their property. The impacts can be similar to the dilemmas of adjacent-land management. Some inholdings are dramatic, such as old mining claims, homesteads, and ranches, which may provide their modern owners prime access to wilderness surroundings. In recent years, there have been proposals to build resorts on private inholdings, such as in Montana's Absaroka-Beartooth Wilderness. Sometimes motorized access is granted to inholders on primitive roads, by aircraft, or by boat. Inholdings may be used by commercial outfitters to provide sites for support facilities and services (e.g., cabins, stock facilities, aircraft landing fields), access (e.g., interior private roads), and visitor facilities (e.g., hiker huts, outfitter camps, private dwellings). The most extensive examples of this type of access are under ANILCA (U.S. Public Law 96-487), which provides for various types of access to private or public land inholdings within Alaskan wilderness areas (Tanner 2004).

Federal agencies try to buy out wilderness inholdings, but they are constrained by their overall acquisition priorities and available funds. In the eastern United States, agencies can even condemn lands under

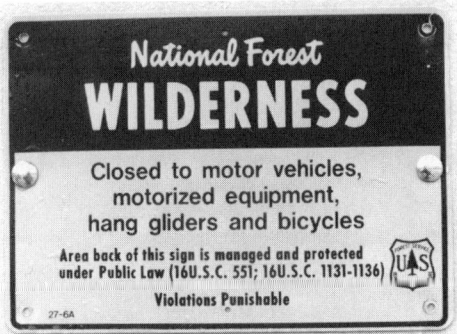

Fig. 13.6. Keeping vehicles out of wilderness is a difficult management task requiring road closures and signs listing prohibited uses (left), wilderness signage to identify entry to wilderness (center), and boundary marking (right). Jim Mahoney, formerly of the BLM Phoenix office, inspects a shotgun-blasted wilderness boundary sign. Installing, maintaining, and monitoring wilderness boundary signs are an important management challenge. Photo courtesy Marilyn Hendee.

authority of the so-called Eastern Wilderness Act (U.S. Public Law 93-622)—although to date it has not been done. A more recent threat to wilderness has been the subdivision and development (or proposal for same) of wilderness inholding lands by developers or speculators who perceive the opportunity to make a profit from the development or may be more likely to inflate the value of the lands and pressure a federal agency to buy them out at a high price. Such situations are active in Colorado's West Elk and Holy Cross Wilderness areas.

Mining

Mineral exploration has been phased out under the Wilderness Act, but mining and extraction from established claims are allowed and are a threat to wilderness resources and values. For example, oil development is being considered in Alaska and silver mining on existing claims near the Cabinet Mountains Wilderness in Montana. Another potential threat may be on wilderness lands purchased by the federal government without subsurface rights to minerals and oil and gas, such as in wilderness areas designated in the eastern United States. Thousands of mining claims exist in BLM wilderness study areas and may impact wilderness in the future if mineral prices rise, leading to pressures for mineral claim development and subsequent mining operations at these locations. Mines can produce acid drainage and siltation that lead to reduced water quality in streams and water bodies. The negative social and ecological impacts of mining on wilderness conditions—naturalness and wildness—are extensive. Even mines that have played out may continue to impact wilderness because of old buildings, many of them eyesores, junk heaps, and roads that continue to erode and which are invitations to vehicle-trespass into wilderness.

Wildland Fire Suppression

Fire prevention and suppression, adjacent to and in wilderness, change ecosystems that are fire-dependent by reducing the natural-fire frequency, leading to changes in ecosystem structure and composition. Allowing natural processes like fire to continue to function as part of wilderness ecosystems and landscapes is now recognized by wilderness managers as important to providing diversity and natural variation (Christensen 1995). The tendency has been for federal agencies to suppress most fires, partially because they fear that the fires would spread to inholdings of other landowners or to adjacent, nonwilderness lands and, in part, because of political pressure to suppress them.

For most of the last century, fires were aggressively attacked, and public vigilance was enlisted in a very successful fire-prevention campaign. This was before it was understood how fires renew ecosystems and provide variation and diversity to fire-dependent ecosystems. The massive stand-replacing fires that occurred in Yellowstone National Park in 1988 and the forest's subsequent regeneration helped educate the general public and wilderness users about the value of fires, although for many people, including some powerful legislators, the fires only confirmed rigid views that fire was bad. A similar process was at work in the intense fire year of 2000, when more than 6 million acres burned in more than 80,000 wildland fires, according to the National Interagency Fire Center (2008). But there is a growing realization that past fire suppression has allowed tremendous fuel loads to build up, and the loads are contributing to the larger and more intense wildfires. The complexity of fire as a natural disturbance process and the natural ecological functions it serves—especially in fire-dependent ecosystems—are becoming more completely understood (see Chapter 11). Although greater understanding is evolving, additional research and scientific inquiry are needed (Parsons 2000).

Although fire suppression, leading to the loss of natural fire regimes and inhibiting fire as a natural process in wilderness, is identified as a problem (Stokes 1996), reversing the trend will be difficult. *Due to the political volatility of any potential "let burn" policies for fire, the managed approach to allow fire in specific management prescriptions and controls will continue even in wilderness. Nature, however, will continue to restore fire whenever conditions permit.*

Air Quality

Polluted air is a threat to wilderness naturalness because of its physical and biological impacts. It reduces visibility that may impact wilderness experiences. Good visibility within a wilderness naturally enhances the enjoyment of that environment and vice versa. For example, the San Gorgonio Wilderness near Los Angeles has severely diminished visibility due to urban air pollution. Bell and others (1985) have reported that visual impairment from pollution can influence visitors to natural areas (e.g., Grand Canyon) to change trip schedules or to choose another location that has better visibility.

Air quality creates impacts on wilderness because of visibility and other indirect biological impacts. In the eastern United States, acid rain from industrial and urban emissions can be especially harmful to high-elevation ecosystems. For example, atmospheric pollutants, sometimes deposited in acid rain or acid fog, can cause changes in vegetation and water quality, and thus affect fish and amphibian populations. Forest fires contribute to air pollution, and managers deciding to allow wildfires to burn in wilderness ecosystems should consider such effects. Even in prescribing wilderness burns, agencies must consider atmospheric effects, but by selecting conditions that minimize impacts on air quality, such managed fires may differ in time, scale, and postfire effects from the natural fire they are supposed to simulate.

The Clean Air Act of 1977 (U.S. Public Law 95-95) provides a complex legal framework that further protects wilderness air resources. In particular, the amendments to this legislation in 1997 required the U.S. Forest Service to monitor eighty-eight (Class I) wilderness areas for visibility and mandated action if air quality deteriorated. All other wilderness areas are under Class II protection, which charges wilderness managers with responsibility to protect air-quality-related values like visibility. In 1996, the USFS notified the State of Washington that visibility in the Alpine Lakes and Goat Rocks Wilderness areas was adversely impacted by a coal-fueled power plant in Centralia, Washington, and subsequently a mediated settlement for air-quality improvement was completed (Stokes 1999). *Thus, wilderness areas, and the air quality in them, serve as an important indicator of overall ambient air quality in the United States.*

Water Projects and Water Quality

Reconstructing and maintaining dams and reservoirs in wilderness for water storage threaten wilderness solitude and the physical landscape; the demand for water produced in wilderness threatens in-stream flows; and wilderness water quality is threatened by some nonconforming wilderness activities, like grazing and mining.

Water storage facilities built before the Wilderness Act was passed continue to be used and maintained. New facilities are allowed under the Wilderness Act if under an executive order from the president. Water storage facilities require maintenance or reconstruction that is very controversial because of the mechanized intrusions that are required. For example, maintenance of dams in the Emigrant Wilderness of California and in the Selway-Bitterroot Wilderness of Montana and Idaho is a contested public issue that has both proponents and opponents. These water storage facilities provide badly needed water for irrigation of agriculture in valleys below the wilderness, especially during late summer months when water is scarce. However, opponents point out that reservoirs behind dams are an unnatural influence in wilderness, and as these dated facilities wear out they should not be repaired, restored, or replaced—nature should be allowed to restore the landscape to its natural condition.

There is also renewed interest and competition in the western United States for water flowing from remote forest and mountain wilderness areas by those wanting more irrigation water and conservation interests wanting to maintain fisheries, aquatic biota, and wildlife. The need to maintain minimum in-stream flows to protect endangered salmon runs in the Northwest is an example.

Fig.13.7. As urban and resort developments advance into wildlands, they may directly adjoin wilderness boundaries. Here a resort development and blacktop road crowd the boundary of Red Rock–Secret Mountain Wilderness (USFS) near Sedona, Arizona. Photo courtesy John C. Hendee.

The quality of water in some wilderness streams may be affected by runoff from grazed areas that can include animal feces and sediments at higher than natural levels. A few streams and rivers that flow through wilderness areas have their origins outside wilderness and may bring water with pollutants and unnatural soil sedimentation levels to these wilderness rivers. Contamination of surface waters in wilderness by such diseases as giardia highlights the insidious and incremental changes in water quality, even in remote wilderness areas. Water quality is also affected by recreation visitors and their activities, such as disposing of food scraps and human waste. *Protection of natural water regimes and water quality in wilderness is an increasingly important challenge.*

Advanced Technology

High-tech clothing and camping equipment are becoming widely used in wilderness, and some fear this protective technology may be partially cutting visitors off from the wilderness experience they seek (Knapp 2000). Certainly modern gear makes wilderness visits easier and may also account for some of the increased visitor use.

The use of GPS equipment, cell phones, radios, beepers, pagers, and other electronic technology by search-and-rescue personnel is generally accepted, much like the use of other high-tech climbing gear. However, the use of electronic technology by wilderness visitors raises questions over whether they are having a primitive-recreation experience with the naturalness and solitude for which wilderness exists (Nash 1996; Shultis 2000). Now that electronic communication is so readily available, it may be an advertised safety feature for commercially guided trips or perhaps even required by insurance companies.

Using aircraft in search-and-rescue operations is somewhat accepted when human life seems at stake, but it is controversial unless really needed because it intrudes on the experiences of other visitors. Aircraft, even where legal landing strips exist in wilderness, impacts the experience and solitude of wilderness visitors who see or hear them (Meyer 1999). The modern, personal transport devices now on the market, such as Jet Skis and snow machines, are further mechanized threats to wilderness.

Modern electronic communication and navigation devices and evolving mechanical transport equipment may give visitors a false sense of security and contribute to irresponsible behaviors based on assumptions that rescue is only a cell phone call and helicopter flight away. Similarly, visitors may tend to rely on technological devices to navigate and not be able to read maps or orient themselves in a wilderness environment, thus robbing themselves of the benefits of wilderness discovery and self-reliance, while also increasing the probability of the need for search-and-rescue activities. The intrusions of advanced technology, at the very least, remind visitors of the society and busy work world that they may be trying to take a break from while in the wilderness. Roderick Nash addressed the question in a 1996 article titled "A Wilderness Ethic for the Age of Cyberspace."

Motorized and Mechanical Equipment Trespass and Legal Use

Motorized equipment or mechanical vehicles trespass, such as all terrain vehicles (ATVs), snowmobiles, or mountain bikes, can dilute wilderness solitude and damage resources. For example, operators of snowmobiles and ATVs can travel cross-country and enter a wilderness area inadvertently or deliberately for convenience, recreation, and exploration. Such trespassing usually goes undiscovered in remote locations. In some areas, vehicles are intentionally taken into or across wilderness areas as a shortcut or used by adjoining landowners or users who do not view their use of a remote area as an intrusion on the wilderness.

Management use of motorized vehicles and mechanical equipment is legal in wilderness where it is the minimum method for accomplishing a legitimate wilderness purpose, and in some areas for certain activities, such as wildlife management and law enforcement in BLM wilderness designated under the California Desert Protection Act, as discussed earlier. Thus, depending on the area or purpose, visitors may see mechanical equipment used by management even in remote wilderness areas, such as helicopters for search and rescue, ATVs for beach patrols, four-wheel-drive vehicles for wildlife-management activities, and chain saws for trail construction or maintenance.

The extent to which motorized and mechanical equipment is used by managers affects a visitor's frame of reference as to what is acceptable in wilderness. *Our view is that motorized and mechanical equipment is justified in wilderness management only when it is absolutely essential as the minimum necessary tool to accomplish the purposes of the Wilderness Act. Very often primitive skills can do the work just as well, economically, and always with less impact on the wilderness environment and experience of wilderness visitors.*

Aircraft Noise and Airspace Reservations

Aircraft overflights of wilderness by commercial and military aircraft cause noise and air pollution, can disturb wildlife and visitors through visual and noise intrusion, and dilute solitude with a reminder of society and modern technology (Tarrant et al. 1995). There are private and public airfields on federal lands that are still in operation within wilderness areas in Montana and Idaho and which are used by private visitors, as well as flights by outfitters moving supplies and customers in and out (Meyer 1999). The Frank Church–River of No Return Wilderness (FC-RONRW) has thirty-one operating airstrips on federal and private lands within the wilderness boundary, and the twelve located on federal land handle about 5,500 landings per year (U.S. Department of Agriculture, Forest Service 1998). The Forest Service has little discretion in managing these airstrips because the Central Idaho Wilderness Act (Public Law 96-312) establishing the FC-RONRW stated that these uses "shall" be permitted, which is different from the Wilderness Act, which stated such uses "may" continue subject to secretary of agriculture regulations. Wilderness areas in Alaska are another example where preexisting use of aircraft, especially floatplanes, is allowed to continue under their wilderness designation legislation, such as ANILCA (U.S. Public Law 96-487).

As aircraft access to wilderness grows, increasingly wilderness managers must determine what management discretion is available to limit access by searching the legislative history establishing the area (Meyer 1999). Then managers likely will face objections from private pilot and wilderness outfitter organizations who do not want such access limited.

Intrusions on wilderness airspace, especially by low-flying military aircraft on training flights in the West or by commercial sightseeing flights in places such as the Grand Canyon, affect wildlife and may offend people who experience them (Tarrant et al. 1995). The California Desert Protection Act contains wording that the designation of sixty-nine wilderness areas by that act would not limit military overflights (U.S. Public Law 103-433). Only in the BWCAW has an airspace reservation been established, limiting overflights in the affected area to above 5,000 feet by presidential order. *Managers with little authority to restrict airspace activity may need to encourage the public to speak out and object to excessive airspace intrusions.*

Urbanization and Encroaching Development

Urban sprawl and development encroach on wilderness boundaries and dilute wilderness with sights, sounds, and diminished remoteness, while increasingly urbanized wilderness visitors are suspected of being more tolerant of crowding and less wild conditions.

Sprawling urban development has dramatically affected wilderness conditions with smog, encroaching roads and ease of access, noise, and casual day use in urban-proximate wilderness areas like San Gorgonio outside Los Angeles. The Pusch Ridge Wilderness near Tucson, Arizona, has suburban backyards abutting its wilderness boundary. Even a remote area such as the Mission Mountains Wilderness in Montana is being affected by development in surrounding valleys, because it is isolating a remnant population of grizzly bears and threatens the bears' viability by interfering with the genetic interchange with other breeding populations (Stokes 1999).

Unfortunately, Congress ruled against buffer zones to insulate wilderness from adjacent nonwilderness activities—first in the Endangered American Wilderness Act (U.S. Public Law 95-237), which encouraged wilderness designation near cities to provide urban populations with much-needed recreation, and then the New Mexico Wilderness act (U.S. Public Law 96-550), which specifically prohibited buffer zones to insulate wilderness from adjacent activities, as did several subsequent wilderness designation acts (see Chapter 5).

The increasingly urbanized and diverse population in the United States may bring different values to wilderness in the future (Hammitt and Schuster 2000; Watson and Landres 1999) that will raise questions for managers, such as are today's urban visitors more likely to tolerate crowding in high-use wilderness areas? The fear is that increasingly urbanized visitors to wilderness may be satisfied with trips to crowded and heavily impacted wilderness due to their lack of previous experience in a more pristine area and a frame of reference more tolerant of crowding and oblivious to impacts. The Forest Service, and presumably the other agencies, have been reluctant to restrict use to achieve solitude standards. They are reluctant because of political pressure and concern that turning some visitors away from high-use areas would displace some of them to other less used and more pristine wilderness sites, thereby spreading impacts to those areas. Unfortunately some visitors who go to these high-use destinations may assume that whatever crowded and impacted conditions they experience are what is intended for wilderness by managers. *Helping visitors to understand what wilderness is, what it is meant to provide, and why management is necessary is an ongoing educational challenge for managers* (Stokes 1999). Creating and implementing management strategies to address increasing wilderness use and impacts will continue to be difficult and controversial because there are many views on what approach is best (Cole 2001a; Oye 2001; Spring 2001; Worf 2001). The wilderness recreation strategy of the Forest Service is an example.

Global Climate Change

National and international panels of experts have acknowledged that the scientific evidence for global climate change is well documented, even though the causes for change are still being debated, such as human-caused versus natural variation (Intergovernmental Panel on Climate Change 2007). The need for some management response to these ongoing climate changes and the potential for subsequent impacts to biological and physical resources were summarized in a report by the U.S. Government Accounting Office (2007) that strongly pointed out the responsibility of federal land-management agencies to give guidance and direction to agency managers on how to address the effects of climate change on federal land and resources. Examples of the effects of climate change are numerous, especially in northern latitudes, where evidence has mounted to show later freeze-up of water bodies, earlier thaw dates, declining glacial ice, changes in permafrost, changes in forest vegetation, more extensive forest insect outbreaks, changes in bird and animal ranges, and other evidence of environmental change (Hinzman et al. 2005). Some situations have caught the attention of the public and the media about how rapidly environmental change is occurring, such as the loss of sea ice in northern latitudes, which has reduced the primary habitat for polar bears and has led to a decline in their abundance. Consequently, the U.S. Department of the Interior proposed listing the polar bear as a threatened species under the U.S. Endangered Species Act of 1973 (U.S. Public Law 93-205) (U.S. Department of the Interior 2007).

Legislation Designating Areas with Compromised Wilderness Conditions

There has been a long-standing concern about the tendency to politically compromise with nonwilderness interests and uses when designating wilderness areas, even going back to 1935 when, upon founding the Wilderness Society, Bob Marshall commented, "in the past far too much good wilderness has been lost by those whose first instinct is to compromise." The Wilderness Act required more than fifty versions before passage in 1964 due to the need to reach a political compromise with what we now refer to as permitted but nonconforming uses like grazing and mining. Many subsequent laws designating wilderness areas have included special provisions and management direction along with permitted and nonconforming uses like aircraft or motorboat access to some areas (see Chapter 5 and Appendix B).

Managing special provisions within wilderness requires a thorough understanding of the Wilderness Act and the intent of an area's designating legislation, along with some information about the values of wilderness held by those for whom a special provision was included (Watson et al. 2004). Even with substantial information and legislative background to guide managers, special provisions can conflict with other visitor and user interests and the overall purpose of the Wilderness Act (Nickas and Proescholdt 2005).

The history of the Wilderness Act and subsequent designation laws suggests that some compromise is necessary in our political system to achieve designation of new areas, and there are still many candidate areas to be considered (Scott 2004). However, the concern is that special provisions and nonconforming uses diminish wilderness character and cause unintended negative consequences for the quality of wilderness in the NWPS. For example, grazing guidelines first put in a Colorado wilderness bill in 1980 set a precedent to include special grazing provisions in subsequent wilderness legislation as well as to allow more motorized access for grazing interests (Nickas and Proescholdt 2005).

Lack of Political and Financial Support for Wilderness Protection and Management

The ideal expressed in the Wilderness Act, to preserve an enduring resource of wilderness, is threatened by a lack of political support sufficient to generate the finances needed for its protection and management. Concern is expressed among federal agencies, such as by former U.S. Forest Service chief Mike Dombeck (1994, 4), "that the resources committed to protect and manage wilderness have not kept pace with our needs," particularly for fieldwork budgets and staff. The evidence of such neglect is wide and deep. Wilderness plans are still in progress, have not been started, or are in need of revision and updating for many areas. Numerous wilderness study areas and roadless areas are still being evaluated to determine if they should be added to the NWPS, and many areas recommended for wilderness years ago have not been acted on by Congress. Without the efforts of volunteers and partnerships with environmental organizations and cooperators, much more wilderness work would be undone.

In 1999, a select panel was formed to examine the quality of management in the wilderness areas after they are added to the NWPS (Brown 2000). The panel's preliminary observations about NWPS management by the four federal agencies included six key questions: (1) Is wilderness being taken seriously in the four federal agencies? (2) Are the organization and level of management within the federal agencies positioned for real

Fig. 13.8. Old mining and homestead sites, such as this one in the Old Woman Mountains Wilderness (BLM) in California, concentrate ruins, trash, and buildings that detract from wilderness conditions and, in addition, attract curious visitors to unsafe situations. Photo courtesy Marilyn Hendee.

strength? (3) Are the funding and staffing resources committed to wilderness management sufficient for the needs? (4) Are all the federal agencies cooperating, collaborating, and being consistent across the NWPS? (5) Are wilderness visitors being well served? (6) Can the training of managers and wilderness research be improved and better focused?

The panel further observed that there was need for more wilderness research, planning, and management training to support current wilderness field staff. Management concerns also included fragmenting administrative responsibilities for wilderness management and multiple jurisdictions for some federal wilderness areas. For example, the Bob Marshall Wilderness administratively includes four national forests and five ranger districts, and the Frank Church–River of No Return Wilderness covers parts of six national forests (but is managed by four of them and six ranger districts).

The threat to wilderness from the lack of adequate resources for protection and management of the NWPS is caused by the lack of political support for this purpose among the larger public—as expressed by Congress. This may be the most serious threat to wilderness in the long run because it is at the root of establishing and maintaining high standards of wilderness naturalness and solitude and funding the people and programs require to implement them. The fact that projections by agencies like the Forest Service actually show recreation as a larger source of revenue than commodity programs, like timber, should help (Mackey 1998), but there is also competition among different kinds of recreation, with wilderness being classified as a small program compared to motorized recreation including car camping and off-road vehicle (ORV) use.

Strong support for wilderness by visitors and environmental-protection groups is key, but even among wilderness users there are divisive forces among commercial and public visitors, educational groups, and organized wilderness programs for personal growth and therapy—not to mention grazing, hunting, and endangered-species interests. Retired Forest Service wilderness coordinator Jerry Stokes (1999) pointed to seven legislative proposals in the 105th Congress and numerous committee hearings that would have weakened wilderness management and diluted the integrity of the Wilderness System. Stokes described the Wilderness System as being under siege. He blamed changes in U.S. society, urbanization, diverse population growth, loss of great statespersons and wilderness champions in Congress, and lack of any cohesive national movement supporting wilderness protection and management (Stokes 1999).

Support by wilderness constituencies for legislative appropriations to the agencies and administrative decisions to allocate more funding and staff resources to wilderness management by the agencies are key to future wilderness protection. However, this too is problematic because the question remains: will visitors support use restrictions on popular wilderness sites so that a higher standard of naturalness and solitude can be achieved? *Expanded public information and education programs are essential components for the future success of wilderness management, and these education programs must reach beyond visitors to the broader public.*

Monitoring Threats to Wilderness

Growing wilderness use and the need for continued and additional research, planning, and monitoring to produce the information necessary to guide management of wilderness resources and values have been reported by numerous authors (Cole 1996b; Cole and Landres 1996; Cole, Watson et al. 1997; Manning and Lime 2000; Cole and Wright 2003). The Forest Service is developing a national framework for monitoring wilderness character based on key words and phrases in the wilderness definition of the 1964 Wilderness Act (Landres et al. 2005). The emphasis on potential threats to wilderness in this chapter highlights the need for planning (Chapters 8 and 9), relates these efforts to resource issues (Chapters 10, 11, and 12), and prepares readers for a discussion of wilderness use, visitor impacts, and management (Chapters 14, 15, and 16).

The threats to wilderness documented in this chapter underscore the need for monitoring wilderness conditions and processes so that baseline conditions can be documented and the changes from internal and external factors can be quantified. The objectives and standards identified in planning and the limits of acceptable change (Chapters 8 and 9) illustrate the quantifiable and objective measures that managers must use for comparison to decide if management actions are working or if new actions are required. Unfortunately, standards outlined in wilderness plans (such as for high-use wilderness sites) have not always been enforced (Oye 2001). This implies (or has encouraged) the acceptance of high levels of use at certain sites that exceed standards—and a new wilderness recreation strategy has been proposed in 2000 by the USFS that would accept higher use at popular sites (Oye 2001). However, others strongly oppose such acceptance and call for rollback of use to prior levels (Worf 2001). Only with well-established and long-term monitoring can we hope to document and understand natural conditions and processes in wilderness to track the degree to which wilderness standards are being maintained or degraded.

A final point, but very important to protecting wilderness against threats, is that U.S. wilderness primarily falls under the jurisdiction of four federal agencies (there is some state wilderness—see Chapter 6): the U.S. Forest Service in the Department of Agriculture and in the U.S. Department of Interior, the National Park Service, the U.S. Fish and Wildlife Service, and the Bureau of Land Management. Wilderness is managed by all of these agencies as part of their larger land-management system, and they all struggle with threats to wilderness and shortage of finances and people (Henry 1996; Jerome 1996; Jarvis 1996; Stokes 1996; Brown 2000). All the agencies managing wilderness need more people and more resources to meet their wilderness management responsibilities.

Summary

One of the greatest values of wilderness to future generations will be "naturalness" because 95 percent of the land will have been heavily managed, manipulated, and developed for human uses. The more than 4.5 percent of the landscape that remains "untrammeled" as wilderness in the United States is a scarce resource and may soon represent the only natural remnants of many ecosystems and landscapes with "wildness." *Threats to wilderness are generally defined as change agents that cause impacts on wilderness resource conditions and values—what causes the impacts and the impacts themselves.*

We described nineteen short-term and long-term, direct and indirect threats to wilderness resources and values that must be addressed to achieve the purposes of the Wilderness Act for present and future generations. Changes that affect wilderness can be considered threats when they are not part of (or weaken) the natural conditions, processes, and variability that were historically part of a wilderness area. Threats to wilderness conditions, attributes, and values are complex, interactive, and both internal and external to wilderness. We provide many examples of both threats and their impacts.

Nineteen categories of internal and external threats to wilderness conditions and values were briefly discussed: (1) fragmentation and isolation of wilderness as ecological islands; (2) threatened and endangered species; (3) increasing commercial and public recreation use; (4) livestock grazing; (5) nonnative species; (6) administrative access, facilities, and intrusive management; (7) adjacent-land management and use; (8) inholdings; (9) mining; (10) wildland fire suppression; (11) air quality; (12) water projects and water quality; (13) advanced technology; (14) motorized and mechanical equipment trespass and legal use; (15) aircraft noise and airspace reservations; (16) urbanization and encroaching development; (17) global climate change; (18) legislation that designates new wilderness areas with compromised wilderness conditions; and (19) lack of political and financial support for wilderness protection and management.

These internal and external threats are both actual and potential threats to wilderness now, with considerable variation among geographic areas of the United States, and between remote and more urban-proximate wilderness areas. Some of these threats and examples are current controversies that may diminish over time, but new controversies will appear. However, this discussion highlights the need for continued planning, management, and monitoring of wilderness conditions and processes if we are to continue to enjoy the values of wilderness now and in the future.

Study Questions

1. Define two threats to wilderness.
2. When are wilderness conditions and values affected by threats?
3. Why is understanding wilderness threats important for wilderness management?
4. What are examples of threats that are internal to wilderness?
5. What are examples of threats that are external to wilderness?
6. How does monitoring contribute to wilderness planning and management?
7. Why are nonindigenous and nonnative species a threat to wilderness naturalness?
8. Why is there a shortage of people and financial resources to manage wilderness?

Acknowledgments

Chad P. Dawson and John C. Hendee were lead authors of this new chapter for the third edition and changes and additions in this fourth edition.

For contributions and review comments on drafts of the third edition of this chapter, we thank the following colleagues. We have benefited from their review comments and input, but we alone bear responsibility for any errors, mistakes, and opinions reported herein: David N. Cole, research biologist, Aldo Leopold Wilderness Research Institute, Missoula, Montana; Ed Krumpe, professor of resource recreation and tourism, University of Idaho, Moscow, Idaho; and Peter Landres, biologist, Aldo Leopold Wilderness Research Institute, Missoula, Montana.

Case Discussion:
Interactions Between Potential Threats to Wilderness

Most of the potential internal and external threats to wilderness are interactive and complex in how they affect wilderness character and attributes. Threats like global climate change are more difficult for managers to address because they were not part of the natural conditions in wilderness and were not included in the HRV for an area. Thus, there is a great deal of uncertainty on how to manage the effects and impacts of this threat.

The U.S. Government Accounting Office conducted a study in 2006 and 2007 to understand how federal land-management agencies were addressing the global climate change threat to federal lands. The report stated (2007, 44–45):

> Climate change has already begun to adversely affect federal resources in a variety of ways. Most experts with whom we spoke believe that these effects will continue—and likely intensify—over the coming decades. Some federal resources, depending on a variety of factors, may be more vulnerable than others. Because this issue is long term, global, and may affect federal resources in a number of ways, it will require foresight on the part of federal agencies to prepare for and minimize the adverse effects of climate change. However, federal resource management agencies have not yet made climate change a high priority. BLM, FS, FWS, NOAA, and NPS are generally authorized, but not specifically required, to address changes in resource conditions resulting from climate change in either their resource management actions or planning efforts. However, none of these agencies have specific guidance in place advising their managers how to address the effects of climate change in either their resource management actions or planning efforts. The resource managers with whom we spoke stated that in the absence of such guidance, they are unsure whether or how to take the effects of climate change into account when carrying out their responsibilities. Such uncertainty may, as unanticipated circumstances arise, force resource managers to set their own priorities, which may be inconsistent with those of the agencies' management and may result in misdirected efforts and wasted resources. Because there is growing evidence that climate change is likely to have wide-ranging consequences for the nation's land and water resources, elevating the importance of the issue in their respective strategies and plans would enable BLM, FS, FWS, NOAA, and NPS to provide effective long-term stewardship of the resources under their purview.
>
> To better enable federal resource management agencies to take into account the existing and potential future effects of climate change on federal resources, we recommend that the Secretaries of Agriculture, Commerce, and the Interior—in consultation with the Director of FS; the Administrator of NOAA; and the Directors of BLM, FWS, and NPS, respectively—develop clear, written communication to resource managers that explains how managers are expected to address the effects of climate change, identifies how managers are to obtain any site-specific information that may be necessary, and reflects best practices shared among the relevant agencies, while also recognizing the unique missions, objectives, and responsibilities of each agency.

Case Discussion Questions:

1. Give three examples *of internal* threats to wilderness that interact with the impacts of global climate change and create situations that threaten wilderness conditions and processes.
2. Give three examples *of external* threats to wilderness that interact with the impacts of global climate change and create situations that threaten wilderness conditions and processes.
3. Why is it necessary to understand the interaction between different threats to wilderness to provide better stewardship for wilderness?
4. When the federal agencies that manage wilderness develop "best management practices" to address global climate change, should those practices include monitoring activities or not? Why or why not?

References

Aplet, G. H.; Keeton, W. S. 1999. Application of historical range of variability concepts to the conservation of biodiversity. In: Baydack, R. K.; Campa, H., III; Haufler, J. B., eds. 1999. *Practical Approaches to the Conservation of Biological Diversity.* Covelo, CA: Island Press, pp. 71–86.

Asher, J. E.; Harmon, D. W. 1995. Invasive exotic plants are destroying the naturalness of U.S. wilderness areas. *International Journal of Wilderness.* 1(2): 35–37.

Bell, P. A.; Malm, W.; Loomis, R. J.; McGlothin, G. E. 1985. Impact of impaired visibility on visitor enjoyment of the Grand Canyon. *Environment and Behavior.* 17(4): 459–474.

Brown, P. 2000. Issues in the quality of U.S. wilderness management. *International Journal of Wilderness.* 6(2): 3–4.

Carter, D. 1997. Maintaining wildlife naturalness in wilderness. *International Journal of Wilderness.* 3(3): 17–21.

Christensen, N. L. 1995. Fire and wilderness. *International Journal of Wilderness.* 1(1): 30–34.

Cole, David, N. 1994. The wilderness threats matrix: A framework for assessing impacts. Research Paper INT-RP-475. Ogden, UT: U.S. Department of Agriculture, Forest Service, Intermountain Research Station.

———. 1996a. Ecological manipulation in wilderness: An emerging management dilemma. *International Journal of Wilderness.* 2(1): 15–18.

———. 1996b. Wilderness recreation in the United States—trends in use, users, and impacts. *International Journal of Wilderness.* 2(3): 14–18.

———. 1997. Recreation management priorities are misplaced— allocate more resources to low-use wilderness. *International Journal of Wilderness.* 3(4): 4–8.

———. 2000. Soul of the wilderness: Natural, wild, uncrowded, or free? Which of these should wilderness be? *International Journal of Wilderness.* 6(2): 5–8.

———. 2001a. Balancing freedom and protection in wilderness recreation use. *International Journal of Wilderness.* 7(1): 12–13.

———. 2001b. Management dilemmas that will shape wilderness in the 21st century. *Journal of Forestry.* 99(1): 4–8.

Cole, David N.; Landres, Peter B. 1996. Threats to wilderness ecosystems: Impacts and research needs. *Ecological Applications.* 6(1): 168–184.

Cole, David N.; McCool, S. F.; Borrie, W. T.; O'Loughlin, J. 2000. *Wilderness science in a time of change conference*, Vol. 5: *Wilderness Ecosystems, Threats, and Management;* May 23–27, 1999; Missoula, MT. RMRS-P-15-Vol-5. Ogden, UT: U.S. Department of Agriculture, Forest Service, Rocky Mountain Research Station.

Cole, David N.; Watson, A. E.; Hall, T. E.; Spildie, D. R. 1997. High use destinations in wilderness: Social and biophysical impacts, visitor responses, and management options. Research Paper INT-RP-496. Ogden, UT: U.S. Department of Agriculture, Forest Service, Intermountain Research Station.

Cole, David N.; Wright, V. 2003. Wilderness visitors and recreation impacts: Baseline data available for twentieth century conditions. General Technical Report RMRS-GTR-117. Ogden, UT: U.S. Department of Agriculture, Forest Service, Rocky Mountain Research Station.

Czech, B.; Krausman, P. R. 1999. Controversial wildlife management issues in southwestern U.S. wilderness. *International Journal of Wilderness.* 5(3): 22–28.

Derlet, R. W.; Carlson, J. R.; and Richards, J. R. 2008. Risk factors for coliform bacteria in Sierra Nevada Mountain wilderness lakes and streams. *International Journal of Wilderness.* 14(1): 28–31.

Dombeck, Mike. 1999. A wilderness agenda and legacy for the U.S. Forest Service. *International Journal of Wilderness.* 5(3): 4–6.

Duff, D. A. 1995. Fish stocking in U.S. federal agency areas: Challenges and opportunities. *International Journal of Wilderness.* 1(1): 17–19.

Hammitt, W. E.; Cole, D. N. 1998. *Wildland Recreation: Ecology and Management.* New York: John Wiley.

Hammitt, W. E.; Schuster, R. M. 2000. Wilderness use in the next 100 years. *International Journal of Wilderness.* 6(2): 12–13.

Henry, W. 1996. Status and prospects for wilderness in the U.S. National Park Service. *International Journal of Wilderness.* 2(3): 19, 47.

Hinzman, L. D.; Bettez, N. D.; Bolton, W. R.; Chapin, F. S.; et al. 2005. Evidence and implications of recent climate change in northern Alaska and other Arctic regions. *Climate Change.* 72: 251–298.

Hunger, D. H.; Christensen, N. A.; Becker, K. G. 1999. Commercial and private boat use on the Salmon River in the Frank Church– River of No Return Wilderness, United States. *International Journal of Wilderness.* 5(2): 31–36.

Intergovernmental Panel on Climate Change. 2007. *Climate Change 2007: The Physical Science Basis; Summary for Policy Makers.* Switzerland, Geneva: Intergovernmental Panel on Climate Change.

Jarvis, J. 1996. Status and prospects for wilderness in the U.S. Bureau of Land Management. *International Journal of Wilderness.* 2(3): 23.

Jerome, P. 1996. Status and prospects for wilderness in the U.S. National Wildlife Refuge System. *International Journal of Wilderness.* 2(3): 22, 47.

Johnson, L. C.; Wallace, G. N.; Mitchell, J. E. 1997. Visitor perceptions of livestock grazing in five U.S. wilderness areas: A preliminary assessment. *International Journal of Wilderness.* 3(2): 14–20.

Kelson, A. R.; Lilieholm, R. J. 1997. The influence of adjacent land activities on wilderness resources: U.S. wilderness manager perceptions. *International Journal of Wilderness.* 3(1): 25–28.

Klyza, C. M., ed. 2001. *Wilderness Comes Home: Rewilding the Northeast.* Middlebury Bicentennial Series in Environmental Studies. Hanover, NH: Middlebury College Press and University Press of New England. 320p.

Knapp, D. 2000. Activities and technology: Technology and wilderness in the 21st century. *International Journal of Wilderness.* 6(2): 20.

Landres, Peter B.; Boutcher, S.; Merigliano, L.; Barns, C.; Davis, D.; Hall, T.; Henry, S.; Hunter, B.; Janiga, P.; Laker, M.; McPherson, A.; Powell, D.; Rowan, M.; Sater, S. 2005. Monitoring selected conditions related to wilderness character: A national framework. General Technical Report RMRS-GTR-151. Fort Collins, CO: U.S. Department of Agriculture, Forest Service, Rocky Mountain Research Station.

Landres, Peter B.; Brunson, M. W.; Merigliano, L.; Sydoriak, C.; Morton, S. 2000. Naturalness and wildness: The dilemma and irony of managing wilderness. In: Cole, D. N.; McCool, S. F.; Borrie, W. T.; O'Loughlin, J., eds. *Wilderness science in a time of change conference*, Vol. 5: *Wilderness Ecosystems, Threats, and Management;* May 23–27, 1999; Missoula, MT. RMRS-P-15-Vol-5. Ogden, UT: U.S. Department of Agriculture, Forest Service, Rocky Mountain Research Station, pp. 377–381.

Landres, Peter B.; Marsh, S.; Merigliano, L.; Ritter, D.; Norman, A. 1998a. Boundary effects on wilderness and other natural areas. In: Knight, R. L.; Landres, P. B., eds. *Stewardship Across Boundaries.* Covelo, CA: Island Press, pp. 117–139.

Landres, Peter B.; Meyer, S. 2000. National wilderness preservation system database: Key attributes and trends, 1964 through 1999. General Technical Report RMRS-GTR-18-Rev. ed. Ogden, UT: U.S. Department of Agriculture, Forest Service, Rocky Mountain Research Station.

Landres, Peter B.; Morgan, P.; Swanson, F. J. 1999. Overview of the use of natural variability concepts in managing ecological systems. *Ecological Applications.* 9(4): 1179–1188.

Landres, Peter B.; White, P. S.; Aplet, G.; Zimmermann, A. 1998b. Naturalness and natural variability: Definitions, concepts, and strategies for wilderness management. In: Kulhavy, D. L.; Legg, M. H., eds. *Wilderness and Natural Areas in Eastern United States: Research, Management and Planning.* Nacogdoches, TX: Stephen F. Austin State University, Arthur Temple College of Forestry, Center for Applied Studies, pp. 41–50.

Liddle, M. J. 1997. *Recreation Ecology: The Ecological Impact of Outdoor Recreation and Ecotourism.* New York: Chapman and Hill.

Machlis, G. E.; Tichnell, D. L. 1985. *The State of the World's Parks.* Boulder, CO: Westview Press. 131p.

Mackey, C. 1998. U.S. wilderness management in the 21st century—politics, policy, and partnerships. *International Journal of Wilderness.* 4(3): 6–11.

Manning, R. E.; Lime, D. W. 2000. Defining and managing the quality of wilderness recreation experiences. In: Cole, D. N.; McCool, S. F.; Borrie, W. T.; O'Loughlin, J., eds. *Proceedings: Wilderness Science in a Time of Change*, Vol. 4: *Wilderness visitors, experiences, and visitor management;* May 23–27, 1999; Missoula, MT. RMRS-P-15. Ogden, UT: U.S. Department of Agriculture, Forest Service, Rocky Mountain Research Station, pp. 13–52.

Manning, Robert E. 1997. Social carrying capacity of parks and outdoor recreation areas. *Parks and Recreation.* 32: 32–38.

Marion, Jeffrey L.; Roggenbuck, Joseph W.; Manning, Robert E. 1993. Problems and practices in backcountry recreation management: A survey of National Park Service Managers. NPS/NRVT/NRR-93/12. Denver, CO: U.S. Department of the Interior, National Park Service. 64p.

Matthews, K. R.; Knapp, R. A. 1999. A study of high mountain lake fish stocking effects in the U.S. Sierra Nevada wilderness. *International Journal of Wilderness*. 5(1): 24–26.

McLaran, M. P. 1990. Livestock in wilderness: A review and forecast. *Environmental Law*. 20(4): 857–889.

Meyer, S. S. 1999. The role of legislative history in agency decision-making: A case study of wilderness airstrip management in the United States. *International Journal of Wilderness*. 5(2): 9–12.

Murray, M. P. 1997. High elevation meadows and grazing: Common past effects and future improvements. *International Journal of Wilderness*. 3(4): 24–28.

Nash, Roderick. 1996. A wilderness ethic for the age of cyberspace. *International Journal of Wilderness*. 2(3): 4–5.

National Interagency Fire Center. 2008. Historical wildland fire information. www.nifc.gov/fire_info/fire_stats.htm; accessed February 1, 2008.

Nickas, G.; Proescholdt, K. 2005. Keeping the wild in wilderness: Minimizing nonconforming uses in the National Wilderness Preservation System. *International Journal of Wilderness*. 11(3): 13–18.

Oye, G. 2001. A new wilderness recreation strategy for National Forest Wilderness. *International Journal of Wilderness*. 7(1): 13–15.

Parsons, David J. 2000. Restoration of natural fire to United States wilderness areas. In: Watson, A. E.; Aplet, G. H.; Hendee, J. C., comps. *Personal, Societal, and Ecological Values of Wilderness: 6th World Wilderness Congress Proceedings on Research, Management, and Allocation*, vol. II; October 24–29, 1998; Bangalore, India. RMRS-P-14. Ogden, UT: U.S. Department of Agriculture, Forest Service, Rocky Mountain Research Station, pp. 42–47.

Posewitz, Jim. 1998. Yellowstone to the Yukon (Y2Y): Enhancing prospects for a conservation initiative. *International Journal of Wilderness*. 4(2): 25–27.

Scott, D. 2004. *The Enduring Wilderness: Protecting our Natural Heritage Through the Wilderness Act*. Golden, CO: Fulcrum.

Shultis, J. 2000. Activities and technology: Gearheads and golems—technology and wilderness recreation in the 21st century. *International Journal of Wilderness*. 6(2): 17–18.

Singh, Jaidev; Jackson, Rodney. 1999. Transfrontier conservation areas: Creating opportunities for conservation, peace and the snow leopard in Central Asia. *International Journal of Wilderness*. 5(3): 7–13.

Spring, Ira. 2001. If we lock people out, who will fight to save wilderness? *International Journal of Wilderness*. 7(1): 17–19.

Stokes, J. 1996. Status and prospects for wilderness in the U.S. Forest Service. *International Journal of Wilderness*. 2(3): 20–21.

———. 1999. Wilderness management priorities in a changing political environment. *International Journal of Wilderness*. 5(1): 4–8.

Tanner, R. J. 2004. Subsistence, inholdings, and ANILCA: The complexity of wilderness stewardship in Alaska. *International Journal of Wilderness*. 10(2): 18–22.

Tarrant, M. A.; Haas, G. E.; Manfredo, M. J. 1995. Factors affecting visitor evaluations of aircraft overflights of wilderness areas. *Society and Natural Resources*. 8: 351–360.

U.S. Department of Agriculture, Forest Service. 1998. Frank Church–River of No Return Wilderness programmatic and operational management plans, Vols. I and II, Draft Environmental Impact Statement. U.S. Department of Agriculture, Forest Service, Intermountain and Northern Regions.

U.S. Department of the Interior. 2007. Endangered and threatened wildlife and plants: 12-month petition finding and proposed rule to list the polar bear (*Ursus maritimus*) as threatened throughout its range. *Federal Register* 72(5): 1064–1099. Washington, DC. U.S. Department of the Interior, U.S. Fish and Wildlife Service, 50 *Code of Federal Regulations*, Part 17.

U.S. Government Accounting Office. 2007. Climate change: Agencies should develop guidance for addressing the effects on federal land and water resources. Washington, DC: U.S. Government Accounting Office.

U.S. Public Law 88-577. The Wilderness Act of September 3, 1964. 78 Stat. 890.

U.S. Public Law 93-205. The Endangered Species Act of December 28, 1973. 87 Stat. 884.

U.S. Public Law 93-622. (Eastern Wilderness) Act of January 3, 1975. 88 Stat. 2096.

U.S. Public Law 95-95. Clean Air Act of August 7, 1977. 42 Stat. 7401.

U.S. Public Law 95-237. Endangered American Wilderness Act of February 24, 1978. 92 Stat. 40.

U.S. Public Law 96-312. Central Idaho Wilderness Act of July 23, 1980. 94 Stat. 948.

U.S. Public Law 96-487. Alaska National Interest Lands Conservation Act (ANILCA) of December 2, 1980. 94 Stat. 2371.

U.S. Public Law 96-550. (New Mexico Wilderness) Act of December 19, 1980. 94 Stat. 3221.

U.S. Public Law 96-560. (Colorado Wilderness) Act of December 22, 1980. 94 Stat. 3266.

U.S. Public Law 103-433. California Desert Protection Act of October 31, 1994. 108 Stat. 4473.

Watson, A.; Landres, P. 1999. Changing wilderness values. In: Cordell, H. K., ed. *Outdoor Recreation in American Life: A National Assessment of Demand and Supply Trends*. Champaign, IL: Sagamore Publishing, pp. 384–388.

Watson, A. E.; Patterson, M.; Christensen, N.; Puttkammer, A.; Meyer, S. 2004. Legislative intent, science and special provisions in wilderness: A process for navigating statutory compromises. *International Journal of Wilderness*. 10(1): 22–26.

Worf, Bill. 1997. Response to "Ecological manipulations in wilderness" by Dr. David N. Cole. *International Journal of Wilderness*. 3(2): 30–31.

———. 2001. The new Forest Service wilderness recreation strategy spells doom for the National Wilderness Preservation System. *International Journal of Wilderness*. 7(1): 15–17.

V. Wilderness Uses and Their Management

Chapter 14

Wilderness Use and User Trends

Introduction 358
The Importance of Understanding Wilderness Use and Users 358
Wilderness-Dependent Uses 358
Wilderness Use: An Overview 359
    Public Recreational Use of Wilderness 359
    Commercial Recreational Use in Wilderness 360
    Indirect Use of Wilderness 361
    Scientific Use of Wilderness 362
    Educational Use of Wilderness 364
    Wilderness Experience Programs for Education, Personal Growth,
        Therapy, and Healing 364
    Commodity Uses 365
        Mining for Minerals, Oil, and Gas 365
        Logging 367
        Water Storage and Use 367
        Grazing 368
Wilderness Recreational Use 369
    Methods for Estimating Wilderness Recreational Use 370
        Direct Counting Methods 370
        Indirect Counting Methods 371
        Units of Measurement 373
        Trends in Recreational Use 374
    Use Characteristics 375
        Length of Stay 375
        Group Size 376
        Method of Wilderness Travel 376
        Recreational Activities 376
        Season and Timing of Use 377
        Type of Group 377
        Visitor Residence 378
    Distribution of Use 378
        Distribution of Use Between Areas 379
        Distribution of Use Within Areas 380
User Characteristics 382
Factors Affecting Wilderness Use Trends and Management 384
    Changing Age and Population Structure 384
    Population Distribution and Diversity 384

Leisure Time 385
Effects of Expanding the Wilderness System 385
Increasing Educational Levels and Environmental Awareness 386
Changing Transportation Cost and Supply 386
Changing Recreational Preferences and Organized Use 386
Constraints to Wilderness Visitation 387
Changing Wilderness Conditions and Regulations 387
Wilderness Recreation Use Projections 387
Summary 389
Study Questions 390
Acknowledgments 390
Case Discussion: Constraints to Wilderness Participation 391
Case Discussion Questions 391
References 391

Introduction

Wilderness is preserved because society believes it has values that justify its protection. Many of these values are based on some kind of direct or indirect human use. This is clear in the Wilderness Act of 1964 (U.S. Public Law 88-577), which states that wilderness is to be preserved for "use and enjoyment" by "the American people of present and future generations." The Wilderness Act defined wilderness as having "outstanding opportunities for solitude or a primitive and unconfined type of recreation… [and] may also contain ecological, geological, or other features of scientific, educational, scenic, or historical value." As use of wilderness has increased and the human effects have been documented by studies, we have learned that primitive recreation and experiencing the naturalness and solitude of wilderness yield important inspirational, physical health, personal growth, and psychological values.

The Importance of Understanding Wilderness Use and Users

Understanding wilderness uses and users is essential to wilderness management. Diverse wilderness uses reflect a range of wilderness values and these uses may overlap or threaten other values, thereby creating conflicts. Most wilderness management is use management, trying to protect the wilderness from use impacts and seeking to minimize conflicts among different wilderness recreation users (e.g., hikers, horseback riders, river rafters, mountain climbers) and among recreation and other uses where allowed (e.g., grazing, commercial outfitting, scientific study).

Wilderness users and uses are inherently complex. Knowledge of user characteristics and the type and extent of use is essential, especially if management seeks to emphasize freedom and minimize visitor regimentation and rely instead on more indirect and light-handed approaches such as information and education to manage use and users.

There are many kinds of wilderness use and they reflect a variety of user values (Driver et al. 1987; Alan E. Watson et al. 1999). Broad-based support exists for the concept of wilderness among the general U.S. population, mostly based on ecological and environmental quality and various values and uses that occur both off-site and on-site (Cordell, Tarrant et al. 1998; Cordell, Murphy et al. 2005). In this chapter, we first identify several different types of wilderness use and discuss the known or estimated amount of each use. Then, because information is limited for all uses except recreation and because recreation is an important use of most wildernesses, the remainder of the chapter reviews the amount, character, and distribution of wilderness recreational use and the characteristics of users. Trends in use and speculation about future use then complete the chapter.

Wilderness-Dependent Uses

Wilderness uses vary in the degree to which they depend on wilderness biological or physical conditions. Based on wording in the Wilderness Act and established knowledge of the concept, when we refer to wilderness conditions we mean the qualities of "naturalness and solitude." Some uses depend on these wilderness conditions but other uses do not. Some activities do not depend on wilderness conditions but can be enhanced by them. Thus, *there is a wilderness-dependence continuum of activities ranging from those dependent on wilderness, to those not dependent but enhanced by wilderness conditions, and to those activities not dependent at all on wilderness conditions* (for example, see the four categories of wildlife dependence on wilderness in Chapter 12).

For example, commodity uses, such as grazing or mining, take place in some wildernesses but do not depend on or require wilderness conditions—they can take place just as well outside of wilderness. Thus, where they are allowed in wilderness they may be called "nonconforming" uses. Certain recreational uses also do not depend on wilderness conditions. Some recreational activities in wilderness—for example, campers playing catch in a wilderness meadow—just take place in a wilderness without depending on the naturalness and solitude qualities of the wilderness. Some activities are enhanced for some people by wilderness conditions, even if the activity is not dependent on naturalness and solitude. For example, observing wildlife or fly-fishing in solitude on a wilderness stream may be enhanced by the wilderness setting. Other activities—for example, observing the results of natural ecological processes on the landscape, experiencing solitude and isolation, and facing the challenges of traveling and living in a large, roadless, undeveloped area—clearly depend on wilderness settings.

Thus, the wilderness-dependence continuum is another concept for planning and management of wilderness activities. Wilderness dependence considers the importance of the setting to the recreation experience, as do the recreation opportunity spectrum (ROS) and wilderness opportunity spectrum (WOS).

The concept of wilderness dependency is basic to managing wilderness use. Chapter 7 sets forth a management principle that wilderness-dependent activities should be favored over those that can be carried on outside wilderness. It is easier to understand this principle if you envision a wilderness area that is approaching the limit of use that it can sustain without unacceptable loss of naturalness and solitude. In such situations, wilderness managers must often make a decision among competing uses, and which of them are more dependent on wilderness conditions. These decisions are always subjective and open to interpretation, but we believe

Fig. 14.1. Educating visitors about appropriate behavior and permitted activities in wilderness areas is becoming increasingly important as more people discover wilderness and should have a day or overnight experience there. Here a U.S. Forest Service volunteer wilderness information specialist informs visitors about Oregon's Three Sisters Wilderness. Photo courtesy Barbara Merlin.

that managers can and must make such judgments in their planning and management processes (see Chapters 8 and 9).

Wilderness Use: An Overview
Public Recreational Use of Wilderness

Recreation is the most obvious and most often reported wilderness use. In 2005, wilderness recreation totaled 10.7 million visits per year to the NWPS and 16.3 on-site days (Bowker et al. 2007). Wilderness use and recreational visitation are often treated as synonyms; however, wilderness use is a broader term that includes *nonrecreational* use, such as grazing. On-site recreational use certainly involves the largest numbers of direct wilderness users, has important localized impacts, and poses major management challenges.

The research information on recreational use and biophysical impacts from recreational use to campsites and trails is very limited. Cole and Wright (2003) reported that only 56 percent of the 625 wilderness units in the NWPS in 2000 had any research information on visitors and their impacts. The types of information available in wilderness units were mostly for campsite impacts (51 percent), visitor use (24 percent), and trail conditions (9 percent). Overall, the FS and NPS had some recreation information for the majority of units in the NWPS they manage (77 percent and 66 percent, respectively), and the BLM and FWS agencies had only a small percentage of areas with any recreation research (17 percent and 11 percent, respectively).

Recreational use of wilderness includes many diverse activities, such as hiking, hunting, fishing, rock climbing, photography, cross-country skiing, snowshoeing, and viewing wildlife. Wilderness visitors take short walks, day hikes, long backpacking trips, and everything in between. Some visitors ride

Fig. 14.2. The 1964 Wilderness Act allows many uses of wilderness, in addition to individual enjoyment. A commercial outfitter has tents for clients, left; scientists at work, second from left; a dam in Colorado's Mount Zirkel Wilderness, third from left, is an example of water development; and a band of sheep in a meadow on the Bridger Wilderness in Wyoming, right, is an example of livestock grazing. Photos (left) courtesy Richard Walker, and others U.S. Forest Service.

horses on day trips, or on long trips with pack animals to carry gear and supplies. Other visitors walk, leading pack animals like horses, burros, mules, or llamas. Some wilderness users float rivers in boats, canoes, kayaks, or rafts. *In some places,* where legislation and regulations permit, some visitors use motorboats, snowmobiles, cruise ships, or airplanes (for example in Alaska and Idaho) for access. People travel in all types and sizes of groups—mostly in small groups with family or friends—and some travel alone. Other visitors travel with commercially outfitted groups composed of friends and strangers, or in groups sponsored by youth, church, environmental, adventure, or educational organizations. Increasingly in recent years, some visit wilderness with organized wilderness programs for personal growth, therapy, and healing.

Commercial Recreational Use in Wilderness

In addition to public recreation use, the Wilderness Act authorizes commercial recreational use of wilderness. *There are two main kinds of commercial-recreation influences on wilderness:* from *outfitting and guiding,* and from *equipment rental.* Outfitters and guides accompany some visitors, adding to and facilitating the use of wilderness. This use usually involves travel by horse or boat, although outfitted and guided backpacking trips are a growing commercial use. Other outfitting businesses involve only equipment rental—the visitors travel without guides using rental equipment, such as canoes, rafts, backpacking gear, cross-country skis, and tents. Rental equipment is now readily available in most major urban areas and cities and towns near wilderness. In a few places, one can rent pack animals like burros or llamas, but few outfitting businesses rent pack stock for do-it-yourself wilderness travel, out of concern for liability issues and proper care of the animals.

Commercial outfitting is a traditional use of wilderness permitted by the Wilderness Act and encouraged by wilderness managers in some areas where use capacity is available and the proposed activities are compatible with wilderness conditions. Outfitters provide camping and food services, but the key services are specialized transportation (horses, rafts, skis, etc.) and specialized equipment (lightweight camping gear, etc.). Outfitting in some areas has drawn criticism for stressing comfort, convenience, and excessive facilities and technology that conflict with wilderness values. Nationwide, the wilderness outfitting business is responding to the need for wilderness use that is light on the land, leaving no trace of use, and responsive to wilderness values and other users. Some outfitters and some wilderness areas reflect this trend more than others. Furthermore, outfitters can serve as educators of wilderness users to enhance visitor experiences and to promote wilderness stewardship.

Commercial outfitters and guides must obtain special-use permits to take groups into wilderness, and this gives managers a chance to influence their operations toward wilderness-compatible activities and the timing and location of use. In many wilderness areas, there is great competition for permits and, as established carrying capacities are reached, commercial outfitter use may be limited in favor of public recreational use. This is a contentious issue, but the underlying problem is the overall increased use of wilderness (Mackey 1998).

Overall estimates of the extent of commercially outfitted and guided wilderness recreation use are not available. Outfitter use varies from none in some wildernesses, such as the smaller wilderness areas in the eastern and western United States, to a large amount of use in others, particularly on whitewater rivers such as the Colorado, Middle Fork of the Salmon, and Selway where specialized equipment and skill are necessary for safe travel. In general, outfitters and guides play a more important role in larger wildernesses where horse travel is common or on wilderness rivers where specialized watercraft and skill are required. However, even in two of the largest wildernesses in the lower forty-eight states, the Bob Marshall Wilderness in Montana and the Selway-Bitterroot in Montana and Idaho, studies in the 1980s showed only a minority of the visitors employ commercial outfitters—about one-fifth in the Bob Marshall Wilderness and one-sixth in the Selway-Bitterroot (Lucas 1985).

Institutional outfitting, which is outfitting and guiding by schools or by nonprofit educational and religious organizations, is a large and growing source of wilderness use. For example, colleges, universities, and schools take classes to wilderness and operate outdoor recreation, education, and adventure programs, and provide equipment-rental services to support such activities. Religious organizations bring recreation groups of all ages to wilderness. Institutional use of wilderness is becoming a contentious issue because of the size of groups, the overall extent of use, and the fact that permits are not required in many wilderness locations. Commercial outfitters complain that, as wilderness-use limits are reached and exceeded, their permits are being restricted while institutional outfitters are not required to obtain permits and are less restricted. Managers, concerned that appropriate agency funding for wilderness management is limited while workloads increase, look to user fees as a more important source of revenues. Institutional outfitters, therefore, could be an important new source of revenues if they are required to obtain and pay for permits (Mackey 1998). This is an important wilderness policy issue and is very contentious.

Indirect Use of Wilderness

Besides direct, on-site recreational use of wilderness, there are several important types of indirect use (Haas et al. 1986; Cordell, Bergstrom, and Bowker 2005). *Millions of people who never set foot in wilderness may nevertheless derive satisfaction from experiencing wilderness indirectly.* This indirect use comes about vicariously through reading; viewing television, photos, and films; and listening to lectures and accounts by others. Indirect use also comes about when people stay at resorts or recreation areas near wilderness where they can view wilderness scenery and experience the excitement in wilderness gateway communities, such as observing trailhead use, outfitter activities, and marketing materials about visiting wilderness areas.

One major study documented greater population growth in western counties having wilderness than in counties without wilderness (Ridzitis 1996). The attraction of small towns and rural areas near wilderness and wildlands for retirees and persons with portable jobs is increasingly noted (Powers 1996). Indirect and vicarious enjoyment is a major use of wilderness. Without wilderness, the values and benefits from these experiences would be diminished. Historical and past accounts of wilderness encounters would remain but would provide only a partial substitute because they lack the possibility that one could actually visit the area. Indirect use produces and is produced by five types of values: option, existence, bequest, pre- and posttrip values, and ecological services.

1. *Option value: Many people value the possibility that they will someday visit wilderness* (Schuster et al. 2005), so keeping open the option to visit wilderness is important to them and, in that sense, they are indirect users. Whether or not they ever actually visit a wilderness, it is worth something to them to know they could. Another option value revolves around the idea that wilderness and other protected areas provide a storehouse of resources that could be used in other ways, such as for commodity production, in the future if needed by society.

2. *Existence value: For many other people, the simple fact that wilderness exists has value even though they may not actually visit it.* Existence values have been defined and measured for more than thirty years (Fisher and Krutilla 1972; Schuster et al. 2005).

3. *Bequest value: Many people want to leave wilderness as an inheritance to later generations.* Schuster et al. (2005) report that a larger percentage of the general public in the United States held bequest values, existence values, and option values than the estimated percentage of the population who directly visited wilderness for recreational use (Bowker et al. 2007).

4. *Pre- and posttrip values: Recreation experiences have not only an on-site phase but also four off-site phases: anticipation, travel to, travel from, and*

recollection. Compared to most other types of recreation, wilderness visits have particularly long and well-developed off-site anticipation and recollection phases. Planning frequently starts well in advance, and experiences are often relived afterward so that even direct on-site recreational use involves substantial indirect use.

5. *Ecological services: By far the largest wilderness value in estimates of economic values of wilderness is ecological services,* such as providing clean air and water, biodiversity, and a place where natural processes and conditions exist relatively unmodified (Loomis and Richardson 2001). Similarly, these ecological service values are the ones reported most frequently by the majority of the U.S. population as important wilderness benefits (Cordell, Murphy et al. 2005).

Some wildernesses or portions thereof are closed to all direct, on-site recreational use, such as some FWS wilderness areas (e.g., small islands). In addition to the indirect or vicarious use these areas provide, they provide breeding grounds that produce many birds and wildlife that are observed and enjoyed outside the wilderness.

Any estimate of the amount of indirect wilderness use would be very difficult to make, and none is available. Sales of wilderness-related books and videotapes are substantial, and films with nature-wilderness themes draw crowds. Numerous popular television programs have been based on nature and wilderness. The clearly positive influence of wilderness on classic North American artists, philosophers, and writers is also well recognized (Mortensen 1997, 1999), and wilderness continues to inspire poetry, prose, and thoughts from less prominent artists and authors. We speculate that the number of indirect users is probably much greater than the number of actual visitors, and this may be part of what is reflected in the broad public support for wilderness.

Scientific Use of Wilderness

One of the major values of wilderness is its potential for scientific use. *Wilderness serves as a living laboratory, particularly for ecological studies looking at interrelationships (e.g., wildlife dynamics and fire-dependent communities) in relatively natural, unmodified conditions and large areas.* Aldo Leopold (1941) wrote of wilderness as a land laboratory or environmental baseline, valuable for learning how healthy ecosystems function and how humans affect the ecosystem.

As the rest of the world becomes more developed and modified by humans, the contrast between wilderness and nonwilderness increases, and the value of wilderness as a natural laboratory will increase. Thus, *scientific use is a substantial wilderness-dependent activity because of the scarcity of unmodified natural ecosystems.* Not only is wilderness valuable for the scientific opportunities it provides (wilderness for science), but science is proving a valuable source of new information to support wilderness and wildland conservation (science for wilderness) (Parsons 2000).

Certain types of environments, such as high mountain and alpine environments and more recently semiarid areas and deserts, are especially well represented in designated wilderness, so studies of them are often located in wilderness. Studies of glaciers are a notable example. Such scientific study is not directly dependent on wilderness; the subject just happens to be there. Other scientific use also depends on objects that happen to be located in wilderness, such as cultural and archeological sites (e.g., prehistoric campsites, dwellings, and artifacts) and paleontological resources (e.g., dinosaur fossils). Scientific activities at such sites often pose management dilemmas because research often requires excavation and removal of fossils and artifacts and protection of exposed sites. These resources are also very vulnerable to removal by visitors. Seeing fossils and artifacts in their natural wilderness setting is also valuable, and the experience is diminished by their removal. Protecting cultural and archeological resources is an important issue, especially in some wilderness areas in the southwestern United States.

Wilderness provides a good setting for some kinds of social and psychological research (Manning 1989). *Isolating small groups of people and their close interdependence in the face of the challenges of wilderness experiences provides valuable behavioral research opportunities.* Numerous research studies on wilderness user experiences, "opportunities for solitude," and tolerance for crowding have been conducted in wilderness areas, wildlands, and wild and scenic river areas (Absher and Lee 1981; Cole, Watson, and Roggenbuck 1995; Graefe et al. 1984; Gramann 1982; Manning 1985). This research on users has evolved from descriptive to more comprehensive studies seeking to better understand the benefits from user experiences in wilderness conditions (Driver et al. 1987; Ewert and MacEvoy 2000). Some studies looked at the reactions of visitors to contacts with other visitors (encounter preferences), the normative behavior of users (Hall and Shelby 1996; Alan Watson et al. 1996; Williams et al. 1991), and the importance and meaning of privacy and solitude (Dawson and Hammitt 1996; Hammitt and Madden 1989). Such research has important implications for wilderness management based on understanding what are reasonable standards and the expectations of visitors. The results of such studies and numerous others (completed in the last several decades), when coupled with the ROS, WOS, and LAC planning approaches (see Chapters 8 and 9), have changed wilderness visitor planning and management.

One of the important scientific values of wilderness is its function as a natural gene pool. In wilderness, the natural genetic diversity of native plants and animals is protected from direct manipulation. Outside wilderness, this diversity is often reduced unnaturally as some species become extinct and as domestic species are selectively bred for greater uniformity, increased yield, and other characteristics. Although the wilderness system represents some types of environments and ecosystems very well, such as alpine areas and deserts, other environments, such as grasslands and low-elevation forests, are not well represented (Davis 1989; Cordell, Murphy et al. 2005). The scientific value of wilderness increases as these yet unrepresented environments and ecosystems are added to the NWPS.

Scientific use of wilderness is pervasive but difficult to measure, and there is some difference of opinion as to its extent (Franklin 1987; Butler and Roberts 1986; Reed et al. 1989; Carr and Hendee 1993). The four federal management agencies tend to focus on what is present in the wilderness resource and what is changing, via inventory and monitoring studies. The USFS, NPS, and colleges and universities focus on studies to better understand ecological relationships in wilderness. The USFS, colleges, and universities also conduct studies focused on assisting managers in their wilderness management efforts.

Science was one of the stated purposes of wilderness from its original definition in the Wilderness Act in 1964. However, how much research is needed to fulfill the mandate of the Wilderness Act that science is one specific purpose of wilderness? The concept of wilderness includes the value of undeveloped lands with minimal human influence that serve as a baseline of environmental integrity and where scientific research can measure natural processes. Because the authors of the Wilderness Act could not have anticipated the current size of the NWPS, the need for scientific information to support management of this system is greater than they originally expected. Furthermore, global environmental conditions have changed at an unexpectedly rapid rate, and this highlights the need for basic science and monitoring within the NWPS to describe changes in natural processes and conditions. During the 1990s, several assessments and articles about wilderness research needs noted that the needs and opportunities for research far exceed the fiscal and staff capabilities of existing wilderness research programs, a theme echoed in several articles (Cordell and Young 1993; Hendee and Ewert 1993; Hendee 1995). As other lands become more impacted by human influences, the value of wilderness as a natural, protected laboratory increases.

A panel that reviewed the USFS research program in wilderness outlined three complementary categories of research (Parsons 2007, 35):

1. Science for wilderness: science that informs effective stewardship and management of wilderness, including the status and trends of ecological conditions
2. Wilderness for landscape sustainability: science that improves understanding of the contributions of wilderness to the ecological processes, services, and integrity of larger landscapes
3. Wilderness for science: science that uses wilderness and similarly managed lands as laboratories to understand the causes and consequences of environmental change, minimally confounded by other influences

These three categories reflect the legislative intent of the Wilderness Act and the place of wilderness on the extreme end of the environmental modification spectrum (described in Chapter 1).

Sharing and disseminating wilderness research findings have taken on several forms to accommodate a worldwide interest in such information. The World Wilderness Congresses, such as the fifth in 1993 at Tromso, Norway; the sixth in 1998 at Bangalore, India;

the seventh in 2001 in South Africa; and the eighth in Alaska in 2005 (Alan Watson, Sproull et al. 2007), have provided opportunities for compiling, comparing, and exchanging information among scientists from several countries through research symposia and published proceedings. The *International Journal of Wilderness* (www.ijw.org) was launched in 1995 and provides a continuing forum for findings of wilderness studies. The 1999 conference on "Wilderness Science in a Time of Change" was one of the most comprehensive gatherings of primarily U.S. wilderness scientists to share studies, perspectives, and dialogue about wilderness research and scientific issues (Cole, McCool, Borrie et al. 2000a, 2000b; Cole, McCool, Freimund et al. 2000; McCool et al. 2000a, 2000b).

Educational Use of Wilderness

Wilderness is used for educational purposes by providing sites for field trips, study areas for student research, and a source of instructional examples. Reed et al. (1989) reported that 39 percent of NWPS wilderness managers claimed that environmental and conservation education programs were being conducted in their wilderness in 1987. Some educational uses based on lessons from large-scale, long-term ecological processes may be dependent on wilderness, but for many topics, other lands are available and may be more suitable and accessible.

Some educational uses are more akin to recreational use: wilderness as a setting for teaching travel and survival skills. Many colleges and universities have field trips that teach wilderness skills. Some youth organizations, such as the Boy Scouts, teach outdoor-living skills, with their application in wilderness as a pinnacle of achievement. Whether such use is really dependent on wilderness is questionable, but trips to wilderness as a test and demonstration of scouting skills are classic. What they really need is large, unroaded areas with wilderness or primitive conditions. Other courses also teach wilderness values as well as low-impact use techniques, and this kind of education would be considered more wilderness-dependent.

Educational use of wilderness is not directly measured, but some indication of its magnitude can be gained from data on wilderness-related courses in colleges and universities and wilderness experience programs. Dawson and Hendee (2004) reported that in 2002 more than 1,500 students took primary wilderness-related courses at colleges and universities in the United States, which was 30 percent lower than what was reported in a 1982–1983 study (Hendee and Roggenbuck 1984). Most courses focused on wilderness appreciation and use, legislation and policy, protection and management, and history. Some university courses use wilderness experiences to teach other subjects, such as environmental education and natural ecosystems. These activities and educational programs can be highly dependent on wilderness.

Studies in the mid-1990s document more than two hundred education-oriented wilderness experience programs (excluding higher education institutions and youth organizations), ranging from several large programs serving hundreds of students (e.g., Teton Science School in Wyoming) to many small programs serving fewer than one hundred per year (Dawson, Tangen-Foster et al. 1998; Friese, Hendee et al. 1998).

Today, schools and higher education institutions may be exempt in some locations from having to secure special-use permits to use wilderness, which means there are no systematic data about the extent of this so-called institutional outfitting. Commercial outfitters would like to see permits and fees for educational use to help share the financial and use quota burden imposed on them (Mackey 1998). *As wilderness enters the twenty-first century, one of the most important and contentious policy issues is whether to, and how to, bring educational use of wilderness into the wilderness use accounting system—and whether to, and how to, charge for such use.*

Wilderness Experience Programs for Education, Personal Growth, Therapy, and Healing

Wilderness experience programs (WEPs) take paying clients into wilderness or comparable lands to develop their human potential through education, personal growth (including leadership and organizational development) or therapy, and healing (Friese, Hendee et al. 1998). Evidence suggests that WEPs' use of wilderness is a very large emerging industry. One nationwide survey of wilderness managers found that, among those reporting WEP use in wilderness they administered, two-thirds said such use was increasing, and more than one-third expressed concern over environmental impacts of use by such groups or conflicts with other users (Gager et al. 1998). Because of this user's size and prospects for continued growth, permitting and dealing with the use of wilderness for education, personal growth, therapy, and healing are a growing management challenge, yet a growing opportunity for enhanced social benefits from wilderness. Many policy and management issues surrounding access to the wilderness affect these organized uses (Ewert et al. 1999).

Wilderness personal growth, therapy, and healing programs use various combinations of challenge, adventure, and reflective activities to help participants get in touch with themselves. Wilderness personal-growth programs seek empowerment; they stretch participants' abilities and determination in wilderness activities to teach them that their capabilities exceed what they imagined and, therefore, they may be self-limiting their performance back home in their daily lives. Wilderness therapy and healing programs seek healing and restored normal functioning; they use the natural consequences of wilderness living and primitive skills to cleanse participants of unwanted stresses and restore normal functioning, with individual and group therapy and solo experiences to help connect participants with their inner selves as the ultimate source of their afflictions—and their recovery. Although wilderness personal growth, therapy, and healing programs serve all age classes, adolescents and young adults are the most frequent participants. Enhanced self-esteem and other variations in empowerment are the most consistently reported outcomes for individuals in studies of wilderness experience programs (Friese, Pitmann et al. 1996).

Surveys in the 1990s (Dawson, Tangen-Foster et al. 1998; Friese, Hendee et al. 1998) identified more than 230 personal-growth programs, ranging from several large programs serving thousands of clients annually—Outward Bound, by far the largest, served about 30,000 clients in 1998—to many small programs serving fewer than 100 clients annually. The cost of those personal-growth programs was about $100 or more per day per participant, so they generate substantial revenues and contribute to a wilderness-related industry.

In 1998, a survey identified thirty-eight wilderness therapy and healing programs; data from five of them projected a total estimate of 12,000 wilderness therapy clients, generating 392,000 wilderness field days and $143 million in revenues. The programs studied ranged from three to eight weeks in length and charged an average of $325 per participant per day, with about 40 percent of clients receiving some degree of copayment from medical insurance (Russell et al. 2000). A higher leader-to-participant ratio in the field and clinical oversight of the process increase the cost of programs for wilderness therapy and healing. The data indicated that such programs were growing, and that increasingly medical insurance companies, social service agencies, judicial authorities, and school officials were turning to outdoor and wilderness treatment programs to help adolescents overcome problem behaviors such as substance abuse, resistant and defiant behavior, emotional adjustment, and psychological problems. Reflecting this growth and its attempt to soften any stigma associated with the word *therapy* and broaden the range of healing programs, a group of highly respected programs introduced the term *outdoor behavioral healthcare* to refer to this use, and organized as an "Outdoor Behavior Health Industry Council" to cooperate in uplifting standards in the industry.

Studies of WEPs and their clients document the attraction of wilderness for personal-growth and healing uses and their partial or complete dependence on wilderness to deliver program outcomes and achieve important social benefits. *Is the use of wilderness for the education, personal growth, therapy, and healing of young people as important as—or more important than—commercial or public recreational use? Increasingly, wilderness managers will face such questions in policy and actions that favor one kind of use over another when limiting use to protect wilderness naturalness and solitude. Moreover, questions of what kinds of groups have to obtain permits and how much they must pay to use wilderness will be more contentious and more frequently debated.*

Commodity Uses

Several on-site commodity uses take place in wilderness, including mining, logging, water storage and use, and grazing. These uses, sometimes referred to as "nonconforming but allowed" uses, are generally limited to where they were established before designation of the area as wilderness.

Mining for Minerals, Oil, and Gas

The 1964 Wilderness Act permits mining under certain prescribed situations but under agency regulations designed to minimize adverse impacts on the wilderness environment (see Chapters 4 and 5 for legislative details). However, as explained in the following discussion, active mining in wilderness is minimal and prospecting for new claims has largely ended. The bigger threat may be on wilderness lands purchased by the federal government without subsurface rights to minerals, oil, and gas, such as areas designated in the eastern United States. However, even in these circumstances, development in or near wilderness would be constrained to help protect the wilderness environment.

Almost all mining claims in wilderness areas are located on national forest and in BLM wildernesses (Browning et al. 1988). Most national parks are closed to all mining, but six were established with special provisions permitting mining. Three of these parks contain wilderness and two have areas proposed for wilderness designation. In 1976, Congress passed the Mining in the Parks Act (U.S. Public Law 94-429), closing these six national parks to new mining claims and placing all existing claims under strict new regulations. The 1964 Wilderness Act permitted continued staking of claims in national forest wilderness only and only until the end of 1983. However, because claims could be staked only on public-domain lands, prospecting was limited to wildernesses in the western United States, which are part of the public domain, and was excluded from wildernesses in the East. In the East, almost all the land in wildernesses, mostly national forests and national parks, was private land before it was acquired by the federal government. However, in many cases the mineral rights on these lands were not acquired by the government, and therefore mining is a possible use but still subject to restrictions to protect the wilderness. These private mineral rights in wilderness areas in the East could be acquired from willing sellers, or even by condemnation (although politically it would be difficult), if funds were available to compensate owners.

Mining in wilderness on the national forests is limited. Mining claims are numerous—just how numerous is hard to determine because historically there was no requirement to notify the USFS when claims were filed on national forestland. Claims were recorded in county courthouses, along with many other claims on other lands. In 1985, it was estimated that there were at least 10,000 mining claims on national forests (Wilkinson and Anderson 1985). In 1984, the Cabinet Mountains Wilderness in Montana on the Idaho border reported more than 700 claims, and in 1998, a proposal was made by a mining company to extract silver from under the Cabinet Mountains Wilderness using a tunnel system operating from just outside the wilderness area. However, reported mineral extraction from any wilderness is limited to nonexistent. This low level of mining is consistent with mineral surveys of wilderness and primitive areas that have failed to turn up major mineral deposits that are economically viable. Restrictions on mining activity to protect the wilderness make mining more difficult and costly and probably have discouraged development. However, Reed et al. (1989) reported that 9 percent of NWPS wilderness managers claimed that there were active surface or subsurface mining claims or maintenance on such claims in their wilderness areas in 1987.

Oil and gas production is under lease inside wilderness in several locations, one example being Indian Mounds Wilderness, managed by the Forest Service in east Texas (Evans 1986). Such use, which operates under different laws than hard-rock mining, is a possibility in many places and is very controversial. Reed et al. (1989) reported that less than 1 percent of NWPS wilderness managers claimed that active producing oil or natural gas wells were in operation in their wilderness in 1987.

The BLM policy for their primitive areas generally prohibited mining and mineral leasing except for valid existing rights established before the area was administratively designated as a primitive area. Under provisions of the Federal Land Policy and Management Act of 1976 (FLPMA) (U.S. Public Law 94-579), mining claims can continue to be filed on lands included in wilderness study areas, provided that the claims or subsequent mining operations do not impair the suitability of such areas for subsequent designation as wilderness. Mining operations in effect on the date of FLPMA can continue, provided they do not cause unnecessary or undue degradation.

When BLM lands are legislatively designated as wilderness, they are withdrawn from all new mining and mineral leasing. BLM regulations for designated wilderness require holders of existing mining claims to file a plan of operations outlining how mining is to be undertaken. Before such plans are approved, the BLM must determine if the mining claim is valid. If it is not, actions are taken to invalidate the claim. Thousands of mining claims exist in BLM WSAs. In future years, these claims may pose a wilderness management problem for the BLM if mineral prices rise, leading to pressures for mineral claim development and subsequent mining operations at these locations. However, although expanding mining seems to be a declining threat, many wilderness areas contain old mining sites littered with old equipment and junk and occasionally abandoned roads that are in visual contrast to the natural landscape and an attractive nuisance for potential vehicle trespass.

Logging

Except for limited timber cutting necessary for mine timbers, logging has been allowed in just one wilderness, the Boundary Waters Canoe Area Wilderness (BWCAW) in Minnesota—an area that has been an exception within the NWPS in a number of ways (Duncan and Proescholdt 1999; Gladden 1990). The Boundary Waters Canoe Area Wilderness Act of 1978 (U.S. Public Law 95-495) was the first amendment to the Wilderness Act, and its purpose was to end logging inside the area. Before logging was ruled out in 1978, the BWCAW locations in which it was permitted had been progressively reduced over the years, with about two-thirds still available for timber harvest before the 1964 Wilderness Act and about 40 percent thereafter.

Another possible exception is cutting trees to control insects or disease, especially if an infestation threatens areas outside the wilderness. This was done in several national forest wildernesses in Texas to control southern pine beetles (Billings 1986), but it was very controversial.

Water Storage and Use

Facilities for water storage, such as dams and reservoirs, are permitted by the 1964 Wilderness Act. New water storage projects require presidential approval, but many small reservoirs built before enactment of the Wilderness Act continue to be used and maintained. It is very controversial when heavy maintenance or reconstruction using mechanical equipment is proposed for these aging facilities, as in the Selway-Bitterroot Wilderness.

Not all wilderness water storage and use depend on dams inside wilderness. Many wildernesses store vast amounts of water in the form of snow and provide a large proportion of the stream flow in the West. *Measuring snowpack to forecast stream flow poses a conflict with wilderness values when snow is measured with the aid of mechanized access, such as helicopters, snowmobiles, or electronic snow-weight recording devices connected to radio transmitting stations.* Wilderness managing agencies have responded somewhat differently to this conflict. In wilderness on the national forests, except where helicopter use was well established before the Wilderness Act, snow measurements are made without mechanized equipment, usually by crews traveling on cross-country skis or snowshoes. Electronic devices that measure and transmit snow data from wilderness locations are not permitted. The U.S. Department of the Interior policy for national park wilderness is to permit existing water-resource-monitoring devices to be retained in wilderness. New devices will be placed in wilderness only if the secretary of the interior decides that essential information cannot be obtained from locations outside the wilderness, and that the proposed device is the minimum necessary to successfully and safely accomplish the objective. Reed et al. (1989) reported that 15 percent of NWPS wilderness managers claimed that weather- or snow-monitoring stations or equipment were being used on a temporary or permanent basis in wilderness they administered in 1987.

Most irrigation water impoundments in wilderness were built long before the Wilderness Act. Typically, the dams are low and constructed of local rock and timber, and shoreline vegetation has had many years to adjust to the higher water level. Thus, many of the reservoirs are not conspicuously unnatural unless water levels fluctuate greatly. *In 1987, about one hundred dams existed in national forest wilderness, with many others in national park wilderness and backcountry, but reservoirs are concentrated in a few areas, so most wildernesses do not have such developments.* Several reservoirs are in the Montana portion of the Selway-Bitterroot Wilderness, which has low dams on thirty to forty lakes, some built in the 1880s and 1890s before the area was even a national forest, let alone a wilderness. A few dams in the BWCAW, built to facilitate log drives, are still in place and maintain lake levels; many more dams have been abandoned. Some wildernesses in the Sierra Nevada of California have small dams to control minimum stream flows so streams will not go dry in late summer with a resulting loss of fish. There are also a few high, concrete hydroelectric dams in the California Sierras, built before the Wilderness Act, with the last such dam constructed on the Rubicon River in the Desolation Wilderness in

1963, only one year before the Wilderness Act became law. Again, as these water storage and control facilities need heavy maintenance, reconstruction, or replacement, this becomes controversial because of perceived conflicts with wilderness values.

Grazing

Grazing by livestock—sheep, cattle, and a few horse herds—is also allowed by the Wilderness Act where it existed before wilderness designation. Wilderness in the national forests and on BLM lands accommodates virtually all of the livestock grazing in wilderness. Commercial grazing (as opposed to grazing of pack stock used for wilderness travel) probably is less common now as a commercial activity than in past decades. Changes in the livestock business generally have reduced the economic attractiveness of extensive grazing in wilderness because of the higher costs of grazing there compared to less remote pastures. Transporting livestock to and from wilderness, maintaining minimal facilities, hiring herders, and alleged losses to predators are some categories of increased costs related to grazing in wilderness compared to other lands. An emphasis on intensive production of forage on more suitable lands and more use of feedlots in the livestock industry further accentuates the cost differences. The fact that fees for grazing public land are low compared to these for private lands probably contributes to maintaining the wilderness grazing demand at current levels.

Nevertheless, grazing takes place on about one-third of the wilderness lands in the NWPS excluding Alaska (Reed et al. 1989). In 1986, about 14 percent of all cattle and sheep grazing in the National Forest System grazed in wilderness at least part of the year. Because grazing seasons are short in many wildernesses, the percentage of Animal Unit Months (AUMs) in wilderness is lower. (One AUM equals one cow or five sheep for one month.) Figures on grazing in BLM primitive areas and potential wilderness are unavailable, but grazing on these lands is substantial; more than 80 percent of BLM land in the western states (excluding Alaska) is subject to grazing (Foster 1976). Grazing often generates pressures for mechanized access to maintain facilities, such as water tanks and for predator control. These issues pose a difficult policy challenge for wilderness managers.

One study of wilderness users in five intermountain western areas found that the proportion of visitors who accepted livestock grazing (43 percent) was similar to the proportion (40 percent) who considered grazing unacceptable (L. Johnson et al. 1997). However, even visitors accepting grazing predicated their approval on proper management, and a majority of all visitors reported that direct encounters and livestock impacts detracted from their wilderness experience (e.g., encounters on trails, manure in campsites) (L. Johnson et al. 1997). Some users reported that some aspects of livestock grazing (e.g., calves with mother cows, cowboys moving cattle) added to their wilderness experience.

Grazing in wilderness will continue where it was an established use of public lands before the Wilderness Act passed in 1964 or in areas designated later. Public-land grazing has a strong and active political constituency in the livestock industry. Without provision for grazing to continue where it was already established, the Wilderness Act could not have passed—so grazing was included as one of the so-called allowed but nonconforming uses, along with mining.

Since 1964, objections to grazing in wilderness and on all public lands have increased, and in response, livestock interests have supported legislation and management direction to protect (and possibly strengthen) their right to graze and manage livestock grazing in wilderness, as the following brief review based on Browning et al. (1988) and McClaran (1990) documents. See also Chapter 5 on grazing.

Livestock grazing and other wilderness management policies in national forest wildernesses were reviewed by U.S. congressional committees in the 95th (1978–1979) and 96th (1980–1981) Congresses. The result was U.S. House of Representatives Committee Report 96-617, which accompanied the Colorado Wilderness Act (U.S. Public Law 96-560) and interpreted and clarified grazing provisions in the Wilderness Act. The Colorado Wilderness Act directed that grazing management in Colorado wildernesses be guided by report 96-617, which stressed its interpretation that the Wilderness Act provided for continuation of existing grazing use; the maintenance and construction of supporting facilities, including "fences, line cabins, water wells and lines, and stock tanks"; and temporary use of motorized equipment to repair facilities and for emergency purposes.

The grazing provisions in House Report 96-617 were extended to apply to other national forest wilderness areas. Several subsequent laws, including the Arizona Wilderness Act (U.S. Public Law 98-406), Utah Wilderness Act (U.S. Public Law 98-428), Wyoming Wilderness Act (U.S. Public Law 98-550), Nebraska Wilderness Act (U.S. Public Law 99-504), and El Malpais National Monument Act (U.S. Public Law 100-225) in New Mexico, also indirectly refer to the guidelines in House Report 96-617. The El Malpais National Monument Act grazing provisions are important because they apply to a BLM wilderness and not a national forest wilderness. In addition, several of these laws provide for administrative review of existing grazing policies to ensure that they are consistent with House Report 96-617. The New Mexico Wilderness Act (U.S. Public Law 96-550) addressed grazing directly in authorizing additional fencing as provided in the grazing allotment management plan for livestock in the Cruces Basin Wilderness.

Although these legislative and committee directions were only directed at BLM wilderness in the case of the El Malpais National Monument Act (U.S. Public Law 100-225), they reflect a collective concern for livestock grazing and management options in wilderness yet to be designated on BLM lands (McClaren 1990). This direction was clearly expressed in the California Desert Protection Act in 1994 (U.S. Public Law 103-433), which designated sixty-nine new wilderness areas in the arid southeastern region of California (J. Watson and Brink 1996). This act ensured not only that grazing in these newly designated wilderness areas could continue, but that established grazing management facilities (fences, watering sites) would be accessible by vehicles for management purposes (by permit holders and agency personnel), and also provided access to watering facilities for wildlife, such as guzzlers.

Livestock grazing is a well-established use of wilderness on national forest and BLM lands, with rights embodied in the Wilderness Act and supported (some would say expanded) in subsequent legislation. These rights must be respected, but we share the concern of other wilderness supporters over the impacts of excessive grazing on biodiversity and naturalness of wilderness ecosystems, and on the aesthetic experience of visitors. Mitigation of these impacts involves good grazing management; grazing levels that leave necessary forage for wildlife; better monitoring; and wider recognition by wilderness managers, grazers, and visitors of potential ecological changes triggered by grazing before they become obvious physical impacts. Murray (1997) described ecological changes and impacts of grazing in high-elevation meadows, the underlying idea being that the more we understand about such impacts, the more vigilant we can be about imminent impacts, and the more likely they will be minimized. *We need more educational information about grazing levels and potential impacts, so wilderness visitors better understand the issues and can work with wilderness managers and the livestock industry to keep grazing within acceptable limits.* Fragile sites, such as high elevations and arid lands, are especially vulnerable and require vigilance by all interests.

Wilderness Recreational Use

Much more is known about recreational use than the other wilderness uses. As an important use, recreation is the focus of a large part of all wilderness management and thus warrants a detailed discussion. However, *wilderness is not just a primitive-recreation area. The Wilderness Act makes it clear that wilderness is established for many purposes, with "primitive and unconfined recreation" only one purpose* (see Chapter 4).

The discussion of wilderness recreational use covers three main points: (1) methods for estimating wilderness recreational use, (2) use characteristics, and (3) distribution of use. *An understanding of the amount, character, and distribution of recreational use is essential to wilderness management because such use is the cause of many impacts, the source of many wilderness values and potential funding.* The amount of wilderness use an area receives is one basis for allocating

Fig. 14.3. A campsite near Bald Knob Lake in the Absaroka-Beartooth Wilderness, Montana, bears out data that campers prefer to be near lakes and streams. Data on the amount, character, and distribution of recreational use are important information for wilderness managers. Photo courtesy U.S. Forest Service.

agency funds among areas for wilderness management, and, as use increases faster than agency funding, user fees may be a larger source of funding support for wilderness management (Mackey 1998; Alan Watson 1998). Therefore, some review of the methods used to measure wilderness use, and their advantages and shortcomings, is necessary background to understand how and where they might be used to determine the extent of use in a wilderness area.

Methods for Estimating Wilderness Recreational Use

Wilderness recreational use is difficult to measure. The typical wilderness has many remote access points with some larger ones, such as the BWCAW and the Selway-Bitterroot Wilderness, having seventy or eighty entries. Even some smaller wilderness areas have more than ten access points. Many access sites are distant from ranger stations or contact points and are, thus, difficult and expensive to monitor.

Compared to use of developed sites, wilderness recreational use is light and variable, and often of low density. This makes it prohibitively expensive to observe all entry points—some would have no use at all to observe on some days. Use is dispersed over such a wide area that it is nearly impossible to make any sort of direct head count, as can be done in developed recreation areas. Therefore, a variety of direct and indirect methods for counting and estimating wilderness use have been devised, including observations, electronic counters, automatic cameras, trail registers, mandatory permits, remote sensing, and guessing. These different methodologies each have their strengths and weaknesses and should be used with some knowledge of the sampling requirements and reliability in estimating total use, information provided in more detail in sources such as the following (Hollenhorst et al. 1992; Hornback and Eagles 1999; Alan Watson, Cole et al. 2000; Yuan et al. 1995).

Direct Counting Methods

Four methods of direct counting are generally used to help estimate use and monitor use patterns: (1) direct observation, (2) field interviews, (3) voluntary self-registration at trailheads, and (4) mandatory permits. Direct observation and field interviews are generally used to determine registration compliance to adjust self-registered use data as a basis to estimate total use, or to obtain more detailed information about users, such as use patterns or user characteristics. Self-registration and mandatory permits are often used because of their greater accuracy in estimating use and lower costs than other direct counting methods.

1. *Direct observation*—Systematically observing visitors or interviewing them (on-site interviews or requesting names and addresses for follow-up mail surveys) at a sample of trailheads on sample days has been used for decades, but the highly variable use pattern makes reliable sampling difficult and costly. Large samples are difficult to obtain. For example, the 74,000-acre Mission Mountains Wilderness in Montana has nineteen access points. If the main use season is about 100 days, this results in 1,900 date/place combinations from which to sample. One person working full-time could sample about 4 percent of these, and then only for about six or seven hours per day because of travel time. Stratified sampling that would concentrate on more often used times and places could increase the efficiency of sampling but would require some reasonably accurate information on use patterns to plan the stratification of the sample.

2. *Field interviews*—Formal or informal interviews may be used to help managers understand user interests, preferences, and needs. Formal interviews are used in scientific studies to estimate visitor use and visitor characteristics, such as when gathering information in a study of visitor-to-visitor encounters and conflicts, or to gather names and addresses for a follow-up mail survey. Sometimes, managers informally interview visitors to get their impressions, learn how many other visitors they have seen, and gather information to help them understand wilderness visitors.

3. *Voluntary self-registration*—Many estimates of wilderness use are based on voluntary self-registration at trail registers. Party size, method of travel, date of entry, length of stay, some data on destination (or itinerary), and visitor residence are usually obtained. The problem, of course, is that some or even many visitors do not register (Lucas 1983; Lucas and Kovalicky 1981; Petersen 1985). As these studies document, some kinds of visitors—especially equestrians, hunters, people making very short visits, and lone individuals—are less

likely to register than others. Thus, the resulting raw registration data not only underestimate use but may also provide biased estimates of its composition. Efforts have been made to develop systems for adjusting trail-register data with adjustment factors applied to raw data from the trail-register cards to compensate for nonregistration. It is necessary to observe a sample of registration behavior to develop the adjustment factors because studies indicate that registration rates may be somewhat to highly variable among users, between wildernesses, and over time. It is also essential to periodically check actual registration rates and recalibrate the adjustment factors employed as a basis for use estimates. Currently, use estimates based on trail registers have a large, but usually unknown, margin of error. If registration rates could be raised, trail registers would be a better basis for use estimates. Low registration rates require large expansion factors and yield undependable estimates. Petersen's 1985 study found that locating trail registers one to three miles from the trailhead and providing a new sign that explained briefly the importance of registering helped raise registration rates. Of course, such a location would miss a casual day hiker who did not travel far into that wilderness.

4. *Mandatory wilderness permits—The most accurate wilderness use data probably come from mandatory visitor-permit systems* (Hendee and Lucas 1973; Hollenhorst et al. 1992; Yuan et al. 1995). In 1980, about 70 percent of NPS wildernesses and about 50 percent of USFS wildernesses required visitors to obtain permits, a practice also common in Canadian wilderness-type areas (Washburne and Cole 1983). Today, in almost all cases, permits must be obtained from the managing agency, but in some wildernesses, permits are issued by cooperators, such as at stores by resort employees, or are self-issued by visitors at trailheads. Permits provide all the information obtained from trail registers, in addition to greater detail on planned routes of travel and the chance for managers to encourage environmentally and socially sensitive behavior, plus answer user questions if the permits are issued in person. An account in 2000 of the required, self-issued permit system in Oregon's Eagle Cap Wilderness described its primary purpose as determining the amount of visitor use in particular areas, but when that purpose was no longer necessary the system was continued because of its observed value to educate visitors about wilderness conduct and stewardship (Carlson 2000).

Some types of visitors (day users, local users, visitors from nearby resorts) are less likely to get permits, just as some visitors do not register at trail registers. Permit compliance varies, although it is usually higher than for trail self-registers and ranges from two-thirds to more than 80 percent of users. Compliance is high in most national parks, in the BWCAW, and in some other areas where the permit system has been in effect for more than twenty years and visitor awareness is high, but compliance might be lower in areas with newer permit requirements. Availability of permits at times and places convenient for visitors will increase compliance, and public information programs and enforcement are necessary. *We strongly recommend an adaptive and comprehensive approach to develop the best permit system for a given area and especially its application only when and where needed as the minimum approach to managing wilderness use.*

Commercial outfitters and guides and nonprofit organizations that bring paying clients to wilderness areas are required to apply for a special-use permit and pay a fee based on numbers of paying clients and revenues from them. This provides a good measure for commercial use of wilderness, although it is a small proportion of total use. Institutional outfitters (e.g., schools, universities, colleges, and churches) usually are not required to obtain permits, yet they may account for substantial use, and there are pressures for policy changes to require them to register their use in some way and pay fees (Mackey 1998).

Indirect Counting Methods

Two general methods of indirect counting are sometimes used to estimate use and monitor use patterns: (1) automatic photography and (2) electronic trail-traffic counters. A third approach, making educated guesses and estimates about visitor use, is more commonly used than is appropriate and produces precise-sounding numbers that can be deceiving, if warnings about the limitations of educated guessing are not made explicit. Other indirect methods, like pressure-plate counters, have been tried on trails but are not widely used due to inaccuracies and maintenance problems.

Table 14.1. Units of Measure for Wilderness Recreational Use

Unit	Definition	Use and Comments
Visitor	1 person who makes 1 or more visits, usually during 1 year	Rarely used (sometimes inaccurately used as a synonym for visit)
Visit (or occasion)	The entry of 1 person into an area, regardless of length of stay (repeat visits counted)	NPS; USFS before late 1960s
Visitor hour	1 person present for 1 hour	NPS occasionally
Visitor day (recreation visitor day or RVD)	1 person present for 12 hours, or equivalent (2 for 6 hours, etc.)	USFS standard unit since 1965
Person day	1 person for 1 calendar day (7 hours or more; 5–7 hours = 0.75; 3 to 5 hours = 0.5; 15 minutes to 3 hours = 0.25; less than 15 minutes = 0)	USFS standard before 1965 (noncomparable to visitor days, but 1.5 visitor days per person day is a rough conversion for wilderness use)
Recreation day	1 person present for any part of a calendar day	Rarely used
People at one time (PAOT)	Total number of people present at one time, usually a day	Some USFS reports, some research
Overnight stay	1 person passing 1 night within an area (could also be called a visitor night)	NPS standard unit; for relation to USFS visitor days, see text
Group (or party) visits	The entry of 1 group (party) into an area	Some research
Group (or party) visitor	As above for 1 group (party)	Rarely used

1. *Automatic photography*—Automatic photography has evolved rapidly in the last thirty years. For example, automatic movie cameras, set to expose one frame at preset intervals, have been used by the National Park Service (Marnell 1977) to estimate use of several wild rivers. Some of these cameras photograph a calendar clock in a corner of each frame to record the date and time of each observation. More recent photographic systems used in some wildernesses employ a digital, slide, or video camera automatically triggered by an electronic traffic counter to film a few frames or images of the visitors. Group size, direction, and method of travel are evident, and day users can be distinguished from overnight backpackers. For protection of visitors' privacy, no identification of individuals is made, and only public areas through which visitors pass are filmed, such as trails, but not campsites or swimming areas. This system has worked well for managers and researchers, with similar results to direct observation but at a much lower cost (Hollenhorst et al. 1992; Lucas and Kovalicky 1981; Petersen 1985; Yuan et al. 1995).

2. *Electronic trail-traffic counters*—The use of automatic electronic trail-traffic counters has been increasing because of newer technology and improved operation. O'Rourke (1994) reports that after rapid advances in the electronics industry, the best type of trail-traffic counters for use by the USFS is active infrared systems because of their accuracy and self-correcting circuitry for some false counting situations. Active infrared systems emit an infrared beam onto a receiver or reflector and register a count whenever there is an interruption by something moving through the beam. Another type of operating unit is recommended for specialized situations, but reduced accuracy is to be expected because of how the system operates. Passive infrared systems emit an infrared beam that

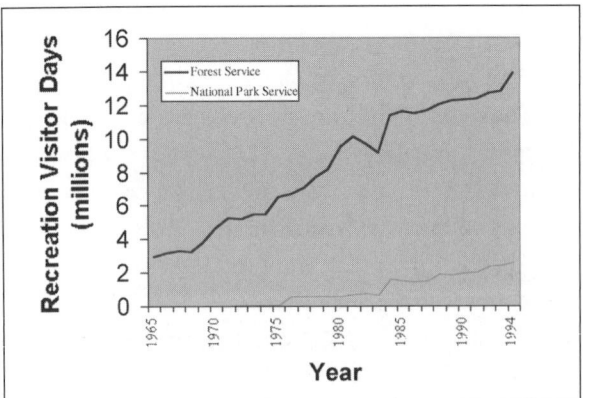

Fig. 14.4. Total recreation visitor days of use on national forest and in national park wilderness areas (Cole 1996).

directly reflects off a passerby and registers a count whenever the reflected light is detected. However, due to lower accuracy with this technology, they should be reserved for areas that require lightweight units that can be quickly installed and removed. At best, automatic electronic trail-traffic counters indicate the number of large moving objects passing during a time interval or date. They cannot identify whether the objects were hikers, packhorses, elk, or cows, or how they clustered into parties or groups. Neither direction (entry or exit) nor route of travel can be obtained from the counters. Nevertheless, automatic counters are being used in a number of wildernesses and provide useful information, especially to estimate trends in use and to estimate compliance with voluntary self-registration at trailheads and mandatory permits (Hollenhorst et al. 1992; O'Rourke 1994; Yuan et al. 1995).

3. *Educated guesses and estimates*—Wilderness use can be estimated by means of an educated guess, hopefully based on previous years' reports and an assumed rate of change. Estimates might be modified according to various observations—how many cars were parked at a particular access this year compared to previous years (but possibly at different times of the season and on different days of the week), changes in the number of guests reported by outfitters, wilderness rangers' impressions, ranger observation, casual field interviews, unusual weather patterns, and so on. Some wildernesses show wide fluctuations in use from year to year. Often this comes from the use of guessing (more educated or less), which may be more sensitive to changes in who does the guessing, or in whose observations are used from year to year, than actual visitation. Educated guessing may be the most commonly used method for estimating wilderness use, especially for smaller or less heavily used areas, or to adjust estimates based on partial data from one of the other use measurement methods. *We recommend that educated guessing to estimate wilderness use be based on as much data and experience as possible.*

Individual wildernesses use a wide variety of techniques to estimate use, and most areas report using a combination of techniques (Washburne and Cole 1983). Accuracy varies from good to very poor. Unfortunately, the methods used and the expected accuracy levels are not reported in annual use estimates. Adjustments for noncompliance on permits or trail registers, if made at all, are too often essentially guesswork based on unsystematic spot checks.

Thus, estimates of wilderness recreational use are based on a mixture of methods, and the resulting level of accuracy is uncertain and probably only fair. Whether errors compensate more than they accumulate is unknown. Comparisons of visitor use over long time periods are particularly unreliable. Older estimates were mainly guesses, and as more accurate methods were adopted, large changes in reported use often developed. Therefore, historical records of trends in use must be treated cautiously.

Certainly, wilderness recreation use data need not be perfect to be useful, but any improvement adds to the management value of the information. The need for accuracy in use data is greater in wilderness areas for which new management plans are being developed or in areas that are monitoring to evaluate the achievement of planning and management standards. Various ways of utilizing wilderness recreational use data are discussed in Chapter 15 on visitor management.

Units of Measurement

Several units of measurement are used to report recreational use, and the units vary between the Forest Service and NPS. The BLM and FWS do not routinely collect visitor data, which leads to further confusion in comparing data between agencies. The most common units of measure are listed in Table 14.1. Unfortunately, the standard units used by the NPS, the overnight stay, and the USFS, recreation visitor day (RVD; twelve-hour period), are not directly comparable. For an individual, overnight stays could be converted approximately into RVDs by multiplying by two and one-half on the assumption that a one-night stay usually is made up of a large part of the first day and the last day; for example, 8 AM to 8 PM, 8 PM to 8 AM, and 8 AM to noon.

This array of units for quantifying and expressing recreation use is confusing. Which of the units is best? As is so often the case, the answer is "that depends." Specifically, it depends on the purpose for which the data are to be used. Often, several different units are relevant to a particular purpose.

For indicating potential impact on camping areas and camper congestion, the overnight stay or the group night provides the most relevant information. With respect to solitude, the visitor day, which reflects how many people are present over time, is generally useful, and people at one time (PAOT) is particularly appropriate. For total use, the number of visitors or RVDs is most useful. Total visits are a partial measure of visitors, are much easier to obtain, and give a general idea of total use tied to certain types of managerial workloads, such as making visitor contacts and issuing permits. However, the great variation in the definitions of use labeled "visits" makes this a difficult unit of measure to interpret. A brief visit of an hour or so counts the same as a two-week stay. Visits can be used with visitor days, if both are reported, to gain an idea of length of stay.

Different wilderness management agencies do not use comparable terms to report use. The USFS is the only agency to report recreational use separately for each wilderness area, and it has used the RVD as the unit of measure since 1965. The NPS reports backcountry camping use, which is defined as camping in minimally developed areas not reached by roads, and has used the overnight stay as a unit of measure since 1971. Thus, aggregate NPS "backcountry overnight stay" includes stays in wildernesses and are the best available information used to represent NPS wilderness camping use, but it does not include wilderness day use, which is often substantial in national parks. The other two federal agencies that manage wilderness areas, the BLM and FWS, do not routinely collect use data, but they host a small proportion of total use in the NWPS.

Because wilderness recreation management often competes with other kinds of outdoor recreation for public funding, and use is one indicator of needed management support, more complete and accurate reports of wilderness use are needed. We recommend that at least visits and visitor days be reported annually for all wildernesses in the NWPS. Overnight stay would be a useful supplemental unit, although calling it a visitor night would clarify its meaning. We recommend use of one or more of the use-measurement methods and less reliance on educated guessing to report use estimates. Sometimes educated guessing to estimate recreational use is better than nothing, but we recommend that it be supported by as much actual data and manager experience as possible.

Trends in Recreational Use

In the following discussion, we summarize and interpret recreation use trends from numerous user studies conducted over the last forty years (Cole 1996; Loomis et al. 1999; Lucas 1980; Lucas and Stankey 1988; Roggenbuck and Lucas 1987; Roggenbuck and Watson 1989).

Wilderness recreation use has increased greatly since 1964. Although use was widely reported to have leveled off and even declined in some areas by 1986, it grew substantially in the 1990s despite predictions that it would not (Cordell, Bergstrom, Hartmann et al. 1990; Cordell and Teasley 1998). Cole (1996) points out that *these "level off or decline" visitor use predictions were incorrect and were detrimental because of their unsupported implications that it might be possible to ease management regulations and promote more wilderness use, and that recreational demand was no longer an appropriate argument for designating additional wilderness areas.*

Although many speculated about the alleged decline in recreational use of wilderness in the 1980s, it turned out to be a temporary setback and applied only in some areas. Seventy-five percent of the reported decline in NPS backcountry use from 1976 through 1989 occurred in five parks: Sequoia–Kings Canyon, Yosemite, Olympic, Shenandoah, and Great Smoky Mountains (Cole 1996). *Recreational use of wilderness continues to grow and is expected to continue. The vast majority of all wilderness RVDs are in USFS areas and secondarily in NPS areas.* The visitation on FWS and BLM lands is small in comparison and has little effect on total NWPS use, but it is expected to grow.

Overall use of the NWPS was estimated by Cole (1996) as about 17 million RVDs in 1994. This was a sixfold increase between 1965 and 1994, an average annual increase of 6.3 percent. He reported that the majority of this increase was due to additional units and acreage being designated as part of the NWPS, which grew from the original fifty-four USFS wildernesses and more than 9 million acres in 1964 to 630 areas and more than 103 million acres by the mid-1990s. Cole further reported that *recreational use in the original fifty-four USFS wilderness areas increased 86 percent from about 3 million RVDs in 1965 to 5.5 million RVDs in 1994; the remaining increase of 11.5 million RVDs was attributed to the addition of 576 wilderness areas designated as part of the NWPS since 1964* (Cole 1996).

Estimates of wilderness use derived from a National Visitor Use Monitoring Project were 10.7

million trips and 16.3 million on-site days (i.e., visits) in 2002 (Bowker et al. 2007). According to Cole (1996) these estimates of wilderness use are difficult to compare with those by other researchers using agency reports and RVD units. The apparently higher numerical estimate by Cole (1996) of 17 million visitor days in 1994 is partially explained by the fact that RVDs are twelve-hour time periods and on-site days are visits from one to twenty-four-hours in length. Thus, the 2002 estimate of use by Bowker et al. (2007) is an actual increase over 1994 even though conversion from one unit of measure to another is, at best, a crude estimation.

The intensity of wilderness use (RVDs per acre) is also an important issue for managers. If the entire NWPS is considered, wilderness use intensity increased from 1965 (0.24 RVD per acre) to 1979 (0.40 RVD per acre) and then declined dramatically to 0.18 RVD per acre (Cole 1996). However, this is misleading because it includes the sparsely used wilderness in Alaska. *If the 56 million acres of designated wilderness in Alaska is excluded from the analysis, then wilderness use intensity grew sharply from 1965 to 1979 and then remained steady at 0.40 visitor day per acre per year until 1994* (Cole 1996).

This type of use-intensity analysis is valuable for assessing concentrations of use across specific areas. Because it can be misleading when the distribution of the use is not known and because use is usually highly concentrated in certain popular locations, it may appear acceptable as an average (RVDs per acre) for an entire wilderness area. For example, across the NWPS one would expect wilderness management needs (and needs for funding) to be greatest for areas with the highest use per acre. *Now that growth in acreage of the NWPS is slowing because most large areas are already designated and use is projected to continue to increase, it is reasonable to expect that use intensities will increase as well—and so will the need for funding of wilderness management.*

Use Characteristics

In the following discussion, we summarize and interpret recreation characteristics from numerous user studies conducted over the last forty years (Cole, Watson, and Roggenbuck 1995; Lucas and Stankey 1988; Roggenbuck and Lucas 1987; Roggenbuck and Watson 1989; Washburne and Cole 1983; Alan Watson, Cole, Friese et al. 1999).

Length of Stay

Most wilderness visits are for short stays, day use, or only one- to two-night stays. Many small or medium-sized wildernesses are predominantly day-use areas. Studies report that the amount of visits that are day use ranges from 7 to 68 percent, but surveys are available for only a few areas. Among the areas for which visitor surveys are available, only the Bob Marshall Wilderness, the Weminuche and Rawah in Colorado, and some wildernesses in Wyoming have little day use. These areas are large, with the wilderness boundary miles beyond the trailhead in many places, and this limits day use. Most large wilderness areas are also far from large population centers, making them more often destinations for vacations and trips of several days, rather than weekend or day hiking trips. The limited information on trends in use suggests that the percentage of day users may be increasing in some areas.

Long wilderness trips of a week or more account for less than one-tenth of all visits, even in the very large areas such as the BWCAW, the Bob Marshall Wilderness, and the Selway-Bitterroot Wilderness. Visits to many areas, especially small to medium-sized wildernesses, averaged two to three days, with wildernesses in the eastern and western United States having similar average lengths of stay (two to three days) and percentage day use (25 to 70 percent). Lengths of stay in most areas having had more than one survey have become a little shorter in more recent years. *Thus, wilderness use trends toward shorter wilderness trips and more day use seem consistent with other sociodemographic trends, such as individuals and families having shorter blocks of leisure time.*

Fig. 14.5. Riders and pack stock travel through a mountain meadow; such trips are a traditional way of visiting a few large western wilderness areas. Hiking is a much more common method of travel. Photo courtesy U.S. Forest Service.

Group Size

Parties of wilderness visitors are generally small; from one-half to three-fourths of the parties at all areas for which data have been collected range from two to four persons. Few persons visit wilderness alone (less than 10 percent of all visitor groups), and parties of more than ten people account for only about 5 percent of all groups.

Party size is declining in areas where long-term studies have been conducted. More wilderness areas have restricted party size, and even organized groups have become smaller because managers and organization leaders have become concerned about the impact of large groups on the wilderness environment and on other visitors.

Method of Wilderness Travel

The most common method of wilderness travel in almost all areas is, with few exceptions, hiking. Most visits to the BWCAW are made in paddled canoes (in 1986, 89 percent of all overnight groups). Some BWCAW visitors use limited horsepower, outboard-motor-powered canoes or boats in specifically designated zones, and a few hike, snowshoe, or ski.

The BWCAW is one of the few areas in the lower forty-eight states where mechanized travel is permitted within a wilderness. Other exceptions include a few large wilderness areas in the West that have wilderness landing fields or airstrips that predate the Wilderness Act or wilderness designation. Such airstrips have remained open to airplanes, such as in the Great Bear Wilderness in Montana and the Selway-Bitterroot Wilderness and Frank Church–River of No Return Wilderness in Montana and Idaho. In these large areas, airstrips are often the starting point of outfitted horseback or float trips. The Frank Church–River of No Return Wilderness includes twelve airstrips on federal land that have about 5,500 total aircraft landings per year, and the USFS is very limited in its ability to close the airstrips based on the statutory history of the wilderness's designation and legislative intent and language, but it is allowed to restrict use levels (Meyer 1999). The BWCAW has a policy not found elsewhere that not only prohibits air access but bans low-altitude flights as a result of Executive Order 10092, signed by President Truman in 1949.

In Alaska, floatplanes and "flight seeing" remain an important method of physical and visual access to vast wilderness areas under ANILCA legislation. Similarly, places like Glacier Bay Wilderness are visited by large cruise ships whose passengers view tidal glaciers and passively experience the rugged mountain ranges in some coastal wilderness areas of Alaska.

Horse use is important in some large western wilderness areas but still accounts for a minority of total use in the NWPS today. Managers of seven wilderness areas in the western United States estimated that more than half of the visitors to these areas rode horses in 1978 (Washburne and Cole 1983), and later studies reported that half of the visitors to the Bob Marshall Wilderness rode horses. Horse use is common in some southwestern wilderness areas, such as the Superstition Wilderness, where most of the visitors riding horses are day users. Horses and pack stock are few or uncommon in wildernesses in the eastern United States and are a small part of the use in most western states.

Recreational Activities

Wilderness visitors participate in a variety of recreational activities in addition to hiking. Surveys of visitors indicate that, on the average, respondents participated in *three major activities: fishing, photography, and nature study.* Studies suggest that in western areas, about one-half or more of the visitors fish, but for many the fishing is somewhat incidental rather than central to their experience (Carpenter and Bowhis 1976; Hendee, Clark et al. 1977). We suggest that the increased wilderness use since these studies were conducted includes a declining percentage of anglers compared to other recreational activities.

Photography and nature study are also major activities in most places, and swimming is common in

Fig. 14.6. A commercial cruise ship enables passengers to visit Glacier Bay Wilderness, a National Park Service–managed area in Alaska. Photo courtesy Chad P. Dawson.

Fig. 14.7. Flight seeing over Denali Wilderness, a National Park Service–managed wilderness, with a commercial outfitter. Federal legislation allows such business in some wilderness area airspace. Photo courtesy Chad P. Dawson.

many southern areas. These are all low-impact, non-consumptive uses that are compatible with wilderness values. Hunting varies from a minor to fairly common use in national forest wildernesses, but it is generally prohibited in national park wildernesses. Mountain climbing is an infrequent use in most areas but common in a few areas, as are other activities that involve challenge and adventure experiences (Ewert and Hollenhorst 1997). The frequency of these activities varies by area, but overall, the percentage of visitors who hunt or fish has declined in recent years while mountain climbers have increased (Cole, Watson, and Roggenbuck 1995).

Season and Timing of Use

Summer is the highest use season in virtually all areas, and that trend continues. A few large western U.S. areas have considerable fall big-game hunting, but in two areas studied (the Bob Marshall and the Selway-Bitterroot), summer visitors still substantially outnumber fall hunters. Many areas in the South, Southwest, and California deserts have substantial winter and spring use (Washburne and Cole 1983). In the North, and at higher elevations in the mountains, winter use is light but much more common than in previous decades, and is growing. Wilderness experience programs aimed at personal growth, education, or therapy operate year-round to achieve program objectives and to take advantage of client demand, weather conditions, and opportunities for solitude.

Many areas experience weekend peaks in use, especially the smaller, more accessible wildernesses near urban population centers, such as the San Gorgonio and San Jacinto Wildernesses near Los Angeles, Pusch Ridge Wilderness near Tucson, Superstition Mountains Wilderness near Phoenix, Alpine Lakes Wilderness near Seattle, and others. Some small wilderness areas in the East do not have extreme weekend peaks in use but do report the majority of annual use occurring during weekends.

Type of Group

In almost all wilderness areas, the majority of the use takes place in small groups of family and close friends. Wilderness use by organization-sponsored groups accounts for a small percentage of use, with one study suggesting about 10 percent and another study indicating that such use may be declining. This refers to organizations such as conservation groups, outdoor recreation clubs, youth groups such as Boy and Girl Scouts, youth camps, and church groups.

Several studies indicate an increase in use by wilderness experience programs aimed at personal growth, education, or therapy and healing (Dawson, Tangen-Foster et al. 1998; Friese et al. 1998, 1999; Russell et al. 2000). One nationwide study of wilderness managers found that two-thirds reported increased use by wilderness experience programs in their areas, with only 1 percent reporting a decline in use by wilderness experience programs (Gager et al. 1998).

Fig. 14.8. Increasingly, visitors use wilderness in winter, such as ski tourers in the Selway-Bitterroot Wilderness, Idaho. Photo courtesy Richard Walker.

Visitor Residence

Some nationally and internationally known wilderness areas draw visitors from all over the nation and other countries. Some internationally attractive wildernesses are the Bob Marshall and Selway-Bitterroot in Montana; John Muir Wilderness, Yosemite, and Sequoia–Kings Canyon National Parks in California; the Great Smoky Mountains in the East; and the BWCAW in Minnesota. Wilderness and backcountry visitors come from many states and foreign countries, but even the majority of visitors to these wildernesses are from the state or region near the wilderness. For example, almost 93 percent of the visits to Sequoia–Kings Canyon backcountry in 1986 were made by Californians, and 60 percent of the visits to the Bob Marshall Wilderness in Montana were made by state residents. The discretionary time and income necessary to travel to a distant wilderness area may limit some wilderness visitation. Numerous studies indicate that, although wilderness is a national resource, *wilderness recreational use typically comes from the home state and region or nearby urban population centers.* This may reflect the constraints of time, expense, and distance as well as residents' preferences for local opportunities. Some implications are that *wilderness is needed fairly close to population centers with nationwide distribution to make wilderness readily accessible to the nation's population.*

Distribution of Use

The geographical distribution of wilderness recreational use is very uneven—there are many people in a few places and only a few in many other wilderness locations. This unevenness of use is a pattern among wildernesses as well as within individual wildernesses. For a variety of reasons, *some wilderness areas, some trailheads, and some popular destinations attract more use than others.* This geographical concentration of use is compounded by the tendency for use to further concentrate during the summer and high-use weekends.

The 5,550-acre Great Gulf Wilderness, located in the White Mountain National Forest of New Hampshire and Maine, reported such an uneven distribution of wilderness recreational use in 1999. Visitation increased through the early season, reaching peaks on July 4, in early August, and on Labor Day weekends, and then decreasing in mid-September. Recreational use on weekends is noticeably greater than that on weekdays and the highest weekend use occurs on holiday weekends like July 4 and Labor Day (Dawson, Simon et al. 2001).

These concentrations of wilderness use pose management challenges and opportunities. Wilderness areas vary in their capacity to absorb the social and environmental impacts of use, and some of the most popular locations (such as high alpine lakes) may be sensitive to physical impacts from use and intrusions on solitude. The tendency of use to concentrate in time and space enables management to focus efforts and resources on these locations and time periods, whether by redirecting, dispersing, or limiting use.

It is clear that *an even distribution—the same amount of use on every acre, or of every mile of the trail system, and of every campsite—is neither possible nor desirable.* It is impossible—barring total regimentation—for several reasons. Trips vary in length; thus, the interior is not as heavily visited as the periphery. In addition, most use is on trails or water routes, which typically branch and diverge, so the main trunks inevitably carry more people than the branches. Furthermore, evenly dispersed use is undesirable for two reasons. *First*, different parts of a wilderness vary in their response to visitor use and in their capacity to sustain use while maintaining the existing environmental conditions. *Second*, with evenly distributed use, persons who really prize solitude and are willing to travel or hike a long way to find it would be frustrated. At the same time, persons not strongly motivated to seek solitude would encounter fewer visitors than they would willingly accept and, in a sense, the greater solitude would not be appreciated by them. The ROS and WOS are management

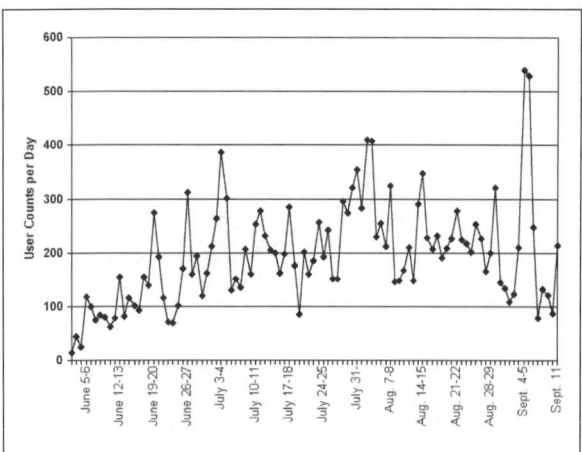

Fig. 14.9. Temporal distribution of recreational use on weekends (labeled above) and weekdays from June 2 through September 11, 1999, in the Great Gulf Wilderness, New Hampshire (Dawson, Simon et al. 2001).

concepts that recognize the desirability of diversity in visitor experiences, and the LAC system (Chapter 9) provides a planning mechanism to provide for it. However, no carrying-capacity formula approach or planning process can exactly determine an optimum distribution of use within a wilderness.

We suspect that in some areas present use is too unevenly distributed and that some redistribution to reduce the extremes is desirable. Badly overused wilderness recreation sites are a common problem. An initiative by the USFS, the so-called wilderness recreation strategy, seeks to allow higher use on some sites than originally planned, allegedly to keep from overusing pristine sites, but this strategy is controversial (Oye 2001; Worf 2001). The issue of optimum-use distributions, and how to encourage them, is considered in many sections of this book—in discussions of principles of management (Chapter 7), planning (Chapter 8), carrying capacity (Chapter 9), impacts of recreation activities and use (Chapter 15), and visitor use management (Chapter 16).

Distribution of Use Between Areas

Reported visitor use shows substantial variation between individual wildernesses in the national parks and national forests. For example, backcountry overnight stays exceeded 105,000 in Yosemite and fewer than 14,000 in Everglades National Park in 1986; visitor days exceeded 450,000 in the John Muir Wilderness (CA) and 17,000 in the Dolly Sods Wilderness (WV) areas of the national forests in 1986. The reasons for such different drawing power are poorly understood, but the most heavily used national forest and national park wildernesses are almost all located relatively close to large population concentrations. California, Minnesota, southern Appalachian, and New England wildernesses receive the most intense recreational-use pressure, but other wildernesses in these same regions are lightly used.

Location near many people is associated with heavy use, but a reputation as an attractive area is also necessary if heavy use is to actually occur. Publicity in national magazines and guidebooks has contributed to the popularity of some wildernesses and some particular trails in specific wildernesses. Some people have speculated that merely designating an area as wilderness attracts extra use, that is, assigning an area a name and identifying it as wilderness may stimulate use. This is often called the "designation effect." However, there is scant evidence that designation results in increased use.

By themselves, visitor totals indicate little about pressure on the resource. For one thing, areas vary in size. For example, although total use in the Collegiate Peaks Wilderness in Colorado and the Bob Marshall Wilderness in Montana was about the same in 1986, the Bob Marshall is about six times as large as the Collegiate Peaks. To aid in comparing areas and use pressure, acreage and use figures can be expressed in terms of visitor days per acre. However, such comparisons do not fully reveal the degree of visitor congestion and the resulting pressure on the soil, vegetation, water, and wildlife because they ignore the length of the use season. For wilderness in the northern Rocky Mountains or Cascade Range of the Pacific Northwest, the main use season might be only two or three month long (because of trails blocked by snow or streams too high to ford). In milder regions, the use season is much longer. Therefore, adding a length-of-season adjustment to annual-use-per-acre figures would show even higher use levels per acre for areas with primarily summer use. For instance, the more than 1 million visitor days of use in 1986 in the BWCAW were not distributed evenly over an entire year—about 90 percent of the visitor days took place from late May through September.

Moreover, *the proportion of usable or effective acreage is not the same in all wildernesses.* Although all wilderness acreage is available in the sense that it provides at least a backdrop and space for isolation (and, of course, is a vital part of the natural ecosystem), only a portion is used directly by visitors. The amount of land available for use is affected by steepness of slope; type of vegetation; and the extent of area in lakes, streams, and wet, boggy soils. In the San Jacinto Wilderness in California, for example, it was estimated that of the 690 acres in one travel zone (management area), approximately 400 acres were unavailable for use because of excessive steepness, type of vegetation, and the presence of excessively wet meadows. Other areas are even more rugged, and the proportion of the usable area might be much smaller. For example, we would estimate that considerably less than 10 percent of the area in the Mission Mountains Wilderness is potentially usable because of steepness.

Usable acreage is also influenced by the degree of access development, number of trails, and travel routes.

Most wilderness use is restricted to travel on the existing trail systems, areas directly adjacent to trails, and overnight camping areas. Some areas have an extensive trail network whereas other areas have only a few trails. Therefore, *for more detailed indices of accessibility and use of an area, the entry points per thousand acres and miles of trail per thousand acres can be calculated for comparisons between different wilderness areas.* Miles of trail per thousand acres and number of entry points per thousand acres relate more closely to use capacity than does gross area, and they vary considerably among wilderness areas. For example, the Great Gulf Wilderness (NH) has about sixty times as many entry points per thousand acres, and more than eight times as dense a trail network (miles of trail per thousand acres), as the Teton Wilderness (WY).

Distribution of Use Within Areas

Use distribution within any particular wilderness can vary as much as use between wilderness areas. For instance, in the BWCAW in 1986, about 61 percent of the user groups entered through only ten of the area's eighty-seven entry points; in fact, two entry points accounted for almost one-fourth of all groups entering the area. A study in Yosemite National Park revealed that 4 percent of trailheads received 68 percent of total use (van Wagtendonk 1981). The 1999 use pattern in the Great Gulf Wilderness in New Hampshire is representative of the uneven spatial distribution of use in many wildernesses. The use data for the Great Gulf Wilderness indicate a pattern of uneven hiking use on some trail segments, particularly those associated with the Appalachian Trail and the summits of the Presidential Range (Dawson, Simon et al. 2001).

Heavy concentration of visitor use along only a few trail miles is common. For example, in the entire Spanish Peaks (CO) trail system the 10 percent of trail miles with the heaviest use accounted for about one-half of all visitor miles in 1970 (Lucas 1980). In 1999, the Great Gulf Wilderness had a concentration of 70 percent of visitor miles of use were on only 40 percent of the available miles of trail, and that is a moderately high concentration index for a small wilderness (Dawson, Simon et al. 2001).

Horse use tends to be more concentrated than hiking use. Horse parties go farther, but they tend to stay on main trails. As a result, areas with heavy horse use generally have more concentrated total use patterns, although horse use accounts for a small portion of use in most areas. The Bob Marshall, the only area studied where a majority of visitors traveled by horse, also showed the most concentrated use. One might expect that users, in an attempt to avoid crowds, would disperse more evenly. *Uneven use seems universal, but those areas for which data for several years are available all show a pattern toward more even distribution of use at access points in more recent years, perhaps partly because more visitors are seeking less crowded, less impacted areas, and partly because managers have taken actions to shift use.*

Short trips, both in distance traveled and duration, are characteristic for most wildernesses. One study of eight Montana and Idaho wildernesses and backcountry areas revealed that only about 14 percent of the visitors traveled more than twenty miles round-trip in the wilderness—a statistic that refutes the notion that wilderness trips are typically long. Only about 2 percent of the visitors traveled more than fifty miles (Lucas 1980).

Distribution of visitor use in many of the arid wildernesses of the southwestern United States is heavily influenced by water availability. Visitor use is concentrated around primary water sources. In the Superstition Wilderness, Arizona, visitors are educated to camp in designated sites that have one maintained fire ring and allow access to water sources, but that are far enough away from those sources to reduce social and biological impacts. All other campsites are rehabilitated and fire rings are removed whenever they occur to keep the majority of the area wild and to provide opportunities for solitude.

Fig. 14.10. Lakes and river attract wilderness visitors, but camping on shorelines and riverbanks is often prohibited to reduce visual and biophysical impacts of visitors. Photo courtesy U.S. Forest Service.

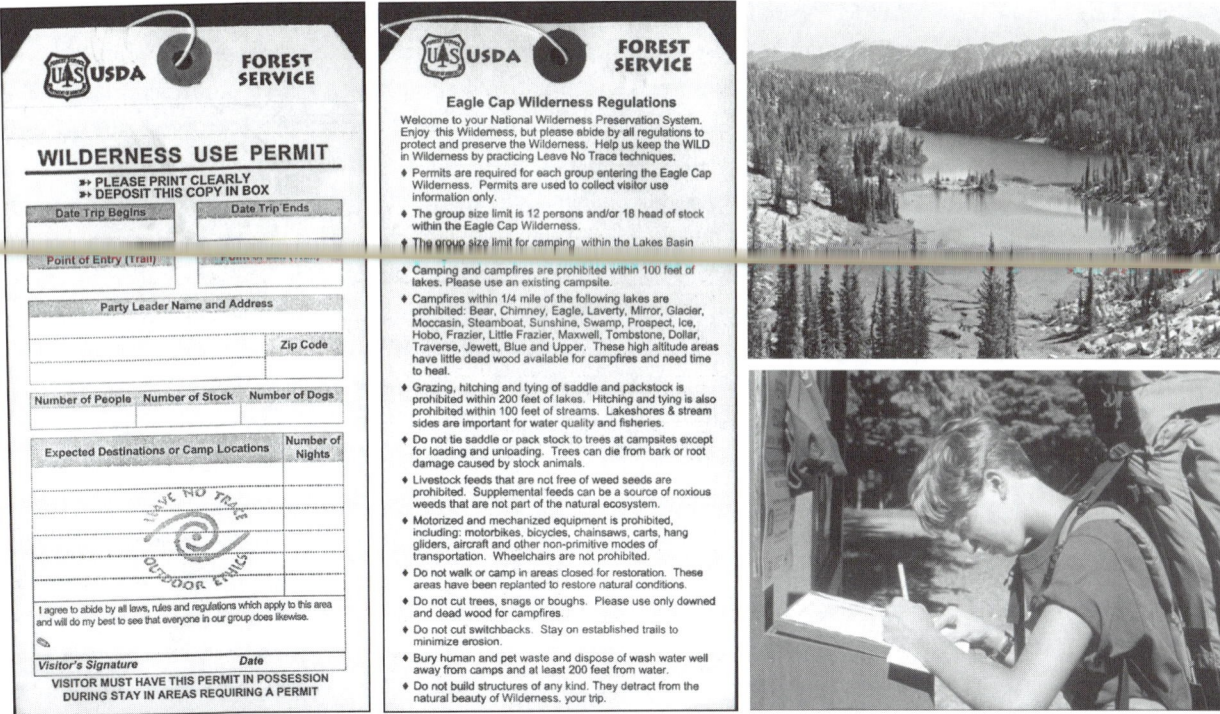

Fig. 14.11. Wilderness visitor permits are required for entry into many wilderness areas. Shown here is an example of a free, "self-issued" permit (left) and some important regulations to educate the visitor (center). Forest Service managers of the Eagle Cap Wilderness in Oregon (top right) require one person from each party to use this type of permit, obtained at a trailhead (lower right). Photos courtesy Tom Carlson.

An effective way of describing wilderness congestion resulting from use concentration is to measure the number of other parties a group could be expected to meet (encounters) per hour or day in a given portion of the wilderness. Six wilderness areas in the Alpine Lakes, Washington; Mount Jefferson, Oregon; and Three Sisters Wildernesses, Oregon, reportedly had extremely high encounter rates in 1991–1992 that exceeded the level preferred by most users (Cole, Watson et al. 1997). For example, weekend visitors to the Alpine Lakes Wilderness typically encountered thirty to forty other groups, whereas weekday visitors encountered ten to fifteen other groups. In the two Oregon wilderness areas, weekend visitors typically encountered twelve to eighteen other groups, whereas weekday visitors encountered five to eleven other groups. Most visitors reported that they expected these levels of encounters and were not overly bothered by them, and only 10 to 23 percent supported management actions to reduce use levels.

Use patterns in most wildernesses are strongly trail-related. The only exceptions are areas where water access predominates, as in coastal southeastern Alaska and the BWCAW. Studies of national forest wilderness areas estimate that fewer than 20 percent of the visitors to most areas travel cross-country and an even smaller percentage of the total distance covered in the wilderness is off-trail travel. Cross-country travel depends on the motives, interests, and experience of the visitors, as well as on the terrain and vegetation. Wilderness visitors are almost locked into trails in steep, heavily forested country, especially in areas with many downed trees, such as forest zones in the northern Rockies, much of the Cascades, and the East. Alpine, plateau-like areas and desert terrain have more cross-country opportunities.

Six wilderness areas in the Alpine Lakes, Mount Jefferson, and Three Sisters Wildernesses had substantial recreational use impacts in 1991–1992, showing on system trails, social trails, campsites, and lakeshores (Cole, Watson et al. 1997). Most visitors reported that they noticed these impacts, that the impacts detracted from their wilderness experience, and that they were supportive of management programs to close trails and campsites or to reestablish vegetation in some areas.

Campsite use tends to be uneven. Extrapolating from research studies and professional experience, we believe the most popular campsites share characteristics of (1) proximity to water, possibly with fishing opportunities; (2) scenic and/or water views; (3) closeness to a trail (several hundred feet); (4) availability of enough level land to camp (4 percent or less slope); and (5) availability of firewood (not as important as it used to be thanks to portable stoves and Leave No Trace awareness). Where wilderness users do not camp so close to water, the impacts on the riparian zone are minimized, and lakeshore access is available to more users. Sites close to water are highly attractive, and getting people to change their selection of campsites may not be easy.

A study of campsites over twelve to sixteen years in portions of Montana's Selway-Bitterroot and Lee Metcalf Wildernesses and Oregon's Eagle Cap Wilderness, showed an increase in the number of campsites by 53 percent, 84 percent, and 123 percent, respectively (Cole 1993). The ecological impacts of locating and managing campsites are discussed in Chapter 15, but the key point here is that campsite impacts increase substantially with campsite proliferation, expansion, and, to a lesser extent, existing campsite deterioration.

User Characteristics

A common stereotype of wilderness visitors pictures them as young, athletic, wealthy, leisured, and urban. This stereotype is commonly reflected in congressional testimony on wilderness designation proposals and in public meetings on proposals in small towns near wilderness. Although there are some distinguishing sociodemographic characteristics among wilderness users, we must be careful to recognize that statistical averages represent some portion of the users but that there are many different types of users with a variety of interests, needs, and expectations. The diversity of experiences sought by visitors is one of the reasons that the WOS is used by wilderness planners to compare the wilderness experience opportunities within each wilderness area and among the wilderness areas of a region. Both stereotypes and diversity of users have been documented in studies over forty years (Roggenbuck and Lucas 1987; Roggenbuck and Watson 1989; Stankey 1971; Alan Watson, Williams et al. 1992). These studies are among the most complete and are the basis for much that follows—we will not cite them repeatedly.

Characteristics of wilderness users are well-known from many studies, and these characteristics are similar from area to area, even in different parts of the nation. In general, wilderness users, compared to the general population, tend to be young but with all age groups represented, predominantly male but with increasing numbers of women, from urban areas but largely near the wilderness area visited, above average in income but rarely wealthy, well educated and in professional or technical occupations or students.

Only a few important changes seem to have occurred in the sociodemographic characteristics of users over the last forty years—higher average education levels, older average age, and a greater percentage of women. Even though there are reported changes in these sociodemographic characteristics, there are no reported shifts in the kinds of trips that users take or their preferences for conditions they encounter (Cole, Watson, and Roggenbuck 1995).

Given the high educational level of users, managers can continue to inform and educate them about wilderness management issues and the need for management actions and programs. Understanding who the wilderness visitors really are is important for both wilderness policy makers and wilderness managers—wilderness visitors are important clients. Policy makers need to know who receives the benefits gained from financial and natural-resource allocations to wilderness designation and management. Managers also need to understand who their clients are so they can communicate with them through information and education programs (e.g., Internet communications for off-site information and education before visitors arrive at the wilderness area).

Age: *Wilderness visitors tend to be younger on average than the general population, yet all age groups are fairly well represented.* Data from about thirty wilderness user studies show that a substantially higher percentage of sixteen- to twenty-five-year-olds and twenty-six- to thirty-five-year-olds are present among wilderness visitors than are in the general population. The thirty-six to forty-five age group is often more highly represented than the general population, and the forty-six to fifty-five age group is present in similar proportion to the general population. The fifty-six and older age groups are underrepresented among wilderness visitors as they are in all types of outdoor recreation. Children under sixteen years of age are substantially underrepresented in most wilderness studies, not because they do not use wilderness, but rather because most studies typically do not survey users less than sixteen years of age. All these age figures refer to individual visitors, but a number of studies have been based on party leaders or persons registering for the entire party, and these persons tend to be older.

A recent study indicates that the average age of visitors to wilderness is increasing, with more older users in the late 1980s and early 1990s (Cole, Watson, and Roggenbuck 1995). Changes in birthrate, the degree to which individual wilderness visitors continue to visit wilderness as they age, and the age groups from which most new visitors are recruited will affect future wilderness use levels. *In summary, wilderness visitors are more likely to be young to middle-aged adults, but with younger and older age classes present as well.*

Gender: *Wilderness use has sometimes been viewed as a male activity* because studies from 1960 through 1983 reported that 70 to 85 percent of users were male. The larger, horse-oriented wildernesses average less female visitation; the smaller hiking areas average a little more. Some recent studies indicate an increasing percentage of women visitors, although their participation rates are still lower than those of male visitors (Cole, Watson, and Roggenbuck 1995).

Residence: *Most wilderness visitors are from large cities and urban areas, as are most Americans.* However, because visitors do not typically travel long distances to visit wilderness, the proportion from urban places depends largely on the degree of urbanization in that region. *Most wilderness users are from the state or region in which the wilderness area is located.* For example, about 60 percent of the 1982 visitors to Montana wildernesses came from urban areas, and 51 percent of the Montana population lived in urban areas in 1980. In highly urbanized southern California, more than 90 percent of wilderness visitors come from urban areas with more than a million people. In general, the percentage of city and urban visitors to wilderness is in proportion to the population of the state or region, and this trend continued to be reported through the early 1990s (Cole, Watson, and Roggenbuck 1995). Although current residence of wilderness visitors has been reported in most studies as overwhelmingly urban, early studies (Hendee, Catton et al. 1968; Lucas 1980; Outdoor Recreation Resources Review Commission [ORRRC] 1962) found more visitors with rural backgrounds. Some people may be attracted to wilderness by fond memories of rural surroundings in childhood contrasted with the pressures of urban living in adulthood. Even urban-born Americans are drawn to the contrast between urban and wilderness. In any case, due to the urbanization of American society in the past forty years, more adults may be visiting wilderness now because they were introduced to the activity by their urban parents, friends, organizational outfitters (such as youth and college outdoor programs), and wilderness experience programs.

Income: *Wilderness visitors are moderately above average in income, as are most types of outdoor recreationists, but all income levels are represented among wilderness visitors.* The above-average income levels reflect the high education levels and high proportion of users with professional occupations among wilderness visitors, and lower-income visitors may actually be student participants. High incomes are not necessary to visit wilderness, although a certain level of investment is required for equipment and transportation to access the wilderness. Thus, wilderness users need some modest level of disposable income. However, because most users visit a wilderness in the state or region near their residence, typical expenditures for wilderness visits are low compared to many other kinds of recreation away from home.

Education: *One of the most distinguishing characteristics of wilderness visitors is high educational levels.* Studies show that with few exceptions, 60 to 85 percent of the visitors to most wildernesses have attended college, and 20 to 40 percent have done graduate study. This higher level of education among wilderness users is related to their above-average income and professional and technical occupations. Why this association between educational attainment and wilderness use? Higher education may sensitize people to the natural world and global concerns about environmental issues such as wilderness. Students at institutions of higher education are more likely to have friends, acquaintances, and organizations that visit wilderness and who may introduce them to such activities.

Occupation: *Studies show that persons in professional or technical occupations and students form the majority of visitors to most wildernesses* and are more prevalent among wilderness visitors than in the general population. Generally, from 20 to 40 percent of the visitors of working age are in professional or technical work, with about one-fourth of the adults and young adults being students. This is related to the high education of wilderness visitors. Homemakers and skilled laborers usually are the next most common in occupation, each comprising about one-tenth of the total in most areas. Other occupational categories are not well represented. Blue-collar workers, other than skilled craftspersons, account for only about 5 percent of all visitors. No changes or trends in this pattern of occupational categories have been recently reported for wilderness users.

Racial and Ethnic Minority Participation: *It is a matter of concern that racial and ethnic minorities are greatly underrepresented among wilderness visitors, although their participation is increasing.* Some researchers have pointed out that the increasing racial and ethnic diversity of the U.S. population is having a small impact on increased wilderness use, but that shifts in user values may occur as more ethnic minorities participate (Cook and Borrie 1995; Alan Watson, Cole, Roggenbuck et al. 1995). Nevertheless, the future of wilderness depends on public support, and wilderness needs the support of the increasingly diverse U.S. population.

Factors Affecting Wilderness Use Trends and Management

Wilderness recreation use has increased greatly over the last fifty years, and overall use of the NWPS is expected to continue. In some areas, visitor use has leveled off and even declined, whereas in other areas there are steady increases. There are numerous factors that affect outdoor recreation and wilderness use (Cordell 1999; Cordell, Tarrant et al. 1998; O'Leary et al. 1989; Bowker et al. 2007), such as sociodemographic changes in the population and other factors (Alan Watson 2000). In the following discussion, we consider several of these factors and their potential impacts on recreational use of wilderness in the future—leading toward a projection of future wilderness use.

Changing Age and Population Structure

Many studies of outdoor-recreation participation have pointed to age as one important predictive measure of future participation, but *the gradual aging of the U.S. population has not led to a decline in wilderness participation as predicted by some researchers.* English and Cordell (1985) found that participation rates among different age cohorts have risen steadily since 1960, a fact that suggests that the expected dampening effects of increased age on recreation participation might be even smaller in the future than predicted in the past. One of the most fundamental changes in U.S. society today is the increasing age of the population. The median age of the U.S. population was 30 in 1980 and was projected to reach 36 by the year 2000 (Dwyer 1994) (U.S. Bureau of the census reported it was 35.3 in 2001).

Although the population is aging, short-term changes are not dramatic in those age classes who visit wilderness the most. The percentage of people in the 18-to-24 age class has declined slightly since 1980; they were 10.4 percent of the population in 1990 but will account for only 8.3 percent by the year 2025 (Dwyer 1994). The percentage of the population that is 65 years of age or older is projected to increase from 12.6 percent in 1990 to 21 percent by the year 2025. Decreases in youth under 18 and increases in the percentage of 65 and over should have little effect on wilderness use because neither age class has previously been a major wilderness user. The 25-to-64-year-old age class, the most inclined to visit wilderness of any age class, will continue to increase as the short-term change is toward a "middle-aged" population. *Thus, the changing age and population structure in the United States should have neither a major dampening nor increasing influence on wilderness use during the first decade of the twenty-first century.*

Population Distribution and Diversity

Changes in the U.S. population distribution might seem to have the potential to impact wilderness use trends in specific regions, but such changes to date are not evident. At times, the relationship is backward. For example, the large migrations to the South and West during the 1970s were not matched by significant growth in wilderness use in those areas.

The U. S. population is projected to grow at a rate of 0.5 percent per year and the majority of that growth (50 million people between 1990 and 2025) is expected to be among racial and ethnic minority groups (Dwyer 1994). Wilderness recreation use is slowly increasing among nontraditional ethnic and racial groups, and this change is most often reported in wilderness areas near urban centers (Cook and Borrie 1995). This trend toward more racial and ethnic diversity is reflected in the U.S. population and recreation participation in general (Dwyer 1994). *Despite the slow increases toward more diverse racial, ethnic, and gender participation in wilderness recreation, this remains a major issue* for wilderness. Whatever the reasons for the underrepresentation of racial and ethnic minorities in wilderness recreation (cultural, economic, or preference), *we need to remove the barriers to participation so the benefits and values of wilderness can be more widely shared—and so wilderness protection will be a relevant goal for our increasingly diverse public.*

Leisure Time

One to several days are necessary to visit most wilderness areas—time for travel to and from the site and the experience. Such blocks of time seem harder to come by for many people in our fast-paced society, despite data that suggest that we have more leisure time. The annual average of hours worked per week, as reported by the U.S. Bureau of Labor Statistics, has stayed relatively constant at 39 to 40 from 1970 to 1995 (Robinson and Godbey 1997). The trend in overall free time increased from 34.8 in 1965 to 39.6 in 1985 and reportedly stabilized at about 40 hours per week in studies conducted in the 1991 through 1995 period (Robinson and Blair 1995). However, many factors, such as age, education, income, race, residence, and marital status, affect the amount of leisure time available to individuals (Robinson and Godbey 1997). For example, today in a typical family with children, both spouses may hold jobs and these dual-income households are on the rise. Although the discretionary income of such households increases, it may be difficult to coordinate household duties, vacations, and free time between spouses so that "getaway" weekends or longer periods are possible. Thus, trends reducing blocks of discretionary time may have a dampening effect on wilderness visitation and encourage shorter trips and use of nearby wilderness areas. However, total recreation use of wilderness is increasing despite a pattern toward shorter trips to nearby wilderness areas.

Effects of Expanding the Wilderness System

One explanation for much of the increased total use of wilderness is the increase in the size of the NWPS. However, expanding the NWPS with new areas has probably had little effect—besides reclassifying the roadless-area-recreation visitors from that category to the wilderness-visitation category. In some cases, the visits were already being counted as wilderness. For example, some of the growth in acres and number of areas in the NWPS resulted from shifts of national forest primitive areas to wilderness or official designation of portions of a national park as wilderness. In both cases, the recreational use was already being counted.

Fig. 14.12. Hundreds of organizations use wilderness as places for personal growth, therapy, education, and leadership development. The Anasazi Foundation, Phoenix, Arizona, takes at-risk youth on personal growth and healing trips in wilderness and deserts of the Southwest. Ezekial Sanchez, one of the organization's leaders holds a stick as he talks with young men. Photo courtesy Jenny Dixon, Anasazi Foundation.

These areas have long been specially designated and widely perceived as wilderness, even though not technically designated.

However, the wilderness recreation market is not fixed. If a new area is designated as wilderness, many of its visitors are the same people who visited the area before it was so designated. The wilderness designation is new, but the land and its recreational attractions were always there; it is "new" only in an official sense. New visitors might indeed show up; perhaps residents from the vicinity or nearby urban areas were attracted by the publicity generated by wilderness designation.

Whatever effect major expansion of the wilderness system may have had on past use of older areas, its future effect probably will be less. The large expansion of the system (Landres and Meyer 1998) in the 1980s (more than 300 areas and 71 million acres added between 1980 and 1985) and the 1990s (69 wilderness areas and more than 4 million acres in the California Desert Protection Act—U.S. Public Law 103-433) is not likely to be duplicated in years to come (J. Watson and Brink 1996). *We anticipate a gradual increase in the NWPS to about 125 million acres during the first three decades of the twenty-first century, some of which may include designation of already-proposed roadless areas in several large national parks. The wilderness movement is strong and resilient, its leaders have committed to a goal of 200 million acres, and citizen efforts are actively inventorying potential wilderness in every region.*

Fig. 14.13. The Aldo Leopold Wilderness Research Institute in Missoula, Montana, administered by the U.S. Forest Service, specializes only in wilderness research. Other federal agencies participating in the institute are the Bureau of Land Management, U.S. Geological Survey, U.S. Fish and Wildlife Service, and National Park Service. Photo courtesy Chad P. Dawson.

Increasing Educational Levels and Environmental Awareness

The most distinguishing socioeconomic characteristic of wilderness users is their high educational levels, and this portion of the population is increasing. The percentage of the population with a college degree or some postsecondary education continued to increase through the 1990s. The introduction of wilderness activities and interests to students through college courses, field trips, outdoor clubs, and other students will likely continue as a factor contributing to future wilderness use.

Some aspects of higher education stimulate environmental concerns in both science-oriented and other students, as global environmental concerns have social, economic, historic, and political influences that are part of a liberal education program. Wilderness is related to these concerns because it is a benchmark and symbol of environmental quality. Therefore, wilderness continues to have broad public recognition as a positive environmental concern in an increasingly environmentally conscious society, and support for wilderness is part of the agenda for numerous environmental and conservation groups.

Changing Transportation Cost and Supply

The impact of gasoline prices on travel for wilderness recreation does not appear to have been great in the past. First, during the 1973–1974 embargo, use of national forest wilderness continued to grow, although the rate of growth slowed. Second, in the early 1980s, increases in the price of gasoline had not kept up with inflation; but despite this, the wilderness use trend was downward for several years in a row. Availability of gasoline has been a problem from time to time, but typically this has been a temporary situation that took place without a long-term impact on wilderness use. *Most wilderness visitors live relatively close to the areas they visit, so it seems unlikely that availability would affect overall wilderness recreation use to any large extent. However, substantial increases in the price of fuel (i.e., increases above the rate of inflation) may decrease wilderness visitation.*

Changing Recreational Preferences and Organized Use

Population surveys of recreation participation suggest that *in recent decades the amount of U.S. citizens participating in wilderness-related activities, such as hiking and backpacking, has remained relatively close to 5 percent.* Cordell and Teasley (1998) reported further that 1.7 percent of the U.S. population claimed to have used a unit of the NWPS during 1994–1995; furthermore, 7.5 percent claimed to have visited a national forest and 10.7 percent claimed to have visited a national park. However, many users probably could not distinguish which type of area they actually visited, and Cordell and Teasley (1998) speculated that some respondents who said they visited a national forest or national park actually visited wilderness. Of course, the reverse may be true as well. Thus, the proportion of the population whose recreational interests might be met in wilderness settings (5 percent), and the proportion who said they visited wilderness (1.7 percent), are consistent with the estimated NWPS visitation level of 17 million visitor days in 1994, especially given that fishing, photography, and nature study are reported as the most popular wilderness recreational activities and are not classified as wilderness-related in this survey.

Numerous changes in the preferences of users for wilderness-dependent activities have increased use among existing and new clientele. Wilderness experience programs (WEPs) offered by organized groups for purposes of personal growth or education have evolved into a growing source of wilderness visitation, often introducing new clientele to the benefits of wilderness use.

Adventure-recreation activities and programs that combine risk and recreation in a wilderness setting are also growing, and challenge both users and managers alike to respond to this growing phenomenon (Ewert and Hollenhorst 1997). For example, a controversial issue in wilderness policy at the turn of the twenty-first century was whether to allow permanently fixed bolts and hangars as anchors to be installed or to remain for mountain climbers in wilderness (Ewert, Hendee et al. 1999). These and other developments in recreation, conservation, environmental group membership, and public information campaigns continually change and shape the public perception of wilderness as a concept and as a place to recreate.

Constraints to Wilderness Visitation

Green et al. (2007) report that there are three categories of wilderness constraints that are impediments or barriers to participation for some people: personal (e.g., don't have enough time because of long work hours or long school hours), structural (e.g., wilderness areas are crowded), and psychological (e.g., concerned for my personal safety). Their study supported the concerns by managers that minorities, women, low-income, elderly, and less educated people had higher probabilities of feeling constrained from participating in recreational activities in wilderness areas than other user groups. Additionally, immigrants felt they were more constrained from participating in recreational activities in wilderness than people born in the United States.

Changing Wilderness Conditions and Regulations

Despite some managerial concerns about declining quality of wilderness conditions and experiences, there is little evidence that user dissatisfaction is negatively influencing wilderness use levels. Most studies of wilderness users report high levels of satisfaction with their experience (Lucas 1985; Van Horne et al. 1985). An analysis of studies comparing satisfaction levels at three wilderness areas with earlier studies—BWCAW (1969 and 1991), Desolation (1972 and 1990), and Shining Rock (1978 and 1990)—indicates that the vast majority of users were very satisfied with their trip or rated its quality as being very good (Cole, Watson, and Roggenbuck 1995). There was no important change in perceived quality except in the case of the Desolation Wilderness (CA) day users who reported increased quality ratings in 1990 compared to 1972.

However, increased wilderness use, new equipment, increased education and information, and more intense management may combine to diminish the free and easy wilderness access of an earlier era when going on a wilderness trip was less complicated and less regulated. Then, solitude and privacy were more likely; regulations, permit requirements, and restrictions less burdensome; firewood more available; and water potable without treatment.

Wilderness conditions do affect visitor use, such as following fires when visitor use declines for a period of time immediately after the fire (Borrie and McCool 2007; Brown et al. 2008). Furthermore, there may be some users who were displaced by some of these changes or who no longer go to wilderness (i.e., dissatisfied users may not be returning to wilderness), but the total extent of such temporal and spatial shifts in use or nonparticipation is not known.

Most wilderness areas now have more regulations. Fires may be prohibited in some areas and stoves required. Giardia infection is now a risk in most wilderness waters and makes boiling water, filters, or purification tablets essential. Historically, visitors might have relaxed around a campfire next to a high mountain lake, maybe sipping clear, cold water dipped from the lake. Today, they might get a ticket from a wilderness ranger for camping too close to the lake, or for the campfire, and they might suffer diarrhea from drinking the water. Today, they might also receive a weather report on their cell phone or from the ranger via radio, thereby reducing uncertainty about what to expect. Today, they are also more likely to encounter more people with whom to share and obtain information. Is wilderness now a different experience?

How will visitors respond to such changes? Although these changes in the wilderness experience may not appeal to some who had a different experience in wilderness decades ago, it is clear that today's wilderness experiences have great appeal to most current users, as witnessed by their high satisfaction ratings reported in studies of wilderness users and growing wilderness use.

Wilderness Recreation Use Projections

We conservatively project that future use will increase at an average of 2 percent annually, based on the following review of use studies and the earlier discussion of influences on future use. *Moreover, we expect that new designations to the NWPS will further increase the size of the system and total use.*

Projections of wilderness recreational use have been limited by basic-use data and by incomplete knowledge of the relationship of wilderness use to causal factors. Probably the earliest projection of wilderness use was made in 1961 as part of the USFS National Forest Recreation Survey project (U.S. Department of Agriculture, Forest Service 1961). That unpublished projection predicted a tripling of wilderness use days by 1976 and a more than eightfold growth by the year 2000—about a 5 percent average annual rate of increase. The actual reported figure for 1976 turned out to be 7,105,600 twelve-hour visitor days, compared to the projection of 5,804,000 use days—a different unit of measure. One cannot convert from one unit to the other with precision, but when adjusted, the projected 1976 use amounts to roughly 8.5 million visitor days. Thus, in 1976, projected use was about 20 percent higher than reported use. The projection technique was based on simple assumptions with no supporting research (there was almost no recreation research at the time) and based on projections of population, income, leisure time, and travel variables that are almost as hard to project as recreation use itself.

At almost the same time, wilderness use was projected as part of the Outdoor Recreation Resources Review Commission (1962) studies, which worked only with USFS data (no wildernesses were managed by any other agency before 1964). Their projected rates of increase were similar to those of the USFS study (about 5 percent per year), which is interesting because their projection procedures were quite different.

Almost twenty years elapsed before other wilderness use projections were developed in response to requirements in the Forest and Rangeland Renewable Resource Planning Act of 1974 (Public Law 93-378). Three projection studies were published in 1982 and 1983, all of them using more advanced statistical techniques and based on more data than the early studies. Jungst and Countryman (1982) developed several models, yielding projected average annual rates of increase of 2.6 percent to 7.2 percent to the year 2020. This wide range reflects the uncertainty of projections. Although the difference between 2.6 and 7.2 may not seem large, in a forty-year period, 2.6 percent results in less than a tripling of use, whereas 7.2 percent results in about a sixteenfold increase in use. We know now that the mid to low range of predicted increase in use was closer to what has happened.

Hof and Kaiser (1983) projected several outdoor-recreation activities that probably parallel wilderness use, such as dispersed primitive camping. Their approach involved estimating per capita participation, using various socioeconomic and supply variables in a regression model, and applying the equation using high, medium, and low projected values for the independent variables. Primitive camping was projected to increase from a 1977 base of 100 to 155, 205, and 311 by the year 2030 in the low, medium, and high scenarios, respectively. These index translate to annual average rates of growth ranging from less than 1 percent to slightly more than 2 percent. Although this has proven to be low, it may have reflected more conservative thinking and a temporary leveling off of wilderness use reported and predicted at that time.

Cordell, Bergstrom et al. (1990) estimated that wilderness demand would be greater than supply by the year 2000 and would continue to increase through the year 2040, based on projection of activities that could occur in wilderness. For example, demand for day hiking was expected to increase beyond 1987 at a rate of 1 percent per year, whereas primitive camping was projected to increase at a rate of 0.5 percent per year (Cordell, Bergstrom et al. 1990). Although this does not provide real numbers for comparison, it reflects the notion of steadily increasing use of a wilderness system that could eventually have demand exceeding supply at acceptable levels of quality.

Overall use of the NWPS was estimated by Cole (1996) as about 17 million RVDs in 1994. This is a sixfold increase between 1965 and 1994 and an average annual increase of 6.3 percent, but increases between 1989 and 1994 averaged only 2.8 percent.

Loomis, Bonetti et al. (1999) forecast wilderness use to increase at 0.5 percent annually in northeastern and Rocky Mountain wilderness areas and 1 percent annually in the southeastern and Pacific coastal states wilderness areas over the next several decades. This is less than the rate of increase (1.4 percent annually) reported by Cole (1996) for national forest areas from 1989 through 1994.

Bowker et al. (2007) estimated NWPS use was 10.7 million trips and 16.3 million on-site days (i.e., visits) in 2002, based on a National Visitor Use Monitoring Project. Their estimates of wilderness use are difficult to compare with other researchers using

RVD units (Cole 1996). The visitor use projections by Bowker et al. (2007) are based on population projections and other factors. They project use to increase from 16.9 million visits in 2010 to 17.3 million visits in 2020 and 17.9 million visits in 2030. Their projection is for the total number of "unique" participants to increase from 2.9 million in 2010 to 3.1 million in 2030. While the total number of participants and total number of visits will increase, the per capita participation will decline over that period.

Wilderness recreation use will remain an important use of the national forests and national parks; it is about 7 percent of total national parks' overnight use and about 5 percent of total recreational use of the national forests. Wilderness managed by the BLM is lightly used but growing as its newly designated wildernesses and wilderness study areas are being discovered. Use of U.S. Fish and Wildlife refuge wilderness is limited in many areas, but accessible areas will be discovered by wilderness visitors seeking less crowded naturalness and solitude.

It is clear that we have limited capabilities to project wilderness use. A handful of studies agree that wilderness use will increase, but they do not agree on the projected rates of increase. All studies have predicted the steady, slow to modest increases seen in the last twenty to forty years. The greatest difficulty in projecting future wilderness use is due to the limited current and past wilderness use information, as past use is the basis for estimating future use.

Wilderness and recreation use trends are complex (Clawson 1985), but *our overall conclusion is that wilderness recreation use will continue to grow steadily at a moderate rate of about 2 percent per year.* This is supported by a combination of factors, previously reviewed in this chapter, including the social and demographic structure of the U.S. population, recreational preferences, and tastes of an increasingly environmentally aware and more highly educated and informed public. Beyond these factors and others discussed in this chapter, *we also believe that the wilderness constituency in the United States and North American population—those who support wilderness protection for reasons in addition to recreation—will continue to grow. Their influences will be measured in ways other than recreation use, such as increased public support for wilderness designation and wildland protection in general.*

Fig. 14.14. As visitor use increased in the popular Three Sisters Wilderness of central Oregon, Les Joslin, above, began a wilderness information specialist program while he was a USFS employee, to advance wilderness education and train volunteers to inform and educate visitors about low-impact hiking and camping, and to assist rangers with gathering visitor statistics and maintaining trailhead facilities. Joslin now runs the Central Oregon Wilderness Education Partnership as a community-based wilderness education partnership with the USFS. Photo courtesy Pat Joslin.

Summary

Wilderness stewardship and management depend on an understanding of all the varied uses wilderness receives, because most wilderness management is management of use of one kind or another, primarily recreation use. Some uses depend on wilderness, but others just take place there and are less dependent on legally mandated wilderness qualities of naturalness and solitude. Recreation is the most obvious wilderness use, but other uses include commercial recreation; indirect, off-site uses like ecological services and benefits from wilderness; scientific uses; educational uses; personal growth, therapeutic, and healing uses; and a variety of direct commodity uses such as mining, water storage, and grazing.

Wilderness recreation use provides quality opportunities and benefits to users, and challenges to wilderness managers. More is known about recreational use than most other wilderness uses, because it has been more widely studied and measured. Many direct- and indirect-use measurement methods are employed, ranging from direct observation and use counts at a sample of access points and days to self-registration and visitor permits. Recreational use is reported in many different statistical units, but visitor days (USFS) and overnight stays (NPS) are most common; unfortunately, they are not easily comparable.

Most wilderness visits are short, with day use predominating in many—maybe most—areas. Parties visiting wilderness usually are small, most often two to four people, representing families or small groups of friends. The most common method of travel is hiking, with a few exceptions. Activities vary, with fishing, photography, nature study, and swimming reported as popular wilderness pursuits. Summer is the main use season, with exceptions mainly in the South and Southwest and at low elevations. Winter use in the North and at high elevations is growing. Many areas have sharp weekend peaks in use. Some visitors to many wildernesses come from all over the nation, but most wilderness visitors live relatively close to the area visited. Wilderness recreational use is distributed very unevenly; some wildernesses, entry points, trails, and campsites are heavily used, while many others have little use. A completely even distribution is not possible and not desirable, but, in many areas, use probably is too uneven and concentrated at attractive locations.

Wilderness visitors are younger than the general population, but all age groups are represented. About one-fourth of wilderness visitors are female, but this proportion has increased over time and is probably increasing total wilderness use. Visitors are only slightly more likely to be urban residents and are moderately above average in income, as are most types of outdoor recreationists. The single most distinguishing socioeconomic characteristic of wilderness visitors is a high level of education, and this is related to their better-than-average incomes and typically professional and technical occupations.

Trends in wilderness recreational use have changed greatly over the years. In the 1940s, 1950s, and 1960s, use increased rapidly, but in recent decades, the rate of growth, though still high, has gradually slowed. In the 1980s and early 1990s, use fluctuated in many wilderness areas but total NWPS use has shown an overall increase. Projections of future use vary widely. Although annual percentage growth is slow, total use in the NWPS has grown dramatically due to the growth in the number of wilderness areas as acreage added to the wilderness system over the last forty years. Understanding recreational use and users helps managers to protect the valuable wilderness resource, manage wilderness conditions, and influence the wilderness user's experience.

Our projections for future recreation use in wilderness are synthesized from several studies and conclude that use will continue to increase at about 2 percent per year, with total use and total number of participants increasing even though the per capita participation rates will likely decline due to a variety of factors.

Study Questions

1. Why is understanding wilderness use important for wilderness management?
2. What are the important nonrecreational uses of wilderness? Which ones are substantially wilderness-dependent?
3. What are the different methods to measure recreation use in wilderness? What strengths and weaknesses does each method have?
4. In what way is wilderness use unevenly distributed? What are the major management implications of that use distribution pattern?
5. What are the major characteristics of wilderness recreation use?
6. What are the major characteristics of wilderness visitors?
7. What are the trends in recreational use of wilderness? What are the major management implications of those trends?

Acknowledgments

Chad P. Dawson and John C. Hendee completed minor editing of this chapter for the fourth edition. The third edition of this chapter was a major revision by Chad P. Dawson and John C. Hendee from the second edition chapter on wilderness use; we recognize and thank our colleague, Robert Lucas, for his seminal efforts on the topic in the first two editions. We thank numerous colleagues for reviews of the third edition, as we have benefited from their review comments, but we alone bear responsibility for any errors and opinions reported herein.

Case Discussion: Constraints to Wilderness Participation

The benefits and values of wilderness are available to all segments of society however, not all segments of the U.S. population visit wilderness or, at least, not at the same participation rates. Green et al. (2007) studied wilderness participation to understand whether different segments of the population did not want to participate or if they did want to participate and perceived some type of constraints that kept them from visiting wilderness. Their study supported the concerns by managers that minorities, women, low-income, elderly, and less educated people had higher probabilities of feeling constrained from participating in recreational activities in wilderness areas than other user groups. Additionally, immigrants felt they were more constrained from participating in recreational activities in wilderness than people born in the United States. The NWPS protects wilderness areas for the use, enjoyment, and benefit of all society, but not all want to or perceive they can take advantage of these direct wilderness experience opportunities. Managers need to be aware that various segments of the public face one or more constraints to their participation in wilderness-and sometimes information and education programs or other management activities can reduce those real or perceived constraints.

Case Discussion Questions:

1. If managers can find ways to reduce the constraints felt by minorities, women, low-income, elderly, and less educated people, how do you think more participation by each of these segments will change the use characteristics (e.g., methods of wilderness travel, season and timing of use) reported in wilderness visitor studies in the future?
2. How will the factors affecting wilderness use trends and management (reported in this chapter) positively or negatively impact those segments of society—minorities, women, low-income, elderly, and less educated people—that are already underrepresented in wilderness visitation?
3. Why is it important for managers to attempt to reduce or remove the real and perceived constraints to participation in recreation activities in wilderness for minorities, women, low-income, elderly, and less educated people?

References

Absher, James D.; Lee, Robert G. 1981. Density as an incomplete cause of crowding in backcountry settings. *Leisure Sciences*. 4(3): 231–247.

Billings, Ronald F. 1986. Coping with forest insect pests in southern wilderness areas, with emphasis on the southern pine beetle. In: Kulhavy, David L.; Conner, Richard N., eds. *Wilderness and Natural Areas in the Eastern United States: A Management Challenge*. Nacogdoches, TX: Stephen F. Austin University, School of Forestry, Center for Applied Studies, pp. 120–125.

Borrie, W. T.; McCool, S. F. 2007. Describing change in visitors and visits to the "Bob." *International Journal of Wilderness*. 13(3): 28–33.

Bowker, J. M.; Murphy, D.; Cordell, H. K.; English, D. B. K.; et al. 2007. Wilderness recreation participation: Projections for the next half century. In: Watson, A.; Sproul, J.; Dean, L. *Science and stewardship to protect and sustain wilderness values: Eighth World Wilderness Congress Symposium*; September 30–October 6, 2005; Anchorage, AK. Proceedings RMRS-P-49. Fort Collins, CO: U.S. Department of Agriculture, Forest Service, Rocky Mountain Research Station, pp. 367–373.

Brown, R. N. K.; Rosenberger, R. S.; Kline, J. D.; Hall, T. E.; Needham, M. D. 2008. Visitor preferences for managing wilderness recreation after wildfire. *Journal of Forestry*. 106(1): 9–16.

Browning, J. A.; Hendee, J. C.; Roggenbuck, J. W. 1988. 103 *Wilderness Laws: Milestones and Management Direction in Wilderness Legislation, 1964–1987*. Bulletin 51. Moscow: University of Idaho, College of Forestry, Wildlife and Range Sciences.

Butler, Lisa Mathis; Roberts, Rebecca S. 1986. Use of wilderness areas for research. In: Lucas, Robert C., comp. *Proceedings: National Wilderness Research Conference*: Current research; July 23–26, 1985; Fort Collins, CO. General Technical Report INT-212. Ogden, UT: U.S. Department of Agriculture, Forest Service, Intermountain Research Station, pp. 398–405.

Carlson, T. 2000. The Eagle Cap Wilderness permit system: A visitor education tool. *International Journal of Wilderness*. 6(2): 27–28.

Carpenter, M. Ralph; Bowhis, Donald R. 1976. Attitudes toward fishing and fisheries management of users in Desolation Wilderness, California. *California Fish and Game*. 62(3): 168–178.

Carr, D.; Hendee, J. C. 1993. An assessment of current research in the National Wilderness Preservation System. In Hendee, J. C.; Martin, V. G., eds. *Proceedings of the 5th World Wilderness Congress Symposium: International wilderness allocation, management and research*; September 24–October 2, 1993; Tromso, Norway. Ojai, CA: The WILD Foundation, pp. 297–303.

Clawson, Marion. 1985. Outdoor recreation: Twenty-five years of history, twenty-five years of projection. *Leisure Sciences*. 7(1): 73–100.

Cole, David N. 1993. Campsites in three western wilderness: Proliferation and change in condition over 12 to 16 years. Research Paper INT-463. Ogden, UT: United States Department of Agriculture, Forest Service, Intermountain Research Station.

———. 1996. Wilderness recreation use trends, 1965 through 1994. Research Paper INT-RP-488. Ogden, UT: U.S. Department of Agriculture, Forest Service, Intermountain Research Station.

Cole, David N.; McCool, S. F.; Borrie, W. T.; O'Loughlin, J. 2000a. *Wilderness science in a time of change conference*, Vol. 4: *Wilderness Visitors, Experiences, and Visitor Management*; May 23–27, 1999; Missoula, MT. RMRS-P-15. Ogden, UT: U.S. Department of Agriculture, Forest Service, Rocky Mountain Research Station.

———. 2000b. *Wilderness science in a time of change conference*, Vol. 5: *Wilderness Ecosystems, Threats, and Management*; May 23–27, 1999; Missoula, MT. Proceedings RMRS-P-15. Ogden, UT: U.S. Department of Agriculture, Forest Service, Rocky Mountain Research Station.

Cole, David N.; McCool, S. F.; Freimund, W.; O'Loughlin, J. 2000. *Wilderness science in a time of change conference*, Vol. 1: *Changing*

Perspectives and Future Directions; May 23–27, 1999; Missoula, MT. RMRS-P-15. Ogden, UT: U.S. Department of Agriculture, Forest Service, Rocky Mountain Research Station.

Cole, David N.; Watson, Alan E.; Hall, T. E.; Spildie, D. R. 1997. High use destinations in wilderness: Social and biophysical impacts, visitor responses, and management options. Research Paper INT-RP-496. Ogden, UT: U.S. Department of Agriculture, Forest Service, Intermountain Research Station.

Cole, David N.; Watson, Alan E.; Roggenbuck, Joseph W. 1995. Trends in wilderness visitors and visits: Boundary Waters Canoe Area, Shining Rock, and Desolation Wilderness. Research Paper INT-RP-483. Ogden, UT: U.S. Department of Agriculture, Forest Service, Intermountain Research Station.

Cole, David N.; Wright, V. 2003. Wilderness visitors and recreation impacts: Baseline data available for twentieth century conditions. General Technical Report RMRS-GTR-117, Ogden, UT: U.S. Department of Agriculture, Forest Service, Rocky Mountain Research Station.

Cook, Barbara; Borrie, William. 1995. Trends in recreation use and management of wilderness. *International Journal of Wilderness*. 1(2): 30–34.

Cordell, H. K. 1999. *Outdoor Recreation in American Life: A National Assessment of Demand and Supply Trends*. Champaign, IL: Sagamore Press.

Cordell, H. K.; Bergstrom, J. C.; Bowker, J. M. 2005. *The Multiple Values of Wilderness*. State College, PA: Venture Publishing.

Cordell, H. K.; Bergstrom, John C.; Hartmann, Lawrence A.; English, Donald B. K. 1990. An analysis of the outdoor recreation and wilderness situation in the United States: 1989–2040. A Technical Document Supporting the 1989 USDA Forest Service RPA Assessment. General Technical Report RM-189. Fort Collins, CO: U.S. Department of Agriculture, Forest Service, Rocky Mountain Forest and Range Experiment Station.

Cordell, H. K.; Murphy, D.; Riitters, K.; Harvard, J. E., III. 2005. The natural ecological value of wilderness. In: Cordell, H. K.; Bergstrom, J. C.; Bowker, J. M. *The Multiple Values of Wilderness*. State College, PA: Venture Publishing, pp. 205–249.

Cordell, H. K.; Tarrant, M. A.; McDonald, B. L.; Bergstrom, J. C. 1998. How the public views wilderness: More results from the USA survey on recreation and the environment. *International Journal of Wilderness*. 4(3): 28–31.

Cordell, H. K.; Teasley, Jeff. 1998. Recreational trips to wilderness: Results from the USA national survey on recreation and the environment. *International Journal of Wilderness*. 4(1): 23–27.

Cordell, H. K.; Young, M. 1993. An evaluation of wilderness research: State of knowledge and research needs for the 90's. Proceedings of the 1992 *Annual Meeting of the Society of American Foresters*; October 25–28, 1992; Richmond, VA: Washington, DC: Society of American Foresters, pp. 196–201.

Davis, George D. 1989. Preservation of natural diversity: The role of ecosystem representation within wilderness. In: Freilich, H. R., comp. *Wilderness benchmark 1988: Proceedings of the National Wilderness Colloquium*; January 13–14, 1988; Tampa, FL. General Technical Report SE-51. Asheville, NC: U.S. Department of Agriculture, Forest Service, Southeastern Forest Experiment Station, pp. 76–82.

Dawson, Chad P.; Hammitt, William E. 1996. Dimensions of wilderness privacy for Adirondack Forest Preserve hikers. *International Journal of Wilderness*. 2(1): 37–41.

Dawson, Chad P.; Hendee, J. C. 2004. Wilderness-related courses in natural resources programs at U.S. colleges and universities. *International Journal of Wilderness*. 10(1): 33–36.

Dawson, Chad P.; Simon, M.; Oreskes, R.; Davis, G. 2001. Great Gulf Wilderness use estimation: Comparisons from 1976, 1979, and 1999. In: Kyle, G., ed. *Proceedings of the 2000 Northeastern Recreation Research Symposium*; April 2–4, 2000; Bolton Landing, NY. General Technical Report NE-276. Newton Square, PA: U.S. Department of Agriculture, Forest Service, Northeastern Research Station, pp. 283–288.

Dawson, Chad P.; Tangen-Foster, Jim; Friese, Gregory T.; Carpenter, Josh. 1998. Defining characteristics of U.S.A. wilderness experience programs. *International Journal of Wilderness*. 4(3): 22–27.

Driver, B. L.; Nash, Roderick; Haas, Glenn. 1987. Wilderness benefits: A state-of-knowledge review. In: Lucas, Robert C., comp. *Proceedings: National Wilderness Research Conference: Issues, State-of-Knowledge, Future Directions*; July 23–26, 1985; Fort Collins, CO. General Technical Report INT-220. Ogden, UT: U.S. Department of Agriculture, Forest Service, Intermountain Research Station, pp. 294–319.

Duncan, Richard A.; Proescholdt, K. 1999. Protecting the Boundary Waters Canoe Area Wilderness: Litigation and legislation. *Denver University Law Review*. 76(2): 621–658.

Dwyer, John F. 1994. Customer diversity and the future demand for outdoor recreation. General Technical Report RM-252. Fort Collins, CO: U.S. Department of Agriculture, Forest Service, Rocky Mountain Forest and Range Experiment Station.

English, Donald B. K.; Cordell, H. Ken. 1985. A cohort-centric analysis of outdoor recreation participation changes. In: Watson, Alan E., ed. *Proceedings: Southeastern Recreation Research Conference*; February 28–March 1, 1985; Myrtle Beach, SC. Statesboro, GA: Georgia Southern College, Department of Recreation and Leisure Services, pp. 93–110.

Evans, Kent E. 1986. Indian Mounds Wilderness Area: Perceived wilderness qualities and impacts of oil and gas development. In: Kulhavy, David L.; Conner, Richard N., eds. *Wilderness and Natural Areas in the Eastern United States: A Management Challenge*. Nacogdoches, TX: Stephen F. Austin University, School of Forestry, Center for Applied Studies, pp. 156–165.

Ewert, A.; MacEvoy, L. 2000. The effects of wilderness settings on organized groups: A state-of-knowledge paper. In: McCool, S. F.; Cole, D. N.; Borrie, W. T.; O'Loughlin, J., comps. *Wilderness science in a time of change conference*, Vol. 3: *Wilderness as a Place of Scientific Inquiry*; May 23–27, 1999; Missoula, MT. RMRS-P-15. Ogden, UT: U.S. Department of Agriculture, Forest Service, Rocky Mountain Research Station, pp. 13–26.

Ewert, Alan W.; Hendee, John C.; Davidson, Sam; Brame, Richard; Mackey, Craig. 1999. Wilderness access issues for education, personal growth and therapeutic use. *International Journal of Wilderness*. 5(3): 13–18.

Ewert, Alan W.; Hollenhorst, Steven J. 1997. Adventure recreation and its implications for wilderness. *International Journal of Wilderness*. 3(2): 21–26.

Fisher, Anthony; Krutilla, John V. 1972. Determination of optimal capacity of resource-based recreation facilities. *Natural Resources Journal*. 12(3): 417–444.

Foster, John D. 1976. Bureau of Land Management primitive areas—are they counterfeit wilderness? *Natural Resources Journal*. 16(3): 621–663.

Franklin, Jerry F. 1987. Wilderness ecosystem research—a scientific perspective. In: Lucas, Robert C., comp. *Proceedings: National Wilderness Research Conference: Issues, State-of-Knowledge, Future Directions*; July 23–26, 1985; Fort Collins, CO. General Technical Report INT-220. Ogden, UT: U.S. Department of Agriculture, Forest Service, Intermountain Research Station, pp. 42–46.

Friese, Gregory T.; Hendee, John C.; Kinziger, Michael L. 1998. Wilderness experience program industry in the United States: Characteristics and dynamics. *Journal Experiential Education*. 21(1): 40–45.

Friese, Gregory T.; Kinziger, Michael L.; Hendee, John C. 1999. History and status of use of wilderness for personal growth. In: Cordell, K., ed. *Outdoor Recreation in American Life: A National Assessment of Demand and Supply Trends*. Champaign, IL: Sagamore Press, pp. 380–384.

Friese, Gregory T.; Pittman, T.; Hendee, John C. 1996. *Studies of the Use of Wilderness for Personal Growth, Therapy, Education and Leadership Development: An Annotation and Evaluation*. Moscow: University of Idaho Wilderness Research Center.

Gager, Dan; Hendee, J. C.; Kinziger, M.; Krumpe, E. 1998. What managers are saying and doing about Wilderness Experience Programs. *Journal of Forestry*. 96(8): 33–37.

Gladden, James N. 1990. *The Boundary Waters Canoe Area: Wilderness Values and Motorized Recreation*. Ames: Iowa State University Press.

Graefe, Alan R.; Donnelly, Maureen P.; Vaske, Jerry J. 1986. *Crowding and specialization: A reexamination of the crowding model: Proceedings of the National Wilderness Research Conference*; July 23–26, 1985; Fort Collins, Colorado. Ogden, UT: U.S. Department of Agriculture, Forest Service, Intermountain Research Station, pp. 333–338.

Gramann, James H. 1982. Toward a behavioral theory of crowding in outdoor recreation: An evaluation and synthesis of research. *Leisure Sciences*. 5(2): 109–126.

Green, G. T.; Bowker, J. M.; Johnson, C. Y.; Cordell, H. K.; Wang, X. 2007. An examination of constraints to wilderness visitation. *International Journal of Wilderness*. 13(2): 26–36.

Haas, Glenn E.; Herman, Eric; Walsh, Richard. 1986. Wilderness values. *Natural Areas Journal*. 6(2): 37–43.

Hall, Troy; Shelby, Bo. 1996. Who cares about encounters? Differences between those with and without norms. *Leisure Sciences*. 18: 7–22.

Hammitt, William E.; Madden, Mark A. 1989. Cognitive dimensions of wilderness privacy: A field test and further explanation. *Leisure Sciences*. 11: 293–301.

Hendee, John C. 1995. Universities must play a larger role in wilderness. *Trends*. 32(1): 22–27.

Hendee, John C.; Catton, William R., Jr.; Marlow, Larry D.; Brockman, C. Frank. 1968. Wilderness users in the Pacific Northwest—their characteristics, values, and management preferences. Research Paper PNW-61. Portland, OR: U.S. Department of Agriculture, Forest Service, Pacific Northwest Forest and Range Experiment Station.

Hendee, John C.; Clark, Roger N.; Dailey, Thomas E. 1977. Fishing and other recreation behavior at high-mountain lakes in Washington State. Research Note. PNW-304. Portland, OR: U.S. Department of Agriculture, Forest Service, Pacific Northwest Forest and Range Experiment Station. 27p.

Hendee, John C.; Ewert, Alan. 1993. Wilderness research: Future needs and directions. *Journal of Forestry*. 91(2): 18–21.

Hendee, John C.; Lucas, Robert C. 1973. Mandatory wilderness permits: A necessary management tool. *Journal of Forestry*. 71(4): 206–209.

Hendee, John C.; Roggenbuck, Joseph W. 1984. Wilderness-related education as a factor increasing demand for wilderness. In: *Proceedings of the International Forest Congress 1984: Forest Resources Management—The Influence of Policy and Law*; August 6–7, 1984; Quebec City, Quebec, Canada. Bethesda, MD: Society of American Foresters, pp. 273–278.

Hof, John G.; Kaiser, H. Fred. 1983. Projections of future forest recreation use. Resources Bulletin W0-2. Washington, DC: U.S. Department of Agriculture, Forest Service.

Hollenhorst, Steven J.; Whisman, Steven A.; Ewert, Alan W. 1992. Monitoring Visitor Use in Backcountry and Wilderness: A Review of Methods. General Technical Report PSW-134. Albany, CA: U.S. Department of Agriculture, Forest Service, Pacific Southwest Research Station.

Hornback, K. E.; Eagles, P. F. J. 1999. *Guidelines for Public Use Measurement and Reporting at Parks and Protected Areas*. Gland, Switzerland, and Cambridge, UK: International Union for Conservation of Nature and Natural Resources (IUCN). 90p.

Johnson, Janet L.; Franklin, Jerry F.; Krebill, Richard G., coords. 1984. *Research natural areas: Baseline monitoring and management*: Proceedings of a symposium; March 21, 1984; Missoula, MT. General Technical Report INT-173. Ogden, UT: U.S. Department of Agriculture, Forest Service, Intermountain Forest and Range Experiment Station.

Johnson, Laura C.; Wallace, George N.; Mitchell, John E. 1997. Visitor perceptions of livestock grazing in five US wilderness areas, a preliminary assessment. *International Journal of Wilderness*. 3(2): 14–20.

Jungst, Steven E.; Countryman, David W. 1982. Two regression models for projecting future wilderness use. *Iowa State Journal of Research*. 57(1): 33–41.

Landres, Peter; Meyer, S. 1998. National Wilderness Preservation System Database: Key attributes and trends, 1964 through 1998. Research Paper RMRS-GTR-18. Ogden, UT: U.S. Department of Agriculture, Forest Service, Rocky Mountain Research Station.

Leopold, Aldo. 1941. Wilderness as a land laboratory. *The Living Wilderness*. 6(3): 3.

Loomis, J.; Bonetti, K.; Echohawk, C. 1999. Demand for and supply of wilderness. In: Cordell, K., ed. *Outdoor Recreation in American Life: A National Assessment of Demand and Supply Trends*. Champaign, IL: Sagamore Publishing, pp. 351–375.

Loomis, J.; Richardson, R. 2001. Economic values of the U.S. wilderness system: Research evidence to date and questions for the future. *International Journal of Wilderness*. 7(1): 31–34.

Lucas, Robert C. 1980. Use patterns and visitor characteristics, attitudes, and preferences in nine wilderness and other roadless areas. Research Paper INT-253. Ogden, UT: United States Department of Agriculture, Forest Service, Intermountain Research Station. 89p.

———. 1983. Low and variable visitor compliance rates at voluntary trail registers. Research Note INT-326. Ogden, UT: U.S. Department of Agriculture, Forest Service, Intermountain Forest and Range Experiment Station.

———. 1985. Visitor characteristics, attitudes, and use patterns in the Bob Marshall Wilderness Complex, 1970–82. Research Paper INT-345. Ogden, UT: U.S. Department of Agriculture, Forest Service, Intermountain Research Station.

Lucas, Robert C.; Kovalicky, Thomas J. 1981. Self-issued wilderness permits as a use measurement system. Research Paper INT-270. Ogden, UT: U.S. Department of Agriculture, Forest Service, Intermountain Forest and Range Experiment Station.

Lucas, Robert C.; Stankey, George H. 1988. *Shifting trends in wilderness recreational use: Proceedings of the National Outdoor Recreation Forum*; January 13–14, 1988; Tampa, FL. General Technical Report SE-52. Asheville, NC: U.S. Department of Agriculture, Forest Service, Southeastern Forest Experiment Station, pp. 357–367.

Mackey, C. 1998. U.S. wilderness management in the 21st century—politics, policy and partnerships. *International Journal of Wilderness*. 4(3): 6–11.

Manning, R. E. 1989. Social research in wilderness: Man in nature. Wilderness Benchmark Conference. In: Freilich, Helen R., comp. *Proceedings of the National Wilderness Colloquium*; January 13–14, 1988; Tampa, FL. General Technical Report SE-51. Asheville, NC: U.S. Department of Agriculture, Forest Service, Southeastern Forest Experiment Station, pp. 120–132.

Manning, Robert E. 1985. Crowding norms in backcountry settings: A review and synthesis. *Journal of Leisure Research*. 17(2): 75–89.

Marnell, Leo F. 1977. Methods for counting river recreation users. In: *Proceedings: River Recreation Management and Research Symposium*; January 24–27, 1977; Minneapolis, MN. General Technical Report NC-28. St. Paul, MN: U.S. Department of Agriculture, Forest Service, North Central Forest Experiment Station, pp. 77–82.

McClaran, M. P. 1990. Livestock in wilderness: A review and forecast. *Environmental Law*. 20(4): 857–889.

McCool, S. F.; Cole, D. N.; Borrie, W. T.; O'Loughlin, J. 2000a. *Wilderness Science in a Time of Change Conference*, Vol. 2: *Wilderness Within the Context of Larger Systems*; May 23–27, 1999; Missoula, MT. RMRS-P-15. Ogden, UT: U.S. Department of Agriculture, Forest Service, Rocky Mountain Research Station.

———. 2000b. *Wilderness Science in a Time of Change Conference*, Vol. 3: *Wilderness as a Place of Scientific Inquiry*; May 23–27, 1999; Missoula, MT. RMRS-P-15. Ogden, UT: U.S. Department of Agriculture, Forest Service, Rocky Mountain Research Station.

Meyer, S. S. 1999. The role of legislative history in agency decision-making: A case study of wilderness airstrip management in the United States. *International Journal of Wilderness*. 5(2): 9–12.

Mortensen, Charles O. 1997. Henry David Thoreau: A lecture and a wilderness legacy. *International Journal of Wilderness*. 3(4): 13–16.

———. 1999. Environmental perception: The influence of wilderness on United States artists, writers and their legacy. In: Watson, A. E.; Aplet, G. H.; Hendee, J. C., comps. *Personal, societal, and ecological values of wilderness: 6th World Wilderness Congress Proceedings on Research, Management, and Allocation*, Vol. II; October 24–29, 1998; Bangalore, India. RMRS-P-14. Ogden, UT: U.S. Department of Agriculture, Forest Service, Rocky Mountain Research Station, pp. 120–122.

Murray, Michael P. 1997. High elevation meadows and grazing: Common past effects and future improvements. *International Journal of Wilderness*. 3(4): 24–28.

O'Leary, Joseph T.; Dottavio, F. D.; McGuire, F. 1989. Social factors in recreation participation and demand: Implications from national surveys; January 13–14, 1988; Tampa, FL. General Technical Report SE-52. Asheville, NC: U.S. Department of Agriculture, Forest Service, Southeastern Forest Experiment Station, pp. 275–289.

O'Rourke, Deb. 1994. *Trail traffic counters for Forest Service trail monitoring*. Missoula, MT: U.S. Department of Agriculture, Forest Service, Technology and Development Program.

Outdoor Recreation Resources Review Commission (ORRRC). 1962. *Wilderness Recreation—A Report on Resources, Values, and Problems*. ORRRC Study Report 3. Washington, DC: U.S. Government Printing Office. 352p.

Oye, Garry. 2001. A new wilderness recreation strategy for national forest wilderness. *International Journal of Wilderness*. 7(1): 13–14.

Parsons, D. J. 2000. Wilderness science and the Aldo Leopold Wilderness Research Institute. *International Journal of Wilderness*. 6(1): 44.

———. 2007. An outside assessment of wilderness research in the Forest Service. *International Journal of Wilderness*. 13(3): 34–35, 39.

Petersen, Margaret E. 1985. Improving voluntary registration through location and design of trail registration stations. Reserach Paper INT-336. Ogden, UT: U.S. Department of Agriculture, Forest Service, Intermountain Forest and Range Experiment Station.

Powers, Thomas M. 1996. Wilderness economics must look at wilderness thru the windshield not the rearview mirror. *International Journal of Wilderness.* 2(1): 5–9.

Reed, Patrick; Haas, Glenn; Beum, Frank; Sherrick, Lois. 1989. Non-recreational uses of the National Wilderness Preservation System. A 1988 telephone survey. In: Freilich, Helen R., comp. *Proceedings of the National Wilderness Colloquium*; January 13–14, 1988; Tampa, FL. General Technical Report SE-51. Asheville, NC: U.S. Department of Agriculture, Forest Service, Southeastern Forest Experiment Station, pp. 220–228.

Ridzitis, Gundars. 1996. *Wilderness and the Changing American West.* New York: John Wiley.

Robinson, John P.; Blair, J. 1995. The national macroenvironmental activity pattern survey (preliminary report). Washington, DC: Environmental Protection Agency.

Robinson, John P.; Godbey, Geoffrey. 1997. *Time for Life: The Surprising Ways Americans Use Their Time.* University Park: Pennsylvania State University Press.

Roggenbuck, Joseph W.; Lucas, Robert C. 1987. Wilderness use and user characteristics: A state-of-knowledge review. In: Lucas, Robert C., comp. *Proceedings: National Wilderness Research Conference: Issues, state-of-knowledge, future directions*; July 23–26, 1985; Fort Collins, CO. General Technical Report INT-220. Ogden, UT: U.S. Department of Agriculture, Forest Service, Intermountain Research Station, pp. 204–245.

Roggenbuck, Joseph W.; Watson, Alan E. 1989. Wilderness recreation use: The current situation. In: Watson, Alan E., comp. *Outdoor recreation benchmark 1988: Proceedings of the National Outdoor Recreation Forum*; January 13–14, 1988; Tampa, FL. General Technical Report SE-52. Asheville, NC: United States Department of Agriculture, Forest Service, Southeastern Forest Experiment Station, pp. 346–356.

Russell, Keith; Hendee, John C.; Phillips-Miller, D. 2000. How wilderness therapy works: An examination of the wilderness therapy process to treat adolescents with behavioral problems and addictions. In: McCool, S. F.; Cole, D. N.; Borrie, W. T.; O'Loughlin, J., comps. *Wilderness Science in a Time of Change Conference*, Vol. 3: *Wilderness as a Place of Scientific Inquiry*; May 23–27, 1999; Missoula, MT. RMRS-P-15-Vol-3. Ogden, UT: U.S. Department of Agriculture, Forest Service, Rocky Mountain Research Station, pp. 207–217.

Schuster, R. M.; Tarrant, M.; Watson, A. 2005. The social value of wilderness. In: Cordell, H. K.; Bergstrom, J. C.; Bowker, J. M. *The multiple values of wilderness.* State College, PA: Venture Publishing, pp. 113–142.

Stankey, George H. 1971. Myths in wilderness decision making. *Journal of Soil and Water Conservation.* 25(5): 183–188.

U.S. Department of Agriculture, Forest Service. 1961. Table G, Summary of National Forest Recreation Survey, Form No. 10, Appendix 18, National Forest Recreation Survey (NFRS). Unpublished report on file at U.S. Department of Agriculture, Forest Service, Intermountain Research Station, Forestry Sciences Laboratory, Missoula, MT.

U.S. House of Representatives. 1979. House Report 96-617. 96th Congress, 1st Session. Washington, DC: U.S. Government Printing Office. 14p.

U.S. Public Law 88-577. The Wilderness Act of September 3, 1964. 78 Stat. 890.

U.S. Public Law 93-378. The Forest and Rangeland Renewable Resources Planning Act (RPA) of 1974. 88 Stat. 476.

U.S. Public Law 94-429. Mining in the Parks Act of September 28, 1976. 90 Stat. 1342.

U.S. Public Law 94-579. (Wilderness Study) Federal Lands Policy and Management Act (FLPMA) of 1976. 90 Stat. 2743.

U.S. Public Law 95-495. (Boundary Waters Canoe Area Wilderness) Act of October 21, 1978. 92 Stat. 1649.

U.S. Public Law 96-560. (Colorado Wilderness) Act of December 2, 1980. 94 Stat. 2371.

U.S. Public Law 96-550. (New Mexico Wilderness) Act of December 19, 1980. 94 Stat. 3221.

U.S. Public Law 98-406. Arizona Wilderness Act of August 28, 1984. 98 Stat. 1485.

U.S. Public Law 98-428. Utah Wilderness Act of September 28, 1984. 98 Stat. 1657.

U.S. Public Law 98-550. Wyoming Wilderness Act of October 30, 1984. 98 Stat. 2807.

U.S. Public Law 99-504. Nebraska Wilderness Act of October 20, 1986. 100 Stat. 1802.

U.S. Public Law 100-225. (El Malpais and Cebolla Wilderness) Act of December 31, 1987. 101 Stat. 1539.

U.S. Public Law 103-433. California Desert Protection Act of October 31, 1994. 108 Stat. 4473.

Van Horne, Merle J.; Szwak, Laura B.; Randall, Sharon A. 1985. Outdoor recreation activity trends—insights from the 1982-83 nationwide recreation survey. In: Wood, James D., Jr., ed. *Proceedings: National Outdoor Recreation Trends Symposium II*; February 25–27, 1985; Myrtle Beach, SC. Atlanta, GA: U.S. Department of the Interior, National Park Service, Southeast Regional Office, pp. 109–130.

van Wagtendonk, Jan W. 1981. The effect of use limits on backcountry visitation trends in Yosemite National Park. *Leisure Sciences.* 4(3): 311–323.

Washburne, Randel F.; Cole, David N. 1983. Problems and practices in wilderness management: A survey of managers. Research Paper INT-304. Ogden, UT: U.S. Department of Agriculture, Forest Service, Intermountain Forest and Range Experiment Station.

Watson, A., Sproull, J.; Dean, L. 2007. *Science and stewardship to protect and sustain wilderness values: Eighth World Wilderness Congress Symposium.* September 30–Oct., 6, 2005; Anchorage, AK. Proceedings RMRS-P-49. Fort Collins, CO: U.S. Department of Agriculture, Forest Service, Rocky Mountain Research Station.

Watson, Alan E. 1998. Sustainable financing of parks and protected areas—are user fees the answer? *International Journal of Wilderness.* 4(3): 5.

Watson, Alan E. 2000. Wilderness use in the year 2000: Societal changes that influence human relationships with wilderness. In: Cole, D. N.; McCool, S. F.; Borrie, W. T.; O'Loughlin, J., comps. *Wilderness Science in a Time of Change Conference*, Vol. 4: *Wilderness Visitors, Experiences, and Visitor Management*; May 23–27, 1999; Missoula, MT. RMRS-P-15. Ogden, UT: U.S. Department of Agriculture, Forest Service, Rocky Mountain Research Station, pp. 53–60.

Watson, Alan E.; Cole, D. N.; Friese, G.T.; et al. 1999. Wilderness uses, users, values and management. In: Cordell, K., ed. *Outdoor Recreation in American Life: A National Assessment of Demand and Supply Trends.* Champaign, IL: Sagamore Publishing, pp. 377–401.

Watson, Alan E.; Cole, D. N.; Roggenbuck, J. W. 1995. Trends in wilderness use characteristics. In: Thompson, J. L.; Lime, D.; Gartner, B.; Samer, W., comps. *Proceedings of the Fourth International Outdoor Recreation and Tourism Trends Symposium*; May 14–17, 1995; St. Paul, MN. St. Paul: University of Minnesota, College of Natural Resources, pp. 68–71.

Watson, Alan E.; Cole, D. N.; Turner, D. L.; Reynolds, P. S. 2000. Wilderness Recreation Use Estimation: A Handbook of Methods and Systems. Reserach Paper RMRS-GTR-56. Ogden, UT: U.S. Department of Agriculture, Forest Service, Rocky Mountain Research Station.

Watson, Alan E.; Hendee, John C.; Zaglauer, Hans P. 1996. Human values and codes of behavior: Changes in Oregon's Eagle Cap Wilderness visitors and their attitudes. *Natural Areas Journal.* 16(2): 89–93.

Watson, Alan E.; Williams, Daniel R.; Roggenbuck, Joseph W.; Daigle, John J. 1992. Visitor characteristics and preferences for three national forest wildernesses in the South. Research Paper INT-455. Ogden, UT: U.S. Department of Agriculture, Forest Service, Intermountain Research Station.

Watson, J.; Brink, P. 1996. The California Desert Protection Act—a time for desert parks and wilderness. *International Journal of Wilderness.* 2(2): 14–17.

Wilkinson, Charles F.; Anderson, H. Michael. 1985. Wilderness. *Oregon Law Review.* 64(1, 2): 334–370.

Williams, Daniel R.; Roggenbuck, Joseph W.; Bange, Steve. 1991. The effect of norm-encounter compatibility on crowding perceptions, experience, and behavior in river recreation settings. *Journal of Leisure Research.* 23(2): 154–172.

Worf, Bill. 2001. The new Forest Service wilderness recreation strategy spells doom for the National Wilderness Preservation System. *International Journal of Wilderness.* 7(1): 15–17.

Yuan, Susan; Maiorano, Brian; Yuan, Michael; Kocis, Susan M.; Hoshide, Gary T. 1995. Techniques and equipment for gathering visitor use data on recreation sites. 2300-Recreation, 9523-2838-MTDC. Missoula, MT: U.S. Department of Agriculture, Forest Service, Technology and Development Program.

Chapter 15
Ecological Impacts of Wilderness Recreation and Their Management

by David N. Cole

Introduction 396
Importance of Wilderness Recreation Impacts 396
Recreational Activities and Associated Impacts 398
    Trampling 398
    Campfires 400
    Trail Construction and Maintenance 402
    Pack-Stock Grazing for Recreational Visits 403
    Wildlife Disturbance 404
    Water Pollution and Disposal of Human Waste 406
Managing Campsite Impacts 407
    Campsite Impacts 407
        Vegetation Change 408
        Changes in Soil Condition 408
        Temporal Patterns of Impact 409
    Factors That Influence Amount of Impact 410
        Amount of Use 410
        Type and Behavior of Site Users 411
        Environmental Conditions and Durability of the Site 412
    Management Strategies and Techniques to Control Impacts 413
        Limiting Amount of Use 413
        Dispersal of Use 414
        Temporary Campsite Closures 414
        Limitations on Length of Stay 415
        Party Size Limits 415
        Encouraging Use of Resistant Sites 415
        Setbacks from Water 415
        Site Hardening and Facilities 416
        Containing Impacts 417
    Monitoring Campsite Conditions 418
    Campsite Restoration 420
        Five Steps for Site Restoration Plans 420
        Guidelines to Implementing Site-Restoration Plans 421
        Visitor Management 422
        Site Stabilization and Preparation 422
        Propagation, Collection, and Planting 422
        Site Maintenance 423
Managing Trail Impacts 423
    Three Common Problems: Erosion, Muddiness, User-Created Trails 424
        Erosion 424
        Muddiness 426
        User-Created Trails 426

| Monitoring Trail Conditions 428
| Trail Rehabilitation 429
| Managing Pack and Saddle Stock 429
| Types of Stock Impact 430
| Management Strategies and Techniques 431
| Limiting Use 431
| Encouraging Less Damaging Behavior 432
| Managing the Timing of Use 433
| Managing Location of Use 433
| The Challenge of Stock Management 434
| Summary 434
| Study Questions 435
| Acknowledgments 435
| Case Discussion: Baseline Wilderness Data 435
| Case Discussion Questions 436
| References 436

Introduction

Four parallel trails gouged into a wildflower-dotted alpine meadow, denuded campsites with severe soil erosion, numerous trees battered and scarred by tethered livestock. Such recreational impacts are all too common in wilderness. Wilderness management to prevent and restore such impacts is a challenge and directly related to managing for quality wilderness experiences.

This chapter begins with a discussion of the significance of recreational impacts, its purpose being to bring recreation impacts into perspective with other wilderness management problems. This discussion is followed by a description of important types of recreational impacts, those caused by trampling, campfires,

Fig. 15.1. Recreation activities are highly concentrated at campsites and frequently result in severe ecological impacts. This site has lost a substantial amount of vegetation, organic matter has eroded, and mineral soil is compacted. Photo courtesy David Cole, U.S. Forest Service.

construction and maintenance of trails, pack animals, wildlife disturbance, and water pollution and disposal of human waste. The remainder of the chapter deals with impacts associated with campsites, trails, pack and saddle stock, and alternative management responses to problems based on ecological impacts.

The examples in this chapter are largely drawn from the author's experience in the western United States. The international literature was consulted to assess the degree to which these examples are typical, and if it was possible to generalize the management implications. Other examples and studies are included where western U.S. results are not broadly applicable.

Importance of Wilderness Recreation Impacts

To evaluate the importance of recreation impacts in wilderness areas, it is important to reexamine the goals of wilderness management. The relevant phrases in the Wilderness Act state that wilderness is an area "which is protected and managed so as to preserve its natural conditions and which *generally* appears to have been affected *primarily* by the forces of nature, with the imprint of man's work *substantially* unnoticeable" (emphasis added). Clearly some recreation impacts will be tolerated, but (1) the integrity of relatively undisturbed wilderness ecosystems should not be substantially compromised, and (2) the evidence of impacts should not be conspicuous. The first of these objectives is concerned with protecting wilderness ecosystems; the second, with protecting the quality of the visitor's experience. Both are important.

What constitutes a substantial compromise of ecosystem integrity? There is no clear answer to this question. *I suggest that the most important impacts are those that seriously disrupt ecosystem composition, structure, and function and that are evident at large spatial scales. Impacts that alter a large proportion of a relatively rare ecosystem are particularly detrimental, as are irreversible changes. But where impacts occur repeatedly, even easily reversible changes are important.*

The ecological impacts of recreation can be severe. The composition, structure, and function of sites that are used for recreation are often nearly completely altered. At the site scale, then, recreation impacts are as profound as any human threat to wilderness. However, these impacts are highly localized. Recreation use is highly concentrated along a few major trails and at a few popular destinations. This leaves the vast majority of most wildernesses essentially unvisited and, therefore, virtually undisturbed by recreation use. Marion and Leung (1997) estimate that only 0.05 percent of the backcountry at Great Smoky Mountains National Park has been directly disturbed by recreation use. At large scales, then—from the perspective of landscapes, watersheds, and regions—recreation impacts are generally not severe. At these scales, far more potent threats to the integrity of wilderness ecosystems exist. Acid rain is altering the basic ecology of lakes throughout uncommon ecosystems in certain areas. A New York state wilderness in the Adirondack Mountains provides an example of an area substantially affected by acid rain; aquatic ecosystems are most heavily impacted. In addition to external threats, such as acid rain, internal threats can be potent. The impacts of fire suppression and livestock grazing, though less severe than recreation at the site scale, are expressed over vast wilderness acreages (Cole and Landres 1996).

Generally, then, recreation use, although causing locally severe impacts, does not substantially compromise wilderness landscapes. There are several important exceptions to this generalization, however. *The first exception is the introduction of fish, often exotic species, into lakes and streams without fish or predator species and the subsequent removal of fish through angling.* In such cases, adding a new component to the food chain has impacted entire aquatic ecosystems. It has reduced, displaced, or eliminated competitors or prey organisms and has stimulated the invasion or expansion of predatory populations. Fish stocking has been implicated as one of the reasons for the decline in amphibian populations in the wilderness of the Sierra Nevada, for example (Knapp 2005). Angling, although partially negating the effect of introductions, has altered the population structure of fisheries and, in extreme cases, has nearly eliminated a major consumer and link in the food chain.

Generally, fishing is allowed throughout wilderness. In a few cases, angling pressure is attenuated by "catch-and-release" regulations. Such regulations are in effect in parts of the Frank Church–River of No Return Wilderness, Idaho, and the Golden Trout Wilderness, California, for example. However, wilderness designation seldom brings such restrictions. A more common (and better) action is to discontinue the artificial stocking of fish. This will allow some areas, where reproduction is poor, to revert to a more natural state.

A second recreation impact that can be important at large spatial scales is disturbance of wildlife because of hunting or unintentional harassment. Hunting is generally allowed in wildernesses, outside of the national parks. Although long-lasting effects are difficult to document, unintentional disturbance has undoubtedly altered the distribution, population structure, and behavior of many wildlife species.

Finally, less common but also important, is physical alteration or pollution of uncommon ecosystems or sites inhabited by rare plants or animals. Disturbance of such places can lead to eliminating rare species or altering most examples of certain ecosystem types. Pollution of lakes, disturbance of meadows by pack stock, and camping impacts along desert riparian strips are examples of such impacts that occur in certain wildernesses.

The effect of recreation impacts on visitors, in contrast, is most important when it is apparent at the site scale—the scale most relevant to human interaction with the natural environment, where people spend the most time. Consequently, recreation use—because it severely impacts ecosystems at the site scale—commonly compromises the goal of avoiding conspicuous evidence of human impact. The importance of such impacts depends on the visitor's sense of aesthetics and sensitivity to change, whether or not the change is perceived as desirable, and the magnitude of the change. Thus, trail impacts in meadows are more troublesome than impacts in forests because they are more obvious and aesthetically displeasing, even though the amount

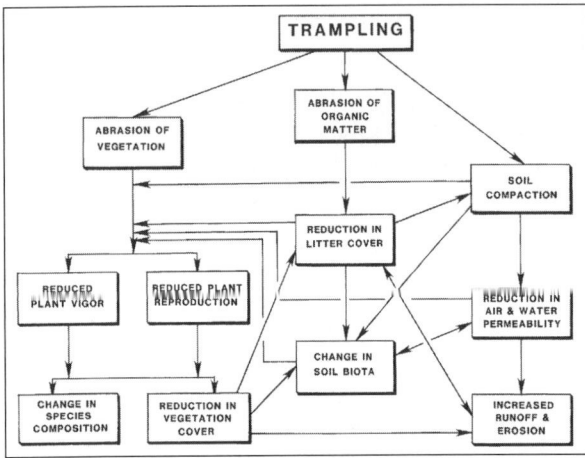

Fig. 15.2. A conceptual model of trampling effects (partially based on Liddle 1979). Note the numerous reciprocal and cyclic relationships between soil and vegetational impacts.

of change in meadows may be lower than in forests. Felling trees and cutting brush may be desirable and appropriate on trails where it makes travel easier but undesirable and inappropriate elsewhere.

Few studies of visitor perceptions of impacts show much relationship between visitor satisfaction and amount of impact. Visitors consistently report that recreation impacts—*if they occurred*—would adversely affect their wilderness experience (Roggenbuck et al. 1993). However, many visitors do not notice ecological impacts that have occurred. Of those who do notice impacts, many do not perceive of these impacts as "damage"—or undesirable change. Finally, most visitors do not change their behavior or have less satisfactory experiences, even when confronted by impacts that they consider undesirable (Farrell et al. 2001). For example, even those who dislike the heavy evidence of horse use in the Bob Marshall Wilderness are likely to continue to camp in the same places and travel the same trails and, on the whole, enjoy it. This suggests that although most wilderness visitors do not like the "idea" of recreation impacts, few visitors are bothered much by even relatively high degrees of alteration.

In dramatic contrast, site impacts are the foremost concern of many managers (Marion, Roggenbuck et al. 1993). Managers are often well aware of such impacts and are charged, as managers, to deal with them despite the apparent indifference of visitors to the impacts around them. Most of the discussion that follows deals more with management to avoid conspicuous evidence of human use than with management to maintain ecosystem integrity.

Recreational Activities and Associated Impacts

Now that we have discussed the importance of recreation impacts in light of wilderness management goals, we will take a more detailed look at the effects of various recreation activities on the wilderness resource.

Trampling

The old adage to "take nothing but pictures and leave nothing but footprints" is outdated. In many places, too many footprints have left an unwanted legacy. The effects of human trampling have been investigated for almost a century, and we now understand many of the general effects of trampling. These can be conveniently displayed in a conceptual model.

Trampling has three initial effects: abrasion of vegetation, abrasion of surface soil organic layers, and soil compaction. Plants can be crushed, sheared off, bruised, and even uprooted by trampling. Any consistent trampling is likely to reduce the vigor and reproductive capacity of all but the most resistant plant species.

Various physiological and morphological changes occur when vegetation is trampled. Changes include reductions in plant height, stem length, and leaf area, as well as in number of plants that flower, number of flower heads per plant, and seed production (Liddle 1997). In Glacier National Park, Montana, Hartley (1999) found reduced carbohydrate reserves in the roots of trampled glacier lilies, presumably a response to a reduction in the ability to photosynthesize after trampling. All of these changes are manifested in reduced vigor and reproduction, leading to less plant biomass and cover. Thus, *most recreation sites exhibit a gradient in which the cover of vegetation decreases from undisturbed vegetation to places where trampling stress is concentrated and no vegetative cover survives. The sharpness of this gradient depends on both the fragility of the vegetation and whether the trampling is concentrated or dispersed.* For example, the gradient from no vegetation to normal cover levels is very narrow along trails, where trampling is concentrated, particularly those trails through fragile vegetation, and quite wide on campsites, where trampling is dispersed, particularly campsites in resistant vegetation (Cole and Monz 2004).

Plants vary in their ability to tolerate trampling. Some plants are even favored by trampling—not as a direct response to abrasion, but in response to reduced competition with other plants and favorable changes in microclimate that result from trampling. For example, trampling often increases light levels and temperature ranges—changes that favor certain species. Many, if not most, of these favored species are introduced plants that are brought into the wilderness by humans and pack stock.

The vegetation that grows in areas of moderate disturbance, then, is very different in composition from undisturbed vegetation. It consists of trampling-tolerant survivors as well as either native or nonnative invading species capable of growing in the local environment and dispersing to the site. Which of these types of plants dominates is highly variable. In Oregon's Eagle Cap Wilderness, dandelion, a Eurasian weed, is the most common plant on lower-elevation campsites. Apparently, it cannot tolerate conditions on campsites in subalpine forests, where the most common plants are native rushes and sedges that were originally on the site and that survive trampling, although at reduced densities (Cole 1982b). Nonnative species are more invasive on low-elevation recreation sites.

The tendency for the original species on a site to decrease in density, creating more favorable conditions for new invading species, explains the observation that species richness—the number of different species occupying a site—often increases with low to moderate levels of trampling before declining to zero as trampling intensifies (Liddle 1997).

Knowledge of which species and types of plants are most tolerant of trampling can be useful in locating sites with resistant vegetation types. Generally, broad-leaved herbs, lichens, low shrubs, and tree seedlings have little tolerance and are quickly eliminated on recreation sites. Eliminating tree seedlings on forested recreation sites portends major, undesirable changes on these sites once the overstory dies. Established trees, because of their size, are little affected by trampling except in a few cases where studies have found reduced growth rates, increased water stress, and damage to root systems, and increased windthrow as a result of erosion.

Cole (1995b) has shown, for groundcover plants, that the ability to resist trampling *(resistance)* decreases with erectness and that broad-leaved herbs are less resistant than grasslike plants and shrubs. Herbs growing in shade are particularly intolerant of trampling because *adaptations to shading*—possession of large, thin leaves and tall stems—make these plants vulnerable when trampled. This explains the common finding that trampling of forested sites generally results in more rapid loss of vegetation than trampling of open woodlands or meadows (Cole 1993b). Low shrubs, such as heather, are relatively resistant to trampling stress, but once damaged they recover slowly. Their *resilience* is low. Grasslike plants are most tolerant of trampling.

Although their size spares them from most trampling damage, large shrubs and trees are affected by a number of associated recreation activities. Vegetation removal during trail construction and for firewood will be discussed later. Trees may be felled for tent poles, hitch rails, or other structures, and shrubs and trees may be removed to create additional tent space. They are also subjected to *deliberate mutilation*—carved initials and ax scars.

Normally, less than one-half of a given volume of soil is solid matter; the rest is pore space that contains air and water. *Trampling compacts the soil—presses together the solid soil particles, filling or compressing many of the pores.* Larger pores—those that permit rapid percolation of water after precipitation and are normally occupied by air—are severely reduced by trampling. This can indirectly affect vegetation and soil microbiota. It can cause oxygen shortages and reduce water availability. Along with the greater difficulty plant roots have in moving through compacted soils, these changes generally reduce plant vigor and retard reproduction and establishment of seedlings. But plants vary greatly in their response to compaction. While some species are adversely affected by any compaction, others respond positively to low levels of compaction (Godefroid and Koedam 2004).

In addition to effects on aboveground vegetation, compaction affects soil biota. Larger organisms, such as earthworms, that help rejuvenate the soil find it more difficult to penetrate dense soils. Microbiota are likely to be adversely affected by lower oxygen concentrations. Of particular concern are adverse effects on mycorrhizal fungi, which improve nutrient uptake and water absorption in plants and may be a limiting factor in revegetating disturbed areas.

Perhaps of greater importance, compaction drastically reduces the rate at which water filters into the soil. Water that does not filter into the soil runs off across the surface. Greater runoff increases erosion potential and decreases the supply of soil water. Although this reduced water supply is unlikely to be a problem in areas with plentiful rainfall, it is likely to increase water stress in arid environments or during dry periods of the year.

Erosion is likely to be a more common and important problem. Deeply eroded trails are unsightly and difficult to use. Erosion on campsites and other sites of concentrated use, by removing the most productive soils on the site, diminishes the potential for vegetation growth on the site. Moreover, the formation of soil is such a slow process that erosion can be considered an irreversible process.

Abrasion and loss of organic matter exacerbate many of these same problems. Normally, thick organic horizons protect mineral soil from much of the direct impact of trampling and decrease surface water runoff. Loss of these layers facilitates increased compaction and runoff. This leads to further loss of organic litter carried off by running water. On canoe-accessible campsites along the Delaware River in Pennsylvania and New Jersey, organic soil horizons were only one-third as thick as on undisturbed sites (Marion and Cole 1996).

Loss of litter directly affects plant and animal populations, both above and below the soil surface. For example, vegetation composition is likely to shift as plants that germinate best on organic media give way to those that are more successful on bare mineral soil. Seeds of most species are unlikely to germinate on a smooth, compacted soil surface devoid of litter without a variety of microenvironments.

Soil biota are a little-studied but important component of ecosystems. Microbial communities develop in response to plant exudates and organic matter. In turn, they contribute to ecosystem functioning by metabolizing nutrients, transforming soil organic matter, producing phytohormones, and contributing to soil food webs. Soil biota are affected by many of the types of impact just described. As soils are compacted, biota requiring larger pores to live in will be eliminated. Populations will decline and shift in composition as primary energy sources—aboveground plants and soil organic matter—are eliminated. Zabinski and Gannon (1997) report substantial reductions in the functional diversity of microbial populations on a backcountry campsite in Montana.

Fig. 15.3. A heavily impacted campsite in the subalpine forest of the Eagle Cap Wilderness, Oregon, illustrates many of the effects of trampling. Note the loss of vegetation and exposed mineral soil and tree roots. Photo courtesy David Cole, U.S. Forest Service.

Campfires

Collecting and burning wood in campfires have their own unique impacts (Fig. 15.4). As with trampling, these impacts are both *aesthetic (fire rings, blackened rocks, and charcoal)* and *ecological (felled trees, deadwood removal, and sterilized soils)*. Ecological impacts result from both the removal of wood, either live or dead, standing or on the ground, from large areas around the campsite and burning the wood in campfires.

Removing firewood and associated trampling greatly enlarge the area affected by camping activities. In Great Smoky Mountains National Park in Tennessee and North Carolina, the area disturbed by firewood collection was typically more than nine times the size of the devegetated area around campsites. In this much larger area, the number of live and dead trees, usually

Fig. 15.4. Remnants of campfires, fire rings and charcoal, have negative aesthetic and ecological consequences. Photo courtesy David Cole, U.S. Forest Service.

the smaller size classes, was reduced, as were woody fuels. Pieces of wood one to three inches in diameter were barely one-third as abundant as on neighboring undisturbed sites (Bratton et al. 1982). T. Hall and Farrell (2001) reported an average 40 percent reduction of woody material thirty to forty-five feet from the center of wilderness campsites in Oregon. Around backcountry campsites in Yellowstone National Park, Taylor (1997) reports that the density of tree saplings is typically reduced for a distance of more than one hundred feet from the center of a campsite.

Because leaves, needles, and twigs—the tree components most critical to nutrient cycling—are little affected by firewood collection, removing downed wood need not adversely affect long-term site productivity. The major source of damage is likely to be the elimination of large (more than three-inch-diameter) woody debris. Decaying wood of this size plays an important role in the ecosystem (Maser et al. 1988). It has unusually high water-holding capacity; accumulates nitrogen, phosphorus, and sometimes calcium and magnesium; and is an important site for nitrogen-fixing microorganisms. It is the preferred substrate for seedling establishment and subsequent growth of certain species. Of particular importance, *ectomycorrhizal fungi*—organisms that develop a symbiotic association with the roots of most higher plants, improving the plants' ability to extract water, nitrogen, and phosphate from less fertile soils—are also concentrated in decayed wood. Consequently, although collection of smaller pieces of wood is unlikely to cause adverse impacts, eliminating large woody debris is likely to reduce site productivity, particularly on droughty and infertile soils.

Loss of large woody debris is likely to adversely affect the populations of invertebrates, small mammals, and birds that use the wood as a food source or living place (Bull et al. 1997). It eliminates sites protected from trampling where seedlings can regenerate and removes natural dams that decrease the potential for soil erosion. *Clearly, loss of large woody debris due to firewood collection is a serious impact where a sizable area is affected. Collection of wood that can be broken by hand is likely to have little effect.*

A wood campfire severely affects the area burned. Fenn et al. (1976) found that a single intense campfire burned 90 percent of the organic matter in the upper inch of soil. *Fires cause pronounced changes in soil chemistry.* Reported fire effects include the loss of nitrogen, sulfur, and phosphorus; increases in pH and many cations (positively charged ions); and reductions in the moisture-holding capacity, infiltration rates, and microbiotic populations of soil. Arocena et al. (2006) report elevated levels of copper in fire pits in backcountry areas of Mount Robson Provincial Park, British Columbia, Canada. Overall, these changes constitute a sterilization of the soil, likely to render the site less hospitable for the growth of vegetation and likely to require at least ten to fifteen years to recover, particularly if the site has been used for some time. Such impacts are particularly pronounced where fires have been built in several places on a single campsite.

Other problems result from carelessness with campfires. Escaped campfires have burned thousands of acres. More common is the destruction of small vegetational coppices by creeping root fires. This has been a serious problem in subalpine tree clumps in the mountains of the Southwest. Here, such clumps are considered beautiful, are valued highly as campsites, and regenerate very slowly, if at all.

Although more research will be necessary before we understand the importance of firewood collection and burning, the preceding review suggests some effective means of minimizing impact. *With firewood collection, the key is not disturbing woody debris larger than about three inches in diameter.* The firewood-collection problem can be minimized by teaching visitors to collect only wood they can break by hand. If campers would leave their axes and saws at home, many impacts—loss of large woody debris, scarring, and felling of trees—would be eliminated. It may also be necessary to prohibit or discourage campfires, or to promote the use of small stoves in areas where wood production is low. If use in such areas is high, firewood will quickly disappear, tempting visitors to use the larger pieces. In fact, campfires should probably be discouraged or prohibited wherever use is very high—regardless of site productivity—to avoid adverse impacts. As of 1991, about one-half of National Park Service backcountry areas prohibited campfires, at least in some places (Marion, Roggenbuck et al. 1993). This prohibition was most common in areas with little firewood—arid, Arctic, and alpine regions. Many areas prohibit fires in zones of low productivity, such as the white bark pine forest, at high elevations in the Sierra Nevada. Such a prohibition at high elevations would also reduce the destruction of tree coppices by escaped campfires.

Fig. 15.5. A bridge crosses a boggy area; such structures, although critical to mitigating resource damage, should be kept to a minimum. Photo courtesy David Cole, U.S. Forest Service.

Minimizing campfire impacts is more complex, involving *choosing between confining impacts to a small total area of concentrated use and dispersing and covering up evidence of use.* Generally, in heavily used areas, campers should be encouraged to use established fire rings so that only a small amount of ground is severely altered. This requires leaving a fire ring at well-used sites, encouraging campers to use these sites, and keeping them clean and attractive. In areas that are infrequently used, it might be better to persuade visitors to use undamaged sites; to build small, less damaging fires; and to camouflage the site when they leave to discourage repeated use of the site and, through the addition of organic matter, initiate recovery of the site. In this case, campers must be willing to select undisturbed sites and be able to leave the site looking undisturbed. *The worst situation is allowing fire rings to move around a site, continually being rebuilt after being removed by rangers, volunteers, or earlier campers, and allowing many fire sites to proliferate at popular destinations.* Unfortunately, this situation is all too common. Further discussion of the pros and cons of dispersal and concentration can be found in the section on campsite management in this chapter and in Reid and Marion (2005).

Trail Construction and Maintenance

Constructing and maintaining trails have pronounced effects on vegetation and soil. Although these impacts are usually deliberate and considered necessary to provide recreation opportunities and to manage visitor traffic, as with all others, they should be kept to a minimum.

The major impacts of trail construction and maintenance stem from opening up tree and brush canopies; building a barren, compacted trail tread that may alter drainage patterns; and creating a variety of new habitats in the process. Common maximum trail clearing widths and heights range from 4 by 8 feet on hiker trails to 8 by 10 feet on trails for stock. This clearance increases light intensity and reduces competition for species that can survive along the trail. These changes can alter the composition of vegetation substantially (Adkison and Jackson 1996, C. Hall and Kuss 1989). Composition along trails also shifts in response to increased trampling and grazing, increased nitrogen from manure and urine, and increased moisture, the result of having fewer trees to intercept precipitation, fewer plants to transpire, and more watershed along the sides of the compacted trail tread. Trail corridors can contribute to the rate of spread of exotic plant species, as has been reported for the Boundary Waters Canoe Area Wilderness, Minnesota (Dickens et al. 2005); Glacier National Park, Montana (Tyser and Worley 1992); and Rocky Mountain National Park, Colorado (Benninger-Truax et al. 1992).

Creating a bare, compacted trail tread and a narrow zone of disturbed vegetation on either side is a dramatic change but is usually accepted by visitors. The dramatic changes are confined to a zone usually no more than eight feet wide, although in meadows compositional changes have been noted more than twenty feet from the center of the trail (Cole 1979). Probably more disturbing are sites, usually in wet soil areas, where the disturbed, barren tread widens, greatly exceeding the recommended tread width of twenty-four inches. This change is caused by use—not construction—although poor location and design during construction may have initiated or contributed to the problem.

Trail construction can also create new habitat by other means. Examples include creating or eliminating rock faces where trails traverse rock outcrops; creating debris slopes where boulders are pushed downslope to build the trail; creating flat, soil-covered surfaces where trails traverse steep talus slopes; and creating wet soil areas where trails impede normal drainages. Again, these changes do not affect large areas and are generally considered to be acceptable; however, they should be recognized as undesirable and kept to a minimum.

Perhaps the two most serious trail impacts are disruption of drainage systems and aesthetic problems resulting from obtrusive trail design or construction. Unfortunately, the solution to one of these problems is likely to aggravate the other. That is, designs to solve drainage problems may be perceived as overengineering, whereas lack of engineering may lead to drainage problems. This trade-off is discussed in more detail in the trail-management section of this chapter. The solution demands a careful balance—enough engineering to avoid disturbing drainage while remaining sensitive to building trails that blend into the natural environment. In each situation, those who construct trails will have to evaluate which is more "natural" and appropriate—a high-standard trail that avoids off-trail disturbance or a low-standard trail that risks the possibility of more resource damage and a less comfortable walking surface.

Fig. 15.6. Grazing by pack and saddle stock has altered vegetation and soil conditions over large proportions of many wilderness areas. Photo courtesy David Cole, U.S. Forest Service.

Pack-Stock Grazing for Recreational Visits

Pack and riding stock trample vegetation and soil along trails and on campsites, as hikers do, leading to the changes noted in the section on trampling impact. Differences between stock and hiker impact are summarized in the section of this chapter on pack-stock management. Here we outline changes occurring on areas grazed by recreational pack stock—meadows and grasslands that are generally unaffected by hikers.

Grazing areas are affected primarily by trampling and grazing (defoliation), although defecation may also cause minor changes. Grazing, by removing leaves, disrupts the ability of plants to manufacture food. Excessive and repetitive defoliation depletes food reserves, reducing plant vigor and reproductive capacity. Numerous studies have illustrated that grazing can reduce current growth; stem, leaf, and seed stalk heights; reproductive activity; basal and foliar cover; and root growth (see McClaran and Cole 1993). Loss of vigor, in turn, makes vegetation more susceptible to trampling damage, particularly penetration of the vegetative mat by stock hooves, and results in a reduction in vegetative cover. In Yosemite National Park, even relatively modest levels of grazing (a few hours per year on picket) reduced the productivity, vegetation cover, basal litter cover, and relative graminoid cover; increased the basal soil cover; and altered the species composition of meadows (Cole, van Wagtendonk et al. 2004).

Trampling causes changes in vegetation and soil conditions, as described earlier. Of particular concern in grazing areas is disturbance of wet meadows. Wet soils, thick organic deposits, and vegetation mats are all susceptible to deformation and disintegration when trampled. Heavy trampling of such sites can lead to a surface of broken sod and hummocks, increased erosion, and even lowering of water tables. In Kings Canyon National Park, California, disturbance by recreational stock, superimposed on the earlier effects of sheep and cattle grazing, led to accelerated rill, channel, and gully erosion of meadows. Gullies up to fourteen feet deep lowered water tables and dried out meadows, promoting invasion of lodgepole pine, before being stabilized through improved meadow management and grazing programs (DeBenedetti and Parsons 1983).

Plants differ in their susceptibility to grazing, much as they differ in susceptibility to trampling. Those capable of growing from buds close to or under the ground are more likely to survive close grazing than those with buds located where they can be removed by grazing. Of even more importance, plants are preferentially grazed so that the least palatable species are most likely to survive grazing. Olson-Rutz et al. (1996a, 1996b) report short-term effects of a pack-stock grazing experiment, including preferential grazing of grasses, decreasing vegetal cover, and reduced subsequent-year production of grass and forb stems. All of these selective forces, along with introducing exotic species in manure, coats, hooves, and supplemental feed, contribute to pronounced changes in species composition and reductions in forage. In the Eagle Cap Wilderness, montane valley-bottom meadows grazed by stock had only about two-thirds the vegetational cover of nearby ungrazed meadows. The grazed meadows were dominated by forbs and had a sizable component of annual and exotic species, whereas the ungrazed meadows were dominated by native grasses and sedges (Cole 1981).

Because it reduces available forage, stock grazing may adversely affect wildlife populations that use the same forage resource. For such competition to occur, there must be overlap in the diets of the stock and wildlife species, and they must be using the same meadows. Of the important ungulates in wilderness areas, competition with elk is most likely. Elk and horses have similar diets and elk commonly use popular grazing areas as winter range. Unpublished range studies in the Bob Marshall Wilderness found that recreation-stock grazing has caused deterioration of forage areas that are used by elk in winter. Competition with bighorn sheep and mountain goats is also possible, but winter range of these species is probably less accessible to livestock.

Wildlife Disturbance

Recreation activities can impact wildlife in four different ways (Knight and Cole 1995b). Animals can be indirectly affected through *habitat modification* or through *pollution* (particularly through leaving trash). They can be directly affected through *exploitation—hunting, fishing, trapping, or collecting*. Finally, wildlife can be *directly disturbed*—either intentionally or unintentionally. We know most about the short-term responses of individuals to disturbance—nest abandonment, elevated heart rates, or flight. We know little about how entire populations or communities of animals are affected or about long-lasting effects (see also Chapter 12).

Firewood collection can affect small mammal and bird populations by altering food sources and living places and eliminating protected sites (Cole and Landres 1995). Similar modifications of habitat are also likely to affect reptile and amphibian populations. Organic trash around campsites also attracts animals, ranging from invertebrates to small rodents; certain birds; and large mammals such as bears. Blakesley and Reese (1988), for example, found seven bird species to be positively associated with campgrounds and seven species to be negatively associated.

Although habitat modification and improper disposal of trash are the major source of impact for smaller wildlife species, most of these changes are highly localized and, with the exception of the attraction of "pest" species to campsites, not evident to most users. Change becomes more important only where a species' entire habitat is disturbed or where the "pest" is a bear. Improper food storage and disposal have frequently been implicated as a causal factor in a chain of events that ultimately ends in the death of bears. Consequently, appropriate food-storage procedures and/or devices have been implemented in many backcountry areas.

Perhaps of more widespread importance to the achievement of wilderness management goals are the impacts resulting from stocking fish, angling, hunting, and unintentionally harassing wildlife. As noted earlier, hunting outside of national parks and angling are legal, accepted practices that are likely to continue in wilderness, despite their considerable alteration of natural conditions (S. Anderson 1995). Fish stocking, which has been shown to adversely affect amphibian populations (Knapp 2005), is a common practice in wildernesses with mountain lakes. Recently, it has been discontinued in a few wildernesses.

Although poorly understood, unintentional harassment, particularly of birds and large mammals, has undoubtedly altered the distribution, structure, and behavior of animal populations (Knight and Cole 1995b). Where harassment affects an entire population (as is likely to be the case with grizzly bear or hunted, localized populations of bighorn sheep and mountain goat) or affects most of a species' habitat, this disturbance is probably much more disruptive to wilderness ecosystems than many of the impacts we have been discussing, such as effects of trampling on trails and campsites.

Harassment of wildlife by recreationists or other wilderness users produces excitement or stress in animals. This may lead to panic, exertion, disruption of essential functions such as breeding or nesting, displacement to other areas, and sometimes death. Animals that are healthy and have ample food and places to escape to are more capable of withstanding harassment than animals that are underfed, highly parasitized, experiencing severe weather, giving birth or nesting, or lacking secure areas for escape. Damage to animals—in terms of increased energy expenditures or radical changes in behavior or distribution—also increases as disturbance becomes more frequent and more unpredictable (Knight and Cole 1995a).

Generalizing about harassment is difficult because of the considerable variability among and within species. Effects on wolves, which are relatively intolerant of disturbance, are much more serious than effects on coyotes. Similarly, effects on eagles, which may not return to feeding sites for several hours after being disturbed, are more serious than effects on jays. Within species, prior experience with humans strongly tempers responses. Some individuals can learn to tolerate at least predictable disturbances (Knight and Temple 1995). Differences between hunted and non-hunted populations can also be profound, because hunted animals have experienced a need to escape. Individuals giving birth or with young are more readily disturbed than others.

Disturbance of several subspecies of bighorn sheep has been widely studied, primarily in California and western Canada. Although a number of studies have implicated harassment as a cause of declining sheep populations, most recent work suggests that sheep can habituate to human intrusion. One Canadian study monitored heart rates and behavioral responses to disturbance. Although largely unaffected by foot traffic approaching from a road below, sheep responded dramatically to the presence of dogs and foot traffic approaching from upslope (an unexpected action that blocks their preferred escape route). The authors conclude that disturbance by visitors can be minimized by confining use to established trail systems and discouraging people from taking dogs (MacArthur et al. 1982).

Wildlife disturbance can be exacerbated by attempts to disperse visitors from popular parts of the wilderness to less visited places. Such dispersal occurs when use levels are limited in popular places, forcing visitors to seek out less popular places. It can also result from educational messages asking visitors to avoid popular places or to seek out places where opportunities for solitude are greater. Where increased dispersal occurs, the frequency of wildlife harassment may increase and the size of secure areas where harassed animals can escape may decrease. Any attempt to alter visitor use distributions should consider the consequences to wildlife.

Off-season use of wilderness can also be particularly problematic. Cross-country skiing and snowshoeing, in particular, can stress populations at a time of year when they are least able to tolerate it. Cassirer et al. (1992) documented disturbance of elk by cross-country skiers in Yellowstone National Park. Ferguson and Keith (1982) have documented a tendency for elk and moose to move away from trails being used by skiers. Importantly, they found that a single skier usually caused the animals to flee; the passage of additional skiers was irrelevant. Possibly a few large parties will cause less disturbance than many widely dispersed small parties. Although little is known about the consequences of such disturbance to reproduction or survival, we do know that flight increases the necessary caloric intake of these animals. Some ungulates adapt to winter conditions by decreasing activity to conserve energy. Disturbance interferes with this adaptation and may increase food demand beyond the supply provided by winter range. Clearly, the severity of such an effect would vary from year to year and from place to place; only through increased monitoring of wildlife populations in relation to disturbance will we be able to measure impacts. Nevertheless, many managers are currently educating winter users about the threats posed by harassment and the need to avoid animals.

Other important problems occur when use is concentrated on limited critical habitat. Recreation use around desert waterholes and salt licks can cause more substantial problems than one would expect from the same amount of total use. Use does not need to coincide with the presence of animals. For example, summer grazing by pack stock can reduce available food sources on critical elk winter range. Managers should identify habitat critical to wildlife at various seasons and develop plans for minimizing disturbance.

Three general approaches to *minimizing problems with wildlife disturbance* can be identified (Gutzwiller and Cole 2005). Of foremost importance in wilderness is *management of people*. Access can be limited, as where overnight use is prohibited or where sensitive places are closed to all visitation—actions that currently are almost entirely confined to FWS and NPS wildernesses. Alternatively, access can be restricted at certain critical times of the year, such as feeding times, nesting times for birds, and postnatal periods for mammals. There is considerable potential to educate users about avoiding wildlife conflict. One of the primary principles of the Leave No Trace educational program is "respect wildlife."

The other potential management strategies involve modifying wildlife behavior and habitats. *Behavioral modification*—habituation to predictable, harmless human activity—is useful where hunting is not allowed. For example, aversive training of "problem" bears has been tried. This can be used to influence

Fig. 15.7. Use of wilderness during winter is increasing in popularity. Such use can stress wildlife populations at a time when they have little tolerance of increased stress. Photo courtesy U.S. Forest Service.

reactions to human disturbance, although the appropriateness of such an approach must be questioned (D. Whittaker and Knight 1998). Finally, *habitats can be modified* to change population distributions or to mitigate disturbance. Again, the appropriateness of such actions in wilderness must be questioned.

Water Pollution and Disposal of Human Waste

Most management concerns with water pollution have centered on the potential for transmitting disease by organisms present in water. Many different organisms are capable of causing illness in humans (Cilimburg et al. 2000). *Three prominent sources of water contamination are (1) visitors, their dogs, and pack stock; (2) domestic livestock; and (3) wildlife.* Even where animal contamination is absent, bacteria and other pathogens can be found in the soil, forest floor, and stream sediment (Silsbee and Larson 1982). Therefore, even so-called pristine areas receiving almost no recreation use can harbor organisms that are harmful to humans.

Water-quality studies in mountainous wilderness in the West have generally found very low levels of bacterial contamination, even in areas of concentrated use. For example, at Rae Lakes, one of the most popular alpine lake basins in Kings Canyon National Park, coliform bacteria counts were usually low enough to allow drinking (Silverman and Erman 1979). Along the Colorado River in Grand Canyon National Park, Arizona, water was unfit for drinking but coliform levels were generally low except when major tributary streams were in flood (Brickler et al. 1983). Here the primary source of contamination appeared to be domestic livestock or wildlife. Springs and streams in Great Smoky Mountains National Park exceeded maximum permissible levels of coliform bacteria, but contamination did not appear related to recreation use (Silsbee and Larson 1982).

Even where contamination is not evident, transmission of disease does occur. Surface waters in many wildernesses are contaminated with giardia (Cilimburg et al. 2000). As with bacterial contamination, most (but not all) experts believe that humans, domestic animals, and wildlife all act as hosts capable of spreading the organism. Beaver have most frequently been implicated as the major source of giardia contamination.

Where level of contamination has been related to amount of recreational use, it is not clear whether areas receiving more recreation use present higher health hazards than lightly used areas do (Cilimburg et al. 2000). In fact, one study of used and unused watersheds in Montana found less contamination in the watershed open to recreation use and a decrease in contamination after the closed watershed was opened to use (Stuart et al. 1971). The authors concluded that the primary source of contamination was wildlife and wildlife populations and, therefore, contamination was reduced by recreation use. In an unpublished study in the Anaconda-Pintler Wilderness, Montana, elevated bacterial counts were most often found just downstream from trail crossings used by horses and pack animals. In the Eagle Cap Wilderness, contamination levels were higher in streams, particularly at mid-elevations in meadows, than in lakes, and coliform counts generally peaked along with runoff after a storm (McDowell 1979). *All this suggests that management of recreation use is likely to do little to reduce health hazards. The most important management action is informing visitors about the prevalence of contamination and the need to treat water.*

Proper control of waste, from both humans and recreational stock, is also important, although this will not eliminate health hazards. Toilets may be appropriate at sites that receive heavy, consistent use throughout the season. However, knowledge is insufficient to provide specific guidance about when toilets are or are not necessary. When provided, toilets must be sensitively located by managers, because their presence—especially when accompanied by a direction sign saying "toilet"—is a reminder of civilization and may not be consistent with the spirit of wilderness.

Where soil is available and use levels are not too high, disposal of human waste in individual "catholes" is recommended. Even this practice poses problems, however. Research in Montana's Bridger Range has shown that significant numbers of intestinal pathogens in feces survived an entire year of burial (Temple et al. 1982). Statements that nature will take care of wastes "in a few days" are misleading and may promote careless disposal. This research shows that the possibility of disease transmission persists for a considerable time. Moreover, depth of burial (two to eight inches) made no difference in survival of pathogens; neither did it matter whether disposal occurred at high or low elevations, in forest or in meadow. This emphasizes the need to promote burial at sufficient depth and far enough away from campsites and water bodies to minimize the chance of direct contact by other users. In some places, particularly on mountain climbs and in narrow desert canyons, wilderness managers are asking visitors to pack out their waste.

Although education campaigns in proper human waste disposal are common, greater emphasis on the potential hazard and the need for careful disposal seems necessary. Pack and riding animals should be kept away from surface water as much as possible, and they should never be confined where manure is likely to contaminate water sources.

Other types of water pollution appear to be more prevalent and more subject to management control. In the Kings Canyon National Park study, where the health hazard was minimal (Silverman and Erman 1979), recreation use was associated with a number of changes in the basic ecology of lakes. The most heavily used lakes had less nitrate, more iron, and more aquatic plants than other lakes (Taylor and Erman 1979). The authors suggest that recreation use—through erosion of trails and campsites, improper waste disposal, destruction of vegetation, and campfires—may cause an increase in trace elements, such as iron, the absence of which formerly limited plant growth. Stimulated plant growth results in increased nitrogen uptake and, therefore, decreased nitrate levels. Insects, aquatic worms, and small clams were more abundant on the bottom of more heavily used lakes (Taylor and Erman 1980).

Although we do not know how common lake eutrophication has become, its effects are felt throughout the food chain. Moreover, such changes are

Fig. 15.8. In many wildernesses, aquatic ecosystems are particularly prone to significant disruption by recreational use. Photo courtesy David Cole, U.S. Forest Service.

long-lasting. In the Kings Canyon National Park study mentioned earlier, changes were still prevalent at Bullfrog Lake, a formerly heavily used lake, sixteen years after it had been closed to camping and grazing. Such changes are very important because they are likely to affect the entire lake ecosystem. This can be a serious problem in wilderness areas with a small number of heavily used lakes. In this situation, recreation use can dramatically alter the structure and functioning of much of that type of ecosystem. This would clearly constitute a serious failure to achieve the wilderness management goal of preserving natural conditions. Aquatic ecosystems may be the wilderness ecosystems most prone to important disruption by recreation use.

Managing Campsite Impacts

In virtually all wilderness areas, the most common and profound impacts of recreation are those associated with campsites and trails. In many areas, the management of pack stock is also an important issue.

Campsite Impacts

Most visitors spend more time at their campsite than anywhere else in wilderness. Unfortunately, this focuses impacts on the very places where visitors spend most of their time. Although natural conditions are desirable, some amount of impact can actually make a campsite more habitable. Some clearing of brush and trees, for example, provides better tent sites, causing many visitors to select sites with some vegetation loss. Studies of wilderness campers show that most campers view small areas of impact as "positive," "Pretty natural, healthy" (Farrell et al. 2001), because they make the site function

Fig. 15.9. In remote, trailless places, impact can be minimized if hikers spread out and disperse their impact. Photo courtesy David Cole, U.S. Forest Service.

well as a temporary human dwelling. Problems arise when damage to campsites becomes extreme or where sites proliferate over entire destination areas, providing constant reminders of the large numbers of people using the area.

Vegetation Change

Many studies have examined changes in vegetation on wilderness campsites. In most cases, trees are mechanically injured and the reproduction is suppressed, and there is profound damage to ground plant cover. Generally, there is little evidence that vigor of large trees is reduced, and aside from outright felling of trees and girdling because of tethering stock, tree mortality is uncommon. An exception to this generalization was documented in the BWCAW, Minnesota, where severe erosion of shallow soil around tree roots has caused high mortality (Marion and Merriam 1985). Ground-level vegetation is more profoundly affected. Plant cover is reduced, usually to bare ground in the central part of the campsite, and plant-species composition changes. Species diversity is usually reduced, and exotic plants often become an important component of the flora.

To illustrate the magnitude of campsite changes, I ask you to consider the median change on twenty-two campsites located in subalpine forests in the Eagle Cap Wilderness (Cole 1982b). On the median campsite, more than two thousand square feet had been obviously disturbed by camping. Almost 90 percent of the ground cover had been lost on the site—as inferred by comparing campsite conditions with those of an undisturbed control site close by. One-half of the site, the central area around the fire ring, was entirely devoid of vegetation. The surviving vegetation was very different in plant-species composition from undisturbed vegetation. Two species, a huckleberry and a heather, contributed almost 40 percent of the cover on undisturbed sites but only 6 percent of the surviving cover on campsites. In contrast, a sedge and a rush that contributed only 8 percent of the cover on undisturbed sites contributed almost 30 percent of the cover on the campsites. Overall, low shrubs and mosses were greatly reduced in cover. Grasses and grasslike plants, although losing some coverage, were less drastically affected, so they became the most abundant type of plant on the campsites.

Essentially all of the trees growing on these sites, 96 percent, had been damaged. Although much of the damage was minor, consisting of broken lower branches and nails driven into the trunk, one-fourth of the trees bore trunk scars from chopping and another one-fourth had been felled. About one-third of the trees had exposed roots, usually a result of tying stock to tree trunks. About 90 percent of the tree seedlings had been eliminated by trampling, which does not bode well for perpetuating forested campsites. Along with the felling of most of the saplings on the site, death of seedlings suggests that overstory trees will not be replaced when they die.

Typical levels of campsite impact vary greatly among different wilderness areas, as well as among campsites within the same area. For example, in the Bob Marshall Wilderness, Montana, campsites are typically much larger (mean of about 3,000 sq ft) than in the Eagle Cap, whereas in Grand Canyon National Park, they are much smaller (mean of about 500 sq ft) (Cole and Hall 1992). Mean campsite size varied between 500 square feet and 800 square feet in four wilderness areas in the south-central United States (McEwen et al. 1996), was about 700 sq ft at Isle Royale National Park (Marion and Farrell 2002), and was about 1,000 square feet at Great Smoky Mountains National Park (Marion and Leung 1997).

Changes in Soil Condition

The changes in soil condition most frequently noted are loss of the organic litter horizon, exposure of bare mineral soil, and compaction of the soil. Various measures of compaction are used, the two most common being bulk density and resistance to penetration. A few studies have also documented decreases in water infiltration rates and changes in organic matter content and soil chemistry.

On the Eagle Cap campsites, the depth of the organic horizons was cut in half (Cole 1982b). In some places, all organic litter was lost. Exposure of bare mineral soil was 1 percent on control plots compared to 31 percent on campsites. Although some of the surface organic matter pulverized by trampling is probably removed by erosion, some evidently moves downward and accumulates in the uppermost mineral horizons because soil organic matter content increased 20 percent on campsites. Similar studies have found both increases and decreases in organic matter on campsites. Loss of vegetation cover and changes in the organic content of soils will have a profound influence on the biota that live in soil. Zabinski et al. (2002) found evidence of dramatic changes in soil biology on Eagle Cap campsites, including reductions in the biomass and respiration of soil microbes, their functional diversity, and the quantity of soil nitrogen available to most plants (potentially available nitrogen). Such changes make it difficult to revegetate disturbed campsites.

Bulk density increased 15 percent and infiltration rates were reduced by about one-third on the Eagle Cap sites. These relatively small changes may result from the sandy, granitic substrate of these campsites, which makes them relatively resistant to compaction. In similar studies, infiltration rates were reduced by two-thirds in the Bob Marshall Wilderness (Cole 1983b) and three-fourths in Grand Canyon National Park (Cole 1986). Finally, several changes in soil chemistry were found. Values of pH increased, soils became less acidic, and there were sizable increases in the concentrations of magnesium, calcium, and sodium. These chemical changes probably reflect input from campfire ashes, excess food, soap, and other substances scattered about the site.

To summarize, almost every parameter examined on the Eagle Cap campsites had been substantially altered by camping. We conclude that "natural conditions" are not being preserved on these wilderness campsites, because these are typical sites—not worst cases or atypical examples.

Despite their severity, campsite impacts are highly concentrated. For the Eagle Cap Wilderness as a whole, less than 0.2 percent of the area has been affected by camping (Cole 1981). Marion and Leung (1997) estimate that less than 0.01 percent of Great Smoky Mountains National Park—where most camping occurs on officially designated campsites—has been affected by camping. In most places, only occasional campsites are encountered. However, large numbers of campsites are concentrated in a few popular destination areas. For example, camping has impacted more than one hundred sites at one popular lake in the Eagle Cap (Cole 1982a). Over one-half of these sites lost more than 25 percent of their vegetation, and most were in sight of the trail. Although this represents little threat to the ecological integrity of the Eagle Cap Wilderness, it does provide conspicuous evidence of human use. Not only is there scant opportunity to camp on an undisturbed site, but pronounced campsite impacts are found on almost every potential campsite area.

Temporal Patterns of Impact

Studies of individual campsites show that they have a typical "life history," moving successively through a "development phase," "dynamic equilibrium phase," and then "recovery phase" (Fig. 15.10).

Impact usually occurs rapidly in the development phase, when a previously unused site is first used as a campsite. On newly established canoe campsites at Delaware Water Gap, for example, most of the impact that occurred over the six years following campsite opening occurred during the first year of use (Marion and Cole 1996). Impact did continue to increase each year for the first three years, but at a decelerating rate. Actual rates of deterioration during the "development" phase vary among kinds of impact. Loss of vegetation occurs rapidly, while exposure and compaction of mineral soil occur more slowly.

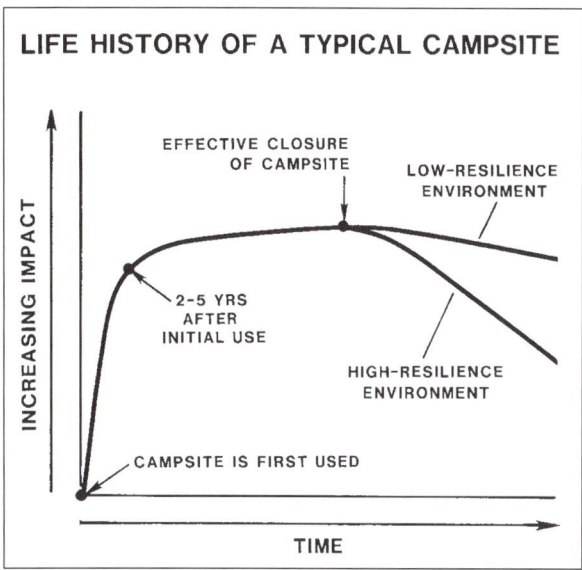

Fig. 15.10. The life history of a typical campsite, illustrating periods of rapid deterioration, relative stability, and slow recovery.

Deterioration occurs more rapidly as amount of use increases and as site durability decreases. For example, one night of camping in a previously unused forest in the Popo Agie Wilderness, Wyoming, eliminated about 60 percent of the vegetation, while four nights eliminated over 80 percent (Cole and Monz 2003). On neighboring meadow sites, four nights of camping had no effect on vegetation cover during the first year of use. After the second year of camping, vegetation cover was reduced significantly, but the magnitude of loss was small (less than 20 percent) compared to the loss that occurred in the forest (greater than 80 percent), and no further loss occurred during the third year of camping.

The "development" phase is followed by a more stable phase of "dynamic equilibrium," where site impacts remain stable with a steady level of use. Where amount and type of use are relatively constant, seasonal and year-to-year fluctuations, dictated largely by climatic variation, may exceed change for many kinds of impact. For example, on long-established campsites in the Eagle Cap Wilderness, mean vegetation cover was 15 percent in 1979, 12 percent in 1984, and 19 percent in 1990 (Cole and Hall 1992). Vegetation cover on these sites might be expected to fluctuate between 10 and 20 percent indefinitely, as long as use characteristics are relatively stable. Devegetated area and soil compaction are impacts that tend to be stable on established campsites. Other impacts tend to deteriorate more steadily because recovery processes are so slow. For example, mineral soil exposure on Eagle Cap campsites increased from 33 percent in 1979 to 41 percent in 1984 and 44 percent in 1990 (Cole and Hall 1992). In these western coniferous forests, the accumulation of new litter cannot keep pace with the rate of litter loss from campsite use. In eastern forests, however, the amount of mineral soil might be more stable because litter production is greater. Tree damage, in particular, tends to be cumulative, worsening over time because recovery is slow or does not occur.

Once a campsite is effectively closed, a "recovery" phase begins, with the rate of recovery highly variable and always slower than site deterioration. Recovery rates vary among kinds of impact and environments, as well as with amount of previous impact. Hartley (1999) reports residual effects thirty years after trampling experiments were conducted in alpine vegetation at Logan Pass in Glacier National Park, Montana. In comparison, most visual evidence of camping on closed riparian campsites at Delaware Water Gap disappeared within six years (Marion and Cole 1996).

Given the same environmental setting, more highly impacted sites will require longer periods to recover. When different environments are compared, however, it is difficult to predict how long recovery will take merely based on how the site is impacted. Some of the environments that are most readily disturbed by camping (such as lush herbaceous vegetation in riparian zones) are capable of relatively rapid recovery once camping stops. There is some evidence that differences in recovery rates, among different environments, may exceed differences in deterioration rates.

Temporal patterns at larger spatial scales are especially important. There is a tendency for impacts to proliferate and spread across the landscape where the distribution of recreation use is not tightly controlled. For example, in two drainages in the Eagle Cap Wilderness, the number of campsites increased from 336 in 1975 to 748 in 1990 (Cole 1993a). *Site proliferation occurs because sites deteriorate rapidly and recover slowly. As use shifts across the landscape, new campsites appear more rapidly than old campsites disappear.*

Factors That Influence Amount of Impact

To better understand how to minimize camping impacts, it is important to understand why some sites are more seriously damaged than others. Total impact is a product of both the intensity of impact at any one place and the areal extent of impacted places. *The major factors that influence how much change occurs on an individual site are (1) the amount and frequency of use the site receives, (2) the type and behavior of its users, and (3) the environmental conditions and durability of the site.* Season of use would be a fourth important factor, except that in most places, campsites are seldom used outside of the main use season. *The areal extent of impact is primarily a result of the spatial distribution of recreation use.*

Amount of Use

The usual assumption has been that the amount of use a site receives is most important in determining its impact, but numerous studies suggest that this assumption is misleading. In the Eagle Cap (Cole 1982b), for example, even campsites used no more than a few nights per year (light-use sites) have been severely altered (Table 15.1). Most overstory trees have been damaged, most

Table 15.1. Relationship Between Selected Campsite Impact Parameters and the Amount of Use a Site Receives

Impact Parameter		Light-Use Sites (N = 6)	Moderate-Use Sites (N = 6)	Heavy-Use Sites (N = 10)	Kendall's Tau (a = 0.05)
			Median		
Camp area	(sq m)	48	224	205	NS
Devegetated area	(sq m)	19	122	93	0.30
Trees with exposed roots	(%)	3	33	39	0.41
Damaged trees	(%)	74	85	97	NS
Seedling loss	(%)	73	92	89	NS
Surviving vegetation cover	(%)	9	6	4	−0.41
Decrease in depth of organic horizons	(%)	3	21	68	0.36
Floristic dissimilarity	(%)	31	60	64	0.33
pH increase	(%)	3	5	11	NS
Decrease in infiltration rates	(%)	8	57	12	NS
Increase in soil organic matter	(%)	19	26	20	NS
Increase in bulk density	(%)	16	11	16	NS

Source: Cole 1982b.

seedlings have been eliminated, most of the vegetation has been lost, soil has been compacted, and soil chemistry has been changed. Sites used an estimated five to ten times more frequently, about one night per week during the main use season (moderate-use sites), differed in the following ways: the disturbed area was usually much larger, as was the devegetated area; exposure of tree roots was pronounced; organic horizons were thinner; and changes in undergrowth species (indicated by the floristic dissimilarity value) were more extreme. Compared with these sites, the only major difference on the most heavily used sites—those used several nights per week—was that organic horizons were even thinner. In an experimental study, one night of camping on previously unused sites caused significant vegetation loss in all four vegetation types studied (Cole 1995a). Four nights of camping caused less than twice the impact of one night.

These and similar results from across the country suggest *there is a general relationship between use and impact similar to that in Fig. 15.11. Only when comparing sites receiving very low levels of use do differences in amount of use make any sizable difference in amount of impact.*

Type and Behavior of Site Users

Certain types of impact on campsites are determined almost entirely by the behavior of campers, for example, damage to trees and "pollution" of the site with campfire ashes, charcoal, food, and so on. Not all parties build campfires or damage trees. A campsite could be heavily used and not suffer tree damage or changes in soil chemistry caused by building campfires or discarding wastes. Other types of impact are little affected by behavior. Even campers who carefully practice low-impact use techniques will still trample vegetation and compact soil.

Three other characteristics of user groups also influence campsite impact—size of the party, length of stay, and whether or not they use pack stock. The effect of party size on campsite impact has never been formally studied. One can assume that large parties will increase the disturbed area of individual campsites. Thus, campsite area and size of the devegetated area would be much larger than on sites used by small parties. However, there is little reason to believe that party size should affect any other characteristic of established campsites. On undisturbed sites, however, large parties will cause impact more rapidly than small parties. It is considerably more difficult for a large party to cause minimal impact when visiting relatively undisturbed places.

Campsites used for long periods of time by the same party tend to be more heavily impacted than other sites. Two factors seem to be at work here. First, use patterns on the site are repeated day after day, leading to severe disturbance of certain parts of the site. For example, places where tents are set up and used for a week or more are likely to be highly altered. Long lengths of stay can be highly damaging on previously unused sites, whereas length of stay may be of little importance on a site where impact levels are already high. The second factor—and this applies even to well-impacted sites—is the natural tendency for people to "improve" and "develop" their campsite the longer they stay.

A final behavioral factor, *the spatial distribution of visitors, also influences the areal extent of impact*. Cole (1992) shows that the factor that most influences amount of campsite impact is the degree to which activities are spatially concentrated. Campsite impacts in the backcountry of Grand Canyon National Park are relatively low—campsites are extremely small—because the rough terrain forces campers to confine their activities to small flat spots, free of rocks and thorny vegetation (Cole 1986). Marion (1995) documents the effects of management actions taken to increase the concentration of use at two spatial scales on canoe campsites at Delaware Water Gap over a five-year period. Intersite concentration of use was increased by reducing the number of designated sites 25 percent. Managers increased the intrasite concentration of use by installing fire grates on each site. This concentrated activities more than in previous years when visitors built fires on many different parts of the site. As a result, mean campsite area decreased more than 30 percent. The effect of these combined actions was to reduce the aggregate area of campsite disturbance 50 percent—from 6.9 acres in 1986 to 3.46 acres in 1991. Although these actions increased the intensity of use on individual campsites, there was no resultant increase in intensity of impact on individual sites (Marion and Cole 1996). Marion and Farrell (2002) found unusually low areal extent of camping disturbance at Isle Royale National Park, which they attribute to a designated camping policy, limitation on site numbers, construction of sites in sloping terrain, use of facilities, and an ongoing program of campsite maintenance.

The effect of pack stock is discussed in more detail later. *Parties with pack stock disturb a larger area than hikers because the campsite includes an area where stock are confined*. Such camps often show more soil disturbance, a result of trampling by heavy, shod animals, and more tree damage, a result of tying horses to trees for extended periods (Cole 1983b).

Environmental Conditions and Durability of the Site

The final factor determining intensity of impact is environmental conditions and the durability of the campsite. Trail condition provides a useful illustration of the importance of location. It is common to find badly eroded or wet sections of trail alternating with better-drained sections that are in good shape, despite the fact that the very same number and type of people are using both good and bad trail segments. Environmental differences such as steepness of slope, soil texture, and moisture content account for most of these differences in condition. *Similarly, on campsites, environmental conditions, including soil characteristics, depth of surface organic horizons, and vegetation type, can greatly influence site durability or the amount of impact it will exhibit in response to use*.

A number of studies have examined, through experimentation, the effects of increasing amounts of trampling on different types of vegetation. Trampling disturbance, particularly loss of vegetation, varies widely among vegetation types. In experimental trampling studies, Cole (1993b) found that as few as twenty hikers can destroy 50 percent of the vegetation cover in some vegetation types; in other vegetation types, 50 percent vegetation cover is lost after 600 hikers have crossed the area. This suggests that campsites in some vegetation types could absorb more than thirty times as much use as campsites in other types, with no more vegetation damage. Clearly, where people camp is

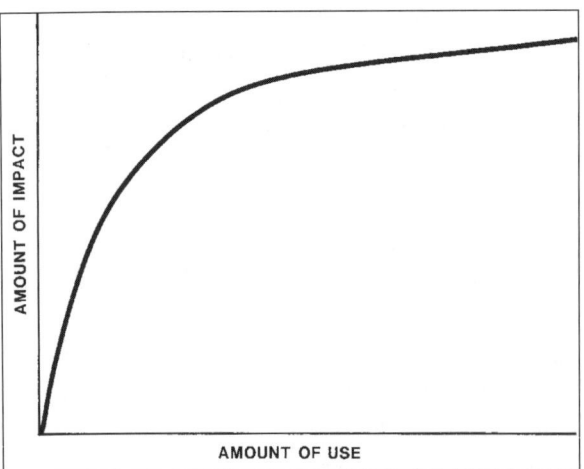

Fig. 15.11. Generalized relationship between amount of use and amount of impact. Only at low-use levels are site impacts likely to reflect different amounts of use.

critical in determining loss of vegetation. Diagrams of the spatial pattern of impact development on experimental campsites in the Popo Agie Wilderness, Wyoming (Cole and Monz 2004), show that impact is determined more by campsite location than by amount of use or number of years of use.

Environmental influences on campsite susceptibility to impacts are complex, and our understanding is still rudimentary. Any environmental setting may be susceptible to one type of impact and resistant to another. There may be little relationship between a site's resistance, its ability to tolerate use without changing, and its resilience, its ability to recover from changes that do occur. Characteristics of durable campsites include (1) either lack of ground-cover vegetation or presence of tolerant vegetation (grasslike plants are most tolerant—short woody plants are least tolerant), (2) an open rather than closed tree canopy, (3) thick organic soil horizons, and (4) a relatively flat but well-drained site.

It is commonly believed that sites located close to lakes and streams are more fragile than sites away from water bodies. When examined in the Eagle Cap Wilderness, this did not prove to be the case (Cole 1982b). Campsites close to lakes were no more highly impacted than sites located away from lakes.

Management Strategies and Techniques to Control Impacts

Each of the factors that influence amount of impact offers a strategy for reducing site impacts. For example, four primary strategies for minimizing impacts involve *(1) limiting amount of use on campsites, (2) changing type of use and behavior in such a way that per capita impact is reduced, (3) shifting use to more durable sites or hardening sites and providing facilities to actively increase site resistance, and (4) containment of impacts—recognizing situations where impacts are inevitable and can best be minimized by limiting their areal spread*. A fifth strategy—site cleanup and rehabilitation—treats symptoms rather than causes (Table 15.2). These strategies can be implemented by various techniques or actions. Effective campsite-management programs call for evaluating objectives, problems, and all potential solutions before selecting a series of coordinated actions, often using several different strategies. Further information on alternative management options can be found in Cole, Peterson et al. (1987) and D. Anderson et al. (1998). Leave No Trace education—one of the most important management techniques—is discussed in Chapter 16.

Limiting Amount of Use

Currently, overnight use is limited in most NPS wildernesses as well as in about twenty USFS wildernesses, and a few BLM and FWS wildernesses. This is little different from the situation in 1980, when overnight use was limited in sixteen classified and fifteen potential wildernesses (Washburne and Cole 1983). Managing campsite impact is just one of many reasons for limiting use. *Because the relationship between amount of use and amount of impact is that a little use causes most of the impact, limiting use, by itself, is likely to do very little to improve the condition of well-established campsites. In most cases, unless all visitation is curtailed on a site, there is little chance for recovery.* The only exception to this rule is where use levels are kept very low and dispersal is practiced (see "Dispersal of Use," next).

In places where use levels are high—the usual case where rationing has been implemented—use limits are likely to be more effective in limiting the *number* of campsites than the *severity* of impacts on individual sites. To limit the number of campsites, however, it is necessary to simultaneously employ the containment strategy (discussion follows) and get campers to use existing or designated sites, rather than make

Table 15.2. Factors That Influence Impact on Campsites.
Each factor defines a strategy and set of specific techniques (of which only one example is provided here) for managing impacts.

Factor	Strategy	Technique (example)
Amount/frequency of use	Reduce use	Institute quotas
Type/behavior of use	Change type/behavior of use	Party size limits
Environmental/site conditions	Increase site resistance	Bridge unavoidable bogs along trails
Use distribution	Contain use	Camping in designated sites only

new sites. In general, then, use limits are likely to be effective only when supporting containment or when they maintain extremely low use levels. Use limits can, however, be effective in dealing with other management problems, particularly crowding.

Dispersal of Use

Because use dispersal has been found to spread campsite impacts as well as use, it is a much less common management strategy today than it was in the past. In 1980, managers were attempting to disperse use in 50 percent of all wildernesses and potential areas to be added to the wilderness system (Washburne and Cole 1983). Goals ranged from shifting use to less frequently visited areas to discouraging camping on impacted campsites. Any dispersal of use will affect the number, distribution, and condition of campsites, and recent research has shown that dispersal is more likely to increase than to decrease impact.

As in the case of use limitation, *use dispersal is unlikely to improve the condition of individual sites unless use levels are very low*. Studies in wilderness areas in the West suggest that even a night or two of use per year can inflict long-term damage (Cole and Monz 2003). We also know that only a slight increase in use will significantly alter previously unused or seldom-used sites. In the Eagle Cap, even the lightly used sites away from trails had typically lost more than 70 percent of their vegetation. Therefore, increased use of little-used areas or sites will increase both the number of impacted sites and their level of impact. The more than 220 campsites around Mirror and Moccasin Lakes in the Eagle Cap (where the average number of parties probably does not exceed ten per night) are partially a result of a management decision, made in the late 1970s and early 1980s, to remove existing rock fire rings and to request visitors not to camp on heavily used sites (Cole 1982a). *At such popular destinations, directing use away from heavily used sites actually spreads campsite damage. Encouraging more use of less popular parts of the wilderness will also increase campsite impacts in these places, with little compensatory improvement in the condition of popular locations*. More recently, Eagle Cap managers have stopped removing all fire rings in places where campfires are allowed, and no longer encourage visitors to avoid heavily used campsites. Recent monitoring of Eagle Cap campsites suggests a trend toward less impact, though the popular long-impacted lakeshore sites remain highly impacted.

Dispersal could improve campsite conditions in lightly used remote parts of wildernesses, if visitors can be encouraged to spread out and camp on undisturbed sites. In this vast majority of wilderness acreage, dispersal can help perpetuate the "ideal" wilderness situation where no sites become heavily impacted. However, if use increases, it may be necessary to impose limits. Ironically, use limits may be more effective in low-use situations than in the high-use places they are most commonly applied. The dispersal strategy must be supported with an intensive educational program wherein campers are taught minimum-impact camping techniques and how to select apparently unimpacted and resistant sites. Indeed, concentrating use and impact in popular places and dispersing use and impact elsewhere is one of the most critical Leave No Trace messages. This allows sites to fully recover before being used again; otherwise sites will deteriorate, and dispersal will merely spread lasting impacts. Because of this possibility, monitoring campsites and their condition is particularly important to a dispersal program.

Temporary Campsite Closures

Rest-rotation systems—where certain heavily impacted sites are temporarily closed to allow recovery before being used again—are likely to be effective only if required recovery periods are short in relation to periods of use and deterioration. This is seldom the case, however. Recovery almost always is slower than deterioration. The effectiveness of a temporary campsite-closure program was monitored around Big Creek Lake in the Selway-Bitterroot Wilderness, Montana, where seven out of fifteen campsites were closed to allow recovery. Eight years after closure, vegetation on closed sites was still only one-third as extensive as that on controls, and mineral soil exposure was 25 percent compared with only 0.1 percent on controls (Cole and Ranz 1983). The most profound change since initiation of the closures was to create seven new campsites, close to the closed sites, on which conditions have rapidly deteriorated. Within eight years of their creation, loss of vegetation and soil exposure were as high on new sites as on long-established sites. The likely effect of a rest-rotation system, then, is an increase in the number and area of impacted sites without substantial improvement in the condition of sites in use.

Limitations on Length of Stay

Limits on length of stay have been placed both on the maximum number of nights allowed in the wilderness and at individual campsites. Imposing area-wide length-of-stay limits is unlikely to have any effect on site impacts. The major benefit of such a regulation is to allow more people to use an area in which total use is limited. Length-of-stay limits for campsites are likely to do little to improve campsite conditions in popular parts of wildernesses or on popular campsites because new parties are likely to occupy sites shortly after they are vacated. Such a limit will help prevent "homesteading," although this is more a social than an ecological problem. It can also prevent serious deterioration of sites that had not been heavily impacted previously and avoid the tendency for sites used for long periods to become developed or improved. However, in these cases *the common fourteen-day limit is too long. To avoid damaging little-used areas, sites should never be used more than a night or two in succession.*

By reducing length of stay to a minimum and sleeping and eating in different places, users can reduce per capita impact in little-disturbed areas. Using this technique, a party traveling through the wilderness will prepare and eat supper in one location, clean up, and travel farther to a good bed site. In the morning, the party gets up and moves on to a good breakfast site. A typical camp is never established. *Generally, the most valuable use of length-of-stay limits is minimizing time spent at little-used sites.* This goal is most effectively accomplished through education, particularly when such limits are not imposed on heavily used sites.

Party Size Limits

Party size limits apply to a majority of wildernesses where the number of people per party is limited, the limits ranging from six to sixty persons, the most common being ten (Monz et al. 2000). As noted earlier, larger parties are likely to disturb larger areas, but in the most highly disturbed part of the campsite, severity of impact is unlikely to be much greater than with small parties. Establishing lower party size limits could reduce the size of campsites and devegetated zones; however, such an action might result in the need for more campsites. *To be effective in maintaining small campsites, limits probably should be ten or fewer, and users should be educated to not spread out on campsites.* Excessively large sites may require partial revegetation and some means of keeping visitors off the periphery.

Party size limits are of most value in lightly used parts of wilderness where dispersal is being practiced. Rate of impact tends to increase with party size, so a small party will find it much easier to leave little trace of their visit than a large party. Again, limits must be quite low and might be most effectively implemented as part of a program to foster appropriate use of places off the beaten track. *Impacts of larger groups can be reduced if party members will spread out during travel and break up into small dispersed camping units.*

Encouraging Use of Resistant Sites

Because certain sites are much more durable than others, encouraging use of resistant sites can minimize impacts by directing use either to resistant sites or away from fragile ones. This can be done through regulation or education. Camping in meadows is often discouraged or prohibited. Such campsites—both when occupied and after use—are much more obvious and aesthetically displeasing than sites set back in forests. Most research suggests, however, that grassland and meadow vegetation, particularly if it is dry, is much more resistant to damage than the forest undergrowth (Cole and Monz 2003). Therefore, in lightly used areas, where the dispersal strategy is being practiced, visitors should be encouraged to camp on meadows and grasslands. Here, encounters with other parties are unlikely, and it is most important to minimize trampling damage. In heavily used areas, where even resistant vegetation will be lost, one should encourage camping in forested areas with thick organic horizons, so impacts and other campers will be screened by trees.

Setbacks from Water

One of the most common management actions is to discourage camping close to streams and lakes. Such setbacks have social and ecological justifications as well as repercussions. Three conditions that might make lakeshores particularly vulnerable are (1) moist soil with great potential for vegetation damage and soil compaction, (2) steeply sloping shores prone to erosion, and (3) potential for water pollution. Soil moisture and slope steepness do not necessarily decrease with distance from water, however. Flat rock outcrops close to shores are undoubtedly much more tolerant of use than moist or sloping sites a considerable distance from the lake.

Fig. 15.12. Dry meadows make durable campsites, provided stays are short and campfires are not built there. Photo courtesy David Cole, U.S. Forest Service.

Most water-quality studies suggest that even in high-use areas pollution from human sources does not present a significant health problem (Silverman and Erman 1979). However, in at least one case, heavy use appears to have altered benthic plant populations and the concentration of certain ions (Taylor and Erman 1979). More monitoring is necessary to determine whether this is a common problem. Around heavily used lakes, particularly in areas where lakes are rare, setbacks may be justified as a means of reducing pollution.

Another justification for setbacks from water is the tendency of visitors to develop trails from campsites to the lake or stream. Social reasons—maintaining public access and the aesthetic qualities of the lakeshore or streamside, plus reducing the visibility of campers—may also justify setbacks. This action will keep visitors from camping where they most like to camp. Moreover, Christensen and Cole (2000) note that visitors report that social justifications for lakeshore setbacks are less persuasive than ecological justifications. Setbacks will increase the area altered by camping, at least for the short term, because visitors will develop a second set of campsites away from the lake. Moreover, a setback will often eliminate most of the potential places to camp near water.

Setbacks should be instituted because they will solve a specific problem. Often much could be accomplished by persuading visitors to prevent water pollution; this would avoid imposition of rigid setbacks. Elsewhere, setbacks may be necessary only in a few heavily used places. Many wildernesses are adopting more flexible setback policies than in the past. This may involve a rigid minimum setback of, say, fifty feet, with the suggestion that camps be at least one hundred feet back if the terrain permits. Where setbacks are established, the old sites closer to water should be actively rehabilitated.

Site Hardening and Facilities

True site hardening, wherein a site's durability is increased through manipulation, such as planting hardy grasses, is almost nonexistent in wilderness. More common is providing facilities that absorb or concentrate impact: fireplaces, tent pads, shelters, stock-holding facilities, toilets, and trash cans. Building facilities is a controversial action. *We support the installation of such facilities to protect resources or for visitor safety, but not for visitor comfort and convenience.* Trails are an example of an almost universally accepted facility that serves to absorb and concentrate impact. The other most common facilities in wilderness are toilets, shelters, constructed fireplaces, tables, and a drinkable water supply.

Facilities should be the exception rather than the rule—a means of dealing with concentrated use, particularly by novice users, in a few places in the wilderness. In Great Smoky Mountains National Park, for example, shelters receive 37 percent of the backcountry use, but because they concentrate impact, they account for only 10 percent of the disturbed area on campsites in the park (Marion and Leung 1997). Although shelters may seem inappropriate to many in wilderness, they are effective in reducing resource impact. We feel that wilderness should offer a range of recreational opportunities, including a few places that must accommodate heavy, localized use. Providing facilities may help prevent excessive deterioration of these places, while their judicious use preserves quality experiences for those who choose to visit such popular locations.

The facilities that can be most readily defended as necessary are constructed fireplaces, stock facilities, and toilets. Stock facilities are discussed in the section on pack-stock management. Fireplaces are most appropriate in areas of high fire hazard, for example, some wildernesses in southern California and the Southwest. However, in a number of other places they are used to confine campfire damage and to designate an acceptable campsite. Generally, this is necessary if visitors will otherwise build new fire rings or disturb new sites.

Toilets are an undesirable but sometimes appropriate facility, particularly where use is so high that the likelihood increases of visitors digging up previously buried fecal material. Much can be accomplished by teaching people to defecate far from high-use camps. However, in some situations, toilets become a necessity. The most common toilet is a wooden box a few feet high. In some areas, however, outhouses are enclosed, and in some wildernesses, composting toilets have been installed.

Containing Impacts

Containing use is a well-developed principle of site management outside of wilderness and, though many consider it inappropriate in wilderness—labeling it the "sacrifice site" concept—it is already being consciously and unconsciously applied within wilderness. The label "sacrifice site" is an unfair one, because it implies that managers have the option of not "sacrificing" sites. This is only possible if recreation use is very low or not allowed. Given that recreation use will be allowed, the choices are between having relatively large and relatively small numbers of impacted ("sacrificed") sites. *The containment strategy is an attempt to keep the number of impacted sites relatively small. For example, a trail contains and concentrates use.* Exhorting visitors to stay on trails and not to shortcut switchbacks is an example of appropriate containment. *Applied to campsites, the same "containment concept" would encourage use of existing sites to avoid rapid deterioration of new sites.* Currently a number of wildernesses, particularly those managed by the NPS, allow camping only on designated sites, at least in certain parts of the wilderness. Others encourage the use of existing campsites.

Most sites deteriorate substantially even when used only a few nights per year. Therefore, at a heavily used destination area, the choice is between a few deteriorated sites—the result of containment—and many deteriorated sites—the result of dispersal. Containment is a better strategy for minimizing impacts, unless management is willing to reduce use to extremely low levels and actively rehabilitate deteriorated sites. Generally, this is neither practical nor desirable. Wilderness can and should accommodate a range of opportunities; having a few popular locations with a handful of well-impacted campsites seems appropriate as long as the vast majority of the area remains largely undisturbed.

Although this conclusion may be disturbing to those who want pristine wilderness, we find it a sensible compromise that allows generous opportunities for recreational use. The finding that heavily used sites show little more impact than sites used a few times per year has its positive side; concentrating use on a few sites will not result in ever-increasing deterioration, provided that sites are well located and inappropriate visitor behavior is discouraged. Moreover, as long as most people want to visit popular areas and use the most heavily impacted sites, site containment will allow natural conditions to be preserved throughout most of the wilderness; opportunities for solitude will be preserved for those who seek it; and the need to manipulate visitor distributions and behavior, which results in loss of freedom, will be minimized.

Containment can be accomplished either through regulation—by allowing camping only on designated sites—or through education—by encouraging the use of existing campsites. Sites can be either clustered or dispersed. Although they are easier to administer and patrol, we feel that clustered sites are usually undesirable because they reduce campsite solitude and exacerbate problems such as bear encounters, waste disposal, and depleting firewood supplies. Where use is so high that unacceptably large numbers of campsites are required, use limits may have to be established.

Where use limits are established, managers often set up a reservation and fixed itinerary system, too. Visitors are required to stay at sites they reserve before entry. With such a system, the number of sites necessary to accommodate a given number of parties is minimized because the need for overflow sites is eliminated. Reservations and a fixed itinerary greatly reduce freedom and spontaneity, however, and are among the most unpopular actions taken in wilderness. *Compliance with fixed-itinerary systems is often low, defeating their efficiency* (Stewart 1989). Such regulations should be applied only if absolutely necessary. Visitor freedom can be maintained by limiting use at trailheads and then allowing free movement within the area. Based on historic use patterns, trailhead quotas can be set to keep use levels in destination areas within acceptable limits most of the time (Parsons et al. 1981).

Containment need not—and usually should not—be practiced throughout an entire wilderness. For example, in many wildernesses managed by the NPS, the wilderness is divided into a number of use areas or zones. In the most heavily used areas, camping is allowed only at designated sites. Elsewhere, visitors can camp wherever they choose. Both dispersal and containment can be practiced in the same wilderness area. Visitors who are properly equipped, skilled, and sensitized to low-impact use can be dispersed, while those who are novices or poorly equipped for low-impact use can be contained on designated sites.

Once a containment strategy is established, with or without use limits, the existing number of campsites can be reduced. This requires keeping people off closed sites and actively rehabilitating them. Usually, closed sites are identified through signing or a string enclosure. Open sites can also be signed, but a less obtrusive tactic might be leaving fire rings only on open sites. It is important to leave open sites where people want to camp and to have a few more sites open than the maximum number of parties anticipated at any time. *Spildie et al. (2000) evaluated the effectiveness of a containment strategy in a subalpine lake basin in the Selway-Bitterroot Wilderness, Idaho.* Actions taken included closing certain campsites to all use, requiring groups with stock to camp in one of nine designated stock sites in the lake basin, and restoring closed sites and portions of sites still available for use. *In just five years, disturbed area decreased 37 percent and bare area decreased 43 percent.* They project that, in a few decades, disturbed and bare area can be reduced to just 36 percent and 24 percent, respectively, of what it had been before imposition of the containment program. A containment strategy adopted in Shenandoah National Park, Virginia, reduced the area of campsite disturbance by 50 percent in just three years (Reid and Marion 2004).

Monitoring Campsite Conditions

Effective campsite management requires detailed information about the location and condition of campsites and trends over time. Sites should be inventoried to identify locations and problems that need attention, and to plan the types of corrective management actions that will be required. If done carefully, the inventory can provide a baseline for identifying trends in campsite number and condition and possibly relating changing conditions to changes in visitor traffic and management actions. Finally, the inventory and monitoring system can be a critical part of the limits of acceptable change (LAC) planning process (Stankey et al. 1985).

Several decades ago, very few wildernesses had any monitoring data. In a recent survey of wildernesses, we found that *campsites are being monitored in a portion of 50 percent of all wildernesses* (Cole and Wright 2003). A variety of different campsite-monitoring techniques have been developed. (Refer to Cole 1989 and Marion 1991 for further detail on available monitoring techniques.) These techniques vary in the quantity and quality of the information produced, as well as in cost. Techniques that produce precise, reliable information on a number of separate impact parameters require more time and cost more than those that produce either relatively little information or less precise information.

One common monitoring method uses photographs. From experience, *photographs have not proven to be reliable substitutes for field measurements* or estimates of parameters, such as vegetation cover or tree damage. Patches of sunlight and shade often make interpretation of ground cover difficult, and it is seldom possible to distinguish features beyond the closest trees. *Nevertheless, as supplementary documentation, photos are indispensable.* They can help identify the site for future measurements, record campsite features not measured in the field, and provide a visual supplement to collected data.

Where it can be afforded, the best monitoring data come from careful measurements in the field, using techniques similar to those employed in campsite impact research studies (for example, Cole 1982b and Marion and Cole 1996). However, such techniques often require spending an hour or more measuring each campsite. Because it is important to collect monitoring data on all campsites—to permit assessments of change in the number and distribution of campsites over time—such techniques are prohibitive except in wildernesses with small numbers of campsites. Most wildernesses have so many sites—Kings Canyon–Sequoia National Parks have more than 7,400 sites—and funding levels are so low that a system using visual estimates will be most practical.

In an effort to increase campsite monitoring in wilderness, the U.S. Forest Service developed a minimum protocol for campsite monitoring—a rapid technique that would provide a minimum standard of useful information. Instructions are as follows. First, visit all places where campsites are likely to be located. At each inventoried campsite, either use a GPS to obtain site coordinates or carefully place a dot on a topographic map and obtain site coordinates from the map. Particularly if a GPS is not used, take a photograph of the campsite to facilitate future relocation of the campsite.

Then, independently assess (1) ground-cover disturbance of the main campsite, (2) impact to standing trees and roots, and (3) size of disturbed area (including satellite tent pads and stock-holding areas). (4) Record disturbance to the ground cover of the central portion of the campsite (disregarding satellite disturbed areas) as one of the following classes.

> Class 1—Ground vegetation flattened but not permanently injured. Minimal physical change except for possibly a simple rock fireplace.
> Class 2—Ground vegetation worn away around fireplace or center of activity.
> Class 3—Ground vegetation lost on most of the site, but humus and litter still present in all but a few areas.
> Class 4—Bare mineral soil widespread over most of the campsite.

This is an adaptation of the condition class assessment technique first developed by Frissell (1978).

Record tree damage as one of the following classes, depending on the number of trees that have been severely damaged. Assess damage off-site as well as on-site, particularly in stock-holding areas associated with the campsite. Include any trees judged to have been damaged as a result of camping activities at the site being monitored. Severely damaged trees are those that (1) have been felled and are at least four inches in diameter where felled (if trees have multiple stems, consider the tree felled if any stem at least four inches in diameter has been cut off), (2) have scarring that exceeds one square foot in total area, or (3) have highly exposed roots (more than three feet of root sticks out at least one inch above the ground surface).

> 0—No more than three severely damaged trees
> 1—four to ten severely damaged trees
> 2—More than ten severely damaged trees

Record disturbed area as one of the following classes, depending on the size of the area disturbed by camping activities, including the main campsite, satellite tent pads, and areas where horses are confined. Where there is a landing area for boats, include this. In most situations, disturbed places are distinguished by obvious vegetation loss (either complete lack of vegetation or sparse vegetation resulting from trampling). Where vegetation is naturally absent, it may be necessary to identify disturbed places on the basis of flattening of soil or litter on the forest floor. When there are multiple separate disturbed parts of the campsite, do not include undisturbed areas in between. For example, if there is a main campsite, two tent pads, and a stock-holding area, assess the size of each of the four areas separately and then sum them. Social trails between separate disturbed areas can be ignored.

> 0—No more than 250 square feet
> 1—251 to 1000 square feet
> 2—More than 1000 square fee

Finally, assign the campsite an overall impact rating from 1 to 8. This is the sum of the groundcover disturbance rating (1–4), the tree damage rating (0–2), and the disturbed area rating (0–2). It should take no more than a minute or two to assign a rating.

This approach is inexpensive, and it produces quite reliable, precise information. Its primary drawbacks are that it is not very sensitive, and the amount of information produced is limited. This system is most helpful in showing the location of campsites, which campsites are most heavily impacted, and how the distribution of sites changes over time. It reveals little about which types of impact are most serious, and by the time campsite conditions have changed enough to be reflected in a changed rating, a profound amount of change will have occurred.

A slightly more time-consuming technique involves quick estimates of a number of different impact parameters. Fig. 15.13 shows one side of a form, based on this system, used to inventory all campsites in the Bob Marshall, Great Bear, and Scapegoat Wildernesses in Montana. *Both the impact index and the individual parameter ratings can be mapped.* Fig. 15.14 shows the location and overall condition (impact index) of campsites in a portion of the Bob Marshall Wilderness. Such a map is valuable for identifying problem areas and specific types of problems. This makes it much easier to tailor a campsite-management plan to the area. Obviously, different management strategies are necessary for Upper Holland Lake, with many highly impacted sites; Koessler Lake, with only one site, but a highly impacted one; and George Lake, with many sites, none of which are highly impacted.

Such a system provides considerable information at relatively low cost; however, the information

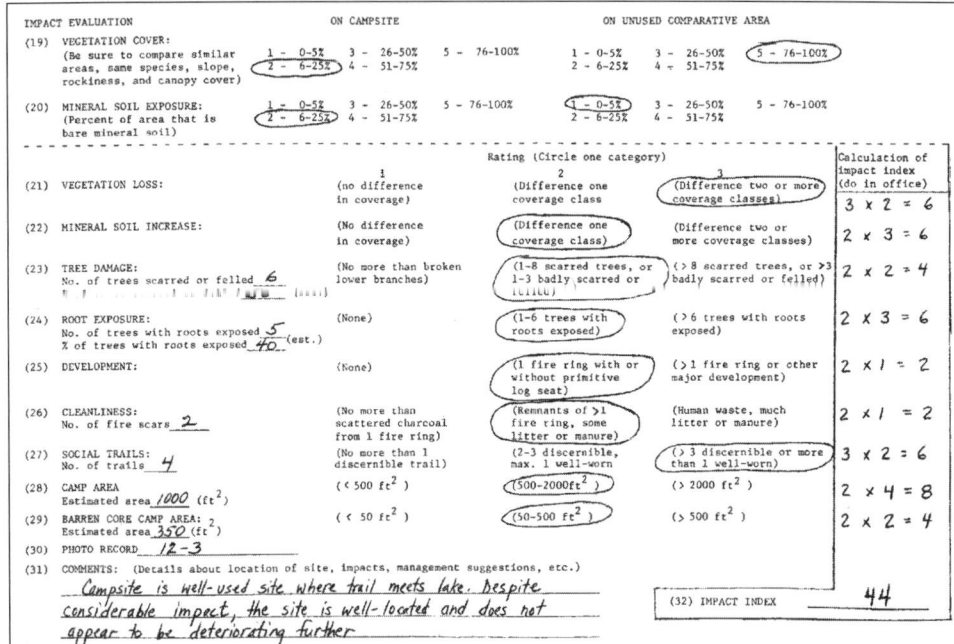

Fig. 15.13. Completed back of form used to inventory campsites in the Bob Marshall Wilderness Complex.

is not very precise. It is not uncommon, for example, for different evaluators to give the same campsite very different ratings for individual parameters. We have found, however, that overall index ratings do not vary greatly among different evaluators. This suggests that the primary limitation to these techniques is in drawing conclusions about changes over time—for separate types of impact, such as amount of vegetation loss—on individual campsites.

Marion (1991) developed a technique that is a compromise (in both cost and precision) between research-level techniques and the rapid assessments used in the Bob Marshall Wilderness. Certain parameters—such as campsite area—are measured carefully, while other parameters are visually estimated. After using this technique in a number of parks in the eastern United States, Marion reports, it takes two trained evaluators about thirty minutes to monitor a campsite. Another compromise approach is to inventory all campsites using a condition-class rating and then to carefully monitor a sample of campsites to identify subtle trends.

Regardless of the technique employed, the wilderness should be reinventoried periodically, perhaps every five years. This will show whether the number of sites is increasing or decreasing and whether conditions on individual sites are improving or deteriorating. If standards have been established for campsite condition, the ratings can be related to specific standards to determine where and what type of management is needed. By relating documentation of existing conditions (through monitoring) to specific objectives (standards) and then carefully choosing solutions from the array of possible management strategies described earlier, campsite management should become more effective and efficient.

Campsite Restoration

As we noted, *temporary closures as part of a campsite rest-rotation system are unlikely to succeed.* However, campsites may be *permanently* closed to enforce setbacks from lakes, to close poorly located sites, or to reduce the number of sites in places where containment is being tried. Closed campsites are common in wilderness. In many cases, closed sites are simply left alone. Increasingly, however, managers are realizing that they need to assist the recovery process because recovery rates are so slow. Most restoration programs are run by highly committed but untrained volunteers. Experimentation, documentation, and communication of what does and does not work are usually lacking. Barriers to restoration success are challenging and poorly understood. Moreover, suitable techniques and materials can be highly area- and site-specific. For all of these reasons, success and the ability to generalize about site rehabilitation have been limited.

Five Steps for Site-Restoration Plans

The likelihood of restoration success can be increased through careful planning. Common problems that can be avoided through careful planning include destruction of restoration work because the cause of the problem was not adequately dealt with, failure of plantings due to inadequate site preparation and plant selection, loss of plantings due to inadequate long-term maintenance of the site, and an inability to learn from successes and failures due to insufficient monitoring. Although there are many appropriate formats for a restoration plan, all are likely to contain the following five steps.

1. *Establish restoration goals and objectives. Goals define the intended result of the project*—for example, whether to return the site to its predisturbance condition or to simply reduce current levels of impact. *Objectives provide specific measurable targets*. It is critical to simultaneously consider goals for both the site and the larger landscape in which the site occurs. For example, in addition to restoring the lakeshore, what are the goals for the entire lake basin? Goals might be to confine camping impacts to places more than 200 feet from lakes or to confine them to other existing campsites. *Without statements of goals for larger areas, restoration programs may do little more than displace impacts to other places in the wilderness.*

2. *Conduct a situational analysis. Describe and analyze the site and its context, answering the following questions*: (a) what is the nature and extent of the problem, (b) what caused the problem, (c) what ecological and visitor use characteristics constrain the restoration effort, and (d) what are the ecological and visitor use characteristics of a fully successful project? It is helpful to place the problem in a historical context. Is the problem recent or has it existed for decades? Is it relatively stable or getting worse? Does the problem result from visitor use, natural events, management actions, or some combination of these? Evaluate the difficulty of restoring the site. Sites can be difficult to restore because they are harsh, unproductive, or have short growing seasons. They can also be difficult because the needs and desires of visitors in the area are such that visitors cannot easily be kept off the site.

3. *Design the site. Lay out the physical and ecological characteristics of the site-to-be*, including accommodations for visitor use and management. The design should prescribe the desired topographic configuration and drainage, as well as spatial arrangements of vegetation and substrate—usually in an attempt to emulate characteristics of the natural site. When designing vegetation planting, both vertical and horizontal structure should be considered.

4. *Develop implementation procedures. Describe the procedures necessary to achieve the site design*. Procedures to be specified, discussed in more detail later, include (a) site stabilization, (b) site preparation, (c) propagation or collection of vegetation or propagules, (d) planting procedures, (e) maintenance procedures, and (f) visitor management.

5. *Develop monitoring procedures*. A monitoring program should (a) define the area to be monitored, (b) decide which parameters should be monitored, (c) prescribe techniques for each parameter, (d) define a sampling/monitoring plan, and (e) include records of planting, maintenance, and management programs. Be sure to note when monitoring should be done, by whom, and how it will be used.

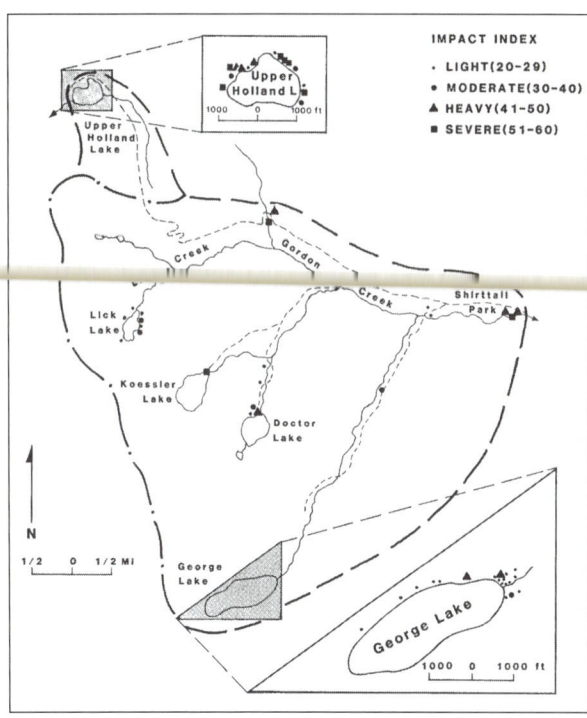

Fig. 15.14. This map of a portion of the Bob Marshall Wilderness displays the location and overall condition (impact index) of each campsite. This provides a graphic overview of campsite conditions at one point in time.

Guidelines to Implementing Site-Restoration Plans

Although generalization is difficult, a few guidelines related to steps in the process of implementing a restoration design can be suggested. Refer to Therrell et al.

(2006) for a recent handbook on site restoration, most relevant to high-elevation lands in the western United States.

Visitor Management

The first step in any rehabilitation program is to effectively close the site to all use. Even day use, where horses are tied to trees or people inadvertently walk across the site, can frustrate a rehabilitation attempt. The most effective approach to closure consists of helping people understand the reasons for closures and letting them know about other desirable places to camp. It is best to get this information to visitors before they enter the area so they can adjust travel plans. In a number of cases, rope or string between stakes or trees has been honored. A sign declaring the site closed (Fig. 15.15), the reason for the closure, and the location of alternative open sites in the vicinity promotes compliance. Use of the site can also be discouraged by partially burying large rocks and logs (iceberging) or planting dead trees and shrubs (vertical mulching).

Site Stabilization and Preparation

Once closed, the site should be cleaned up, eliminating fire rings, charcoal, excess firewood, and trash. Erosion is usually not a serious problem on campsites; however, if erosion is occurring it must be dealt with, usually through drainage manipulation. Then the soil needs to be prepared. Compacted soil should be cultivated (scarified) to a depth of at least four inches to facilitate seed germination, root and plant growth, and water infiltration. Clods need to be broken up, leaving a crumb texture. Long-established campsites usually have soils lacking in organic matter and a healthy soil biota. Scarification may have little effect if organic matter is not either incorporated into the soil or spread on top of the soil. Locally collected raw organic matter—such as from well-rotted logs—can be mixed with the scarified soil. This may not be desirable in grasslands because grasses generally prefer more neutral soils. In a recent study on subalpine campsites in the Eagle Cap Wilderness, revegetation success was increased by adding organic matter and compost to soils (Cole and Spildie 2007). Other types of organic fertilizers—which largely increase the organic content of soil while slowly adding low levels of limiting nutrients—are also available, but have not been evaluated on wilderness campsites. Inorganic fertilizers can be added at this stage, but they have seldom improved revegetational success and they tend to favor exotic species.

Propagation, Collection, and Planting

Under some favorable circumstances, natural revegetation may occur without much assistance within a short period of time, but elsewhere may require decades or centuries. In the West, at least, natural revegetation is likely to be most rapid at lower elevations, on more productive soils, and in areas that receive plenty of light and moisture. In slow recovery places, revegetation can be facilitated by transplanting whole plants or plant cuttings, or by seeding.

Transplanting, a technique used frequently and successfully, is time-consuming and can disturb adjacent areas from which plants are removed. Consequently, in a number of areas, plants grown in nurseries, from seed, cuttings, or root divisions obtained close to the restoration site, are transported to the backcountry for transplanting (Rochefort and Gibbons 1992). *Transplanting is more successful than seeding in certain environments.* Transplants can provide an early seed source, "nurse" plants that shelter and facilitate the germination and survival of seedlings, and may be the only effective way to get late successional plants established on sites. Experience in the Pacific Northwest suggests the following procedure for transplanting:

1. Select plant species adapted to grow on the site. Species that naturally colonize disturbed sites are good choices, as are plants that reproduce vegetatively. Choose relatively short plants with healthy-looking foliage.

Fig. 15.15. A sign placed on a rope circling a closed campsite can be effective enough in keeping people off the site to allow recovery to occur. Photo courtesy David Cole, U.S. Forest Service.

2. Water both the plants to be transplanted and the area to be transplanted one day before transplanting.
3. Place the plant upright in a hole slightly larger than the root ball. Make certain that roots are not doubled over on themselves. Fill in the excess space with organic matter and soil. When tamped down firmly, the top of the root ball should be slightly below the ground to facilitate watering and to reduce the risk of damage from frost heaving.
4. Water thoroughly. If it is warm or dry, it may be necessary to water the plants daily or to shade them. Where this is not feasible, survival rates can be increased by pruning some flowers, leaves, and branch tips and by providing large root balls.
5. Add a one-inch layer of mulch over the transplanted area and around the base of the transplants. Mulch can be leaves, pebbles, commercial mulch, decaying wood, grass, or any other material that insulates yet allows free movement of air and moisture. Lightweight mulches may have to be anchored by limbs, stones, or similar objects.

Fig. 15.16. Forest Service volunteers loosen compacted soil and add organic materials onto a damaged campsite to speed its recovery. Highly rotten wood was collected and incorporated into the soil of this campsite, temporarily closed in the Sawtooth Wilderness, Idaho. Photo courtesy David Cole, U.S. Forest Service.

Seeding is less complicated. Again, seed should be gathered close to the restoration site (to maintain genetic integrity) from plants adapted to the site. In some cases, nonnative species have been used because seed is more readily available, and it was thought that such plants would be replaced by native species. In Mount Rainier National Park, for example, red fescue, a nonnative species, was planted as a cover crop to reduce erosion, soil temperatures, and frost action, and to increase the organic matter in disturbed areas. In four years, native plants did not invade red fescue sites; bare areas that were not planted with red fescue showed better revegetation (Van Horn 1979). *To us, seeding with nonindigenous plants is an unacceptable alteration of natural conditions in wilderness—the finding that it is frequently unsuccessful confirms this belief.*

When seeding, it is important to become familiar with special germination requirements, such as scarification or stratification, of the species used. Seeds should either be scattered over the prepared soil and covered with one-half inch of soil or dropped into holes one-half inch deep. Soil should be tamped, mulched, and watered. From experiments in the Eagle Cap Wilderness (Cole and Spildie 2007), we have learned that some species are best for transplanting, while others grow best from seed and some colonize rapidly on their own if sites are adequately prepared.

Site Maintenance

It may be necessary to water plantings for years, if conditions are droughty. It will also be necessary to maintain ropes and signs that keep visitors off sites.

Even with all this effort, restoration is likely to require long time periods. Transplants on road cuts in the alpine zone of Rocky Mountain National Park were surviving after forty years, but they had not spread significantly (Stevens 1979). Three years after being planted, only 19 percent of transplants on closed subalpine campsites in Yosemite National Park were alive and total vegetation cover had increased only 1 percent (Moritsch and Muir 1993). *Clearly, it is much better to avoid damage than to try to fix it after it occurs.*

Site rehabilitation is an appropriate means for correcting past abuses, but it should be used judiciously. In addition to being costly, it interjects horticulture and landscaping into the wilderness. *As a general principle, we believe rehabilitation should be used to restore wilderness campsites that could then be protected through a new management program. Rehabilitation should not be used to bandage sore spots where no other change in management is implemented.*

Managing Trail Impacts

Impacts on and along trails result from the trampling of hikers and pack stock and the effects of trail construction and maintenance. As discussed in more detail earlier in this chapter, *these impacts include loss of vegetation and shifts in plant-species composition, exposure of bare mineral soil, soil compaction, and changes in microhabitats, including changes in drainage and erosion.*

Where trail construction is carefully planned, most of these changes are of little concern; although pronounced, most changes are localized and deliberate. Most wilderness trails were originally constructed to provide administrative access, particularly for firefighters, but currently are maintained primarily for recreational purposes. Most trail impacts only warrant concern when they provide obtrusive evidence of human use, become difficult to use, or require large amounts of money and labor to maintain. Although trail problems are usually highly localized, maintaining and relocating trails often costs more than any other aspect of wilderness management.

Three Common Problems: Erosion, Muddiness, User-Created Trails

The most common problems with trails are (1) excessive erosion, (2) muddy stretches in areas of water-saturated soils, and (3) development of informal, user-created trails, either adjacent to existing trails or in areas where no trails were planned. The first two problems make the trail difficult to use; all three suggest either "overuse" or improper use to visitors (Table 15.3). Two other problems result from attempts to correct the first three problems: excessive engineering and the proliferation of open and closed trails where trails have been frequently relocated. Both of these situations provide abundant evidence of human use and manipulation of the resource—evidence that cannot be eliminated but should be kept to a minimum.

Erosion

Although erosion can be significant on parts of a trail system, studies of trails in the Selway-Bitterroot Wilderness, Montana, and Guadalupe National Park, Texas, showed that little erosion is occurring over these two trail systems (Cole 1991; Tinsley and Fish 1985). Material eroded from trail "banks" or the tread itself is usually deposited elsewhere on the trail. *Soil is lost from the trail system only where water drains off the trail,* and much of that can be compensated for by sediment washed onto the trail from above by overland flow. Although trail troughs often change—either deepening through erosion or being filled in through deposition—trail systems as a whole usually exhibit a relatively steady state.

What is critical, however, are those stretches where erosion is pronounced. Severe erosion does occur, and it can make a trail difficult to use, either because it is too deep and narrow or because exposed roots and rocks make footing difficult. This tempts people to leave the trail and make a new trail. Deeply rutted trails also exacerbate their own erosion problems by more effectively channeling water.

Although trampling of vegetation can cause limited amounts of erosion, the primary effect of trampling is to make a trail susceptible to erosion by loosening up the soil (DeLuca et al. 1998), reducing infiltration rates, and removing vegetation. *Running water is the principal agent of erosion.* Streams, snowmelt water, and water from springs all cause erosion when channeled down a trail. In some places, rainfall can also be intense enough to erode trails. For this reason, the main factors that determine degree of erosion damage are trail grade, orientation (what Marion and Leung 2001 refer to as slope alignment angle) and drainage—factors that affect the channeling and erosive force of water in the trail—and soil texture, the primary factor determining how readily the soil is detached and carried away. Leung and Marion (1996) provide a useful review of what is known about the influence of environmental factors on trail degradation.

Amount and type of use are generally less important to causing erosion than trail location and design features. Several studies have found that trails were not substantially deeper where use levels were higher (Cole 1991; Marion and Leung 2001). This is not to say that

Fig. 15.17. The effect of a horse's hoof is very different from the effect of a hiker's boot. Hikers tend to compact the soil; horses can punch holes in the soil, leaving soil detached and easily eroded. David Cole, U.S. Forest Service.

Table 15.3. Common Trail-Impact Problems and Strategies and Techniques for Mitigating Each Problem

Problem	Strategy	Technique (example)
Erosion	Improve location and/or design	Build water bars
Muddiness	Improve location and/or design	Route trails around boggy areas
Multiple trails	Improve location	Relocate trails
Shortcutting switchbacks	Change user behavior	Convince visitors to stay on existing trails
Informal trail systems	Reduce use	Reduce use quotas

a thousand hikers per day would not cause more erosion than only one hiker per day. Rather, it means that beyond a low threshold of use—a substantial amount of use is required to eliminate vegetation and render a trail vulnerable to water erosion—location and design are more important determinants of erosion than amount of use. Therefore, to prevent erosion, use levels would have to be so low that bare trails do not form. Virtually eliminating use to avoid trail problems is inefficient—not to mention unpopular with the public—given that trails can be designed to accommodate heavy use.

It has been suggested that type of footwear is an important determinant of amount of damage. Studies indicate, however, that lug soles are not substantially more destructive than other types of footwear. Kuss (1983) found no significant difference in the volume of soil eroded from a stretch of trail after being trampled by lug-soled and corrugated rubber-soled boots. This lack of difference was found despite increases in soil yield after 600 and 2,400 trampling passes, amounting to 1.4 and 1.7 times the yield of undisturbed trail, respectively.

Whether trail use is by hikers or by parties with pack and saddle stock is an important indicator of potential erosion. A small bearing surface carrying heavy weight, a horse's hoof can generate pressures of up to 1,500 pounds per square inch. These pressures, along with sharp shoes, cause stock to break up, not compact, the trail surface. Detached soil is more easily eroded and makes trails dustier when dry and muddier when wet. In an experimental study, DeLuca et al. (1998) found that horse traffic resulted in much higher sediment yields (an indicator of erosion potential) from established trails than either hikers or llamas. Hooves also tend to punch holes through meadow turf and disrupt wet soils. Use of stock on frequently used, properly located, and well-maintained trails is unlikely to aggravate problems. However, *on little-used trails that are steep, pass through wet meadows, and are seldom maintained, stock use can be much more damaging than hiker use.*

For most trails, the most effective solution to erosion problems lies in locating the trail where it is resistant to erosion and, where this is not possible, designing it to minimize erosion. Studies have found that erosion is most severe where trails are located in soils of homogeneous texture and that lack rocks. Erosion-prone soils consist primarily of sand, silt, or clay, rather than a combination of these different particle sizes; fine sand and silt soils are particularly prone to erosion (Bryan 1977). Such soils are frequently encountered in glacial deposits, particularly in valley bottoms and enclosed basins. Meadows also often have homogeneous, fine-textured soils. Meadow soils can also be highly organic, and organic soils are particularly prone to deterioration (Stewart and Cameron 1992). Steep slopes and places where drainage or snowmelt runs down the trail are prone to erosion (Leung and Marion 1999a). Streambanks with steep slopes, abundant moisture, and, in many places, fine-textured soils can frequently experience excessive erosion.

For these reasons, *in the mountainous West, erosion problems are frequently avoided by locating trails on ridges, talus slopes, and bedrock, away from alluvial plains and the glacial deposits of valley bottoms.* South- and west-facing slopes are often preferred locations because snow melts earlier there. *Where trails receive regular use, trails are best located in forests outside of meadows.* Although the vegetation in meadows is relatively resistant (Cole 1993b), regular trail-traffic use will eliminate even resistant vegetation and expose soils that are highly prone to erosion.

Several trail-design features can help minimize erosion by getting water off the trail. It is important to avoid steep grades by locating trails on side hills and by providing switchbacks where necessary. To divert water off the tread, trails are usually outsloped and often incorporate dips and rises (often called a rolling grade) rather than long, continuous downslope stretches. Water bars—logs, boards, timbers, or rocks installed across a trail, usually on an angle and sloping out—are common means of directing water off the tread and minimizing erosion. Water bars must be spaced closely enough so that water cannot build up excessive speed and erosive power. They must be securely anchored and large enough to keep water from running around or over them, forming destructive little waterfalls and they must be maintained frequently because they cease to function when dislodged or buried in the sediment deposited behind them. Finally, all of these techniques need to be part of the original trail design; once a deep trough has eroded, none of these techniques will be effective. Birkby (1996) and Hesselbarth and Vachowski (1996) provide good how-to instructions on these techniques.

It is also necessary to keep water from flowing onto the trail. The most common devices used are cross ditches (rock- or log-armored ditches crossing the trail), culverts (wood, metal, or fiberglass drainages buried underneath the trail), and parallel ditches (depressions that carry water adjacent to but lower than the trail tread). All of these devices must be carefully placed and maintained or they can aggravate problems and appear unnatural.

Erosion of stream banks can be minimized by locating stream crossings where banks are low, gentle, and stable. Where this is not possible, angling the trail across rather than directly up the bank and incorporating drainage devices can help minimize damage.

Muddiness

Muddy stretches are particularly disturbing to recreationists and both hikers and pack stock balk at walking through muddy, wet areas. *In an attempt to avoid muddy sections, hikers and stock may travel parallel to the trail and enlarge the muddy area until it can be hundreds of yards long and unnecessarily wide.* These muddy areas usually result from trampling while soils are water-saturated. Again, amount and type of use are of little importance because it takes only a little trampling to do most of the damage; however, damage is much more rapid with heavy use and with stock use (Leung and Marion 1999a). Muddiness can be a season-long problem in places where the water table is always close to the surface, or it can be temporary, occurring during snowmelt or when heavy rains fall on trails that have been churned to dust. *The solution to trail muddiness is to either locate trails on dry soils or shield the wet soils from trampling through trail engineering. Relocation is preferred, if a better location exists.*

Snowfields that do not melt until late in the season should be identified, and trails should be located away from meltwater channels. Identifying areas where the water table is close to the surface is more difficult. By noting the plant species and plant communities that grow in places along existing trails where muddiness is a problem, one can identify reliable vegetation indicators of potential problems. In the Selway-Bitterroot Wilderness, for example, more than two-thirds of the muddiness problems in one trail system were found in one vegetation type, which, along with vigorous growth of four individual species, can be used to identify sites to avoid (Cole 1983a).

Where trails must be built through water-saturated soils and it is not possible to improve the drainage, some sort of bridging is necessary to shield the vulnerable soil from trampling. Stepping stones sometimes provide a simple solution. A common type of bridging—*corduroy*—consists of three or more logs laid on the ground as stringers and bound together with wire or nails. It is notorious for not lasting long and should only be a temporary measure (Hesselbarth and Vachowski 1996). More permanent solutions include *puncheon*, a deck or flooring placed on stringers to elevate the trail, and *turnpiking*, an elevated trail of earth or gravel fill supported either by logs or flat rocks. Turnpikes may or may not be ditched on either side or may use culverts to facilitate drainage. Although such a trail is a permanent improvement and shows a "substantially noticeable imprint of man's work," huge muddy areas are also noticeable and undesirable. Therefore, *it is our opinion that trail engineering in wilderness is appropriate where necessary to cope with mud as long as (1) relocation is not feasible, (2) the design fits the environment, and (3) the practice is not carried to the extreme where every small wet place is bridged.*

User-Created Trails

The three most common types of undesirable impromptu trails are (1) multiple, parallel, or braided trails; (2) shortcuts on switchbacks; and (3) informal trail systems that traverse popular trailless areas or that fan out across popular destination areas. Each of these situations is the result of a unique set of circumstances; consequently, very different management approaches are required for each. The one element they have in common is that they are caused by people leaving the existing trail system—out of either dissatisfaction with the trail itself or a desire to go someplace else. Therefore, understanding visitors behavior and possibly accommodating their desires, as well as informing them of the damage they are doing, are common means of dealing with these three situations.

Multiple trails result from people avoiding surfaces that provide difficult footing—usually because they are slippery and muddy or because the trail is deep and narrow. Such surfaces are common in areas that are poorly drained or have homogeneous fine-grained textures, and in open areas, such as meadows, where it is easier for visitors to spread out. In either case, hikers and horses walk beside the trail, forming a new one. Other than educating people to stay on existing trails, the best solution to multiple trails is trail relocation. There has been a strong trend to get trails out of meadows into adjacent forested areas. However, trails abandoned in meadows usually need to be actively rehabilitated or they may continue to erode and may take centuries to recover.

In some places, such as Yosemite National Park, it has been difficult to get people to stop using trails in the meadows or along other preferred routes. In these cases, engineering solutions such as turnpiking have been employed to overcome the problems that encouraged users to form multiple trails. Although engineering may be preferable to doing nothing, we feel it should be a last option. Widening the trail trough can be a means of making it easier for visitors to stay on the trail tread. Armoring the trailside with rock—by making it uncomfortable to leave the tread—can also be helpful.

Shortcutting switchbacks is a common problem that is usually dealt with by trying to change visitor behavior and through trail design. Whether through education or regulation, visitors are asked not to shortcut switchbacks. In some places signs have been erected along the trail; this is an undesirable intrusion and should not be used unless absolutely necessary. Designs that can effectively reduce shortcutting include screening one switchback from another, building barriers of rock or vegetation, avoiding the use of numerous short switchbacks, and using wide turns.

Fig. 15.18. Trails through areas of water-saturated soils develop into ever larger bogs. In many cases the only solution is to bridge the area. It is best to bridge the entire bog, rather than just part of it—the problem with this trail. Photo courtesy U.S. Forest Service.

Informal trail systems indicate too much use of an area intended to be kept trailless. They are difficult to control. In popular destination areas, managers should consider developing an "official" trail system based on existing use patterns and encouraging its use. This strategy should control the size of the informal trail network and keep trails out of fragile environments. When laying out an official trail it is critical to provide access to the places visitors seek to go. If this is not done, the informal system will continue to be used, frustrating this management strategy.

Informal trails can also be controlled by reducing use, closing and rehabilitating trails, and persuading visitors to spread out. This is probably only realistic in remote, lightly used parts of the wilderness, where management objectives are to keep the area trailless. Once trails begin to develop, use must be reduced or an official trail should be designated. Although it may seem inappropriate to reduce use in lightly used places—particularly if use is not limited in much more heavily used destination areas—this may be the only means of meeting the objective of avoiding trails. Several actions can be taken at an early stage to make such an action less likely. Users of such areas should be educated about the need to travel in small parties, to spread out rather than follow in each other's footsteps, and to avoid using incipient trails. Of particular importance, management should not attempt to disperse use from heavily used destination areas to these lightly used areas unless they are willing to accept trails in these places.

Fig. 15.19. Multiple trailing is a common problem, particularly in meadows. Such trails tend to form where the trail tread is difficult to use (note the water flowing down the main tread of this trail). The most common solution has been to relocate trails in the forest or at the edge of the meadow where soils are drier and have less potential to erode. Photo courtesy Randel Washburne, U.S. Forest Service.

As much of the preceding discussion concludes, next to proper trail location, engineering—surfacing, bridging, ditching, and so on—is the most effective means of avoiding trail problems. Problems that cannot be reduced or avoided by trail relocation will generally require an engineering solution. Although engineering is an appropriate means of providing for use without excessive resource damage, it is counter to the wilderness ideal, and therefore should be kept to a minimum.

Excessive relocation should be avoided because frequent relocation can produce a maze of closed trails. Managers must exercise restraint when relocating trails. This problem can be reduced by actively rehabilitating trails. A good rule of thumb, however, is not to relocate a trail unless (1) the new section is in a more resistant environment or is better designed, and (2) hikers can be kept off the old section of trail. Unless both these conditions are met, relocation will only compound problems by disturbing new sites.

Monitoring Trail Conditions

Monitoring of trails can be used to ascertain trends in trail condition, where and how to locate and design trails, whether trail-management programs are working, and where and how trails need to be fixed.

For assessing trends in condition, rapid survey techniques are quick and easy and usually provide adequate precision. One approach is to sample conditions along trails at a systematic specified distance, such as every 328 feet (100 m). Using such an approach on a trail system in Montana, Cole (1991) showed that mean trail width (the zone disturbed by trampling) increased from 40 inches (100 cm) to 50 inches (125 cm) between 1980 and 1989; however, bare width (the zone without any vegetation) and trail depth did not change significantly. Such data can also be used to assess the portion of a trail that exceeds certain depth or width standards or that contains certain "detracting features" (such as root exposure or muddiness). Leung and Marion (1999b) assessed the influence of sampling interval on the accuracy of estimates of the lineal extent of "detracting features" on trails at Great Smoky Mountain National Park. They report that a sampling interval of less than 328 feet (100 m) is best, although an interval of 328 feet (100 m) to 1,640 feet (500 m) is probably acceptable. They advocate using a sampling approach to assess the lineal extent of problems but advise against using a sampling approach to estimate the frequency of occurrence of problems.

Frequency of problems—and also lineal extent—are most accurately and usefully assessed with a census of the trail system. The first step here is to define exactly what is or is not considered a problem. As trails are walked, the number and length of problems are recorded and their location is mapped. Such information is useful not only in assessing trend, but also in budgeting for trail maintenance and in allocating resources to certain trail segments. At Great Smoky Mountains National Park, Leung and Marion (1999a) were able to show that (1) soil erosion and wet soil were the most extensive problems; (2) water bars were more effective than drainage dips in diverting water off trails; and (3) although serious impact problems were fairly well distributed through the park, trails with wet, muddy treads were concentrated in places with high horse use.

For assessing the effectiveness of management programs or maintenance techniques on specific trail segments, the expense of more detailed replicable measurements may be justified. Typically, these techniques provide accurate measures of trail cross-sectional area at fixed locations that can be precisely relocated. Increases in trail cross-sectional area indicate erosion has occurred. Marion, Leung et al. (2006) describe new techniques for measuring cross-sectional area (based on a variable interval), as well as improved ways to interpret maximum incision measures.

Only recently, methods for monitoring networks of user-created trails were developed. Cole, Watson et al. (1997) mapped trails and classified segments according to their level of disturbance:

1. Disturbed but retain at least 20 percent vegetation cover
2. Trails with less than 20 percent vegetation cover but less than 1.5 feet wide
3. trails with less than 20 percent vegetation cover and more than 1.5 feet wide

This approach has been adapted and applied to user-built trail assessment work being conducted in a number of places (Marion, Leung et al. 2006), including Yosemite National Park (Bacon et al. 2006).

Trail Rehabilitation

Much of the information about rehabilitating campsites also applies to trails, particularly the techniques available for reestablishing vegetation and the need to eliminate all use. The major difference is the need, in many places, to stop erosion and to replace the soil lost by erosion. It also may be more difficult to keep people from using the trail if it leads where they want to go.

One must first provide a desirable alternative route to minimize use of a trail. This may require observing use patterns and visitor behavior, and even some questioning of users about their itinerary and what alternative routes might be acceptable. Once the alternative has been provided, one should try to de-emphasize the old trail. Careful selection of a starting point for the relocation can make it easier to hide the old trail. If it cannot be hidden, block it with logs, rocks, or brush. Finally, if this does not work, it will be necessary to erect signs such as "Please stay on the trail to prevent damage."

Avoiding further trail erosion starts with looking for the erosion source. Sometimes runoff has been directed down the trail, and it must be diverted elsewhere. Ditches and water bars across the trail can be used to keep water off the trail. Where trails are deeply eroded, it may be necessary to place rock or log check dams in the trail to reduce water velocity and to allow backfilling by sediment (Rochefort and Gibbons 1992). Material used to fill in trail troughs—other than what is deposited behind check dams—must be judiciously selected. Where the trail follows the contour, it is often possible to move material deposited below the trail, because of construction, back into the tread. Other good sources of material are soil from streambeds and rock from talus slopes or other trail work. It is best not to remove too much material from any single place and to make certain the source area is blended into its surroundings and hidden from view. Regardless of where the material comes from, revegetation will be most effective if the trough is filled to grade with soil.

Fig. 15.20. Informal trail networks at popular destination areas can be unsightly. Managers often try to confine use and impacts to certain trails, while closing and restoring other trails. Photo courtesy Alan Watson, U.S. Forest Service.

Multiple trails have been successfully restored in Tuolumne Meadows, Yosemite National Park, by cutting off the sod ridges between the multiple trails at the level of the trail tread and stacking them in the shade. The soil beneath both the trails and the ridges was dug up to eliminate compaction, and sand was added to bring the trail up to the level of the surrounding meadow. Finally, the sod was divided into transplant plugs and planted. Although this technique will work only in certain environments, it is well suited to multiple trails in meadows—a very common problem in many areas. Eagen et al. (2000) report on the continued success of this basic technique in Tuolumne Meadows. Bay and Ebersole (2006) documented successful transplanting of alpine turf onto closed trails in the Sangre de Cristo Wilderness, Colorado. Rock walls were built across trails, backfilled with rock from talus slopes and raw soil, and planted with pieces of turf cut for the new trail.

Managing Pack and Saddle Stock

Although the backpacking "boom" of the late 1960s and 1970s relegated pack and saddle stock to a minority use in all but a few wildernesses, *stock use (both riding and pack animals) is still an important and accepted tradition.*

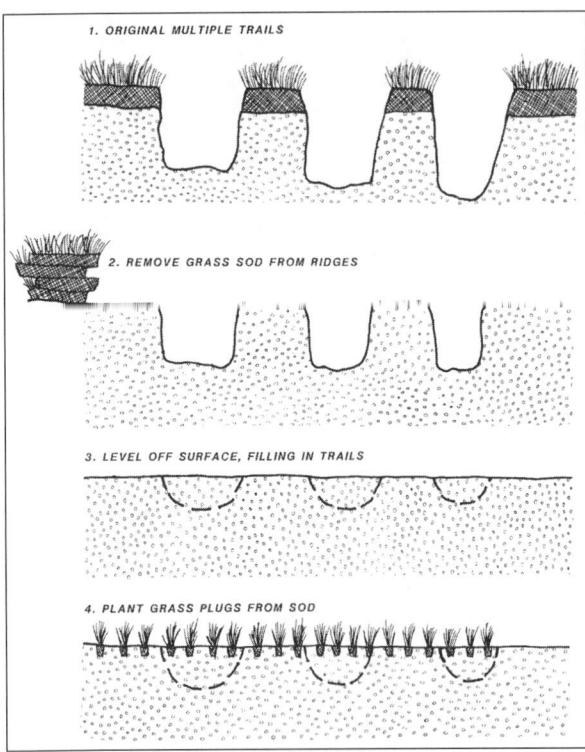

Fig. 15.21. Multiple trails can be revegetated by removing sod, leveling the soil surface, and transplanting sod plugs. Photo courtesy U.S. Forest Service.

This is in contrast to some other nations, such as Australia, where use of stock in wilderness is not allowed. As of 1990, stock use occurred in about one-half of all wilderness areas (McClaran and Cole 1993). *About one of every nine parties entering wilderness in 1990 traveled with stock, and about one-third of all nonadministrative use of stock is commercial, as opposed to private.* This situation is probably not much different today.

Despite the prevalence of such problems, the published literature on pack-stock impacts and their management is sparse. Certain impacts are similar to those caused by hikers and can be managed similarly; others are qualitatively different and require very different sorts of management techniques.

Types of Stock Impact

Generally the impact of stock on trails is similar to that caused by hikers except that it is more pronounced. Weaver and Dale (1978) examined the effects of controlled amounts of use by horses and hikers on trail width and depth, percentage of bare ground, and soil compaction (bulk density). Trails produced by 1,000 horse passes were 2 to 3 times as wide and 1.5 to 7 times as deep as trails produced by 1,000 hiker passes. Compaction increased about 1.5 to 2 times as rapidly on horse trails as on hiker trails. Finally, one-half of the vegetation was lost after 1,000 hiker and 600 horse passes on a grassland, and after 300 hiker and only 50 horse passes in a forest. *In a more recent experimental trampling study in Montana, Cole and Spildie (1998) report that just 25 horse passes cause more loss of vegetation cover than 150 passes by hikers or llamas.* In a Tasmanian shrubland, Whinam and Chilcott (1999) report that broken plant material, as a percentage of plant biomass, was 0.1 percent after trampling by hikers, compared to 39.2 percent after trampling by horses. These experimental results suggest that *creating multiple trails and new trails in trailless areas will occur much more rapidly with stock use than with hiker use. The trails created will also be wider, deeper, more compacted, and less vegetated.*

The effect of stock on existing trails, which can differ from their effect on undisturbed sites, was examined experimentally in Great Smoky Mountains National Park (P. Whittaker 1978). Again, horse use caused more pronounced increases in trail width, trail depth, and litter loss than hiker use. In Montana, horse traffic on existing trails resulted in more than double the sediment yield caused by a comparable amount of hiker traffic (DeLuca et al. 1998). Although hiker use generally tends to increase soil compaction on the trail, horse use loosens the soil, making it more susceptible to erosion (Fig. 15.7). *Trail widening is accentuated by the tendency for stock to walk on the downslope side of the trail.* This breaks down the outer edge of the trail so that new soil must be brought in to rebuild the trail. The result is a wide trail, a much wider area of disturbance, and an ongoing trail-maintenance problem.

Equestrian trails require considerably more maintenance than hiking trails. They must be "brushed out" to a greater height and width. Fallen trees must be quickly removed or detour trails will rapidly develop. Stock often break or dislodge drainage devices such as water bars. Stock disturbance of muddy trail sections can be particularly severe and can be corrected only with some type of bridging more elaborate than that required for foot travel.

At campsites, the magnitude of impact caused by stock parties as opposed to hiking parties is even more pronounced than on trails. Moreover, stock parties cause a number of impact types that other parties do not. Unfortunately, very little data on differences between these two types of use are available. Unpublished data (archived at the Aldo Leopold Wilderness

Research Institute, Missoula, MT) from Dr. Sidney Frissell's work in the early 1970s in what is now the Lee Metcalf Wilderness indicate that campsites used by stock parties were, on the average, ten times as large and had seven times as much exposed mineral soil as sites used primarily by backpackers.

Stock parties, generally being larger than backpacker parties, consequently disturb a larger area, to which is added an adjacent area where stock is kept. The animals are often tied to trees, resulting in a large number of damaged trunks and exposed roots. Because stock parties usually carry saws and axes to clean windfalls from trails, trees are often felled for tent poles to support large canvas tents and for firewood.

Although stock are usually kept adjacent to the camp, they occasionally are brought into the central camp area. Here, the action of their shod hooves can cause rapid deterioration of the site. Loss of organic soil horizons, increased compaction, and decreased infiltration rates, typical of all campsites, are particularly pronounced on sites used by stock parties. Seeds of exotic plants contained in horse feed readily germinate and grow on such disturbed sites. On Bob Marshall campsites, exotic species accounted for just 5 percent of the vegetation cover on backpacker sites, compared with 43 percent of the cover on stock sites (Cole 1983b).

In areas where stock are grazed or confined for the night, impacts result from both trampling and defoliation of plants. These impacts, described in the earlier section on grazing impacts, are unique to parties with stock and often affect a much larger area than all other recreational impacts combined. In a portion of the Eagle Cap Wilderness, an estimated 1.8 percent of the area had been significantly altered by recreational use. About three-fourths of this disturbed area consisted of areas used only by stock for grazing (Cole 1981).

Not all pack animals cause similar impacts. Studies show that llamas cause substantially less impact than horses both to existing trails (DeLuca et al. 1998) and to vegetation (Cole and Spildie 1998).

Management Strategies and Techniques

Despite a strong tradition of stock use in some wilderness areas, it clear that social, environmental, and administrative costs associated with stock use are much more pronounced than with a comparable amount of hiker use. As with campsite and trail impact, a number of factors influence the severity of stock impacts. The most important are frequency and amount of use, party size and behavior, time of use, and location of use. These define four primary management strategies: (1) limiting or reducing use, (2) encouraging less damaging behavior, (3) managing the timing of use—discouraging use during times of the year when the potential for damage is high—and (4) managing location of use—encouraging use of particularly resistant environments and containing use are particularly valuable in controlling the spread of stock impact (Table 15.4).

Limiting Use

Stock impacts on trails and campsites are unlikely to be greatly diminished merely by reducing use, unless use levels are cut to almost nothing. Because stock use causes impact even more rapidly than hiker use, this conclusion has even more serious implications for managing stock use. Unless all stock use is eliminated, there will be few situations where reducing stock use will produce substantial benefits in improved trail and campsite conditions. Reducing use can help minimize social and aesthetic impacts (horse user–hiker conflicts, manure, and so on), but other less drastic actions can also be taken. Except in cases where all

Table 15.4. Factors That Influence Pack-Stock Impacts
Each factor defines a strategy and set of specific techniques (of which only one example is provided here) for managing impacts.

Factor	Strategy	Technique (example)
Amount/frequency of use	Reduce use	Close overgrazed meadows
Party size/behavior	Change behavior	Promote low-impact use
Season of use	Change timing of use	Prohibit use during spring
Environmental/site conditions	Increase site resistance	Reinforce trail with log cribbing
Use distribution	Contain use	Prohibit off-trail travel with stock

Fig. 15.22. Stock-holding areas adjacent to campsites greatly enlarge the area of disturbance. Such areas often experience serious tree damage, loss of vegetation, and soil erosion. Photo courtesy U.S. Forest Service.

recreational use must be limited, or where management objectives indicate that all stock use and impact are inappropriate—and all stock use is prohibited—there seems to be little justification for limiting the amount of stock use. Better solutions to stock-caused problems exist.

One important exception to this is limits on stock use in specific meadows to avoid overgrazing. Managers of Sequoia–Kings Canyon National Parks, for example, have implemented a management plan with specific limits for use of certain meadows with a history of being overgrazed. When the limit is reached, the meadow is temporarily closed. The need for closure is evaluated by monitoring meadow condition or by determining use levels from posttrip itineraries that each stock party is required to submit. Some of these meadows have length-of-stay limits so that more users can share the limited resource.

However, just as it is inequitable to ration stock use when hiker use is unlimited, it is also inequitable, where rationing is instituted, not to recognize that stock cause considerably more damage than hikers. *In terms of comparable impact, more hikers than horse users can use an area.* As the earlier review of impact studies shows, *trampling by a horse typically causes four to eight times as much impact as trampling by a hiker.* Thus, a typical group of three horse riders with two pack animals is likely to cause about seven to fourteen times the impact of a group of three backpackers (everything else being equal). *When the impacts of grazing and confining stock are factored in, a group of three, choosing to travel with stock, might typically cause about twenty times the impact of a comparable group of three choosing to travel on foot.* Consequently, where use is rationed, an equitable allocation (based on relative impact) could give a hiker twenty times the odds of a horse rider of obtaining a permit.

Encouraging Less Damaging Behavior

Minimum-impact stock practices can do much to reduce impacts, particularly in and around campsites. Outside of malicious or thoughtless chopping of trees, most tree damage—a particularly severe problem on stock sites—results from felling trees for firewood or tent poles and tying horses to trees. Available techniques and equipment (Aaland 1993; Hampton and Cole 2003) can largely eliminate these practices and their resultant impacts. Moreover, stock do not need to be in camp areas after unloading; if stock were kept off campsites, soil disturbance and horse manure would also be greatly reduced. If these ideas were adopted by stock users, campsites used by stock parties would suffer no more damage than sites used by comparably sized groups of backpackers.

Unfortunately, it may be more difficult to persuade stock users to adopt Leave No Trace techniques. In a study of attention to low-impact messages posted on trailside bulletin boards—the most common of educational media—Cole, Hammond et al. (1997) found that horse riders' knowledge of appropriate low-impact techniques was much lower than hikers' knowledge. Moreover, fewer horse riders (27 percent) than hikers (71 percent) stopped to read the messages; the horse riders who did read the messages gave them less attention; and the horse riders remembered less of the message after their trip. Given that stock groups have so much more potential to cause impact than hiking groups, these findings are disturbing. Hopefully, there are educational media other than trailside bulletin boards that can be effective in reaching horse riders. Rather than rely on education, many areas have established regulations designed to prevent inappropriate behavior—such as prohibiting the practice of tying stock to trees.

Although a limit on party size is currently the most common pack-stock management technique in wilderness—about one-half of wildernesses with stock use have a limit—the number allowed ranges from five to thirty-five animals per party, with twenty-five the most common limit (Monz et al. 2000). *Such high limits will have very little beneficial effect. Moreover, there is little equity in the common situation where the maximum allowable number of stock exceeds the allowable number of hikers. Limits based on number of "heartbeats" are more equitable.* Using this approach, it is permissible for any group to have a given maximum number of heartbeats, say ten, which can consist of any combination of people, stock, or dogs.

Stock impacts could also be reduced by requiring parties to carry stock feed, thus virtually eliminating grazing damage. A number of wildernesses with stock use, mostly NPS areas, have taken this action. Many more recommend packing in feed. In many places, parties bring only enough feed to supplement grazing. Where deterioration is not yet critical, this approach may be more acceptable to users and may postpone or avoid the need for regulations. There is, of course, a trade-off in that additional animals are needed if feed is packed in. Also, it is important that the feed not contain exotic weeds. Certified weed-free hay is now required in many wilderness areas. Other possibilities include the prohibition of grazing where areas have been overgrazed.

Managing the Timing of Use

Although only a few wildernesses prohibit stock use during certain seasons of the year (McClaran and Cole 1993), the impact of both grazing and trampling is highly dependent on seasonal variables, particularly the phenology or stage in the annual cycle of development of plants, and the moisture content of the soil. *Generally, plants and soil are most vulnerable to disturbance during spring when plants are using stored nutrients for growth and when soils are water-saturated.* This has led Sequoia–Kings Canyon National Parks—the backcountry area with the longest history of research and management of pack-stock impact—to make their primary management tool a system of opening dates that are determined from information on the type of hydrologic year and the vegetation and soils. A research program examined the composition and distribution of the parks' major forage areas as well as their susceptibility to differing intensities, frequencies, and times of use (DeBenedetti and Parsons 1983). Opening dates were prescribed for three types of hydrological years (wet, normal, and dry), allowing the user to predict when use will be allowed.

Managing Location of Use

Among the most effective way to manage the impact of stock is to control where stock use is and is not allowed. A common action is to keep stock away from stream banks and lakeshores, where trampling can be particularly destructive and, in some cases, causes accelerated erosion. This will also reduce the risk of water pollution. In most NPS wilderness, stock use is only permitted on established trails (McClaran and Cole 1993). Moreover, some parks—like Glacier National Park—only allow overnight stock use in selected areas judged to be resistant to stock impact. Less restrictive alternatives exist, although resultant impact will be greater. One approach is to permit stock use on mainline trails and selected trails into destination areas. For example, imagine a series of lakes up a tributary valley. Stock use might be permitted as far as a stock camp at the first lake above the valley bottom. However, foot traffic only is allowed beyond this first lake. Through this compromise, horse riders are provided with access to the tributary valley without opening up the entire string of lakes to the impacts typically associated with stock traffic. Where certain trails are closed to stock use, trail-maintenance costs can be substantially reduced.

Sequoia–Kings Canyon National Parks have an even less restrictive policy. They prohibit stock use only in places that have never received regular stock use and in certain meadows being maintained as representative examples of pristine ecosystems. Such a policy provides opportunities for horse users to enjoy much of the wilderness while avoiding degradation of currently undamaged areas and providing protection for some representative examples of meadow types that might be entirely altered by unrestricted grazing. It also provides places for hikers to go where they know they will not see stock parties.

Confinement of stock impacts (limiting where they can go) offers a reasonable compromise between providing horse riders with access to wilderness and confining the heavy impacts typically associated with stock use. Spildie et al. (2000) evaluated the effectiveness of a confinement strategy designed to reduce the impact of stock in a subalpine lake basin in the Selway-Bitterroot Wilderness. Stock users were permitted to camp overnight at any of the lakes in the lake basin but were required to stay in a designated stock site provided at each lake. They were required to tie their stock to a hitch line that was to be strung between two designated trees, confining the impacts of stock containment to a very small area. They also were not allowed to graze their stock in the lake basin. The benefits of this program—in reducing impact—have been substantial. In just five years,

Fig. 15.23. Hitch rails, like corrals, serve to concentrate the impact of stock use in a small area. Photo courtesy David Cole, U.S. Forest Service.

disturbed area decreased 37 percent and bare area decreased 43 percent. Disturbed and bare areas in the lake basin should eventually decline to just 36 percent and 24 percent, respectively, of what they were in the early 1990s. Moreover, stock users retain the ability to access the entire lake basin, as long as they use specific campsites and either graze their stock outside of the lake basin or pack in feed.

About one-half of NPS wilderness areas require stock parties to camp at certain sites specifically designated for their use; this confines damage to one site in a particularly resistant area. In Yellowstone National Park, Wyoming, grazing is allowed only in designated meadows that are periodically closed to allow recovery. Corrals, drift fences, and hitch rails can also be used to confine the spread of impact. Where such measures have been studied, most visitors support both restricting stock use in certain areas and requiring stock parties to use certain campsites.

The Challenge of Stock Management

The damage caused by stock is generally much greater than that caused by hikers; many types of impact can be attributed solely to stock; and in most areas only a very small proportion of visitors benefit from and are associated with this impact. Many hikers consider this an inappropriate impact on the wilderness resource, and this, along with other inconveniences such as exposure to horse manure and moving off the trail to let horse users pass, creates conflicts between backpackers and horse users. These conflicts appear to be growing. Complaints about stock are the most common complaint from visitors in many wilderness areas.

On the other hand, stock use is a generally accepted, traditional type of wilderness use. Managers must grapple with the questions of how much of this impact is acceptable and where it should be allowed. An important question is whether to allow stock use in places now suffering from too much use. Stock impacts are likely to be less severe here than elsewhere because impacts are already so serious that further impact is unlikely. However, because such places receive heavy hiking use too, horse-hiker conflicts are likely to also be severe in such locations. Where impacts are excessive, managers can attempt to persuade stock users to use low-impact camping techniques, or they can confine stock use to certain trails and campsites. The most effective management programs will use a variety of techniques to reduce damage resulting from use of stock.

Summary

This chapter has identified the various environmental effects of wilderness recreation use and described many techniques for managing them *based on ecological impacts*. Several general points summarize this material.

First, impact is inevitable wherever wilderness recreation use is allowed. Therefore, consistent with the goal of providing recreation opportunities, management can only limit impact, not prevent it. Nevertheless, to prevent impact from increasing incrementally, with little ability to keep track of cumulative impacts, it is imperative to set specific objectives and standards that will place a limit on impact. Then, through monitoring of conditions, managers will be able to identify more clearly when specific impacts have become so pronounced as to demand management attention.

Second, many strategies and techniques are available to help managers deal with each type of impact. Too often, managers try only one technique—often the one they are most familiar with or one the neighboring manager is using. In most cases, however, using several techniques simultaneously will be much more effective than using just one. An appropriate process for selecting a suite of management techniques is to identify the source of the problem, formulate actions that can eliminate the problem at its source, implement a preferred set of actions, and, most important, monitor results.

Managers have been preoccupied with "too much use" as the major cause of problems, and limiting use to a *"carrying capacity" as the principal solution.* Amount of use is just one of many factors influencing amount of impact; often it is one of the less influential factors. Likewise, limiting use to a carrying capacity is only one of many alternative management techniques, and often it is not very effective. However, there clearly are situations where use must be limited. *More consideration should be given to limiting use of wilderness that is still relatively lightly used and impacted—to see that it remains in that state.* Where use is limited, other actions—such as confinement of impact—are often necessary if use limits are to be beneficial.

Finally, managers must show equal concern for both quality of wilderness visitors' experience (see Chapter 16) and environmental impacts. The two are inextricably bound together; actions that affect one will affect the other—sometimes in positive and sometimes in negative ways. Therefore, managers must clearly define problems and how alternative actions will deal with them (see Chapter 16). Managers must also consider how an action to correct a problem in one specific place will affect other places and other wilderness conditions. Only through more effective integration of ecological and visitor experience concerns, both in research and management, can we develop the more holistic approach that will make wilderness management more effective.

Study Questions

1. Select two recreation impact "situations" (type of impact, environmental type, and location) that represent substantial alterations of natural conditions and two recreation impact situations that may *not* represent substantial alterations. Defend your choices.
2. Provide a detailed description and one example of (a) direct, (b) indirect, and (c) cyclic effects of trampling on vegetation and soil.
3. Suggest two actions that managers might take to influence the amount of use that campsites receive and thereby reduce impact. Contrast the appropriateness of each action on campsites in heavily used and popular destination areas with that on campsites in lightly used, remote places in the wilderness.
4. Compare and contrast the magnitude and nature of impacts caused by (a) small parties and large parties and (b) hiker parties and stock parties. Describe where impacts are similar and where they differ.
5. What do studies suggest about the relative impact (magnitude) of horses versus hikers? What about llamas? Suggest and defend management actions that might be taken to minimize differences.
6. Discuss the dilemma between dispersing use and dispersing impacts. How would this dilemma affect decisions about tearing down rock fire circles (rings) at a popular site or closing heavily used campsites?

Acknowledgments

The chapter author, David Cole, expresses his thanks to numerous field assistants and colleagues, to Bob Lucas for giving him the opportunity to work on wilderness, and to wildlands for their inspiration.

Case Discussion:
Baseline Wilderness Data

The research information on recreational use and biophysical impacts from recreational use on campsites and trails is very limited. Cole and Wright (2003) reported that only 56 percent of the 625 wilderness units in the NWPS in 2000 had any research information on visitors and their impacts. The types of baseline information available in wilderness units were mostly for campsite impacts (51 percent), visitor use (24 percent), and trail conditions (9 percent). Monitoring campsite and trail conditions is essential to understanding what management strategies and techniques are needed to control impacts.

> **Case Discussion Questions:**
> 1. What specific information would you seek in *current resource conditions* and baseline studies of campsites regarding vegetation change, changes in soil conditions, temporal patterns of impact, and user-created campsites in order to make decisions about management strategies and techniques needed to control visitor impacts on *campsites*?
> 2. What information about amount, type, season, and location of *visitor use* do you need to decide how to manage use and control impacts on *campsites*?
> 3. What specific information would you seek in *current resource conditions* and baseline studies of trails regarding erosion, muddiness, and user-created trails in order to make decisions about management strategies and techniques needed to control visitor impacts on *trails*?
> 4. What information about amount, type, season, and location of *visitor use* do you need to decide how to manage use and control visitor impacts on *trails*?

References

Aaland, Dan. 1993. *Treading Lightly with Pack Animals.* Missoula, MT: Mountain Press.

Adkison, Gregory P.; Jackson, Marion T. 1996. Changes in ground-layer vegetation near trails in midwestern U.S. forests. *Natural Areas Journal.* 16: 14–23.

Anderson Dorothy H.; Lime, David W.; Wang, Theresa L. 1998. *Maintaining the Quality of Park Resources and Visitor Experiences: A Handbook for Managers.* St. Paul: University of Minnesota, Department of Forest Resources, Cooperative Park Studies Unit.

Anderson, S. H. 1995. Recreational disturbance and wildlife populations. In: Knight, Richard L.; Gutzwiller, Kevin J., eds. *Wildlife and Recreationists: Coexistence Through Management and Research.* Washington, DC: Island Press, pp. 157–168.

Arocena, Joselito M.; Nepal, Sanjay K.; Rutherford, Michael. 2006. Visitor-induced changes in the chemical composition of soils in backcountry areas of Mt Robson Provincial Park, British Columbia, Canada. *Journal of Environmental Management.* 79: 10–19.

Bacon, James; Roche, James; Elliot, Crystal; Nicholas, Niki. 2006. VERP: Putting principles into practice in Yosemite National Park. *The George Wright Forum.* 23(2): 73–83.

Bay, Robin F.; Ebersole, James J. 2006. Success of turf transplants in restoring alpine trails, Colorado, U.S.A. *Arctic, Antarctic, and Alpine Research.* 38: 173–178.

Benninger-Truax, Mary; Vankat, John L.; Schaefer, Robert L. 1992. Trail corridors as habitat and conduits for movement of plant species in Rocky Mountain National Park, Colorado, USA. *Landscape Ecology.* 6: 269–278.

Birkby, Robert C. 1996. *Lightly on the Land: The SCA Trail-Building and Maintenance Manual.* Seattle, WA: The Mountaineers.

Blakesley, J. A.; Reese, K. P. 1988. Avian use of campground and non-campground sites in riparian zones. *Journal of Wildlife Management.* 52: 399–402.

Bratton, Susan P.; Stromberg, Linda L.; Harmon, Mark E. 1982. Firewood-gathering impacts in backcountry campsites in Great Smoky Mountains National Park. *Environmental Management.* 6: 63–71.

Brickler, Stan; Tunnicliff, Brock; Utter, Jack. 1983. Use and quality of wildland water: The case of the Colorado River corridor in the Grand Canyon. *Western Wildlands.* 9(2): 20–25.

Bryan, Rorke B. 1977. The influence of soil properties on degradation of mountain hiking trails at Grovelsjon. *Geografiska Annaler.* 59A(1–2): 49–65.

Bull, E. L.; Parks, C. G.; Torgersen, T. R. 1997. Trees and logs important to wildlife in the Interior Columbia River Basin. General Technical Report PNW-GTR-391. Portland, OR: U.S. Department of Agriculture, Forest Service, Pacific Northwest Research Station.

Cassirer, E. Frances; Freddy, David J.; Ables, Ernest D. 1992. Elk responses to disturbance by cross-country skiers in Yellowstone National Park. *Wildlife Society Bulletin.* 20: 375–381.

Christensen, Neal A.; Cole, David N. 2000. Leave-no-trace practices: Behaviors and preferences of wilderness visitors regarding use of cookstoves and camping away from lakes. In: Cole, D. N.; McCool, S. F.; Borrie, W. T.; O'Loughlin, J., comps. *Wilderness Science in a Time of Change,* Vol. 4: *Wilderness Visitors, Experiences, and Visitor Management;* May 23–27, 1999; Missoula, MT. RMRS-P-15. Ogden, UT: U.S. Department of Agriculture, Forest Service, Rocky Mountain Research Station, pp. 77–85.

Cilimburg, Amy; Monz, Christopher; Kehoe, Sharon. 2000. Wildland recreation and human waste: A review of problems, practices, and concerns. *Environmental Management.* 25: 587–598.

Cole, David N. 1979. Reducing the impact of hikers on vegetation: An application of analytical research methods. In: Ittner, R.; Potter, D.; Agee, J.; Anschell, S., eds. *Recreational impact on wildlands; conference proceedings;* October 27–29, 1998; Seattle, WA. R-6-001-1979. Portland, OR: U.S. Department of Agriculture, Forest Service, Pacific Northwest Region, pp. 71–78.

———. 1981. Vegetational changes associated with recreational use and fire suppression in the Eagle Cap Wilderness, Oregon: Some management implications. *Biological Conservation.* 20: 247–270.

———. 1982a. Controlling the spread of campsites at popular wilderness destinations. *Journal of Soil and Water Conservation.* 37: 291–295.

———. 1982b. Wilderness campsite impacts: Effect of amount of use. Research Paper INT-284. Ogden, UT: U.S. Department of Agriculture, Forest Service, Intermountain Forest and Range Experiment Station.

———. 1983a. Assessing and monitoring backcountry trail conditions. Research Paper INT-303. Ogden, UT: U.S. Department of Agriculture, Forest Service, Intermountain Forest and Range Experiment Station.

———. 1983b. Campsite conditions in the Bob Marshall Wilderness, Montana. Research Paper INT-312. Ogden, UT: U.S. Department of Agriculture, Forest Service, Intermountain Forest and Range Experiment Station.

———. 1986. Recreational impacts on backcountry campsites in Grand Canyon National Park, Arizona, USA. *Environmental Management.* 10: 651–659.

———. 1989. Wilderness campsite monitoring methods: A sourcebook. General Technical Report INT-259. Ogden, UT: U.S. Department of Agriculture, Forest Service, Intermountain Research Station.

———. 1991. Changes on trails in the Selway-Bitterroot Wilderness, Montana, 1978–89. Research Paper INT-450. Ogden, UT: U.S. Department of Agriculture, Forest Service, Intermountain Research Station.

———. 1992. Modeling wilderness campsites: Factors that influence amount of impact. *Environmental Management.* 16: 255–264.

———. 1993a. Campsites in three western wildernesses: Proliferation and changes in condition over 12 to 16 years. Research Paper INT-463. Ogden, UT: U.S. Department of Agriculture, Forest Service, Intermountain Research Station.

———. 1993b. Trampling effects on mountain vegetation in Washington, Colorado, New Hampshire, and North Carolina. Research Paper INT-464. Ogden, UT: U.S. Department of Agriculture, Forest Service, Intermountain Research Station.

———. 1995a. Disturbance of natural vegetation by camping; experimental applications of low level stress. *Environmental Management.* 19: 405–416.

———. 1995b. Experimental trampling of vegetation, II: Predictors of resistance and resilience. *Journal of Applied Ecology*. 32: 215–224.

Cole, David N.; Hall, Troy E. 1992. Trends in campsite condition: Eagle Cap Wilderness, Bob Marshall Wilderness and Grand Canyon National Park. Research Paper INT-453. Ogden, UT: U.S. Department of Agriculture, Forest Service, Intermountain Research Station.

Cole, David N.; Hammond, Timothy P.; McCool, Stephen F. 1997. Information quantity and communication effectiveness: Low-impact messages on wilderness trailside bulletin boards. *Leisure Sciences*. 19: 59–72.

Cole, David N.; Landres, Peter B. 1995. Indirect effects of recreationists on wildlife. In: Knight, R. L.; Gutzwiller, K. J., eds. *Wildlife and Recreationists: Coexistence Through Management and Research*. Washington, DC: Island Press, pp. 183–202.

———. 1996. Threats to wilderness ecosystems: Impacts and research needs. *Ecological Applications*. 6: 168–184.

Cole, David N.; Monz, Christopher A. 2003. Impacts of camping on vegetation: Response and recovery following acute and chronic disturbance. *Environmental Management*. 32: 693–705.

———. 2004. Spatial patterns of recreational impact on experimental campsites. *Journal of Environmental Management*. 70: 73–84.

Cole, David N.; Petersen, Margaret E.; Lucas, Robert C. 1987. Managing wilderness recreation use: Common problems and potential solutions. General Technical Report INT-259. Ogden, UT: U.S. Department of Agriculture, Forest Service, Intermountain Research Station.

Cole, David N.; Ranz, Beth. 1983. Temporary campsite closures in the Selway-Bitterroot Wilderness. *Journal of Forestry*. 81: 729–732.

Cole, David N.; Spildie, David R. 1998. Hiker, horse, and llama trampling effects on native vegetation in Montana, USA. *Journal of Environmental Management*. 53: 61–71.

———. 2007. Vegetation and soil restoration on highly impacted campsites in the Eagle Cap Wilderness, Oregon. General Technical Report RMRS-GTR-185. Fort Collins, CO: U.S. Department of Agriculture, Forest Service, Rocky Mountain Research Station.

Cole, David N.; van Wagtendonk, Jan W.; McClaran, Mitchel P.; Moore, Peggy E.; McDougall, Neil K. 2004. Response of mountain meadows to grazing by recreational pack stock. *Journal of Range Management*. 57: 153–160.

Cole, David N.; Watson, Alan E.; Hall, Troy E.; Spildie, David R. 1997. High-use destinations in wilderness: Social and biophysical impacts, visitor responses, and management options. Research Paper INT-RP-496. U.S. Department of Agriculture, Forest Service.

Cole, David N.; Wright, Vita. 2003. Wilderness visitors and recreation impacts: Baseline data available for twentieth century conditions. RMRS-GTR-117. Ogden, UT: U.S. Department of Agriculture, Forest Service, Rocky Mountain Research Station.

DeBenedetti, Steven H.; Parsons, David J. 1983. Protecting mountain meadows: A grazing management plan. *Parks*. 8(3): 11–13.

DeLuca, T. H.; Patterson, W. A., IV; Freimund, W. A.; Cole, D. N. 1998. Influence of llamas, horses, and hikers on soil erosion from established recreation trails in western Montana, USA. *Environmental Management*. 22: 255–262.

Dickens, Sara Jo M.; Gerhardt, Fritz; Collinge, Sharon K. 2005. Recreational portage trails as corridors facilitating non-native plant invasions of the Boundary Waters Canoe Area Wilderness (U.S.A.). *Conservation Biology*. 19: 1553–1657.

Eagen, Sean; Newman, Peter; Fritzke, Susan; Johnson, Louise. 2000. Restoration of multiple-rut trails in the Tuolumne Meadows of Yosemite National Park. In: Cole, D. N.; McCool, S. F.; Borrie, W. T.; O'Loughlin, J., comps. *Wilderness Science in a Time of Change, Vol. 5: Wilderness ecosystems, threats, and management;* May 23–27, 1999; Missoula, MT. RMRS-P-15. Ogden, UT: U.S. Department of Agriculture, Forest Service, Rocky Mountain Research Station, pp. 188–192.

Farrell, Tracy A.; Hall, Troy E.; White, Dave D. 2001. Wilderness campers' perception and evaluation of campsite impacts. *Journal of Leisure Research*. 33: 229–250.

Fenn, Dennis B.; Gogue, G. Jay; Burge, Raymond E. 1976. Effects of campfires on soil properties. Ecological Service Bulletin 5. Washington, DC: U.S. Department of the Interior, National Park Service.

Ferguson, Michael A. D.; Keith, Lloyd B. 1982. Influence of Nordic skiing on distribution of moose and elk in Elk Island National Park, Alberta. *Canadian Field-Naturalist*. 96: 69–78.

Frissell, Sidney S. 1978. Judging recreation impacts on wilderness campsites. *Journal of Forestry*. 76: 481–483.

Godefroid, S.; Koedam, N. 2004. Interspecific variation in soil compaction sensitivity among forest floor species. *Biological Conservation*. 119: 207–217.

Gutzwiller, Kevin J.; Cole, David N. 2005. Assessment and management of wildland recreation disturbance. In: Braun, C. E., ed. *Techniques for wildlife investigations and management,* 6th ed. Bethesda, MD: The Wildlife Society, pp. 779–796.

Hall, Christine N.; Kuss, Fred R. 1989. Vegetation alteration along trails in Shenandoah National Park, Virginia. *Biological Conservation*. 48: 211–227.

Hall, Troy E.; Farrell, Tracy A. 2001. Fuelwood depletion at wilderness campsites: Extent and potential ecological significance. *Environmental Conservation*. 28: 241–247.

Hampton, Bruce; Cole, David. 2003. *Soft Paths,* 3rd ed. Mechanicsburg, PA: Stackpole Books. 225p.

Hartley, Ernest A. 1999. Visitor impact at Logan Pass, Glacier National Park: A thirty-year vegetation study. In: Harmon, D., ed. *On the Frontiers of Conservation: Proceedings of the 10th George Wright Society Biennial Conference on Research and Resource Management in Parks and Public Lands;* March 22–26, 1999; Asheville, NC. Hancock, MI: The George Wright Society, pp. 297–305.

Hesselbarth, Woody; Vachowski, Brian. 1996. Trail Construction and Maintenance Notebook. 9623-2833-MTDC. Missoula, MT: U.S. Department of Agriculture, Forest Service, Missoula Technology and Development Center.

Knapp, Roland A.; 2005. Effects of non-native fish and habitat characteristics on lentic herpetofauna in Yosemite National Park, USA. *Biological Conservation*. 121: 265–279.

Knight, Richard L.; Cole, David N. 1995a. Factors that influence wildlife responses to recreationists. In: Knight, R. L.; Gutzwiller, K. J., eds. *Wildlife and Recreationists: Coexistence Through Management and Research*. Washington, DC: Island Press, pp. 71–80.

———. 1995b. Wildlife responses to recreationists. In: Knight, Richard L.; Gutzwiller, Kevin J., eds. *Wildlife and Recreationists: Coexistence Through Management and Research*. Washington, DC: Island Press, pp. 51–70.

Knight, Richard L.; Temple, Stanley A. 1995. Origin of wildlife responses to recreationists. In: Knight, R. L.; Gutzwiller, K. J., eds. *Wildlife and Recreationists: Coexistence Through Management and Research*. Washington, DC: Island Press, pp. 81–91.

Kuss, Fred R. 1983. Hiking boot impacts on woodland trails. *Journal of Soil and Water Conservation*. 38: 119–121.

Leung, Yu-Fai; Marion, Jeffrey L. 1996. Trail degradation as influenced by environmental factors: A state-of-the-knowledge review. *Journal of Soil and Water Conservation*. 51: 130–136.

———. 1999a. Assessing trail conditions in protected areas: An application of a problem-assessment method in Great Smoky Mountains National Park, USA. *Environmental Conservation*. 26: 270–279.

———. 1999b. The influence of sampling interval on the accuracy of trail impact assessment. *Landscape and Urban Planning*. 43: 167–179.

Liddle, Michael J. 1997. *Recreation Ecology: The Ecological Impact of Outdoor Recreation and Ecotourism*. London: Chapman and Hall.

MacArthur, Robert A.; Geist, Valerius; Johnston, Ronald H. 1982. Cardiac and behavioral responses of mountain sheep to human disturbance. *Journal of Wildlife Management*. 46: 351–358.

Marion, Jeffrey L. 1991. Developing a natural resource inventory and monitoring program for visitor impacts on recreation sites: A procedural manual. NPS/NRVT/NRR-91/06. U.S. Department of the Interior, National Park Service.

———. 1995. Capabilities and management utility of recreation impact monitoring programs. *Environmental Management*. 19: 763–771.

Marion, Jeffrey L.; Cole, David N. 1996. Spatial and temporal variation in soil and vegetation impacts on campsites. *Ecological Applications*. 6: 520–530.

Marion, Jeffrey L.; Farrell, Tracy A. 2002. Management practices that concentrate visitor activities: Camping impact management at Isle Royale National Park, USA. *Journal of Environmental Management*. 66: 201–212.

Marion, Jeffrey L.; Leung, Yu-Fai. 1997. An assessment of campsite conditions in Great Smoky Mountains National Park Research/Resources Management Report. Atlanta, GA: U.S. Department of the Interior, National Park Service, Southeast Regional Office.

———. 2001. Trail resource impacts and an examination of alternative assessment techniques. *Journal of Park and Recreation Administration*. 19: 17–37.

Marion, Jeffrey L.; Leung, Yu-Fai; Nepal, Sanjay K. 2006. Monitoring trail conditions: New methodological considerations. *The George Wright Forum*. 23(2): 36–49.

Marion, Jeffrey L.; Merriam, L. C. 1985. Recreational impacts on well-established campsites in the Boundary Waters Canoe Area Wilderness Bulletin. AD-SB-2502. St. Paul: University of Minnesota, Agricultural Experiment Station. 16p.

Marion, Jeffrey L.; Roggenbuck, Joseph W.; Manning, Robert E. 1993. Problems and practices in backcountry recreation management: A survey of National Park Service managers. NPS/NRVT/NRR-92/12. Denver, CO: U.S. Department of the Interior, National Park Service.

Maser, C.; Tarrant, R. F.; Trappe, J. M.; Franklin, J. F. 1988. From the forest to the sea: A story of fallen trees. General Technical Report PNW-GTR-229. Portland, OR: U.S. Department of Agriculture, Forest Service, Pacific Northwest Research Station.

McClaran, Mitchel P.; Cole, David N. 1993. Pack stock in wilderness: Use, impacts, monitoring, and management. General Technical Report INT-301. Ogden, UT: U.S. Department of Agriculture, Forest Service, Intermountain Research Station.

McDowell, Theodore R. 1979. *Geographic variations in water quality and recreational use along the Upper Wallowa River and selected tributaries*. Dissertation. Corvallis: Oregon State University.

McEwen, Douglas; Cole, David N.; Simon, Mark. 1996. Campsite impact in wildernesses in the south central United States. Research Paper INT-RP-490. Ogden, UT: U.S. Department of Agriculture, Forest Service, Intermountain Research Station. 12p.

Monz, Christopher; Roggenbuck, Joseph; Cole, David; Brame, Richard; Yoder, Andrew. 2000. Wilderness party size regulations: Implications for management and a decision-making framework. In: Cole, D. N.; McCool, S. F.; Borrie, W. T.; O'Loughlin, J., comps. *Wilderness Science in a Time of Change*, Vol. 4: *Wilderness Visitors, Experiences, and Visitor Management*; May 23–27, 1999; Missoula, MT. RMRS-P-15. Ogden, UT: U.S. Department of Agriculture, Forest Service, Rocky Mountain Research Station, pp. 265–273.

Moritsch, Barbara J.; Muir, Patricia S. 1993. Subalpine revegetation in Yosemite National Park, California: Changes in vegetation after three years. *Natural Areas Journal*. 13: 155–163.

Olson-Rutz, K. M.; Marlow, C. B.; Hansen, K.; Gagnon, L. C.; Rossi, R. J. 1996a. Pack horse grazing behavior and immediate impact on a timberline meadow. *Journal of Range Management*. 49: 546–550.

———. 1996b. Recovery of a high elevation plant community after pack horse grazing. *Journal of Range Management*. 49: 541–545.

Parsons, David J.; Stohlgren, Thomas J.; Fodor, Paul A. 1981. Establishing backcountry use quotas: An example from Mineral King, California. *Environmental Management*. 5: 335–340.

Reid, Scott E.; Marion, Jeffrey L. 2004. Effectiveness of a confinement strategy for reducing campsite impacts in Shenandoah National Park. *Environmental Conservation*. 31: 274–282.

———. 2005. A comparison of campfire impacts and policies in seven protected areas. *Environmental Management*. 36: 48–58.

Rochefort, Regina M.; Gibbons, Stephen T. 1992. Mending the meadow: High-altitude meadow restoration in Mount Rainier National Park. *Restoration & Management Notes*. 10(2): 120–126.

Roggenbuck, Joseph W.; Williams, Daniel R.; Watson, Alan E. 1993. Defining acceptable conditions in wilderness. *Environmental Management*. 17: 187–197.

Silsbee, David G.; Larson, Gary L. 1982. Bacterial water quality: Springs and streams in the Great Smoky Mountains National Park. *Environmental Management*. 6: 353–359.

Silverman, G.; Erman, D. C. 1979. Alpine lakes in Kings Canyon National Park, California: Baseline conditions and possible effects of visitor use. *Journal of Environmental Management*. 8: 73–87.

Spildie, David R.; Cole, David N.; Walker, Sarah C. 2000. Effectiveness of a confinement strategy in reducing pack stock impacts at campsites in the Selway-Bitterroot Wilderness, Idaho. In: Cole, D. N.; McCool, S. F.; Borrie, W. T.; O'Loughlin, J., comps. *Wilderness Science in a Time of Change*, Vol. 5: Wilderness ecosystems, threats, and management; May 23–27, 1999; Missoula, MT. RMRS-P-15. Ogden,

UT: U.S. Department of Agriculture, Forest Service, Rocky Mountain Research Station, pp. 199–208.

Stankey, George H.; Cole, David N.; Lucas, Robert C.; Petersen, Margaret E.; Frissell, Sidney S. 1985. The limits of acceptable change (LAC) system for wilderness planning. General Technical Report INT-176. Ogden, UT: U.S. Department of Agriculture, Forest Service, Intermountain Research Station.

Stevens, D. R. 1979. Problems in revegetation of alpine tundra. In: Linn, R., ed. *Proceedings, First Conference on Scientific Research in the National Parks*; November 9–12, 1976; New Orleans, LA. Series No. 5. Washington, DC: U.S. Department of the Interior, National Park Service, pp. 241–245.

Stewart, D. P. C.; Cameron, K. C. 1992. Effect of trampling on the soils of the St. James Walkway, New Zealand. *Soil Use and Management*. 8: 30–36.

Stewart, William P. 1989. Fixed itinerary systems in backcountry management. *Journal of Environmental Management*. 29: 163–171.

Stuart, D. G.; Bissonnette, G. K.; Goodrich, T. D.; Walter, W. G. 1971. Effects of multiple use on water quality of high-mountain watersheds: Biological investigations of mountain streams. *Applied Microbiology*. 22: 1048–1054.

Taylor, J. 1997. Leave only footprints? How backcountry campsite use affects forest structure. *Yellowstone Science*. 5(1): 14–17.

Taylor, T. P.; Erman, D. C. 1979. The response of benthic plants to past levels of human use in high mountain lakes in Kings Canyon National Park, California, USA. *Journal of Environmental Management*. 9: 271–278.

Taylor, T. P.; Erman, Don C. 1980. The littoral bottom flora of high elevation lakes in Kings Canyon National Park. *California Fish and Game*. 66: 112–119.

Temple, Kenneth L.; Camper, Anne K.; Lucas, Robert C. 1982. Potential health hazard from human waste in wilderness. *Journal of Soil and Water Conservation*. 37: 357–359.

Therrell, Lisa; Cole, David; Claassen, Victor; Ryan, Chris; Davies, Mary Ann. 2006. *Wilderness and Backcountry Site Restoration Guide*. 0623 2815. Missoula, MT: U.S. Department of Agriculture, Forest Service, Missoula Technology and Development Center.

Tinsley, Bradford E.; Fish, Ernest B. 1985. Evaluation of trail erosion in Guadalupe Mountains National Park, Texas. *Landscape Planning*. 12: 29–47.

Tyser, R. W.; Worley, C. A. 1992. Alien flora in grasslands adjacent to road and trail corridors in Glacier National Park, Montana (U.S.A.). *Conservation Biology*. 6: 253–262.

Van Horn, Joseph C. 1979. Soil and vegetation restoration at the Sunrise Developed Area, Mt. Rainier National Park. In: Ittner, R.; Potter, D.; Agee, J.; Anschell, S., eds. *Recreational Impact on Wildlands: Conference Proceedings*; October 27–29, 1978; Seattle, WA. R-6-001-1979. Portland, OR: U.S. Department of Agriculture, Forest Service, Pacific Northwest Region, pp. 286–291.

Washburne, Randel F.; Cole, David N. 1983. Problems and practices in wilderness management: A survey of managers. Research Paper INT-304. Ogden, UT: U.S. Department of Agriculture, Forest Service, Intermountain Forest and Range Experiment Station.

Weaver, T.; Dale, D. 1978. Trampling effects of hikers, motorcycles and horses in meadows and forests. *Journal of Applied Ecology*. 15: 451–457.

Whinam, J.; Chilcott, N. 1999. Impacts of trampling on alpine environments in central Tasmania. *Journal of Environmental Management*. 57: 205–220.

Whittaker, Doug; Knight, Richard L. 1998. Understanding wildlife responses to humans. *Wildlife Society Bulletin*. 26: 312–317.

Whittaker, Paul L. 1978. Comparison of surface impact by hiking and horseback riding in the Great Smoky Mountains National Park. Management Report 24. Gatlinburg, TN: U.S. Department of the Interior, National Park Service, Southeast Region. 32p.

Zabinski, Catherine A.; DeLuca, Thomas H.; Cole, David N.; Moynahan, October S. 2002. Restoration of highly impacted subalpine campsites in the Eagle Cap Wilderness, Oregon. *Restoration Ecology*. 10: 275–281.

Zabinski, Catherine A.; Gannon, J. E. 1997. Effects of recreational impacts on soil microbial communities. *Environmental Management*. 21: 233–238.

Chapter 16
Wilderness Visitor Management: Stewardship for Quality Experiences

by Chad P. Dawson, John C. Hendee, and Rudy M. Schuster

Introduction	440
Visitor Management Considerations	441
Five Types of Visitor Impacts	442
Type 1. Illegal Actions	442
Type 2. Careless or Thoughtless Violations of Regulations	442
Type 3. Unskilled Actions Result from a Lack of Skills or Knowledge	442
Type 4. Uninformed Actions	442
Type 5. Unavoidable Impacts	443
Managers' and Visitors' Perceptions of Problems	443
Visitor Impacts on Social Conditions	445
Satisfactions and Benefits	445
Normative Behavior	446
Visitor Density and Perceived Crowding	446
Stress and Coping with Dissatisfying Conditions	446
Sense of Place and Relationships with Place	447
Visitor Conflict	447
Elements of Visitor Use Subject to Management	448
Amount and Distribution of Use	448
Redistributing Visitor Use	449
Method of Travel	450
Party Size	450
Length of Stay	450
Visitor Behavior	450
Visitor Effects on the Resource	450
Managing for Wilderness Experiences	451
Naturalness—Natural Conditions and Wilderness Experiences	451
Solitude—Social Conditions and Wilderness Experiences	452
Managerial Presence—Influences on Wilderness Experiences	453
Combined Influences on Wilderness Experiences	454
Wilderness Visitor Management Approaches	454
Direct Versus Indirect Visitor Management	454
Regulation—Costs and Benefits	454
Acceptance Does Not Equal Preference	456
Use of Regulations by Managers	457
Guidelines for Regulation	457
Location of Management	459
Indirect and Direct Management Techniques	460
Indirect Management Techniques	460

Design of the Physical Setting	460
Information and Education Programs	463
Information to Redistribute Use	465
Minimum-Impact Education Programs	466
Entry Requirements	468
Fees	469
Direct Management Techniques	470
Group-Size Limits	471
Length-of-Stay Limits	471
Campfire Restrictions	472
Prohibiting Certain Types of Use	472
Camping Setbacks	473
Rationing Use	473
Restricting Activities	475
Other Management Tools and Techniques	475
Wilderness Rangers	475
Litter Control	476
Monitoring Wilderness Experience Quality	476
Encounter Levels on Trails and at Campsites	476
User-to-User and Activity Conflicts	477
Preferences and Dislikes	478
Use Simulation Models	478
Summary	480
Study Questions	482
Acknowledgments	482
Case Discussion: Relationships with Wilderness	483
Case Discussion Questions	483
References	484

Introduction

Wilderness visitor management is guided by two objectives of the Wilderness Act (U.S. Public Law 88-577): (1) maintaining the natural setting and (2) providing "outstanding opportunities for solitude or a primitive and unconfined type of recreation." Wilderness recreation use can impact a relatively pristine natural setting, and too many visitor contacts diminish solitude, cause conflicts, and affect naturalness. *The objectives of wilderness management—maintaining naturalness and solitude—must be achieved without diminishing the wilderness character of an area.* Fortunately, most benefits reported in studies of wilderness use—benefits that can be categorized as relating to development of self, development of community or group social connections, and spiritual development—are related to wilderness naturalness and solitude. We call the primeval qualities of wilderness the "primal hypothesis of wilderness experience benefits" because they attribute development of self, development of community, and spiritual development to the experience of naturalness and solitude (White and Hendee 2000).

The Wilderness Act generally, and the spirit of the wilderness idea, restrict developing facilities for visitors, such as paving trails and providing campsite facilities, because that would reduce naturalness and interfere with primitive and unconfined experiences. Wilderness managers must rely primarily on means other than physical site development to control unwanted impacts and to manage visitor use. Furthermore, because the public holds many different values for wilderness (Cordell et al. 1998) and these values are sometimes in conflict with one another or wilderness (Manning and Valliere 1996), it is necessary for managers to specify the standards of naturalness and solitude for each wilderness area.

The Wilderness Act (with a few specific exceptions) prohibits structures or installations, commercial enterprises, permanent or temporary roads, motorboats, motorized equipment, aircraft, or mechanical transport (such as bicycles). Established nonconforming uses, such as mechanized access to operate and maintain existing dams or airplane access, have been authorized by legislation in certain areas. The Wilderness Act also provides for some specific, limited exceptions to these constraints, *for some wilderness areas*, including existing private rights; emergencies involving the health and safety of people within a wilderness; fire, insect, and disease control; water control structures; access to private inholdings, valid mining claims, and occupancies; commercial operations by outfitters and guides; continuing motorboat and aircraft use established before an area was designated wilderness; and livestock grazing. Many of these exceptions do not apply to wilderness in the national parks where, conceptually, park-resource protection is more comprehensive due to other NPS legislative mandates.

Some wilderness designation laws expand some of the exceptions to the Wilderness Act, compromises that were necessary to their passage. These wilderness designation laws and their exceptions are covered in Chapter 5. For example, the Alaska National Interest Lands Conservation Act of 1980 (ANILCA, U.S. Public Law 96-487) added 56 million acres to the wilderness system and specifically incorporates the original Wilderness Act's direction, "except as otherwise expressly provided for." However, because of special conditions in Alaska, ANILCA provides for a number of activities and facilities not permitted in wilderness elsewhere, including continued traditional access by airplanes (mostly floatplanes), motorboats, and snowmobiles; continued use of public and private cabins and possible construction of new public cabins; continued subsistence uses of fish, wildlife, and plants; and provision for fish production facilities such as fish ladders and hatcheries.

Critically important for wilderness visitor management, the Wilderness Act grants exceptions to most of the prohibitions (except commercial enterprises and permanent roads) for administrative activities, but limited to management action *"necessary to meet minimum requirements [emphasis added] for the administration of the area for the purpose of this Act."* This does not give managers a free hand. Management actions must pass stringent tests: the action must be the minimum necessary to meet the requirements for managing the area as wilderness under the Wilderness Act. This minimum requirement concept gives rise to management directives in the Forest Service that managers use the "minimum tools" necessary to achieve specific wilderness management objectives. In general, the act's purposes are to "assure that an increasing population, accompanied by expanding settlement and growing mechanization, does not occupy and modify all areas within the United States…to secure…an enduring resource of wilderness" and to "preserve its wilderness character" while devoting areas to recreational, scenic, scientific, educational, conservation, and historical use consistent with wilderness.

Management actions cannot be justified because they are convenient or economical or because they achieve nonwilderness goals, such as increasing stream flows, increasing fish and wildlife populations, or controlling predators or other natural processes such as fire, unless they threaten resources outside the wilderness. Determining whether a decision or proposed action is consistent with the act or qualifies for an exception is an important challenge for wilderness managers, and many have agonized over decisions as to whether specific actions are defensible under the act or qualify for the exception (Worf 1987). See Chapter 5 for a summary of management direction and clarification contained in the wilderness designation and related laws passed since the Wilderness Act.

Visitor Management Considerations

Wilderness visitor management is one of the most compelling issues facing wilderness managers today. *Two*

Fig. 16.1. The quality of a visitor's wilderness experience is influenced by many biological, social, and managerial conditions. Managers can, to some extent, control many of these conditions to maximize a visitor's opportunities for a high-quality experience. Photo courtesy U.S. Forest Service.

main considerations of wilderness visitor management are (1) to provide visitors opportunities for quality wilderness experiences and (2) to limit impacts on resources caused by visitor use. These two considerations are interrelated. Aspects of them are sometimes subtle and easily overlooked, but they are becoming increasingly important for managers who must implement the seemingly conflicting goals of providing quality experiences while limiting site impacts (Anderson et al. 1998; Manning and Lime 2000). For example, many resource impacts (such as heavily compacted soil in campsites) also affect visitor experiences, so reducing impacts might improve visitor experiences. However, management to control impacts (such as hardening camping sites) could adversely affect visitors' wilderness experiences by reducing naturalness. Often a management solution to one problem has side effects creating another problem, so evaluating alternatives and consequences is critical. A careful assessment of likely management impacts on both visitors and wilderness resources is always required.

Five Types of Visitor Impacts

Wilderness managers must consider five types of undesirable visitor actions and their associated effects and appropriate responses. Table 16.1 outlines the basis of classifying visitors' awareness and motivation, and how they relate to laws and regulations. The five types of visitor actions to be managed are (1) illegal actions, (2) careless actions, (3) unskilled actions, (4) uninformed actions, and (5) unavoidable impacts.

Type 1. Illegal Actions

Illegal actions violate agency regulations established to implement the Wilderness Act or subsequent wilderness designation legislation. Examples are the illegal use of chainsaws or mechanical and motorized vehicles in wilderness, such as mountain bikes, motorbikes, ATVs, or snowmobiles, that disrupt other visitors' experiences and damage wildlife, soil, and vegetation. The managerial response would be law enforcement but also communicate legal restrictions so users will know the rules. Managers understand that some illegal actions may result from ignorance and can be treated more like Type 3 (unskilled) or Type 4 (uninformed) behaviors. Mechanized violations, for example, can be addressed by informing visitors of appropriate nonwilderness areas for semiprimitive, motorized recreation.

Type 2. Careless or Thoughtless Violations of Regulations

Examples of such behavior are littering, parties larger than specified in area restrictions, camping in closed areas, and building wood fires where prohibited. Managers can try to motivate visitors to alter such behavior by persuasion, by making it easier to behave appropriately, and by discouraging inappropriate behavior. For example, managers might provide litterbags or encourage litter pickup through direct appeals by rangers, or physically block old campsites during vegetation recovery. Providing information through effective communication will usually prevent careless actions or thoughtless violations. We must assume that visitors want to do the right action—if they know the rules and the reasons for the rules.

Type 3. Unskilled Actions Result from a Lack of Skills or Knowledge

These actions include such things as digging a drainage ditch around a tent or making a bed or shelter of boughs cut from live trees. Many once-recommended camping practices—such as bough beds, pole shelters, and burying garbage—are now inappropriate. Such practices lead to accumulated impacts, and better equipment and practices are now available. Educating visitors about desirable practices can reduce impacts from unskilled actions, and, when necessary, enforcing rules against such actions—but always trying to educate visitors about why such rules are necessary. Wilderness education is making a difference, such as documented positive changes in wilderness visitor norms in Oregon's Eagle Cap Wilderness over a thirty-year period (Carlson 2000; Watson 2000; Watson, Hendee et al. 1996). The Leave No Trace (LNT) education effort by public recreation agencies and many private organizations is an attempt to combat unskilled actions (Swain 1996).

Type 4. Uninformed Actions

These actions reflect behavior that might have been different if certain information had been communicated to visitors. Uninformed behavior can intensify some types of use impacts, such as by large numbers of visitors who enter a wilderness at a few well-known access points when they might have used other access points or nonwilderness lands if they had known about use patterns and alternative places. Wilderness recreation use is often poorly distributed and

Table 16.1. Five Types of Problem Visitor Actions, Examples, and General Management Responses

Type of Visitor Action	Example	Management Response
1. Illegal actions	Mechanized trespass (e.g., motorbikes or snowmobiles)	Direct regulations supported by law enforcement
2. Careless actions	Littering, nuisance activity (e.g., shouting)	Persuasion, education about impacts, rule enforcement
3. Unskilled actions	Ditching around a tent or burning garbage	Primarily education about low-impact practices, some rule enforcement
4. Uninformed actions	Concentrated visitor use	Education-information programs
5. Unavoidable impacts	Physical impacts from careful use unavoidable impacts; relocation of use to more durable site	Reduction of use levels to limit

concentrated in a few places, which leads to impacts no matter how well visitors try to behave. Use wilderness education to provide visitors with prior information about alternatives available, or as a last resort, limit access at popular entry points.

Type 5. Unavoidable Impacts

These impacts reflect the reality that every party visiting a wilderness, by their very presence, causes some minimum, unavoidable impacts to other users (e.g., visitor crowding and conflicts) and to the resources (e.g., human and pack-stock impacts on trails, shoreline, campsites, and water conditions) (Washburne and Cole 1983). For example, a party arrives on-site, another party is there and notices them, and the solitude of both parties is reduced. Vegetation under a tent or sleeping bag is unavoidably compressed and damaged to some extent. Trails will be impacted by travel, especially in wet conditions. LNT and minimum-impact educational programs have helped reduce visitor impacts that previously were thought to be unavoidable by teaching about appropriate behaviors and equipment (Swain 1996). However, even with all the skill and knowledge possible, there remains a minimum, unavoidable level of impact. If, after all actions to reduce visitor impacts have been taken (e.g., visitor education, trail relocation to more durable sites), the accumulation of unavoidable impacts is excessive, managers can still limit use after other options have been tried. Although this may be considered by some to be a theoretical "last resort" for management to protect wilderness resources and visitor experiences, others believe that rolling back use levels may be politically difficult to impossible to achieve—hence, they suggest that use limits should be evaluated along with other options and not held back as the last alternative.

Managers' and Visitors' Perceptions of Problems

Visitors often see wilderness problems differently than managers, and managers must understand this when defining problems and seeking solutions. Visitors are looking for an experience, and they may not be aware of their impacts, whether careless, unskilled, or uninformed. Trying to change visitor behavior is difficult at best, but it becomes almost impossible if visitors do not perceive their actions and impacts as undesirable, no matter how inappropriate managers may believe it to be. Thus, *education must be a key component of*

Fig. 16.2. Most wilderness stewards regard vegetation damage at campsites, such as trampling and loss of ground vegetation, as seen here in Oregon's Eagle Cap Wilderness, as a serious problem. Many visitors, however, are bothered more by crowding and conflicts with other visitors. Photo courtesy U.S. Forest Service.

Fig. 16.3. Most wilderness visitors tolerate fire rings and campfire remains, but many managers find such visual and physical evidence of use unacceptable. Photo courtesy U.S. Forest Service.

visitor management, and to be good and effective wilderness educators, managers must try to see situations from a visitor's point of view.

Washburne and Cole (1983) reported that managers of different wilderness areas usually indicated similar visitor and environmental problems and usually saw these problems as localized rather than widespread. *Managers considered human impacts on campsite vegetation and trail conditions to be the most serious problem* (Manning, Ballinger et al. 1996; Marion et al. 1993; Washburne and Cole 1983). *Human impacts on lakeshore vegetation rated second and littering was the third most serious problem. Human impacts on trails, packstock impacts on vegetation at campsites, and visitor crowding came next, all with a similar level of rating as a problem. Most other problems, except disposal of human feces, were not generally considered serious.* Typically, a manager who reported these problems described them as affecting "a few places." Visitor impacts to campsite vegetation were the only problem perceived as affecting "many places" in the wildernesses studied. The managers surveyed perceived ecological impacts as a problem more often than visitor crowding or conflicts among users. However, Manning and Lime (2000, 41) compared four studies of National Park Service wilderness managers from 1979 to 1993 and concluded that *crowding and visitor conflicts are increasingly reported by managers and that carrying capacity "has become a pervasive but largely unresolved issue."*

When reading about visitor impact studies in the following paragraphs, it is important to look for the relationships that stand out across several studies rather than one particular area or individual study. For example, a survey of managers who reported use by organized wilderness experience programs (Gager et al. 1998) in the areas they administered revealed their concerns about such use: site impacts (42 percent), lack of wilderness stewardship skills and knowledge (42 percent), large group size (41 percent), overuse in areas already crowded (40 percent), conflicts with other users (36 percent), and establishing new trails and campsites (34 percent).

Most surveys of wilderness visitors indicate more concern with social conditions, such as crowding, conflict between visitor groups, and littering, than with resource conditions, such as campsite and trail impacts (Cole, Watson, and Roggenbuck 1995). Visitors react particularly negatively to littering.

Many visitors do not seem to be very aware or concerned about campsite impacts. Rock fire rings and ashes, unless extreme, seem to be acceptable to most visitors (Shelby and Harris 1985), whereas managers often react very negatively to campfire impacts.

Trail conditions concern visitors in more heavily impacted areas. A study from 1970 to 1982 in the Bob Marshall Wilderness Complex (BMWC) showed a sharp increase in complaints about trails (Lucas 1985), as trail conditions can make travel particularly difficult, especially large mudholes. Some visitors report the desirability of facilities like trails and bridges (Cole, Watson, and Roggenbuck 1995). Some potential rules and regulations are supported by users (e.g., limit party size) and some produce strong negative reactions (e.g., assign campsites) (Cole, Watson, and Roggenbuck 1995). *Generally managers are more concerned about protecting wilderness resources from physical and biological impacts, whereas visitors are more concerned about their experience, including access and freedom from intrusions by other users.* These perspectives are understandable as managers are responsible for protecting the resource and visitors want to experience it. A survey of agency-training needs revealed that managers are well aware that they need to be better trained and educated themselves to effectively and efficiently meet the present and future demands of visitor management in wilderness (Conrad 1997).

Educating visitors is a key management strategy; introducing such concepts as LNT helps visitors minimize their impacts. Other wilderness ethics programs are equally important to foster stewardship and protection of wilderness resources. Hansen (1990) cautioned wilderness managers that information and education programs must be consistent from the front desk at visitor centers and ranger stations to the information dispersed at trailheads and in the backcountry.

Because the supply of wilderness is ultimately limited, we urge visitor management approaches (e.g., information and education) that make it easier—not harder—for visitors to experience wilderness, including redirecting visitor recreation use to a more appropriate activity site or wilderness with less use, if necessary. Extensive evidence shows that wilderness settings lead to wilderness-related experiences that are good for people because they provide numerous human benefits.

Balancing wilderness visitor use against maintaining naturalness and solitude is the essence of the wilderness stewardship challenge before managers today and into the future, especially in view of the need to restore ecological conditions and processes in some wilderness areas. Thus, wilderness management should focus on providing opportunities for visitor experiences and encouraging appropriate wilderness use within the sustainable capacity of the resource—such ideals must be the hallmark of stewardship in the next century.

Visitor Impacts on Social Conditions

The condition of the natural environment in wilderness influences the social setting and the overall experience of visitors while on their trip in wilderness. The social setting is additionally influenced by each visitor's behavior, by visitor interactions within their group, and by between-group interactions during their various activities on a wilderness trip. Visitors arrive in wilderness settings having made choices and have expectations of what they will see and experience. Ultimately, visitors seek some level of satisfaction with their trip and will cope with various environmental and social conditions to arrive at overall trip satisfaction.

In addition to directly accounting for social settings, managers must be aware of how actions and decisions that modify management and resource settings affect the social setting. Managers need to develop a perspective on the natural and social settings important to visitor experiences when making decisions. This section briefly presents basic elements of the social experience and how visitors handle social interactions with each other during on-site experiences and provides a selected summary of recreation management research as it applies to wilderness.

Satisfactions and Benefits

Visitor satisfaction with recreational experiences has been measured using overall trip satisfaction as well as breaking the experience into component types of satisfactions (so-called multiple satisfaction approach). At the most basic level, trip satisfaction measures ask the visitor to indicate satisfaction with the overall experience, and, while this is helpful, it may hide some dissatisfactions that visitors have with certain aspects of the trip that would be helpful for managers to know. Overall trip satisfaction measures do not easily translate to management applications. Multiple satisfaction measures are more complex and target specific attributes associated with the setting and experience that are important to visitors and management. The multiple satisfaction approach can target aspects of the resource that management can control (e.g., trail condition, campsite condition, educational materials, etc.) and the social experiences, such as solitude, encounters, and crowding. Managers can use this approach to determine if the on-the-ground condition matches the objectives and standards defined in the management plan.

Driver and Brown (1983) used expectancy theory to develop the concept that people engage in particular activities and choose particular resource and social settings in order to fulfill personal needs and outcomes from recreational experiences. Needs were defined as a preexisting condition that developed prior to the recreational activity. The visitor chooses the activity, resource, and social setting with the expectation that the combination will satisfy their needs. The visitor may have a need to escape industrial areas, for family bonding, excitement, adventure, social interaction, learning, stress reduction, and so on. The desire to fulfill a need is what motivates people to engage in wilderness recreation.

Satisfactions and benefits are the outcomes of the recreational experience in wilderness. An individual's satisfaction leads to personal benefits such as improved health, reduced stress, and improved capacity to perform his or her work. The combined benefits to individuals form the overall social benefits to society, such as preserving cultural heritage, family cohesion, reduced need for medical services, and others. Driver et al. (1987) proposed that these societal benefits are the important outcomes from recreational use of such resource and social settings as found in the NWPS. The benefits-based approach to management is one of the recent recreation management frameworks that carefully considers the visitors and their attributes (e.g., previous experience, expectations, and behavioral norms) as receiving an experience based on the natural, social, and managerial settings (McCool, Clark et al. 2007).

Normative Behavior

In outdoor recreation management, social norms are defined as shared beliefs about how people ought to behave in certain settings (Heywood 1996) and about standards for what is acceptable and not in certain settings (Hall and Shelby 1996). Normative beliefs guide human behavior during the recreation experience. In wilderness settings normative beliefs concerning appropriate camping practices guide how backpackers establish campsites and interact within the camping group. Norms also govern interactions among groups.

Social norms evolve over time and vary among settings. Normative beliefs concerning standards describe the ability of wilderness visitors to accept on-site conditions, including physical impacts, management regimentation, and crowding. Impacts or crowding beyond the level accepted by visitors adversely affect their wilderness experience. Managers need to be aware that social norms may exist among the people using a wilderness. Social norms can be used to judge the acceptability of management decisions affecting the resource as well as to assess the impact of management decisions on the wilderness experience.

Visitors influence each other's behavior. Information and education programs have a multiplier effect because what is learned and shared directly or indirectly (for example, observations by novice visitors of the behavior of more experienced visitors) and can either reinforce or undermine management actions and programs. Managers can influence normative behavior by modeling, informing, and educating visitors about appropriate behaviors in various wilderness settings (for example, small wood fires where allowed, LNT practices for camping, and selecting campsites within approved regulations).

Visitor Density and Perceived Crowding

Density is an objective measure that counts the number of people in the system using methods such as number of backcountry permits issued, trail counters, or parking lot vehicle counts. Management can control density by limiting the number of permits issued, restricting parking, or charging an access fee.

Three assumptions concerning visitor density were stated by Freimund and Cole (2001, 5): "(1) relatively low use densities are a fundamental desirable attribute of wilderness, (2) use limits are needed at least in some portions of the wilderness system, and (3) science can contribute to better decisions about use limits." The question in terms of the visitor experience becomes, when is visitor density so high that the experience is undermined?

Crowding is defined as a negative and subjective evaluation of visitor density that is made by the visitor (Kuss et al. 1990). Solitude is an important component of the wilderness experience, and some limited social interactions are often expected in wilderness. Thus, solitude, crowding, and high or low density are all relative terms. Whether a wilderness visitor experiences solitude or crowding in a given situation is relative to the individual's perception of that setting.

Stress and Coping with Dissatisfying Conditions

Crowding, solitude, and density are relative terms and management application of them is complex. Contributing to the complexity of these terms is the fact that visitors often initiate a coping process in response to dissatisfying situations (Schuster, Hammit, and Moore 2006). The coping process is when an individual appraises a social or resource situation as problematic and then chooses coping mechanisms that are thought to be able to help solve the problem. These coping mechanisms may be problem-focused and directed at modifying the problem environment or emotion-focused and directed at reducing the internal and personal effect of the problem.

Common coping mechanisms employed in responding to social interactions include rationalization, product shift, displacement, avoidance, confrontation, and problem solving (Anderson and Brown 1984; Schuster, Hammitt, and Moore 2006). Rationalization occurs when the individual justifies the difference between on-site conditions and expectations to maintain cognitive consistency. Product shift is a change in the definition of the recreation opportunity after the on-site experience does not match expected conditions. Displacement is catagorized as intersite displacement (leaving the area altogether), intrasite displacement (participating in the same activity at a different location on the same area), and temporal displacement (participating in the same activity at the same location at a different time of day, week, or year). Avoidance is when a hiker uses subtle behaviors to avoid contact with other wilderness users (e.g., change hiking pace, leave a view spot early, or avoid conversation). Confrontation occurs when the individual actually interacts with the party responsible for causing the problems. Finally, problem solving occurs when the individual makes a plan of action that is designed to modify the environment or reduce and remove the source of the problem.

These issues are relevant to wilderness management because people cope with situations that are not perfect and it is difficult for management to assess experience quality from the visitors' perspective. For example, a wilderness visitor may experience a problem, cope with it, come to a personal resolution for the issue, and not accurately report to management that there was a problem requiring attention. Thus, increased visitor density does not necessarily lead to increased perception of crowding or to conflict, and visitor satisfaction measures as indicators of experience quality may be misleading if only overall trip satisfaction is measured. Most management plans require that social conditions be maintained within certain standards for encounter levels and perceptions of crowding. A report of high trip satisfaction or low perceptions of crowding are considered inaccurate by themselves, if the wilderness visitor had to employ coping mechanisms to achieve the perception of high trip satisfaction.

Sense of Place and Relationships with Place

Attachment to a place is the extent to which an individual values or identifies with a specific natural or cultural setting (Brooks et al. 2006; Farnum et al. 2005). Place dependence is relative to the importance of a resource or location in providing the social and natural setting necessary to participate in a given activity. The wilderness includes a collection of attributes that make it functional for specific purposes. Value is seen by the visitor as the degree to which a place provides satisfying hiking or kayaking locations, opportunities for solitude, or other amenities. Place identity refers to the dimensions of the individual's personality that are defined by the wilderness area, and is a subjective and cognitive combination of the ideas, beliefs, experiences, values, goals, and preferences that the individual holds in relation to that place.

The type and strength of attachment to the place influence the visitor's perception of acceptable resource, social, and management conditions. Managers should be aware that management plans and actions will be viewed differently depending on the place attachment of visitors. Place attachment also has been associated with the local public's reaction to local management actions and broad policy changes.

Visitor Conflict

Recreation conflict has been defined as goal interference attributable to another's behavior (Jacob and Schreyer 1980). Recreation conflict can be categorized into four types: (1) intra-activity conflicts, such as between hikers—maybe two different types of hikers; (2) inter-activity conflicts, such as between backpacking hikers and horse users; (3) conflict between recreation visitors and other permitted users of wilderness, such as between hikers and livestock grazing; and (4) conflict between recreational visitors and management activities and approaches, such as campsite or trail closures in a wilderness with limited alternatives (Schreyer 1990), or perhaps visitors annoyed by mechanical access for wilderness management purposes, such as servicing wildlife water sources in California desert wilderness where such access for wildlife management is allowed.

Conflict between visitors is often a greater problem in the visitors' opinions than sheer numbers of other visitors or crowding (Manning 1985; Stankey 1973). Conflict is usually between different types of visitors (e.g., horse users and backpackers, large and small groups), but sometimes it is caused by objectionable behavior by similar visitors. Large parties are a source of dissatisfaction to most visitors (Cole, Watson, and Roggenbuck 1995). Some conflict stems from different methods of travel, particularly where motorized use is permitted as an exception; for example, outboard motors in parts of the BWCAW (Cole, Watson, and Roggenbuck 1995). Horse/hiker conflicts tend to affect hikers more than horse users (Lucas 1985).

The concept of conflict provides an organizational framework for a manager to anticipate potential issues among wilderness visitors. Conflict is often defined as asymmetrical, meaning that one group experiences conflict while the other does not. For example, wilderness horse rider have expressed concern over llama pack-stock users while llama packers do not perceive problems with horse riders (Blahna et al. 1995). Graefe and Thapa (2004) note that the true nature of conflict is more complicated than the broad goal definition applies or the four categories allow. Asymmetry should not be assumed; a manager should fully investigate conflicts as they arise. Conflict is just as likely to arise within groups as between groups.

The goal interference definition implies that there is social interaction between groups sharing the wilderness. Defining conflict as a discrepancy in social values provides another approach to understanding and anticipating conflict. Vaske et al. (1995) found that even though hikers on Mount Evans in Colorado did not see hunters, they expressed conflict with the group based on differing social values. When asked a hypothetical question concerning potential hiker-hunter interactions in wilderness, hikers indicated that they would consider seeing hunters as problematic (Schuster, Hammitt, Moore, and Schneider 2006).

Activity style or technology also has been used as a categorization variable for conflict. For example, hikers have been found to perceive conflict with mountain bikers due to a perceived difference in social values (Carothers et al. 2001), but the conflict was asymmetric; mountain bikers and dual-sport participants were less likely to perceive conflict with hikers.

Conflict may occur within an individual. Problems may arise during the trip that have no bearing other than on the individual or blame cannot be placed on another. For example, wilderness hikers have been found to experience problems due to getting lost, poor planning, or undertaking an endeavor that is beyond their physical ability. While a manager may not have control over these situations, being aware of them is useful in the larger process of assessing visitor experiences and managing the overall social setting.

Elements of Visitor Use Subject to Management

Management cannot control or shape all the things that influence wilderness visitors' experience. Those aspects that can be controlled or influenced by managers are called management parameters (Shelby and Heberlein 1984). Two aspects of visitor behavior potentially influenced by management parameters are (1) effects on the resource and (2) effects on the experiences of others. How and to what extent can these different aspects of use and their impacts be modified? How can management of each aspect reduce ecological impacts and improve the quality of the visitor's experience? What general issues should managers consider in dealing with each aspect of use? We address these questions in considering seven use characteristics that are subject to management: amount and distribution of use, redistribution of visitor use, method of travel, party size, length of stay, visitor behavior, and visitor effects on the resource.

Amount and Distribution of Use

Managers can directly restrict the numbers of visitors, if that is the minimum requirement (or minimum-tool method) necessary for keeping impacts within the physical or social carrying capacity of the area. Restricting use would be a last resort after other approaches have failed and impacts cannot be reduced sufficiently in other ways. For example, use can be managed *indirectly*, through education, by closing roads, by lengthening trails, or by changing signs; or *directly*, by limiting the number of use permits issued or redirecting use with a ranger on-site. However, direct management of use should be a last resort after educational and indirect approaches have been exhausted. One must weigh the results of limiting the number of visitors entering an area against tighter regulation of use inside the wilderness, which may reduce the quality of visitor experiences. A balance is needed, and the optimum solution may vary from area to area. For example, in some areas, it may be less obtrusive to limit use (with other indirect management techniques) while studying what is an appropriate use level, and then to set higher use limits later, if the study results support such actions.

Wilderness use distributions are typically uneven both spatially and temporally, some of which result from uninformed actions by users who possess limited knowledge of alternative opportunities. However, sometimes concentrated use reflects easily accessible trails, differences in the attractiveness of locations, or proximity to population centers. In any case, a manager's first tool for altering the distribution of use is providing information and education to users. If that does not solve problems, limits might be imposed at entry points, or in certain travel zones, but always with information on alternatives made available to visitors.

Redistributing Visitor Use

Efforts to redistribute recreational use need to be linked to well-defined management objectives to avoid simply spreading problems more widely. Wilderness managers may have five objectives for seeking to shift some use:

1. *Redistribute some use from wilderness to nonwilderness locations*—Many wilderness visitors probably would be as well or better satisfied in areas having more facilities and managed more intensively for dispersed recreation than wilderness can be. These users are seeking relatively primitive, roadless recreation experiences but do not require or necessarily want the completely undeveloped conditions provided in designated wilderness. Managers, by providing information about roadless, nonwilderness locations, might redirect users to desired experiences, thereby relieving pressures on wilderness while at the same time better meeting some visitors' desires. Similarly, information about wilderness needs to stress its special character and de-emphasize recreation opportunities that are not wilderness-dependent. The recreation opportunity spectrum (ROS) provides a particularly useful framework for identifying opportunities to redistribute use to nonwilderness primitive and nonmotorized semiprimitive settings in a balanced, comprehensive way (see Chapter 8).

2. *Redistribute some use from heavily used to less used wilderness areas, or nonwilderness and roadless areas as appropriate*—Use can vary greatly from wilderness to wilderness, and sometimes a lightly used wilderness, capable of absorbing more use, can handle redistributed use from a heavily used area. Through information and education, managers can encourage those visitors who are particularly concerned with solitude and relatively undisturbed conditions to visit more lightly used areas, thereby diverting visitation from heavily used areas. If this approach does not shift use as much as desired, it might need to be supplemented by more direct controls. *Shifting visitors among wilderness areas needs to be carefully planned*, because it could result in merely exporting problems from area to another or reducing diversity within the wilderness system—the concept of a wilderness opportunity spectrum (WOS) suggests intensity-use variation from unused or lightly used areas to more heavily used areas, but still within desired wilderness conditions.

3. *Redistribute use within a wilderness to less used entry points, trails, and campsites*—Several access points, a few miles of trail, and a few campsites usually account for a large proportion of all use in any wilderness area. Providing information about less used locations might help guide visitors to them. If that doesn't work, more direct redistribution of use could be imposed through wilderness permits or on-site dispersal by wilderness rangers. Again, these direct measures should be imposed as last-resort actions, with care not to just redistribute the impacts.

4. *Redistribute some use to more appropriate locations or time periods better able to accommodate the use*—For example, horse users might be encouraged to go to places with more abundant forage and durable sites to reduce physical impacts, and to places where hikers do not usually go to reduce hiker–horse user conflicts. In another example, visitors might be diverted at specific times from certain areas where sensitive wildlife would be adversely affected by disturbance or where the presence of dangerous wildlife, such as grizzly bears with cubs, could present a serious threat to visitors. Most wilderness areas have short use seasons and many have sharp weekend and holiday peaks. Managers might shift use from peak times to low-use periods, first by providing information about peak and off-peak use levels, such as weekdays versus weekends and summer versus fall seasons. Finally, the minimum tool to accomplish the redistribution may be to restrict peak use, while at the same time providing advance information about anticipated peak-use situations and alternative areas.

5. *Concentrate heavy use in more popular locations that are durable and already impacted*—For example, in some wilderness areas recreation visitors might be encouraged to go to already popular places with durable sites to reduce physical impacts to other areas in the wilderness and allow some visitors who seek solitude to have that experience in the rest of the wilderness area. If visitors are appropriately informed about the different experiences possible in other areas, it may improve their satisfaction and reduce user conflicts. However, this strategy can have two serious

drawbacks: some visitors may become accustomed to the higher-impacted sites and crowded conditions in those areas and assume that these conditions are generally allowable in wilderness, and the impacts on these high-use recreation sites may become unacceptable in wilderness. This strategy may require a rest and rotation of sites to allow heavily impacted sites to recover, if research shows that such an approach is ecologically appropriate in a reasonable time frame. If visitor impacts become unacceptable, it may require managers to limit use, especially during peak-use times.

Method of Travel

In most wildernesses, the majority of visitors hike, and horse travel is a distant second. In several areas boats and canoes are used (some with motors in the BWCAW); kayaks and rafts are common on wilderness rivers, and floatplanes and bush planes are used to access certain areas in Alaska, Idaho, and Montana. The method of travel may provoke conflicts and increase both environmental and social impacts. Managers can use information and education to influence behavior of certain types of users to reduce impacts and conflicts (e.g., whitewater outfitters, horse users, and backcountry pilots). Some types of travel may be limited to certain portions of a wilderness and make it possible to separate different travel methods by giving each type of travel its own use zone or certain time of use. Commercial operators of cruise ships in coastal wilderness and flight seeing in Alaskan wilderness areas such as Glacier Bay and Denali require special considerations and management under ANILCA.

Party Size

Large parties are not common in most wildernesses, but when encountered they diminish other visitors' experiences (Cole, Watson, and Roggenbuck 1995) *and, almost inevitably, impact a larger area than smaller parties* (Cole 1986). Managers already limit party size in many places (Washburne and Cole 1983), but this technique needs to be supplemented with education and information. Frequently, large parties are sponsored by organizations that are easier to contact than independent users, and they may be willing to cooperate with wilderness managers.

The question of what limits to place on party size is controversial and becoming more so all the time. Should group size be limited to six, nine, twelve, or fifteen? There is no perfect and completely defensible answer. However, commercial outfitters complicate the question, as do wilderness experience programs that require some reasonable group size to remain cost-effective and economically viable as a business or program (Ewert et al. 1999). Good information on visitor use, user impacts, and carrying capacities, plus some common sense, are needed to make these decisions. The deciding factors on party size will vary from area to area and are derived by using wilderness planning processes (see Chapters 8 and 9).

Length of Stay

Length of stay can be regulated, but generally it does not contribute much to overuse because few parties stay in the wilderness very long (Cole, Watson, and Roggenbuck 1995). Nevertheless, when a party stays for a long time at a popular campsite, it unfairly monopolizes the site and increases impacts. For these reasons, some limit on length of stay at any one campsite seems desirable, and some areas have limits, usually seven to fourteen days. At locations where long stays are likely to occur, some reasonable limit on length of stay may permit more people to visit the area. Here again, good information and some common sense are needed to make decisions on limiting the length of stay.

Visitor Behavior

Visitor behavior includes many actions: staying on trails or shortcutting them, choosing a spot to camp, whether a campfire is built and how big, how human wastes are handled, how noisy a group is, consideration of other visitors, littering, and so on. *Visitor management to influence behavior of wilderness users is key, and it can reduce or eliminate the need to regulate and control visitor use.* All undesirable behavior is not inevitable—it can be influenced through information and education programs. People can adopt new practices and, in fact, research has shown that visitors' understanding of appropriate behavior has changed and improved over time (Lucas 1985; Watson, Hendee et al. 1996).

Visitor Effects on the Resource

Resource impacts can be dramatically increased or reduced, depending on visitor behavior. "Illegal" and "careless" behaviors (Types 1 and 2) may be more serious than "inappropriate behavior by well-meaning but unskilled visitors" (Type 3), but unskilled behavior probably is most common.

Low-impact camping involves using appropriate equipment and learning camping and traveling skills that can reduce resource impacts. For example, the LNT program teaches visitors to use campfires only where wood is abundant, keep them small, and leave little evidence of the fire.

Horse use requires special skill and knowledge because of its potential for damaging soils and vegetation. Most horses weigh more than a thousand pounds, and their weight is supported on small hooves with iron shoes. They are often tied up, which concentrates their impacts. A horse can eat twenty, thirty, or more pounds of grass and other forage each day in a wilderness. Excellent educational materials are available on how to minimize impacts while using horses. In addition, new, light equipment reduces the number of pack animals needed to transport camp equipment, thereby further reducing impacts and probably reducing visitor conflicts as well. In some areas, visitors are using llamas as pack animals, and they cause fewer impacts than horses or mules because they are lighter and have small, unshod hooves. Llama use is still relatively new and information on their impacts is scarce, but growing (Watson, Christiansen et al. 1998).

Managing for Wilderness Experiences

Wilderness experiences can be affected by many factors, some of which managers can control or influence. The Wilderness Act's statement of policy is that wilderness areas "shall be administered for the use and enjoyment of the American people in such manner as will leave them unimpaired for future use and enjoyment as wilderness." The act's definition of wilderness includes "has outstanding opportunities for solitude or a primitive and unconfined type of recreation" and a setting that "generally appears to have been affected primarily by the forces of nature." Thus, wilderness experiences must be characterized by naturalness and solitude. In the following, we discuss the influence of (1) natural conditions (e.g., naturalness), (2) social conditions (e.g., solitude), (3) managerial presence, and (4) combined influences on wilderness experiences.

Naturalness—Natural Conditions and Wilderness Experiences

Wilderness naturalness is diminished and/or enhanced by (1) recreational impacts, (2) nonconforming but allowed uses, and (3) natural ecological processes.

1. *Recreational impacts can diminish naturalness through evidence of recreation use*, such as devegetated campsites, eroded shortcuts across trail switchbacks, social trails between campsites, depleted firewood and littering, and impact on wildlife. Recreational pack-stock use can also diminish naturalness from trampling and overgrazing in points of confinement such as picket lines or tether trees. It is the "apparent naturalness" perceived by visitors that is important under the Wilderness Act and also to visitors' experience, as most are not professionally trained to perceive actual naturalness. Some of the more severe types of impacts (e.g., trail erosion) probably detract from visitor experiences, but more common impacts (e.g., loss of ground-cover vegetation at campsites or lower branches on trees near campsites broken off for firewood) likely have less of an effect, positive or negative, on visitors' experiences. Some evidence shows that visitors may prefer low or modest levels of some campsite impacts rather than totally unmodified conditions (Shelby and Harris 1985, 1986).

2. *Nonconforming but allowed uses (see Chapter 14) have great potential for diminishing the apparent naturalness of wilderness*. Cattle and sheep grazing presents an agricultural rather than wild appearance, and livestock may also consume and trample areas of ground vegetation and create muddy conditions at watering holes and points of concentration. Old mining roads and mine tailings and debris are an obvious intrusion on naturalness, just as are buildings from "inholding" landowners. Even vehicle tracks from administrative access by wilderness managers, where allowed, will diminish apparent naturalness, and managers must set a positive example by using such access only for emergencies.

3. *Natural ecological processes like wildfire, insect infestations, wind, avalanche danger, and fish and wildlife are apparent signs of naturalness that visitors can identify*. The types of wildlife observed, the

Fig. 16.4. For many visitors, seeing wildlife is an important part of their wilderness experience. Photo courtesy U.S. Forest Service.

frequency of observation, and the behavior of animals observed are all expressions of the degree of naturalness. Truly natural conditions may not facilitate the degree of wildlife or scenic visibility most visitors desire. For example, timber harvest areas may improve visibility and concentrate wildlife more than dense forests. Furthermore, wilderness wildlife includes many animal species besides deer, elk, moose, eagles, and other charismatic fauna that most people particularly enjoy seeing. In certain wildernesses, some natural species may be considered dangerous to visitors, especially grizzly bears but also black bears, bison, moose, and poisonous snakes and scorpions. Visitors vary in their appreciation of dangerous animals and in their willingness to share the wilderness with them. Natural wildlife populations often will not be of an optimum size for visitor preferences. For example, deer and elk will usually not be as numerous under wilderness conditions as they might be if the area were managed to maximize their numbers. Some wild animals, such as wolves, wolverines, and cougars, are secretive and difficult to observe. The behavior of some wildlife may also reflect diminished naturalness, such as jays and chipmunks (or even bears) hanging around campsites where they have been attracted to food scraps left by campers or encouraged to beg for food.

Wildfire is a natural process that can dramatically affect "apparent naturalness" and, in addition, insects, diseases, windstorms, floods, landslides, volcanic eruptions, and avalanches can alter vegetation and the landscape over large areas. The type of vegetation, its patterns, and wildlife populations are strongly influenced by fire in most wildernesses. Most visitors will not recognize unnatural successional changes, such as meadows invaded by trees or aspen being replaced by spruce, as the result of wildfire prevention and control.

Visitation generally declines for several years immediately following a fire (Borrie and McCool 2007; R. Brown et al. 2008). However, in many cases, visitors prefer conditions that occur several years after fire has played its natural role, when the landscape is more open and lower vegetation is conducive to larger populations of deer, elk, grouse, songbirds, and other wildlife. On the other hand, when a wilderness fire is actually burning, it can be a threat to visitors, temporarily close some trails, and generate smoke that burns one's eyes and throat and impairs visibility. However, many knowledgeable wilderness visitors support policies to allow fires to burn in fire-dependent ecological communities, and support for fire's natural role has grown over time as a result of education and experience (McCool and Stankey 1986; Taylor and Mutch 1986; Watson, Hendee et al. 1996).

Natural and wild conditions are not automatically recognized and appreciated by visitors during their experiences. However, "apparent naturalness," meaning "generally appears to have been affected primarily by the forces of nature," is required by the Wilderness Act. *One challenge for managers is to educate visitors more fully to understand natural processes so they can better appreciate experiencing naturalness in wilderness.* Besides, we believe the wildness of wilderness is closely linked to the benefits of experiencing wilderness. This is the "primal hypothesis" (White and Hendee 2000). Put another way, there is something about being in natural conditions that helps people come into balance as human beings, and this is the basis for increasing use of wilderness for personal growth, therapy, and healing.

Solitude—Social Conditions and Wilderness Experiences

Visitors' experiences are strongly affected by other visitors and their actions; generally, social conditions affect experiences more than natural conditions. Outstanding opportunities for solitude or a primitive and unconfined type of recreation are affected by the presence of other visitors, conflicts with other visitors, and visitor behavior.

Crowding as a management concern stems from the Wilderness Act's provisions for "outstanding opportunities for solitude," and also from common conceptions about wilderness as a place away from people. Research on wilderness visitors' reasons for visiting wilderness supports the importance of uncrowded conditions and solitude to many visitors. "Solitude" is something of a misnomer, as a common definition of solitude is "the state of being alone." Although some individuals go alone into wilderness, most go with a small group and think of solitude as their group being separate from other groups. Solitude in wilderness generally refers to a group of visitors meeting relatively few other groups of visitors. Meeting no other visitors at all is not as desirable to many visitors as meeting just a few other parties (Cole, Watson, and Roggenbuck 1995). Solitude is certainly not the only appeal of wilderness, and for many visitors it is not the most important, but it is a required condition and largely subject to control by managers who may establish acceptable use levels to protect solitude and naturalness.

Wilderness solitude is a relative term. What is considered crowding or solitude depends on crowding norms held by visitors and those they encounter and is influenced by the situation or location in which encounters occur (Hammitt and Patterson 1991; Manning 1985, 1986; Patterson and Hammitt 1990). For example, studies of canoeists in the BWCAW in Minnesota reported that they preferred much lower encounter levels than did motor canoeists or motorboaters (Cole, Watson, and Roggenbuck 1995). Visitors also have been found to attach greater importance to solitude at campsites compared to when hiking on the trail or during daytime travel by canoe (Stankey 1973; Watson 1995a). However, there is some concern that newer visitors to wilderness are experiencing more crowded conditions than historically were present, and thus, some speculate that newer visitors may be more tolerant of crowding, as that is how they first experienced wilderness.

Solitude is also impacted by encounters with nonrecreational uses, such as grazing and uses outside wilderness that create noise, air pollution, or visual distractions, including aircraft overflights. Encounters with different types of visitors (e.g., large groups), with different styles of use (e.g., horse users), or for different purposes (e.g., wilderness therapy or personal growth) may affect visitors' experiences more than sheer numbers of other parties. Such unwanted encounters often are one-sided impacts to one type of visitor (Marion et al. 1993). Where motorized use is allowed (e.g., outboard motor use in parts of the BWCAW), visitors not using motorized means of travel prefer not to meet those who are (Cole, Watson, and Roggenbuck 1995), but the motorized visitors do not object to the nonmotorized visitors (Adelman et al. 1982). Horse-hiker conflict is usually not as severe, but it also tends to be one-sided, with the hikers complaining more. Large parties are objectionable to persons in the typical smaller groups. Visitors with outfitters and those accessing the wilderness on their own may object to each other, and so may hunters and nonhunters. Reducing such visitor conflicts through wilderness management helps protect solitude.

Some kinds of visitor behavior directly impact user experiences, such as hikers failing to yield the right-of-way to parties with pack stock, inconsiderate visitors shooting guns, or visitors with uncontrolled dogs. Other behavior affects experiences indirectly through environmental impacts such as littering or leaving a messy campsite. All these intrusions impact the solitude of wilderness experiences and may be an unwelcome reminder of similar intrusions that must be dealt with in everyday life.

Managerial Presence—Influences on Wilderness Experiences

How managers operate and behave, especially their attitudes and presence, can greatly affect wilderness visitors' experience. Experiences can be damaged by management actions intended to solve other problems, poorly designed actions, or implementation that has not been well thought out. Direct types of management have more potential for adverse effects than indirect approaches (this is discussed in more detail in the next section). Excessive regulations, in particular, can detract from the sense of "primitive and unconfined recreation" called for in the Wilderness Act. Managerial presence, if too prominent or regulatory in attitude, can diminish the sense of experiencing wilderness and facing challenges on one's own. Managerial contact with visitors is usually preferable outside rather than inside the wilderness, although wilderness rangers usually report a pleasant reception to low-key or less formal encounters with visitors in the wilderness.

Combined Influences on Wilderness Experiences

All three influences on wilderness experiences—naturalness, solitude, and managerial—will vary from wilderness to wilderness and within each wilderness, based on variation in use levels as affected by differences in ease of access, attractions, and natural conditions.

Such variation is desirable, as long as all conditions meet basic wilderness definitions and objectives. Total quality in wilderness experiences will be greater by providing a diverse set of opportunities so that people, who vary in their desires and abilities, can find what they want (i.e., a spectrum of opportunities among wilderness areas). Different portions of each wilderness also vary in capacity, durability, and suitability for various types of use; some areas are reached by easy trails close to a trailhead; and some places are very remote with no trails. Thus, there is a wilderness opportunity spectrum (WOS) among most wilderness areas and even within an individual wilderness area (e.g., trailless zones or no-campfire zones) (see Chapters 8 and 9).

Wilderness Visitor Management Approaches

Several important issues to consider in wilderness visitor management include direct versus indirect management, the role and use of regulations, location of management, and multiple combinations of management methods.

Direct Versus Indirect Visitor Management

Direct management emphasizes regulating behavior by restricting individual choice as managers exert control over visitors (e.g., limiting use, permit requirements). Indirect management emphasizes influencing or modifying behavior by managing factors that influence visitors' decisions, such as information and education or making access difficult by limiting parking space or road maintenance. Individual visitors retain freedom to choose, as managers try to influence visitors while allowing more variation in use and behavior. Direct and indirect visitor management approaches are discussed in more detail later in this chapter and are illustrated in Table 16.2.

Usually, indirect management should be the first choice, with direct management used only when indirect means cannot achieve management objectives. Most wilderness users prefer indirect management, and the concept of wilderness as an undeveloped, open, and unconfined place accentuates the desirability of a management philosophy that is as indirect, unobtrusive, and subtle as possible. Trying indirect management first agrees with the concept of minimum requirements and the minimum-tool rule that calls for the minimum necessary regulation, action, or force to solve a problem or meet a management objective (see Chapter 7). Education is the key to most indirect management, and it is the main managerial response to careless, unskilled, and uninformed actions (Types 2, 3, and 4 in Table 16.1).

Direct controls are sometimes necessary to address specific problems, but they should be applied in ways that allow as much visitor freedom as possible. Respecting visitor freedom is a high priority in wilderness visitor management. Visitor regulation is a common and important tool for managing both ecological impacts and visitor experiences, but it should be applied with restraint and only when and where it is the minimum necessary approach to solve the problem.

There is some controversy about whether indirect management approaches are as effective as direct management (Shindler and Shelby 1993); however, there are too few studies to conclusively define when one is more effective than the other and under what conditions (Manning and Lime 2000). Some wilderness researchers suggest that indirect and direct management approaches should not be thought of as alternatives to each other, but rather that they form a continuum of management options (McCool and Christiansen 1996) and many of these techniques can be used effectively together (Cole, Watson, Hall et al. 1997).

Employing several approaches and techniques in a balanced, coordinated manner is needed in wilderness visitor management. Relying on only one approach is less effective than an integrated combination of approaches—neither regulations, education, design, nor wilderness rangers alone are the complete answer. *No single approach is so powerful, versatile, or acceptable in terms of costs to visitors or managers that it constitutes a panacea.* Judgment and creativity are essential in wilderness visitor and resource management.

Regulation—Costs and Benefits

Regulations can help solve some problems, but solutions are not free. For managers, the costs of regulation are in terms of employees' time needed to explain and enforce them, plus potential political costs from angry

Table 16.2. Direct and Indirect Techniques for Managing the Type and Amount of Wilderness Use
Modified from Manning and Lime 2000

Type of Management	Method	Specific Techniques
INDIRECT—Emphasis on influencing or modifying use and/or behavior. Individual retains freedom to choose. Control less complete. More variation in use possible.	1. Physical design and alterations	Improve, maintain, or neglect access roads. Improve, maintain, or neglect campsites. Make trails more or less difficult. Build trails or leave areas trailless. Improve fish or wildlife populations or take no action (stock or allow depletion or elimination).
	2. Information and education programs	Information to redistribute use. Advertise recreation opportunities in surrounding area, outside wilderness. Minimum-impact education programs (e.g., LNT). Advertise underused areas and patterns of use.
	3. Entry and eligibility requirements	Charge constant visitor fee. Charge differential fees by trail zone, season, entry point. Require proof of wilderness knowledge and/or skills (or group permits).
DIRECT—Emphasis on regulation of behavior. Individual choice restricted. High in degree of control.	1. Increased enforcement	Impose fines. Increase surveillance of area.
	2. Zoning	Separate incompatible uses (hiker-only zones, areas with horse use). Prohibit uses at times of high damage potential (no horse use in high meadows until soil moisture declines). Limit camping to setbacks from water or other features.
	3. Rationing use	Rotate use (open or close access points, trails, campsites). Require reservations. Assign campsites and/or travel routes to each camper group. Limit usage via access point. Group or party size limits. Limit camping to designated campsites only. Limit length of stay in area.
	4. Restrictions on activities	Prohibit certain types of use. Restrict building campfires. Restrict certain recreation activities.

letter writers and e-mailers, phone callers, or even campaigns led by people offended by government regulations. For visitors, the costs are in terms of hassles, oppressive reminders of civilization and government, lost opportunities, and diminished enjoyment from wilderness experiences that are not "primitive" or "unconfined," as called for in the Wilderness Act.

Costs to visitors stem in part from differences between the basic definitions of "recreation" and "regulation." Recreation is usually defined as a type of human experience that results from self-rewarding physical or mental engagements, based on personal free choice during leisure time. This definition stresses internal control and free choice for personal reasons that vary as much as do people. Regulation, in contrast, is defined as a means to govern and to produce order or uniformity. Thus, instead of the self-direction and diversity inherent in recreation, regulation invokes external direction and uniformity.

Tension between recreation and regulation intensifies in wilderness management because the Wilderness Act defines wilderness as an area that provides "outstanding opportunities for solitude or a primitive and *unconfined* type of recreation" (emphasis added). However, the act also requires managers to protect and manage wilderness "so as to preserve its natural conditions"—an almost impossible mandate if uncontrolled recreational use is allowed.

The appropriateness of wilderness recreation regulations depends largely on the benefits and costs of implementing a specific regulation, compared to nonregulatory alternatives for solving a particular problem. Furthermore, the effectiveness of wilderness recreation regulations depends on information and education programs combined with enforcement to support the regulations.

Research suggests that regulations can diminish the quality of recreational experiences. Visitors to the Rawah Wilderness in Colorado said autonomy strongly added to their satisfaction (Brown and Haas 1980). Some specific expressions of autonomy in the psychological scales used in the study were "doing things your own way," "feeling free from society's restrictions," "freedom of choice," and "traveling where you desire." These notions of autonomy, solitude, and privacy have also been found to be highly valued in other studies of wilderness visitors (Dawson and Hammitt 1996; Hammitt and Madden 1989; Roggenbuck 1980).

The criteria for a *primitive setting* within the ROS system include minimal evidence of management restrictions and controls (U.S. Department of Agriculture, Forest Service 1982). *Semiprimitive settings* have a few more controls and restrictions; more highly developed settings have even more.

Regulation can be beneficial, and, of course, recreation in many sports and games requires rules. Eliminating some freedoms in outdoor recreation can create other, perhaps more valuable, freedoms. Safety is a case in point. For example, removing power boaters' freedom to operate in swimming areas greatly increases swimmers' freedom to swim safely. Even the benefits of safety can have costs if safety regulations detract from high adventure, such as preventing whitewater kayaking in difficult rapids. However, where safety is not an issue, regulations can also protect experiences for some by limiting the behavior of others—for example, regulations requiring quiet in campgrounds after 10 PM.

A key, positive role for regulation is allocating recreational opportunities. Thus, managers can separate conflicting types of recreation such as snowmobiling and skiing. Regulations can be established from planning objectives for an area based on their relative position on the ROS or within the opportunity classes defined in the LAC process (see Chapters 8 and 9).

Research suggests that wilderness visitors often accept regulations fairly well, particularly if the necessity is explained, and once they become accustomed to them. Studies and literature about wilderness regulations go back to the 1970s and 1980s when the need for regulations began to emerge. For example, 82 percent of the applicants for the limited number of permits to visit California's San Gorgonio and San Jacinto Wildernesses supported restricted use of the area; even those turned away without a permit supported rationing through permits by three to one (Stankey 1979). Only 10 to 12 percent of visitors sampled in nine wilderness and backcountry areas believed that restricting use was undesirable when an area was being used beyond capacity (Lucas 1980). Similar acceptance of regulations has been reported in other wilderness areas (Cole, Watson, and Roggenbuck 1995) and for whitewater rivers (McCool and Utter 1982). Overall, recreational visitors and managers share a sense of wilderness stewardship that is necessary to work as allies for wilderness protection—even when user regulations are necessary.

Acceptance Does Not Equal Preference

Accepting restrictive regulations, as described earlier, should not be misinterpreted as indicating that visitors like such regulations. Surveys of visitor attitudes toward regulations show a strong preference for less authoritarian styles of wilderness management (Cole, Watson, and Roggenbuck 1995). Two-thirds of the campers in eight wildernesses studied in Idaho and Montana opposed the idea of a regulation prohibiting wood fires, and horse users rejected by three to one a possible requirement that they carry in all feed for their stock (Lucas 1980). In many cases, conscientious visitors accept restrictions because they see them as necessary to sustain desired wilderness conditions. On the other hand, some people do object to regulations. For example, Behan (1974) strenuously protested what he called "police state wilderness," in response to an article favoring mandatory permits and visitor regulation (Hendee and Lucas 1973). The debate over how much regulation should be used in wilderness management has been going on for several decades, with indirect—less regulatory and information-based—approaches now finding more acceptance by managers as well as visitors.

Visitor acceptance can be a false indicator that all is well if people avoid an area because of what they view as restrictive regulations. These displaced visitors are succeeded or replaced by those who accept the regulation. This can give managers a false sense of success. Research on this invasion and displacement of one kind of user with another due to regulation avoidance is difficult and therefore scarce, but one of the few studies done found that "too many use controls" was one of the top three reasons given by boaters who had stopped using the lower St. Croix and upper Mississippi Rivers (Denburg 1982).

Use of Regulations by Managers

Several surveys of wilderness management practices show that regulations are common. A 1980 study of all wildernesses under all agencies showed that managers of 54 percent of these areas regulated camping, half had mandatory use permits, 15 percent rationed use, and 69 percent of the areas with significant horse use had regulations affecting such use (Washburne and Cole 1983). When asked what they considered the most effective management technique, the majority of managers listed educational contacts; however, a large minority also listed regulations, such as designated campsites and lakeshore camp restrictions, as the most effective techniques.

National Park Service managers reported that they used backcountry overnight permits (68 percent) and group-size restrictions (62 percent) most frequently and then a wide combination of direct and indirect measures to manage or influence user behavior, such as a minimum-impact backcountry-use education program (Manning, Ballinger et al. 1996; Manning and Lime 2000; Marion et al. 1993). Similarly, most state wilderness areas restrict overnight use by requiring that only designated campsites be used, and then they use a broad array of direct and indirect measures to manage or influence user behavior (Peterson 1996).

Guidelines for Regulation

Some general guidelines seem reasonable for managers trying to fit regulation into its appropriate place in recreation management, whether related to managing impacts or visitor experiences, or both.

1. *Do not regulate if effective nonregulatory alternatives exist*—Establishing a regulation, by itself, achieves nothing, although it may provide a sense of satisfaction that something is being done. This may be an illusion, as *the potential effectiveness of a regulation can be tested with three questions.*

Fig. 16.5. Camping near lakes and streams, as in the Selway-Bitterroot Wilderness in Idaho, is often prohibited. Nevertheless, campers often claim they are unaware of the regulation or that they were unable to find a campsite farther from shores and banks. Photo courtesy Richard Behan.

First, is it possible to inform visitors of each regulation? Managers have a responsibility to inform visitors before citing them for a violation. The old principle "ignorance of the law is no excuse" seems out of place in wilderness recreation management. Limited knowledge of recreation regulations is common. For example, the main reason given by violators of a regulation prohibiting camping too close to water in Forest Service–administered wilderness in Colorado was they were unaware of the regulation (Swain 1986).

Second, if visitors are informed about a regulation and understand it, will their behavior change enough to solve the problem? The main underlying question here is whether visitors' motives, knowledge, attitudes, and behavior are understood well enough to estimate their response to a regulation. Visitor behavior can be changed by regulations, but drastic changes in behavior are usually difficult to achieve. For example, Swain (1986) found the main reason Colorado wilderness visitors violated a regulation requiring dogs to be kept on a leash was that they rejected the need for such a regulation, and many said they intended to continue to violate the regulation, even if it meant paying a fine.

Third, if the regulation changes visitor behavior as intended, will it really help achieve management objectives? For example, if visitors do not camp at three existing campsites within 200 feet of an alpine lake, three new campsites may be established beyond 200 feet. Furthermore, parties eating lunch or just resting at the old campsites may trample the vegetation enough to prevent

Fig. 16.6. Managers use trail design to indirectly influence a visitor's experience. Fording a stream (upper left) affects hikers differently than when crossing on a bridge (upper right). Removing trails and roads is an important management option as well. The Bureau of Land Management in California planned to remove hundreds of miles of old four-wheel-drive roads (lower left) following the 1994 passage of the California Desert Protection Act. In partnership with the Student Conservation Association (SCA), restoration began (lower right). Photos courtesy (upper left, upper right) U.S. Forest Service and (lower right, lower left) Dave and Katie Wash, BLM.

substantial recovery, resulting in six impacted sites where once there were only three.

2. *Try to develop effective nonregulatory (indirect) visitor management*—Some wilderness managers are too quick to assume that regulations are the only or the most effective way to achieve objectives, and thereby sell indirect, nonregulatory approaches short. For example, a wilderness ranger reported that about one-half of the campers in his area were camping within 200 feet of shorelines even after several years of intensive efforts to enforce a regulation prohibiting camping within 200 feet of lakes and streams. An information program that described the problem and explained the impacts of campsite location on the environment, on other visitors, and on sanitation might have changed specific campsite selection by visitors, at least as much as the regulation with its poor compliance. Wilderness visitors seem particularly well suited for management programs based on education and information, as one of the most distinguishing socioeconomic characteristics of wilderness visitors is a high level of education. Furthermore, wilderness visitors usually attach high personal importance to wilderness, which suggests many would respond to information about how to protect these areas.

3. *Explain why regulations are necessary*—An explanation of necessary regulations should improve compliance, especially in ways that relate to the most important aspects of the problem. Good explanations could reduce the costs to visitors by reducing perceptions of regulations as arbitrary and capricious and help make them seem more reasonable and necessary. Consistently explaining why regulations are necessary, from public information at the front desk to ranger contacts at the wilderness trailhead, helps support the rationale linking the regulation to a management problem. If a regulation is challenged, clear justification will be needed to enforce or defend it.

4. *Provide the minimum regulation needed to solve the problem*—Regulations span a continuum from severe to relatively mild. Avoid regulations that are more strict or sweeping than needed or that restrict visitor behavior that is not part of the problem. Target the specific behavior as precisely as possible. For example, if campfires are causing unacceptable impacts at some locations, prohibiting fires would deal with the problem more precisely than banning camping in that area.

5. *Regulate outside wilderness rather than regulate activity inside wilderness*—In wilderness and backcountry recreation, more freedom and spontaneity can be preserved if most regulations are applied outside the area; those admitted to an area would be substantially free to travel and

camp with little regulation. Limits on number of visitors admitted, party size, and method of travel are examples of "outside" regulations; assigned campsites or prescheduled travel itineraries are examples of "inside" regulations.

6. *Monitor the effects of the management actions on target problems*—Monitoring of some sort is essential because most management actions are taken with only a limited understanding of their likely consequences. Monitoring is the only way to determine what the management actions really accomplish. Managers must monitor effectively so they can use the information to minimize regulations if possible, to make necessary regulations more effective through progressive fine-tuning, and to develop other approaches to solving problems. Management activities yield very little new knowledge without monitoring because what is done is not documented and results are not objectively observed and tallied. This is especially true when the person judging was also involved in the decision and implementing the management action.

7. *When considering recreation regulations, remember that visitor use and enjoyment are one of the reasons why wilderness exists*—The Wilderness Act in section 2(a) established that wilderness was designated primarily to preserve natural ecosystems while also offering opportunities for wilderness experiences. However, support for wilderness depends also on the support of people who benefit from experiencing it. The visitors are interested in wilderness and should be viewed as being on the same team of wilderness supporters and advocates as the managers. *The goal of wilderness visitor management should be to allow and facilitate as much use and enjoyment of wilderness as is consistent with protecting wilderness naturalness and solitude—not to keep people out of wilderness.* The only precautionary note is that the first priority has to be preserving wilderness character, conditions, and processes.

Location of Management

Applying regulations outside of wilderness instead of inside is part of a more general wilderness visitor management issue. Contact with visitors, in person, through signs, and with printed materials, to present information about suggested practices, regulations, closures, or alternative places to visit is usually best delivered outside the wilderness. To do otherwise could detract from certain dimensions of the wilderness experience: the sense of isolation, solitude, freedom, and being in an undeveloped setting. Pretrip contact usually is effective in achieving the desired change in behavior because the information comes early enough in the decision process to be used. Most visitors react positively to meeting rangers inside the wilderness, but there, wilderness rangers are concentrating on tasks that cannot be done outside the wilderness: monitoring conditions, dealing with violations of regulations, maybe checking permits, educating visitors engaging in inappropriate behavior, and correcting specific on-site problems such as litter cleanup or maintaining trails. Of course, many wilderness management tasks are performed outside wilderness, such as educational contacts, planning, and data analysis.

Fig. 16.7. A national park backcountry directional sign without mileage, such as this (left), is appropriate and welcomed by most wilderness visitors. A cairn, a pile of rocks, marks a trail project (right) in a BLM-managed desert wilderness near Phoenix; cairns can mark routes across sandy, rocky terrain where the tread may not be obvious. Such direct, low-key management also keeps visitors from trampling sensitive vegetation. Photo courtesy Chad P. Dawson.

Indirect and Direct Management Techniques

A variety of potential management tools and techniques are available to deal with specific recreation use situations, including indirect techniques and direct methods (Table 16.2). This section explains indirect and then direct methods, evaluates their effectiveness, discusses visitor attitudes about each, and the use of each method by wilderness managers.

Indirect Management Techniques

Indirect management techniques address factors that influence visitors, rather than directly controlling their behavior. Visitors are still free to choose. The main methods and techniques of indirect management include (1) design of the physical setting, (2) information and education programs, (3) information to redistribute use, (4) minimum-impact education programs, (5) entry requirement, and (6) visitor fees.

Design of the Physical Setting

Design of access, trails, and facilities can indirectly influence visitor behavior. All such designed works need to be developed sensitively to avoid conflicting with the objective of wilderness as an essentially undeveloped area that appears to be substantially natural.

Modifying access to wilderness entry points can alter the amount and type of use by making it harder or easier to visit an area. Because it takes place outside wilderness, many legal constraints on development are lifted. Closing the last section of road and thus making the trail longer to the wilderness boundary would reduce the total amount of use entering the wilderness. Such access modification would tend to reduce day use more than overnight use and reduce hiking more than horse use in areas with substantial levels of horse use. Reducing use can lessen impacts and increase opportunities for solitude for those who visit. Less day use inside wilderness may be desirable to protect naturalness and solitude, and indirect methods may be the minimum necessary action to achieve that objective.

Improving or extending roads closer to the wilderness boundary would likely have the opposite effect and would increase total use and raise the proportion of day users. This might direct some use away from other high-use areas where overuse problems exist, especially if combined with actions to make access more difficult to these problem areas. However, access should be improved only with great care. Improved access might attract new and more users, thereby creating impacts and displacing some visitors with few offsetting gains elsewhere. Modifications in road access to wilderness entry points do occur, although often the road changes are for reasons other than modifying recreational use, such as road development and improvement for logging near wilderness trailheads.

The authors observed the striking effect of road access on an area's use near a large southwestern city; one area typically attracted 600 visitors per year over a bone-jarring four-wheel-drive road to the trailhead, whereas another area the same distance from the urban center drew 10,000 or more visitors per year to its easily accessed trailheads. Of course, other factors played a part in total use, but ease of access to trailheads was a major influence.

In addition, *design and providing facilities at access points can increase or decrease particular types of use and total use.* Horse facilities, such as unloading ramps, corrals, and hitch rails, encourage horse use. In water-oriented areas, such as the BWCAW, ramps for motorboats with trailers result in more boat use, and use by larger boats, than would a launch area requiring a carry-in portage from the parking area. The portage arrangement will screen out large boats and reduce boat use more than canoe use, with this effect more pronounced as the portage lengthens. The size of the parking area may have some effect on amount of use, although people determined to use that entry will usually park somewhere, like down the road, even if the parking lot is full. Conversely, parking should not be expanded where increased use is not desired, because it invites more use.

Visitor acceptance of access modifications has been rarely studied. Stankey (1973) reported about 40 percent of the visitors in three western wildernesses opposed closing parts of roads, but more visitors supported it seven years later (Stankey 1980). Some visitors returning to the same access point they had used before would object to having a longer hike to their destination. New visitors would usually be unaware of the modification if it was done skillfully.

Trail design and maintenance are potentially powerful indirect management tools. Trails are an acceptable, unobtrusive way to influence use patterns. Some areas deliberately left without trails will be lightly used. Trails can be built or rebuilt either to be easy or to be steep and difficult; trail length from point A to B can usually be varied considerably; and bridges can be built to facilitate use or trails can require fording streams, which will reduce or divert use and may stop all travel during high water. Some trails are too rough, rocky, or poorly cleared of blown-down trees for horses and pack animals, but hikers can scramble over and through them, thus separating these types of use.

Trail design can markedly influence the quality of the experience for the people who hike or ride horses. A trail can lead travelers to changing vistas or travel straight through dense forests. Trails can lead to varied vegetation, rock outcrops, glimpses of water, and perhaps increase chances of seeing wildlife. Trail design offers a unique opportunity for managers to subtly diversify and enhance visitors' experiences. The sequence of visual experiences and level of challenge are strongly influenced by trail design.

A trail can pass right by popular campsites, thus encouraging their use, or swing around through trees or behind a ridge to shift camping use elsewhere. This can also reduce feelings of crowding by limiting campers' observation of trail traffic, and vice versa. A winding, up-and-down trail cuts down on observation of other travelers just as long, straight, flat stretches increase sightings. Carefully thought-out trail loops or alternate routes can reduce visual and direct encounters among users. Trail design and maintenance also can facilitate or discourage off-trail travel and exploration, such as a rocky, steep trail section with a few strategic natural barriers left in place near the trailhead.

Despite the potential power of trail design and maintenance, its practical use is constrained. Most trails were built more than fifty years ago, before areas were designated as wilderness, primarily as an administrative transportation system, especially for fire control. In the West, early trails were intended mainly for horse travel, not hikers. Most trails were designed to get crews and supplies to fires quickly or to service fire lookouts. Neither environmental protection nor visitor experiences were important objectives. Budgets for trail work have been low for many years and maintenance of existing trails (often by volunteers working for expenses) consumes most of the available funds. Some trails are not maintained. Most new trail construction is replacement of poorly located short sections of old trail. Washburne and Cole (1983) report that 22 percent of all wildernesses reported some new trail construction and 37 percent had upgraded some existing trails. Only 8 percent reported closing any trails, and 14 percent cut back maintenance on some trails to reduce use. Trailless zones were designated in 35 percent of the wildernesses. Now, twenty-five years later, visitor use pressures are greater and there may be even less money available for trail construction and upgrading.

Visitor acceptance of trail design changes is virtually unstudied. The sort of redesign discussed here would probably be well accepted by visitors; probably most would be unaware of it (one measure of success). This unobtrusiveness is a key characteristic of indirect management. If trails are laid out skillfully, visitors' experiences should be improved substantially, and the wilderness environment better protected.

In most wildernesses studied, visitors agree that trails are appropriate and desirable (Stankey and Schreyer 1987). Support is stronger for simple, narrow, winding, low-standard trails than for high-standard trails and strong for leaving the trail system about as it is. Horse users prefer higher-standard trails than hikers (Lucas 1980). However, support for simple, low-standard trails and existing trail systems does not mean that badly eroded trails or numerous large mudholes are acceptable. A study of trends in recreational use of the Bob Marshall Wilderness Complex (BMWC) in Montana, made up of three adjacent wildernesses, showed a sixfold increase in complaints about trail conditions from 1970 to 1982 (Lucas 1985). Visitors desire maintenance, not expansion, of trail systems. Support for bridges, especially over large streams, is strong and widespread (Stankey and Schreyer 1987; Cole, Watson, and Roggenbuck 1995).

Signs within the wilderness can indirectly affect visitor experiences in several ways. Directional signs, usually at trail junctions, make travel easier and perhaps a bit safer. They also are an obvious work of people and reduce the visitor's sense of adventure and reliance on map-reading and orienteering skills. Most wildernesses have trail signs, but most managers strive for a balance, installing only limited, simple signs, particularly at confusing junctions. Wilderness signs might provide only one destination and not give distances or only give a directional arrow and a trail name. Interpretive signs, explaining natural features or historic events, are rarely used in wilderness, but such information is often shared via guidebooks and printed materials. Visitors generally support the concept of directional signs and some

Fig. 16.8. A variety of toilet designs address a basic human impact in wilderness. They range from a toilet made from wood in the BWCAW (left), to nonnatural material design (center) to a box toilet (right) commonly used at higher elevations in the Selway-Bitterroot Wilderness. Photos courtesy U.S. Forest Service and Chad P. Dawson.

complain if they are scarce (Lucas 1985), although most studies show a larger proportion are neutral about signs (Stankey and Schreyer 1987).

Trail systems and associated bridges and signs are the main recreational facilities in most wildernesses. Some wildernesses have other facilities at campsites, such as outhouses, corrals, and hitch rails, and a few areas have shelters, all to protect wilderness resources. *The use of facilities should be carefully limited in wilderness to avoid changing visitor experiences to something other than primitive. Facilities should not be provided for the comfort and convenience of visitors, but rather to protect wilderness resources and values, including wilderness experiences.* For example, bridges are often out of place in wilderness, and we suggest their limited use or removal unless they are a minimum tool to provide safety or have major historical significance.

Generally, visitors seem to support little or no development of campsites and little increase in numbers of campsites. Research suggests that user preferences are generally for more primitive-type campsites (Stankey and Schreyer 1987). Visitors tend to reject fireplaces but loose, rock fire rings are generally acceptable. Minimum-impact and LNT camping programs are influencing visitor behavior toward small campfires or no fires in environmentally sensitive locations and the use of campstoves to reduce resource and visual impacts.

Outhouses vary from enclosed, roofed buildings such as the pit toilets found in developed campgrounds to simple boxes that rise only a few feet above the ground. About 15 percent of all wildernesses had some enclosed pit toilets in 1980 (Washburne and Cole 1983), and 19 percent had some open box-type pit toilets. Wilderness toilets are found in many parts of the country, but the greatest concentration seems to be in national forests in the Pacific coast states and in national parks. In heavily used areas, toilets concentrate human waste and may reduce the chance of visitor contact with it, although decomposition is slower than with the individual (cat hole) burial system. In places with limited soil near campsites, as in the BWCAW, toilets must be carefully placed to reduce the chance of water pollution. In the few studies conducted, visitor opinion tends to be neutral or negative toward outhouses in wilderness (Lucas 1985; Cole, Watson, and Roggenbuck 1995). Attitudes about toilets seem to vary in relation to the perceived need in each area—thus, there is support for them in the BWCAW—and to local management practices where their use is established and they are accepted (Stankey and Schreyer 1987). *Toilets, and any obvious maintenance programs, are an obvious sign of humans that intrudes on wilderness conditions; they should be used only where they are clearly essential—the minimum tool to solve the sanitation problem of human waste disposal.*

Corrals for holding horses near camps are found in some western wildernesses, particularly at outfitter camps. They confine the impacts of horses to a selected area, which is usually devoid of vegetation, and minimize impacts elsewhere. Corrals are probably a good example of a "sacrifice area" as well as a work of humans, but may be the best solution to limiting impacts where parties with large numbers of stock camp, especially for long periods, such as during hunting seasons. Visitors' experiences are affected by the presence of a structure, associated soil and vegetation impacts, and, for horse users, a way to avoid the challenge of keeping stock nearby under primitive conditions. Horse users generally favor corrals; hikers oppose them, as demonstrated in several studies over a long period of time (Hendee, Catton et al. 1968; Lucas 1980, 1985). Hitch posts or rails are less obtrusive facilities, useful for holding stock for shorter periods of time than corrals, with similar effects on visitor experiences.

Somewhat equivalent facilities in the BWCAW would be docks and canoe rests, most of which are being removed. Both are a convenience and are obtrusive. Most visitors rejected docks in an early study in the BWCAW (Stankey 1973).

Shelters, usually open-fronted Adirondack types, are found in some areas, mostly in national forests in the Pacific coast states, the Northeast, and national parks. Enclosed public cabins are found in national forest wilderness in Alaska and are specifically authorized there by law. *Shelters concentrate impacts and, though they might have some advantages for that purpose, they also serve as an attractive nuisance and draw curious visitors.* Visitor attitudes about shelters are almost unstudied. Shelters reduce the primitive character of the visitor's experience and can reduce solitude because a number of parties often share a shelter overnight. Shelters function primarily as comfort and convenience facilities, with little or no wilderness protection role except in Alaska, where severe weather and bear danger confront visitors, and in Great Smoky Mountains National Park, where open shelters can be closed to keep out bears. *Shelters are out of place in wilderness, and we suggest their gradual elimination unless they are a minimum tool to provide safety or have major historical significance.*

Information and Education Programs

Informing and educating wilderness visitors and prospective visitors are a highly acceptable indirect management action—maybe the most important of all wilderness management techniques (Manning and Lime 2000). Information and education do not alter the wilderness resource directly, as do facilities, and they do not regulate or control visitors. Visitors retain freedom to choose, but their choices are made with more information. Good information and education programs enable visitors to have more rewarding experiences and help them build appreciation for appropriate wilderness behavior, experiences, and values; come closer to achieving the type of experience they are seeking; and develop a deeper appreciation for an area. Such programs also help managers achieve objectives by influencing where visitors go, what they do, and how they do it (Roggenbuck and Ham 1986; Roggenbuck 1992). Effective programs require consistent messages and information from initial contacts with personnel at the front desk, trailhead message boards, and contacts with rangers at trailheads or on patrol in the wilderness.

Wilderness visitors provide a particularly good audience for information and education programs because one distinguishing socioeconomic characteristic of wilderness visitors is high education levels. Most wilderness visitors also place a high personal value on wilderness, suggesting that most want to use it carefully. Information and education are now widely used by wilderness managers, and many are quite creative. Visitor acceptance of educational programs is rarely a concern, but effectiveness often is a question. More research is needed on the effectiveness of wilderness education on wilderness behavior. Some studies do document effects on the behavior of wilderness visitors, such as a study of the use of bulletin board messages at trailheads to increase visitor knowledge of low-impact wilderness travel practices (Cole, Hammond et al. 1997). Another study documented improvement in wilderness behavior norms over a twenty-eight-year period, perhaps reflecting that wilderness education efforts are working (Watson, Hendee et al. 1996).

Roggenbuck and Manfredo (1990) list five factors that influence information program success: (1) timing of the message, (2) message content, (3) recipient characteristics, (4) message-source characteristics, and (5) communication methods.

Fig. 16.9. Corrals, like hitch posts, somewhat mitigate the impact of horses at campsites in some western wilderness areas, such as this one at an outfitter camp in the Bob Marshall Wilderness. Photo courtesy U.S. Forest Service.

1. Time the message's delivery appropriately. Messages seeking visitor redistribution and reduced impacts are probably most effective during the trip planning phase. For example, if a gas campstove, trowel for disposal of human waste, high lines for tying horses, or feed for stock are to be recommended, a message at the trailhead or in the wilderness has little influence on new visitors who do not have the recommended equipment.
2. Message content is important. If managers use persuasion, the message should provide arguments that are strong, relevant, novel, and simple enough to comprehend. Ending arguments with questions rather than statements can increase the likelihood that recipients will think about the message. Repetition of messages is likely to increase comprehension and acceptance. Carefully organize and design the message. We must assume that fully explaining undesirable situations, and the rationale for particular management actions, may cause visitors to modify their use patterns and behavior. An appropriate middle ground may exist for the right level of detail—enough to adequately explain and motivate, but not so much that attention is lost (Cole, Hammond et al. 1997). Messages need to be matched to settings; a simple, standardized message is appropriate for some types of behavior (e.g., littering), but some recommended actions need more detail (e.g., firebuilding procedures or campsite selection).
3. Identify and understand the audience; make messages relevant specifically to them. Consider the types of behavior and potential for problems from each audience group, their motivations, background, and relevant experience. Visitor characteristics also influence success. Visitors are more receptive to messages if they (a) think of themselves as being a part of the problem, (b) have relatively low levels of prior knowledge and experience, and (c) are part of small groups. Group leaders are likely to be more receptive than group members, unless they are highly experienced.
4. Message-source characteristics are most important when persuasion is used or when the visitor is not very motivated to listen to or think about messages. In these kinds of situations—"learning situations where the recipient is in a hurry, in a distracting environment, is tired, is part of a large group, or is in a situation where the flow of complex information is forced and fast paced (as in some video programs)" (Roggenbuck and Manfredo 1990, 110), agencies should seek out attractive or well-respected individuals to deliver messages. Home, club meeting place, school, ranger station, trailhead, and the interior of the wilderness may all be appropriate, depending on the message and audience. Sporting goods and outdoor-recreation-equipment stores may be good contact points for wilderness information and education, and they may have their own education programs about wilderness activities and travel into which other messages can be integrated. The increasing use of electronic communication such as e-mail and the World Wide Web suggests that these approaches should be considered as potentially effective during the trip planning phase (i.e., information obtained prior to the trip allows for more informed visitor decision making and area and activity selection) (Queen et al. 1998). Information for trip planning, appropriate visitor behavior (such as using LNT techniques), and regulatory and permitting processes for access to and use of wilderness are especially effective on websites that are well maintained and updated regularly.
5. Select communication methods that fit the audience, message, and situation. A variety of methods will probably be most effective, including video, slides, posters, presentations, printed materials, and electronic communication. A survey of wilderness managers reported that six communication methods were used most often: brochures, personnel at agency offices, maps, signs, personnel in the backcountry, and message displays at trailheads (Doucette and Cole 1993). The more the visitors are involved with the educational program, the more they will retain and understand the information. For example, interactive teaching techniques are often considered more effective than traditional lectures. It is generally believed that personal communication is superior to printed material (Doucette and Cole 1993), but this may vary. For example, one study found that printed matter could be as effective as face-to-face communication for dispersing wilderness visitors (Roggenbuck and Berrier 1982). Written materials need to be delivered in situations where visitors have the time to process the information. A variety

of channels should be used to help ensure visitors get the message. Because the educational process is so complex, it is important to focus on a few messages rather than try a shotgun approach. For example, Cole, Hammond et al. (1997) studied visitor attention to low-impact messages posted on a trailside bulletin board and the extent to which visitors gained new knowledge from these messages. They found that visitors' knowledge was increased by exposure to messages. However, knowledge gain was as great when just two messages were posted as when eight messages were posted. They concluded that, in this situation, most visitors were only willing to allocate enough attention or able to process about two bits of information.

Manning and Lime (2000) developed eleven guidelines from a review of published literature for effective use of information and education programs in recreation management:

- Personal contact by wilderness field staff can be an effective approach to delivering information and education programs.
- Use of multiple techniques is more effective than single techniques.
- Information and education programs are more effective with younger, less experienced, and less knowledgeable visitors.
- Brochures, personal messages, and audiovisual programs are generally considered more effective than the same information on signs.
- Messages may be more effective for visitors during their pretrip planning or early in their trip.
- Messages from highly credible sources are more effective.
- Computer-based information dissemination can be an effective technique.
- Volunteers, outfitters, and commercial guides are effective and efficient communication partners.
- Information on the impacts, costs, and consequences of inappropriate visitor behavior can be an effective approach to reinforcing communication messages.
- Role modeling of appropriate behavior by wilderness field staff can be an effective approach to reinforcing communication messages.
- Messages should be targeted to specific visitor audiences when possible.

Wilderness managers should also try to educate visitors and the general public about wilderness values, purposes, and philosophy. Such information can deepen their appreciation for wilderness and help improve the quality of their wilderness experience.

Information to Redistribute Use

In the early 1980s, managers of most wildernesses said they were providing information to try to redistribute use (Washburne and Cole 1983), usually within a particular wilderness, and much less often to other areas outside the wilderness. Most efforts to redistribute use within a wilderness focused on shifting use among access points, but some tried to alter visitors' choices of areas to camp (Lucas 1981).

Today, wilderness information and education efforts are even more widespread, and more managers seek to shift use by making visitors more aware of alternative destinations and by providing descriptive information that makes some places more attractive and others less so. Such use of information to influence use distributions can have desirable effects even if there is little change in overall distributions, because the information can help visitors better match their desires to the characteristics of alternative places. For example, it may make it possible for those who prefer lower levels of encounters to visit lightly used areas offering such opportunities, and those who better tolerate higher levels of contact with other visitors go to areas where more encounters are likely. Information can help visitors develop realistic expectations, and this will improve their satisfaction, at least when satisfaction is defined as the congruence between expectations and reality of the experience.

Using information to shift use within wilderness has met with varied success in three places studied: Yellowstone National Park (Krumpe and Brown 1982), Shining Rock Wilderness in North Carolina (Roggenbuck and Berrier 1982), and the BWCAW (Lime and Lucas 1977). In each of these areas, about one-third of the visitors receiving information used it to pick entry points or campsites in line with managers' objectives.

Information campaigns must be geared to management objectives. Managers must decide if their objectives call for bringing about a general redistribution (say, from heavy- to light-use areas) or site-specific redistribution, such as from an impacted area to another location (probably a more appropriate objective), or help visitors match their desires and experiences better, such as suggesting new trails and other wilderness or nonwilderness locations (a very appropriate objective and probably the easiest to achieve). Objectives should also guide the design and conduct of the information materials, media, and campaign strategy.

The information must be available to visitors in the planning stage of their trip. After people have arrived at an access point, it usually is too late to influence that trip, although later trips might be affected. If information does not reach a large proportion of visitors early, its potential for improving experiences is severely reduced.

Information should cover a variety of attributes of the environmental, use, and managerial conditions. Different visitors have different objectives and will respond to various types of information in different ways. More complete and detailed information will aid visitors in choosing places to visit best suited for the experience they are seeking and may improve the credibility of information. However, managers must be sensitive to providing too much detailed information, thus taking away the sense of exploration and discovery that contributes to wilderness experiences for many people and supports the idea of wilderness. For example, an aggressive outreach education program was developed for the Superstition Wilderness in Arizona as a way to contact urban visitors before they entered the wilderness itself. The program includes telephone and personal contacts at the office, formal education presentations, and interactions at trailheads. The program is intended to inform and educate visitors before their trip starts and then to allow them to seek a more "unconfined" recreation experience with fewer ranger contacts within the wilderness.

Information and education are important tools for managing the distribution of wilderness use, but they are not a panacea for use distribution problems. In fact, unless used with care (e.g., providing consistent messages for all types of contacts), information could stimulate use and create problems that would not otherwise occur. Shifting use is a two-edged sword, and managers need to be aware and careful to monitor the degree to which they are solving overuse problems or merely redistributing such problems.

Minimum-Impact Education Programs

By 1980, only a few wildernesses did not have a minimum-impact education program (Washburne and Cole 1983). Since then, emphasis on such programs has increased and substantial staff and budget are devoted to them, such as "pack it in and pack it out," or Leave No Trace. One study found that three-fourths of all backcountry areas in national parks had a minimum-impact educational program (Marion et al. 1993).

Minimum-impact education programs reach the public through a variety of contacts: at federal agency information centers and ranger stations, at agency websites, at schools and colleges, through organizations such as the Boy Scouts, at sporting goods and outdoor-recreation equipment stores, at wilderness access points, and on wilderness trails and campsites through wilderness rangers. Some organizations have developed their own low-impact education programs, such as the Boy Scouts, Outward Bound, the National Outdoor Leadership School, and the Wilderness Education Association. The content of such programs has varied because some minimum-impact practices were based on judgment and assumptions and have only recently been based on field research (Cole 1989). The LNT program is gaining support throughout the NWPS and is designed to develop outdoor skills and ethics among wilderness users and managers (Swain 1996). The LNT educational program focuses on camping behavior and stresses resource impacts and social impacts on other visitors' experiences. Reducing resource impacts includes concerns about soil compaction, vegetation trampling, campfires, and litter. Reducing social impacts includes concerns about party size, trail etiquette, human noise, and group solitude experiences. With the LNT program, consistency in low-impact education has emerged, even among the different federal land-managing agencies. This consistency is most apparent in *the following seven principles that are the crux of the LNT program (Leave No Trace Center for Outdoor Ethics 2008):*

1. *Plan ahead and prepare for your wilderness trip*—Take time to learn about the area you plan to visit so you know what to expect. Choose the area you visit based on an evaluation of your group's ability and interest in minimizing impact. Travel in small groups and take appropriate equipment. Tents with lightweight poles eliminate the need to cut trees for poles or to build lean-tos. Foam pads replace bough beds and waterproof tent floors make drainage ditches unnecessary. Stoves either eliminate or reduce firewood gathering and campfire impact.

On backcountry trips particularly, repackage food so that potential trash is reduced. Finally, if traveling in bear country, be prepared and knowledgeable.

2. *Travel and camp on durable surfaces*—This is a complex principle derived to a great extent from research about the relationship between amount of use and amount of impact. Numerous studies show that a little use causes a lot of impact. Consequently, in popular high-use areas it may be best to concentrate use and impact. Stay on established trails and select a campsite that is already impacted. Select a site that is large enough to accommodate your group. Set up tents and the "kitchen" in places that have already been disturbed. Leave your site clean and attractive so the next group will want to camp there. In remote and relatively pristine places, disperse use and impact. Spread out while hiking and select a campsite with no evidence of previous use. Try to select travel routes and campsites that are durable. Disperse tents, activities, and traffic routes when camped in a pristine area and naturalize the site when you leave—so the next group that happens by will not recognize it as a campsite. Keep stays short in pristine areas. Finally, stay off lightly impacted trails and campsites. Lightly impacted places are in a state of flux. If they continue to be used, they are likely to deteriorate rapidly and substantially. However, if left alone, they usually can restore the natural conditions.

3. *Dispose of waste properly*—Pack out litter and waste food. Be particularly careful in bear country. Dispose of human waste appropriately, in toilet facilities if they are provided or in cat holes at least 200 feet away from water, trails, and campsites. Do all washing away from camp and never directly in streams, lakes, or springs.

4. *Leave what you find*—Never blaze trees, leave flagging, or build cairns along trails. Never make trenches around tent sites or build campsite "improvements." Avoid damaging live trees and plants and leave natural objects and cultural artifacts.

5. *Minimize campfire impacts*—Cook on stoves and minimize the use of campfires. If having a fire in a high-use area, use an existing fire ring. Keep fires small. Use only dead and downed wood that can be broken by hand. Burn the fire until only ash or small coals are left. Be sure the fire is out and scatter ashes, leaving a clean and attractive fire ring. In remote areas, select a durable fire site. Use a fire pan, build the fire on a mound of mineral soil, or build it in a shallow pit in mineral soil. Do not line the fire with rocks. Naturalize the fire site when you leave.

6. *Respect wildlife*—Travel quietly and give animals the space they need to feel secure. Avoid approaching animals too closely, feeding them, or touching them. Learn about locations and times of the year when disturbance of animals is particularly problematic. Leave pets at home or keep them under control.

7. *Be considerate of other visitors*—Travel and camp quietly. Minimize undesired contact between groups by taking breaks and camping away from the trail, from public attractions (such as lakeshores), and from other groups.

The LNT program recognizes that educational programs need to be tailored to individual places and user groups. Behavior that may be appropriate in one place may be disastrous in another place. This is most obvious in principle 2, "Travel and camp on durable surfaces." Recommended behaviors in popular high-use areas (where use concentration is appropriate) are precisely the opposite of recommended behaviors in remote areas (where use dispersal is appropriate). More subtle differences also exist. For this reason, the LNT program has developed different outdoor-skill booklets for different ecoregions in the United States, as well as for specialized recreational activities (such as river floating, horse use, rock climbing, snow camping, caving, and sea kayaking). Visitors must accumulate knowledge and the wisdom to consider the many variables that determine the most appropriate behavior for any given situation. Then they must use their own judgment.

Relatively little research has examined the effectiveness of educational programs in reducing recreation impact problems. The only inappropriate behavior that has been studied extensively is littering. Numerous studies have shown that littering can be reduced with persuasive communication techniques. Successful programs have been based on rewards, punishment, and environmental cues such as trash cans, written appeals about the need to keep places free of litter, and demonstrations in which role models pick up litter (Roggenbuck 1992). Written messages are often least effective. Punishment-oriented themes are often most effective. At Mount Rainier National Park, the mere presence of a uniformed ranger was the most effective of various techniques designed to keep visitors on established trails in meadows (Swearingen and Johnson 1995).

The following six principles of low-impact education (not just LNT) have been compiled from both experience and research:

1. *Educational programs should be guided by specific objectives*—It is important to identify specific problems and the users that are the primary cause of the problems. Messages should be targeted to these specific users and problems. Certain behaviors and user groups will be more amenable to change than others. Oset's (1990) two-pronged approach to education is to (a) deal with specific problems caused by specific users and (b) invest in the education of future users (such as through grade school programs).

2. *Messages should be clear, concise, and consistent*—Visitors may be overloaded with information and, consequently, miss the most important points if managers do not target their messages. Target a few specific behaviors. Short statements that demonstrate the desired behavior and why it is important are most likely to be effective. When more than one educational channel is used, or when different personnel are involved, the message should be consistent. Otherwise, visitors are likely to be confused. Unfortunately, the complexity of certain judgments essential for low-impact recreation (such as where to camp) often defies concise, consistent answers.

3. *Timing of educational messages is important*—To steer visitors away from heavily used areas, or to influence the equipment they take, messages must be sent to prospective visitors when they are planning their trip. Timing of messages should be varied for maximum effectiveness.

4. *A combination of techniques is likely to be most effective*—A combination of techniques allows messages to be repeated. It also makes it more likely that most visitors will be contacted. Different messages and media may be needed for different user groups. There is no reason, for example, to burden backpackers with all of the details of low-impact stock use. However, backpackers are more effectively contacted through messages on bulletin boards than stock users are (Cole, Hammond et al. 1997).

5. *Messages should be presented in a professional manner*—Text and dialogue should be accurate and easy to understand. High-quality equipment and materials should be used for graphics and productions. The options for effective communication increase if celebrities or respected individuals deliver the messages.

6. *Personnel must be personable, well trained, and committed*—Personnel who contact the public should be cheerful, polite, outgoing, and have a positive attitude. The ability to speak well, give accurate directions, and listen attentively to the visitors' experiences is helpful. Personnel need to be trained in communication skills and low-impact techniques. They should have personal knowledge of the area. Finally, they must believe in the value of minimizing recreation impact and the importance of education in achieving that goal.

Entry Requirements

To drive an auto, a person must demonstrate knowledge and skill to obtain a license. In most states, youngsters must pass a course in firearms safety to obtain a hunting license. Although human life is not so directly at stake as in driving or hunting, a test or schooling in wilderness skills might be required to certify knowledge and skills of wilderness visitors. The course might be a few hours long and be conducted in a classroom, using films or video and presenting general principles, ethics, and outdoor skills, or it could be longer and take place in the outdoors. Several educational programs are already designed to teach people how to use wilderness safely and sensitively; some are weeks long; such as the National Outdoor Leadership School (NOLS), Lander, Wyoming, and some are college-level wilderness skills courses. The Wilderness Education Association, Bloomington, Indiana, has set accreditation standards for some college wilderness education curricula. This is not a new idea, as the notion of "certified outdoorsmen" was proposed more than six decades ago (Wagar 1940).

Some professional guides, but not the general public, must pass tests, except for mountain climbers or whitewater boaters in some places. A license requirement might be imposed only for heavily used areas, or only for those areas where use is actually rationed. In such places, wilderness permits might be issued only to visitors with licenses or to groups with a licensed leader, and others could gain experience by visiting other less used areas. Obviously, the details could be complicated and administrative costs could be high. Costs for visitors would vary with the length and demands of a course, but could be substantial. Training would have to be readily available; otherwise, eligibility requirements could foster elitism—such as making some wilderness experiences only available to those with the time and money to take the training.

Although it may seem elitist to make wilderness entry dependent on certification of knowledge, there are some segments of wilderness visitors who are already going in that direction for trip leaders. For example, to be employed by most wilderness experience programs (WEPs) guides are usually required to have some level of first aid training, ranging from basic first aid to wilderness first-responder certification (Tangen-Foster and Dawson 1999). Different levels of training and certification are required to keep participants in the program safe and healthy; however, WEPs are required to obtain permits to use federal lands, and this requires proof of insurance. The insurance companies base the program's eligibility and rates on several factors, including staff training (usually requiring first aid) and experience.

Certainly, *if skills, knowledge, and sensitivity to wilderness values could be raised, people could better enjoy wilderness without destroying it, and the need for direct controls would be reduced.* Increased skills and a deeper understanding of wilderness could raise the quality of visitor experiences. Such courses could reduce visitor conflict and resource impacts and, thus, further improve experiences.

Fees

Fees are another technique for potentially modifying use and are classified here as an indirect technique because they do not directly control what visitors do. Even if a cost is imposed for the privilege of visiting an area, the choice of paying or not is up to the visitor. To date, it is not clear that any wilderness fees have been charged at a sufficient level to reduce or shift use, but the potential exists to do that with increased entry fees or parking fees, and would likely have the following effects.

Constant fees—that is, the same charge at all times and places—would tend to reduce use, with the degree of use reduction a function of the level of the entry fee. However, if the recreation experience is unique (e.g., running the Colorado River through the Grand Canyon), then even moderate fees may not reduce use. Overall, a flat entry fee would probably reduce short stays, but the longer stays with a per-day fee would probably not have this effect. Ideally, a fee would encourage people who place a low value on wilderness or who have other sites or activities to seek an alternative for which no fee or a lower fee is charged. Unfortunately, persons who value wilderness but whose incomes are low would be discouraged. This is true, of course, for anything bought and sold, but the validity of this income effect for public goods and services is questioned by many people. It raises the issue of pricing some people out of the market. This income effect could be offset by free or lower-priced entry permits for certain types of organizations or low-income visitors.

Variable fees could have a much more focused effect on the distribution and timing of use. Fees could vary among places, high at heavily used wildernesses and at overused entry points, and low or none at seldom-visited areas. Fees could be raised at peak periods and lowered or dropped entirely during the off-season. Perhaps ideally, persons who had qualified for a "wilderness license" could pay a reduced fee on the assumption they would have less impact on the wilderness.

Use of wilderness for recreation is managed to maintain both the quality of the resource and the experience for the visitors. The related planning, management, and enforcement activities to support recreational use of wilderness are costly in terms of staff and funding. Recreation managers have been increasingly challenged to provide quality opportunities with limited or declining budgets.

Administrative costs, especially for variable fees, could be moderate to high. Most wilderness visitors object to fees, and visitor studies usually show charging is one of the least popular methods of controlling use (Stankey and Schreyer 1987). Yet, *visitor surveys show that fees at any reasonable level would not be beyond the incomes of most visitors, and fee levels proposed to date are still a small fraction of the travel expense of a wilderness trip.* One study showed greater support for fees that would be used to protect and manage the area visited, rather than going into the general treasury (Martin 1986). Fees are increasingly used for river management where entry permits are required (Welsh 2000).

The concept of allocating fees for use in or near the areas where the fees are collected is one premise for a major "recreation fee demonstration program" (the so-called fee demo) in the Forest Service that was authorized by the U.S. Congress in 1996 to help shift the cost of providing federal recreation and wilderness opportunities and programs to those who use them. These types of programs have been met with some user acceptance and some strong user resistance; for example, environmental groups sponsored national fee-demo-resistance days in 1999 and 2000. The resistance is partially related to past wilderness user experiences and whether or not such fees have been traditionally collected in a given wilderness area for public access (Richer and Christensen 1999; Williams et al. 1999; Winter et al. 1999). For example, there was a parking fee charged at the two main access areas to the Superstition Wilderness in 1997 through 2000, and the funds supported continuing a resident volunteer program at those sites to inform and educate visitors. Some believe that this approach is the way of the future (Mackey 1998). Because wilderness management needs to grow and limited budgets continue, the concept of wilderness user fees will seem more appropriate, and fees will play a larger role in wilderness management in the future.

A fee approach to supporting wilderness management is controversial for numerous reasons, such as the social equity of charging user fees to the public when some segments of the public may not have the same financial resources as others. Others object to charging the public for using "their public land." Probably one of the more troubling aspects of this approach is the tendency for recreational users who pay a use fee to be viewed as customers, and with that perspective comes the associated expectation to satisfy customer needs and wants, which may run contrary to the resource-protection purposes of the Wilderness Act.

Direct Management Techniques

Direct management techniques regulate and restrict individual choices, such as rationing use with permits. Management may be able to exercise a high degree of control based on its regulatory authority, but generally this has high administrative costs, and occasionally political costs from offended users who formally object. To some extent, direct management is a less sensitive approach and requires less knowledge about visitors, their behavior, and likely responses to management actions than does indirect management. Some knowledge of use is needed, but if use is to be directly controlled to fit management objectives, less real understanding of it is required. Like all wilderness management efforts, direct regulation should only be used when it is the minimum tool to meet the objective or solve the problem (Lucas 1982).

This means that direct management may be somewhat easier than indirect management and perhaps demands less professional skill. Direct management does, however, have more potential for confrontations, conflict, controversy, and political backlash than indirect techniques, and it carries higher costs for visitors. Direct techniques should be applied only as a last resort, after indirect methods have been used as effectively as possible and have accomplished as much as they can to solve management problems, thereby leaving direct control as the minimum tool that will work.

Direct management can also include "strong suggestions" to discourage certain behaviors, as well as regulating and rationing use. Many of the specific techniques discussed next and shown in Table 16.2 can take the form of either regulations or recommended practice. The distinction is basically whether or not noncompliance is considered a punishable violation. This distinction is probably significant to visitors. A recommendation or suggestion, no matter how strong, imposes less on the quality of visitors' experiences than an enforceable regulation. Unfortunately, suggestions are more of a burden for conscientious visitors than for indifferent or careless visitors, but most wilderness visitors do care about their impacts. A strong suggestion costs less for managers to implement than a formal regulation requiring field enforcement and follow-up. Managers should strive to encourage visitors to adopt desired behavior as a social norm, and to the extent this occurs, social pressure can become a substitute for agency regulations.

All regulations can either be uniform within a wilderness or vary from location to location. A number of types of visitor behavior can be managed directly. Although there are numerous direct management actions that might be applied (Table 16.2), only *seven techniques of direct management found to be effective* are discussed here: (1) group or party size limits, (2) length-of-stay limits, (3) campfire restrictions, (4) prohibiting certain types of use, (5) camping setbacks, (6) rationing use, and (7) restrictions on activities.

Group-Size Limits

Setting a maximum group or party size has both social and environmental benefits. The social impacts are that *most visitors feel large parties are out of place in wilderness and quality of experience suffers* (Cole, Watson, and Roggenbuck 1995; Marion et al. 1993). Quality of experiences will be raised if contacts with large parties are reduced or eliminated. Environmental impacts are that *unusually large parties expand the area impacted at campsites and probably intensify impacts in the core of the campsite.* Slow recovery of soil and vegetation means that the effects of a large group persist for a long time, probably for years, continuing to degrade visitor experiences (Cole 1987).

Maximum group sizes are usually set based on typical group sizes and the estimated resource and social impacts of large groups. About one-half of all wildernesses areas limited their group sizes in the early 1980s (Washburne and Cole 1983). Maximum group size tends to vary with the character of an area. For example, large groups may be appropriate where large rafts are needed to run rough, whitewater rivers. In 1980, one-fourth of all wildernesses had a limit on numbers of horses or other stock, ranging from five to fifty head (Washburne and Cole 1983). Large groups and parties with large numbers of stock are not common, so most visitors are not inconvenienced. Research has found somewhat mixed attitudes about regulating group size, but a majority of visitors in most wildernesses studied supported a twelve-person limit (Lucas 1980, 1985) and strongly supported party size limits, particularly in more heavily used areas (Cole, Watson, and Roggenbuck 1995; Stankey and Schreyer 1987).

Monz et al. (2000) reported that use of group-size limits has been relatively stable within the federal agencies over fifteen years, and by 1999 the USFS (73 percent) and NPS (68 percent) wilderness areas were more likely to use group-size restrictions than the FWS (22 percent) and the BLM (17 percent) wilderness areas. The main reasons for using group-size restrictions were to reduce environmental impacts of recreational use and conflicts between groups. *The most common group size for people is ten (with a median of twelve people) in federal wilderness areas that use such a restriction.* The most common group size for horses and pack stock is twenty-five (with a median of fifteen people) in federal wilderness areas that use such a restriction.

There seems to be a consensus that group-size limits, perhaps six to twelve persons, are reasonable in most places. However, party size limits for commercial outfitters and wilderness experience programs promise to be extremely controversial because these organizations must take enough clients at a time to be economically viable. Perhaps party size limits for public use could be smaller than for such organizations running wilderness trips under special-use permits that may also require other restrictions. Consistent application of regulations is best when and where it is possible and appropriate.

Length-of-Stay Limits

Although it can be regulated, length of stay usually does not contribute much to overuse. Few parties stay very long, and those that do usually remain in more remote, less used, less impacted areas and have well-experienced visitors able to minimize impacts. An exception may be commercial outfitter and big-game hunter camps, which may feature rustic convenience facilities to ease hardship for paying guests or make bad weather tolerable. However, an extended wilderness adventure to the most remote possible location seems particularly consistent with the purposes of the Wilderness Act and with the philosophy of wilderness. Publicizing length-of-stay limits may diminish feelings of wilderness, even for visitors planning short trips.

On the other hand, when a party stays for a long time at one popular campsite, it is unfair to other visitors and hard on the site. Therefore, a reasonable limit on length of stay at any one campsite seems useful. Time spent at a particular location—not total time in the wilderness—seems to be the critical issue.

General length-of-stay limits were in effect in 28 percent of all wildernesses in 1980 (Washburne and Cole 1983), with almost one-half of all wildlife refuge wildernesses and a few wildernesses managed by other agencies permitting only day use. Stay limits are now common in national parks, with an average of seven to nine days (Marion et al. 1993). Enforcement is difficult without a permit system.

Limits on time allowed at any one wilderness campsite are not as common as general limits on visits to an entire area. This is true despite the fact that general stay limits are harder to enforce and, in our view, less useful than campsite limits. Seven to fourteen days are the most common campsite limits, especially in national forest wilderness. Hunting camps, usually set up in one place for a week or two in national forest wildernesses, may account for keeping some limits longer. National park limits usually range from one to three nights. Short limits, probably in the three- to five-day range, may be best for quality of experience and also may be desirable for minimizing ecological impacts.

Campfire Restrictions

Campfires can (and should) be prohibited or discouraged throughout a wilderness or at certain places, such as near timberline where trees are scarce and slow-growing and near water sources. Campfires can be restricted to officially designated sites, sometimes to designated fire rings or grates. Campfires have substantial environmental impacts (Cole and Dalle-Molle 1982) because fire destroys vegetation where it is built, and it sterilizes soil. Fuel-wood gathering can cause more widespread and often severe impacts. Campfires and horse use often produce the most conspicuous impacts in a wilderness.

Campfires are a traditional part of wilderness experiences and a treasured part of a camping trip for most people. Four decades ago, three-fourths of the visitors to three wildernesses in Washington and Oregon agreed that camping was not complete without experiencing an evening campfire (Hendee, Catton et al. 1968). More than twenty-five years ago, about 90 percent of the visitors to nine areas in the northern Rockies and California had campfires, often in addition to a campstove (Lucas 1980, 1985).

Educational efforts to encourage limited use of fires, small fires, minimum-impact fire-building and fuel-collection methods, and use of gas campstoves for most cooking must be a major part of any wilderness management program. Timberline or subalpine areas may require special education efforts or, in some cases, prohibit campfires. Extreme fire danger may require temporary or seasonal prohibitions on camp- or cooking fires.

Campfire policies vary among wilderness managing agencies. Twenty-five years ago, national park managers often prohibited campfires everywhere (43 percent of all areas), often required that fires be built only at designated sites (22 percent), and sometimes merely discouraged campfires (16 percent) (Washburne and Cole 1983). In contrast, at that time less than 1 percent of national forest wildernesses had overall fire prohibitions, just 6 percent limited fires to designated sites (almost all in California and the BWCAW), 4 percent prohibited fires in subalpine areas, and 22 percent merely discouraged fires. Today, restrictions on the use of fires by wilderness visitors are common and widespread.

The presence of rock campfire rings also affects visitor experiences. In many wildernesses, especially in national forests, managers encourage visitors to remove all fire rings and "naturalize" the site to remove all evidence of fire. The objective is to eliminate the visual impact of campfires, but the ecological effects may remain. Wilderness rangers may then remove the fire rings visitors do not obliterate, often over and over again. This approach does not always make sense because of the cost of such work and the potential spread of ecological impacts to more areas rather than a few. Further, surveys indicate that visitors are not unduly disturbed by one reasonably sized fire ring (Lucas 1985; Shelby and Harris 1985). A better approach is the common policy of leaving one or a few well-located, small fire rings per site, or removing rings only where fires or all camping is prohibited or discouraged (e.g., Superstition Wilderness). This seems in line with visitor desires, requires less work by wilderness rangers, and probably limits environmental impacts.

Prohibiting Certain Types of Use

Some types of recreational uses conflict with each other, and the impacts are often greater on one type of user than on others. Hikers tend to dislike encountering horse users and their impacts (Shew et al. 1986), and occasionally hikers bother horse users. Many people in small groups dislike meeting large parties. Canoeists usually resent motorboats in the few wildernesses where some motorized use is allowed. Prohibiting one type of use may improve the quality of experience for other visitors, but it obviously eliminates or diminishes experiences for those whose use is prohibited. Enforcement is relatively easy when prohibited uses are conspicuous, but confrontations may result.

Twenty-five years ago, prohibiting certain uses in some sections of a wilderness was already common, such as camping near or at some alpine lakes, with almost one-half of the NPS and about one-fifth of USFS wildernesses report having portions closed to horses (Washburne and Cole 1983). Now many areas have prohibitions on horses near lakes and streams or on some trails, and a few have seasonal prohibitions on horses. Much of the limited motorized use allowed under special exceptions is also zoned; for example, outboard-motor use in the BWCAW is allowed only on certain waterways.

Excluding some uses from certain sections of a wilderness can be a valuable management tool, but it should be done with restraint. Associated education efforts to reduce objectionable behavior and to increase mutual understanding and acceptance should be tried first, and if successful, may avoid the need for prohibitions or limit the extent of the area to which they need to be applied.

Camping Setbacks

In many wildernesses, camping within a certain distance from water, trails, and other campsites is discouraged or prohibited. A 100-foot or 300-foot no-camping zone is most common. These setbacks are encouraged or required to minimize resource impacts and social impacts such as crowding and visual intrusions that may be reduced by encouraging people to select secluded sites that are screened from view by topography and vegetation, rather than imposing a specific setback (Marion et al. 1993).

Some water setback restrictions refer only to lakes, some to lakes and streams, occasionally only to named streams. The various reasons given for setbacks include reduced environmental impacts, but shorelines are not necessarily more ecologically fragile, and water pollution depends on where wastes are disposed, not where camp is located. Social reasons are just as important, and benefits to quality of experience include a reduced sense of crowding, keeping camps out of the foreground of the prime visual attraction, and keeping the lakeshore or streamside open to all visitors without disturbing the privacy of someone's camp.

Restricting camping near water goes against a natural preference of users. Water is a powerful attraction, and most people want to be close to it, mainly for aesthetic reasons. Locations of existing campsites, developed by visitors' free choices over many years, clearly reflect a preference for certain locations near water, trails, and other users. Besides reducing quality by forcing people out of preferred sites, the supply of potential places to camp is greatly reduced, and in steep, mountainous country, there may be no legal places to camp near many destinations because reasonably level ground is often only available near water.

Setback enforcement is difficult. In most places with setback regulations, the majority of violation notices are issued for camping too close to water (Swain 1986). Visitors feel harassed, which is unpleasant for them and for wilderness rangers. Furthermore, Robertson (1986) found that wilderness campers were poor judges of distance from shorelines.

Many wildernesses also prohibit camping within a certain distance of trails. Trail and water no-camping zones often overlap, and this closes large areas to camping. If areas suitable for camping do exist, many new campsites will develop behind the setback, while the old ones closer to water typically persist, increasing total impacts considerably (Cole 1989).

Although setback regulations are useful, they may have negligible effects in reducing impacts and a negative effect on visitor experiences. Benefits are not clear and may not justify the full-scale enforcement campaign that is required. Educating visitors on where to camp and why to camp there is the first course of action. The advantages of not camping too close to water should be explained, with specific distances from water, trails, and other campers suggested as the minimum setback wherever possible. Reasons for picking a site screened from view should be explained. The relationship of impacts to campsite choices could be explained too so that visitors can make reasonably good choices under varying conditions.

Rationing Use

Rationing use can have major potential impacts on quality of experiences. Those permitted to enter enjoy heightened solitude and quality of experience, but the experience is eliminated for those whose wilderness entry is denied. For example, a study of the BWCAW in 1991 concluded that overnight visitors who reported having difficulty finding a campsite would benefit from reduced limits of visitor use (Watson 1995b). How use is rationed can further impact experiences. For example, several techniques can be used to limit or redistribute use. A daily limit can be set for the number of parties or individuals at all or certain entry points, with the itinerary up to each party admitted; they can make it up as they go. Alternately, limits could also be placed on the number of persons staying overnight in the entire area, in specific zones, or at individual campsites—in this case, numbers admitted would depend on the itineraries and planned lengths of stay of parties already in the area. If use were at capacity, the numbers leaving would determine numbers to be admitted. Numerous approaches exist for allocating use within an area and among user groups, but not all approaches are equally acceptable to each user group (Cable and Watson 1998).

Limited-entry permits are administered in a variety of ways. Some or all are usually available for advance reservations. Some proportions of the available permits often are held back for last-minute drop-ins on a first-come, first-served basis (sometimes called queuing, from the similarity to waiting in line to be served). A mixture of reservation and first-come, first-served permits seems to be fair. People who carefully plan far ahead, perhaps because of their need to schedule vacations and travel far from their home to the wilderness, need some advance assurance they will be able to visit the wilderness. Other people cannot or at least prefer not to plan so much, and operate in a more spontaneous, spur-of-the-moment fashion. If they live close to the wilderness—and most visits are from persons in the same state or region—they may want to react to last-minute, personal schedule or weather changes by going to the wilderness or not. A mixed system gives each type of person a chance at a permit.

Limited permits could also be issued through a lottery, as is done for some big-game hunting, such as bighorn sheep permits or some river-running permits like the Colorado River in the Grand Canyon. Handling such a lottery for entries on various dates is more cumbersome than assigning hunting permits for clearly defined hunting seasons, but it is done regularly for river-floating permits. However, few wilderness areas use a lottery approach, and visitors surveyed twenty-five years ago did not like the lottery concept. In fact, it was the least favored alternative for limiting use, except on rivers where visitors had become familiar with lotteries and accept them (Stankey and Baden 1977). At the turn of the twenty-first century, lottery systems with call-ins for cancellations and unallocated launch dates are used on fourteen rivers where use is rationed and applications are required by management (Welsh 2000).

All of these and other approaches to rationing use—reservations; lottery; first-come, first-served; fees; and skill/knowledge—have different combinations of advantages and disadvantages. Each system benefits certain types of visitors at the expense of others. Visitor acceptance varies from low to high, with some uncertainty. Difficulty of administration varies, but only queuing (first-come, first-served) seems to be easy.

Stankey and Baden (1977) proposed five guidelines for managing wilderness rationing that are now more than thirty years old but still make sense, and are supplemented with our comments (in parentheses):

1. Start with an accurate base of knowledge about use, users, and impacts (in the target area or zone).
2. Use direct rationing only after less restrictive measures have failed to solve the problems (that is, when rationing really is the minimum tool that will solve the problem).
3. Combine rationing techniques to minimize and equalize costs to managers and users (e.g., entry quotas, site quotas, reservation schedules, etc.).
4. Ration so the people to whom the experience is most valuable are more likely to get permits (the visitors whose desired experiences are most wilderness-dependent).
5. Monitor all rationing programs so their effectiveness can be objectively evaluated (in terms of physical impacts and social impacts such as crowding and encounters).

With any reservation system, there are usually no-shows, especially with free reservations. Generally, if a reservation has not been picked up and used by a certain time, such as 10 AM, it is released for use by drop-in visitors. Some people might have applied for a permit "just in case we decide we want to go," and others may have to unavoidably change plans. Successful applicants for very scarce, hard-to-get permits might become ineligible for another permit for a time to give other people a chance. If permits are issued at several locations, a communication system is needed to know when the limit has been reached. A computer reservation system, similar to that used for airline ticket sales, is one potential solution. Generally, public acceptance of controlled entry seems to have been good, especially when the reasons have been explained.

The psychological impacts of rationing systems and ethical burden on managers warrant comment. For example, whether a party is in the "wrong" campsite by choice or accident, confronting them is likely to be an uncomfortable experience for everyone. Apprehensions about whether they are in compliance could nag visitors. Will a ranger come along? Will the ranger check their permit? Will the ranger reprimand them, make them move, issue a violation notice, or what? If the camp quota is not full anyway, will the ranger take this into account? If the ranger ignores the violation, is this fair to the party that reluctantly passed a lovely, empty campsite late in the day and pushed on with aching feet until sunset to reach the campsite listed on their permit? Rationing systems can cause a loss of the sense of wilderness freedom and exploration, sometimes for a fairly small gain, and place the wilderness rangers in a particularly uncomfortable enforcement role. *Rationing is needed in many wildernesses and wilderness locations, but we urge their implementation as a last resort,* when they truly are the minimum tool to solve overuse problems and their application is the most light-handed approach possible.

One way to lessen the impacts of controlled itineraries is to inform and educate visitors, before their arrival at the wilderness entry points, about the need for and use of rationing. Wilderness travel information can indicate the alternatives and opportunities, such as the likelihood that a particular campsite or entry through a particular trailhead will be available. Such information can be sent by mail or accessed by the visitor via the Internet to help facilitate travel planning (Queen et al. 1998). This is an example where the rapid expansion of electronic communication can facilitate wilderness information and education for better visitor management.

Restricting Activities

Many specific activities can be prohibited or discouraged to reduce or eliminate adverse impacts on the environment and visitor experiences. Whatever restrictions managers impose, they must monitor results to justify whether to continue the restrictions, drop them, or modify them. For example, target shooting might be restricted to protect solitude. Hunting and fishing regulations can also affect wilderness use and activities, can contribute or detract from experiences, and alter the wildlife and fish presence and behavior in the ecosystem. Naturalness, of course, is the goal. Fish and game management, however, is a state responsibility except in national parks and wildlife refuges. State fish and game departments may object to special regulations for wilderness hunting or fishing because they feel they unfairly discriminate against these activities (see Chapter 11).

Other Management Tools and Techniques

Several other management tools and techniques also provide means of implementing either direct or indirect management. One of the most important of these is wilderness ranger contact with visitors.

Wilderness Rangers

Wilderness rangers carry out many of the management actions discussed above. Most patrol the wilderness during its high-use season, usually summer or fall, but some are stationed at busy trailheads to contact entering visitors and are sometimes called wilderness information specialists. A few wilderness specialists are year-round employees who compile and analyze wilderness data and conduct public education programs during the off-season. Wilderness rangers can serve many functions, such as gathering field data on resource conditions, use, and visitor actions, provided they receive some of the training that a manager would receive (Conrad 1997). They can influence visitor behavior by suggesting, advising, and informing; enforcing regulations; modeling appropriate behavior; performing emergency trail repairs and minor maintenance; rehabilitating campsites; planning and directing or doing some cleanup; and giving emergency assistance. As field managers, wilderness rangers can help solve site-specific problems by providing visitors with information and advice, thereby reducing both the need for area-wide regulations and more direct controls.

Almost all wildernesses and backcountry areas have wilderness rangers, although sometimes the large area patrolled by one person permits only superficial coverage. In some areas, great importance is placed on rangers as a key wilderness management tool. From all available evidence, wilderness rangers have proven to be very useful, and visitor surveys also indicate they are well accepted by the public. Because of their importance, rangers should be selected carefully for their interest and ability to work with people as well as their knowledge and concern for wilderness (Dawson 2007). Wilderness rangers are, in effect, people managers, and they are an agency's prime contact with the wilderness-using public. Most are volunteers or

Fig. 16.10. Controlling litter depends on cooperation between land managers and the public. While the "pack it in, pack it out" regulation generally works for most visitors to wilderness, occasionally it doesn't. Here, Sierra Club volunteers clean up litter at a campsite in the Selway-Bitterroot Wilderness. Photo courtesy Richard Walker.

seasonal employees, but this does not diminish the importance of their work and the need for their continued training.

Litter Control

Litter cleanup and efforts to further reduce littering, usually through the "pack-it-in, pack-it-out" regulation, are essential in wilderness. Research indicates that litter detracts seriously from wilderness experiences and is a common user complaint identified in visitor studies (Cole, Watson, and Roggenbuck 1995). Litter is an intrusion from civilization on the wilderness experience. Many believe that a clean site is more likely to be kept clean by users; at least it improves their experience and provides no excuse to leave it littered because it was found that way.

Litter does not all have to be removed by agency employees. In field tests in developed campgrounds, rangers contacted families with children and solicited the help of children to clean the campground in return for fire-prevention- and environmental-education-type rewards—badges, comic and coloring books, and so on. An adaptation of the incentive system was developed for wilderness and backcountry (Muth and Clark 1978). It is actually an "appeal system" because it was not feasible for wilderness rangers to carry a supply of incentive rewards, children are not as numerous, and many wilderness users objected to the notion of rewards—an appeal for visitors' help is all that is necessary. This is because the "pack-it, pack-it-out" norm has become accepted, and in some areas visitors are not permitted to take bottles or cans with them (reusable containers are permitted). This regulation is reported to have greatly reduced littering.

Monitoring Wilderness Experience Quality

Wilderness visitor management requires monitoring visitor experiences just as resource conditions must be monitored. Wilderness management affects visitor experiences, but unfortunately, monitoring methods for social conditions and experiences are not as well developed and tested as for resource conditions.

If the LAC system (Chapter 9) is being used, monitoring of all established indicators is essential. Some of these indicators will relate to visitor experiences, such as the intensity of use and presence of impacts. In any wilderness, whatever management system is being used, identifying problems requiring management attention comes best from systematic, objective monitoring. *The results of such monitoring indicate the effectiveness of the management actions and whether or not additional or different effort is needed.*

To be most useful, information gathered by managers in the field needs to be collected systematically and accurately and carefully recorded. Three important aspects of wilderness experience quality to monitor are (1) encounter levels on trails and at campsites, (2) user-to-user and activity conflicts, and (3) visitor preferences and dislikes. The three aspects are reviewed next, but numerous other aspects can be monitored depending on the variables and standards specified in the LAC opportunity classes (see Chapter 9).

Encounter Levels on Trails and at Campsites

The number of parties met per day by a group while traveling on trails or waterways is a measure of the degree of solitude or crowding and is one measure of wilderness experience quality (Cole, Watson, and Roggenbuck 1995; Hammitt and Rutlin 1995). Encounter levels have been used as an LAC indicator. In one approach, wilderness rangers or volunteers tally how many individuals and groups they meet each day. To reasonably represent visitors' encounter experiences, the agency people must travel at about the same speed and for about the same distance as typical visitors. Tallies might be adjusted slightly to compensate for unusually fast, long travel, or for coverage of only part of a typical day's travel. It is essential to keep formal notes, preferably on a standard form, and to record the trail segments covered. Later these data can be compiled, compared, and evaluated for patterns and trends.

A second approach is for fieldworkers to ask visitors about their encounter experiences. "How many other individuals or groups did you meet yesterday?" (or "today" if late in the day). The visitors' route and answers would be noted. More encounter information can be obtained from visitors than in the first approach, and thus results would be more valid. Answers should be recorded systematically in a notebook with date and location of travel noted, along with weather, which might influence the data analysis.

Similarly, most visitors place particular value on a campsite isolated from the sight and sound of other campers. How successful they are in achieving isolation significantly influences their quality of experience. Therefore, managers should ask about and monitor campsite encounters, and LAC systems will usually include it as an indicator.

At least two approaches can be used to monitor campsite solitude: direct observation and asking visitors. Because of their comparability, both could be used and the results compared and combined to improve accuracy. In direct observation, a wilderness ranger records how many other occupied camps were observable from each occupied camp, and how much noise was audible. This must be done in the evening or early morning, when most people are in camp, which limits the number of observations possible.

If asking visitors about their experiences, when an individual or group is contacted in camp or on the trail, it would be common to ask: "Did you camp here last night?" or "Where did you camp last night?" Then, as appropriate: "Was anyone else camped nearby?" "How many groups?" "Was their noise or presence an intrusion for you?"

All the data need to be carefully recorded and the campsite identified as precisely as possible. Campsite environmental conditions should also be monitored, and a system of numbering campsites should be established to facilitate data recording and tabulation. Locating camps mentioned by groups on the trail may involve going over a map with them, and sometimes it will be imprecise, but a general area and good information can usually be recorded.

Fig. 16.11. Managing recreational use of wilderness requires combining a range of activities, such as designing and maintaining trails, providing wilderness rangers, conducting information and education programs, and writing regulations. Here a trail crew repairs a trail. Photo courtesy John C. Hendee.

User-to-User and Activity Conflicts

The behavior of all visitors affects every user, to some extent, either positively (e.g., good social exchanges, trail information) or negatively (e.g., noise, dogs, litter, camping too close together). The degree to which visitors are impacted by others' behavior depends on their expectations and the degree to which the behavior is considered intrusive or not "normal" (Jacob and Schreyer 1980). Other users consider some activities intrusive because they diminish the solitude or naturalness experience they expect and seek in the wilderness. For example, campfire smoke may annoy other users, the sound of gunshots by hunters or shooters may not be appreciated by nonhunters or nonshooters, and pack-stock users may expect hikers to give them the right-of-way when they meet on the trail, which may not be acceptable to the hikers. These kinds of behaviors can become conflicts depending on the sensitivity of those who believe their experience has been intruded on or the expectations of visitors who believe such activities are unnecessary or inappropriate in wilderness (Watson 1994, 1995a).

Fig. 16.12. Charging fees for permits or for parking at heavily used wilderness areas may be a source of funds to support management and possibly discourage some use. While some visitors don't mind, others object to paying for their use of public lands and find such direct management intrudes on their sense of freedom. Here signs approaching the Superstition Wilderness (USFS) near Phoenix remind users they must purchase a permit (left), the fee per vehicle per day (middle), and a vending machine (cash or credit cards accepted) at the trailhead parking to collect the fee (right). Photos courtesy Marilyn Hendee and John C. Hendee.

Preferences and Dislikes

Managers need some feedback from visitors on the things they prefer or dislike, whether as part of a formal monitoring program or not. Reactions to trail conditions, campsite conditions, litter, specific management policies or regulations, and information provided by the agency are all possible topics managers might want to monitor. There is no way to do this except by talking with or surveying visitors. The same questions or topics should be raised with all visitors. Some of these topics are commonly discussed with visitors anyway, especially trail conditions. In these cases, it is just a matter of making careful notes after the conversation. To be most useful, the information should be collected in a standardized way. Recording spontaneous complaints and compliments is fine, but mixing such responses with answers to specific questions such as "What do you think about our closure of camping at Summit Lake?" will produce a misleading, noncomparable mixture of information.

Visitors' preferences and dislikes (satisfactions and dissatisfactions) are a measure of the quality of the wilderness experience. Some factors affecting visitors' experiences include the number and type of user encounters (especially compared to prevailing visitor norms and expectations), perceptions of crowding, actual use levels, and trail and campground conditions (Manning 1999; Patterson and Hammitt 1990; Watson, Asp et al. 1997). These factors are complicated because visitors may adjust their expectations toward actual conditions, such as use levels after experiencing the area, or adopt other "coping" behaviors or rationales to achieve satisfaction (Hammitt and Patterson 1991).

Use Simulation Models

In some wildernesses, visitor use must be reduced and/or redistributed to achieve specific management objectives. We have discussed a variety of techniques that can lower use levels and redistribute visitors away from critical sites. However, the specific consequences of redistributing use are not entirely predictable. For example, if a trailhead quota system is implemented, how will that affect the actual number of users present at a certain lake or on a certain trail within the wilderness? How will it affect the number of trail encounters between groups on trails or at campsites? The complexity of travel routes, which typically overlap, and the variability of travelers'

Fig. 16.13. Maintaining naturalness and solitude even in the vast reaches of Alaska, here in the USFS-managed Stikine-LeConte Wilderness, is difficult because the Alaska National Interest Lands Conservation Act (1980) permits activities such as aircraft access usually prohibited elsewhere. Photo courtesy John C. Hendee.

decisions usually are so great that neither intuition nor simple calculations can accurately answer these questions (i.e., predict on-site use levels and distribution). Rigid itineraries provide a more determinate result, at least for use of camp areas and associated encounters between camping parties, but not for encounters between parties while traveling on the trails. However, for many reasons not all parties adhere to their itinerary, so results are not as predictable as they might seem. If use patterns and encounters resulting from any given total-use level and entry-point distribution cannot be predicted, then trial-and-error experimentation might be used. However, trial and error is very time-consuming, and managers would have to try a policy for at least a year to see how it worked. Results for any one year could be influenced by uncontrolled outside factors such as weather. Detailed information on use patterns and encounters would be available only with special, costly studies.

When systems are too complex for analytic solutions and not suited to real-world experimentation, simulation modeling may provide answers. Simulation models are simplified replicas of a particular real-world system, usually formulated so that a computer program can represent them. One wilderness travel simulation model was developed thirty years ago to provide a better way to formulate and evaluate use-management policies (Lucas and Shechter 1977; Shechter and Lucas 1978; Smith and Krutilla 1976). This simulation model provided an experimental way for managers to represent use patterns and quickly "simulate" the probable effects of different use-management policies. Variability in visitor behavior is incorporated in the model, but in just a few minutes, use can be simulated for a number of seasons to average out variations. Because the experimentation takes place in the computer instead of the real world, the high resource and social costs and expense of field data collection are avoided. Even the most extreme patterns can be tested inexpensively without damage to sensitive resources.

This wilderness use simulation model includes a replica of an area's travel system (entry points, trails, water or cross-country routes, and campsites) and provides for different types of groups (several sizes and methods of travel). The computer program for the model generates visiting parties of different kinds who arrive at the area at various simulated dates and clock times, enter at particular access points, select routes of travel, and move along them. The simulated parties may overtake and pass slower parties moving in the same direction (overtaking encounters), pass parties moving in the opposite direction (meeting encounters), or pass by parties camped in areas visible from trails or other travel routes such as rivers (visual encounters). Parties that stay overnight select places to camp, which they may share with other camping parties (camp encounters). On a later day, camping parties leave the campsite, continue on their chosen routes, and eventually leave the area.

To make the model operational, data are needed on the area and its use. The travel network must be known and also something about how different types of visitors behave within it—their patterns of arrival, various routes followed and relative popularity of each, travel speeds, and so on. This information is supplied to the model in probabilistic terms; for example, there might be one chance in ten that a party entering at Deer Creek would select a route to Arrow Lake with a two-night stay there.

Fig. 16.14. Designing entrances to wilderness parking and trailheads requires managers to balance between providing necessary facilities and information and being intrusive. Informational signs at trailheads (left) use brief, directed messages (left center), and more permanent displays educate visitors about the wilderness experience and low-impact recreation practices (right center). While permit systems can bring visitors and managers together for a face-to-face exchange at permit-issuing offices, even a self-issuing permit system (right) is an opportunity to educate visitors and collect information about them. Photos courtesy Marilyn Hendee and Chad P. Dawson.

Management objectives are critical in this process. The simulator will determine use distribution and encounter frequency for any use pattern, but whether this is acceptable or unacceptable is always a managerial decision based on management objectives. Which technique is chosen to shift the use pattern is also a managerial responsibility. Use might be rationed, a new trail built or another closed, or information supplied to visitors to change their location choices. The simulator is not concerned with how use patterns are altered but only with the consequences of the alteration. The simulator is an aid to managers in decision making, but it does not make the decisions. It is a way to try out or simulate a potential management plan before implementing it, to evaluate the results to help decide whether to proceed or modify the planned action. Both the actions to be tested and the evaluation of results depend on the managers' simulated decisions—their professional skill—not the computer. By testing use-management actions before implementing them, the simulation model could help put use modification plans on a sounder, more justifiable basis.

The wilderness use simulation model has been applied to test areas in the Spanish Peaks Primitive Area (Smith and Krutilla 1976), the Desolation Wilderness (Shechter and Lucas 1978), and the Green and Yampa Rivers in Dinosaur National Monument, in Colorado and Utah (Lime et al. 1978); a sixty-three-mile section of the Appalachian Trail (Manning and Potter 1984); and the Colorado River in Grand Canyon National Park in Arizona (Underhill et al. 1986). Results of these applications indicate that when the current existing-use situation was simulated, the resulting use patterns and encounter experiences generally agreed with data from visitor surveys, indicating that the model was valid. Where there were discrepancies, they stemmed primarily from oversimplification of certain data describing typical use (e.g., including too few different travel routes to adequately reflect the variability of visitor movements), rather than inherent flaws in the simulation model.

In the Desolation Wilderness and Dinosaur National Monument in Colorado, many simulation scenarios were tested. Use was increased and decreased by varying amounts, and uneven distributions were made more even by shifting use from popular entries to less used access points, and from heavily used weekends to weekdays. Some clear relationships, not all expected, emerged. Changing the timing of visitor entries had little effect on use pressures or encounter levels. Changes in total use (all other things remaining the same) produced proportionate results. For example, if total use doubles, use of any specific location doubles, on the average. This predictable, proportional relationship provides a basis for comparing results of more complex scenarios in which use is redistributed against the same amount of total use but without any change in its distribution. Almost all use-redistribution scenarios produced lower average daily encounters for visitors than the same total use without redistribution. Some use redistributions were much more effective in reducing average encounter levels than equivalent across-the-board decreases in total use.

Average encounters do not tell the whole story, however. The frequency of extreme encounter levels, including both high levels and low levels, changed substantially for different scenarios. A manager probably would be more concerned about reducing or eliminating experiences of unsatisfactory quality than about just altering averages. In addition, changes at key trouble spots were even more pronounced, another result more relevant to a manager's perspective than overall averages.

Computer-based simulation models are still an experimental tool in wilderness management and are being rediscovered and developed for wilderness and river-recreation management situations, such as on the Colorado River in Grand Canyon National Park (Daniel and Gimblett 2000; Gimblett et al. 2000). Simulation models will be needed in the future as use increases and managers are challenged to maintain experience quality, visitor independence, flexibility, and spontaneity as well as protecting resources in the face of growing demands on limited wilderness resources.

Summary

Visitor management is a vital part of wilderness stewardship. It is guided by two principal objectives: providing a natural setting (naturalness) and a special type of quality visitor experience (solitude and primitive and unconfined recreation). Both objectives are important, but sometimes achieving them results in conflict.

Five types of undesirable visitor actions exist, varying in severity, each with different kinds of appropriate management responses: illegal actions, careless actions, unskilled actions, uninformed actions, and unavoidable impacts. *Managers and visitors differ in their perceptions of problems;* managers tend to perceive resource impacts as more important than social problems, whereas visitors tend to take the opposite point of view. These divergent perceptions reduce compliance with regulations and acceptance of recommendations. Managers need to be aware of visitor perceptions and work to narrow gaps between visitors' views and their own perceptions of problems and conditions.

Many aspects of wilderness recreation use are subject to management: amount of use, distribution of use, timing of use, party size, length of stay, and visitor behavior. Professional management of these aspects of wilderness recreation to protect natural conditions and opportunities for appropriate wilderness experiences requires adequate data about visitors and resources, much of which currently are of poor quality or are completely lacking.

Some wilderness visitor management approaches are direct, imposing control on visitors' actions, whereas other approaches are indirect, altering factors that influence visitors' choices. In general, indirect approaches are preferable, being more consistent with wilderness concepts. Regulation is the prime example of a direct approach and should be used with restraint and only when it is the minimum tool to achieve the objective, and with recognition of the limitations on effectiveness of regulations. *Monitoring results of management actions is essential to improve future management, and much improvement is needed.* Management actions implemented from outside a wilderness are generally preferable to management inside, especially in regulations and educational contacts. No single management approach is effective enough or acceptable enough to be the sole choice; a blend of approaches is essential.

Many important values of wilderness depend on the quality of visitor experiences. Managers can influence the factors that affect experience quality: natural, social, and managerial conditions. Natural conditions depend on management of user impacts and allowing natural processes such as natural wildfires to shape wilderness ecosystems. Social conditions depend on management of solitude/crowding, user conflicts, and certain types of problem behavior. Management conditions influence visitor experiences through style and tone, as well as specific actions. Diversity in all three conditions is desirable.

Indirect management involves less restrictive visitor management techniques. Design can include access, trails, and facilities to modify the amount, type, and distribution of recreational use. Most specific facilities, other than trails and bridges, play only a limited role in wilderness, receive limited support from visitors, and need to be used with restraint. Information and education are essential tools for managing visitors and their experience quality. These tools can be used to redistribute use, to expand visitors' understanding of wilderness values, to reduce visitor conflict, and to increase visitors' skills in minimum-impact use.

Direct management is a necessary tool in some situations. Regulations and use rationing are the most common direct management approaches, but strong recommendations are a less authoritarian alternative. Some types of regulations are more often appropriate than others. Group-size limits, length-of-stay limits at campsites, campfire restrictions in some situations, prohibiting some types of use in parts of areas, and sometimes rationing use all have their place in the manager's tool kit. Some management techniques are harder to justify, such as total prohibition of types of use allowed under the Wilderness Act and fixed itineraries under some permitting approaches. Rationing, when necessary, can be done in many ways, and each way has different advantages and disadvantages.

The importance of monitoring visitor experience quality and resource conditions cannot be overstressed. Monitoring visitor-to-visitor encounters on trails and at campsites, and satisfactions and dissatisfactions have been reported as some of the important visitor experience indicators. Use of simulation models can provide a way to estimate results of alternative actions and might be considered as "monitoring possible future conditions."

Wilderness management should stress visitor education and information more, and regulation less, except where direct regulation is the minimum method to achieve planning standards and objectives. Visitor understanding of "wilderness ethics" and minimum-impact use techniques is increasing. Benefits are apparent in many places: litter is less common, groups are smaller; and so on. Continuing emphasis on education can further improve visitor behavior and further diminish the need to rely on regulation. Managers need to evaluate their educational efforts and the results of education programs to guide their improvement.

Content of messages directed at visitors will tend toward more specific information geared to variations in the situations visitors encounter. How to communicate such messages most effectively and efficiently and how to reach key audiences that are the source of problem behaviors are challenges managers will need to meet successfully. Management budgets will remain limited, and programs must be cost-effective to produce the results needed to protect the resource while providing recreation opportunities in wilderness.

> ## Study Questions
> 1. What two considerations are stressed in wilderness visitor management?
> 2. How can wilderness recreation use problems be classified? How does this relate to appropriate management responses to those problems?
> 3. How do managers' and visitors' perceptions of problems in wilderness compare?
> 4. What are some of the important concepts in understanding visitor impacts on social conditions in wilderness?
> 5. What aspects of wilderness recreation use can managers control or influence?
> 6. What is the nature of visitor experiences in wilderness? What are the three main impacts on those experiences?
> 7. What characterizes direct and indirect visitor management? What are the strengths and weaknesses of each?
> 8. What are the main types of indirect visitor management? How would each affect experiences? What are the pros and cons of each?
> 9. What factors must be dealt with if wilderness recreation use must be rationed?
> 10. What would a wilderness manager monitor to keep track of experience quality?
> 11. How could simulation models of visitor use help management of the quality of the wilderness experience?

Acknowledgments

This chapter in the third edition was a major revision and synthesis by Chad P. Dawson and John C. Hendee of Chapters 15 and 17 in the second edition and we recognize and thank our colleague, Robert C. Lucas, for his seminal efforts on these topics in the first two editions.

We recognize and thank our colleague David N. Cole, research biologist at the Aldo Leopold Wilderness Research Institute, for his written contributions to two indirect management topics in this chapter: information and education programs and minimum-impact education programs.

For review comments and inputs on drafts of this chapter in the third edition, we thank the following colleagues; we have benefited from their review comments, but we alone bear responsibility for any errors, mistakes, and opinions reported herein: Dr. John J. Daigle, College of Natural Sciences, Forestry and Agriculture, Department of Forest Management, The University of Maine, Orono; Dr. John Shultis, Resource Recreation and Tourism Program, University of Northern British Columbia, Prince George; Gregory Hansen, wilderness manager, Mesa Ranger District, Tonto National Forest, USDA, Forest Service, Mesa, Arizona; James Mahoney, wilderness manager, BLM Field Office, Phoenix, Arizona; Kim Crumbo, wilderness specialist (retired), Grand Canyon National Park; and Dr. Alan Watson, social research scientist, Aldo Leopold Wilderness Research Institute, Missoula, Montana.

The fourth edition includes contributions from Dr. Rudy Schuster on the social aspects of visitors and their management and additional updating throughout the chapter by Chad P. Dawson and John C. Hendee.

Case Discussion:
Relationships with Wilderness

Managers continually work at understanding visitors and the interest of various public groups in using wilderness directly or indirectly. Over time, wilderness managers have had to become students of sociology, recreation management, and other social sciences, in addition to the natural sciences and management, to be good stewards of the NWPS. Much of wilderness management work is people management—both direct and indirect management of people.

Recently, the concepts of the importance of place and relationships with wilderness have become part of the research literature. One research publication (Dvorak and Borrie 2007, 14–15) summarizes three reasons for taking a relational approach to understanding visitors:

> First, by understanding how visitors conceptualize their relationship with wilderness and the variety of cultural and social forces that influence these relationships, wilderness managers and researchers may be able to develop new indicators and standards to guide management. These relationship indicators and standards could be used to facilitate opportunities for quality wilderness experiences based on various concepts (e.g., experience use history, life stage, affinity for technology) of an individual's relationship. For example, wilderness recreation opportunities could be assessed to determine how they provide experiences for families with young children or for individuals considered as "veterans" in that area. While developing such indicators and standards may be challenging, the process represents an evolution in thinking about protected areas and an attempt to find new ways to address experience quality.
>
> Second, a relationship framework integrates with the responsibility of managers to preserve wilderness character for future generations, but also current generations "in the future." By acknowledging that wilderness is an enduring resource with ongoing significance, a relationship framework posits the examination and understanding of management actions in the context of an individual's lifetime. It moves from documenting visitor experiences as snapshots of the individual or consumer-oriented one-time transactions, to attempting to understand how experience and forces of change affect relationships over time. This shift in focus provides managers with information as they make difficult, value-based decisions about what desired wilderness conditions should be and mean for future generations.
>
> Finally, acknowledging changing relationships could provide more latitude in future decision-making. It focuses greater attention on the temporal and dynamic aspects of the interactions individuals have with an area. It places greater emphasis on the examination of both current visitor trends and possible future changes that occur in the general population. Such foresight may allow managers to be more proactive in decision making, in contrast to a reactive reliance on satisfaction or singular outcome-based approaches to understanding visitor experiences.

Case Discussion Questions:

1. What are some of the research challenges of implementing this type of approach to understanding visitors?
2. How does this relationship approach help managers to understand the reasons for conflicts that visitors have with some management decisions, such as access issues that institute visitor use fees and allow increased commercial access to wilderness?
3. Most visitor research is a snapshot in time and space and is a static look at visitor characteristics, use, preferences, needs, and satisfactions. The relationship approach is suggesting longer-term monitoring and consideration of broader societal trends such as demographic change. How can the current system of visitor studies be improved by some additional research on visitor relationships with wilderness?

References

Adelman, Bonnie J. E.; Heberlein, Thomas; Bonnicksen, Thomas M. 1982. Social psychological explanations for the persistence of a conflict between paddling canoeists and motor craft users in the Boundary Waters Canoe Area. *Leisure Sciences.* 5(1): 45–61.

Anderson, D. H.; Brown, P. J. 1984. The displacement process in recreation. *Journal of Leisure Research.* 16: 61–73.

Anderson, D. H.; Lime, D. W.; Wang, T. L. 1998. *Maintaining the Quality of Park Resources and Visitor Experiences: A Handbook for Managers.* St. Paul: University of Minnesota, Department of Forest Resources, Cooperative Park Studies Unit.

Behan, R. W. 1974. Police state wilderness: A comment on mandatory wilderness permits. *Journal of Forestry.* 72(2): 98–99.

Blahna, D. J.; Smith, K. S.; Anderson, J. A. 1995. Backcountry llama packing: Visitor perceptions of acceptability and conflict. *Leisure Sciences.* 17:185–204.

Borrie, W. T.; McCool, S. F. 2007. Describing change in visitors and visits to the "Bob." *International Journal of Wilderness.* 13(3): 28–33.

Brooks, J. J.; Wallace, G. N.; Williams, D. R. 2006. Place as relationship partner: An alternative metaphor for understanding the quality of visitor experience in a backcountry setting. *Leisure Sciences.* 28: 331–349.

Brown, Perry J.; Haas, Glenn E. 1980. Wilderness recreation experiences: The Rawah case. *Journal of Leisure Research.* 12(3): 229–241.

Brown, R. N. K.; Rosenberger, R. S.; Kline, J. D.; Hall, T. E.; Needham, M. D. 2008. Visitor preferences for managing wilderness recreation after wildfire. *Journal of Forestry.* 106(1): 9–16.

Cable, S.; Watson, A. E. 1998. Recreation use allocation: Alternative approaches for the Bob Marshall Wilderness Complex. RMRS-RN-1. Ogden, UT: U.S. Department of Agriculture, Forest Service, Rocky Mountain Research Station.

Carlson, T. 2000. The Eagle Cap Wilderness permit system: A visitor education tool. *International Journal of Wilderness.* 6(2): 27–28.

Carothers, P.; Vaske, J. J.; Donnelly, M. P. 2001. Social values versus interpersonal conflict among hikers and mountain bikers. *Leisure Sciences.* 23: 47–61.

Cole, David N. 1986. Ecological changes on campsites in the Eagle Cap Wilderness, 1979 to 1984. Research Paper INT-368. Ogden, UT: U.S. Department of Agriculture, Forest Service, Intermountain Research Station.

———. 1987. Research on soil and vegetation in wilderness: A state-of-knowledge review. In: Lucas, Robert C., comp. *Proceedings: National Wilderness Research Conference: Issues, state-of-knowledge, future directions*; July 23–26, 1985; Fort Collins, CO. General Technical Report INT-220. Ogden, UT: U.S. Department of Agriculture, Forest Service, Intermountain Research Station, pp. 135–177.

———. 1989. Low impact recreational practice for wilderness and backcountry. General Technical Report INT-265. Ogden, UT: U.S. Department of Agriculture, Forest Service, Intermountain Research Station.

Cole, David N.; Dalle-Molle, John. 1982. Managing campfire impacts in the backcountry. General Technical Report INT-135. Ogden, UT: U.S. Department of Agriculture, Forest Service, Intermountain Forest and Range Experiment Station.

Cole, David N.; Hammond, T. P.; McCool, S. F. 1997. Information quantity and communication effectiveness: Low-impact messages on wilderness trailside bulletin boards. *Leisure Sciences.* 19: 59–72.

Cole, David N.; Watson, Alan; Hall, Troy; Spildie, David. 1997. High-use destinations in wilderness: Social and biophysical impacts, visitor responses, and management options. Research Paper INT-RP-496. Ogden, UT: U.S. Department of Agriculture, Forest Service, Intermountain Research Station.

Cole, David N.; Watson, Alan E.; Roggenbuck, Joseph W. 1995. Trends in wilderness visitors and visits: Boundary Waters Canoe Area, Shining Rock, and Desolation Wilderness. Research Paper INT-RP-483. Ogden, UT: U.S. Department of Agriculture, Forest Service, Intermountain Research Station.

Conrad, Richard. 1997. National survey highlights agency training needs in the United States. *International Journal of Wilderness.* 3(4): 9–12.

Cordell, H. K.; Tarrant, M. A.; McDonald, B. L.; Bergstrom, J. C. 1998. How the public views wilderness: More results from the USA survey on recreation and the environment. *International Journal of Wilderness.* 4(2): 28–31.

Daniel, T. C.; Gimblett, H. R. 2000. Autonomous agents in the park: An introduction to the Grand Canyon River trip simulation model. *International Journal of Wilderness.* 6(3): 39–43.

Dawson, C. P. 2007. New opportunities for educating future wilderness and wildland managers in a changing technological world. *International Journal of Wilderness.* 13(3): 36–39.

Dawson, Chad P.; Hammitt, William E. 1996. Dimensions of wilderness privacy for Adirondack Forest Preserve hikers. *International Journal of Wilderness.* 2(1): 37–41.

Denburg, Ronald F. 1982. *Crowding and social displacement on the lower St. Croix and upper Mississippi Rivers.* Thesis. Madison: University of Wisconsin.

Doucette, J.; Cole, David N. 1993. Wilderness visitor education: Information about alternative techniques. Technical Report INT-295. Ogden, UT: U.S. Department of Agriculture, Forest Service, Intermountain Research Station.

Driver, B.; Brown, P. 1983. *Contributions of Behavioral Scientists to Recreation Resource Management.* New York: Plenum.

Driver, B.; Nash, R.; Haas, G. 1987. Wilderness benefits: A state of knowledge review. In: *National Wilderness Research Conference: Issues, State of Knowledge, Future Directions.* GTR-INT-220. Ogden, UT: U.S. Department of Agriculture, Forest Service Intermountain Research Station, pp. 294–319

Dvorak, R. G.; Borrie, W. T. 2007. Changing relationships with wilderness: A new focus for research and stewardship. *International Journal of Wilderness.* 13(3): 12–15.

Ewert, A.; Hendee, J. C.; Davidson, S.; Brame, R.; and Mackey, C. 1999. Wilderness access issues for education, personal growth, and therapeutic use: A U.S. panel summary. *International Journal of Wilderness.* 5(3): 13–18.

Farnum, J.; Hall, T.; Kruger, L. E. 2005. Sense of place in natural resource recreation and tourism: An evaluation and assessment of research findings. General Technical Report PNW-GTR-660. Portland, OR: U.S. Department of Agriculture, Forest Service, Pacific Northwest Research Station.

Freimund, W. A.; Cole, D. N. 2001. Use density, visitor experience, and limiting recreational use in wilderness: Progress to date and research needs. In: Freimund, W. A.; Cole, D. N., eds. *Visitor Use Density and Wilderness Experience.* Missoula, MT: U.S. Department of Agriculture, Forest Service, Rocky Mountain Research Station, pp. 3–8.

Gager, Dan; Hendee, J. C.; Kinziger, M.; Krumpe, E. 1998. What managers are saying and doing about Wilderness Experience Programs. *Journal of Forestry.* 96(8): 33–37.

Gimblett, H. R.; Itami, R. M.; Richards, M. 2000. Simulating wildland recreation use and conflicting spatial interactions using rule-driven intelligent agents. In: Gimblett, H. R., ed. *Integrating GIS and Agent-Based Modeling Techniques for Understanding Social and Ecological Processes.* New York: Oxford University Press.

Graefe, A. R.; Thapa, B. 2004. Conflict in natural resource recreation. In: Manfredo, M. J.; Vaske, J. J.; Field, D.; Brown, P.; Bruyere, B., eds. *Society and Natural Resources: A Summary of Knowledge.* Jefferson City, MO: Modern Litho, pp. 209–224.

Hall, T.; Shelby, B. 1996. Who cares about encounters? Differences between those with and without norms. *Leisure Sciences.* 18: 7–22.

Hammitt, William E.; Madden, Mark A. 1989. Cognitive dimensions of wilderness privacy: A field test and further explanation. *Leisure Sciences.* 11: 293–301.

Hammitt, William E.; Patterson, M. E. 1991. Coping behavior to avoid visitor encounters: Its relationship to wild land privacy. *Journal of Leisure Research.* 23(3): 225–237.

Hammitt, William E.; Rutlin, William M. 1995. Use encounter standards and curves for achieved privacy in wilderness. *Leisure Sciences.* 17(4): 245–262.

Hansen, Greg. 1990. Teaching a wilderness ethic—reaching beyond the forest. In: Lime, David, ed. *Managing America's Enduring Wilderness Resource.* St. Paul: University of Minnesota Extension Service, pp. 123–130.

Hendee, John C.; Catton, William R., Jr.; Marlow, Larry D.; Brockman, C. Frank. 1968. Wilderness users in the Pacific Northwest—their characteristics, values, and management preferences. Research Paper PNW-61. Portland, OR: U.S. Department of Agriculture, Forest Service, Pacific Northwest Forest and Range Experiment Station.

Hendee, John C.; Lucas, Robert C. 1973. Mandatory wilderness permits: A necessary management tool. *Journal of Forestry.* 71(4): 206–209.

Heywood, J. L. 1996. Social regularities in outdoor recreation. *Leisure Sciences.* 18: 23–37.

Jacob, Gerald R.; Schreyer, Richard. 1980. Conflict in outdoor recreation: A theoretical perspective. *Journal of Leisure Research.* 12(4): 368–380.

Krumpe, Edwin E.; Brown, Perry J. 1982. Redistributing backcountry use through information related to recreational experiences. *Journal of Forestry*. 80(6): 360–362.

Kuss, F. R.; Graefe, A.; Vaske, J. J. 1990. *Visitor Impact Management: A Review of Literature*. Washington, DC: National Parks and Conservation Association.

Leave No Trace Center for Outdoor Ethics. 2008. Leave No Trace Program Principles. Boulder, CO. www.lnt.org/programs/principles.php; accessed February 1, 2008.

Lime, David W.; Anderson, Dorothy H.; McCool, Stephen F. 1978. An application of the simulator to a river recreation setting. In: Shechter, Mordechai; Lucas, Robert C., eds. *Simulation of recreational use for park and wilderness management*. Baltimore: Johns Hopkins University Press, pp. 153–174.

Lime, David W.; Lucas, Robert C. 1977. Good information improves the wilderness experience. *Naturalist*. 28(4): 18–20.

Lucas, Robert C. 1980. Use patterns and visitor characteristics, attitudes, and preferences in nine wilderness and other roadless areas. Research Paper INT-253. Ogden, UT: U.S. Department of Agriculture, Forest Service, Intermountain Forest and Range Experiment Station.

———. 1981. Redistributing wilderness use through information supplied to visitors. Research Paper INT-277. Ogden, UT: U.S. Department of Agriculture, Forest Service, Intermountain Forest and Range Experiment Station.

———. 1982. Recreation regulations—when are they needed? *Journal of Forestry*. 80(3): 148–151.

———. 1985. Visitor characteristics, attitudes, and use patterns in the Bob Marshall Wilderness Complex, 1970–1982. Research Paper INT-345. Ogden, UT: U.S. Department of Agriculture, Forest Service, Intermountain Forest and Range Experiment Station.

Lucas, Robert C.; Shechter, Mordechai. 1977. A recreational visitor travel simulation model as an aid to management planning. *Simulation and Games*. 8(3): 375–384.

Mackey, C. 1998. U.S. wilderness management in the 21st century—politics, policy, and partnerships. *International Journal of Wilderness*. 4(3): 6–11.

Manning, Robert E. 1985. Crowding norms in backcountry settings: A review and synthesis. *Journal of Leisure Research*. 17(2): 75–89.

———. 1986. Density and crowding in wilderness: Search and research for satisfaction. In: Lucas, Robert C., comp. *Proceedings: National Wilderness Research Conference: Current Research*; July 23–26, 1985; Fort Collins, CO. General Technical Report INT-212. Ogden, UT: U.S. Department of Agriculture, Forest Service, Intermountain Research Station, pp. 440–448.

———. 1999. *Studies in Outdoor Recreation: Search and Research for Satisfaction*, 2nd ed. Corvallis: Oregon State University.

Manning, Robert E.; Ballinger, N.; Marion, Jeffrey L.; Roggenbuck, Joseph W. 1996. Recreation management in natural areas: Problems and practices, status and trends. *Natural Areas Journal*. 16: 142–146.

Manning, R. E.; Lime, D. W. 2000. Defining and managing the quality of wilderness recreation experiences. In: Cole, D. N.; McCool, S. F.; Borrie, W. T.; O'Loughlin, J., comps. *Proceedings: Wilderness Science in a Time of Change*, Vol. 4: *Wilderness Visitors, Experiences, and Visitor Management*; May 23–27, 1999; Missoula, MT. RMRS-P-15. Ogden, UT: U.S. Department of Agriculture, Forest Service, Rocky Mountain Research Station, pp. 13–52.

Manning, Robert E.; Potter, Fletcher I. 1984. Computer simulation as a tool in teaching park and wilderness management. *Journal of Environmental Education*. 15(3): 3–9.

Manning, Robert E.; Valliere, William A. 1996. Environmental values, environmental ethics, and wilderness management—an empirical study. *International Journal of Wilderness*. 2(2): 27–32.

Marion, Jeffrey L.; Roggenbuck, Joseph W.; Manning, Robert E. 1993. Problems and practices in backcountry recreation management: A survey of National Park Service managers. NPS/NRVT/NRR-93/12. Denver: U.S. Department of the Interior, National Park Service.

Martin, Burnham H. 1986. Hikers' opinions about fees for backcountry recreation. In: Lucas, Robert C., comp. *Proceedings: National Wilderness Research Conference: Current Research*; July 23–26, 1985; Fort Collins, CO. General Technical Report INT-212. Ogden, UT: U.S. Department of Agriculture, Forest Service, Intermountain Research Station, pp. 483–488.

McCool, Stephen F.; Christensen, N. 1996. Alleviating congestion in parks and recreation areas through direct management of visitor behavior. In: *Crowding and Congestion in the National Park System: Guidelines for Management and Research*. St. Paul: University of Minnesota, Agricultural Experiment Station, pp. 67–83.

McCool, Stephen F.; Clark, R. N.; Stankey, G. H. 2007. An assessment of frameworks useful for public land recreation planning. General Technical Report PNW-GTR-705. Portland, OR: U.S. Department of Agriculture, Forest Service, Pacific Northwest Research Station. 125p.

McCool, Stephen F.; Stankey, George H. 1986. Visitor Attitudes Toward Wilderness Fire Management Policy: 1971–1984. Research Paper INT-357. Ogden, UT: U.S. Department of Agriculture, Forest Service, Intermountain Research Station.

McCool, Stephen F.; Utter, Jack. 1982. Recreation use lotteries: Outcomes and management. *Journal of Forestry*. 80(1): 10–11, 29.

Monz, C.; Roggenbuck, J.; Cole, D.; Brame, R.; Yoder, A. 2000. Wilderness party size regulations: Implications for management and a decision-making framework. In: Cole, D. N.; McCool, S. F.; Borrie, W. T.; O'Loughlin, J., comps. *Proceedings: Wilderness Science in a Time of Change*, Vol. 4: *Wilderness Visitors, Experiences, and Visitor Management*; May 23–27, 1999; Missoula, MT. RMRS-P-15. Ogden, UT: U.S. Department of Agriculture, Forest Service, Rocky Mountain Research Station, pp. 265–273.

Muth, Robert M.; Clark, Roger N. 1978. Public participation in wilderness and backcountry litter control: A review of research and management experience. General Technical Report PNW-75. Portland, OR: U.S. Department of Agriculture, Forest Service, Pacific Northwest Forest and Range Experiment Station.

Oset, Robert. 1990. Administrative resuscitation: New life for wilderness education. In: Lime, David W., ed. *Managing America's Enduring Wilderness Resource: Proceedings of the Conference*; September 11–17, 1989; Minneapolis, MN. St. Paul: University of Minnesota, Extension Service, pp. 113–117.

Patterson, M.; Hammitt, W. E. 1990. Backcountry encounter norms, actual reported encounters, and their relationship to wilderness solitude. *Journal of Leisure Research*. 22(3): 259–275.

Peterson, M. R. 1996. Wilderness by state mandate: A survey of state-designated wilderness areas. *Natural Areas Journal*. 16(3): 192–197.

Queen, L; Freimund, W.; Peel, S. 1998. Enhancing the potential for wilderness electronic communication. In: Kulhavy, D. L.; Legg, M. H., eds. *Wilderness and Natural Areas in Eastern North America: Research, Management and Planning*. Nacogdoches, TX: Stephen F. Austin State University, Arthur Temple College of Forestry, Center for Applied Studies, pp. 180–184.

Richer, Jerrell R.; Christensen, Neal A. 1999. Appropriate fees for wilderness day use: Pricing decisions for recreation on public lands. *Journal of Leisure Research*. 31(3): 269–280.

Robertson, Rachel D. 1986. Actual versus self-reported wilderness visitor behavior. In: Lucas, Robert C., comp. *Proceedings: National Wilderness Research Conference: Current Research*; July 23–26, 1985; Fort Collins, CO. General Technical Report INT-212. Ogden, UT: U.S. Department of Agriculture, Forest Service, Intermountain Research Station, pp. 326–332.

Roggenbuck, Joseph W. 1980. Wilderness user preferences: Eastern and western areas. In: *Proceedings: Wilderness Management Symposium*; November 13–15, 1980; Knoxville, TN. Atlanta, GA: U.S. Department of Agriculture, Forest Service, Southern Region, pp. 103–146.

———. 1992. Use of persuasion to reduce resource impacts and visitor conflicts. In: Manfredo, J. J., ed. *Influencing Human Behavior: Theory and Applications in Recreation, Tourism, and Natural Resources*. Champaign, IL: Sagamore Publishing, pp. 149–208.

Roggenbuck, Joseph W.; Berrier, Deborah L. 1982. A comparison of the effectiveness of two communication strategies in dispersing wilderness campers. *Journal of Leisure Research*. 14(1): 77–89.

Roggenbuck, Joseph W.; Ham, Sam. 1986. Use of information and education in recreation management. In: *A Literature Review: The President's Commission on Americans Outdoors*. Report M-59-M-71. Washington, DC: U.S. Government Printing Office.

Roggenbuck, Joseph W.; Manfredo, Michael J. 1990. Choosing the right route to wilderness education. In: Lime, David W., ed. *Managing America's Enduring Wilderness Resource*; September 11–17, 1989; St. Paul, MN. St. Paul: University of Minnesota, Extension Service, pp. 103–112.

Schreyer, R. 1990. Conflict in outdoor recreation: The scope of the challenge to resource planning and management. In: Vining, J., ed. *Social Science and Natural Resources Management*. Boulder, CO: Westview Press, pp. 13–31.

Schuster, R. M.; Cole, D.; Hall, T.; Baker, J.; Oreskes, R. 2006. Appraisal of and response to social conditions in the Great Gulf Wilderness: Relationships among perceived crowding, rationalization, product shift, satisfaction, and future behavioral intentions. In: Burns, R.; Robinson, K. *Proceedings of the 2006 Northeastern Recreation Research Symposium;* April 9–11, 2006; Bolton Landing, NY. General Technical Report NRS-P-14; Newton Square, PA: U.S. Department of Agriculture, Forest Service, Northern Research Station, pp. 488–496.

Schuster, R. M.; Hammitt, W. E.; Moore, D. 2006. Stress appraisal and coping response to hassles experienced in outdoor recreation settings. *Leisure Sciences.* 28: 97–113.

Schuster, R. M.; Hammitt, W. E.; Moore, D.; Schneider, I. E. 2006. Coping with stress resulting from social value conflict: Nonhunters' response to anticipated social interaction with hunters. *Human Dimensions of Wildlife Management.* 11: 101–113.

Shechter, Mordechai; Lucas, Robert C. 1978. *Simulation of Recreational Use for Park and Wilderness Management.* Baltimore: Johns Hopkins University Press.

Shelby, Bo; Harris, Richard. 1985. Comparing methods for determining visitor evaluations of ecological impacts: Site visits, photographs, and written descriptions. *Journal of Leisure Research.* 17(1): 57–67.

———. 1986. User standards for ecological impacts at wilderness campsites. In: Lucas, Robert C., comp. *Proceedings: National Wilderness Research Conference: Current Research*; July 23–26, 1985; Fort Collins, CO. General Technical Report INT-212. Ogden, UT: U.S. Department of Agriculture, Forest Service, Intermountain Research Station, pp. 166–171.

Shelby, Bo; Heberlein, Thomas A. 1984. A conceptual framework for carrying capacity determination. *Leisure Sciences.* 6(4): 433–451.

Shew, Richard L.; Saunders, Paul R.; Ford, Joseph D. 1986. Wilderness managers' perceptions of recreational horse use in the northwestern United States. In: Lucas, Robert C., comp. *Proceedings: National Wilderness Research Conference: Current Research*; July 23–26, 1985; Fort Collins, CO. General Technical Report INT-212. Ogden, UT: U.S. Department of Agriculture, Forest Service, Intermountain Research Station, pp. 320–325.

Shindler, B.; Shelby, Bo. 1993. Regulating wilderness use: An investigation of user group support. *Journal of Forestry.* 91: 41–44.

Smith, V. Kerry; Krutilla, John V. 1976. *Structure and Properties of a Wilderness Travel Simulator: An Application to the Spanish Peaks Area.* Baltimore: Johns Hopkins University Press.

Stankey, George H. 1973. Visitor perception of wilderness recreation carrying capacity. Research Paper INT-142. Ogden, UT: U.S. Department of Agriculture, Forest Service, Intermountain Forest and Range Experiment Station.

———. 1979. Use rationing in two southern California wildernesses. *Journal of Forestry.* 77(6): 347–349.

———. 1980. A comparison of carrying capacity perceptions among visitors to two wildernesses. Research Paper INT-242. Ogden, UT: U.S. Department of Agriculture, Forest Service, Intermountain Forest and Range Experiment Station.

Stankey, George H.; Baden, John. 1977. Rationing wilderness use. General Technical Report INT-198. Ogden, UT: U.S. Department of Agriculture, Forest Service, Intermountain Forest and Range Experiment Station.

Stankey, George H.; Schreyer, Richard. 1987. Attitudes toward wilderness and factors affecting visitor behavior: A state-of-knowledge review. In: Lucas, Robert C., comp. *Proceedings: National Wilderness Research Conference: Issues, State-of-Knowledge, Future Directions*; July 23–26, 1985; Fort Collins, CO. General Technical Report INT-220. Ogden, UT: U.S. Department of Agriculture, Forest Service, Intermountain Research Station, pp. 246–293.

Swain, Ralph W. 1986. *Colorado wilderness violators: Who they are and why they violate.* Thesis. Fort Collins: Colorado State University.

———. 1996. Leave No Trace (LNT)—Outdoor skills and ethics program. *International Journal of Wilderness.* 2(3): 24–26.

Swearingen, T. C.; Johnson, D. R. 1995. Visitors' responses to uniformed park employees. *Journal of Park and Recreation Administration.* 13: 73–85.

Tangen-Foster, J.; Dawson, C. P. 1999. Risk management programs in wilderness experience programs. *International Journal of Wilderness.* 5(3): 29–34.

Taylor, Jonathon G.; Mutch, Robert W. 1986. Fire in wilderness: Public knowledge, acceptance, and perceptions. In: Lucas, Robert C., comp. *Proceedings: National Wilderness Research Conference: Current Research*; July 23–26, 1985; Fort Collins, CO. General Technical Report INT-212. Ogden, UT: U.S. Department of Agriculture, Forest Service, Intermountain Research Station, pp. 49–59.

Underhill, A. Heaton; Xaba, A. Busa; Borkan, Ronald E. 1986. The wilderness use simulation model applied to Colorado River boating in Grand Canyon National Park, USA. *Environmental Management.* 10(3): 367–374.

U.S. Department of Agriculture, Forest Service. 1982. ROS user's guide. Washington, DC: U.S. Department of Agriculture, Forest Service.

U.S. Public Law 88-577. The Wilderness Act of September 3, 1964. 78 Stat. 890.

U.S. Public Law 96-487. Alaska National Interest Lands Conservation Act (ANILCA) of December 2, 1980. 94 Stat. 2371.

Vaske, J. J.; Donnelly, M. P.; Wittmann, K.; Laidlaw, S. 1995. Interpersonal versus social-values conflict. *Leisure Sciences.* 17: 205–222.

Wagar, J. V. K. 1940. Certified outdoorsmen. *American Forests.* 46(11): 490–492, 524–525.

Washburne, Randel F.; Cole, David N. 1983. Problems and Practices in Wilderness Management: A Survey of Managers. Research Paper INT-304. Ogden, UT: U.S. Department of Agriculture, Forest Service, Intermountain Forest and Range Experiment Station.

Watson, Alan E. 1994. The nature of conflict between hikers and recreational stock users in the John Muir Wilderness. *Journal of Leisure Research.* 26(4): 372–385.

———. 1995a. An analysis of recent progress in recreation conflict research and perceptions of future challenges and opportunities. *Leisure Science.* 17(3): 235–238.

———. 1995b. Opportunities for solitude in the Boundary Waters Canoe Area Wilderness. *Journal of Applied Forestry.* 12(1): 12–18.

———. 2000. Wilderness use in the year 2000: Societal changes that influence human relationships with wilderness. In: Cole, D. N.; McCool, S. F; Borrie, W. T.; O'Loughlin, J, comps. *Proceedings: Wilderness Science in a Time of Change:* Vol. 4: *Wilderness Visitors, Experiences, and Visitor Management;* May 23–27, 1999; Missoula, MT. RMRS-P-15. Ogden, UT: U.S. Department of Agriculture, Forest Service, Rocky Mountain Research Station, pp. 53–60.

Watson, Alan E.; Asp, C.; Walsh, J.; Kulla, A. 1997. The contribution of research to managing conflict among national forest users. *Trends.* 34(3): 29–35.

Watson, Alan E.; Christensen, N. A.; Blahna, D. J.; Archibold, K. S. 1998. Comparing manager and visitor perceptions of llama use in wilderness. Research Paper RMRS-RP-10. Ogden, UT: U.S. Department of Agriculture, Forest Service, Rocky Mountain Research Station.

Watson, Alan E.; Hendee, John C.; Zaglauer, Hans P. 1996. Human values and codes of behavior: Changes in Oregon's Eagle Cap Wilderness visitors and their attitudes. *Natural Areas Journal.* 16(2): 89–93.

Welsh, R. T. 2000. A comparison of strategies for rationing and managing use on selected rivers in the United States in 1986 and 1998. In: Watson, A. E.; Aplet, G. H.; Hendee, J. C., comps. *Personal, Societal, and Ecological Values of Wilderness: Sixth World Wilderness Congress Proceedings on Research, Management, and Allocation,* Vol. II; October 24–29, 1998; Bangalore, India. Proceedings RMRS-P-14; Ogden, UT: U.S. Department of Agriculture, Forest Service, Rocky Mountain Research Station, pp. 25–34.

White, David; Hendee, J. C. 2000. Primal hypotheses: The relationship of naturalness and solitude to the wilderness experience benefits of self, development of community, and spiritual development. In: Cole, D. N.; McCool, S. F; Borrie, W. T.; O'Loughlin, J., comps. *Proceedings: Wilderness Science in a Time of Change:* Vol. 3: *Wilderness Visitors, Experiences, and Visitor Management;* May 23–27, 1999; Missoula, MT. RMRS-P-15. Ogden, UT: U.S. Department of Agriculture, Forest Service, Rocky Mountain Research Station, pp. 223–227.

Williams, Daniel R.; Vogt, Christine A.; Vitterso, Joar. 1999. Structural equation modeling of users' response to wilderness recreation fees. *Journal of Leisure Research.* 31(3): 245–268.

Winter, Patricia L.; Palucki, Laura J.; Burkhardt, Rachel L. 1999. Anticipated responses to a fee program: The key is trust. *Journal of Leisure Research.* 31(3): 207–226.

Worf, William A. 1987. Introduction to wilderness research needs panel discussion. In: Lucas, Robert C., comp. *Proceedings: National Wilderness Research Conference: Issues, State-of-Knowledge, Future Directions;* July 23–26, 1985; Fort Collins, CO. General Technical Report INT-220. Ogden, UT: U.S. Department of Agriculture, Forest Service, Intermountain Research Station, pp. 351–352.

VI. The Future

Chapter 17

Future Issues and Challenges in Wilderness Stewardship

Wilderness—Past, Present, and Future	487
A Growing Importance of Wilderness	488
Trends in Wilderness Use	490
Management Funding and Resources	491
An Expanding—but Ultimately Limited—Wilderness System	492
Conclusion	493
References	494

Wilderness—Past, Present, and Future

We have traced the development of the wilderness concept and discussed its various resources and values, threats to the resource, and management actions that might be employed to enhance and protect it. Wilderness, as a land-use concept, has evolved a great deal over the past century, particularly since passage of the Wilderness Act in 1964, as has our capacity to manage this valuable and dwindling resource.

Aldo Leopold (1970, 264, 270) noted that "wilderness is the raw material out of which man has hammered the artifact called civilization"; he goes on to point out that only remnants of it remain, so that "*wilderness is a resource which can shrink but not grow* [emphasis added]. Invasions can be arrested or modified in a manner to keep an area usable either for recreation, or for science, or for wildlife, but the creation of new wilderness in the full sense of the word is impossible."

What about the future for wilderness? What we do today or fail to do today largely shapes the kind of wilderness future we will deal with tomorrow. The implications for thinking about wilderness as we begin the twenty-first century are clear. The kind of wilderness system we will have tomorrow—a generation or even a century from now—will depend on what we do today. This kind of thinking underlies *the concern of many for designating as much wilderness as possible now, to save it for tomorrow,* otherwise many options will be gone. In addition, the implications are just as clear for protecting wilderness already designated; if stewardship is lax, wilderness qualities and conditions will diminish or vanish.

In the U.S. experience with wilderness, defining, protecting, and managing it have derived from a particular set of environmental and sociopolitical conditions. To what extent will these conditions prevail in the future, and, to the extent that they change, how will the new conditions affect wilderness uses and values to society and needs for its management? Land at the primitive end of the environmental modification spectrum has always had a tendency to move toward more modification, for many societal reasons. What Robert Marshall noted in 1937 (240) is still true today:

Fig. 17.1. Stewardship of the NWPS is receiving increased emphasis as more visitors and other users compete for access. Some of the primary objectives will be to maintain the qualities of natural ecosystems in settings such as the Spanish Peaks area of the Lee Metcalf Wilderness, Montana. Photo courtesy U.S. Forest Service.

The world is full of conflicts between genuine values. Often these conflicts are resolved entirely from the standpoint of one of the competing values, and thus whole categories of human enjoyment may be needlessly swept away. It is far more conducive to human happiness to attempt some rational balance that will make possible for the immensely different types of people the varied values they crave. Emphatically this is true of the conflict between the values created by the modification of Nature and the values of the primitive…The fate of unmodified Nature rests in the activity of its friends.

We contend that the quality of the National Wilderness Preservation System in the future depends substantially on the steps we take now to manage existing areas and on the high standards set for its stewardship. As we noted in the opening chapter, *failure to implement adequate and appropriate management and stewardship today will lead to diminished or diluted wilderness tomorrow.*

A Growing Importance of Wilderness

We see wilderness resources and values increasing in importance for many reasons—for direct recreational values; for therapy and healing, including respite for all users from the accelerating pace of today's world; for increased scientific understanding of natural systems and the world around us; as a source of material values to humankind, such as for medicines and genetic stock for agriculture and forestry; and as a protected source of ecological services such as clean air, water, soil, and natural systems for climate regulation. Although economic estimates can place dollar amounts on some of these values, and the current economic estimates are impressive, they are but a partial indicator of the total value of wilderness. We see wilderness remaining as an important, even essential, component of a spectrum of land uses for a variety of important societal opportunities and values. Nearly a half century ago, Howard Zahniser (1956, 41) eloquently stated some of those values:

> In the areas of wilderness that are still relatively unmodified by man it is, however, possible for a human being, adult or child, to sense and see his own humble, dependent relationship to all of life. In these areas, thus, are the opportunities for so important, so neglected a part of our education—gaining of the true understanding of our past, ourselves, and our world which will enable us to enjoy the conveniences and liberties of our urbanized, industrialized, mechanized civilization and yet not sacrifice an awareness of our human existence as spiritual creatures nurtured and sustained by and from the great community of life that comprises the wildness of the universe, of which we ourselves are a part.

We are optimistic about the future of wilderness. There is little evidence that public interest is waning, that wilderness is just a fad. To the contrary, wilderness has shown remarkable tenacity as an item on the public agenda, as evidenced by 171 pieces of federal legislation strengthening and expanding the NWPS since 1964. We think public interest and advocacy will continue, empowered by modern communication methods such as the Internet. We also see growing international interest and support for wilderness as a measure of the imposing strength and resiliency of the wilderness idea.

Fig. 17.2. Wilderness in Alaska is relatively uncrowded and unaffected by human activities, but even here choices must be made between naturalness and wildness, lack of crowding and freedom. Photo courtesy David Cole.

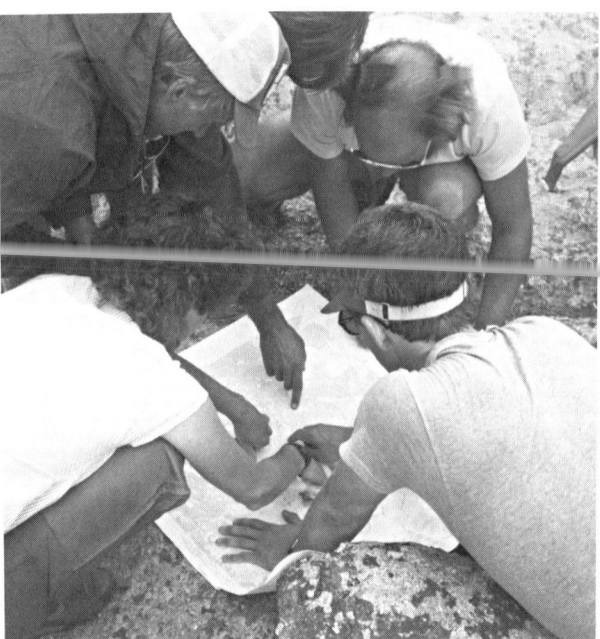

Fig. 17.3. Wilderness provides a unique setting for educational, personal growth, and human development programs. A group supervised by the National Outdoor Leadership School plans a cross-country hike in the Popo Agie Wilderness, Wyoming. Photo courtesy National Outdoor Leadership School, Lander, Wyoming.

Whatever problems exist with the wilderness program in the United States, it nevertheless stands as an achievement that many other nations envy and hold up as a model to which they aspire—namely a special category to protect lands with natural and wild values. For example, nine nations now have wilderness systems by law or administrative mandate, and several others provide some protection for comparable areas. Wilderness protection is on the international agenda as another means of protecting planet Earth, with a growing compilation of information and dialogue on international possibilities in proceedings of seven World Wilderness Congresses (Martin 2001).

Some wildernesses have experienced extremely high-use pressures, necessitating intensive management, including rationing use. Undoubtedly, such measures will continue to be needed, and in more areas. Nevertheless, we do not foresee a future in which people seeking a wilderness experience must always book a time and route months ahead, where rangers lurk behind every tree, and where every step is monitored by sensors and satellites. Many areas will be open and accessible without heavy-handed direct restrictions. Although the past four decades have seen a generally steady increase in the level of recreational use of wilderness, the Wilderness System has also expanded. Unfortunately, the continuing growth and interest in wilderness leave managers constantly struggling to catch up. This is unlikely to change because personnel and financial support are always issues, and management adjustments to increased visitation and impacts are too often made after the fact. Yet, the challenge and mandate for wilderness stewardship remain the same—to try to plan by looking ahead rather than continually reacting to the latest crisis, and to resist the temptation to come down hard on every issue, focusing instead on light-handed, indirect approaches to management, which most managers and users prefer.

Some authors over the last fifty years, like Howard Zahniser—chief architect of the Wilderness Act and former head of The Wilderness Society—warned that "we must not only protect the wilderness from exploitation. We must also see that we do not ourselves destroy its wilderness character in our own management programs. We must remember that the essential quality of the wilderness is its wildness" (Zahniser 1992, 52). The concern for management impacts on wilderness conditions is not a new issue. Cole (2000) puts forward the concept that wilderness can be managed for naturalness, wildness, uncrowded conditions (outstanding opportunities for solitude), or a minimum of management restrictions (unconfined type of recreation), but all four of these attributes of wilderness cannot be simultaneously maximized. Consequently, we support the idea of a wilderness opportunity spectrum whereby management and stewardship approaches seek to maintain diversity within and among wilderness areas—striving always to maintain minimum standards within the guidelines

Fig. 17.4. When a road through the remote Izembeck National Wildlife Refuge on the Alaska Peninsula was proposed, the wilderness character of the area provided the strongest argument against development. Photo courtesy U.S. Fish and Wildlife Service.

of the Wilderness Act and to maintain higher standards of wildness where they existed when an area was designated.

Trends in Wilderness Use

Wilderness recreation use provides quality experience opportunities and benefits to users and challenges to wilderness managers. More is known about recreational use than most other wilderness uses, because it has been more widely studied and measured. From the 1940s through the 1960s, recreation use increased rapidly, but in recent decades, the rate of growth, though still high, has gradually slowed. In the 1980s and early 1990s, recreation use fluctuated in many individual wilderness areas, but total NWPS recreation use increased.

Projections of future use vary widely. The annual percentage of growth in recreation use of wilderness has been small, but total use in the NWPS has grown dramatically due to the growth in the size of the Wilderness System as new areas have been added over the last twenty years. Understanding wilderness recreation use and users is essential to protecting wilderness resources and values while managing conditions that influence wilderness users' experiences.

We expect recreational use of wilderness will continue to grow in the twenty-first century. We expect more demand for day use and easy access; more demand for information to support pretrip planning and area selection; more demand for challenge, adventure, and risk-taking activities; and greater diversity among users (Chavez 2000; Hammitt and Schuster 2000). We are concerned about the trends toward the use of more electronic technology and high-tech equipment that may insulate visitors from truly experiencing wilderness conditions, and about the demand for more "packaged" and service-oriented commercial experiences, requiring less understanding of and connection with nature (Roggenbuck 2000; Shultis 2000).

Fig. 17.5. Units of the National Wilderness Preservation System constitute about 4 percent of the land area of the United States, protect a range of ecosytems, and provide access to wild and varied landscapes. Everglades National Park Wilderness, Florida (upper left); Boundary Waters Canoe Area, Minnesota (upper right); Steens Mountain Wilderness, Oregon (lower left); and Aravaipa Canyon Wilderness, Arizona (lower right). Photos courtesy (upper left) National Park Service, (upper right) U.S. Forest Service, (lower left) Oregon Bureau of Land Management, (lower right) Arizona Bureau of Land Management.

Fig. 17.6. Millions of acres of Bureau of Land Management lands in the Southwest received wilderness designation in the 1990s. Opportunities for wilderness solitude abound (left). Stewardship challenges include removing old mining roads (center) through rehabilitation prohibiting vehicle intrusions (right). Photos courtesy Marilyn Hendee.

Some social benefits are at stake in the kinds of use facilitated or stymied by wilderness management policies as we approach use limits in a growing number of areas. For example, will special permits for organized user groups favor commercial parties, science-education groups, organized youth groups, or wilderness therapy and personal growth groups? Increasingly, such difficult choices are directly influenced by wilderness planners and managers.

Although we are concerned about these pressures that may dilute the natural qualities and conditions of the wilderness experience and may have substantial negative impacts on wilderness, we hope that such pressures will be a force in helping define and strengthen a more biocentric approach to wilderness management and stewardship (Haas and Wells 2000). In the past, the case for more wilderness designation and management was often based on the need to set aside areas because of the increasing recreational demands placed on such natural areas. This argument will surely continue; however, we believe it is time to expand the concern for wilderness, and the kind of wilderness system we need, by placing a greater emphasis on the full spectrum of environmental, scientific, and social values a Wilderness System can deliver. Examples of such values include expanding ecosystem representation; protecting genetic diversity; habitat for wilderness-dependent species, like grizzly bear; and the education, inspiration, and healing experiences that wilderness can provide.

During the twentieth century, increasing recreational use of wilderness provided an important and immediately obvious justification for its protection; it has also been the source of many of the most pressing management challenges. However, the case for wilderness is justified by much more than its recreation use, and more emphasis needs to be placed on its multiple values. While we, as wilderness stewards, are often preoccupied with direct use of wilderness, the public has broader interests in wilderness. Cordell and Stokes (2000) report that in national surveys the top five values of wilderness, reported by 75 percent or more of the populations, are protection for water quality, for wildlife habitat, for air quality, for endangered species, and for future generations.

Good wilderness management depends on an understanding of all the varied uses it receives, because most wilderness stewardship is management of use of one kind or another, but primarily recreational use. Some uses depend on wilderness, but others just take place there and are less dependent on wilderness attributes and conditions. Public recreation is the most obvious wilderness use, but other uses include commercial recreation; indirect, off-site uses like ecological services and benefits from wilderness; scientific uses; education uses; personal growth, therapeutic, and healing uses; and a variety of limited but legally allowable, direct commodity uses such as mining, water storage, and grazing.

Management Funding and Resources

Government financial constraints and demand for cost-effectiveness and efficient delivery of services will undoubtedly continue to increase, along with recognition that wilderness protection is possible only with adequate management and stewardship. We think this will require managers to continue to seek new ways of accomplishing their work, such as forming alliances among different levels of government and among the public, private, and volunteer sectors of society. Many of these approaches are already well under way.

Fig. 17.7. Backpackers gather at information panels at a wilderness trailhead (top); untrammeled wilderness spreads out in front of backpackers in Alaska (middle); a canoeist paddles in Minnesota's Boundary Waters (bottom). Photos courtesy (top) Alan Ewart, (middle) David Cole, (bottom) John Roggenbuck.

(Joslin 2000); and interagency programs have trained scout leaders and leaders of specialized groups such as horse users to educate their constituents in LNT and minimum-impact skills (Miranda 2000). Because partnerships and volunteers will be even more essential in the future, wilderness managers will need more and better new skills in managing and organizing people. Effective work by volunteers and other partners requires strong planning, facilitation, and guidance by professional wilderness managers. Volunteers and partnerships must supplement professional wilderness management, not substitute for it, or wilderness quality will diminish. At the same time, involving partners and volunteers in wilderness work yields important benefits to these workers, as well as to the wilderness they serve. We believe that the strength of support for and appreciation of wilderness in the future will depend on making wilderness experiences and wilderness work opportunities available to as wide a spectrum of people as possible.

An Expanding—but Ultimately Limited—Wilderness System

When the first edition of this book was published in 1978, fewer than 10 million acres of wilderness existed. Now 107.4 million acres, the NWPS will likely grow beyond that. Few people, even among the most dedicated wilderness enthusiasts, foresaw such an outcome. It is further evidence, in our view, of the strength and breadth of support for wilderness in the United States.

Ironically, the rapid expansion of wilderness, coupled with the many laws Congress has passed to create such a system, has raised fears about a gradual deterioration in standards and perhaps a slow, insidious creep of more lenient, less demanding qualities for areas admitted to the NWPS. Yet, by and large, the demanding standards of wilderness management described by the act have been consistently endorsed by the words and actions of Congress. Moreover, although some areas admitted to the NWPS currently do not meet all standards for wilderness, they provide an opportunity for managers to begin the process of restoring wilderness qualities. As the late senator Frank Church (1977) noted thirty years ago, this opportunity to restore wilderness qualities represents the "great promise of the Wilderness Act."

Threats to wilderness integrity and conditions are numerous and will require the vigilance of wilderness

In recent years, many crucial wilderness management jobs—trail construction and maintenance, campsite rehabilitation, visitor contact—have been possible only because of volunteer or partnership participation. For example, trails have been built by conservation groups and local activity clubs (Wash and Wash 2001); volunteers have served as wilderness information specialists providing information to visitors and have collected basic resource inventory data managers, rangers, users, volunteers, outfitters, and the general public. As Robert Marshall (1930, 52) commented more than seven decades ago: "There is just one hope of repulsing the tyrannical ambition of civilization to conquer every niche on the whole earth. That hope is the organization of spirited people who will fight for the freedom of the wilderness." Examples of threats to wilderness abound. Just a few examples include an Alaska Department of Transportation proposal to build a seven-mile-long section of road across the Izembek National Wildlife Refuge through a portion of a designated wilderness area (Clark 2000), proposals to extract oil reserves from under the Arctic National Wildlife Refuge, proposals to dig the Rock Creek mine for silver and copper under the Cabinet Mountains Wilderness in Montana, a proposal to undesignate a portion of the Brigantine Wilderness in New Jersey so off-road vehicles can drive down the beach, and proposals to allow private landowners to use motor vehicles to cross BLM wilderness lands in the Southwest to access their private inholdings. Scores of wilderness plans are contentious and may only be resolved through compromise; unfortunately, the compromises will likely diminish wilderness qualities in some way.

With a growing national and international population, the need and demand for limited resources of all types will only increase in the United States and throughout the world. One of the greatest values of wilderness to future generations will be its relative "naturalness," because 96 percent of the land will have been heavily managed, manipulated, and developed for human uses. *Wilderness in the United States will be increasingly scarce, and the remnants of functioning ecosystems and natural landscapes that legislation protects will be stressed to provide places for escape from escalating pressures of modern life.*

Protecting wilderness from threats to dilute or diminish its naturalness and solitude is the central mission of wilderness management and stewardship. The threats are many, and they can be internal or external to any given wilderness area. We discuss nineteen threats in Chapter 13, too many to summarize here. However, anticipating and managing the short-term and long-term threats to wilderness resources and values are the most important challenge to achieving the purposes of the Wilderness Act for present and future generations. This is the challenge of wilderness management and stewardship.

Conclusion

The future will further affirm the important symbolic values underlying preserving wilderness. In the last two hundred years, the United States has been settled and modified virtually from shore to shore and border to border. At 107.4 million acres, the Wilderness System is, in one sense, vast. Yet, we must remember it constitutes slightly more than 4 percent of the nation's total land area and less than 2 percent of the lower forty-eight states. How much more will be added in future wilderness designations—perhaps another 1 percent, or about 25 million acres, of the United States? Thus, the wilderness values available to our society—historical, ecological, recreational, spiritual, inspirational—are ultimately finite. The availability of naturalness and solitude will depend on what natural land areas we have been able to preserve, and how well we have preserved them. The values of wilderness to present and future generations will change and shift with societal influences, but those values will not diminish. The challenge to provide adequate and appropriate management and stewardship of these protected natural land areas will increase dramatically in proportion to their scarcity.

The Wilderness Act is a historic and important landmark in U.S. history, but the concept of wilderness as a symbol of nature is not fixed in time or legislation—rather it is fixed in the human consciousness. As Olaus Murie (1961, n.p.) noted about the Arctic National Wildlife Range (now the Arctic National Wildlife Refuge): "Certainly a wilderness area, a little portion of our planet left alone, undeveloped, will furnish us with a number of very important uses… We have only begun to understand the basic energies which through the ages have made this planet habitable. If we are wise, we will cherish what we have left of such places in our land." *The ultimate goal of this book is to help ensure that there are some wild places, some wildness, some naturalness, some solitude—some wilderness—for us to enjoy and for future generations to experience.*

References

Chavez, Deborah J. 2000. Wilderness visitors in the 21st century: Diversity, day use, and preferences. *International Journal of Wilderness*. 6(2): 10–11.

Church, Frank. 1977. Wilderness in a balanced land use framework. Wilderness Resource Distinguished Lectureship No. 1. Moscow: University of Idaho Wilderness Research Center.

Clark, Jamie Rappaport. 2000. The value of wilderness to the U.S. National Refuge System. *International Journal of Wilderness*. 6(3): 7–11.

Cole, David N. 2000. Natural, wild, uncrowded, or free? Which of these should our wilderness be? *International Journal of Wilderness*. 6(2): 5–8.

Cordell, Ken; Stokes, Jerry. 2000. The social value of wilderness: A Forest Service perspective. *International Journal of Wilderness*. 6(2): 23–24.

Haas, Glenn; Wells, Marcella. 2000. A more pristine wilderness. *International Journal of Wilderness*. 6(2): 21–22.

Hammitt, William E.; Schuster, Rudy M. 2000. Wilderness use in the next 100 years. *International Journal of Wilderness*. 6(2): 12–13.

Joslin, Les. 2000. A community-based wilderness education partnership in central Oregon. *International Journal of Wilderness*. 6(3): 27–30.

Leopold, Aldo. 1970. *A Sand County Almanac with Essays on Conservation from Round River*. New York: Sierra Club and Ballantine Books. 295p.

Marshall, Robert. 1930. The problems of the wilderness. *The Scientific Monthly*. 43–52.

———. 1937. The universe of the wilderness is vanishing. *Nature Magazine*. 29(4): 235–240.

Martin, Vance. 2001. The World Wilderness Congress. *International Journal of Wilderness*. 7(1): 31–34.

Miranda, Barb. 2000. The central and southern Sierra wilderness education project: An outreach program that works. *International Journal of Wilderness*. 6(3): 35–37.

Murie, Olaus. 1961. Wilderness philosophy, science, and the Arctic National Wildlife Range. In: Dahlgren, G., Jr., ed. *Proceedings: Twelfth Alaskan Science Conference, American Association for the Advancement of Science; August 28–September 1, 1961, Fairbanks.* University of Alaska.

Roggenbuck, Joseph W. 2000. Meanings of wilderness experiences in the 21st century. *International Journal of Wilderness*. 6(2): 14–17.

Shultis, John. 2000. Gearheads and golums: Technology and wilderness recreation in the 21st century. *International Journal of Wilderness*. 6(2): 17–18.

Wash, Dave; Wash, Katie. 2001. The BLM partnership with the Student Conservation Association: Restoring wilderness in the California desert. *International Journal of Wilderness*. 7(1): 30.

Zahniser, Howard. 1956. The need for wilderness areas. *The Living Wilderness*. 21(59): 37–43.

Zahniser, Edward. 1992. *Where Wilderness Preservation Began: Adirondack Writings of Howard Zahniser*. Utica, NY: North Country Books.

Appendix A

The Wilderness Act

Act of September 3, 1964

(Public Law 88-577, 78 Stat. 890; 16 U.S. Code 1121 [note], 1131-1136)

Sec 1. This Act may be cited as the "Wilderness Act" (16 U.S. Code 1121 [note])

WILDERNESS SYSTEM ESTABLISHED STATEMENT OF POLICY

Sec. 2. (a) In order to assure that an increasing population, accompanied by expanding settlement and growing mechanization, does not occupy and modify all areas within the United States and its possessions, leaving no lands designated for preservation and protection in their natural condition, it is hereby declared to be the policy of the Congress to secure for the American people of present and future generations the benefits of an enduring resource of wilderness. For this purpose there is hereby established a National Wilderness Preservation System to be composed of federally owned areas designated by Congress as "wilderness areas," and these shall be administered for the use and enjoyment of the American people in such manner as will leave them unimpaired for future use and enjoyment as wilderness, and so as to provide for the protection of these areas, the preservation of their wilderness character, and for the gathering and dissemination of information regarding their use and enjoyment as wilderness; and no Federal lands shall be designated as "wilderness areas" except as provided for in this chapter or by a subsequent Act.

(b) The inclusion of an area in the National Wilderness Preservation System notwithstanding, the area shall continue to be managed by the Department and agency having jurisdiction thereover immediately before its inclusion in the National Wilderness Preservation System unless otherwise provided by Act of Congress. No appropriation shall be available for the payment of expenses or salaries for the administration of the National Wilderness Preservation System as a separate unit nor shall any appropriations be available for additional personnel stated as being required solely for the purpose of managing or administering areas solely because they are included within the National Wilderness Preservation System.

(c) A wilderness, in contrast with those areas where man and his own works dominate the landscape, is hereby recognized as an area where the earth and its community of life are untrammeled by man, where man himself is a visitor who does not remain. An area of wilderness is further defined to mean in this chapter an area of undeveloped Federal land retaining its primeval character and influence, without permanent improvements or human habitation, which is protected and managed so as to preserve its natural conditions and which (1) generally appears to have been affected primarily by the forces of nature, with the imprint of man's work substantially unnoticeable; (2) has outstanding opportunities for solitude or a primitive and unconfined type of recreation; (3) has at least five thousand acres of land or is of sufficient size as to make practicable its preservation and use in an unimpaired condition; and (4) may also contain ecological, geological, or other features of scientific, educational, scenic, or historical value.

Previously Classified Areas

Sec. 3. (a) All areas within the national forests classified at least 30 days before the effective date of this Act by the Secretary of Agriculture or the Chief of the Forest Service as "wilderness," "wild," or "canoe" are hereby designated as wilderness areas. The Secretary of Agriculture shall—

(1) Within one year after the effective date of this Act file a map and legal description of each wilderness area with the Interior and Insular Affairs Committees of the United States Senate and the House of Representatives, and such descriptions shall have the same force and effect as if included in this Act: Provided however, that, correction of clerical and typographical errors in such legal descriptions and maps may be made.

(2) Maintain, available to the public, records pertaining to said wilderness areas, including maps and legal

descriptions, copies of regulations governing them, copies of public notices of, and reports submitted to Congress regarding pending additions, eliminations, or modifications. Maps, legal descriptions, and regulations pertaining to wilderness areas within their respective jurisdictions also shall be available to the public in the offices of regional foresters, national forest supervisors, and forest rangers.

(b) The Secretary of Agriculture shall, within ten years after the enactment of this Act, review, as to its suitability or nonsuitability for preservation as wilderness, each area in the national forests classified on the effective date of this Act by the Secretary of Agriculture or the Chief of the Forest Service as "primitive" and report his findings to the President. The President shall advise the United States Senate and House of Representatives of his recommendations with respect to the designation as "wilderness" or other reclassification of each area on which review has been completed, together with maps and a definition of boundaries. Such advice shall be given with respect to not less than one-third of all the areas now classified as "primitive" within three years after the enactment of this Act, not less than two-thirds within seven years after the enactment of this Act, and the remaining areas within ten years after the enactment of this Act. Each recommendation of the President for designation as "wilderness" shall become effective only if so provided by an Act of Congress. Areas classified as "primitive" on the enactment of this Act shall continue to be administered under the rules and regulations affecting such areas on the enactment of this Act until Congress has determined otherwise. Any such area may be increased in size by the President at the time he submits his recommendations to the Congress by not more than five thousand acres with no more than one thousand two hundred and eighty acres of such increase in any one compact unit; if it is proposed to increase the size of any such area by more than five thousand acres or by more than one thousand two hundred and eighty acres in any one compact unit the increase in size shall not become effective until acted upon by Congress. Nothing herein contained shall limit the President in proposing, as part of his recommendations to Congress, the alteration of existing boundaries of primitive areas or recommending the addition of any contiguous area of national forest lands predominantly of wilderness value. Notwithstanding any other provisions of this chapter, the Secretary of Agriculture may complete his review and delete such area as may be necessary, but not to exceed seven thousand acres, from the southern tip of the Gore Range–Eagles Nest Primitive Area, Colorado, if the Secretary determines that such action is in the public interest.

(c) Within ten years after the effective date of this Act the Secretary of the Interior shall review every roadless area of five thousand contiguous acres or more in the national parks, monuments and other units of the national park system and every such area of, and every roadless island within the national wildlife refuges and game ranges, under his jurisdiction on the effective date of this Act and shall report to the President his recommendation as to the suitability or nonsuitability of each such area or island for preservation as wilderness. The President shall advise the President of the Senate and the Speaker of the House of Representatives of his recommendation with respect to the designation as wilderness of each such area or island on which review has been completed, together with a map thereof and a definition of its boundaries. Such advice shall be given with respect to not less than one-third of the areas and islands to be reviewed under this subsection within three years after enactment of this Act, not less than two-thirds within seven years of enactment of this Act. A recommendation of the President for designation as wilderness shall become effective only if so provided by an Act of Congress. Nothing contained herein shall, by implication or otherwise, be construed to lessen the present statutory authority of the Secretary of the Interior with respect to the maintenance of roadless areas within units of the national park system.

(d) (1) The Secretary of Agriculture and the Secretary of the Interior shall, prior to submitting any recommendations to the President with respect to the suitability of any area for preservation as wilderness—

>(A) give such public notice of the proposed action as they deem appropriate, including publication in the Federal Register and in a newspaper having general circulation in the area or areas in the vicinity of the affected land;

>(B) hold a public hearing or hearings at a location or locations convenient to the area affected. The hearings shall be announced through such means as the respective Secretaries involved deem appropriate, including notices in the Federal Register and in newspapers of general circulation in the area: *Provided*, that if the lands involved are located in more than one State, at least one hearing shall be held in each State in which a portion of the land lies;

>(C) at least thirty days before the date of a hearing advise the Governor of each State and the governing board of each county, or in Alaska the borough, in which the lands are located, and Federal departments and agencies concerned, and invite such officials and Federal agencies to submit their views on the proposed action at the hearing or by no later than thirty days following the date of the hearing.

(2) Any views submitted to the appropriate Secretary under the provisions of (1) of this subsection with respect to any area shall be included with any recommendations to the President and to Congress with respect to such area.

(e) Any modification or adjustment of boundaries of any wilderness area shall be recommended by the appropriate Secretary after public notice of such proposal and public hearing or hearings as provided in subsection (d) of this section. The proposed modification or adjustment shall then be recommended with map and description thereof to the President. The President shall advise the United States Senate and the House of Representatives of his recommendations with respect to such modification or adjustment and such recommendations shall become effective only in the same manner as provided for in subsections (b) and (c) of this section.

LIMITATION OF USE AND ACTIVITIES

Sec. 4. (a) The purposes of this chapter are hereby declared to be within and supplemental to the purposes for which national forests and units of the national park and national wildlife refuge systems are established and administered and—

(1) Nothing in this Act shall be deemed to be in interference with the purpose for which national forests are established as set forth in the Act of June 4, 1897 (30 Stat. 11), and the Multiple-Use Sustained-Yield Act of June 12, 1960 (74 Stat. 215).

(2) Nothing in this Act shall modify the restrictions and provisions of the Shipstead-Nolan Act (Public Law 539, Seventy-first Congress, July 10, 1930; 46 Stat. 1020), the Thye-Blatnik Act (Public Law 733, Eightieth Congress, June 22, 1948; 62 Stat. 568), and the Humphrey-Thye-Blatnik-Andresen Act (Public Law 607, Eighty-fourth Congress, June 22, 1956; 70 Stat. 326), as applying to the Superior National Forest or the regulations of the Secretary of Agriculture.

(3) Nothing in this Act shall modify the statutory authority under which units of the national park system are created. Further, the designation of any area of any park, monument, or other unit of the national park system as a wilderness area pursuant to this Act shall in no manner lower the standards evolved for the use and preservation of such park, monument, or other unit of the national park system in accordance with the Act of August 25, 1916, the statutory authority under which the area was created, or any other Act of Congress which might pertain to or affect such area, including, but not limited to, the Act of June 8, 1906 (34 Stat. 225; 16 U.S.C. 432 et seq.); section 3(2) of the Federal Power Act (16 U.S.C. 796[2]); and the Act of August 21, 1935 (49 Stat. 666; 16 U.S.C. 461 et seq.).

(b) Except as otherwise provided in this Act, each agency administering any area designated as wilderness shall be responsible for preserving the wilderness character of the area and shall so administer such area for such other purposes for which it may have been established as also to preserve its wilderness character. Except as otherwise provided in this Act, wilderness areas shall be devoted to the public purposes of recreational, scenic, scientific, educational, conservation, and historical use.

(c) Except as specifically provided for in this Act, and subject to existing private rights, there shall be no commercial enterprise and no permanent road within any wilderness area designated by this Act and, except as necessary to meet minimum requirements for the administration of the area for the purpose of this Act (including measures required in emergencies involving the health and safety of persons within the area), there shall be no temporary road, no use of motor vehicles, motorized equipment or motorboats, no landing of aircraft, no other form of mechanical transport, and no structure or installation within any such area.

(d) The following special provisions are hereby made:

(1) Within wilderness areas designated by this Act the use of aircraft or motorboats, where these uses have already become established, may be permitted to continue subject to such restrictions as the Secretary of Agriculture deems desirable. In addition, such measures may be taken as may be necessary in the control of fire, insects, and diseases, subject to such conditions as the Secretary deems desirable.

(2) Nothing in this Act shall prevent within national forest wilderness areas any activity, including prospecting, for the purpose of gathering information about mineral or other resources, if such activity is carried on in a manner compatible with the preservation of the wilderness environment. Furthermore, in accordance with such program as the Secretary of the Interior shall develop and conduct in consultation with the Secretary of Agriculture, such areas shall be surveyed on a planned, recurring basis consistent with the concept of wilderness preservation by the United States Geological Survey and the United States Bureau of Mines to determine the mineral values, if any, that may be present; and the results of such surveys shall be made available to the public and submitted to the President and Congress.

(3) Notwithstanding any other provisions of this Act, until midnight December 31, 1983, the United

States mining laws and all laws pertaining to mineral leasing shall, to the same extent as applicable prior to the effective date of this Act, extend to those national forest lands designated by this Act as "wilderness areas"; subject, however, to such reasonable regulations governing ingress and egress as may be prescribed by the Secretary of Agriculture consistent with the use of the land for mineral location and development and exploration, drilling, and production, and use of land for transmission lines, waterlines, telephone lines, or facilities necessary in exploring, drilling, producing, mining, and processing operations, including where essential the use of mechanized ground or air equipment and restoration as near as practicable of the surface of the land disturbed in performing prospecting, location, and, in oil and gas leasing, discovery work, exploration, drilling, and production, as soon as they have served their purpose. Mining locations lying within the boundaries of said wilderness areas shall be held and used solely for mining or processing operations and uses reasonably incident thereto; and hereafter, subject to valid existing rights, all patents issued under the mining laws of the United States affecting national forest lands designated by this Act as wilderness areas shall convey title to the mineral deposits within the claim, together with the right to cut and use so much of the mature timber therefrom as may be needed in the extraction, removal, and beneficiation of the mineral deposits, if needed timber is not otherwise reasonably available, and if the timber is cut under sound principles of forest management as defined by the national forest rules and regulations, but each such patent shall reserve to the United States all title in or to the surface of the lands and products thereof, and no use of the surface of the claim or the resources therefrom not reasonably required for carrying on mining or prospecting shall be allowed except as otherwise expressly provided in this Act: Provided, that, unless hereafter specifically authorized, no patent within wilderness areas designated by this Act shall issue after December 31, 1983, except for the valid claims existing on or before December 31, 1983. Mining claims located after the effective date of this Act within the boundaries of wilderness areas designated by this Act shall create no rights in excess of those rights which may be patented under the provisions of this subsection. Mineral leases, permits, and licenses covering lands within national forest wilderness areas designated by this Act shall contain such reasonable stipulations as may be prescribed by the Secretary of Agriculture for the protection of the wilderness character of the lands consistent with the use of the lands for the purposes for which they are leased, permitted, or licensed. Subject to valid rights then existing, effective January 1, 1984, the minerals in lands designated by this Act as wilderness areas are withdrawn from all forms of appropriation under the mining laws and from disposition under all laws pertaining to mineral leasing and all amendments thereto.

(4) Within wilderness areas in the national forests designated by this Act, (1) the President may, within a specific area and in accordance with such regulations as he may deem desirable, authorize prospecting for water resources, the establishment and maintenance of reservoirs, water-conservation works, power projects, transmission lines, and other facilities needed in the public interest, including the road construction and maintenance essential to development and use thereof, upon his determination that such use or uses in the specific area will better serve the interests of the United States and the people thereof than will its denial; and (2) the grazing of livestock, where established prior to the effective date of this Act shall be permitted to continue subject to such reasonable regulations as are deemed necessary by the Secretary of Agriculture.

(5) Other provisions of this Act to the contrary notwithstanding, the management of the Boundary Waters Canoe Area, formerly designated as the Superior, Little Indian Sioux and Caribou Roadless Areas, in the Superior National Forest, Minnesota, shall be in accordance with regulations established by the Secretary of Agriculture in accordance with the general purpose of maintaining, without unnecessary restrictions on other uses, including that of timber, the primitive character of the area, particularly in the vicinity of the lakes, streams, and portages: Provided, that nothing in this Act shall preclude the continuance within the area of any already established use of motorboats.

(6) Commercial services may be performed within the wilderness areas designated by this Act to the extent necessary for activities which are proper for realizing the recreational or other wilderness purposes of the areas.

(7) Nothing in this Act shall constitute an express or implied claim or denial on the part of the Federal Government as to exemption from State water laws.

(8) Nothing in this chapter shall be construed as affecting the jurisdiction or responsibilities of the several States with respect to wildlife and fish in the national forests. (16 U.S.C. 1133)

RIGHTS OF NON-FOREST LANDS OWNERSHIP

Sec. 5. (a) In any case where State-owned or privately

owned land is completely surrounded by national forest lands within areas designated by this Act as wilderness, such State or private owner shall be given such rights as may be necessary to assure adequate access to such State-owned or privately owned land by such State or private owner and their successors in interest, or the State-owned land or privately owned land shall be exchanged for federally owned land in the same State of approximately equal value under authorities available to the Secretary of Agriculture: *Provided, however,* that the United States shall not transfer to a State or private owner any mineral interests unless the State or private owner relinquishes or causes to be relinquished to the United States the mineral interest in the surrounded land.

(b) In any case where valid mining claims or other valid occupancies are wholly within a designated national forest wilderness area, the Secretary of Agriculture shall, by reasonable regulations consistent with the preservation of the area as wilderness, permit ingress and egress to such surrounded areas by means which have been or are being customarily enjoyed with respect to other such areas similarly situated.

(c) Subject to the appropriation of funds by Congress, the Secretary of Agriculture is authorized to acquire privately owned land within the perimeter of any area designated by this chapter as wilderness if (1) the owner concurs in such acquisition or (2) the acquisition is specifically authorized by Congress. (16 U.S.C. 1134)

Sec. 6. (a) The Secretary of Agriculture may accept gifts or bequests of land within wilderness areas designated by this Act for preservation as wilderness. The Secretary of Agriculture may also accept gifts or bequests of land adjacent to wilderness areas designated by this Act for preservation as wilderness if he has given sixty days' advance notice thereof to the President of the Senate and the Speaker of the House of Representatives. Land accepted by the Secretary of Agriculture under this section shall become part of the wilderness area involved. Regulations with regard to any such land may be in accordance with such agreements, consistent with the policy of this Act, as are made at the time of such gift, or such conditions, consistent with such policy, as may be included in, and accepted with, such bequest.

(b) The Secretary of Agriculture or the Secretary of the Interior is authorized to accept private contributions and gifts to be used to further the purposes of this Act. (16 U.S.C. 1135)

REPORT TO CONGRESS

Sec. 7. At the opening of each session of Congress, the Secretaries of Agriculture and Interior shall jointly report to the President for transmission to Congress on the status of the wilderness system, including a list and descriptions of the areas in the system, regulations in effect, and other pertinent information, together with any recommendations they may care to make. (16 U.S.C. 1136)

APPROVED SEPTEMBER 3, 1964

Legislative History:

House Reports: No. 1538 accompanying H.R. 9070 (Committee on Interior & Insular Affairs) and No. 1829 (Committee of Conference).

Senate Report: No. 109 (Committee on Interior & Insular Affairs). Congressional Record: Vol. 109 (1963):

April 4, 8, considered in Senate.

April 9, considered and passed Senate.

Vol. 110 (1964): July 28, considered in House.

July 30, considered and passed House, amended, in lieu of H.R. 9070.

August 20, House and Senate agreed to conference report.

Appendix B
Chronological Listing of Wilderness Designation and Wilderness-Related Laws

All 171 public laws designating and affecting wilderness are listed in chronological order from 1964 through December 2007. Because many of the acts do not have titles, clarifications regarding wilderness or other designations are indicated in parentheses. Each law is followed by a list of up to four potential items, separated by a (•): states affected by that law; management agencies designated by that law; the number of Wilderness Study Areas (WSAs) designated, if any; and whether special provisions apply. The term *special provisions*, as used here, refers to specific management directions that apply only to the wildernesses designated by the particular act. These usually reflect unique circumstances and may be exceptions to provisions of the Wilderness Act of 1964. If one of these items is not included, it does not apply to that law. For example, the first law listed, the Wilderness Act, affected twelve states, designated the USDA Forest Service as the managing agency for all these wildernesses, designated no WSAs, and included special management provisions. (Adapted and expanded from: [1] Peter Landres and Shannon Meyer. 2000. National Wilderness Preservation System Database: Key Attributes and Trends, 1994–1999, 2nd ed. rev. RMRS-GTR-18, Fort Collins, CO: U.S. Department of Agriculture, Forest Service, Rocky Mountain Research Station; and [2] James A. Browning, John C. Hendee, and Joe W. Roggenbuck. 1988. 103 Wilderness Laws: Milestones and Management Direction, in Wilderness Legislation, 1964–1987; Bulletin 51. Moscow: University of Idaho, Idaho Forest, Wildlife and Range Experiment.)

U.S. Public Law 88-577. The Wilderness Act of September 3, 1964. 78 Stat. 890. *AZ, CA, CO, ID, MT, NV, NH, NM, NC, OR, WA, WY • FS • Special Provisions*

U.S. Public Law 90-271. (San Rafael Wilderness) Act of March 21, 1968. 82 Stat. 51. *CA • FS*

U.S. Public Law 90-318. (San Gabriel Wilderness) Act of May 24, 1968. 82 Stat. 131. *CA • FS*

U.S. Public Law 90-532. (Great Swamp National Wildlife Refuge Wilderness) Act of September 28, 1968. 82 Stat. 883. *NJ • FWS*

U.S. Public Law 90-544. (North Cascades National Park, Ross Lake and Lake Chelan NRAs, Paseyten Wilderness, and Glacier Park Wilderness Modification) Act of October 2, 1968. 82 Stat. 930. *WA • FS • 1 WSA*

U.S. Public Law 90-548. (Mount Jefferson Wilderness) Act of October 2, 1968. 82 Stat. 936. *OR • FS*

U.S. Public Law 91-58. (Ventana Wilderness) Act of August 16, 1969. 83 Stat. 101. *CA • FS*

U.S. Public Law 91-82. (Desolation Wilderness) Act of October 10, 1969. 83 Stat. 131. *CA FS • Special Provisions*

U.S. Public Law 91-479. (Sleeping Bear Dunes National Lakeshore Designation and Wilderness Study) Act of October 21, 1970. 84 Stat. 1075. *MI • NPS*

U.S. Public Law 91-504. (Omnibus Wildlife Refuge, National Park, and National Forest Wilderness) Act of October 23, 1970. 84 Stat. 1104. *AK, FL, MA, ME, MI, NM, OK, OR, WA, WI • FS, FWS, NPS*

U.S. Public Law 91-660. (Gulf Islands National Seashore Designation and Wilderness Study) Act of January 8, 1971. 84 Stat. 1967. *FL, MS • NPS*

U.S. Public Law 91-661. (Voyageurs National Park Designation and Wilderness Study) Act of January 8, 1971. 84 Stat. 1970. *MN • NPS*

U.S. Public Law 92-154. (Canyonlands National Park Boundary Adjustment and Wilderness Study) Act of November 12, 1971. 85 Stat. 421. *UT • NPS*

U.S. Public Law 92-155. (Arches National Park Designation and Wilderness Study) Act of November 12, 1971. 85 Stat. 422. *UT • NPS*

U.S. Public Law 92-207. (Capitol Reef National Park Designation and Wilderness Study) Act of December 18, 1971. 85 Stat. 739. *UT • NPS*

U.S. Public Law 92-230. (Pine Mount Wilderness) Act of February 15, 1972. 86 Stat. 38. *AZ • FS*

U.S. Public Law 92-237. (Buffalo National River Designation and Wilderness Study) Act of March 1, 1972. 86 Stat. 44. *AK • NPS*

U.S. Public Law 92-241. (Sycamore Canyon Wilderness) Act of March 6, 1972. 86 Stat. 48. *AZ • FS*

U.S. Public Law 92-260. (Oregon Dunes NRA Designation and Wilderness Study) Act of March 23, 1972. 86 Stat. 99. *OR • FS*

U.S. Public Law 92-364. (Cedar Keys Wilderness) Act of August 7, 1972. 86 Stat. 505. *FL • FWS*

U.S. Public Law 92-395. (Scapegoat Wilderness) Act of August 20, 1972. 86 Stat. 578. *MT • FS*

U.S. Public Law 92-400. (Sawtooth NRA Wilderness Designation and Wilderness Study) Act of August 22, 1972. 86 Stat. 612. *ID • FS • 1 WSA • Special Provisions*

U.S. Public Law 92-476. (South Absaroka to Washakie Wilderness Name Change and Stratified Primitive Area Inclusion) Act of October 9, 1972. 86 Stat. 792. *WY • FS • Special Provisions*

U.S. Public Law 92-493. (Lava Beds Wilderness Designation) Act of October 13, 1972. 86 Stat. 811. *CA • NPS*

U.S. Public Law 92-510. (Lassen Volcanic Wilderness) Act of October 19, 1972. 86 Stat. 918. *CA • NPS • Special Provisions*

U.S. Public Law 92-521. (Eagle Cap Wilderness Addition and Wilderness Study) Act of October 21, 1972. 86 Stat. 1026. *OR • FS • 1 WSA*

U.S. Public Law 92-528. (Indian Peaks Wilderness Study) Act of October 21, 1972. 86 Stat. 1050. *CO • FS*

U.S. Public Law 92-536. (Cumberland Island National Seashore Designation and Wilderness Study) Act of October 23, 1972. 86 Stat. 1066. *GA • NPS*

U.S. Public Law 92-593. (Glen Canyon NRA and Wilderness Study) Act of October 27, 1972. 86 Stat. 1311. *UT, AZ • NPS*

U.S. Public Law 93-429. (Okefenokee Wilderness) Act of October 1, 1974. 88 Stat. 1179. *GA • FWS • Special Provisions*

U.S. Public Law 93-439. (Big Thicket National Preserve and Wilderness Study) Act of October 11, 1974. 88 Stat. 1254. *TX • NPS*

U.S. Public Law 93-440. (Big Cypress National Preserve and Wilderness Study) Act of October 11, 1974. 88 Stat. 1257. *FL • NPS*

U.S. Public Law 93-477. (NPS Adjustments and Cape Lookout National Seashore Wilderness Study) Act of October 26, 1974. 88 Stat. 1445. *NC • NPS*

U.S. Public Law 93-550. (Farallon Islands Wilderness Designation and Point Reyes National Seashore Addition) Act of December 26, 1974. 88 Stat. 1744. *CA • FWS*

U.S. Public Law 93-620. The Grand Canyon National Park Enlargement (and Wilderness Study) Act of January 3, 1975. 88 Stat. 2089. *AZ • NPS*

U.S. Public Law 93-622. (Eastern Wilderness) Act of January 3, 1975. 88 Stat. 2096. *AL, AK, FL, KT, NH, NC, TN, SC, GA, VT, VA, WV, WI • FS • 17 WSAs • Special Provisions*

U.S. Public Law 93-626. (Cape Canaveral National Seashore Designation and Wilderness Study) Act of January 3, 1975. 88 Stat. 2121. *FL • NPS*

U.S. Public Law 93-632. (National Wildlife Refuge and National Forest Omnibus Wilderness Designation and Emigrant Basin Wilderness Study) Act of January 3, 1975. 88 Stat. 2154. *AK, CA, FL, GA, LA, ME, MT, ND, NJ, NM, OH, SC • FWS, FS • 1 WSA*

U.S. Public Law 94-31. (Grand Canyon National Park Enlargement Act Amendment for Wilderness Study) Act of June 10, 1975. 89 Stat. 172. *AZ • NPS*

U.S. Public Law 94-146. (Flat Tops Wilderness) Act of December 12, 1975. 89 Stat. 802. *CO • FS*

U.S. Public Law 94-199. (Hells Canyon NRA and Wilderness Designation and Wilderness Study) Act of December 31, 1975. 89 Stat. 1117. *OR, ID • FS • Special Provisions*

U.S. Public Law 94-268. (Bristol Cliffs Wilderness boundary modification) Act of April 16, 1976. 90 Stat. 370. *VT • FS*

U.S. Public Law 94-352. (Eagles Nest Wilderness) Act of July 12, 1976. 90 Stat. 870. *CO • FS*

U.S. Public Law 94-357. Alpine Lakes Area Management (and Wilderness Designation) Act of 1976. July 12, 1976. 90 Stat. 2949. *WA • FS • Special Provisions*

U.S. Public Law 94-544. (Point Reyes Wilderness) Act of October 18, 1976. 90 Stat. 2515. *CA • NPS*

U.S. Public Law 94-557. (National Wildlife Refuge and National Forest Omnibus Wilderness Designation and Wilderness Study) Act of October 19, 1976. 90 Stat. 2633. *AK, AR, FL, IL, LA, MN, MO, MT, NE, NC, WA, WY • FS, FWS • 8 WSAs*

U.S. Public Law 94-567. (National Park and Monument Omnibus Wilderness Designation and Coronado NF Wilderness Study) Act of October 20, 1976. 90 Stat. 2693. *AZ, CA, CO, HI, MI, NM, SD, VA, WY • FS, NPS • 1 WSA • Special Provisions*

U.S. Public Law 94-579. (Wilderness Study) The Federal Land Policy and Management Act (FLPMA) of 1976. 90 Stat. 2743.

U.S. Public Law 95-150. The Montana Wilderness Study Act of November 1, 1977. 91 Stat. 1243. *MT • FS*

U.S. Public Law 95-237. Endangered American Wilderness Act of 1978. February 24, 1978. 92 Stat. 40. *AZ, CA, CO, ID, MT, NM, OR, UT, WA, WY • FS • 1 WSA • Special Provisions*

U.S. Public Law 95-249. (Absaraka-Beartooth Wilderness) Act of March 27, 1978. 92 Stat. 162. *MT • FS*

U.S. Public Law 95-450. The Indian Peaks Wilderness Area, the Arapaho National Recreation Area, and the Oregon Islands Wilderness Area Act of October 11, 1978. 92 Stat. 1095. *CO, OR • FS, FWS • Special Provisions*

U.S. Public Law 95-494. (Blackjack Springs and Whisker Lake Wilderness) Act of October 21, 1978. 92 Stat. 1648. *WI • FS*

U.S. Public Law 95-495. (Boundary Waters Canoe Area Wilderness) Act of October 21, 1978. 92 Stat. 1649. *MN • FS • Special Provisions*

U.S. Public Law 95-546. (Great Bear Wilderness and Bob

Marshall Wilderness Addition) Act of October 28, 1978. 92 Stat. 2062. *MT • FS*

U.S. Public Law 95-614. (Sandia Mountain Wilderness addition on Cibola National Forest) Act of November 8, 1978. 92 Stat. 3095. *NM • FS*

U.S. Public Law 95-625. National Parks and Recreation Act of November 10, 1978. 92 Stat. 3489. *AR, AZ, FL, HI, MS, ND, NM, TX • NPS • 1 WSA • Special Provisions*

U.S. Public Law 96-248. (Sandia Mountain Wilderness Purchase Addition) Act of May 23, 1980. 94 Stat. 355. *NM • FS • Special Provisions*

U.S. Public Law 96-312. Central Idaho Wilderness Act of July 23, 1980. 94 Stat. 2371. *ID • FS • Special Provisions*

U.S. Public Law 96-476. Rattlesnake National Recreation Area and Wilderness Act of October 19, 1980. 94 Stat. 2271. *MT • FS • Special Provisions*

U.S. Public Law 96-550. (New Mexico Wilderness) Act of December 19, 1980. 94 Stat. 3221. *NM • FS • 6 WSAs*

U.S. Public Law 96-487. Alaska National Interest Lands Conservation Act (ANILCA) of December 20, 1980. 94 Stat. 2371. *AK • FS, FWS, NPS • Special Provisions*

U.S. Public Law 96-560. (Colorado Wilderness) Act of December 22, 1980. 94 Stat. 3266. *CO, LA, MO, SC, SD • FS • 10 WSAs • Special Provisions*

U.S. Public Law 96-585. (Fire Island Wilderness) Act of December 23, 1980. 94 Stat. 3379. *NY • NPS • Special Provisions*

U.S. Public Law 97-211. (Florida Keys Wilderness Addition and Deletion) Act of June 30, 1982. 96 Stat. 141. *FL • FWS*

U.S. Public Law 97-250. (Cumberland Island Wilderness) Act of September 8, 1982. 96 Stat. 709. *GA • NPS*

U.S. Public Law 97-283. (Sandia Mountain Wilderness Boundary Adjustment and Addition) Act of October 5, 1982. 96 Stat. 1215. *NM • FS*

U.S. Public Law 97-384. (Charles C. Dean Wilderness) Act of December 22, 1982. 96 Stat. 1942. *IN • FS • Special Provisions*

U.S. Public Law 97-407. Paddy Creek Wilderness Act of January 3, 1983. 96 Stat. 2033. *MO • FS • Special Provisions*

U.S. Public Law 97-411. Cheaha Wilderness Act of January 3, 1983. 96 Stat. 2046. *AL • FS*

U.S. Public Law 97-466. (West Virginia Wilderness) Act of January 13, 1983. 96 Stat. 2538. *WY • FS • Special Provisions*

U.S. Public Law 98-140. Lee Metcalf Wilderness and Management Act of October 31, 1983. 97 Stat. 903. *MT • BLM, FS, FWS • Special Provisions*

U.S. Public Law 98-141. Public Lands and National Parks (and Colorado Wilderness Study Areas) Act of October 31, 1983. 97 Stat. 909.

U.S. Public Law 98-231. (To Add Frank Church to the River of No Return Wilderness Name) Act of March 14, 1984. 98 Stat. 60. *ID • FS*

U.S. Public Law 98-289. Irish Wilderness Act of May 21, 1984. 98 Stat. 199. *MO • FS*

U.S. Public Law 98-321. Wisconsin Wilderness Act of June 19, 1984. 98 Stat. 250. *WI • FS • Special Provisions*

U.S. Public Law 98-322. Vermont Wilderness Act of June 19, 1984. 98 Stat. 253. *VT • FS • Special Provisions*

U.S. Public Law 98-323. New Hampshire Wilderness Act of June 19, 1984. 98 Stat. 259. *NH • FS*

U.S. Public Law 98-324. North Carolina Wilderness Act of June 19, 1984. 98 Stat. 263. *NC • FS • 5 WSAs*

U.S. Public Law 98-328. Oregon Wilderness Act of June 26, 1984. 98 Stat. 272. *OR • BLM, FS • Special Provisions*

U.S. Public Law 98-339. Washington State Wilderness Act of July 3, 1984. 98 Stat. 299. *WA • BLM, FS • Special Provisions*

U.S. Public Law 98-406. Arizona Wilderness Act of August 28, 1984. 98 Stat. 1485. *AZ • BLM, FS • 3 WSAs • Special Provisions*

U.S. Public Law 98-425. California Wilderness Act of September 28, 1984. 98 Stat. 1619. *CA • FS, NPS • 3 WSAs • Special Provisions*

U.S. Public Law 98-428. Utah Wilderness Act of September 28, 1984. 98 Stat. 1657. *UT • FS • Special Provisions*

U.S. Public Law 98-430. Florida Wilderness Act of September 28, 1984. 98 Stat. 1665. *FL • FS • 2 WSAs • Special Provisions*

U.S. Public Law 98-508. Arkansas Wilderness Act of October 19, 1984. 98 Stat. 2349. *AR • FS • Special Provisions*

U.S. Public Law 98-514. Georgia Wilderness Act of October 19, 1984. 98 Stat. 2416. *GA • FS*

U.S. Public Law 98-515. Mississippi National Forest Wilderness Act of October 19, 1984. 98 Stat. 2420. *MS • FS*

U.S. Public Law 98-550. Wyoming Wilderness Act of October 30, 1984. 98 Stat. 2807. *WY • FS • Special Provisions*

U.S. Public Law 98-574. Texas Wilderness Act of October 30, 1984. 98 Stat. 3051. *TX • FS • Special Provisions*

U.S. Public Law 98-578. Tennessee Wilderness Act of October 30, 1984. 98 Stat. 3088. *TN • FS • 2 WSAs*

U.S. Public Law 98-585. Pennsylvania Wilderness Act of October 30, 1984. 98 Stat. 3100. *PA • FS • Special Provisions*

U.S. Public Law 98-586. Virginia Wilderness Act of October 30, 1984. 98 Stat. 3105. *VA • FS • 4 WSAs • Special Provisions*

U.S. Public Law 98-603. San Juan Basin Wilderness Protection Act of October 30, 1984. 98 Stat. 3155. *NM • BLM, FS • Special Provisions*

U.S. Public Law 99-68. (To Designate the Wilderness in Point Reyes National Seashore as the Philip Burton

Wilderness) Act of July 19, 1985. 99 Stat. 166. *CA • NPS*

U.S. Public Law 99-197. Kentucky Wilderness Act of December 23, 1985. 99 Stat. 1351. *KY • FS*

U.S. Public Law 99-490. Tennessee Wilderness Act of October 16, 1986. 100 Stat. 1235. *TN • FS*

U.S. Public Law 99-504. Nebraska Wilderness Act of October 20, 1986. 100 Stat. 1802. *NE • FS • Special Provisions*

U.S. Public Law 99-555. Georgia Wilderness Act of October 27, 1986. 100 Stat. 3129. *GA • FS*

U.S. Public Law 99-584. (Boundary Adjustments) Texas Wilderness Act Amendments of October 29, 1986. 100 Stat. 3322. *TX • FS*

U.S. Public Law 99-635. (Olympic National Park Wilderness Boundary Adjustments/Deletions) Act of November 7, 1986. 100 Stat. 3528. *WA • FS*

U.S. Public Law 100-184. Michigan Wilderness Act of December 8, 1987. 101 Stat. 1275. *MI • FS • Special Provisions*

U.S. Public Law 100-225. (El Malpais and Cebolla Wilderness) Act of December 31, 1987. 101 Stat. 1539. *NM • BLM • 1 WSA • Special Provisions*

U.S. Public Law 100-326. (Rough Mountain, Rich Hole, Barbours Creek, Shawvers Run, Lewis Fork, and Mountain Lake Wildernesses) Act of June 7, 1988. 102 Stat. 584. *VA, WV • FS*

U.S. Public Law 100-499. Winding Stair Mountain National Recreation and Wilderness Area Act of October 18, 1988. 102 Stat. 2491. *OK • FS • Special Provisions*

U.S. Public Law 100-524. Congaree Swamp National Monument Expansion and Wilderness Act of October 24, 1988. 102 Stat. 2606. *SC • NPS*

U.S. Public Law 100-547. Sipsey Wild and Scenic River and Alabama Addition Act of October 28, 1988. 102 Stat. 2736. *AL • FS • Special Provisions*

U.S. Public Law 100-668. Washington Park Wilderness Act of November 16, 1988. 102 Stat. 3961. *WA • NPS • Special Provisions*

U.S. Public Law 101-85. (To commemorate the twenty-fifth anniversary of the Wilderness Act of 1964, which established the National Wilderness Preservation System) Act of August 14, 1989. 103 Stat. 592. *US*

U.S. Public Law 101-184. (To commemorate the contributions of Senator Clinton P. Anderson to the establishment of the National Wilderness Preservation System, and for other purposes) Act of November 28, 1989. 103 Stat. 1334. *US*

U.S. Public Law 101-195. Nevada Wilderness Protection Act of December 5, 1989. 103 Stat. 1784. *NV • FS • Special Provisions*

U.S. Public Law 101-336. Americans with Disabilities Act of 1990 of July 26, 1990. 104 Stat. 328. *US • FS, BLM, FWS, NPS • Special Provisions*

U.S. Public Law 101-378. Admiralty Island National Monument Land Management Act of August 17, 1990. 104 Stat. 470. *AK • FS*

U.S. Public Law 101-401. Maine Wilderness Act of September 28, 1990. 104 Stat. 863. *ME • FS • Special Provisions*

U.S. Public Law 101-512. Department of the Interior and Related Agencies Appropriations Act, 1991 of November 5, 1990. 104 Stat. 1915. *WV • FS • Special Provisions*

U.S. Public Law 101-612. Smith River National Recreation Act of November 16, 1990. 104 Stat. 3209. *CA • FS*

U.S. Public Law 101-626. Tongass Timber Reform Act of November 28, 1990. 104 Stat. 4426. *AK • FS*

U.S. Public Law 101-628. Arizona Desert Wilderness Act of November 28, 1990. 104 Stat. 4471. *AZ • BLM, FWS • Special Provisions*

U.S. Public Law 101-633. Illinois Wilderness Act of November 28, 1990. 104 Stat. 4577. *IL • FS • Special Provisions*

U.S. Public Law 102-217. Chattahoochee National Forest Protection Act of December 11, 1991. 105 Stat. 1667. *GA • FS*

U.S. Public Law 102-301. Los Padres Condor Range and River Protection Act of June 19, 1992. 106 Stat. 242. *CA • FS • Special Provisions*

U.S. Public Law 103-77. Colorado Wilderness Act of August 13, 1993. 107 Stat. 756. *CO • BLM, FS • Special Provisions*

U.S. Public Law 103-433. California Desert Protection Act of October 31, 1994. 108 Stat. 4473. *CA • BLM, FWS, NPS • Special Provisions*

U.S. Public Law 104-167. Mollie Beattie Wilderness Area Act of July 29, 1996. 110 Stat. 1451. *AK • FWS*

U.S. Public Law 104-208. Omnibus Consolidated Appropriations Act, 1997; Oregon Resource Conservation Act of September 30, 1996. 110 Stat. 3009. *OR • FS, FWS*

U.S. Public Law 104-333. Omnibus Parks and Public Lands Management Act of November 12, 1996. 110 Stat. 4093. *AK, NM, OR • BLM, FWS, NPS • Special Provisions*

U.S. Public Law 105-60. Hood Bay Land Exchange Act of October 10, 1997. 111 Stat. 1269. *AK • FS*

U.S. Public Law 105-74. (To require the Secretary of the Interior to exchange certain lands located in Hinsdale County, Colorado) Act of November 12, 1997. 111 Stat. 1460. *CO • BLM*

U.S. Public Law 105-75. Slate Creek Addition to Eagles Nest Wilderness, Arapaho and White River National Forests, Colorado Act of November 12, 1997. 111 Stat. 1462. *CO • FS*

U.S. Public Law 105-76. Boundary Adjustment and Land Conveyance, Raggeds Wilderness, White River National Forest, Colorado Act of November 12, 1997. 111 Stat. 1463. *CO • FS*

U.S. Public Law 105-82. Marjory Stoneman Douglas Wilderness and Earnest F. Coe Visitor Center Designation Act of November 13, 1997. 111 Stat. 1540. *FL • NPS*

U.S. Public Law 106-65. National Defense Authorization Act for Fiscal Year 2000 (Military Lands Withdrawal Act of 1999) Act of October 5, 1999. 113 Stat. 885. *AZ • FWS • Special Provisions*

U.S. Public Law 106-76. Black Canyon of the Gunnison National Park and Gunnison Gorge National Conservation Act of October 21, 1999. 113 Stat. 1126. *CO • BLM, NPS*

U.S. Public Law 106-145. Otay Mountain Wilderness Act of December 9, 1999. 113 Stat. 1711. *CA • BLM • Special Provisions*

U.S. Public Law 106-156. Dugger Mountain Wilderness Act of December 9, 1999. 113 Stat. 1741. *AL • FS • Special Provisions*

U.S. Public Law 106-176. Omnibus Parks Technical Corrections Act of March 10, 2000. 114 Stat. 23. *NM, OR • BLM, FS*

U.S. Public Law 106-291. Department of the Interior and Related Agencies Appropriations Act, 2001 (exclusion from Argus Range Wilderness) Act of October 11, 2000. 114 Stat. 922. Sec. 137. *CA • BLM • Special Provisions*

U.S. Public Law 106-301. Utah West Desert Land Exchange Act of October 13, 2000. 114 Stat. 1059. *UT • BLM*

U.S. Public Law 106-324. (To dedicate the Big South Trail in the Comanche Peak Wilderness Area of the Roosevelt National Forest in Colorado to the legacy of Jaryd Atadero) Act of October 19, 2000. 114 Stat. 1291. *CO • FS*

U.S. Public Law 106-353. Colorado Canyons National Conservation Area and Black Ridge Canyons Wilderness Act of October 24, 2000. 114 Stat. 1374. *CO, UT • BLM • Special Provisions*

U.S. Public Law 106-399. (Steens Mountain) Act of October 30, 2000. 114 Stat. 1655. *OR • BLM • Special Provisions*

U.S. Public Law 106-456. (Spanish Peaks Wilderness) Act of November 7, 2000. 114 Stat. 1955. *CO • FS • Special Provisions*

U.S. Public Law 106-471. (Priest and Three Ridges Wilderness) Act of November 9, 2000. 114 Stat. 2057. *VA • FS*

U.S. Public Law 106-530. Great Sand Dunes National Park and Preserve Act of November 22, 2000. 114 Stat. 2527. *CO • NPS*

U.S. Public Law 106-541. Water Resources Development Act of December 11, 2000. 114 Stat. 2572. *MN • FS*

U.S. Public Law 106-554. Black Rock Desert–High Rock Canyon Emigrant Trails National Conservation Area Act of December 21, 2000. 114 Stat. 2763. *NV • BLM • Special Provisions*

U.S. Public Law 107-20. Supplemental Appropriations Act, 2001 (Apostle Islands National Lakeshore Wilderness Study) Act of July 24, 2001. 115 Stat. 155 Sec. 2601. *WI • NPS*

U.S. Public Law 107-63. Department of the Interior and Related Agencies Appropriations Act, 2002 (Amendment to Black Rock Desert–High Rock Canyon Emigrant Trails National Conservation Area Act of 2000) Act of November 5, 2001. 115 Stat. 414. Sec. 135. *NV • BLM • Special Provisions*

U.S. Public Law 107-107. National Defense Authorization Act for Fiscal Year 2002; including the Fort Irwin Military Withdraw Act of 2001 and (Wilderness Designation on Vieques Island) Act of December 28, 2001. 115 Stat. 1012. Sec. 1504 and Division B–Title XXIX. *Puerto Rico, CA • FWS • Special Provisions*

U.S. Public Law 107-206. 2002 Supplemental Appropriations Act for the Further Recovery from and Response to Terrorist Attacks on the United States (addition to Black Elk Wilderness) Act of August 2, 2002. 116 Stat. 820. Sec 706. *SD • FS*

U.S. Public Law 107-216. James Peak Wilderness and Protection Area Act of August 21, 2002. 116 Stat. 1055. *CO • FS, NPS • Special Provisions*

U.S. Public Law 107-282. Clark County Conservation of Public Land and Natural Resources Act (Omnibus Wilderness Designation) Act of November 6, 2002. 116 Stat. 1994. *NV • FS, NPS, BLM • Special Provisions*

U.S. Public Law 107-334. Mount Nebo Wilderness Boundary Adjustment Act of December 16, 2002. 116 Stat. 2876. *UT • FS • Special Provisions*

U.S. Public Law 107-350. (Quail Springs Wilderness Study Area) Act of December 17, 2002. 116 Stat. 2975. *NV • BLM • Special Provisions*

U.S. Public Law 107-361. (Sand Mountains Wilderness Study Area) Act of December 17, 2002. 116 Stat. 3020. *ID • BLM • Special Provisions*

U.S. Public Law 107-370. Big Sur Wilderness and Conservation Act of December 19, 2002. 116 Stat. 3071. *CA • FS, NPS, BLM • Special Provisions*

U.S. Public Law 108-7. Consolidated Appropriations Resolution, 2003; including T'uf Shur Bien Preservation Trust Area Act (Sandia Mountain Wilderness) Act of February 20, 2003. 117 Stat. 11. Division F–Title IV. *NM • FS*

U.S. Public Law 108-95. Mount Naomi Wilderness Boundary Adjustment Act of October 3, 2003. 117 Stat. 1165. *UT • FS*

U.S. Public Law 108-199. Consolidated Appropriations Act, 2004 (Designation of Congaree National Park Wilderness) Act of January 23, 2004. 118 Stat. 3. Division H–Sec. 139. *SC • NPS*

U.S. Public Law 108-424. Lincoln County Conservation, Recreation, and Development Act (Omnibus Wilderness Designation) Act of November 30, 2004. 118 Stat. 2403. *NV • BLM • Special Provisions*

U.S. Public Law 108-447. Consolidated Appropriations Act, 2005; including Gaylord A. Nelson Apostle Islands National Lakeshore Wilderness Act and Cumberland Island Wilderness Boundary Adjustment Act of December 8, 2004. 118 Stat. 2809. Division E–Title I. Sec. 140 and Sec. 145. *GA, WI • NPS • Special Provisions*

U.S. Public Law 109-94. Ojito Wilderness Act of October 26, 2005. 119 Stat. 2106. *NM • BLM • Special Provisions*

U.S. Public Law 109-97. Agriculture, Rural Development, Food and Drug Administration, and Related Agencies Appropriations Act, 2006 (Gaylord A. Nelson Wilderness Redesignation) Act of November 10, 2005. 119 Stat. 2120 Sec. 440. *WI • NPS*

U.S. Public Law 109-115. Transportation, Treasury, Housing and Urban Development, the Judiciary, the District of Columbia, and Independent Agencies Appropriations Act, 2006 (North McCullough Mountains Wilderness) Act of November 30, 2005. 119 Stat. 2396. Sec. 180. *NV • BLM*

U.S. Public Law 109-118. Caribbean National Forest Act of December 1, 2005. 119 Stat. 2527. *Puerto Rico • FS • Special Provisions*

U.S. Public Law 109-163. National Defense Authorization Act for Fiscal Year 2006 (Cedar Mountain Wilderness Area) Act of January 6, 2006. 119 Stat. 3136 Title I–Subtitle H. *UT • BLM • Special Provisions*

U.S. Public Law 109-309. Amend Ojito Wilderness Act of October 6, 2006. 120 Stat. 1727. *NM • BLM • Technical Amendment*

U.S. Public Law 109-362. Northern California Coastal Wild Heritage Wilderness Act of October 17, 2006. 120 Stat. 2064. *CA • FS, BLM • 1 WSA • Special Provisions*

U.S. Public Law 109-372. Idaho Land Enhancement Act of November 27, 2006. 120 Stat. 2645. *ID • FS, BLM*

U.S. Public Law 109-382. New England Wilderness Act of December 1, 2006. 120 Stat. 2673. *NH, VT • FS • Special Provisions*

U.S. Public Law 109-432. Tax Relief and Health Care Act of December 20, 2006, including the Pam White Wilderness Act of 2006. 120 Stat. 2922. *NV • FS, BLM • Special Provisions*

Glossary

abiotic environment. Includes climatic conditions such as temperature and moisture regimes, and inorganic substances supplied by mineral soil.

acid-neutralizing capacity (ANC). The scientific indicator of buffering capacity (see *buffering capacity* and Chapter 15).

acid rain. A phenomenon that occurs when sulfur dioxide and nitrogen oxides are chemically transformed into acidic sulfates and nitrates during atmospheric transport and are subsequently deposited downwind as acid precipitation (either rain or snow), acid fog, or acidic particles (see Chapter 15).

activity plan. Used by the BLM. A Wilderness Management Plan prepared for a specific wilderness area or two or more closely related areas. It helps implement the BLM's wilderness management policy (see Chapter 8).

ANC. See *acid-neutralizing capacity*.

anthropocentric. A philosophical viewpoint that sees wilderness primarily from a human-oriented perspective. The naturalness of the wilderness is less important than facilitating human use and convenience. Programs that would alter the physical and biological environment to produce desired settings are encouraged (see Chapter 1).

autotrophs. Organisms that are green plants and provide an ecosystem's entire energy base by fixing solar energy and by using simple inorganic substances to build complex organic substances.

Bailey-Küchler System. A system using Robert G. Bailey's ecoregion concept and Küchler's system of potential natural vegetation for classifying ecosystems in the United States and Puerto Rico. The system considers both physical environment factors, such as climate and soil, and biological environment factors, such as vegetation.

bequest value. Protection of wilderness for future generations.

biocentric. A philosophical viewpoint that emphasizes maintaining natural systems at the expense of recreational and other human uses, if necessary, because wilderness values depend on naturalness and solitude. The goal of this philosophy is to permit natural ecological processes to operate as freely as possible, because wilderness values for society ultimately depend on retaining naturalness and solitude (see Chapter 1).

biological oxygen demand. The oxygen used in meeting the metabolic needs of aerobic microorganisms in water rich in organic matter (as in water polluted with sewage).

biome. A major type of terrestrial community, for example, tropical rainforests, deserts, etc., that is recognizable by the characteristic structure of the dominant vegetation in that community.

biosphere reserves. Designated areas in an international network of representative protected areas. Biosphere reserves are representative examples of the world's major biomes, which help complete a portrait of the world's ecosystems, and they often contain areas where environmental conditions are actively managed and accommodate human activity, as well as natural areas. Ideally, biosphere reserves contain a natural or core zone where minimal human impact is present and surrounding zones with human activity and natural resource use (see Chapter 3).

biota. Fauna and flora together.

biotic environment. Biological members of an organism's habitat that interact with it, including competitors, predators, and parasites.

buffering capacity. Soil's ability to neutralize acids without a change in pH (see Chapter 15).

bulk density. The weight of soil packed into a given volume (see Chapter 15).

carrying capacity. The maximum level of use an area can sustain without exceeding the limits of acceptable change (LAC) in social and environmental conditions. As applied to recreational use of wilderness, often includes the effects of such use on experience quality due to crowding and conflict (see Chapters 8 and 9).

classified wilderness. Areas formally protected by the Wilderness Act of 1964 and its extension to eastern lands by the so-called Eastern Wilderness Act, to public lands by the Federal Land Policy and Management Act of 1976, and to Alaska by the Alaska National Interest Lands Conservation Act of 1980.

common wildlife. In wilderness, species that happen to be found in wilderness, but that also live in more modified environments. Their relationship to wilderness is incidental (see Chapter 12).

de facto wilderness. Public lands that are wilderness in the general sense of the term, roadless and undeveloped, but which as wilderness have not been designated by Congress. Lands potentially available for wilderness classification.

de jure wilderness. Wilderness areas officially designated and protected under provisions of the Wilderness Act of 1964, and belonging to the NWPS.

direct management. Management that emphasizes regulating people's behaviors; individual choice is restricted, and managers aim at directly controlling visitor behavior with regulations and use requirements (see Chapter 16). Contrasts with indirect management.

ecosystem. Includes all the organisms of an area, their environment, and the linkages or interactions among them; all parts of an ecosystem are interrelated. The fundamental unit in ecology, containing both organisms and abiotic environments, each influencing the properties of the other and both necessary for the maintenance of life (see Chapters 10 and 11).

ecotone. A habitat created by the juxtaposition of distinctly different habitats; a transitional zone between habitat types, containing species of both habitats.

ecotype. A genetically differentiated subpopulation that is restricted to a specific habitat.

EIS. See *environmental impact statement*.

environmental impact statement (EIS). A required report for all federal actions that will lead to significant effects on the quality of the human environment. The report must be systematic and interdisciplinary, integrating the natural and social sciences as well as the design arts in planning and in decision making. The report must identify (1) the environmental impact of the proposed action, (2) any adverse environmental effects that cannot be avoided should the proposal be implemented, (3) alternatives to the proposed action, (4) the relationship between local short-term uses of human environment and the maintenance and enhancement of long-term productivity, and (5) any irreversible and irretrievable commitments of resources that would be involved in the proposed action should it be implemented.

environmental modification spectrum. A concept that describes a continuum of settings that range from the totally modified landscape of a modern city to those remote and pristine reaches of a country (see Chapters 1 and 7). Related to the recreation opportunity spectrum (ROS) and wilderness opportunity spectrum (WOS) (see Chapter 8).

existence value. Protection of wilderness just to know there are such places relatively unaffected by humans.

exotic pathogens. Nonindigenous pathogenic organisms introduced into an ecosystem that may have potentially profound effects on otherwise pristine landscapes. Because ecological interactions are unknown, and host species may lack mechanisms for resistance, effects may be permanent (see Chapter 10).

exotic species. Includes species introduced into a wilderness that may have adapted to wilderness ecosystems and compete with resident native (indigenous) wildlife species. (The term *alien species* is also used because "exotic" may connote something fascinating or wondrous.) (See Chapter 10.)

feral. Having returned to a wild state from domestication, such as feral horses, goats, pigs, and dogs.

fire-dependent. An ecosystem evolving under periodic perturbations by fire and that consequently depends on periodic fires for normal ecosystem functioning (see Chapter 11).

fire regime. The kind of fire activity (frequency and intensity) that characterizes a specific region (see Chapter 11).

Forest Land and Resource Management Plan. Also known as the Land Management Plan (LMP) or Forest Plan used by the USFS. These plans translate national and regional direction into forest and wilderness goals, a description of the lands where prescriptions apply, a statement of the desired future condition, and a set of standards and guidelines for meeting the goals (see Chapter 8).

forest structure. Often divided into four conceptual aspects: age structure, species composition, horizontal structure or mosaic pattern, and vertical structure or fuel ladders (see Chapter 11).

gene. Generally, a unit of genetic inheritance. In biochemistry, gene refers to the part of the DNA molecule that encodes a single enzyme of structural protein unit.

gene pool. All of the genes present in a population of organisms. A population includes all members of a species living in a particular area and breeding together.

General Management Plan (GMP). Required for each national park under the National Parks and

Recreation Act of 1978 (U.S. Public Law 95-625). A GMP describes the basic management philosophy for the entire national park and provides the strategies for addressing issues and achieving identified management objectives over a five- to ten-year period and sometimes longer (see Chapter 8).

giardiasis. An intestinal disease caused by the protozoan pathogen *Giardia lamblia*, which has become common in wilderness waters over the past thirty years (see Chapter 15). It is commonly called giardia.

GMP. See *General Management Plan*.

grazing. Foraging for food by domestic livestock (sheep, cattle, horses) and allowed in wilderness under grazing permits.

habitat. Place where an animal or plant normally lives, often characterized by a dominant plant form or physical characteristic, for example, a forest habitat.

hard release. Lands permanently released for other multiple-use purposes, including resource development, following their consideration and nonselection for wilderness classification.

heterotrophs. Organisms that use the complex substances produced by autotrophs as a food base and rearrange or decompose them. For example, heterotrophs include animals, microbes, and fungi. They may be either ingestors or decomposers (see Chapter 10).

historic range of variability (HRV). The "naturally bounded" (climate, organism pool, soil parent material, topography, disturbances, etc.) behavior of an ecosystem over time.

indigenous. Living or occurring naturally in an area; native, endemic people, flora, or fauna (see Chapter 3).

indirect management. Management that emphasizes modifying people's behaviors by managing factors and situations that influence their decisions; visitors retain their freedom to choose; a "light-handed" approach to wilderness management (see Chapter 16). An example would be retaining a low standard access road to help limit use of a popular wilderness trailhead. Contrasts with direct management.

inholding. A tract of land under private ownership within public lands, such as a wilderness area.

interim management. The management of areas under study or consideration for wilderness classification during their review period, in a way that does not degrade their wilderness qualities or preclude their being classified by Congress as units of the NWPS. In other words, the decision to study an area for possible wilderness designation is also a decision to temporarily manage that area as a wilderness.

intrinsic processes. Amelioration or preparation of severe sites by pioneering species and their eventual elimination by the less hardy climax species through competition, invasion, succession, and displacement (see Chapter 10).

L-20 Regulation. The first systematic program of wilderness protection promulgated by the USFS in 1929. It was primarily a list of permitted and prohibited uses for designated roadless areas on national forests (see Chapter 2).

LAC. See *limits of acceptable change*.

limits of acceptable change (LAC). A planning framework that establishes explicit measures of the acceptable and appropriate resource and social conditions in recreation settings as well as the appropriate management strategies for maintaining and/or achieving those conditions.

minimum tool rule. Apply only the minimum-impact policy, device, force, regulation, instrument, or practice to bring about a desired result (see Chapter 7). Achieve results using the most "light-handed" approach.

monitoring. Systematic gathering, comparing, and evaluation of data (see Chapter 8).

National Environmental Policy Act of 1969 (NEPA). A federal law establishing as national policy that all federal agencies give full consideration to environmental effects in planning their programs. To implement this policy, the act requires specific "action-forcing" procedures that agencies must observe (see *environmental impact statement*).

NEPA. See *National Environmental Policy Act of 1969*.

nondegradation. A concept that calls for the maintenance of existing environmental conditions if they equal or exceed minimum standards, and for the restoration of conditions below minimum levels (see Chapter 7).

omnibus bill. A wilderness classification act designating several wilderness areas or WSAs.

opportunity class. Designates the availability of a particular quality or kind of experience that is appropriate to the conditions.

organic acts. A law providing original or basic direction to an agency. For example, the Federal Land and Policy Management Act of 1976 is regarded as an organic act for the BLM because it gives basic direction to the agency.

pH. A measure of how acidic or alkaline (basic) a solution is on a scale of 0–14, with 0 being *very acidic*, 14 being *very alkaline*, and 7 being *neutral*. The abbreviation stands for potential of Hydrogen (see Chapter 15).

potential of Hydrogen. See *pH*.

primary succession. A process usually slower than secondary succession because it involves amelioration (improvement) of extreme site conditions by gradual alterations brought about by the organisms (see Chapter 10).

primitive areas. (1) USFS areas set aside under administrative regulation L-20 between 1929 and 1939. Nearly 14 million acres were set aside under the L-20 Regulation, but the lack of strict enforcement and its permissive acceptance of other forest-management activities, including logging, led to its replacement in 1939 by the U regulations. (2) An administrative designation formerly used by the BLM to manage lands for wilderness preservation purposes. It was superseded by the FLPMA in 1976. (3) Refers to an administrative category of protected areas on the national forests before the Wilderness Act of 1964, and either classified as wilderness or identified for subsequent wilderness consideration by that act.

Ramsar Sites. A designation for wetland areas of international importance, such as the Everglades in Florida. As with biosphere reserves and world heritage sites, areas on the Ramsar list have no legal standing because of such listing, but the international recognition may support their protection.

RARE. See *Roadless Area Review and Evaluation*.

RARE II. See *Roadless Area Review and Evaluation II*.

recreation opportunity spectrum (ROS). A planning approach identifying a range of recreational environments across a spectrum: urban recreation areas, rural countryside, highly developed campgrounds, intensively managed multiple-use forests, national parks, recreation and scenic areas, roadless wildlands, and wilderness. The ROS defines six classes: primitive, semiprimitive nonmotorized, semiprimitive motorized, roaded natural, rural, and urban (see Chapters 1 and 8).

Refuge Management Plans. Used by the FWS, these plans are used for each refuge unit, which can include separate plans for management of hunting and fishing, grazing, public use, fire control, and other important activities (see Chapter 8).

Renewable Resources Assessment and Program. Required for the USFS by the Forest and Rangeland Renewable Resources Planning Act (RPA), as amended by the National Forest Management Act. The RPA calls for (1) preparation of an assessment of the supply and demand for the nation's forest and rangeland resources, and (2) development of a management program for national forests that considers alternative management directions and the role of the national forests (see Chapter 8).

research natural area (RNA). Area set aside to preserve representative ecosystems for scientific study and educational purposes (see Chapter 6).

Resource Management Plan (RMP). Used by the BLM, a type of land-use plan designed to guide the management of an entire resource area (see Chapter 8).

RMP. See *Resource Management Plan*.

RNA. See *research natural area*.

Roadless Area Review and Evaluation (RARE). A USFS effort in the early 1970s to systematically inventory, review, and evaluate the relative values for future uses of existing roadless areas. The RARE process identified the extent of roadless lands remaining on the national forests and recommended each area for wilderness consideration, further study, or release for other multiple use (see Chapter 5).

Roadless Area Review and Evaluation II (RARE II). The second Roadless Area Review and Evaluation, in 1977–1979, incorporated new roadless area criteria and the requirements of the National Forest Management Act. The final RARE II inventory list contained 2,919 roadless areas, including 62 million acres in national forests and national grasslands in 38 states and Puerto Rico. Each area's resources were estimated, site-specific information was reviewed, and potential for uses was assessed (e.g., potential timber, harvesting, grazing, and mineral extraction). RARE II also sought to assess how each area might contribute to qualities of the Wilderness System, such as ecological diversity (see Chapter 5).

ROS. See *recreation opportunity spectrum*.

secondary succession. Fairly rapid changes following cutting or burning of a forest or removal of grazing animals from a depleted range (see Chapter 10).

silviculture. The cultivation of forest trees; forestry.

simulation model. Simplified replica of a real-world system, usually described so that it can be represented by a computer program, and the effects of changes simulated (see Chapter 16).

soft release. Also known as "Colorado release." First included in the Colorado Wilderness Act, soft release provides for the release of areas considered for but not classified as wilderness, and the same treatment as other lands in the national forest planning process. Thus, soft-released areas may be considered again as to their wilderness potential during the next cycle of forest planning and ultimately classified as wilderness (see *hard release*).

successional changes. Long-term, predictable trends of an ecosystem, as opposed to short-term cyclical changes. An orderly process of community development involving changes in species structure and community processes with time.

sustainable development. Natural resource utilization that does not exceed the permanent productive capacity of the ecosystem. Thus, a level of development use that can be permanently sustained (see Chapter 3).

synecology. The relationship of organisms and populations to biotic factors in the environment.

trampling. Walking on vegetation and soil by humans and pack stock that may cause abrasion of vegetation, abrasion of surface soil organic layers, and compaction of soils (see Chapter 15).

transactive planning. A strategy used by wilderness managers that incorporates the concept of societal guidance, wherein people are willing and able to construct and guide their own future if given the opportunity to do so. It recognizes the importance of social interaction among those affected by a decision. Small working groups of citizens are the basis of the transactive planning process, and they interact with a planner. Transactive planning has been used in coordination with the LAC process (see Chapter 9).

untrammeled. Not subject to human controls and manipulations that hamper the free play of natural forces. A word describing desired wilderness conditions used in the Wilderness Act.

U regulations. Regulations U-1, U-2, and U-3(a) replaced the L-20 Regulation in 1939 as the authority under which roadless lands in the national forests were administratively protected. Regulation U-1 established wilderness areas, tracts of land of not less than 100,000 acres. Regulation U-2 defined wild areas as tracts of land between 5,000 and 100,000 acres that could be protected, modified, or eliminated by the chief of the USFS. Regulation U-3(a) established roadless areas for recreational use under natural conditions (see Chapter 4).

visibility. An air quality–related value Congress has singled out for protection under the Clean Air Act.

WARS. See *Wilderness Attribute Rating System*.

wilderness. The legal definition is found in the Wilderness Act of 1964, section 2(c) (U.S. Public Law 88-577): "A wilderness, in contrast with those areas where man and his own works dominate the landscape, is hereby recognized as an area where the earth and its community of life are untrammeled by man, where man himself is a visitor who does not remain." This legal definition places wilderness on the "untrammeled" or "primeval" end of the environmental modification spectrum. Wilderness is roadless lands, legally classified as component areas of the NWPS, and managed to protect its qualities of naturalness, solitude, and opportunity for primitive types of recreation (see Chapter 1).

wilderness-associated wildlife. Includes species commonly associated with wildland habitat characteristic of wilderness (see Chapter 12).

Wilderness Attribute Rating System (WARS). A system used in RARE II to document and assess the wilderness attributes of roadless areas (see Chapter 5).

Wilderness Classification Act. Legislation that designates an area or areas as part of the NWPS.

wilderness-dependent. Dependent on the wilderness conditions of naturalness and solitude (see Chapter 7).

wilderness-dependent wildlife. Species dependent on conditions of naturalness and solitude and thus species whose continued existence is dependent on and/or reflective of wilderness conditions (see Chapter 12).

wilderness designation (or wilderness classification). Includes all processes and activities of government agencies and interested publics to identify candidate areas and secure their permanent protection under the U.S. Wilderness Act of 1964 or other legal statute in the case of states or other countries (see Chapter 1).

wilderness management. Government and citizen activity to identify—within the constraints of the Wilderness Act—goals and objectives for classified wildernesses and the planning, implementation, and administration of policies and management actions to achieve them. Involves the application of guidelines and principles to achieve established goals and objectives, including management of human use and influences to preserve naturalness and solitude (see Chapter 1).

wilderness opportunity spectrum (WOS). A spectrum of wilderness conditions including finer gradations of naturalness and solitude, that is, primitive conditions. The WOS includes, for example, pristine, primitive, and portal designations, indicating decreasing degrees of naturalness and solitude. Like the ROS, WOS is a kind of zoning, which delineates particular areas where different management prescriptions or restrictions on visitor behavior apply (see Chapter 8).

wilderness study area (WSA). Area found to possess candidate characteristics for wilderness designation and designated for formal study to assess its suitability for wilderness classification. WSAs were designated during the BLM's wilderness review process (during the inventory phase) begun in 1978 and ending in 1980. WSAs have also been identified in wilderness legislation that mandates their study (see Chapter 5).

world heritage sites. An international classification system to recognize and designate areas that represent a major stage of the earth's evolutionary history; significant ongoing geological processes, biological evolution, and human interaction with the natural environment; superlative natural phenomena, formations, or features; and with natural habitats where threatened or endangered species of animals or plants of outstanding universal value can survive (see Chapter 3).

WOS. See *wilderness opportunity spectrum*.

WSA. See *wilderness study area*.

Abbreviations and Acronyms

ANCSA. Alaska Native Claims Settlement Act of 1971
ANILCA. Alaska National Interest Lands Conservation Act of 1980
BLM. Bureau of Land Management, U.S. Department of the Interior
BMWC. Bob Marshall Wilderness Complex, Montana
BWCAW. Boundary Waters Canoe Area Wilderness, Minnesota
CAMPFIRE. Communal Areas Management Programme for Indigenous Resources, Zimbabwe
CBD. Convention on Biological Diversity, 1992
CEQ. Council of Environmental Quality
CITES. Convention on International Trade in Endangered Species
CNPPA. The Commission on National Parks and Protected Areas
CONCOM. Council of Nature Conservation Ministers, Australia
CPS. Canadian Park Service
DOC. Department of Conservation, New Zealand
EA. Environmental Assessment
EIS. Environmental Impact Statement
EPA. Environmental Protection Agency
FAO. Food and Agricultural Organization
FLPMA. Federal Land Policy and Management Act of 1976
FMC. Federated Mountain Clubs, New Zealand
FWS. U.S. Fish and Wildlife Service, U.S. Department of the Interior
IJW. International Journal of Wilderness
IUCN. International Union for Conservation of Nature and Natural Resources
IWLF. International Wilderness Leadership Foundation
LAC. Limits of Acceptable Change
MAB. Man and Biosphere Program
NAAQS. National Ambient Air Quality Standards
NEM:PAA. National Environmental Management: Protected Areas Act, 1969 (South Africa)
NFMA. National Forest Management Act of 1976
NGO. Nongovernmental organization
NOAA. National Oceanic and Atmospheric Administration
NPS. National Park Service, U.S. Department of the Interior
NWPS. National Wilderness Preservation System
NZFS. New Zealand Forest Service
ORRRC. Outdoor Recreation Resources Review Commission
P.L. Public Law
RAWS. Remote Automated Weather Station
RIM. Recreation Information Management (system)
RNA. Research Natural Area
ROS. Recreation Opportunity Spectrum
RPA. Forest and Rangeland Renewable Resources Planning Act of 1974
SAF. Society of American Foresters
TNC. The Nature Conservancy, a nonprofit organization
UNEP. United Nations Environmental Program
UNESCO. United Nations Educational, Scientific, and Cultural Organization
USFS. U.S. Forest Service, U.S. Department of Agriculture
VIM. Visitor Impact Management
WAG. Wilderness Action Group (WAG)
WARS. Wilderness Attribute Rating System (RARE II)
WCPA. World Commission on Protected Areas
WOS. Wilderness Opportunity Spectrum
WSA. Wilderness Study Area
WWC. World Wilderness Congress

Index

A

Abiotic environment, 252
Absaroka-Beartooth Wilderness (Montana), 115, 180, 259, 345
Access modification, visitor acceptance of, 460
Ackerman, Danie, 66
Actions, 204
Activity plan, 203
Adirondack Forest Preserve, 173
Adjacent-lands
    management and use of, 345
    managing wilderness in relation to, 191–93
Administrative access, facilities, and intrusive management, 344–45
Administrative Procedures Act (APA) (1946), 155–56
Admiralty Island Wilderness, 175
Advanced technology, 348
Affirmative congressional action, 95
Africa. See also South Africa
    protected areas in, 55
    wilderness areas in, 74, 75, 76
African Cape buffalo, 310
African elephants, 310, 312
African lions, 309, 310
Agriculture, U.S. Department of Regulation, 36, 200
Agrupación Sierra Madre, 77
Agua Tibia Wilderness (California), 107
Ahwahnee Hotel, 37
Aircraft noise and airspace reservations, 349
Air quality, 293, 347
    monitoring of, 201
Alaska
    special provisions for wilderness in, 2–3
    wilderness legislation in, 152
    wilderness travel in, 376
Alaska Coalition, 141
Alaska National Interest Lands Conservation Act (ANILCA) (1980), 3, 212
    access to private or public land inholdings and, 345
    amendment of, 202
    construction and maintenance in wilderness, 150
    designation of national parks by, 152
    expansion of Fish and Wildlife Service and, 167
    passage of, 145
    reaffirmation of Congress intent of, 149
    soft release language in wilderness laws of, 151
    wilderness areas in, 107, 141–42, 164, 320, 441
    wilderness study initiatives in, 135
Alaska Native Claims Settlement Act (ANCSA) (1971), 140–41
Alaskan public domain, disposition of, 139–41
Alaskan Statehood Act (1958), 139
Albright, Horace M., 38
Aleutian Islands Wilderness (Alaska), 58
Allagash River (Maine), 170
Alligators, 313
Alpine Lakes Area Management Act (1976), 152
Alpine Lakes Coalition Society, 13
Alpine Lakes Preservation Society, 13
Alpine Lakes Wilderness (Washington), 347
    recreational use of, 377, 381

Alpine Lakes Wilderness, cont.
    wilderness proposal for, 12–13
Amazon Basin, 82
American chestnut, 262–63
Anderson, Clinton, 95, 96
Andrus, Cecil, 141
Angling, encouraging styles with wilderness experiences, 328–29
Animals
    carnivores, 310, 312
    diseases in wilderness, 320
    elimination of predatory, 261
    grazing by domestic, 260–61
    herbivores, 314
    introduced, 262
    loss of species, 261
Animal Unit Months (AUMs), 368
Antarctica, wilderness areas in, 79–80
Antarctic Treaty System (ATS), 79
Antelopes, 324
Anthropocentric position, 19–21, 28
Anthropogenic ecosystem changes, 265
Antipoacher patrols, 314
Antiquities Act (1906), 141
Anza-Borrego Desert State Park, 173
Apartheid, 67
Appalachian Mountains, 262
    fire-dependent ecosystem in, 302
Appalachian Trail, 150, 168–69
    recreational use of, 380
    use of simulation models at, 480
Arabian oryx, 323
Arabuko-Sokoke Forest (Kenya), 317
Arctic National Wildlife Refuge (Alaska), 259
Arizona Desert Wilderness Act (1990)
    Bureau of Land Management recommendations of, 153
    designation of desert under, 212
    designation of wilderness areas by, 242
    facilities and structure in, 150
    grazing and, 369
    military overflights and, 151
    motorized use in, 148
    National Wilderness Preservation System and, 164
    prohibiting buffer zones in, 149
    wilderness legislation and, 146
Arkansas Wilderness Act, 149
Asiatic elephants, 321
Aspinall, Wayne, 95, 96
Assisted trout migration into wilderness, 193–94
Audubon Society, 317
Ausoni Wilderness Area, 75
Australia
    protected areas in, 55
    wilderness in, 60–62
Australian Park Authority, 82
Automatic photography in estimating wilderness recreational use, 372
Availability, wilderness designation and, 119
Avalanche Pass, 174
Avalon Wilderness Reserve, 64

B

Backpacking, evolution of, 36
Badlands National Park (South Dakota), 314
Badlands Wilderness, 314
Bailey-Küchler mapping system, 123–24, 315
Bailey's Province-level Ecosystem classification system, 166–67
Bald eagles, 310, 330
Bandelier National Monument (New Mexico), 295
    1977 wildfire at, 294
Banff, Canada, 63, 321
Barguzin Zapovednik, 69
Base datum of normality, 252
Baseline wilderness data, 435
Bavarian Forest National Park, 52
Bears, 310, 321
    black, 322
    brown, 310, 312
    grizzly, 252, 263, 309, 310, 312, 313, 316, 322, 326
    polar, 310, 312, 313, 326
Bear Trap Canyon Unit, 205, 206, 211
Beluga whales, 312
Benai Biosphere Reserve, 78
Benchmark wilderness laws, 143–46
Bergland, Bob, 141
Big Bend National Park, 152
Bighorn sheep, protection of, 149, 243, 252, 309, 310, 325
Biocentric approach to wilderness management, 19–25, 28, 276
Biodiversity, protecting, 80–81
Biophilia Hypothesis of E. O. Wilson, 2
Biosphere Reserve Project, 58, 174, 175
Biosphere reserves, 56, 340
Biotic community, 252
Bison, 309, 322, 330
Bisti/De-Na-Zin Wilderness, 153
Bitterroot National Forest, Blodgett Canyon in, 128
Black bears, 322
Blitzen River (Oregon), 171
Bonneville Power Administration (BPA), 104
Boone and Crockett Club, 317
Boreal forest region, fire-dependent ecosystems in, 301–2
Borrego Desert State Park, 173
Borrego Palm Canyon wilderness, 173
Boundary effects, 264
Boundary Waters Canoe Area Wilderness (BWCAW) (Minnesota)
    borders of, 81
    camping at, 32
    fire in, 276
    litigation and, 156–57
    logging in, 367
    management by Forest Service and, 45
    military overflights and, 151
    mining protection and, 106, 147
    recovery of natural conditions in, 328
    recreational use of, 375, 378, 379, 387
    Superior National Forest consolidation with, 92
    use patterns of, 381
    water and marine concept and, 82

Boundary Waters Canoe Area Wilderness, cont.
    wilderness canoeing in, 63
    wilderness travel in, 376
Boundary Waters Canoe Area Wilderness Act (1978), 146, 147, 151, 156–57
Boundary Waters–Voyageur Waterway and Canadian Heritage River, 85
Branff National Park, 64
Bridger, Jim, 7, 32
British Columbia Forest Service, 64
Brooks Range (Alaska), 42
Brower, David, 32
Brown bears, 310, 312
Brucellosis, 320
Buffalo
    African Cape, 310
    cape, 314, 322
Buffer zones, 148–49, 192
Bulldozers, 289
Burton, Phil, 127
Bush, George W.
    administration of, 153
    roadless area system and, 168
Buthelezi, Mangosutho, 67
Butterflies, 310

C

Cabeza Prieta Wilderness (Arizona), 318, 341
Cabinet Mountains Wilderness (Montana), 103, 346, 366
Caledonian Forest, 53
California, University of, Wild Land Research Center at, 93
California condors, 319
California Desert Protection Act (1994)
    adding of acres to system, 164
    administrative direction and, 319
    Bureau of Land Management wilderness study areas and, 153
    designation of wilderness areas in, 369
    mechanized access for wildlife management, 149, 344
    military overflights and, 151, 349
    passage of, 145–46
    provision for motorized use of, 148
California-Oregon Coast Ranges region, fire-dependent ecosystems in, 298
California Park and Recreation Commission, 172
California State Wilderness System, 172–73
California v. Bergland, 125–26, 127, 129
California v. Block, 150
California Wilderness Act, 173
California Wilderness Preservation System, 172–73
Campfires, 400–402
    restrictions of, 472
Camping setbacks, 473
Campsite indicators on White Mountain National Forest, 248
Campsites, 41
    impacts of, 407–10
    restoration of, 420–23
    revegetation of, 266
    stock impact and, 430–31
    use of, 381–82
Canada. *See also specific parks*
    National Parks Act in, 62–63
    wilderness in, 85–86
    wilderness protection in, 62
Canada lynx, 310
Canaima National Park (Venezuela), 77
Canyonlands National Park, 152
Cape buffalo, 314, 322
Cape Romain National Wildlife Refuge, 319
Capitan Wilderness (New Mexico), 128
Carhart, Arthur, National Wilderness Training Center (Missoula, Montana), xx, 46, 153

Carhart, Arthur H., 7–8, 16, 40–41, 43, 90
Caribou, 252, 309, 310
    mountain, 310
Caribou-Speckled Mountain, 205
Carnivores, 310, 312
Carpathian Mountains, 53
Carrying capacities, 32, 39
    setting, as necessary to prevent unnatural change, 187–88
Carson, Kit, 7
Carter, Jimmy, 122–23, 141
Cascades Range (Oregon and Washington) fire-dependent ecosystem in, 299
Catlin, George, 6
Catskill Forest Preserve, 173
Cave Creek Complex Fire, 279
Cedar Fire (2003), 294
CEMEX, 77
Central America, 55
Central Idaho Wilderness Act (1980), 107, 147, 149, 150, 349
Central Suriname Nature Reserve, 81
Cerro Grande Fire (2000), 296
Chapman, H. H., 92
Cheetahs, 309, 314
Cherrystems, 243
Chestnut blight, 262–63
Chihuahuan Desert, 77
Chimpanzees, 316
Chitwan National Park, 309
Christianity, wilderness and, 5
Chugach National Forest, 141–42
Chumash Indian rock paintings, 116, 117
Church, Frank, 25, 96, 107, 122, 154
Ciénga de Zapata, 57–58
Civilian Conservation Corps, 42
Civilization, 31
Clark, Kendall, 28
Class I wilderness, 59–71
Class II wilderness, 59, 71–75
Class III wilderness, 59, 75–83
Clean Air Act (1977), 200, 293, 319, 347
Cleveland, Treadwell, Jr., 40
Clifty Wilderness (Kentucky), 164
Climatic factors, fire management and, 278–79
Climax state, 254
Clinton, Bill, 153
    roadless area system and, 168
Coal deposits, 147
Coast Redwoods region (California), 298–99
Code of Federal Regulations, 319
Cole, David, 341
Collaborative decision-making processes, 198
Collegiate Peaks Wilderness (Colorado), recreational use of, 379
Colorado release, 128
Colorado River Management Plan (CRMP), 33, 35, 177
Colorado Wilderness Act (1980)
    administrative review of current policies and, 149
    grazing management and, 13, 104–5, 145, 146, 319, 343, 368
    soft release language in wilderness laws, 151
    wilderness ecosystems and, 258
Colorado Wilderness Act (1993), 153
Commercial outfitting, 360–61, 364
Commercial recreational use, 360–61
    increasing, 341–43
Commodity uses, 359, 365–69
Common native wildlife, 311
Commonsense policy, need for, 25
Communal Areas Management Programme for Indigenous Resources (CAMPFIRE) (Zimbabwe), 74

Communications technology, 47
Complementary conservation areas, 168
Composition, ecosystem, 253
Comprehensive conservation plans (CCPs)
    for Kenai National Wildlife Refuge (Alaska), 202, 215
Computer-based simulation models, 480
Conditional fire zone, 297
Condors, California, 319
Confederated Salish and Kootenai Tribes, 68
Conservation Act (New Zealand, 1987), 71, 73
Conservation biology, 81
Conservation International (CI), 78, 81
    wilderness assessment, 52
Conservation NGOs, 82
Conservation's Wilderness Institute, xx
Constant change, 254
Consultative Forum for the Environment, 67
Continental Divide National Scenic Trail, 169
Convention on Biological Diversity (CBD), 80
Convention on International Trade in Endangered Species (CITES), 314
Copeland Report, 43
Coronado National Memorial (Arizona), 103
Corrals for holding horses, 462–63
Cottonwood Point Wilderness (Arizona), 139
Cougars, 310, 321, 322
Council of Nature Conservation Ministers, 61
Coweeta Experimental Forest (North Carolina), 175
Crater Lake National Park (Oregon), 103, 287
Craters of the Moon National Monument (Idaho), 129, 257
Crocodile Lake National Wildlife Refuge, 176
Crocodiles, 310, 312, 313, 322
Cross-country travel, 381
Crowell, John, 126
Crown fires, 277, 279
Cutler, Rupert, 16, 122, 123
Cutthroat trout, 310

D

Daintree area (Queensland, Australia), 57
Damaging activities, focusing management on, 188
Deam, Charles C., Wilderness Act (Indiana), 151
Death Valley National Monument, 103
Death Valley National Park (California), 132
    expansion of, 146
Death Valley Wilderness (California), 165
Debt-for-nature swap, 77–78
Deer
    red, 311, 314, 321
De facto wilderness lands, 119
Defenders of Wildlife (2008), 322
Degradation, 266
De Grazia, Sebastian, 25
Denali National Park and Preserve (Alaska), 103, 211–12, 322
Denevan, William, 260–62
Depredation, 320–22
Desert tortoise, 310
Desolation Wilderness, 150, 367–68
    recreational use of, 387
    use of simulation models at, 480
Determining historical dynamics (HRV), 268
Developing countries
    wilderness recognition in, 53–54
    wilderness wildlife in, 317
Development Opportunity Rating System (DORS), 123
Dikaya mestnost (wild place), 69
Dinosaur National Monument (Colorado), 44–45
    use of simulation models at, 480
Direct counting in estimating wilderness recreational use, 370–71
Direct human impact, 262

Direct management techniques, 470–75
Direct observation in estimating wilderness recreational use, 370
Direct visitor management, limits of acceptable change process and, 246
Diseases
　control of, in wilderness, 102–3, 149
　interaction of fire with, 283
　in wilderness animals, 320
Distemper, 320
Disturbance, 254–55
Diversity, effect of fire on, 283
Dods, Dolly, Wilderness (WV), recreational use of, 379
Domestic animals, grazing by, 260–61
Drakensberg Mountains, 66
Drakensberg wildernesses, 66–67
Drive-through sequoia trees, 38
Drought episodes, 279
Dry lightning storms, 298

E

Eagle Cap Wilderness in the Wallowa-Whitman National Forest (Oregon), 46, 288, 371
　recreational use of, 382
　stock use in, 431
　vegetation damage at, 443
Eagles
　bald, 310, 330
　golden, 310
Early American scene, wilderness and, 5–6
Earth Summit (Rio de Janeiro, Brazil, 1992), 80
Eastern Deciduous Forest, fire-dependent ecosystem in, 302
Eastern Wilderness Act (1975), 121–22
　maintaining naturalness and solitude and, 15, 23
　wilderness areas in, xix, 3, 108, 182, 257
　wilderness study areas in, 144
Echo Park Dam controversy, 44
Echo Park in Dinosaur National Monument, 93
Ecological impacts of wilderness recreation, 395–435
　campsite impact management, 407
　campsite impacts, 407–10
　factors that influence amount of, 410–13
　monitoring conditions, 418–20
　restoration, 420–23
　strategies and techniques to control, 413–18
　importance of impacts, 396–98
　pack and saddle stock management, 429–34
　recreational activities, 398
　　campfires, 400–402
　　pack-stock grazing for visits, 403–4
　　trail construction and maintenance, 402–3
　　trampling, 398–400
　　water pollution and disposal of human waste, 406–7
　　wildlife disturbance, 404–6
　trail impact management, 423–24
　　erosion, 424–26
　　monitoring conditions, 428–29
　　muddiness, 426
　　rehabilitation, 429
　　user-created trails, 426–28
Ecological islands, fragmentation and isolation of wilderness as, 340
Ecological restoration in wilderness, 265–67
Ecologists Union, 175
Ecology, 8
Economic considerations, wilderness fire-management programs and, 292
Economic values of wilderness, 10–11
Economic values of wilderness wildlife, 316–17
Ecosystems
　approach to planning, 198
　human influences on, 20

Educated guesses in estimating wilderness recreational use, 373
Educational programs, 364, 383
　low-impact, 468
　minimum-impact, 466–68
Education programs, 463–65
Eland, 314
El Carmen–Big Bend Conservation Corridor, 77, 81
El Carmen Wilderness Area, 77
Electrical power infrastructure in assessing wilderness, 52
Electronic trail-traffic counters in estimating wilderness recreational use, 372–73
Elephants, 310, 313, 314, 315, 322. See also Pachyderms
　African, 310, 312
　Asoatoc, 321
Elk, 252, 313
El Malpais and Cebolla Wilderness Act, 148, 152
El Malpais National Monument Act, grazing and, 369
Emerson, Ralph Waldo, 7
Emigrant Basin Primitive Area (California), 108
Emigrant Wilderness (California), 347
Endangered American Wilderness Act (1978)
　buffer zones and, 192
　controlling fires, insects, and disease, 102
　debate in, 16
　fishery management in, 106–7
　passage of, 138, 144–45
　quality of wilderness values and, 137
　urbanization and encroaching development and, 349–50
　wilderness study areas and, 122
Endangered species, 341
　protection for, 319
Endangered Species Act (1973), 318–19
　reasons for enactment of, 19
Endangered Wilderness Act, 149, 150
Endangered wildlife, 184
Energy flows, effect of fire on, 282
Entry requirements, 468–69
Environmental assessment requirements, 199
Environmental groups, wilderness legislation and, 116–18
Environmental impact statements (EIS), 199
Environmental modification spectrum (EMS), 3, 22, 163
　managing wilderness as most pristine extreme on, 180–81
　Environmental Protocol to the Antarctic Treaty (Madrid Protocol) (Antarctica, 1991), 79–80
Equestrian trails, stock impact and, 430
Erosion, 424–26
Etosha Pan (Namibia), 75
Europe
　protected areas in, 55
　reclaiming wilderness in, 52–53
Evans, Dan, 133–34
Everglades National Park (Florida), 57, 58, 175
　establishment of, 39
　recreational use of, 379
Everglades National Park Act (1934), 39
Everglades National Park Wilderness (Florida), 58, 313
Executive Order 10092 (1949), 376
Exotic pathogens, introduction of, 261
Exotic weeds, 261–62
Experimental forests, 42
Expert Outdoorsman, 44
External threats, 338
Ezemvelo KZN Wildlife (South Africa), 56

F

Facilities in wilderness, 150

Featherstone, Alan Watson, 53
Federal Committee on Research Natural Areas, 174
Federal Forest Service, 70
Federal Job Corps, 11
Federal Land Policy and Management Act (1976), 3, 98, 104, 134, 144, 153, 366
Federated Mountain Clubs (FMCs), 72
Feely, Jim, 65–66
Fees, 469–70
Fern Ridge Preserve, 176
Ferrets, 314
Field interviews in estimating wilderness recreational use, 370
Finland, legal protection of wilderness in, 59–60
Finnish Forest and Park Service, 59–60
Fire(s), 184, 276–304
　as boundary effect, 264
　camp, 400–402, 472
　control of, in wilderness, 102–3, 149, 260, 263
　crown, 277, 279
　effect of, on energy flows, 282
　effect of, on succession, 280–81, 283
　fire-dependent ecosystems in, 298–302
　　Boreal Forest region, 301–2
　　California-Oregon Coast Ranges region, 298
　　Cascades Range of Oregon and Washington, 299
　　Coast Redwoods region, California, 298–99
　　Eastern Deciduous Forest, Appalachian, and Gulf Region, 302
　　Intermountain region and Southwest, 299
　　Lake Superior region, 301
　　Rocky Mountain region, 299–300
　　Sierra Nevada region of California, 298
　fire-management planning constraints, 291–94
　　economic considerations and, 292
　　fire behavior prediction, 294
　　interagency planning and, 292
　　limiting suppression impacts, 294–95
　history and evolution of fire-management programs, 296–98
　ignition as challenge in, 304
　lightning-caused, 281, 284, 290, 296, 298–99
　management of, 18
　natural (historical) role of, 280–83
　objectives of management, 285–89
　occurrence and behavior, 277
　　ignition and spread determinants, 277–78
　　major drought episodes, 279
　　regimes, 279–80
　　seasonal and climatic factors, 278–79
　　size and intensity, 279
　periodic, 276
　policy alternatives in, 289–91
　predictions on behavior of, 294
　prescribed, 192, 285, 290
　prevention of, 18
　programs and management, 303
　size and intensity of, 279
　stand-replacement, 300
　suppression of, 18, 283–85
　　limiting impacts, 294–95
　　wildland, 346–47
　trends and future needs, 302–3
　weather and, 277
　wild, 452
Fire-dependent ecosystems in American wilderness regions, 298–302
Fire-dependent plant communities, 280
Fire exclusion, 289, 293
Fire regimes, 276, 279–80
Fire roads, 42
Firewood collection, 404
Fish
　introduced, 262

Fish, cont.
   nonnative, 321
   protection of, in wilderness, 149
   stocking, 321
Fish and Wildlife Service, xiii, 3, 94
   in designating Department of Interior Roadless Lands, 129–34
   in developing roadless area review criteria, 130–31
   wilderness management planning in, 202
   wilderness study process and, 131–32
Fitz Creek Fire (1973), 297
Fitzpatrick Wilderness, 149
Flathead Indian nation, wilderness protection and, 68–69
Florida Wilderness Act (1984), 147
Foley, Tom, 127
Food and Agricultural Organization (FAO), 55
Ford, Gerald, 122
Forest Act (New Zealand, 1949), 73
Forest Act (South Australia, 1968), 64–65, 66
Forest and Rangeland Renewable Resources Planning Act (1974), 123, 200, 201
Forest Land and Resource Management Plan, 201
Forest management, 17
Forest products, meaning of, 40
Forest Service, U.S, xiii, xv, 90
   inventory of potential wilderness additions, 119
   launch of Operation Outdoors, 45
   management of Indian Mounds Wilderness by, 366
   roadless lands review, 16, 115–19, 121
   soft or hard release, 128–29
   statewide wilderness bills, 127
   sufficiency-release language, 127–28
   wilderness designation in 1980s, 126–29
   wilderness fire management and, 286
   wilderness management planning in, 200–202
   on wilderness remaining in national forests, 90
   in wilderness struggles, 43
Forest structure, 284–85
Forest wilderness, inventorying, 41
Forgotten lands, 134
Fosso del Capanno Wilderness Area, 74–75
Foxes, 320
Frank Church–River of No Return Wilderness (Montana)
   aircraft use in, 107, 349
   cougar-ungulate interaction and, 330
   mining management zone in, 147, 149
   national forests in, 351
   wilderness areas in, 165
   wilderness travel in, 376
   wolves and, 316
Fuels
   effect of fire on accumulation of, 282
   fire and, 277
   manipulating, without fire, 290
Future issues and challenges in wilderness stewardship, 487–93

G

Gallatin National Forest, 313
Garden of Versailles, 5
Gas, mining for, 365–67
Gazelle, 311, 323
   Mongolian, 309, 310, 311, 312
General management plans (GMPS), 199–200
Germany, national parks in, 52
Ghost River Wilderness, 64
Gil, Patricio Robles, 77
Gila National Forest (Mimbres, New Mexico), 28, 40, 41
Gila Wilderness Reserve, 257
   establishing, 41
Gilligan, James P., 92–93
Glacier Bay–Admiralty Island Biosphere Reserve, 175
Glacier Bay National Monument (Alaska), 103, 152
Glacier Bay National Park (Alaska), 20
Glacier Bay Wilderness, 20, 175
Glacier National Park (Montana), 325
   Hidden Lake area of, 130
   stock use in, 433
Glacier Peak Wilderness (Washington), 186, 269
Glacier Point, 38
Glen Canyon Dam, completion of, 45
Global climate change, 265, 267, 350
   effects of, on annual climatic regimes, 279
Global Conservation Fund (GCF), 81
Global Environmental Facility of the World Bank, 80, 83
Goal-achievement framework in writing wilderness management plans, 198, 204–6
Goat Rocks Wilderness (Washington), 347
   recreational use of, 377, 381
Goats
   mountain, 309, 310, 344
Golden eagles, 310
Gondwana, 79
Gore Range–Eagles Nest Primitive Area (Colorado), 116, 157
   litigation and, 118–19
Gorillas
   mountain, 316
Gospel-Hump Wilderness (Idaho), 149
Grand Canyon National Park (Arizona), 32, 45, 102, 132, 188
   fire in, 276
   float trip in, 34
   popularity of, 33
   recreational use on Colorado River in, 177
   use of simulation models at, 480
Grand Teton National Park (Wyoming)
   establishment of, 38–39
   fire management program in, 297
Grand Teton National Park Act (1929), 39
Grassroots environmental organizations on soft versus hard release, 128
Grassy Lake, 270
Graves, Forester, 40
Graves, Henry, 40
Grazing, 41, 319, 359, 403
   by domestic animals, 260–61
   livestock, 243–44, 323–24, 343, 368–69
   public-land, 368
   in wilderness, 104–5
Great Arctic Reserve, 70
Great Barrier Reef Marine Park, 82
Great Bear Wilderness (Montana), wilderness travel in, 376
Great Britain, wilderness in, 52–53
Great Dividing Range, 60–62
Greater ecosystem concept, 264–65
Greater St. Lucia Wetland Park, 65, 82
Greater Yellowstone Area fires (1988), 279, 287, 291, 292, 295, 300
Greater Yellowstone Ecosystem, 264–65
Great Gulf Wilderness (NH), 169, 205
   recreational use of, 378, 380
Great Smoky Mountains National Park (North Carolina), 132, 152, 175, 262, 313, 316, 328
   recreational use of, 374, 378
   trail condition in, 428
Great Swamp National Wildlife Refuge (New Jersey), 98, 129
Greeley, William B., 40, 41
Green Lake, 36
Greenland, wilderness in, 55
Grizzly bears
   economic land use and, 309

Grizzly bears, cont.
   elimination of, from many wilderness ecosystems, 263
   injuries by, 322
   management of, 252
   need of tracts of undisturbed habitat of, 316
   as wilderness-dependent wildlife, 310, 312, 313
   in Yellowstone ecosystem, 326
Grizzly Country (Russell), 312
Group-size limits, 471
Guanaco, 314
Guatopo National Park (Venezuela), 77
Gulf region, fire-dependent ecosystem in, 302

H

Habitat management, 21
Hansen, Greg, 18
Hanson, Rich, 47
Hard release, 126–27, 128–29, 150–51
Hartzog, George, 130
Hatfield, Mark, 15, 129
Hayakawa, Samuel, 128
Hayman Fire (Colorado) (2002), 294
Heart of the West, 81
Hellgate Treaty (1855), 68
Hells Canyon Wilderness Act, 151
Helman criteria, 61
Helms, Jesse, 128
Hemlocks, protecting, in southeast, 271–72
Herbivores, 314
Hetch Hetchy Valley, 37
Hickel, Walter J., 140
High-density recreational facilities, 192
High Peaks Wilderness, 183
High Sierra Hikers Assn., et al. v. Bernie Weingardt, et al., 158
Hildebrand, Joel H., 43
Hildebrand Committee, 43
Hillary, Edmund, 78
Himalayan thar, 321
Historical range of variability (HRV), 252, 255, 264, 267, 337
   fire management and, 285
   human disturbances effect of on, 255–56
Historical roots of wilderness management, 31–48
   destruction by popularity, 32–35
   intellectual dilemma, 31–32
Historic Preservation Act (1966), 319–20
Homo sapiens, 31
Hoopa Valley Indian Reservation, 170
Hoover Wilderness, 108
Horses
   corrals for holding, 462–63
   uses of, 376, 380
   wild, 309
Horton Plains, 68
Hubbard Brook Experimental Forest (New Hampshire), 175
Human access points in assessing wilderness, 52
Human-caused change, 196–97, 325
Human disturbances, effect of, on historical range of variability, 255–56
Human ecology view of outdoor recreation, 22
Human Influence Index rating, 52
Human influences
   on ecosystems, 20
   transmission of, 326
   wilderness protection and, 183–84
Humankind's historical role in wilderness, 258–59
Human safety, 322
Human waste, disposal of, 406–7
Humphrey, Hubert, 96
   wilderness legislation and, 94
Hunting, 324–25, 377
   encouraging styles with wilderness experiences, 328–29

I

Ickes, Harold L., 92
Ignition and spread determinants, 277–78
   as challenges to managing fires in wilderness, 304
Ignition source, fire and, 277
Ilmen'ski Zapovednik, 69
Imbewu, 67
Independence Fire (1979), 297
Indian Affairs, Bureau of (BIA), 68–69, 94
   exclusion of lands from wilderness, 133–34
Indian Mounds Wilderness, 147–48, 366
Indian Peaks Act, 152
Indian Peaks Wilderness (Colorado) (1978), 150
Indirect counting methods in estimating wilderness recreational use, 371–73
Indirect management techniques, 460–70
Indirect use of wilderness, 361–62
Indonesia, wildland watershed in, 78–79
Information programs, 463–65
Information to redistribute use, 465–66
Inholdings, 345–46
Insects
   controlling, 102–3, 149
   infestations and disease of, 18
   interaction of fire with, 283
Institutional outfitting, 361
Intensity of wilderness use, 375
Intentionally large ecological effects, 266
Interagency planning, wilderness fires and, 292
Interagency Wilderness Fire Management Program, 292
Interim management, 13–14
Interior Roadless Lands, Department of, designating, 129–34
Intermountain region, fire-dependent ecosystem in, 299
Internal threats, 338
International Biological Program (IBP), 174
International Convention for the Protection of the World Cultural and Natural Heritage, 175
International Journal of Wilderness (IJW), xv, 46, 51, 83, 154, 364
International League of Conservation Photographers, 83
Internationally Adjoining Protected Areas (IAPA), 81
International Union for the Conservation of Nature and Natural Resources (IUCN), 52, 54–56
   Commission on National Parks and Protected Areas, 83
   Wilderness Task Force of, 154
International Waterfowl and Wetlands Research Bureau, 57
International wilderness, 50–86
   in developing countries, 53–54
   forces of change and wilderness protection in, 51–58
   implications of recognition for, 58
   Ramsar sites, 57–58
   reclaiming, in Europe, 52–53
   related protection of nature, 54
International Wilderness Leadership (WILD) Foundation, xv, 56, 66, 67, 75, 83
Introduced fish and animals, 262
Invasive plants, 261–62
Inyo National Forest (California), 32, 33
Irrigation water impoundments, 367
Isle Royale, 313
Italy, wilderness areas in, 74–75

J

Jaguars, 310, 313, 314
Japan, wilderness areas in, 75
Jasper National Park, 63, 64, 300
Johnson, Harold "Bizz," 127
Johnson, Huey, 126
Johnson, Lyndon B., 90, 94
   increase of primitive area and, 117
   passage of Wilderness bill and, 95
Joshua Tree National Park, expansion of, 146

K

Karlton, Lawrence, 129
Katun River, 70
Katun Zapovednik, 70
Kenai National Wildlife Refuge (Alaska), comprehensive conservation plans (CCPs) for, 215
Kennedy, John F., 96
Keystone species, 313
Kgalagadi Transfrontier Park (South Africa), 67
Kiang, 314
Kiger Creek (Oregon), 171
Kings Canyon Wilderness National Park, 39, 284, 321
   fire management in, 291
Klamath River, 170
Kluane National Park, 322
Kneipp, L. F., 40, 41–42
Kofa Wildlife Refuge Wilderness (Arizona), 182, 190
Komodo dragons, 310, 313
Kootenai Indians, 68
Kootenay, Canada, 63
Kosciusko National Park (New South Wales), 60
Kozacek, Susan, 28
Kruger National Park (South Africa), 316
Kuchel, Thomas, 117
Küchler's mapping, 166
KwaZulu-Natal (South Africa), 82

L

Laguna Atascosa National Wildlife Refuge (Texas), 132
Lake Colden, 174
Lake Crescent, 134
Lake Ozette, 134
Lake St. Lucia (South Africa), 65, 82
Lake Superior region, fire-dependent ecosystems in, 301
Lake trout, 326
La Michilla Biosphere Reserve (Mexico), 56
Lamprey River, 171
Land allocation, 198
Land health, study of, 9
Land Management, Bureau of (BLM), xiii, 2, 3, 134–39
   primitive area policies, 366–67
   public domain lands and, 167
   roadless area review process, 135
      inventory, 135–36
      reporting wilderness or nonwilderness recommendations, 139
      wilderness study, 136–39
   wilderness management planning in, 202–3
   Wilderness Study areas, 153
Landscape ecology, 268
Landscape-scale intervention, 266
Land transformation in assessing wilderness, 52
Lane, Franklin K., 38
Langford, Nathaniel P., 37
Lassen Volcanic National Park (California), interagency fire planning and, 292
Latin America, wilderness areas in, 76–78
Leafy spurge, 184
Leave No Trace program (LNT), 43, 204
LeConte, Joseph N., 36
Legal wilderness, 4
Legislative Act on Wilderness Reserves (Finland, 1991), 59
Legislative management direction for specific issues, 146–50

Leishmaniasis, 320
Length-of-stay limits, 471–72
Leonard, Richard M., 44
Leopards, 309, 310, 313, 316
   snow, 314, 321
Leopold, A. Starker, 309, 312
Leopold, Aldo, 487
   administrative zoning of wilderness and, 72
   advocate for wilderness protection, 41
   decision making for national forest by, 43
   as early leader of wilderness movement, 16
   ecological values of wilderness and, 255
   in Forest Service, 7, 8
   L-20 Regulation and, 90
   recreational values of, 12
   scientific use of wilderness and, 362
   U regulations and, 91
   vision of, 4
   wilderness ecosystems and, 252
   wilderness wildlife and, 309, 315–16
Leopold, Aldo, Wilderness Research Center Institute (Missoula, Montana), xx, 46, 153, 302, 337
Leopold report of 1963, 286, 296
"Let burn" policies, 192
Lewis and Clark Trail, 336
Light-handed management techniques, 32
Lightning, 277
Lightning-caused fires, 281, 284, 296, 298–99
   management of, 290
Light pollution, 263
Limits of acceptable change (LAC) process, 187, 190–91, 198, 218–49
   acceptable conditions in, 226–28
   campsite conditions and, 418
   carrying-capacity concept, 219–22
   concept of, 222–33
   defining goals and desired conditions, 224–25
   evaluating and selecting preferred alternative, 232
   identifying alternative opportunity class allocations, 231
   identifying area issues, concerns, and threats, 225–26
   identifying management actions for each alternative, 231–32
   implementing, 234–45
   implementing actions and monitor conditions, 232–33
   inventorying existing resource and social conditions, 229–30
   managing visitor use and impact and, 219
   positive benefits to wilderness from, 233–34
   selecting indicators of resource and social conditions, 228–29
   specifying standards for resource and social indicators for each opportunity class, 230–31
   timing in using, 223
   for wilderness planning, 206–7
Lions, 312, 313, 314, 316
   African, 309, 310
Litigation, wilderness legislation and, 155–56
Litter control, 476
Livestock grazing, 243–44, 323–24, 343, 368–69
Lodgepole pine, 269
Logging, 41, 91, 263
Lone Peak, 122
Louisiana Pacific Timber Co., 126
Lower Delaware River (New Jersey and Pennsylvania), 171
Low-impact education, principles of, 468
L-20 Regulation, 41–42, 90–91, 98, 109–10, 196
Luquillo Experimental Forest (Puerto Rico), 175
Lynx, Canada, 310

M

Mack Lake Burns, 295–96
Maha Eliya Montane Forest, 68
Manageability, 137–38
Management-by-reaction, 197
Management directions, special provisions in wilderness legislation, 146
Management funding and resources, 491–92
Management paradox, 16–17
Management problems
 designation limiting uses and, 16
 designation of areas with unusual, 15–16
Managerial presence, 453
Man and Biosphere (MAB) Coordinating Council, 56
Man and Biosphere (MAB) Program, 56
Mandatory wilderness permits in estimating wilderness recreational use, 371
Mange, 320
Marble Mountain Wilderness (California), 101
Maricopa Complex Wilderness, 212
 implementing limits of acceptable change process for management plan, 242–44
Marine Protected Areas, 55
Marine wilderness, 82
Marion Lake (Oregon), 15
Mariposa Grove, Yosemite National Park (California), 278
Marshall, Bob, Wilderness of Montana and Washakie Wilderness (Wyoming), xix
 acreage of, 351
 consolidation and reclassification of wilderness areas in, 92
 as grizzly country, 309
 mineral leases and, 128
 precipitation levels and patterns in, 104
 primitive areas of, 91
 recreational use in, 361, 375, 378, 379, 380
Marshall, George, 2
Marshall, Robert
 career of, 42–43
 as chief of Division of Recreation and Lands, 91
 favoring of wilderness-dependency, 186
 founding of Wilderness Society, 350
 ideas for establishing wilderness areas, 92
 vision of, 4
 as wilderness advocate, 2
 wilderness designations and, 96
 as wilderness movement leader, 16
 writings of, 7
Martin, Vance, 56
Mather, Stephen T., 38, 41, 91
Mavuradonna Wilderness Area, 74
McArdle, Richard, 91
McCaskill, Lance, 72
McCloskey, Maxine, 82
McCloskey, Michael, 5, 6, 11, 12, 81–82
McClure, James, 128
McCollum, Ian, 66
McNeil River Falls (Alaska), 322
Measurement, units of, in estimating wilderness recreational use, 373–74
Mechanized access, 344
Mednyi Island arctic fox, 320
Megaherbivores, 310, 312
Melcher, John, 122
Meso-predator release, 261
Metcalf, Lee, 96
Metcalf, Lee, Wilderness (Montana), 205, 206
 recreational use of, 382
Michigan Wilderness Act, 149
Migratory ungulates, 312
Migratory waterfowl, 310
Military overflights in wilderness, 151
Millers Reach Fire (Alaska) (1996), 294

Milton, John, 312
Minerals
 leasing, 147
 mining for, 365–67
Minimum-impact camping, 43
Minimum-impact education programs, 466–68
Minimum-impact stock practices, 432–33
Minimum-impact wilderness skills, 44
Minimum Requirements Decision Guide (MRDG), 267, 271
Mining, 346
 for minerals, oil, and gas, 365–67
 potential for, 15–16
 in wilderness, 103–4, 147–48
Mining in the Parks Act (1976), 366
Mission, 66, 45
Mission Mountains Tribal Wilderness Area (Montana), 68, 69, 107
 recreational use of, 379
Mkhize, Vance, 56
Mojave National Preserve, creation of, 146
Mongolian gazelle, 309, 310, 311, 312
Monongahela National Forest (West Virginia), 259
Montana, University of, College of Forestry, xx
Moore, Gordon and Betty, Foundation, 81
Moose, 309
Mormon Pioneer National Historic Trail, 169
Morton, Rogers C. B., 140
Morton National Park (New South Wales, Australia), 61
Moths, 326
Motorized and mechanical equipment trespass and legal use, 348–49
Motorized use in wilderness, 148
Mountain caribou, 310
Mountain climbing, 377
Mountain goats, 309, 310
 translocation of, 344
Mountain gorillas, 316
Mountain Travel, 78
Mount Angeles Roadless Area, 133
Mount Baker–Snoqualmie National Forest, wilderness study on, 12
Mount Belukha, 70
Mount Everest, 78
Mount Jefferson Wilderness, 15
Mount Kosciusko, 60
Mount Logan Wilderness, 100
Mount McKinley National Park, 103
Mount Rainier, 134
Mount San Jacinto Wilderness, 172
Mount Trumbull Wilderness, 265
Mount Washington (New Hampshire), 336
Mount Whitney (California), 32–33, 188
 visitation to, 33
Muddiness, 426
Muir, Andrew, 67
Muir, John, 7, 8, 11–12, 16, 36, 37
Muir, John, Trail, 33
Muir, John, Wilderness (California), 2, 33, 45, 208, 270, 321
 recreational use of, 378, 379
Multiple-Use, Sustained-Yield Act (1960), 97, 100, 119, 200
Murie, Margaret, 312
Murie, Olaus, 72–73, 312
Musk ox, 309

N

Namib Desert (Namibia), 75
Namibia, wilderness areas in, 75
Nash, Roderick, 44
National Ambient Air Quality Standards (NAAQS), 293
National Environmental Management: Protected Areas Act No. 57 (South Australia, 2003), 64

National Environmental Policy Act (NEPA) (1970), 120, 190
 wilderness planning under, 198–99
National Forest Management Act (NFMA) (1976), 123, 200
National forest roadless lands, 152–53
 review of, 115–19
National forests, 94, 167
 evolution of wilderness in, 39–43
National historic trails, 169
National Interagency Fire Center (2008), 346
National Oceanic and Atmospheric Administration, 82
National Outdoor Leadership School (NOLS), 8
National Park Organic Act, 100
National parks, 167
 evolution of wilderness in, 36–39
National Parks Act (Canada, 1988), 62–63
National Parks Act (New Zealand, 1977), 73
National Parks Act (New Zealand, 1980), 71, 73
National Parks and Recreation Act (1978), 199
National Parks Association, wilderness legislation and, 94
National Park Service (NPS), xiii, 3
 in designating Department of Interior Roadless Lands, 129–34
 in developing roadless area review criteria, 130–31
 fire management and, 277
 launching of Mission 66, 45
 Program Standards, 199
 wilderness fire management and, 286
 wilderness management planning in, 199–200
 wilderness study process and, 131–32
National Park Service Act (1916), passage of, 37–38
National Park System, 94
 wilderness in, 152
National Plan for American Forestry, 43
National recreation trails, 168
National scenic trails, 168–69
National Survey on Recreation and the Environment (NSRE), 48
National Trails System, 168–69
National Visitor Use Monitoring Project, 374–75
National Wild and Scenic River System, 140
National Wilderness Federation, 318
National Wilderness Preservation Council, 94
 elimination of, 96
 standards from, 144
National Wilderness Preservation System, xiii, xvi, xix, 2, 3, 48, 97
 complementary conservation areas and, 163–78
 ecological and landscape diversity within, 166–67
 establishment of, 45
 federal agency management in, 164–68
 growth of, 164–68, 336
 overall use of, 388
 size of, xvii, 165, 385
 statistical sources for information concerning, xix–xx
National Wildlife Federation, 317
 wilderness legislation and, 94
National Wildlife Refuges and Game Range System, 94
National Wildlife Refuge System Administration Act (1966), 202
Native Americans
 role in shaping wilderness landscapes of North America, 258
 use of fire, 277, 284, 291
Native Lands and Wilderness Council, 80, 83
Natural Heritage Wilderness Act (Sri Lanka, 1988), 68
Naturalness, 451–52

Naturalness, cont.
　　anthropocentric approach and, 20
　　assessing, for wilderness planning, 267–68
　　defining, in wilderness fire management, 287–89
　　incremental loss of, 336–37
　　as relative concept, 338
Natural recovery, 266
Natural variability, 338
Natural water catchment, 218
Nature, past disruption of, 18
Nature Conservancy (TNC), 163–64, 175–76
Nature Conservation Acts (South Australia, 1968), 64–65
Nature Conservation Law (Japan, 1972), 75
Nature study, 376–77
Nebraska Wilderness Act, 148
　　grazing and, 369
Need, wilderness designation and, 119
Nellie Juan–College Fiord WSA (Chugach National Forest), 141–42
Nelson, Michael, 51
Nepal, wilderness in, 78
Neuberger, Richard L., 92
Nevada Wilderness Act, 146
New England Wilderness Act (2006), 199
New Mexico Wilderness Act, 150, 151, 350
　　grazing and, 369
New South Wales (NSW), 61
New wilderness knowledge and legislative direction, 153–54
New Wilderness Recreation Strategy, 181, 183, 189
New York, wilderness preservation in, 173
New York State Adirondack Park, 183
New Zealand
　　Forest Service in, 73
　　National Park circles in, 72
　　protected areas in, 55
　　wilderness areas in, 71–74
Nilgiri tahr, 314
Nixon, Richard, 122
Noatak and Gates of the Arctic Wilderness Complex, 165
Noatak National Preserve, 58
Noatak Wilderness, 58
Nonconforming but allowed use, 343
Nonconforming uses, 359
Nondegradation, 14–15
　　concept of, 182–83
Nonnative fish, 321
Nonnative species, 311, 322, 343–44
Nonrecreational use, 359
Nonreserved national forest, 118–19
　　designating, 119–22
Norgay, Tenzing, 78
Normative behavior, visitor impact and, 446
North America, protected areas in, 55
North American desert bighorn sheep, 323
North Cascades National Park (Washington), 36, 134, 152
Northern Sierra Madre, 81
Northern Sierra Madre Wilderness Foundation, 76
North Fork of Lone Pine Creek Portal, 33
North Maricopa Mountains Wilderness (Arizona), 169, 212
Northwestern Hawaiian Islands Marine National Monument, 82
Northwest Forest Planning process, 265
Norton v. Southern Utah Wilderness Alliance, 156
No-take management, 82
Ntombela, Magqubu, 67
Nutrient cycles, effect of fire on, 282

O

Oakland Hills Fire (1991), 294
Oelschlaeger, hypothesis by, 5
Oglala Sioux, 53
Oil, mining for, 365–67
Okanogan-Wenatchee National Forest, wilderness study on, 12
Okavango Delta (Botswana), 58
Okefenokee National Wildlife Refuge, 58
Okefenokee Swamp Act (1974), 148
Old Faithful, 37, 38
Old Woman Mountains Wilderness, 24, 190
Olson, Sigurd, 10, 312
Olympic National Park (Washington)
　　mountain goats in, 311, 321
　　recreational use of, 374
　　roadless areas of, 152
　　study of aquatic habitats at, 268
　　transfer of national forest wilderness and, 91
　　wilderness review of, 132–34
　　wilderness study area of, 160
Olympic Primitive Area, 91
Operation Outdoors, 45
Opinion Leaders Programme, 67
Orangutans, 311
Orcas, 313
Oregon Natural Resources Council, 129
Oregon Wilderness Act, 149
Organic Act (1897), 40
Organisms, 252
Organ Pipe Cactus National Monument (Arizona), 103
Original wilderness character, 38
Osprey, 310
Outdoor recreation, human ecology view of, 22
Outdoor Recreation Resources Review Commission (ORRRC), 10, 93, 383
　　land classes developed by, 132–33
　　studies of, 388
Outfitting, 106–8
　　commercial, 360–61
　　institutional, 361
Outhouses, 462
Outside influences, 262
Outward Bound, 8, 46
Ouzel Fire (1978), 295, 296
Overflights, 102
Overgrazing, 261

P

Pachyderms, 315, 324. *See also* Elephants
Pacific Crest Trail, 168–69
Pack stock
　　grazing for recreational visits, 403–4
　　managing, 429–34
Paddy Creek Wilderness Act (Missouri), 151
Palanan Wilderness Area, 76
Pantanal jaguars, 310
Panther, 324
Papahānaumokuākea Marine National Monument, 82
Paradigm shift in ecology, 254
Parker v. United States, 157
The Park Planning Source Book, 200
Parks Canada, 62–63
Passenger pigeons, 309
Patagonian puma, 311
Peace Parks Foundation, 81
Peace River, 321
Peak Wilderness, 68
Pehrson, Jan, 34
Pemigewasset, 205
Pend d'Oreille Indians, 68
Pennsylvania Wilderness Act, 149
Peregrine falcon, 310
Periodic fires, 276
Permits, 32
Personal wilderness values, 26–27
Peterson, R. Max, 129

Petrified Forest National Park (Arizona), 129
Pharmaceutical industry, drugs originating in wild areas, 9
Philippine National Integrated Protected Areas System Act (1992), 75–76
Philippines, wilderness areas in, 75–76
Photography, 376–77
Pigeons, passenger, 309
Pinchot, Gifford, 39–40
Plants
　　fire and, 276, 280–81
　　invasive, 261–62
　　loss of species, 261
Player, Ian, 65–66, 67
Pleasuring grounds, 39
Poaching, 324–25
Polar bears, 310, 312, 313, 326
Political value of wildlife, 317–18
Pollutants, 262
Population density in assessing wilderness, 52
Population growth, 17
Potlatch Corp. v. United States, 158
Powell, John Wesley, 32
Prairie dogs, 313
Precedents, wilderness management and, 14
Predatory animals, elimination of, 261
Prescribed fires, 285
　　igniting, 290
　　policies on, 192
Presidential Range–Dry River Wilderness (New Hampshire), 205, 257, 336
　　recreational use of, 380
President's Council on Environmental Quality, 56–57
Primates, 324
Primeval influence, 265
Primitive area, 41
　　early appearance of as term, 60
Primitive simplicity, 42
Primitiveness, 61
Pristine myth, 260–62
Private land protection efforts, 175–76
Protected Area Network (PAN), 53
　　Parks Foundation (PPF) of, 53
Provincial Parks and Conservation Reserves Act (Canada, 2006), 63, 64
Przewalski's horse, 323
Public domain lands, 167
Public involvement
　　plan-review process and, 210–11
　　in wilderness management, 189–90
Public-land grazing, 368
Public Land Law Review Commission, proposed legislation to create, 95
Public Land Order Number 4582, 140
Public recreational use, increasing, 341–43
Public recreational use of wilderness, 359–60
Pumas, 310, 313
　　Patagonian, 311
Purity doctrine, 14, 107
Pusch Ridge Wilderness (Arizona), 122, 349–50
　　recreational use of, 377
Putin, Vladimir, 70

Q

Quality standards for wilderness study, 138
Quetico Provincial Wilderness Park (Canada), 63, 81, 85–86, 106
Quotas, 32

R

Rabies, 320
Raccoons, 319
Ramsar Convention on Wetlands of International Importance, 57, 58
Ramsar sites, 57–58
RARE. *See* Roadless Area Review and Evaluation

(RARE)
Rationing use, 473-75
Reagan, Ronald, 128
Recreational developments, 40
  in primitive areas, 42
Recreational rivers, 170
Recreational saturation point, 39
Recreational use, 197, 398
  aesthetics and, 316
  campfires, 400-402
  of Colorado River in Grand Canyon National Park, 177
  pack-stock grazing for visits, 403-1
  trail construction and maintenance, 402-3
  trampling, 398-400
  water pollution and disposal of human waste, 406-7
  of wilderness, 21
  wildlife disturbance, 404-6
Recreation opportunity spectrum (ROS), 378-79
  adding system of roadless recreation areas, 168
  allocating of land uses, 192
  guidelines for land-use planning for recreation, 180
  inclusion of urban and recreation areas, 3
  maintaining diversity among wildland settings, 22
  redistribution of vision use and, 449
  wilderness-dependence and, 359
  zoning and, 209
Red deer, 311, 314, 321
Reed, Nathaniel, 130, 131
Reindeer husbandry, 60
Reintroductions, 323
  controversy over wolf, 331-32
Religion, wilderness and, 5
Remoteness, 61
Renewable Resources Assessment Program, 201
Research natural areas (RNAs), 163, 174
Reserves Act (New Zealand, 1977), 71, 73
Residential development, 17
Resort concept, 38
Resource Management Plan (RMP), 203
Restoration, 266
Rewilding, 53
Rhinos, 309, 310, 312, 315, 316
Richardson's blue grouse, 310
Rinderpest, 320
Rio Tinto Zinc, 82
Roadless Area Review and Evaluation (RARE)
  principal objectives of, 120
  Sierra Club and, 122
Roadless Area Review and Evaluation (RARE I), 16, 119-20, 123
Roadless Area Review and Evaluation (RARE II), 9, 119, 122-26, 166, 257
  California v. Bergland, 125-26
  evaluation of areas, 124-25
  inventory, 123-24
  recommendations from, 125
Roadless areas, 167-68
  evaluation of, for wilderness suitability, 115
  national forest, 152-53
  review criteria for, 130-31
  review of Forest Service, 122-26
  review of national forest, 115-19
Roberts, James O. M. "Jimmy," 78
Rocks and Islands Wilderness (California), 165
Rocky Mountains, 269
  fire-dependent ecosystem in, 299-300
  National Parks in, 321
  World Heritage Site in, 63
Roosevelt, Teddy, 317
Russell, Andy, 312
Russia, wilderness areas in, 69-71

S

Saddle stock, managing, 429-34
Sagarmatha National Park, 78
St. Francis of Assisi, 5
St. Regis Canoe Area, 173
Salamanders, 310
Salish Indians, 68
Salmon, 326
Sami culture, 59
San Bernardino Mountains (California), 262
A Sand County Almanac and Sketches Here and There (Leopold), 7
Sandia Mountains, 122
Sandwich Range, 205
San Gorgonio Wilderness (California), 45, 95, 192, 347
  management controls in, 15
  recreational use of, 377
San Jacinto Wilderness (California), recreational use of, 377, 379
San Juan-Rio Grande National Forests Wilderness, 259
  implementing limits of acceptable change process for, 238-40
San Rafael Primitive Area (California), 15, 116-18
Santa Ana or Foehn winds, 277
Santa Rosa Mountains Wilderness Area, 172
Sarawak, wildlands in, 78
Sawtooth Wilderness Act, 151
Saylor, John, 22, 96
  wilderness legislation and, 94
Scale in wilderness ecosystems, 256
Scapegoat Wilderness, part of the Bob Marshall Wilderness Complex (Montana), 280
Scenic rivers, 170
Schulz, Florian, 81
Science of land health, 252
Scientific Committee for Antarctic Research and Council of Managers of National Antarctic Programs (COMNAP), Committee on Environmental Protection, 80
Scientific theme, 8-10
Scientific values of wilderness, 20-21, 362-64
Scotland, wilderness in, 53
Sea otters, 313
Seasonal factors, fire management and, 278-79
Sediment, effect of fire on, 283
Seeding, 423
Selective restriction, 188
Self-willed land, 31
Selous Game Reserve (Tanzania), 317
Selway-Bitterroot Wilderness (Idaho and Montana), 15, 165, 196, 213, 347, 361, 367
  fire management program in, 297
  recreational use of, 375, 378, 382
  wilderness travel in, 376
Selway River (Idaho), 170
Sequoia-Kings Canyon National Park (California), 33, 378
  establishment of, 39
  fire in, 276
  fire management program in, 291, 296-97
  recreational use of, 374
  stock use in, 433
Sequoia trees, drive-through, 38
Sequoia Wilderness, 33
Serengeti National Park (Tanzania), 74, 320
Shasta-Trinity National Forest, controversy over, 125-26
Shelters, 463
Shenandoah National Park (Virginia), 121, 328
  recreational use of, 374
Shining Rock, recreational use of, 387
Shivalik Ranges (India), 321
Siberian tigers, 309, 310

Sieker, John H., 91
Sierra Club, 51
  founding of, 7
  1951 Biennial Wilderness Conference of, 93
  organized outings of, 36
  Outing Committee of, 44
  RARE process and, 121, 122, 135
  sponsorship of High Sierra Wilderness Conference, 44
  wilderness legislation and, 94
  wilderness management and, 43, 44
Sierra Club v. Lyng, 158
Sierra Club v. Watt, 135
Sierra Estrella, 212
Sierra Nevada Mountains (California), 2, 262, 269, 367
  fire-dependent ecosystems in, 298
Siffleur Wilderness Areas, 64
Silver Pass, 2
Silviculture, 21
Simulation models
  use of, 478-80
Sinhgangan (Rain) Forest, 68
Skeleton Coast National Parks (Namibia), 75
Sleeping sickness, 320
Smith, Mike, 82
Smithsonian Institution, 94
Smoke, impacts of, on visitors, 287
Snowfields, 426
Snow leopards, 314, 321
Snow measurements, 367
Social conditions, visitor impacts on, 445-48
Sociological wilderness, 4
Soft release, 127, 128-29, 151
Soil condition, changes in, 408-9
Solitude, 452-53
Sonoran Desert, 77
South Africa, wilderness protection in, 64-68
Southeast Asia, wilderness areas in, 78-79
Southern Rockies, 81
South Fork Meadows Trail, 192
South Fork Primitive Area, 91
South Maricopa Mountains, 212
Southwest, fire-dependent ecosystem in, 299
Soviet National Park System, 70
Spanish Peaks Primitive Area, use of simulation models at, 480
Spanish Peaks (Colorado) trail system, recreational use of, 380
Special-interest species, 269
Special provisions and congressional direction, 150-52
Spectacled bear, 311
Spiritual development, 10
Spotted knapweed, 184
Sri Lanka, wilderness protection in, 68
Stahl, Carl J., 8
Standards, 204
Standing Bear, 53-54
Stand-replacement fires, 300
State Committee on Environmental Protection, 70
Statewide wilderness bills, 127
State wilderness systems, 171-73
Steens Mountain Wilderness, 155
Stegner, Wallace, 10
Step-down management plans, 202
Stikine-LeConte Wilderness (Alaska), 107, 141, 191
Stock management, challenges of, 434
Stokes, Jerry, 352
Stolby Zapovednik, 70
Structures
  ecosystem, 253
  in wilderness, 150
Subalpine meadows, 268-69
Succession, 254-55

Succession, cont.
   effect of fire on, 280–81, 283
Sufficiency-release language, 127–28, 150–51
Suitability, wilderness designation and, 119
Sumner, Lowell, 39, 42, 44
Sundarbans (India-Bangladesh border), 58
Superior National Forest (Minnesota), 32, 41, 92, 105
Superstition Mountains Wilderness (Arizona)
   implementing limits of acceptable change process for management plan, 244–45
   recreational use of, 377, 380
Sustainable development, 54
Switzerland, lack of wilderness in, 52
Sycamore Canyon Wilderness (Arizona), 184

T

Table Top, 212
Tanzania, wilderness areas in, 74
Tapirs, 324
Taylor Complex Fire, 279
Taymyr Zapovednik, 70
Temporary wilderness status, 95
Termites, 313
Terrestrial wilderness, 82
Teton Wilderness Plan, 297
Texas Wilderness Act (1984), 147
Thahansi kale, 68
Thompson, Tom L., 201
Thoreau, Henry David, 6
Threatened sites, focusing management on, 188
Threatened species, 341
Threats
   external, 338
   internal, 338
   wilderness conditions and values affected by, 337–39
Three Sisters Primitive Area (Oregon), 92
   recreational use of, 381
Three Sisters Wilderness (Oregon), 56, 58, 164, 175, 189
Tibetan antelope, 309, 311, 312, 323
Tigers, 309, 310, 312, 313, 321, 322
   Siberian, 309, 310
Tongariro National Park, 71
Tongass National Forests (Alaska), 125
Tonto National Forest, 244
Topography/landscape features, fire and, 277–78
Tourist-development plans, 40
Trail(s)
   construction of, 402–3
   design of, 460–61
   maintenance of, 266, 402–3, 460–61
   managing impacts of, 423–24
   monitoring conditions, 428–29
   rehabilitation of, 429
   standards for, 45
Train, Russell, 56
Trampling, 398–400, 403
Transboundary Conservation, 81
Transboundary effects, 326–27, 340
Transfrontier conservation, 81
Trappers Lake (Colorado)
   development of, 40
   proposal for, 8
Tribal Council, 69
Tribal Ordinance 79A, 69
Tropical Wilderness Protection Fund (TWPF), 81
Trosper, Thurman, 69
Truman, Harry S., signing of Executive Order 10092 (1949) by, 376
Trypanosomiasis, 320
Tsarmitunturi, 59
Tsetse fly, 320

U

Udall, Morris, 122
Udall, Stewart, 140
Ulaanbatarr-Beijing railroad, construction of, 310–11
Umbrella effects, 312
Umfolozi Game Reserve, 65
Ungulates, 321, 322
United Nations Educational, Scientific & Cultural Organization (UNESCO), 55, 57, 175
United Nations Environmental Program (UNEP), 51, 55
UN List of National Parks and Protected Areas (2003), 55
Upper Colorado River Storage Project, 93
Urbanization and encroaching development, 349–50
U regulations, 42–43, 91–94, 98, 110, 196
Use allocation, 198
Use-intensity analysis, 375
User-created trails, 426–28
Utah Wilderness Act, 148, 149, 150
   grazing and, 369

V

Values, recent research in, 12
Vegetation
   change in, 408
   influence of fire on, 281–82
   manipulating, without fire, 290
Vermont Wilderness Act, 150
Virginia Wilderness Act, 149, 152
Virgin Islands National Park, 175
Visitor Experience and Resource Protection (VERP), 200
Visitor impacts on social conditions, 445–48
Visitor management considerations, 441–42
Voluntary self-registration in estimating wilderness recreational use, 370–71
Volunteer worker programs, 45
Voyageurs National Park, 106

W

Wagar, J. V. K., 43–44
Wallowa-Whitman National Forest (Enterprise, Oregon), 28
Walruses, 312
Wapiti (elk), 309
Wapiti Wilderness (Murie and Murie), 312
Washington Mount Jefferson Wilderness (Oregon), recreational use of, 381
Washington Park Wilderness Act (1988), 134
Washington Wilderness Act, 149
Water and marine wilderness, 81–83
Water pollution and disposal of human waste, 406–7
Water projects and water quality, 347–48
Water quality, effect of fire on, 283
Water resource development, 104
Water storage and use, 367–68
Water yields, effect of fire on, 283
Watt, James, 135
Waugh, Frank A., 40
Weather, fire and, 277
Weaver, James, 16
Weeds, 184
   eradication of new infestations, 266
   exotic, 261–62
Wekiva River (Florida), 171
West Virginia Wilderness Act (1983), 147, 151
Wetlands
   defined, 58
   wilderness values of, 58
Whales, Beluga, 312
White Cap fire-management area, 297
White Clay Creek (Delaware and Pennsylvania), 171
White-eared kob, 312
White Goat, 64

White Mountain National Forest (New Hampshire and Maine), 169, 205
   campsite indicators on, 248
   implementing limits of acceptable change process for management plan of, 240–42
   management plan for, 23, 199
White-tailed ptarmigan, 310
Whitney Portal Trailhead, 33
Wild and Scenic Rivers Act (1968), 157, 169–71
Wild areas
   defined, 91–92
Wild dogs, 314
Wildebeest, 309, 310, 312
Wilderness
   access to, 46–47
   activities in, xvii
   appreciation of, 36
   assessment of, 52
   carrying capacity for, 19
   categories of benefits, 11
   conservation of
      evolution of, 66
      reasons for, 51
   constraints to participation, 391
   defined, xix, 4–5, 22, 51, 318
   in developing countries, 53–54
   early American scene and, 5–6
   ecological restoration in, 265–67
   etymology of, 31
   evaluation of roadless lands for suitability, 115
   evolution of, 90
      in national forests, 39–43
      in national parks, 36–39
   expansion of, 492–93
   fragmentation and isolation of, as ecological islands, 340
   growing importance of, 488–90
   herbivores in, 321
   historical origins of, 5
   historical role of humans in, 258–59
   historical themes and values, 6–12
      economics, 10–11
      experiential, 7–8
      scientific, 8–10
      symbolic and spiritual, 10
   international recognition for, 58
   legal, 4
   legislation designating areas with compromised conditions, 350–51
   modern-day American views on, 48
   Native Americans role in shaping, 258
   native wildlife found in, 311
   naturalness and solitude in, 22–24
   nonnative species found in, 311
   personal values in, 26–27
   philosophy in, 19
   places in, xvii
   progress in preservation of, 2
   public interest in, 2
   reclaiming, in Europe, 52–53
   relating themes and values to management, 11–12
   relationships with, 483
   in Rocky Mountain National Park, 152
   scientific values of, 20–21
   shaping habitat in with natural processes, 327
   sociological, 4
   as special place, xiii
   statutory protection for, 92–94
   threats to, 17, 337–39, 339–52, 354
   wildlife in, 312–13
Wilderness Act (1987) (NSW 1987), 61
Wilderness Act, U.S. (1964), 2, 83, 89–111, 133–34, 200

Wilderness Act, cont.
   changes and compromises to pass, 96
   comparing 1956 Bill to, 111
   Congressional debate over, 6
   defined, 257–58
   defining of wilderness in, 22
   designation criteria and, 15
   evolution of wilderness protection, 109–10
   hierarchy of wilderness management and, 319
   intent of, 20
   as landmark legislation, 493
   legal wilderness as defined by, 4
   legislative history, 94–96
   L-20 Regulation and, 90–91
   management implications of, 143–44
   multiple agencies and missions and, 318
   passage of, 6, 7
   purposes of, 336
   reasons for enactment of, 19
   Section 1—title, 97
   Section 2—establishes wilderness system, 97–99
   Section 3—National Wilderness Preservation System, U.S. (NWPS) —extent of system, 99–100
   Section 4—use of wilderness areas, 100–108
      actions to control disease, 102–3
      actions to control fire, 102–3
      actions to control insects, 102–3
      Boundary Waters Canoe Area, 105–6
      grazing in wilderness, 104–5
      mining in wilderness, 103–4
      outfitting and guiding, wildlife in wilderness, 106–8
      overflights, 102
      water resource development, 104
   Section 5—state and private lands within wilderness areas, 108
   Section 6—gifts, bequests, and contributions, 108
   Section 7—annual reports, 108–9
   some exceptions and ambiguities in wilderness legislation, 109
   statutory protection for wilderness, 92–94
   test of time for, xiii
   text of, 96, Appendix A
   U regulations and, 91–94
   wilderness in, 257, 337
   wilderness use in, 358–59
   wilderness visitor management and, 440–41, 442
Wilderness Action Group (WAG) (Southern Africa), 67, 73, 75
Wilderness and Ecological Reserves Act (Newfoundland and Labrador, 1980), 64
Wilderness and Scenic Rivers Act, reasons for enactment of, 19
Wilderness areas
   management of legally designated, 3
   objectives in, 189
   use of, 100–108
Wilderness Areas Act (Alberta, 1971), 64
Wilderness Areas Act (Ontario, 1959), 63
Wilderness Areas Protection Act (Nova Scotia, 1998), 64
Wilderness-associated wildlife, 310–11
Wilderness Associazione Italiana per la Wilderness (AIW) (Italy), 74
Wilderness Bill (1956), 94
Wilderness character, 258
   wildlife as measure of, 312
Wilderness Conference, 44
Wilderness-dependent uses, 358–59
   favoring, 185–86
Wilderness-dependent wildlife, 310

Wilderness designation
   evolution of, 115–43
   giving meaning to, 18
   wilderness management and, 12–16
Wilderness Designation Acts, xix
   management implications, 143–55
Wilderness ecosystems, 252–72
   assessing naturalness for planning, 267–68
   changes in, 259–64
      direct human impact, 262
      exotic weeds, 261–62
      introduced fish and animals, 262
      loss of animal species, 261
      loss of plant species, 261
      outside influences, 262
      Pristine myth, 260–62
   components of, 252–54
   defined, 257–58
   dynamics of, 254–56
   fire in (See Fire(s))
   humankind's historical role in, 258–59
   importance of scale, 256
   managing, 264
      boundary effects, 264
      ecological restoration in wilderness, 265–67
   monitoring and adaptive management, 268–70
   profound versus cosmetic change, 262–64
Wilderness experience, benefit of, 9–10
Wilderness experience programs (WEPs), 386
Wilderness fire management
   controversy and support for programs in, 297–98
   natural in, 287–89
Wilderness fires, monitoring and evaluating, 295
Wilderness Foundation (South Africa), 67
Wilderness Leadership School (WLS), 66, 67
Wilderness legislation
   exceptional and ambiguities in, 109
   litigation and, 155–56
   management direction and special provisions in, 146
Wilderness license, 44
Wilderness litigation
   extent of, 157
   themes and management implications, 157–58
Wilderness management
   alternatives to, 18
   effects of wilderness designation on, 13–16
   evolution of, 45–47
   historical roots of (See Historical roots of wilderness management)
   need for, 16–18
   objectives and guidelines for wildlife, 327–29
   origin of, 43–45
   principles of, 180–94
      applying minimum tools, regulations, or force to achieve objectives, 189
      favoring wilderness-dependent activities, 185–86
      focusing management on threatened sites and damaging activities, 188
      guiding management with written plans with specific area objectives, 186–87
      involving public as key to success of, 189–90
      managing human influences, 183–84
      managing wilderness and sites within, following concept of nodegradation, 182–83
      managing wilderness as most pristine extreme on environmental modification spectrum, 180
      managing wilderness biocentrically to produce human values and benefits, 184–85
      managing wilderness comprehensively, 181–82
      managing wilderness in relation to its adjacent lands, 191–93

Wilderness management, cont.
      monitoring wilderness conditions and experience opportunities to guide long-term wilderness stewardship, 190–91
      setting carrying capacities to prevent unnatural change, 187–88
   as relatively new, 2–3
   wilderness designation and, 12–13
   wilderness philosophy and, 19
   writing plan for, 207–12
Wilderness management planning, 196–215
   in Bureau of Land Management, 202–3
   criteria for evaluating, 212–14
   in federal agencies, U.S., 199–203
   in Fish and Wildlife Service, 202
   in Forest Service, 200–202
   goal-achievement framework for writing, 204–6
   limits of acceptable change in, 206–7
   need for, 196–97
   preparing, 207–12
   tools and techniques, 203–6
   trends and evolution of, 197–99
      plan flexibility and effectiveness, 198
      wilderness planning under National Environmental Policy Act (NEPA) (1970), 198–99
Wilderness managers, professionalism of, 45–46
Wilderness opportunity spectrum (WOS), 209–10, 359, 378–79
Wilderness planning
   assessing naturalness for, 267–68
   limits of acceptable change for, 206–7
Wilderness protection
   classes of, 58–83
      Class I wilderness, 59–71
      Class II wilderness, 59, 71–75
      Class III wilderness, 59, 75–83
   evolution of, 109–10
   forces of change in 21st century, 51–58
   lack of political and financial support for, 351–52
Wilderness Protection Act (South Australia, 1992), 61
Wilderness Provincial Park Act (Alberta, 1959), 64
Wilderness rangers, 475–76
Wilderness recreational use, 369–70
   characteristics, 382–84
      age, 382–83
      education, 383
      gender, 383
      income, 383
      occupation, 383–84
      racial and ethnic minority participation, 384
      residence, 383
   distribution, 378–79
      of use between areas, 379–80
      of use within areas, 380–82
   estimating, 370–75
      direct counting, 370–71
      indirect counting, 371–73
      trends in, 374–75
      units of measurement, 373–74
   use characteristics, 375
      activities, 376–77
      group size, 376
      length of stay, 375
      method of travel, 376
      season and timing of use, 377
      type of group, 377
      visitor residence, 378
Wilderness-related international protection of nature, 54
Wilderness Society, 65, 93
   founding of, 7, 9, 43, 91
Wilderness Society v. U.S. Fish and Wildlife Service, 158

Wilderness species, 309
Wilderness stewardship, xiii
   appeal of term, 17
   future issues and challenges in, 487–93
   wildlife-related problems in, 318–27
Wilderness Stewardship Challenge, 201–2
Wilderness study areas (WSAs), 16
   quality standards for, 138
Wilderness study process, 131–32
Wilderness Task Force (WTF), 56
Wilderness use and user trends, 357–91, 490–91
   factors affecting, 384
      changing age and population structure, 384
      changing conditions and regulations, 387
      changing recreational preferences and organized use, 386–87
      changing transportation cost and supply, 386
      constraints to visitation, 387
      effects of expanding system, 385
      increasing educational levels and environmental awareness, 386
      leisure time, 385
      population distribution and diversity, 384–85
   importance of understanding, 358
   recreation use projections, 387–89
   wilderness-dependent uses, 358–59
   wilderness use
   commercial recreational, 360–61
   commodity, 365–69
   education, personal growth, therapy, and healing programs, 364–65
   indirect, 361–62
   public recreational, 359–60
   scientific, 362–64
Wilderness-user safety, 293–94
Wilderness values, xvii, 28
   increase in appreciation of, 5
   quality of, 136–37
Wilderness visitor management, 440–83
   approaches, 454
      accepting restrictive regulations, 456–57
      direct versus indirect visitor management, 454
      guidelines for regulation, 457–59
      location of management, 459
      regulation, 454–56
      use of regulations by managers, 457
   considerations, 441–42
      managers' and visitors' perceptions of problems, 443–45
      types of visitor impacts, 442–43
      visitor impacts on social conditions, 445–48
   direct techniques, 470–75
   elements of visitor use subject to management, 448
      amount and distribution of use, 448
      behavior, 450
      effects on resource, 450–51
      length of stay, 450
      method of travel, 450
      party size, 450
      redistributing use, 449
   indirect techniques, 460–70
   litter control, 476
   managing for wilderness experiences, 451
      combined influences on, 454

Wilderness visitor management, cont.
   managerial presence, 453
   naturalness, 451–52
   solitude, 452–53
   monitoring quality, 476–78
   rangers, 475–76
   use of simulation models, 478–80
Wilderness Watch v. *Mainella*, 158
Wilderness wildlife
   categories of, 310–11
   as environmental and laboratory, 315–16
Wildfires, 452
Wildhorse Creek (Oregon)
Wild horses, 309
Wildland fire suppression, 346–47
The Wildlands Project, 81
Wildlife, 309–32
   categories of, 310–11
   disturbance of, 404–6
   effect of fire on habitat of, 282–83
   as measure of wilderness character, 312
   movement of, 264
   policies toward, 39
   preservation of, 313–15
   protection of, in wilderness, 149
   relationships in, 311–18
   resources in, 309–18
   viewing, 215
   in wilderness, 106–8, 312–13
   wilderness-associated, 310–11
   wilderness-dependent, 310
   in wilderness stewardship, 318–27
Wildlife Conservation Society, 68
   criteria in assessing wilderness, 52
   electrical power infrastructure, 52
   human access points, 52
   land transformation, 52
   population density, 52
Wildlife Management Institute, 318
   wilderness legislation and, 94
Wildness, 338
Wild pigs, 313
Wild rivers, 82, 170
Willmore Wilderness Park, 64
Willmore Wilderness Park Act (Alberta, 1962), 64
Wilson, Edward O., 9, 310
   Biophilla Hypothesis of, 2
Wilson Creek (North Carolina), 171
Wirth, Conrad L., 93
Wolf Peak, 180
Wolverines, 310
Wolves, 252, 309, 310, 312, 317, 321
   elimination of, from many wilderness ecosystems, 263
   reintroduction of, 321–22, 331–32
Wood Buffalo National Park (Canada), 64, 320
World Bank, 51
   international recognition for wilderness, 58
World Commission on Protected Areas (WCPA), 55
   Guidelines for Protected Areas, 56
World Conservation Monitoring Centre (WCMC), 55
World Heritage Convention, 56, 58
World heritage sites, 56–57, 174–75
World Parks Congress, 55, 56
World Resources Institute, 51

World Wilderness Congresses (WWC), xv, 74
   holding of wilderness policy agency meetings and, 154
   introduction of first world wilderness inventory, 51
   new wilderness categories under, 55
   organization of first, 66
   as project of International Wilderness Leadership (WILD) Foundation, 83
   providing opportunities for compiling, comparing, and exchanging information, 363–64
   providing wilderness focus in diverse places, 46
Wrangell-Saint Elias Wilderness (Alaska), 165
Wright, George M., 39
Wright, George M., Society, 39
Wyoming Wilderness Act, 148, 149
   grazing and, 369

Y

Yaks, 323
Yard, Robert Sterling, 38
Yellow-legged frog, 321
Yellow star thistle, 184
Yellowstone Act, 37
Yellowstone ecosystem, 326
Yellowstone Falls, 37
Yellowstone Game Protection Act (1894), 317
Yellowstone National Park (Wyoming)
   as biosphere reserve, 56, 175
   bison range at, 330
   Congressional wilderness proposal of, 132
   creation of, 6, 37
   fire in, 276
   fire management program in, 297
   1988 fire season in, 279, 287, 291, 292, 295, 300
   as first national park, 36–37
   nonnative fish in, 321
   protecting wilderness values in, 90
   replacing of fire-fall with chicken fall, 38
   roadless areas and, 152
   stand-replacing fires in, 346
   wildlife diseases and, 320
   wildlife movement and, 264
   wildlife preservation and, 315
   as world heritage site, 175
Yellowstone to Yukon: Freedom to Roam (Schulz), 81
Yoho, Canada, 63
Yosemite National Park (California), 37, 38, 57, 90, 175
   fire in, 276
   fire management program in, 296–97
   recreational use of, 374, 378, 379
Yosemite Valley, granting to California, 6

Z

Zahniser, Howard, 65, 93–94, 96, 98
Zapata Swamp, 57–58
Zapovednik Management, 70–71
Zapovedniks (forbidden areas), 69
Zimbabwe, wilderness areas in, 74
Zoning, 209–10
Zululand reserves, 65
Zunino, Franco, 74

About the Authors

Chad P. Dawson is professor and former chair of the Department of Forest and Natural Resources Management, College of Environmental Science and Forestry at the State University of New York, and managing editor of the *International Journal of Wilderness*.

John C. Hendee is professor emeritus and retired dean of the College of Natural Resources at the University of Idaho, is a founder and editor-in-chief of the *International Journal of Wilderness*, and is the WILD Foundation vice president for science and education.